W9-ACZ-916

INSTRUCTOR RESOURCES

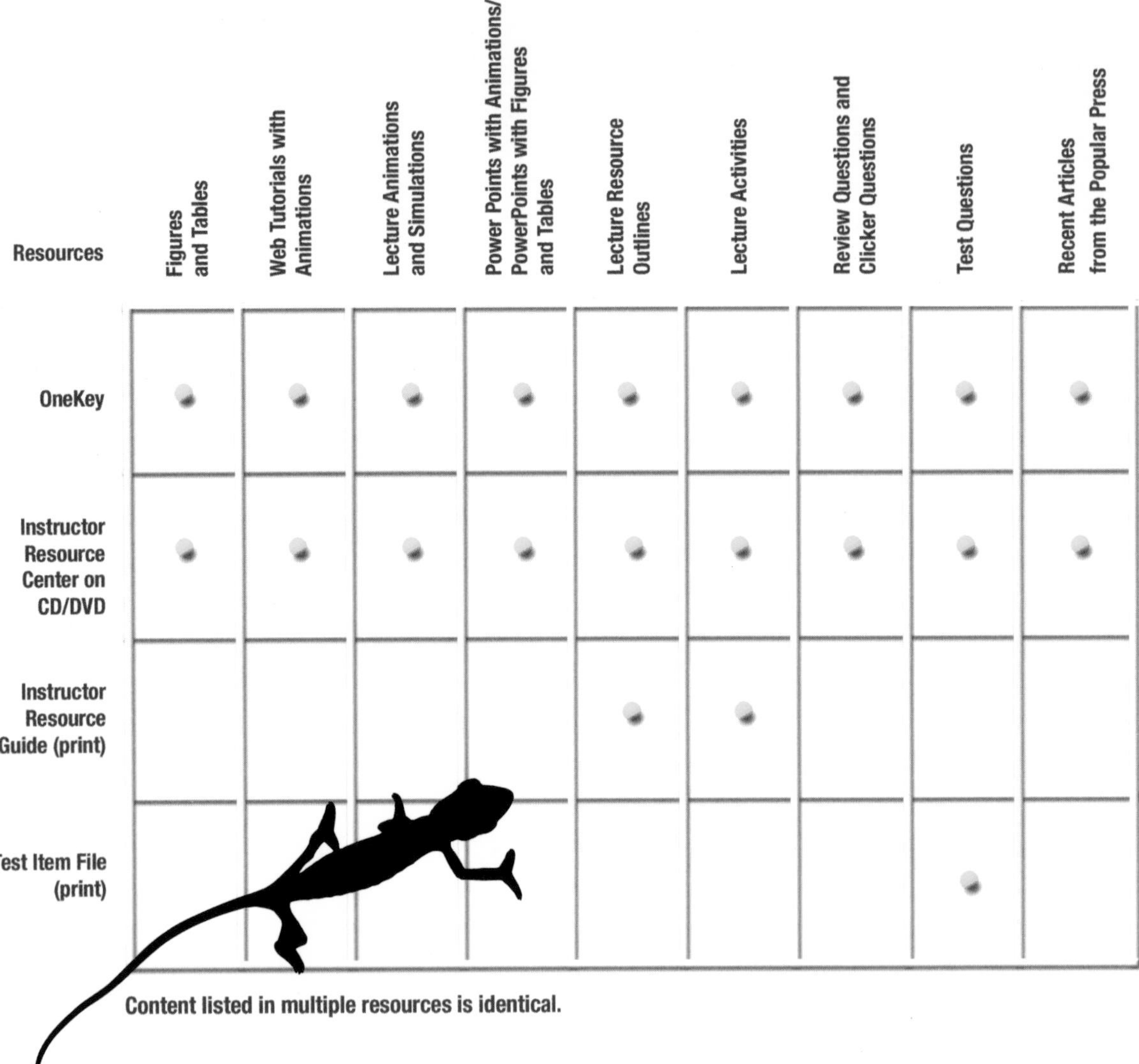

Resources	Figures and Tables	Web Tutorials with Animations	Lecture Animations and Simulations	Power Points with Animations/ PowerPoints with Figures and Tables	Lecture Resource Outlines	Lecture Activities	Review Questions and Clicker Questions	Test Questions	Recent Articles from the Popular Press
OneKey	●	●	●	●	●	●	●	●	●
Instructor Resource Center on CD/DVD	●	●	●	●	●	●	●	●	●
Instructor Resource Guide (print)					●	●			
Test Item File (print)								●	

Content listed in multiple resources is identical.

OneKey

OneKey offers the best teaching and learning resources all in one place. It is all you need to plan and administer your course. Conveniently organized by textbook chapter, these compiled resources help save you time. OneKey is available in CourseCompass, WebCT, and Blackboard platforms.

- OneKey— *in CourseCompass* (0-13-185275-2) — request an instructor access code online (www.coursecompass.com) or from your Prentice Hall representative to get started.
- OneKey— *in Blackboard* (0-13-149534-8) — available as a downloadable course cartridge, with resources ready to incorporate into an existing course or to be used to construct a new one.
- OneKey— *in WebCT* (0-13-149533-X) — available as a downloadable course cartridge, with resources ready to incorporate into an existing course or to be used to construct a new one.

The following digital resources, including electronic versions of all the instructor's print resources, are available in OneKey:

- *Figures and Tables* — more than 700 labeled and unlabeled illustrations, tables, and photos from the text, formatted for large lecture hall presentations.
- *Web Tutorials* — interactive activities with animations.
- *Lecture Animations and Simulations* — high-quality animations and simulations illustrating key biology concepts, specifically designed to be shown during your lectures.
- *Prepared PowerPoint® Shows* — presentations for each chapter available in two formats: with Web Tutorial animations embedded, and with book figures and tables embedded. Labels for figures and tables are all editable in PowerPoint.
- *Lecture Resource Outlines* — detailed lists of all the resources we provide to incorporate into your lecture.
- *Lecture Activities* — ideas for presentations and assessments that promote active learning, including discussion questions, demonstrations, and group and collaborative activities.
- *Review Questions* — all review questions, including Self Test questions, Essay Challenge questions, end-of-chapter questions, and figure caption questions from the text.
- *Clicker Questions* — PowerPoint® slides containing multiple-choice questions for use with classroom response systems. Ask your Prentice Hall representative and visit our website (www.prenhall.com/crs) for more information about using a wireless polling system in your classroom.
- *Test Questions* — electronic version of the Test Item File available as editable Word® files and formatted for import to WebCT or Blackboard courses. We also provide our TestGen software to help you more easily create and manage tests, including multiple versions of each test, and tests for online delivery.

Instructor Resource Center on CD/DVD (0-13-149528-3)

All of OneKey's digital instructor resources located on CD and DVD, organized for easy use.

Instructor Resource Guide (0-13-149522-4)

The print alternative to some of the digital instructor resources provides:

- *Lecture Resource Outlines* — detailed lists of all the resources we provide to incorporate into your lecture.
- *Lecture Activities* — ideas for presentations and assessments that promote active learning, including discussion questions, demonstrations, and group and collaborative activities.
- *Review Questions and Answers* — solutions to the end-of-chapter questions and figure caption questions in the text.

Test Item File (0-13-149521-6)

The print version of the electronic test bank offers:

- More than 2700 questions compiled and reviewed by a carefully selected team of non-majors biology educators.
- Three question types—factual, conceptual, and applied—to test your students' abilities from recall to understanding of core principles to the application of facts to novel situations.
- Questions with a variety of difficulty levels.

Transparency Pack (0-13-149524-0)

More than 250 four-color acetates of figures and tables from the text.

STUDENT RESOURCES

Resources	Self Test Questions	Essay Challenge Questions	Web Investigations	Issues in Biology	Bizarre Facts	Hints and Feedback for All Answers	Web Tutorials	Links	Concept Maps	Recent Articles from the Popular Press	Figures and Tables
OneKey	●	●	●	●	●	●	●	●		●	
Student Study Companion	●	●	●	●	●				●		
Accelerator CD							●				
Student Lecture Notebook											●
Research Navigator with Guide to Evaluating Online Resource										●	

Content listed in multiple resources is identical.

Laboratory Manuals

Explorations in Basic Biology, Tenth Edition (0-13-145312-2)

by Stanley Gunstream

This best-selling laboratory manual can accompany one- or two-semester introductory biology courses for both non-biology majors and mixed biology majors/non-majors. It includes 41 self-contained, easy-to-understand exercises that blend traditional experiments with investigative exercises.

- **Instructor's Manual to Explorations in Basic Biology, Tenth Edition (0-13-145313-0)**

Biological Explorations: A Human Approach, Fifth Edition (0-13-145314-9)

by Stanley Gunstream

Specifically designed for courses in general biology where the human organism is emphasized—and for courses in human biology— this lab manual contains 33 outstanding exercises by the author of Explorations in Basic Biology, Tenth Edition.

- **Instructor's Manual to Biological Explorations: A Human Approach, Fifth Edition (0-13-147096-5)**

OneKey

OneKey is all students need for anywhere-anytime access to course materials. Conveniently organized by textbook chapter, these materials reinforce and apply what they have learned in class. OneKey contains:

- *Self Test* — multiple choice, labeling, and fill-in-the-blank questions that test students' mastery of the major chapter concepts, as well as their understanding of the connections these biological concepts have to their lives. All questions include helpful tips and immediate feedback to all responses
- *Essay Challenge* — essay questions that challenge students to think conceptually about biology and apply it to their own lives.
- *Web Investigations* — activities related to the chapter-opening stories that lead students to internet sites to answer a series of investigative questions.
- *Issues in Biology* — short articles and investigative questions that help students explore significant issues in contemporary biology.
- *Bizarre Facts* — fun, short articles about the oddities of the biological world, including links that enable students to further explore these unusual phenomena.
- *Web Tutorials* — activities that offer students a visual view of concepts and test their knowledge using animations and simulations.
- *Links* — links to web sites on topics addressed in the chapter.

Accelerator CD

Included in every copy of the text, the Accelerator CD is intended to speed up students' experience with OneKey. This CD contains high-bandwidth files (such as movies and animations). When pages that require these files load in OneKey, the Web site will first attempt to find them on the CD. If the Accelerator CD is loaded in the CD drive of the student's computer, the file will load much more quickly. This is especially helpful for students who use a dial-up connection to the internet.

Student Study Companion (0-13-185277-9)

The print alternative to OneKey provides:

- *Self Test questions*
- *Essay Challenge questions*
- *Web Investigations*
- *Issues in Biology*
- *Bizarre Facts*
- *Concept Maps for students to complete*

Student Lecture Notebook (0-13-149525-9)

This portable lecture companion provides figures and tables from the textbook with space for taking notes. Designed to help students take better notes in this visual discipline while still listening to the lecture.

Research Navigator (Available alone or automatically with OneKey)

This tool equips students with the means to start a research assignment or paper or to access full-text articles. It includes extensive help on the online research process as well as three exclusive databases of well-known and reliable source material, including the EBSCO Academic Journal and Abstract Database, The New York Times Search by Subject Archive, and "Best of the Web" Link Library. It enables students to efficiently and effectively make the most of their research time and stay up-to-date on the issues.

Thinking About Biology: An Introductory Laboratory Manual, Second Edition (0-13-145820-5)

by Mimi Bres and Arnold Weisshaar

This lab manual is designed for a one-semester, non-majors introductory biology laboratory course with a human focus. Its 20 easy-to-stage labs can be adjusted for use in 3-hour, 2-hour, or 90-minute lab sessions. This manual specifically develops problem-solving skills and connects biology topics to timely subjects and real-life experiences.

- **Instructor's Manual to Thinking About Biology, Second Edition (0-13-145821-3)**

Symbiosis (www.prenhall.com/symbiosis)

With Symbiosis, Prentice Hall's Custom Laboratory Program for Biology, you can build a lab manual that exactly matches your teaching style, content needs, equipment availability, and course organization. You select the labs you want (including all those from the lab manuals listed to the left), choose the sequence, and can even add your own material with our online book-building system. Please ask your Prentice Hall representative for more details, or email biology_service@prenhall.com.

CUSTOM OPTIONS

The Custom Core Edition is designed to provide you with additional options to meet the content needs of your course. The Custom Core Edition consists of chapters 1–16 from *Life on Earth, Fourth Edition* and is supported by the complete teaching and learning package.

You can further customize the Custom Core Edition by adding your choice of chapters 17–30 from the full text. You can select any combination of chapters to address only the topics that you cover in your course. The chapters available for customization are highlighted in the Table of Contents.

For more information on customization, contact your local Prentice Hall representative or your Pearson Custom Editor.

Life on Earth

Life on Earth

FOURTH EDITION

Teresa Audesirk
Gerald Audesirk
University of Colorado at Denver

Bruce E. Byers
University of Massachusetts, Amherst

CUSTOM CORE EDITION

Upper Saddle River, New Jersey 07458

Executive Editor: Teresa Chung
Editor in Chief, Science: John Challice
Development Editor: Anne Reid
Production Editor: Tim Flem/PublishWare
Executive Managing Editor: Kathleen Schiaparelli
Assistant Editors: Colleen Lee, Andrew Sobel
Media Editor: Travis Moses-Westphal
Marketing Manager: Andrew Gilfillan
Art Editor: Sean Hogan
Managing Editor, Audio/Video Assets: Patricia Burns
Director of Creative Services: Paul Belfanti
Art Director: John Christiana
Page Composition: PublishWare
Manager of Formatting: Allyson Graesser
Manufacturing Buyer: Alan Fischer
Editor in Chief of Development: Carol Trueheart
Assistant Managing Editor, Science Media: Nicole Jackson
Media Production Editor: Tyler Suydam
Assistant Managing Editor, Science Supplements: Becca Richter
Editorial Assistants: Marilyn Coco; Nancy Bauer
Cover and Interior Designer: Susan Anderson
Illustrators: Imagineering; Rolando Corujo; Hudson River Studios; Howard S. Friedman; David Mascaro; Edmund Alexander; Roberto Osti
Director, Image Resource Center: Melinda Reo
Manager, Rights and Permissions: Zina Arabia
Interior Image Specialist: Beth Boyd-Brenzel
Cover Specialist: Karen Sanatar
Image Permission Coordinator: Debbie Latronica
Photo Researcher: Yvonne Gerin
Cover Photograph: Ingo Arndt Wildlife Photography

Pearson Prentice Hall
Pearson Education, Inc.
Upper Saddle River, NJ 07458

Printed in the United States of America

10 9 8 7 6 5 4 3 2 1

ISBN 0-13-149506-2 (Custom Core Edition)

ISBN 0-13-149512-7 (Unbound Edition)

Pearson Education Ltd., *London*
Pearson Education Australia Pty., Limited, *Sydney*
Pearson Education Singapore, Pte. Ltd.
Pearson Education North Asia Ltd., *Hong Kong*
Pearson Education Canada, Ltd., *Toronto*
Pearson Educación de Mexico, S.A. de C.V.
Pearson Education—Japan, *Tokyo*
Pearson Education Malaysia, Pte. Ltd.

Brief Contents

About the Authors

Terry and Gerry Audesirk grew up in New Jersey, where they met as undergraduates. After marrying in 1970, they moved to California, where Terry earned her doctorate in marine ecology at the University of Southern California and Gerry earned his doctorate in neurobiology at the California Institute of Technology. As postdoctoral students at the University of Washington's marine laboratories, they worked together on the neural bases of behavior, using a marine mollusk as a model system.

Terry and Gerry are now professors of biology at the University of Colorado at Denver, where they have taught introductory biology and neurobiology since 1982. In their research lab, funded by the National Institutes of Health, they investigate the mechanisms by which neurons are harmed by low levels of environmental pollutants.

Terry and Gerry share a deep appreciation of nature and of the outdoors. They enjoy hiking in the Rockies, running near their home in the foothills west of Denver, and attempting to garden at 7000 feet in the presence of hungry deer and elk. They are long-time members of many conservation organizations. Their daughter, Heather, has added another focus to their lives.

Bruce E. Byers, a midwesterner transplanted to the hills of western Massachusetts, is a professor in the biology department at the University of Massachusetts, Amherst. He has been a member of the faculty at UMass (where he also completed his doctoral degree) since 1993. Bruce teaches introductory biology courses for both nonmajors and majors; he also teaches courses in ornithology and animal behavior.

A lifelong fascination with birds ultimately led Bruce to scientific exploration of avian biology. His current research focuses on the behavioral ecology of birds, especially on the function and evolution of the vocal signals that birds use to communicate. The pursuit of vocalizations often takes Bruce outdoors, where he can be found before dawn, tape recorder in hand, awaiting the first songs of a new day.

To Heather, Jack, and Lori and in memory of Eve and Joe

T. A. & G. A.

To Bob and Ruth, with gratitude

B. E. B.

Contributors: Instructor and Student Resources

Carole Browne

Wake Forest University

I enjoy teaching humanities students, many of whom are convinced that biology is too difficult or is not relevant to their daily lives. I focus on the biology that they need to understand current issues, with the hope that they will continue to follow these issues even after they are no longer enrolled in biology.

Anne Galbraith

University of Wisconsin – La Crosse

Nonmajors have a refreshing perspective on biology because they are coming from so many different types of disciplines. I really know I'm doing a good job when a non-science major tells me that biology is their favorite class that semester.

James A. Hewlett

Finger Lakes Community College

My true love is teaching a mixed-majors undergraduate biology class. The mention of recombinant DNA brings out clamors for detailed protocols from majors and concerns regarding the ethical and social aspects of such a practice from liberal arts students. Art students become scientific illustrators as part of a class project, while business majors are enthused over the marriage of biology and the NASDAQ. These combinations make teaching undergraduates enjoyable.

Sandra Johnson

Middle Tennessee State University

Teaching undergraduate biology gives me an opportunity to make new connections with established ideas. When students need a different perspective to connect with a biological concept, I'm challenged to communicate the old idea in a new way. Even more exciting, students may then develop their own interpretations, leading to better understanding for all. These opportunities convey the excitement and adventure of biology—of expanding new knowledge. That's special!

Stephen Kilpatrick

University of Pittsburgh, Johnstown

Teaching nonmajors is a fun challenge. The natural world provides a lot of examples of amazing organisms and processes, so I can first get students' attention and then sneak in some basic biology and help the students understand how it applies to their lives.

Nancy Pencoe

State University of West Georgia

Nonmajors! Though many a professor cringes at the thought of having to teach nonmajors biology, I embrace the challenge of teaching students with a wide range of experiences and viewpoints. It is this diversity that makes the course both enjoyable and memorable for the student as well as the instructor.

Kelli Prior

Finger Lakes Community College

I enjoy generating interest in biology's applications outside the classroom, from the food we eat to what we hear about on the news. I see to it that students realize that they need not become biologists to appreciate this relevance to their everyday lives.

Greg Pryor

Francis Marion University

Living in a multimedia world where megabytes of information and misinformation are available at the click of a button, the typical non-science major is excited and curious—yet generally confused—about biology. With this in mind, I aim to captivate these students, facilitate their learning, and encourage their critical thinking. I also endeavor to teach students to be cautious about accepting simple "sound bite" versions of the complex biological issues that affect their lives.

Connie Russell

Angelo State University

I like teaching introductory biology for two reasons. First, I get to learn new ways of connecting seemingly unrelated concepts together, and then I get to break them down into their component parts so that I can initiate them to novice biologists. Secondly, many non-majors come to my class with a distinct dislike for science because they don't see the relevance of it to their own lives. I get the chance to help them understand that everyone needs to think like a scientist to be a successful citizen.

Andrew Storfer

Washington State University

I enjoy teaching nonmajors biology because it is extremely important to teach non-science majors about the importance of biology and the influence biological research and discovery has on their everyday lives. From our improved knowledge of human disease causes and treatment, to implications of cloning and stem cell research, to conservation of biodiversity, understanding biology is critically important. In addition, I enjoy the opportunity to enhance students' critical thinking skills.

Essays

Earth Watch

Health Watch

Scientific Inquiry

Biotechnology Watch

Evolutionary Connections

Contents

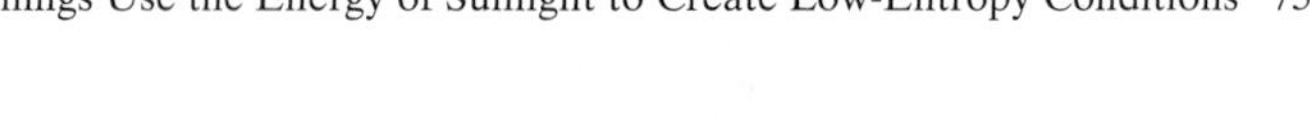

LONG BEACH

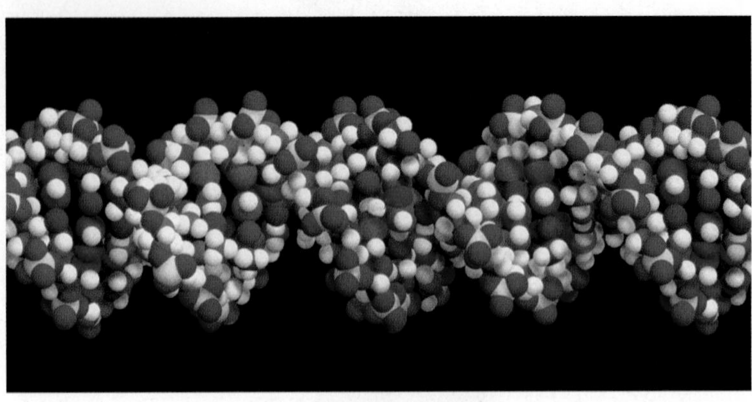

UNIT TWO

Inheritance *111*

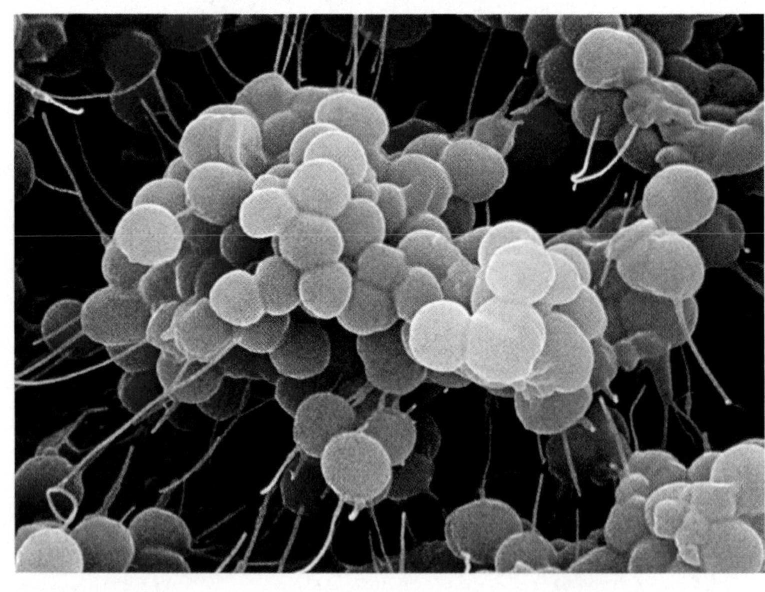

THE FOLLOWING CHAPTERS ARE AVAILABLE AS CUSTOM CHAPTERS:

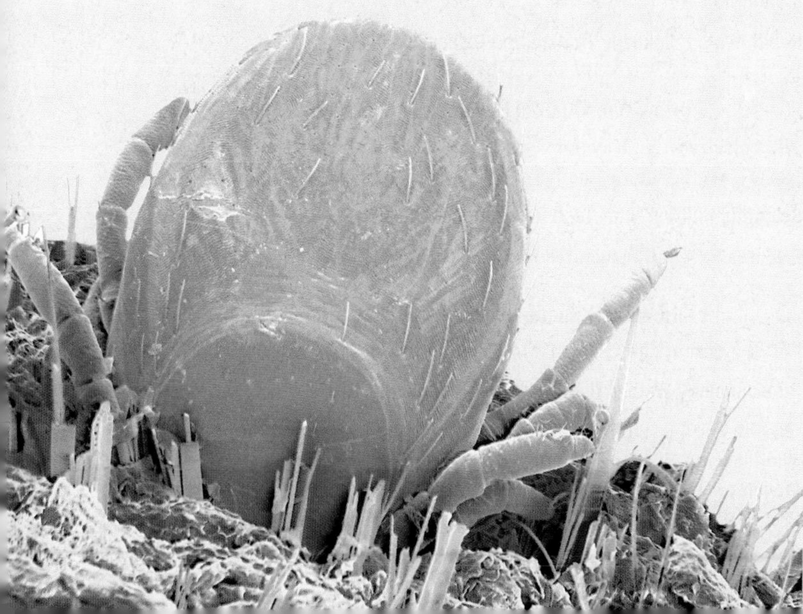

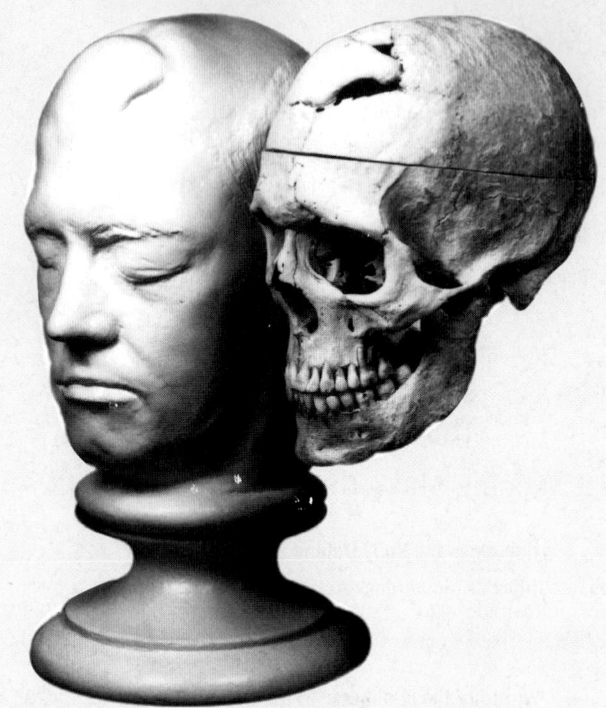

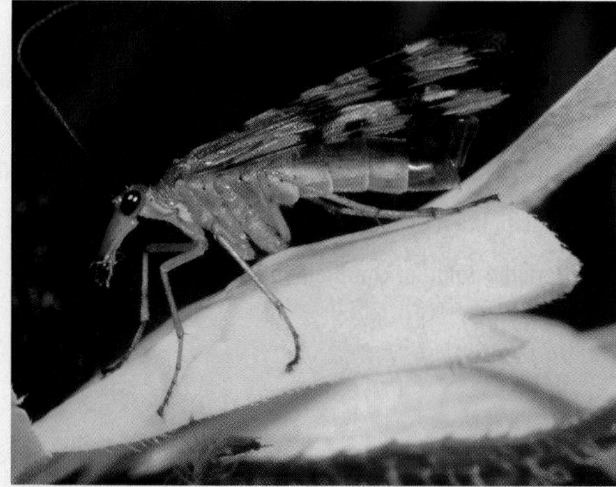

Preface

Are recent claims about cloned human babies true? Are genetically engineered crops safe? Are people causing climate change? How can we stop the spread of AIDS? Will physicians soon be transplanting pig hearts into people routinely? Why are antibiotic medicines becoming less effective? Many of today's most important and controversial social, medical, environmental, and ethical issues are related to biology. The need for voters, jurors, and citizens in general to understand the basic concepts of biology has never been more urgent.

Many of the students who will use this text are enrolled in a course that will provide both their first and final exposure to biology before they leave formal education. We hope that they will emerge from the course prepared to ask intelligent questions, make informed choices, and scrutinize science articles in the popular press with an educated and critical eye. We also hope that students will better understand and appreciate their own bodies, the other organisms with which we share Earth, the evolutionary forces that mold all life-forms, and how complex interactions within ecosystems sustain us and all other life on our planet. Perhaps most of all, we hope that students will develop a fascination with life that will inspire them to keep learning science. To help instructors achieve these teaching goals, we offer this revision of *Life on Earth*. Now in its fourth edition, *Life on Earth* helps students effectively manage a wealth of scientific information and motivates their learning.

Helping Students Manage Information and Get Motivated

The fourth edition of *Life on Earth*—which is not just a textbook, but rather a complete package of teaching aids for the instructor and learning aids for students—has been revised with three specific goals:

- To help instructors **manage the presentation of the wealth of biological information** with the goal of producing scientifically literate students
- To help students build familiarity with the **process of science** through an engaging storyline
- To help students **relate** this information to their own lives so as to understand its importance and relevance

Life on Earth

... Is Organized Clearly and Uniformly to Maximize Conceptual Understanding

Throughout each chapter, students will find aids that help them navigate through the large amount of information they face in this course. We have organized this text to help students see the forest and the trees, without getting lost in thickets of detail.

- "At a Glance" at the start of each chapter brings together the chapter's major subheadings and now includes the titles of essays as well. Instructors can easily assign—and students can easily locate—key topics within each chapter.
- Major sections are introduced as questions to which the student will find answers in the section, while minor subheadings are presented as summary statements that reflect content. A crucial outgrowth of this organizational scheme is that it imparts an understanding of biology as a hierarchy of interrelated concepts, rather than a set of isolated, independent subjects. It also reminds students of the importance in science of asking questions.
- The "Summary of Key Concepts" pulls together the important concepts. To be clear and consistent, we use the same major subheadings in both the "At a Glance" and the chapters themselves, allowing instructors and students to move efficiently among the different components within a chapter.
- In cases where animations or online practice may help make a point more clearly, we have added Media Activity tabs within each chapter. These direct the students to OneKey, which contains relevant activities, animations, and practice tests; the Media Activity numbers in the book correspond with OneKey for easy navigation. Detailed descriptions of each Media Activity are found at the end of each chapter.

... Actively Engages and Motivates Students

Scientific literacy cannot be imposed on students; they must actively participate in acquiring both the core facts and the way of thinking like a scientist. This job is a lot easier if students recognize that biology is about their personal lives as well as the life all around them. To help engage and motivate students, this new edition incorporates the following:

- **Openers/Closers**. Each chapter opens with a strikingly illustrated, brief essay. The opener essays are based on recent news items, on situations in which students might find themselves, or on particularly fascinating biological topics. For example, students will investigate blood doping by elite athletes (p. 99); contemplate the use of DNA to solve historical mysteries (p. 113) and crimes (p. 193); see cloned, genetically engineered pigs that are being raised as potential organ donors for humans (p. 379); and follow along as a scientist estimates the number of species on Earth (p. 283). Each opener ends with a cliff-hanger, seducing students to read more. At the end of each chapter, we revisit the story, allowing students to explore the topic a bit further in light of what they have just learned and, often, to find answers to questions raised in the initial piece. The "revisited" essays conclude with a new "Consider This" segment that poses an open-ended question to encourage deeper thinking about the topic. If you want to encourage students to read the openers, you can assign these questions.
- **Links to Life**. A new feature in the fourth edition, "Links to Life" ends each chapter on a relevant note. These short, informally written segments are related to subjects that are both very familiar to the student and relevant to the chapter.

- **Caption Queries**. New in this edition, selected figure captions in each chapter include questions designed to encourage readers to review and extend their new knowledge of the pictured structure or process. These help link the graphics to the content, a helpful exercise for today's more visual learners.
- **Bioethics**. Many topics explored in the text have ethical implications. New in the fourth edition, these topics are now identified by an icon that alerts students and teachers to the possibility of further discussion and exploration.
- **Essays**. We retain our two original essay libraries, including "Earth Watch," environmental essays that explore issues such as the loss of biodiversity, the growing ozone hole, and invasions of exotic species, and our medically related "Health Watch" essays, which investigate topics such as sexually transmitted diseases, the dangers of artificial steroids, and how smoking damages the lungs. New in this edition, "Biotechnology Watch" essays examine the impact of new technologies such as cloning, *in vitro* fertilization, and genetic modification of organisms.

... Emphasizes Scientific Reasoning

In order to make sound decisions at the voting booth or to evaluate assertions made by the media, students need to think critically. To help students develop scientific reasoning skills, we have added the following:

- **Openers and Closers**. In addition to the engaging, relevant biological storyline, the opener essays build students' familiarity with the process of science and demonstrate how scientists use the process to reach conclusions. The opening and the closing essays incorporate many examples of questions posed, hypotheses stated, predictions made, and experiments performed. Our aim is to show by example how scientists gather objective evidence about interesting questions relevant to students' lives. We hope that student readers of *Life on Earth* will begin to think like scientists. Nothing would please us more than a student who, upon hearing an assertion on television or reading a claim on the Internet, instinctively asks, "What is the evidence and how was it gathered?"
- **Is This Science?** We have added new "Is This Science?" questions to the "Applying the Concepts" critical-thinking questions at the end of each chapter. The new questions are designed to help students practice their scientific reasoning and critical-thinking skills.

... Contains Superior Illustrations for Greater Clarity and Consistency and to Inspire Reader Interest

Benefiting from the advice of reviewers, a talented biological illustrator, and careful scrutiny by the authors and developmental editor, we have extensively revised the illustration program. For the fourth edition, we have:

- Expanded the consistent use of color. We have been vigilant in tracking the use of color to provide consistency in illustrating specific atoms, structures, and processes. We have also made the colors more vibrant to better distinguish individual parts of a figure, to help engage the modern readers' interest, and to focus attention on the most important aspects of the illustration.
- Improved overall quality. We have redrawn the more diagrammatic figures for greater interest and accuracy.
- Enhanced label clarity. We have revised the size, placement, and font of figure labels for more consistency and readability.
- Organized content more efficiently. We have modified the placement of parts of multipart figures for easier navigation through the figure.
- Explained figure content more clearly. Through the judicious use of "talking boxes," we have placed more explanatory statements within figures for greater clarity and ease of reading.
- Modified figure captions to enhance function. Our figure titles now summarize the content; we have made the captions more concise and, to stimulate critical thinking, have added thought-provoking questions to several captions within each chapter.

... Provides Print and Media Resources That Aid User Exploration

For Instructors

- **OneKey:** OneKey for *Life on Earth, Fourth Edition,* supplies anywhere, anytime access to conveniently organized course materials. OneKey provides instructors and students with a single location for our entire collection of superb teaching and learning resources. OneKey also includes everything instructors need to plan and administer courses. All instructor resources are consolidated in one location to maximize your effectiveness and efficiency.
  - All of the figures from the text, both labeled and unlabeled, specially formatted for large lecture hall presentation.
  - The new Biology Lecture Animations & Simulations library, designed for use in class. This stunning collection of animations and simulations illustrates over 50 key concepts identified for nonmajors as the most difficult to visualize by a panel of biology instructors. Unlike animations built for student tutorials, these are designed specifically for use in the lecture environment and provide you with optimal flexibility and control.
  - All the student Web Tutorials from the Student Accelerator CD.
  - Two prebuilt PowerPoint® shows, one containing all art from each chapter, and another containing all animations, simulations, and videos.
  - Lecture resource outlines, which map all our resources by chapter and section, so you know what we have available as you prepare your class.
  - Lecture activities, which are presentation and assessment ideas for use during class to promote active learning.
  - Clicker questions, which are ready to go in PowerPoint® format and can be used with any Classroom Response System, including those from InterWrite (formerly EduCue) and H-ITT.
  - Test-Gen EQ software provides electronic versions (both Mac and Windows) of the complete test bank for the text, including over 2700 additional questions beyond those already found in the textbook or Student Study Companion. OneKey is available for CourseCompass, BlackBoard, and WebCT.
- **Instructor Resource Center on CD/DVD**: This CD/DVD set contains all the digital resources listed above in an easy-to-use, easy-to-search, portable format. The material can be browsed by chapter or resource type and is easily exported to your local drive.
- **Instructor Resource Guide**: This booklet is the print companion to the Instructor's Resource Center on CD. It includes the Lecture Activities, Lecture Resources, Key Terms, end-of-chapter questions and answers, and figure caption questions and answers.

- **Transparency Pack**: This includes 300 four-color acetates of figures and tables from the text.
- **Biology in the News**: This first volume of 34 two-minute video clips from all areas of biology will enrich your classroom teaching and motivate your students by connecting biology to their everyday lives.

For the Student

- **OneKey**: OneKey for *Life on Earth* offers a single location for anywhere, anytime access to students who want extra practice or to access to the many amazing digital resources that come with this text. New to this edition: a perpetual gradebook for students, tracking the results (and averages) of all the quizzes they have taken over the term. www.prenhall.com/audesirk
- **Accelerator CD**: Each copy of this text includes an Accelerator CD. The CD works with OneKey to provide a faster Internet experience. The Accelerator CD stores some OneKey content, so that even dial-up connections can provide snappy performance.
- **Student Lecture Notebook**: Includes key pieces of art from the textbook, with space to take notes.
- **Student Study Companion**: This printed Companion provides questions and review material to students without access to the Internet.

Acknowledgments

Life on Earth is truly a group effort. To meet the dauntingly complex challenge of putting together a text and supplement package of this magnitude, Prentice Hall has assembled an experienced and skilled development team. The text benefited considerably from the thoughtful suggestions of Developmental Editor Anne Reid. She helped us keep the text clear, consistent, and student friendly. Tim Flem, our Production Editor, coordinated the efforts of the photo researcher, copy editor, art studio, and authors. He skillfully brought the art, photos, and manuscript together into a seamless whole while dealing good naturedly with last-minute improvements. Formatting this book is not an easy endeavor, but Tim applied his expertise with great attention to detail. Photo Researcher Yvonne Gerin tracked down excellent photos. Margo Quinto tackled the job of copyediting with exceptional skill.

We also thank Art Director John Christiana for guiding the text and cover design with flair and talent and Art Editor Sean Hogan for coordinating such an immense art program.

Travis Moses-Westphal, our Media Editor, deftly carried the vision for the media program and oversaw the seamless integration of the text, media, and print supplements. Assistant Editors Colleen Lee and Andrew Sobel provided invaluable support and assistance. Editorial Assistant Marilyn Coco cheerfully and efficiently trafficked the manuscript.

Andrew Gilfillan, our Marketing Manager, oversees a large and dedicated salesforce with energy, talent, and enthusiasm. Andrew provides inspired marketing concepts, shares success stories, and makes sure that users' comments always get through to the authors. We thank Paul Corey, now president of the Engineering, Science, and Mathematics division of Prentice Hall, for his confidence and support through this and the past three editions. Finally, but most importantly, our editors: Editor in Chief John Challice has supported us from the very beginning. Executive Editor Teresa Ryu Chung has assumed leadership of the team with talent and zeal, combined with a clear sense of where the project should be going and how to get it there without killing the authors. Her total commitment to the project, her organizational ability, and her sensitivity to all the people involved have been crucial to its success.

So here we acknowledge, with deep appreciation, our "coach" and all our teammates!

Terry and Gerry Audesirk
Bruce E. Byers

Comments and questions are welcome from students and faculty. Please email us at Biology_Service@prenhall.com.

Fourth Edition Reviewers

Marilyn C. Baguinon, *Kutztown University of Pennsylvania*
Gail F. Baker, *LaGuardia Community College*
Kathleen L. Bishop, *University of North Texas*
Edward M. Brecker, *Palm Beach Community College*
Sharon K. Bullock, *Virginia Commonwealth University*
Dave Cox, *Lincoln Land Community College*
Sandi Gardner, *Triton College*
Gail E. Gasparich, *Towson University*
Dale Holen, *Pennsylvania State University, Worthington Scranton Campus*
Tom Langen, *Clarkson University*
Elizabeth E. LeClair, *DePaul University*
Kimberly G. Lyle-Ippolito, *Anderson University*
Amy McMillan, *Buffalo State College*
Lance Myler, *State University of New York Canton*
Kim Cleary Sadler, *Middle Tennessee State University*
Mark Sandheinrich, *University of Wisconsin–La Crosse*
Ruth Sporer, *Rutgers University*
Jeff Travis, *University of Albany*
Jennifer Woodhead, *Brunswick Community College*
Mark L. Wygoda, *McNeese State University*
Samuel J. Zeakes, *Radford University*

Fourth Edition Contributors

Carole Browne, *Wake Forest University*
Anne Galbraith, *University of Wisconsin–La Crosse*
James A. Hewlett, *Finger Lakes Community College*
Patricia Johnson, *Palm Beach Community College*
Sandra Johnson, *Middle Tennessee State University*
Stephen Kilpatrick, *University of Pittsburgh, Johnstown*
Nancy Pencoe, *State University of West Georgia*
Kelli Prior, *Finger Lakes Community College*
Greg Pryor, *Francis Marion University*
Connie Russell, *Angelo State University*
Andrew Storfer, *Washington State University*

Previous Edition Reviewers

W. Sylvester Allred, *Northern Arizona University*
Judith Keller Amand, *Delaware County Community College*
William Anderson, *Abraham Baldwin Agriculture College*
Steve Arch, *Reed College*
Kerri Lynn Armstrong, *Community College of Philadelphia*
G. D. Aumann, *University of Houston*
Vernon Avila, *San Diego State University*
J. Wesley Bahorik, *Kutztown University of Pennsylvania*
Michelle Baker, *Utah State University*
Neil Baker, *Ohio State University*
Bill Barstow, *University of Georgia, Athens*
Colleen Belk, *University of Minnesota, Duluth*
Michael C. Bell, *Richland College*
Gerald Bergtrom, *University of Wisconsin*
Arlene Billock, *University of Southwestern Louisiana*
Brenda C. Blackwelder, *Central Piedmont Community College*
Raymond Bower, *University of Arkansas*
Robert Boyd, *Auburn University*
Marilyn Brady, *Centennial College of Applied Arts & Technology*
Neil Buckley, *State University of New York, Plattsburgh*
Virginia Buckner, *Johnson County Community College*
Arthur L. Buikema, Jr., *Virginia Polytechnic Institute*
William F. Burke, *University of Hawaii*
Robert Burkholter, *Louisiana State University*
Kathleen Burt-Utley, *University of New Orleans*
Linda Butler, *University of Texas, Austin*
W. Barkley Butler, *Indiana University of Pennsylvania*
Bruce E. Byers, *University of Massachusetts, Amherst*
Sara Chambers, *Long Island University*
Nora L. Chee, *Chaminade University*
Joseph P. Chinnici, *Virginia Commonwealth University*
Dan Chiras, *University of Colorado, Denver*
Bob Coburn, *Middlesex Community College*
Martin Cohen, *University of Hartford*
Mary U. Connell, *Appalachian State University*
Joyce Corban, *Wright State University*
Ethel Cornforth, *San Jacinto College, South*
David J. Cotter, *Georgia College*
Lee Couch, *Albuquerque Technical Vocational Institute*
Donald C. Cox, *Miami University of Ohio*
Patricia B. Cox, *University of Tennessee*
Peter Crowcroft, *University of Texas, Austin*
Carol Crowder, *North Harris Montgomery College*
Donald E. Culwell, *University of Central Arkansas*
Robert A. Cunningham, *Erie Community College, North*
Karen Dalton, *Community College of Baltimore County–Catonsville Campus*
David H. Davis, *Asheville-Buncombe Technical Community College*
Jerry Davis, *University of Wisconsin, LaCrosse*
Douglas M. Deardon, *University of Minnesota*
Lewis Deaton, *University of Southwestern Louisiana*
Fred Delcomyn, *University of Illinois, Urbana*
Lorren Denney, *Southwest Missouri State University*
Katherine J. Denniston, *Towson State University*
Charles F. Denny, *University of South Carolina, Sumter*
Jean DeSaix, *University of North Carolina, Chapel Hill*
Ed DeWalt, *Louisiana State University*
Daniel F. Doak, *University of California, Santa Cruz*
Matthew M. Douglas, *University of Kansas*
Ronald J. Downey, *Ohio University*
Ernest Dubrul, *University of Toledo*
Michael Dufresne, *University of Windsor*
Susan A. Dunford, *University of Cincinnati*
Mary Durant, *North Harris College*
Ronald Edwards, *University of Florida*
Rosemarie Elizondo, *Reedley College*
George Ellmore, *Tufts University*
Joanne T. Ellzey, *University of Texas, El Paso*
Wayne Elmore, *Marshall University*
Carl Estrella, *Merced College*
Nancy Eyster-Smith, *Bentley College*
Deborah A. Fahey, *Wheaton College*
Gerald Farr, *Southwest Texas State University*
Rita Farrar, *Louisiana State University*
Marianne Feaver, *North Carolina State University*
Linnea Fletcher, *Austin Community College, Northridge*
Charles V. Foltz, *Rhode Island College*
Matthew Fountain, *State University of New York, Fredonia*
Douglas Fratianne, *Ohio State University*
Scott Freeman, *University of Washington*
Donald P. French, *Oklahoma State University*
Don Fritsch, *Virginia Commonwealth University*
Teresa Lane Fulcher, *Pellissippi State Technical Community College*
Michael Gaines, *University of Kansas*
Irja Galvan, *Western Oregon University*
Gail E. Gasparich, *Towson University*
Farooka Gauhari, *University of Nebraska, Omaha*
George W. Gilchrist, *University of Washington*
David Glenn-Lewin, *Iowa State University*
Elmer Gless, *Montana College of Mineral Sciences*
Charles W. Good, *Ohio State University, Lima*
Margaret Green, *Broward Community College*
Carole Griffiths, *Long Island University*
Lonnie J. Guralnick, *Western Oregon University*
Martin E. Hahn, *William Paterson College*
Madeline Hall, *Cleveland State University*
Georgia Ann Hammond, *Radford University*
Blanche C. Haning, *North Carolina State University*
Helen B. Hanten, *University of Minnesota*
John P. Harley, *Eastern Kentucky University*
Stephen Hedman, *University of Minnesota*
Jean Helgeson, *Collins County Community College*
Alexander Henderson, *Millersville University*
James Hewlett, *Finger Lakes Community College*
Alison G. Hoffman, *University of Tennessee, Chattanooga*
Leland N. Holland, *Paso-Hernando Community College*
Laura Mays Hoopes, *Occidental College*
Michael D. Hudgins, *Alabama State University*
Donald A. Ingold, *East Texas State University*
Jon W. Jacklet, *State University of New York, Albany*
Rebecca M. Jessen, *Bowling Green State University*
Florence Juillerat, *Indiana University–Purdue University at Indianapolis*

Thomas W. Jurik, *Iowa State University*
Arnold Karpoff, *University of Louisville*
L. Kavaljian, *California State University*
Hendrick J. Ketellapper, *University of California, Davis*
Kate Lajtha, *Oregon State University*
Elizabeth E. LeClair, *DePaul University*
Patricia Lee-Robinson, *Chaminade University of Honolulu*
William H. Leonard, *Clemson University*
Edward Levri, *Indiana University of Pennsylvania*
Graeme Lindbeck, *University of Central Florida*
Jerri K. Lindsey, *Tarrant County Junior College, Northeast*
John Logue, *University of South Carolina, Sumter*
William Lowen, *Suffolk Community College*
Ann S. Lumsden, *Florida State University*
Steele R. Lunt, *University of Nebraska, Omaha*
Douglas Lyng, *Indiana University Purdue University, Fort Wayne*
Daniel D. Magoulick, *The University of Central Arkansas*
Paul Mangum, *Midland College*
Michael Martin, *University of Michigan*
Linda Martin-Morris, *University of Washington*
Kenneth A. Mason, *University of Kansas*
Margaret May, *Virginia Commonwealth University*
D. J. McWhinnie, *De Paul University*
Gary L. Meeker, *California State University, Sacramento*
Thoyd Melton, *North Carolina State University*
Joseph R. Mendelson III, *Utah State University*
Karen E. Messley, *Rockvalley College*
Timothy Metz, *Campbell University*
Glendon R. Miller, *Wichita State University*
Neil Miller, *Memphis State University*
Jack E. Mobley, *University of Central Arkansas*
John W. Moon, *Harding University*
Richard Mortenson, *Albion College*
Gisele Muller-Parker, *Western Washington University*
James Murphy, *Monroe Community College*
Kathleen Murray, *University of Maine*
Lance Myler, *State University of New York, Canton*
Robert Neill, *University of Texas*
Harry Nickla, *Creighton University*
Daniel Nickrent, *Southern Illinois University*
Jane Noble-Harvey, *University of Delaware*
David J. O'Neill, *Community College of Baltimore County, Dundalk Campus*
James T. Oris, *Miami University, Ohio*
Marcy Osgood, *University of Michigan*
C. O. Patterson, *Texas A & M University*
Fred Peabody, *University of South Dakota*
Harry Peery, *Tompkins–Cortland Community College*
Rhoda E. Perozzi, *Virginia Commonwealth University*
Gary Pettibone, *Buffalo State College*
Bill Pfitsch, *Hamilton College*
Ronald Pfohl, *Miami University, Ohio*
Bernard Possident, *Skidmore College*
Elsa C. Price, *Wallace State Community College*
James A. Raines, *North Harris College*
Karen Raines, *Colorado State University*
Mark Richter, *University of Kansas*
Todd Rimkus, *Marymount University*
Robert Robbins, *Michigan State University*
William D. Rogers, *Ball State University*
Paul Rosenbloom, *Southwest Texas State University*
K. Ross, *University of Delaware*
Mary Lou Rottman, *University of Colorado, Denver*
Albert Ruesink, *Indiana University*
Christopher F. Sacchi, *Kutztown University*
Alan Schoenherr, *Fullerton College*
Edna Seaman, *University of Massachusetts, Boston*
Linda Simpson, *University of North Carolina, Charlotte*
Anu Singh-Cundy, *Western Washington University*
Russel V. Skavaril, *Ohio State University*
John Smarelli, *Loyola University*
Shari Snitovsky, *Skyline College*
Jim Sorenson, *Radford University*
Mary Spratt, *University of Missouri, Kansas City*
Benjamin Stark, *Illinois Institute of Technology*
William Stark, *Saint Louis University*
Kathleen M. Steinert, *Bellevue Community College*
Barbara Stotler, *Southern Illinois University*
Gerald Summers, *University of Missouri, Columbia*
Marshall Sundberg, *Louisiana State University*
Bill Surver, *Clemson University*
Eldon Sutton, *University of Texas, Austin*
Dan Tallman, *Northern State University*
David Thorndill, *Essex Community College*
William Thwaites, *San Diego State University*
Professor Tobiessen, *Union College*
Richard Tolman, *Brigham Young University*
Dennis Trelka, *Washington & Jefferson College*
Sharon Tucker, *University of Delaware*
Gail Turner, *Virginia Commonwealth University*
Glyn Turnipseed, *Arkansas Technical University*
Lloyd W. Turtinen, *University of Wisconsin, Eau Claire*
Robert Tyser, *University of Wisconsin, La Crosse*
Robin W. Tyser, *University of Wisconsin, LaCrosse*
Kristin Uthus, *Virginia Commonwealth University*
F. Daniel Vogt, *State University of New York, Plattsburgh*
Nancy Wade, *Old Dominion University*
Susan M. Wadkowski, *Lakeland Community College*
Jyoti R. Wagle, *Houston Community College, Central*
Michael Weis, *University of Windsor*
DeLoris Wenzel, *University of Georgia*
Jerry Wermuth, *Purdue University, Calumet*
Jacob Wiebers, *Purdue University*
Carolyn Wilczynski, *Binghamton University*
P. Kelly Williams, *University of Dayton*
Roberta Williams, *University of Nevada, Las Vegas*
Sonya J. Williams, *Langston University*
Sandra Winicur, *Indiana University, South Bend*
Bill Wischusen, *Louisiana State University*
Chris Wolfe, *North Virginia Community College*
Colleen Wong, *Wilbur Wright College*
Wade Worthen, *Furman University*
Robin Wright, *University of Washington*
Brenda L. Young, *Daemen College*
Cal Young, *Fullerton College*
Tim Young, *Mercer University*

Chapter 1

An Introduction to Life on Earth

Seen from space, Earth provides few hints of the abundant life on its surface.

AT A GLANCE

1.1 Why Study Biology?

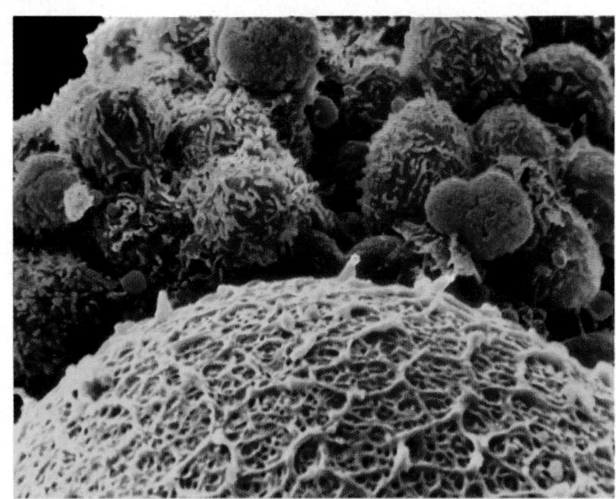

Seen through a microscope, Earth's life offers a view of spectacular beauty and diversity. Here, a human egg rests among other cells that support and protect it.

Why study biology, the science of life? Perhaps you are already fascinated by the study of life and need no further encouragement, but even if you have not yet discovered the thrill of biological inquiry, there are still good reasons to explore biology.

Biology Helps You Understand Your Body

Much of biology is devoted to understanding how bodies work and how disruptions and diseases can prevent them from working properly. Studying biology will help you understand more about the inner workings of your own body and about how different diseases do their damage. This kind of knowledge can be empowering. When you discuss your health with your physician, you will have a much better understanding of the subject matter. Biological knowledge helps you become an informed consumer of health care, and helps you take charge of your own health.

Biology Helps You Become an Informed Citizen

Our society currently faces a host of complex social and ethical issues, many of which will require difficult decisions. As a member of society, you will have a voice in those decisions and a responsibility to make your views known (and to elect representatives who share your views). Many of today's most controversial problems involve biology: Do we need to worry about the impact of pollution on our health? Should we do anything to halt the extinction of endangered species? Are genetically modified foods helpful or harmful? Should cloning of humans be legal? Should scientists be allowed to use cells from human embryos to develop cures for diseases? Is population control a good idea? The more you know about the basic biology that underlies these and other issues, the more likely you will be to have an informed opinion and to be an effective advocate for your point of view. And the more you know about how science works, the better equipped you will be to make your own, independent evaluation, based on evidence, of the claims you encounter in the media or in discussions with other people.

(a)

(b)

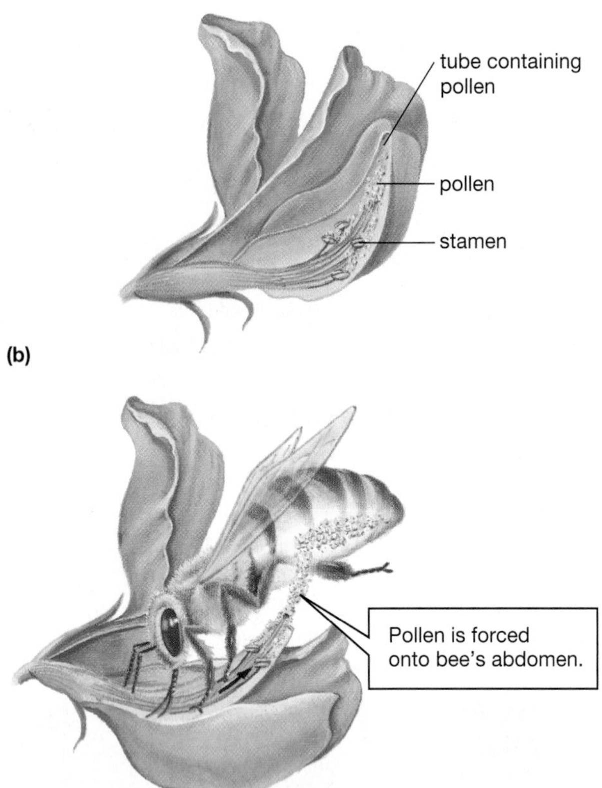

Figure 1-1 The structure of lupine flowers helps ensure pollen transfer ***(a)*** The lower petals of lupine flowers form a tube within which the stamens fit snugly. The stamens shed pollen within the tube. ***(b)*** When the weight of a foraging bee pushes on the lower petals, the stamens are thrust forward, and pollen is ejected onto the bee's abdomen.

Figure 1-2 Wild lupines and fir trees Thousands of people visit Hurricane Ridge in Washington State's Olympic National Park each summer to gaze in awe at Mt. Olympus, but there are also wonders at their feet.

Biology Can Open Career Opportunities

Biology is relevant to many of today's careers. Are you considering a career in biotechnology, or in the pharmaceutical industry, or in the medical equipment business? Do you plan to be a nurse, physician's assistant, physicial therapist, medical technologist, or some other type of health care professional? Perhaps your plans include a job as a wildlife manager, forester, environmental consultant, veterinarian, or zookeeper. Or maybe you have set your sights on becoming a lawyer, with a specialization in enviromental law or patent law. Perhaps a career in biomedical research, or bioinformatics, or genetic counseling appeals to you. And the nation always seems to need qualified science teachers for its schools, colleges, and universities.

A working knowledge of biology will increase your chances of entering any of those careers, and many more. Biological knowledge is an important prerequisite for a whole host of jobs, including many of today's fastest-growing professions.

Biology Can Enrich Your Appreciation of the World

Some people feel that science promotes a cold, clinical view of life, and that scientific explanations of the natural world rob us of a sense of wonder and awe. Nothing could be further from the truth. Biological knowledge only deepens our appreciation of nature's majesty.

Consider, for example, a bee searching for food. Several years ago, we watched such a bee at a patch of lupine flowers. When we looked closely at one of the flowers, we saw that the lower petals formed a tube around the reproductive structures—both the pollen-laden male structures (*stamens*) and the sticky pollen-capturing female structure (*stigma*). We had recently learned that, in young lupine flowers (Fig. 1-1a), the weight of a bee compresses the tube, pushing pollen onto the bee's abdomen (Fig. 1-1b). When a pollen-dusted bee later visits an older flower, a few grains of pollen may drop off onto the stigma (not shown in Fig. 1-1), which sticks out through the lower petals of more mature flowers. The pollen grains carry sperm that can fertilize the eggs that lie below the stigma, beginning the production of new lupine seeds.

Did our new knowledge about lupine flowers make us appreciate them less? Far from it. Rather, we looked on lupines with new delight, thrilled with our fresh insight into how the flower's form was connected to its function, and how the lives of bee and plant were intertwined. Later that summer, we were on Hurricane Ridge in Olympic National Park, enjoying the colorful display of the alpine meadows there (Fig. 1-2). As we crouched beside another lupine plant, an elderly man stopped to ask what we were looking at so intently. He listened with interest as we explained the flower's structure, and then he went off to another patch of lupines to watch the bees foraging. He too felt the increased sense of wonder that comes with understanding.

We hope that, as you read this text, you will share this dual sense of understanding and wonder. We hope you *won't* think of biology as just another course to take, just another set of facts to memorize. Biology can be much more than that. It can be a pathway to a new understanding of yourself and of the life on Earth around you. Join us in a journey of biological discovery.

1.2 How Do Biologists Study Life?

A biologist's job is to answer questions about life. What causes cancer? Why are frog populations shrinking worldwide? What happens when a sperm and egg meet? How does HIV cause AIDS? When did the earliest mammal appear? How do bees fly? The list of questions is both endless and endlessly fascinating.

Anyone can ask an interesting question about life. A biologist, however, is distinguished by the manner in which he or she goes about finding answers. A biologist is a scientist and accepts only answers that are supported by evidence, and then only by a certain kind of evidence. Scientific evidence consists of observations or measurements that are easily shared with others and that can be repeated by anyone who has the appropriate tools. The process by which this kind of evidence is gathered is known as the **scientific method**.

The Scientific Method Is the Basis for Scientific Inquiry

The scientific method proceeds step by step (Fig. 1-3). It begins when someone makes an **observation** of an interesting pattern or phenomenon. The observation, in turn, stimulates the observer to ask a **question** about what was observed. Then, after a flash of insight or a period of long, hard thought, the person proposes an answer to the question, an explanation for the observation. This proposed explanation is a **hypothesis**. A good hypothesis leads to a **prediction**, typcially expressed in "if ... then" language. The prediction is tested with observations or **experiments**. These experiments produce results that either support or refute the hypothesis, and a **conclusion** is drawn about it. A single experiment is never an adequate basis for a conclusion; the experiment must be repeated not only by the original experimenter but also by others.

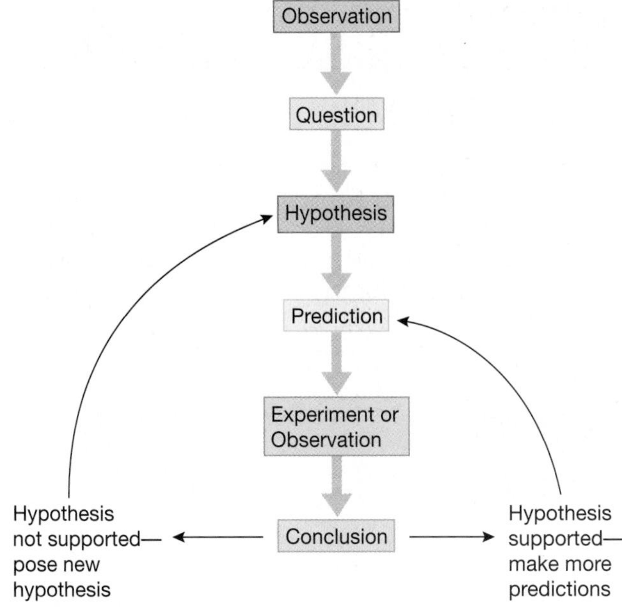

Figure 1-3 The scientific method

The Scientific Method Is Useful in Everyday Life

You may find it easier to visualize the steps of the scientific method through an example from everyday life (Fig. 1-4). Imagine this scenario: Late to an appointment, you rush to your car and make the *observation* that it won't start. This observation leads directly to a *question*: Why won't the car start? You quickly form a *hypothesis*: The battery is dead. Your hypothesis leads in turn to an if-then *prediction*: If the battery is dead, then a new battery will cause the car to start. Next, you design an *experiment* to test your prediction: You replace your battery with the battery from your roommate's new car and try to start your car again. Your car starts immediately, and you reach the *conclusion* that your dead battery hypothesis is correct.

Figure 1-4 The scientific method in everyday life

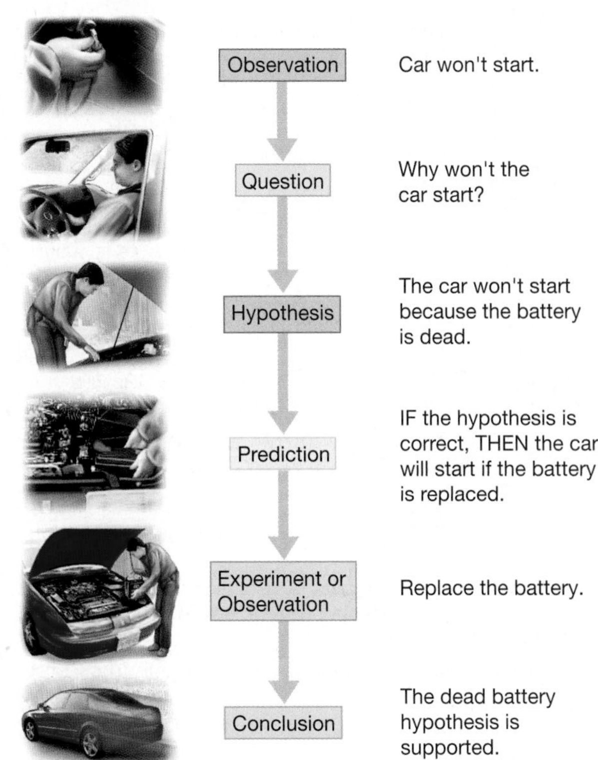

Well-Designed Experiments Incorporate Controls

Simple experiments (like our car battery experiment) test the hypothesis that a single factor, or **variable**, is the cause of an observed phenomenon. The most straightforward test is usually an experiment in which only a single variable is changed. To be scientifically valid, however, the experiment must also rule out *other* possible variables that might have caused the phenomenon. So experiments usually also include **controls**, sections of the experiment in which no variable is changed. The results from the control sections can then be compared with those from the experimental sections. For two classic examples of well-designed experiments, see "Scientific Inquiry: Controlled Experiments, Then and Now."

With the importance of controls in mind, let's revisit our imaginary car battery experiment. Can we really be confident that the old battery was dead? Perhaps the battery was fine all along, and you just needed to try to start the car again. Or perhaps the battery cable was loose and simply needed to be tightened. To control for these other variables, you might re-install your old battery, making sure the cables are secured tightly, and attempt to restart the car. If your car repeatedly refuses to start with the old battery but repeatedly starts immediately with your roommate's new battery, you have isolated a single variable, the battery. And (although you may have missed your appointment) you can now safely draw the conclusion that your old battery is dead.

The scientific method is powerful, but it is important to recognize its limitations. In particular, scientists can seldom be sure that they have controlled *all* the variables other than the one they are trying to study. For this reason, scientific

SCIENTIFIC INQUIRY

CONTROLLED EXPERIMENTS, THEN AND NOW

A classic experiment by the Italian physician Francesco Redi (1621–1697) beautifully demonstrates the scientific method. Redi investigated why maggots appear on spoiled meat. When he began this work, the appearance of maggots was considered to be evidence of *spontaneous generation*, the production of living things from nonliving matter.

Redi *observed* that flies swarm around fresh meat and that maggots appear on meat left out for a few days. He formed a testable *hypothesis*: The flies produce the maggots. In his *experiment*, Redi wanted to test just one variable: the access of flies to the meat. Therefore, he took two clean jars and filled them with similar pieces of meat. He left one jar open (the *control* jar) and covered the other with gauze to keep out flies (the *experimental* jar). He did his best to keep all the other variables the same (for example, the type of jar, the type of meat, and the temperature). After a few days, he observed maggots on the meat in the open jar, but saw none on the meat in the covered jar. Redi *concluded* that his hypothesis was correct and that maggots are produced by flies, not by the nonliving meat (Fig. E1-1). Only through controlled experiments could the age-old hypothesis of spontaneous generation be laid to rest.

More than 300 years after Redi's experiment, today's scientists still use the same approach to design their experiments. Consider the experiment that Malte Andersson designed to investigate the long tails of male widowbirds. Andersson *observed* that male, but not female, widowbirds have extravagantly long tails, which they display while flying across African grasslands (Fig. E1-2). This observation led Andersson to ask the *question*: Why do the males, and only the males, have such long tails? His *hypothesis* was that males have long tails because females prefer to mate with long-tailed males, which therefore have more offspring than shorter-tailed males. From this hypothesis, Andersson *predicted* that if his hypothesis were true, then more females would build nests on the territories of males with artificially lengthened tails than would build nests on the territories of males with artificially shortened tails. He then captured

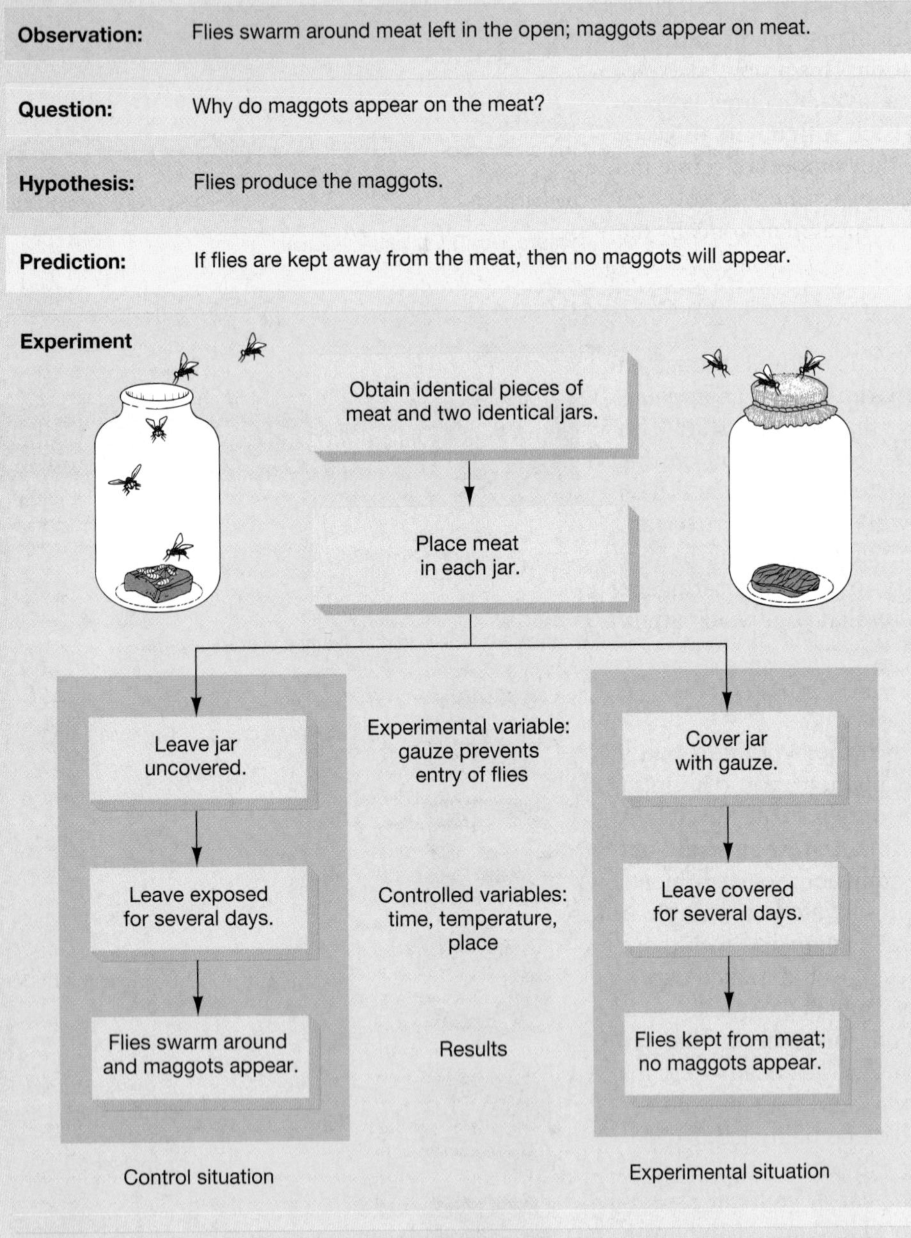

Figure E1-1 The experiments of Francesco Redi QUESTION: *Redi showed that maggots don't appear by spontaneous generation, but did his experiment conclusively demonstrate that flies cause maggots? What kind of follow-up questions might be asked, and what kind of experiment would be necessary if Redi really wanted to determine the source of maggots?*

Figure E1-2 A male widowbird

some males, trimmed their tails to about half their original length, and released them (*experimental* group 1). Another group of males had the tail feathers that had been removed from the first group glued on as tail extensions (*experimental* group 2). Finally, Andersson had two *control* groups. In one, the tail was cut and then glued back in place (to control for the effects of capturing the birds and manipulating their feathers). In the other, the birds were simply captured and released. The experimenter was doing his best to make sure that tail length was the only variable that was changed. After a few days, Andersson counted the number of nests that females had built on each male's territory. He found that males with lengthened tails had the most nests on their territories, males with shortened tails had the fewest, and control males (with normal-length tails) had an intermediate number (Fig. E1-3). Andersson *concluded* that his hypothesis was correct, and that female widowbirds prefer to mate with males that have long tails.

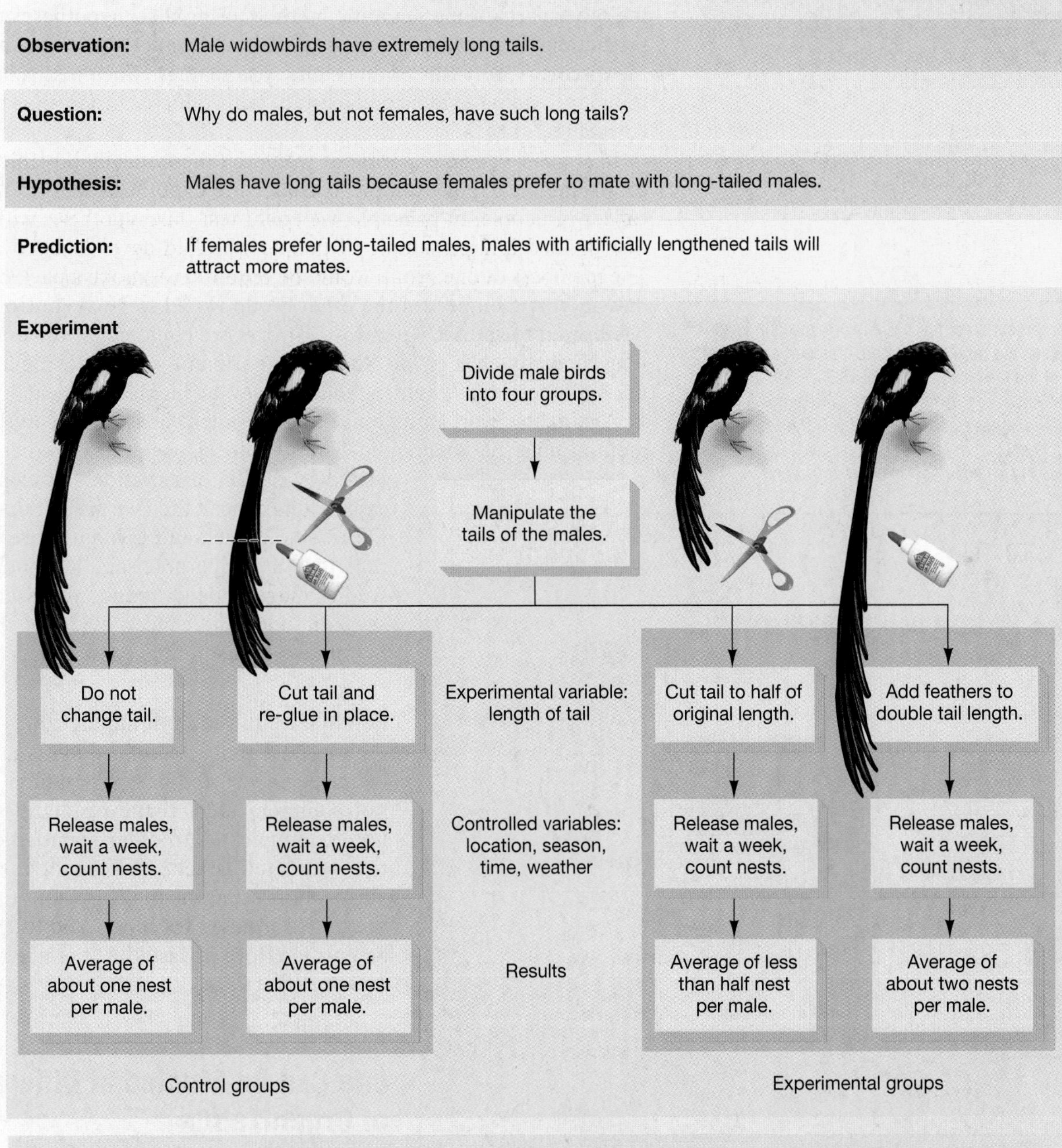

Figure E1-3 The experiments of Malte Andersson

Figure 1-5 A feathered dinosaur Note the impressions of feathers on the forelimbs and tail of this fossil *Caudipteryx* dinosaur.

conclusions remain tentative and are subject to revision if new observations or experiments demand it. In any case, nearly every conclusion immediately raises further questions that lead to further hypotheses and more experiments (why did your battery die?). Science is a never-ending quest for knowledge.

Experiments Are Not Always Possible

A well-designed experiment is usually the most convincing way to test a hypothesis, but biology includes many hypotheses that are not suited to experimental tests. For example, evolutionary biologists often ask questions about events from the historical past. Consider, for example, the hypothesis that the ancestors of today's birds were dinosaurs. These hypothesized ancestors went extinct long ago, of course, and there is no experiment that can demonstrate how they evolved millions of years ago. Nonetheless, the biologists who study this question do use the other parts of the scientific method, using their hypotheses to make testable predictions. For example, if dinosaurs were the ancestors of birds, then we predict the discovery of fossils of dinosaurs with feathers. Such fossils have indeed been found, providing evidence consistent with the conclusion that the hypothesis is correct (Fig. 1-5).

In some cases, an experiment would be theoretically possible but is impractical or unethical. For example, consider the hypothesis that smoking causes lung cancer in people. In principle, we could test this hypothesis with an experiment that divided a large sample of people who had never smoked into two groups. The members of one group would be required to smoke a pack of cigarettes each day, and the members of the other group would serve as controls and would not be allowed to smoke. After, say, 20 years, we could count the number of cases of lung cancer in each group. Such an experiment would provide a powerful test of the hypothesis but would, needless to say, be highly unethical.

Again, however, an inability to experiment does not mean that the scientific method must be abandoned. The hypothesis generates predictions that can be tested by careful observation. For example, if smoking causes lung cancer, then we predict that a random sample of smokers should contain more lung cancer victims than a comparable sample of nonsmokers. Many studies have indeed found an association between smoking and lung cancer, evidence that supports the hypothesis that smoking causes cancer.

Figure 1-6 Levels of organization of life Life is organized in levels of organization, some of which are represented here. Items at each level are the building blocks of the next level. **EXERCISE:** *Think of a scientific question that can be answered by investigating at the cell level, but that would be impossible to answer at the tissue level. Then think of one answerable at the tissue level but not at the cell level. Repeat the process for two other pairs of adjacent levels of organization.*

cells — nerve cell
tissues — nervous tissue
organs — the brain
organisms — pronghorn antelope
populations — herd of pronghorn antelope
communities — snakes, antelopes, hawks, bushes, grass

Science Requires Communication

Finally, note that *communication* is an important part of science. Even the best experiment is useless if it is not communicated. If the results of experiments are not communicated to other scientists in enough detail to be repeated, conclusions cannot be verified. Without verification, scientific findings cannot be safely used as the basis for new hypotheses and further experiments. Scientific effort is wasted if it is not reported in clear and accurate detail.

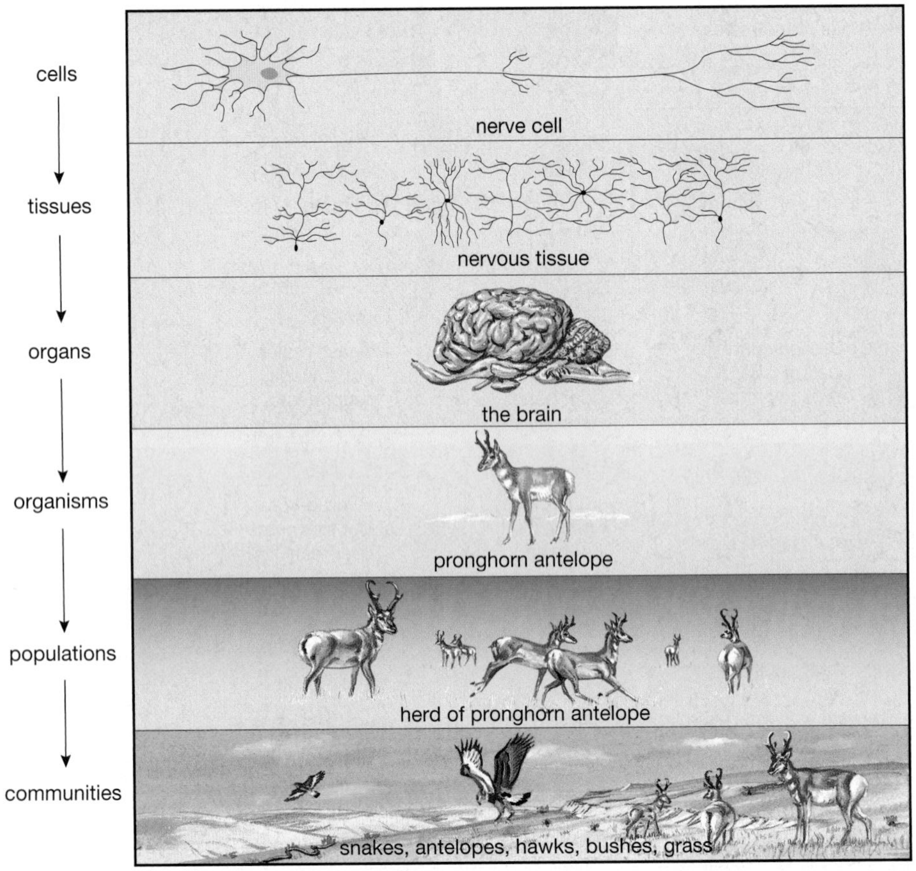

Life Can Be Studied at Different Levels of Organization

As they investigate questions about life, biologists often view the living world as a series of *levels of organization,* with each level providing the building blocks for the next level (Fig. 1-6). For example, the **cell** is the smallest unit of life (Fig. 1-7), but in many life-forms, cells of similar type combine to form structures known as **tissues**. Different tissues, in turn, can combine to form **organs** (for example, a heart or a kidney), and a collection of different organs forms the basis of an **organism**. A collection of organisms of the same type constitutes a **population**, and a collection of different populations makes up a **community**. Note that each level of organization incorporates multiple members of the previous level; a community contains many populations, a population contains many organisms, and so on. (Also note that we could describe many more levels of organization than we have mentioned here.)

Biologists choose to work at an appropriate level of organization. One of the first decisions a biologist must make when designing an experiment is to choose a level of organization at which to study a problem. This decision is ordinarily based on the question to be answered. For example, if you wanted to know how frogs make croaking sounds, you would study frog organs (structures within the frog body). The question of how frogs croak would be impossible to answer if you focused on frog cells or frog communities. On the other hand, if you want to know whether global warming is reducing the number of frogs in the world, it would do you no good to study frog organs. To answer that question, you would have to study frog populations. It is important for scientists to recognize and choose the level of organization that is most appropriate to the question at hand.

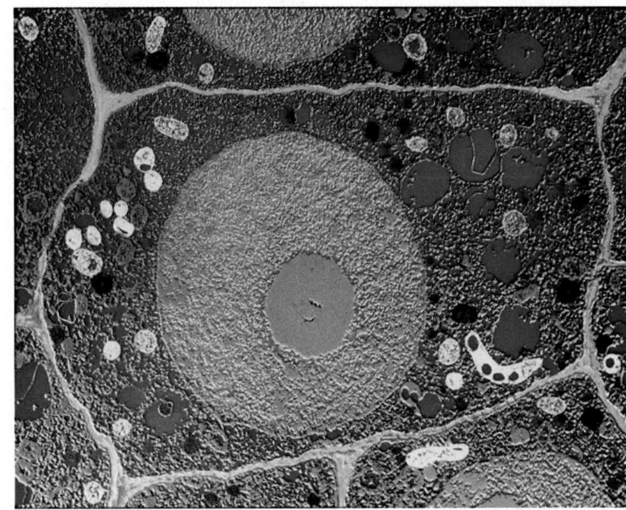

Figure 1-7 The smallest unit of life The boundary of this plant cell is formed by a cell wall, which is green in this colorized micrograph.

Science Is a Human Endeavor

Scientists are real people. They are driven by the same ambitions, pride, and fears as other people, and they sometimes make mistakes. As you will read in Chapter 8, ambition played an important role in the discovery by James Watson and Francis Crick of the structure of DNA. Accidents, lucky guesses, controversies with competing scientists, and the intellectual powers of individual scientists contribute greatly to scientific advances. To illustrate what we might call "real science," let's consider an actual case.

To study bacteria, microbiologists must use pure *cultures*—that is, plates containing colonies of a single type of bacteria free from contamination by other bacteria, molds, and so on. Only by studying a single type at a time can we learn about that particular bacterium. Consequently, at the first sign of contamination, a culture is normally thrown out. On one occasion, however, Scottish bacteriologist Alexander Fleming turned a ruined culture into one of the greatest medical advances in history.

In the 1920s, one of Fleming's bacterial cultures became contaminated with a patch of a mold called *Penicillium*. As he was about to throw out the culture dish, Fleming observed that no bacteria were growing near the mold. Why not? Fleming hypothesized that perhaps *Penicillium* had released a substance that killed off the bacteria growing nearby. To test this hypothesis, Fleming grew some pure *Penicillium* in a liquid nutrient broth. He then filtered out the *Penicillium* mold and applied the liquid in which the mold had grown to an uncontaminated bacterial culture. Sure enough, something in the liquid killed the bacteria. Further research into these mold extracts resulted in the production of the first *antibiotic*—penicillin, a bacteria-killing substance that has since saved millions of lives (**Fig. 1-8**).

Fleming used the scientific method. His experiment began with an observation and proceeded to a question, a hypothesis, and a prediction, followed by experimental tests, which led to a conclusion. But the scientific method alone would have been useless without the lucky combination of accident and a brilliant scientific mind. Had Fleming been a "perfect" microbiologist, he wouldn't have had any contaminated cultures. Had he been less observant and less curious, he would have thrown out the spoiled culture dish. Instead, he turned the contaminated dish into the beginning of antibiotic therapy for bacterial diseases. As French microbiologist Louis Pasteur said, "Chance favors the prepared mind."

Figure 1-8 Penicillin kills bacteria **QUESTION:** *Why, do you think, do some molds produce substances that are toxic to bacteria?*

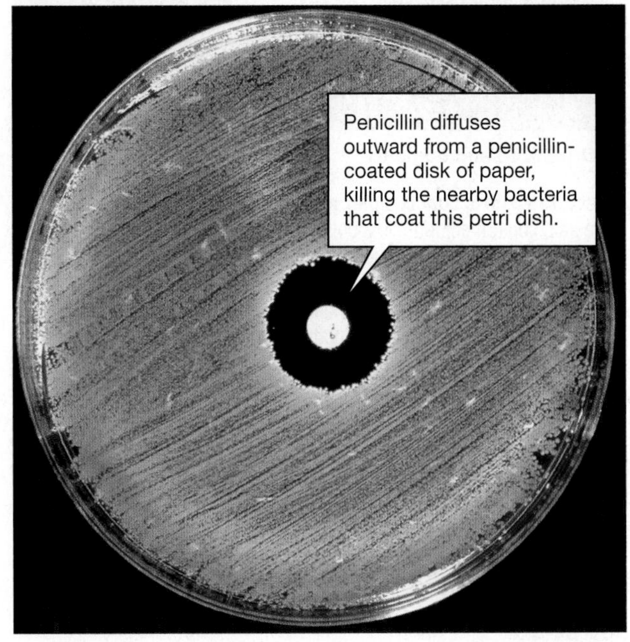

(a) Organized

(b) Complex

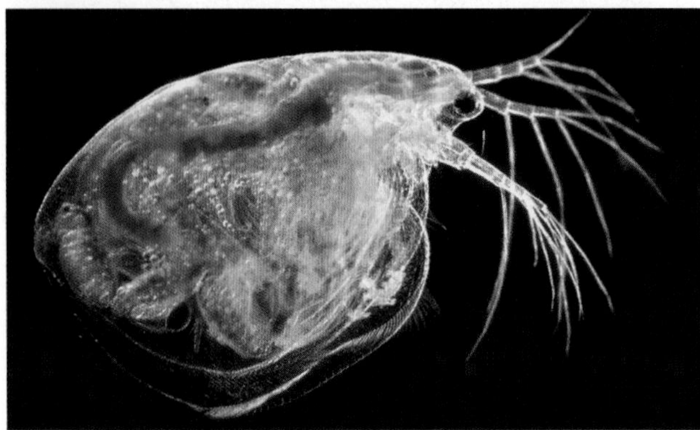

(c) Organized and complex

Figure 1-9 Life is both complex and organized ***(a)*** Each crystal of table salt, sodium chloride, is a cube, showing great organization but minimal complexity. ***(b)*** The water and dissolved materials in the ocean have complexity but very little organization. ***(c)*** The waterflea is only 1 millimeter long (1/1000 meter; smaller than the letter *i*), yet it has legs, a mouth, a digestive tract, reproductive organs, light-sensing eyes, and even a rather impressive brain in relation to its size.

Scientific Theories Have Been Thoroughly Tested

Scientists use the word "theory" in a way that is different from its everyday usage. If Dr. Watson were to ask Sherlock Holmes, "Do you have a theory as to the perpetrator of this foul deed?" he would be asking for what a scientist would call a hypothesis—an educated guess based on observable evidence, or clues. A **scientific theory** is far more general and more reliable than a hypothesis. Far from being an educated guess, a scientific theory is a general explanation of important natural phenomena, developed through extensive and reproducible tests. For example, the theory of gravity, which states that objects exert attraction for one another, is fundamental to the science of physics. Similarly, the *cell theory*, which states that all living things are composed of cells, is fundamental to the study of biology.

Perhaps the most important theory in biology is the *theory of evolution*. Since its formulation in the mid-1800s by Charles Darwin and Alfred Russel Wallace, the theory of evolution has been supported by an overwhelming accumulation of evidence, including fossil finds, geological studies, radioactive dating of rocks, breeding experiments, and research results in genetics, molecular biology, and biochemistry. People who say that evolution is "just a theory" profoundly misunderstand what scientists mean by the word "theory."

1.3 What Is Life?

What is life? This short, simple question does not have a short, simple answer. Although each of us has an intuitive understanding of what it means to say that something is alive, that intuition cannot be easily translated to a precise definition. Life is so diverse and complex that it has proved impossible to devise a definition that neatly divides the living from the nonliving. Standard dictionary definitions are of little help, because they typically use phrases such as "the quality that distinguishes living organisms from dead organisms," without providing much insight as to what that mysterious "quality" might be.

Because we cannot define life precisely, we must instead build our definition bit by bit, by describing a series of different features of living things. In fact, this entire textbook is really an extended attempt to define life. As you read the book and attend your biology classes, you will learn about many different aspects of the living world. Our hope is that, as you proceed, a picture of life will emerge, in much the same way that a painted image gradually takes shape from the patches of color an artist applies to a canvas. We begin with a brief discussion of some properties that are shared by living things. Taken together, these properties form a combination of characteristics that is not found in nonliving objects.

Living Things Are Both Complex and Organized

Compared with nonliving matter of similar size, living things are highly complex and organized. A nonliving crystal of table salt consists of just two chemical elements, sodium and chlorine, arranged in a precise way; the salt crystal is *organized* but simple (Fig. 1-9a). The nonliving water of an ocean contains atoms of all the naturally occurring elements, but these atoms are randomly distributed; the oceans are *complex* but not organized (Fig. 1-9b). In contrast, even the simplest organisms contain dozens of different elements linked together in thousands of specific combinations to form cells. Cells are both complex *and* organized, and every living thing consists of at least one cell. In many organisms, cells are further organized into larger and more complex assemblies such as eyes, legs, digestive tracts, and brains (Fig. 1-9c).

Figure 1-10 Living things reproduce As they grow, these polar bear cubs will resemble, but not be identical to, their parents. The similarity and variability of offspring are crucial to the evolution of life.

Living Things Grow and Reproduce

At some time in its life cycle, every organism becomes larger—that is, it *grows*. This characteristic is most obvious in plants and animals, which tend to start out very small and grow tremendously during their lives. Even single-celled bacteria, however, grow to about double their original size before they divide.

Organisms are also able to reproduce, giving rise to offspring of the same type (Fig. 1-10). Methods for producing offspring vary quite a bit among different kinds of organisms, but the result is always the same—the production of new individuals.

Living Things Respond to Stimuli

Organisms detect and respond to stimuli in their environments. For example, plants grow toward a source of light. Some kinds of bacteria can move to get away from a poisonous chemical. Animals have sensory organs and muscular systems that allow them to detect and respond to light, sound, chemicals, and many other stimuli in their surroundings. Animals can also respond to stimuli inside their bodies. For example, when you feel hungry, you sense the contractions of your empty stomach and the low levels of sugars and fats in your blood. You then respond by finding some food and eating.

Living Things Acquire and Use Materials and Energy

To grow, reproduce, and maintain their high level of organized complexity, organisms must obtain materials and energy from the environment (Fig. 1-11). The materials and energy power an organism's **metabolism**, the sum total of all the chemical reactions needed to sustain life.

Organisms obtain **energy**—the ability to do work, such as carrying out chemical reactions, growing leaves in the spring, or contracting a muscle—in one of two ways. Plants and some single-celled organisms capture the energy of sunlight and store it in sugar molecules, a process called **photosynthesis**. Most remaining organisms, which are not capable of photosynthesis, must consume energy-rich molecules contained in the bodies of other organisms.

Figure 1-11 Living things acquire energy and materials from the environment The green plants seen here capture energy from the sun and materials from the air, water, and soil. The insect gains both energy and materials from the plants, while the toad extracts energy and materials from its insect prey.

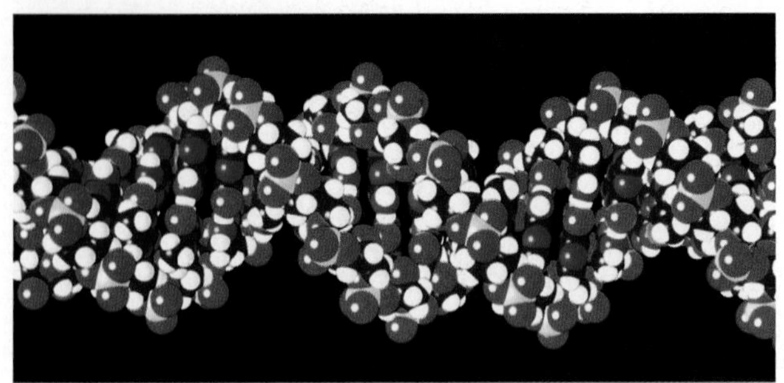

Figure 1-12 DNA A computer-generated model of DNA, the molecule of heredity.

Living Things Use DNA to Store Information

All known forms of life use a molecule called **deoxyribonucleic acid**, or **DNA**, to store information (Fig. 1-12). Each organism's DNA acts as an instruction manual, a blueprint that guides the construction and operation of the organism's body. Each segment of DNA that contains a set of instructions is a **gene**, and when an organism reproduces, it passes a copy of its DNA to its offspring. This transmission of genes to offspring is the basis of heredity, the means by which offspring inherit the characteristics of parents.

1.4 Why Is Life So Diverse?

One of the most striking characteristics of living things is that they all contain DNA. The DNA molecule is a rather complex structure, consisting of a distinctive arrangement of a specific set of parts. How could this particular unique structure have ended up in the bodies of every living thing from the smallest bacterium to the largest whale? The answer to this question lies in the common ancestry of all life. That is, every organism has descended from the same ancestor, and that common ancestor used DNA to store information. DNA has been passed down from generation to generation (remember that when an organism reproduces, it passes a copy of its DNA to its offspring). Every organism living today carries the legacy of a shared ancient heritage.

If modern organisms descended from earlier organisms, and if all life therefore shares such fundamental qualities as the presence of DNA, why are living things so diverse? If an oak tree is related to a butterfly, why are they so different?

Evolution Accounts for Both Life's Unity and Its Diversity

Over long periods of time, organisms change. This kind of change is not the kind that occurs within an individual during its lifetime, the kind that makes you look different now than you did when you were a small child. Instead, we're talking about the change that takes place within groups of organisms, from generation to generation. Each generation is slightly different from the one before it, so, after many generations, a typical member of the group may bear little resemblance to the typical group members of earlier times. This kind of accumulating generation-to-generation change within a group is known as **evolution**.

Evolutionary change has led to the amazing variety of life-forms on Earth. At the same time, evolution accounts for the presence of features that demonstrate life's unity, because evolutionary change has taken place along the branches of a single, gigantic "family tree" that formed as ancestors gave rise to new groups of modified descendants. All the biological structures, mechanisms, systems, and interactions that we will discuss in this book arose through evolution. The idea that evolution gave rise to all of life's key features and characteristics is the guiding principle that unifies the study of biology.

Natural Selection Causes Evolution

The most important process by which evolution occurs is **natural selection**. Natural selection occurs because the characteristics of the different individuals in a group vary, and some individuals possess characteristics that help them survive and reproduce more successfully than do others that lack those traits. The individuals with these favorable traits tend to have a greater number of offspring, and

bio ETHICS **EARTH WATCH**

WHY PRESERVE BIODIVERSITY?

The loss of species is the folly our descendants are least likely to forgive us.

E. O. Wilson, Professor, Harvard University

Scientists have estimated that there are 8 million to 10 million species on Earth today. The vast majority of these species live in the Tropics. Unfortunately, tropical habitats are being rapidly destroyed and disrupted by human activities. For example, a recent analysis of thousands of satellite photos concluded that worldwide tropical rainforest cover has decreased by about 23,000 square miles per year for the past decade. Most of lost forest had been destroyed by logging or to clear land for agriculture. Similarly, a worldwide survey of coral reefs revealed that about 30% of Earth's reef area is severely damaged, again mostly as the result of human influences such as pollution.

The rapid destruction of habitats in the Tropics is causing many species to go extinct, as their homes disappear (Fig. E1-4). The biologist E. O. Wilson has calculated that current rates of habitat loss result in the extinction each year of as many as 25,000 species, most of them disappearing before they are even discovered and named. If Wilson's estimate is reasonably accurate, the current pace of extinctions is massively high, and the Earth of the future will have many fewer species than are currently present. Does it matter? Is there any reason for us to try to slow the loss of biodiversity?

One possible reason to protect Earth's biodiversity is an ethical one. As the only species on the planet with the power to destroy other species, perhaps we have an ethical obligation to protect them from extinction. But even if you do not agree with this ethical argument, there may be compelling practical reasons to save other species from destruction. Our ecological self-interest may be at stake.

For example, Earth's species form communities, highly complex webs of interdependent life-forms whose interactions sustain one another. These communities play a crucial role in processes that purify the air we breathe and the water we drink, build the rich topsoil in which we grow our crops, provide the bounty of food that we harvest from the oceans, and decompose and detoxify our waste. We depend entirely on these "ecosystem services." When our activities cause species to disappear from communities, we take a big risk. If we remove too many species, or remove some especially crucial species, we may disrupt the finely tuned processes of the community and undermine its ability to sustain us.

Figure E1-4 Biodiversity threatened Destruction of tropical rain forests by indiscriminant logging threatens Earth's greatest storehouse of biological diversity.

those offspring tend to inherit the favorable traits from their parents. Those traits thus become more common in the group.

For example, consider how natural selection might have influenced the evolution of beaver teeth. Beavers with larger teeth might have been able to chew down trees more efficiently, build bigger dams and lodges, and eat more bark than "ordinary" beavers. Because these big-toothed beavers obtained more food and better shelter than their smaller-toothed relatives, they probably raised more offspring. The offspring inherited their parents' genes for larger teeth. Over time, less-successful, smaller-toothed beavers became increasingly scarce. After many generations, all beavers had large teeth.

Life's Diversity Is Currently Threatened

One outcome of evolution is that Earth is now home to a tremendous variety of **species**, or types of organisms. In many places, these species have evolved complex interrelationships with one another and with their surroundings. The word **biodiversity** is often used to describe this wealth of species and the complex interrelationships that sustain them. In recent decades, a single species, *Homo sapiens* (modern humans), has drastically increased the rate at which the environment is changing. Unfortunately, many other species have been unable to cope with this rapid change. In the habitats most affected by humans, many species are being driven to extinction. This problem is further explored in "Earth Watch: Why Preserve Biodiversity?"

LINKS TO LIFE: The Life Around Us

The next time you walk across campus, take a moment to observe the astonishing array of creatures thriving in a place as domesticated as a college campus. As you walk, observing life, think about the "why" behind what you see. For example, one scene you may encounter is a flowerbed with honeybees and butterflies flitting among the flowers, gathering the sweet nectar that powers their flight. There's a lot of biology in this scene.

The plants' green color is due to a unique molecule, chlorophyll, that traps specific wavelengths of solar energy and uses them to power the life of the plant and to synthesize the sugar in the nectar gathered by the bees and butterflies. Showy flowers evolved to entice insects to the energy-rich nectar. Why? If you look carefully at a bee, you may see yellow pollen clinging to its legs or to the hairs coating its body. The plants "use" the insects to fertilize each other, and both plants and insects benefit.

Inside the flowers, the sugar in nectar is assembled by chemical reactions that combine carbon dioxide and water, releasing oxygen as a waste product. So as you breathe out air rich in carbon dioxide, you are nourishing the plants with your "waste gas." Conversely, with each breath you take, you are inhaling the life-sustaining "waste gas" from the plants around you: oxygen. Wherever you look, if you look in the right way, you'll see evidence of the interdependence of living things, and you will never take life on Earth for granted.

Chapter Review

Summary of Key Concepts

1.1 Why Study Biology?

Practical benefits include greater understanding of health and disease, new career options, and more informed opinions on environmental and bioethical controversies. Most important, the more you know about living things, the more fascinating they become.

1.2 How Do Biologists Study Life?

Biology is the science of life. Knowledge in biology is acquired through the scientific method. First, an observation is made, which leads to a question. Then a hypothesis is formulated that suggests a possible answer to the question. The hypothesis is used to predict the outcome of further observations or experiments. A conclusion is then drawn about the hypothesis. Conclusions are based only on results that can be shared, verified, and repeated. A scientific theory is a general explanation of natural phenomena, developed through extensive and reproducible experiments and observations.

1.3 What Is Life?

Organisms possess the following characteristics: Their structure is complex and organized; they have cells; they grow; they reproduce; they acquire energy and materials from the environment; they respond to stimuli; and they use DNA to store information.

1.4 Why Is Life So Diverse?

Evolution by natural selection has brought about both life's unity (shared features) and its diversity (a huge number of incredibly different forms). Evolution is the theory that modern organisms descended, with modification, from preexisting life-forms. Species evolve as a consequence of (1) variation among members of a population, (2) inheritance of those variations by offspring, and (3) natural selection of the variations that best help an organism to survive and reproduce in its environment.

Key Terms

biodiversity *p. 11*
cell *p. 6*
community *p. 7*
conclusion *p. 3*
control *p. 3*
deoxyribonucleic acid (DNA) *p. 10*
energy *p. 9*
evolution *p. 10*
experiment *p. 3*
gene *p. 10*
hypothesis *p. 3*
metabolism *p. 9*
natural selection *p. 10*
observation *p. 3*
organ *p. 7*
organism *p. 7*
photosynthesis *p. 9*
population *p. 7*
prediction *p. 3*
question *p. 3*
scientific method *p. 3*
scientific theory *p. 8*
species *p. 11*
tissue *p. 7*
variable *p. 3*

Thinking Through the Concepts

Multiple Choice

1. *Which of the following is paired incorrectly?*
   a. organ—a structure formed of tissues
   b. cell—the smallest unit of life
   c. genes—units of heredity
   d. population—a group of tissues of the same type
   e. tissue—a structure formed of cells
2. *Fleming's discovery of penicillin demonstrated that*
   a. a methodical approach is the best way to get new ideas
   b. ambition motivates the best scientists
   c. accidents and chance occurrences can be important in science
   d. a scientific theory is similar to an educated guess
   e. molds generally kill bacteria
3. *Choose the answer that best describes the scientific method.*
   a. observation, hypothesis, experiment, absolute proof
   b. guess, hypothesis, experiment, conclusion
   c. observation, hypothesis, experiment, conclusion
   d. hypothesis, experiment, observation, conclusion
   e. experiment, observation, hypothesis, conclusion
4. *Which of the following are characteristics of living things?*
   a. They reproduce.
   b. They respond to stimuli.
   c. They are complex and organized.
   d. They acquire energy.
   e. all of the above
5. *All organisms contain DNA because*
   a. organisms with favorable traits leave more offspring
   b. the characteristics of individuals within a population vary
   c. DNA is abundant in the environment and all organisms absorb it
   d. DNA is a simple molecule that has formed by chance in many different organisms
   e. all life has descended from a common ancestor

Review Questions

1. Describe the scientific method. In what ways do you use the scientific method in everyday life?
2. Starting with the cell, list the levels of organization of life, briefly explaining each level.
3. What is the difference between a scientific theory and a hypothesis? Explain how each is used by scientists.
4. What are the differences between a salt crystal and a tree? Which is living? How do you know? How would you test your knowledge? What controls would you use?
5. What is evolution? Briefly describe how evolution occurs.

Applying the Concepts

1. Design an experiment to test the effects of a new dog food, "Super Dog," on the thickness and water-shedding properties of the coats of golden retrievers. Include all the parts of a scientific experiment. Design objective methods to assess coat thickness and water-shedding ability.
2. Think of two different types of organisms that you have seen interacting, for example, a caterpillar on a milkweed plant or a beetle in a flower. What questions do you have about this interaction? Choose one of your questions and devise a single, simple hypothesis about its answer. Use the scientific method and your imagination to design an experiment that tests this hypothesis. Be sure to identify variables and control for them.

For More Information

Attenborough, D. *Life on Earth*. Boston: Little, Brown, 1979. Gorgeously illustrated and beautifully written introduction to the diversity of life on Earth; the inspiration for our title.

Dawkins, R. *The Blind Watchmaker*. New York: W. W. Norton, 1986. An engagingly written description of the process of evolution, which Dawkins compares to a blind watchmaker.

Ehrlich, P. R. *The Machinery of Nature*. New York: Simon & Schuster, 1986. Using layperson's terms, an eminent ecologist and author explains the science of ecology and the biological rationale for environmental concern.

Leopold, A. *A Sand County Almanac*. New York: Oxford University Press, 1949 (reprinted in 1989). A classic by a natural philosopher; provides an eloquent foundation for the conservation ethic.

Swain, R. B. *Earthly Pleasures*. New York: Charles Scribner's Sons, 1981. Insightful essays stress the interrelatedness and diversity of life.

Thomas, L. *The Medusa and the Snail*. New York: Bantam Books, 1980. The late physician, researcher, and philosopher Lewis Thomas shares his awe of the living world in a series of delightful essays.

Wilson, E. O. *The Diversity of Life*. New York: W. W. Norton, 1992. A celebration of the diversity of life, how that diversity evolved, and how humans are affecting it.

WEB TUTORIAL

To access a Web Tutorial visit http://www.prenhall.com/audesirk4. *Log in to the Web site selected by your instructor, navigate to this chapter, and select the appropriate Web Activity number.*

1.1 Experimental Design

Estimated time: 10 minutes

This activity will introduce you to the scientific method, which is at the center of all scientific studies.

1.2 What Is Life?

Estimated time: 15 minutes

Explore the characteristics of life and the way we use the scientific method to ask questions about life.

1.3 Web Investigation: The Life Around Us

Estimated time: 10 minutes

The Internet can be a wonderful source of information for the novice scientist. Unfortunately, not all of the information on the World Wide Web is correct. In this exercise you will learn how to evaluate a Web site.

UNIT 1

The Life of a Cell

Single cells can be complex, independent organisms, such as these two protists. A large *Euplotes* prepares to eat a much smaller *Paramecium*.

Chapter 2

Atoms, Molecules, and Life

Many of today's most popular foods contain artificially modified molecules. Olestra, a new molecule that mimics fat, has proved to be an especially controversial addition to food.

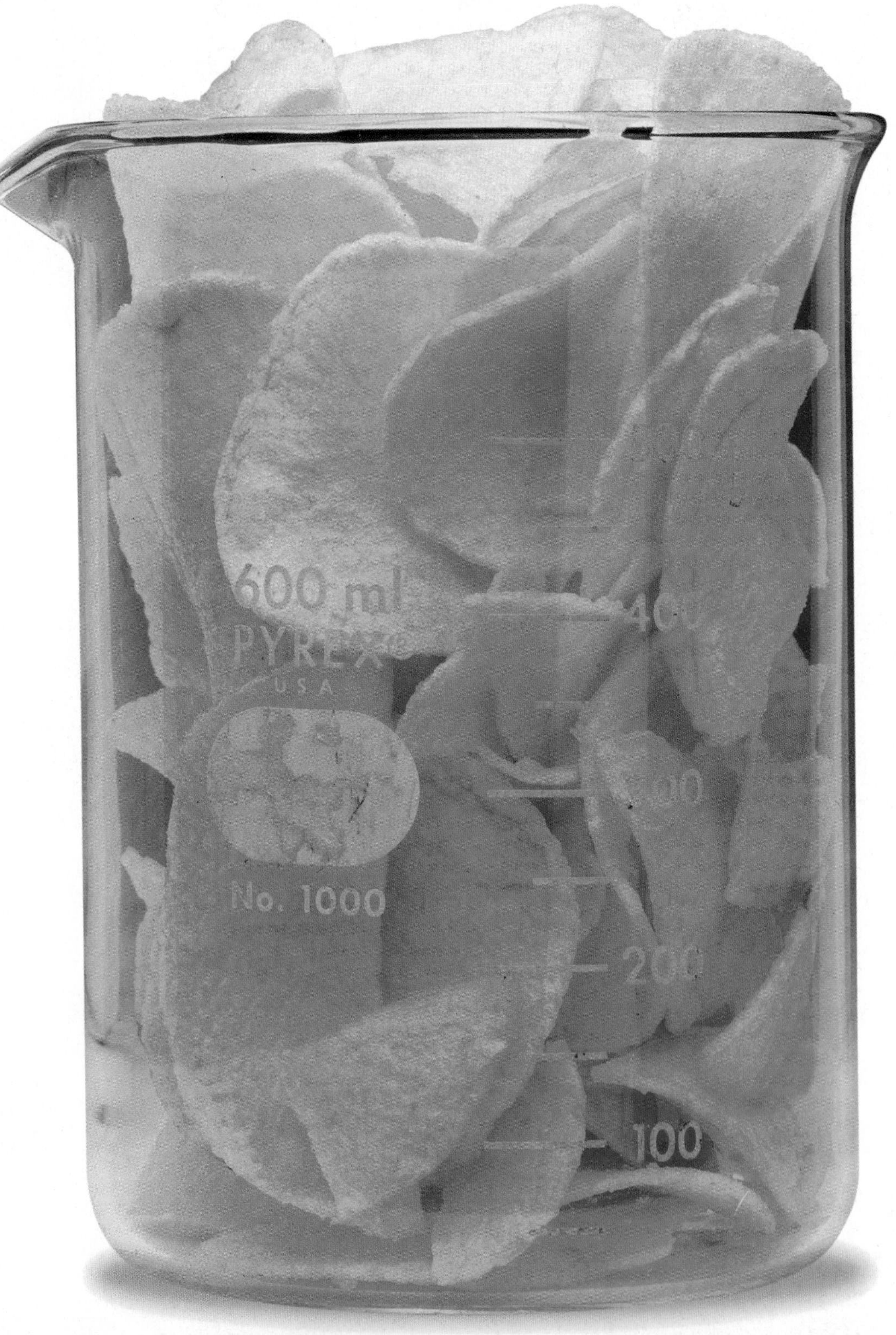

AT A GLANCE

Improving on Nature?

Obesity is a major public health problem, and today's consumers demand foods that help them lose weight. One response to this demand is the artificial fat known as olestra. Like real fat, olestra adds a pleasing texture and taste to foods. Our bodies, however, are not able to extract the energy stored in olestra's molecules. Olestra is completely indigestible and passes through the body unchanged. This indigestibility ensures that potato chips made with olestra have far fewer calories than normal chips.

Olestra could be a valuable ally in the battle against obesity, but is it safe to eat? Consumer advocates have raised concerns about its safety. For example, during test marketing, one organization reported receiving more than 1000 consumer complaints of digestive distress after eating foods made with olestra.

Olestra's inventors protested that counting complaints was not a scientifically rigorous method for testing olestra's safety. After all, can we really assume that callers' illnesses had actually been caused by eating olestra, and not by some other cause? A more rigorous approach was neccesary to test the hypothesis that olestra causes digestive distress.

In one such test, a group of researchers gathered more than 1000 volunteers. Each volunteer received a bag of potato chips to eat while watching a movie. Half the subjects received chips made with olestra; the other half got chips made with normal vegetable oil. Subjects were not told whether their chips contained olestra. Two days after the movie, each subject was questioned about his or her digestive symptoms. The result: There was no significant difference between the two groups; 15.6% of olestra-chip eaters and 17.6% of regular-chip eaters experienced digestive distress in the 48 hours following the movie. The researchers concluded that olestra did not cause illness.

Can fake foods help treat the obesity epidemic? Or are they unsafe to eat?

Skeptics were not persuaded by the movie theater study and called for more research. We will describe some of this additonal research at the close of this chapter, after an introduction to the chemistry of biological molecules.

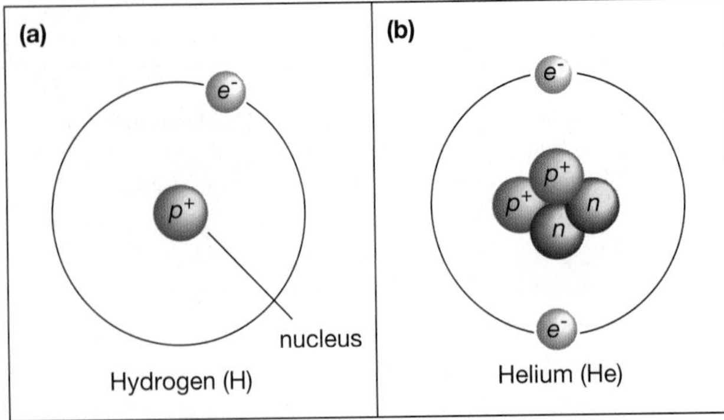

Figure 2-1 Atomic models The two smallest atoms, hydrogen and helium. In these simplified models, the electrons are represented as miniature planets, circling around a nucleus. The nucleus of hydrogen consists of a proton. The nucleus of helium consists of two protons and two neutrons.

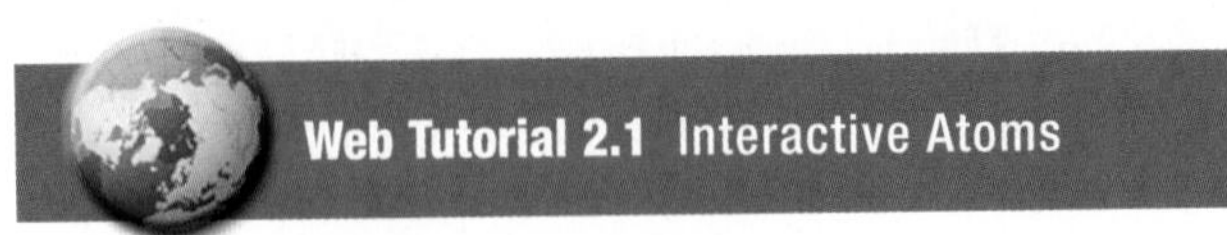

Web Tutorial 2.1 Interactive Atoms

2.1 What Are Atoms?

An **element** is a substance that can neither be broken down nor converted to other substances by ordinary chemical means. For example, carbon is an element, so if you took a diamond (a form of carbon) and cut it into pieces, each piece would still be carbon. If you could make finer and finer divisions, you would eventually produce a pile of carbon **atoms**. Atoms are the basic structural units of matter.

Atoms Are Composed of Even Smaller Particles

Atoms are made up of even smaller units called subatomic particles. Physicists have discovered dozens of different subatomic particles, but we will discuss only three important types: protons, neutrons, and electrons. In each atom, positively charged **protons** and uncharged **neutrons** together form a central **atomic nucleus**. **Electrons**, which are lighter than neutrons and protons and are negatively charged, orbit the atomic nucleus (Fig. 2-1). An atom by itself has an equal number of electrons and protons and is therefore electrically neutral.

There are 92 types of atoms and, therefore, 92 elements in nature. Each type of atom is the basic unit of a different element. The number of protons in the nucleus, called the **atomic number**, is unique for each element. For example, a hydrogen atom has one proton in its nucleus, a carbon atom has six protons, and an oxygen atom has eight. Each element also has other chemical properties. For example, some elements, such as oxygen and hydrogen, are gases at room temperature; others, such as lead and iron, are solids.

Atoms of the same element may have different numbers of neutrons; when they do, the differing atoms are called **isotopes** of the element. Some, but not all, isotopes are **radioactive**; that is, they spontaneously break apart, forming different types of atoms and releasing energy in the process. For example, radioactive isotopes of uranium decay to form lead. Some scientists use radioactive isotopes to determine the age of fossils (see "Scientific Inquiry: How Do We Know How Old a Fossil Is?" on page 264).

Electrons Orbit the Nucleus, Forming Electron Shells

As you may know from experimenting with a magnet, like poles repel each other and opposite poles attract each other. In a similar way, negatively charged electrons repel one another but are drawn to the positively charged protons of the nucleus. This attraction holds an atom together as its electrons orbit around the nucleus. Because of their mutual repulsion, only a limited number of electrons can be at the same distance from the nucleus. Electrons orbit at only a few different distances from the nucleus, and these different distances are called **electron shells** (Fig. 2-2). Electrons in different shells have different amounts of energy. Electrons in the innermost shell have the lowest energy, electrons in the second shell have more energy, and so on through additional shells. In some circumstances, electrons can move from one shell to another, with corresponding changes in energy level. An electron must absorb energy to move from an inner to an outer shell, and an electron moving in the opposite direction would release energy.

The electrons in an atom normally first fill the shell closest to the nucleus and then begin to occupy the next shell. The electron shell closest to the atomic nucleus can hold only two electrons. The second shell can hold up to eight electrons. Thus, a carbon atom, with six electrons, has two electrons in the first shell, closest to the nucleus, and four electrons in its second shell.

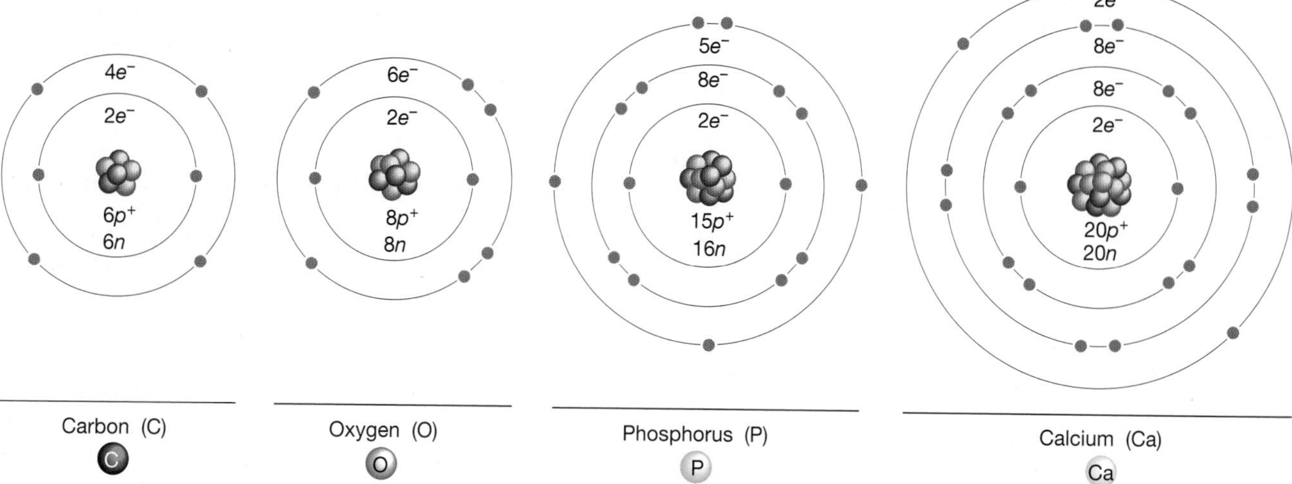

Figure 2-2 Electron shells in atoms Most biologically important atoms have at least two shells of electrons. The first shell, closest to the nucleus, can hold only two electrons, and the next shell can hold a maximum of eight electrons. More distant shells can also hold eight electrons each. **QUESTION:** *Why do biologically active atoms tend to be ones whose outer shells are not full?*

2.2 How Do Atoms Form Molecules?

A **molecule** consists of two or more atoms, which can be of the same or different elements. The atoms are held together by interactions among their outermost electron shells.

Atoms Interact When There Are Vacancies in Their Outermost Electron Shells

Atoms interact with one another according to two basic principles:

- An atom will not react with other atoms when its outermost electron shell is completely full or empty. Such an atom is described as being *inert*.
- An atom will react with other atoms when its outermost electron shell is only partially full. Such atoms are described as *reactive*.

To demonstrate these principles, consider three atoms: helium, hydrogen, and oxygen (see Figs. 2-1 and 2-2). Helium has two protons in its nucleus, and two electrons fill its single electron shell. With its full outer shell, helium is inert. In contrast, hydrogen and oxygen have partially empty outer shells. Hydrogen has one proton in its nucleus and one electron in its single (and therefore outermost) electron shell, which can hold up to two electrons. The oxygen atom has six electrons in its outer shell, which can hold eight. Therefore, we might predict that hydrogen and oxygen atoms will be reactive. We might further predict that oxygen atoms could fill their outer shells by reacting with hydrogen atoms. And, in fact, oxygen reacts readily with hydrogen to form water (H_2O; the 2 subscript after the H indicates that the molecule contains two atoms of hydrogen). In this reaction, the outer shell of an oxygen atom is filled by the electrons from two hydrogen atoms. The resulting water molecule is comparatively unlikely to undergo further reactions. In other words, the water molecule is more stable than the hydrogen and oxygen atoms that gave rise to it.

An atom with an outermost electron shell that is partially full can gain stability in three ways. It can lose electrons (to empty the shell), gain electrons (to fill the shell), or share electrons with another atom (with both atoms behaving as though they had full outer shells, as in the case of water). The results of losing, gaining, and sharing electrons are **chemical bonds**, attractive forces that hold atoms together in molecules. Each element has chemical bonding properties that arise from the arrangement of electrons in its outer shell.

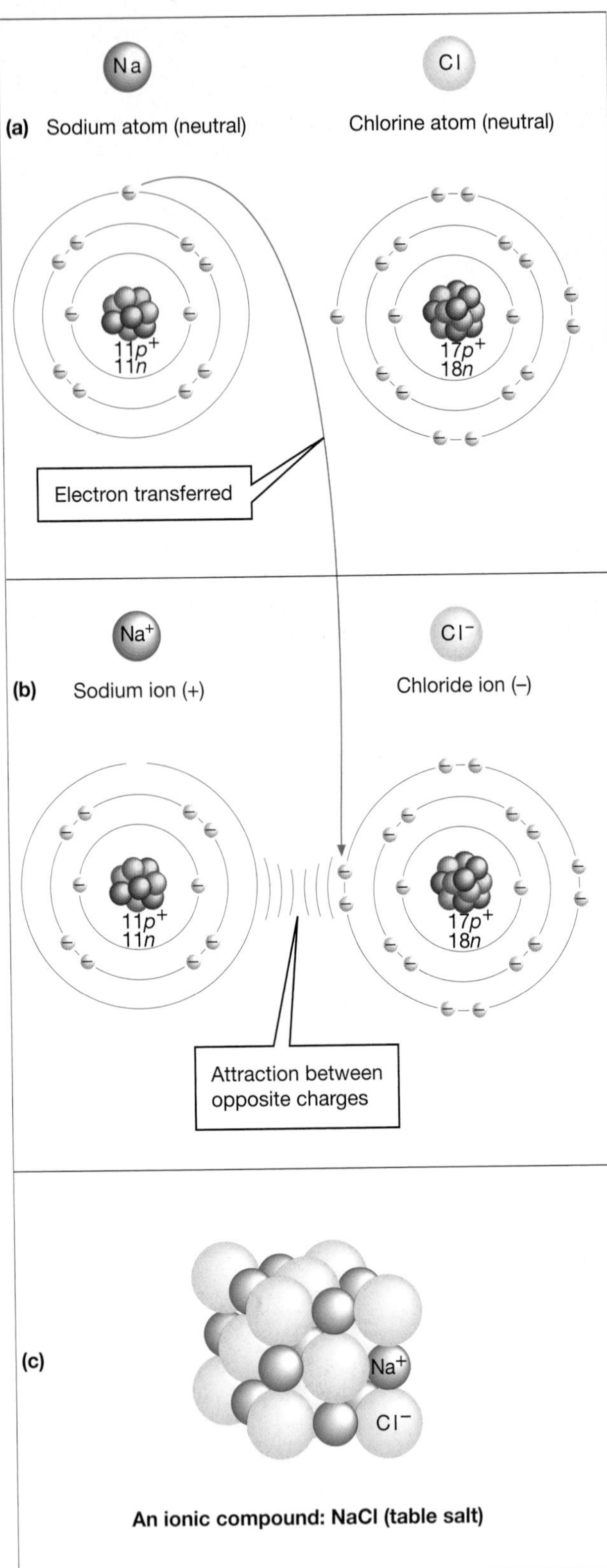

Figure 2-3 The formation of ions and ionic bonds

Chemical reactions make and break chemical bonds to form new substances. Such reactions are essential for the maintenance of life. Whether chemical reactions occur in a plant cell as it captures solar energy, in your brain as it forms new memories, or in your car's engine as it guzzles gas, the reactions make new chemical bonds, or break existing ones, or both.

Charged Atoms Interact to Form Ionic Bonds

The stability that atoms gain by emptying or filling their outermost shells is demonstrated by the formation of table salt (sodium chloride). Sodium (Na) has only one electron in its outermost electron shell, and chlorine (Cl) has seven electrons in its outer shell—one electron short of being full (Fig. 2-3a). Sodium, therefore, can become stable by losing an electron to chlorine (leaving its outer shell empty), and chlorine can fill its outer shell by gaining that electron. Once sodium loses an electron, the protons in the atom outnumber the electrons, so the sodium atom becomes positively charged (Na^+). Similarly, when chlorine picks up an electron, it becomes negatively charged (Cl^-). Positively or negatively charged atoms are called **ions** (Fig. 2-3b).

Opposite charges attract, so sodium ions and chloride ions tend to stay near one another. They form crystals that contain repeating orderly arrangements of the two ions (Fig. 2-3c). The electrical attraction between oppositely charged ions that holds them together in crystals is called an **ionic bond**. Ionic bonds are weak and easily broken, as occurs when salt is dissolved in water (Table 2-1).

Uncharged Atoms Share Electrons to Form Covalent Bonds

An atom with a partially full outermost electron shell can also become stable by sharing electrons with another atom, forming a **covalent bond**. Consider the hydrogen atom, which has one electron in a shell built for two. A hydrogen atom can become stable if it shares its single electron with another hydrogen atom, forming a molecule of hydrogen gas, H_2. Because the two hydrogen atoms are identical, neither nucleus can exert more attraction and capture the other's electron. So each electron's orbiting time is divided equally betwen the two nuclei, forming a single covalent bond. Each hydrogen atom behaves almost as if it had two electrons in its shell. Covalent bonds are stronger than ionic bonds.

Most Biological Molecules Use Covalent Bonding

The molecules in proteins, sugars, carbohydrates, fats, and virtually every other biological molecule are formed of atoms held together by covalent bonds. The atoms most commonly found in biological molecules (hydrogen, carbon, oxygen,

TABLE 2-1 Chemical Bonds

Type of Bond	Bond Forms:
Weak bonds: allow interactions between individual atoms or molecules	
Ionic bonds	Between positive and negative ions
Hydrogen bonds	Between a hydrogen atom in a polar covalent bond and another atom in a polar covalent bond
Strong bonds: hold atoms together within molecules	
Covalent bonds	By the sharing of electron pairs: equal sharing produces nonpolar covalent bonds; unequal sharing produces polar covalent bonds

nitrogen, phosphorus, and sulfur) all need at least two electrons to fill their outermost electron shell and can share electrons with one or more other atoms. Hydrogen can form a covalent bond with one other atom; oxygen and sulfur with two other atoms; nitrogen with three; and phosphorus and carbon with up to four. This diversity of bonding possibilities permits biological molecules to be constructed in almost infinite variety and complexity.

Covalent Bonds Are Either Nonpolar or Polar

In hydrogen gas, the two nuclei are identical, and the shared electrons spend equal time near each nucleus. Therefore, not only is the molecule as a whole electrically neutral, but each end, or *pole*, of the molecule is also electrically neutral. Such an electrically symmetrical bond is called a *nonpolar* covalent bond, and a molecule (such as H_2) formed with such bonds is a **nonpolar molecule** (Fig. 2-4a).

Electron sharing in covalent bonds, however, is not always equal. In many molecules, one nucleus may have a larger positive charge (due to more protons) and therefore attract the electrons more strongly than does the other nucleus. This situation produces a *polar* covalent bond. Although the polar molecule as a whole is electrically neutral, it has charged parts. The atom that attracts the electrons more strongly picks up a slightly negative charge and is thus the negative pole of the molecule. The other atom has a slightly positive charge and is the positive pole. In water, for example, oxygen attracts electrons more strongly than does hydrogen, so the oxygen end of a water molecule is negative and each hydrogen is positive (Fig. 2-4b). Water, with its charged poles, is a **polar molecule**.

Hydrogen Bonds Form between Molecules with Polar Covalent Bonds

Because their covalent bonds are polar, water molecules attract one another. The partially negatively charged oxygen atoms of water molecules attract the partially positively charged hydrogen atoms of other nearby water molecules. This electrical attraction is called a **hydrogen bond** and is weaker than ionic and covalent

(a) Nonpolar covalent bonding

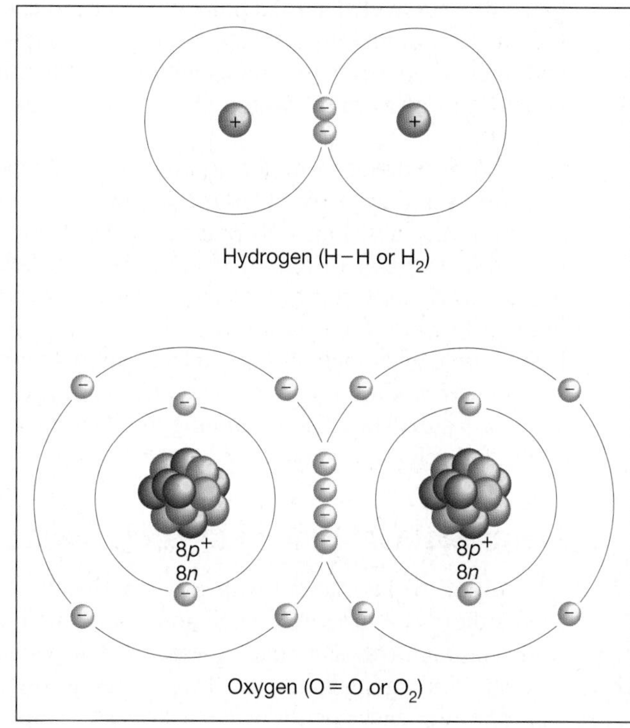

(b) Polar covalent bonding

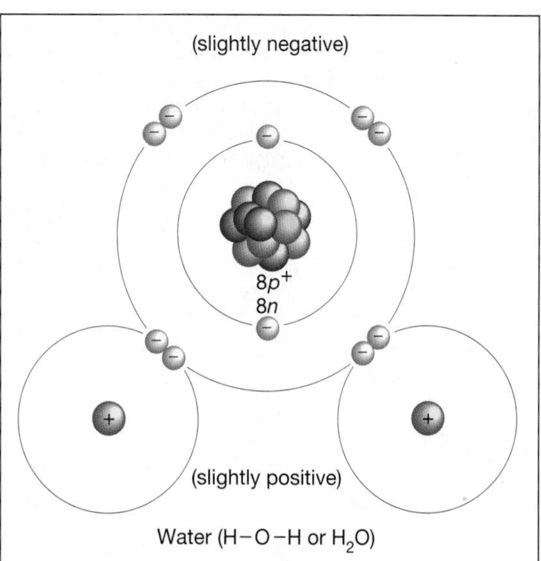

Figure 2-4 Covalent bonds Shared electrons form covalent bonds. ***(a)*** In hydrogen gas, one electron from each hydrogen atom is shared, forming a single covalent bond. In oxygen gas, two oxygen atoms share four electrons, forming a double covalent bond. ***(b)*** Oxygen, lacking two electrons to fill its outer shell, makes one bond with each of two hydrogen atoms to form water. **QUESTION:** *In water's polar bonds, why is oxygen's pull on electrons stronger than hydrogen's?*

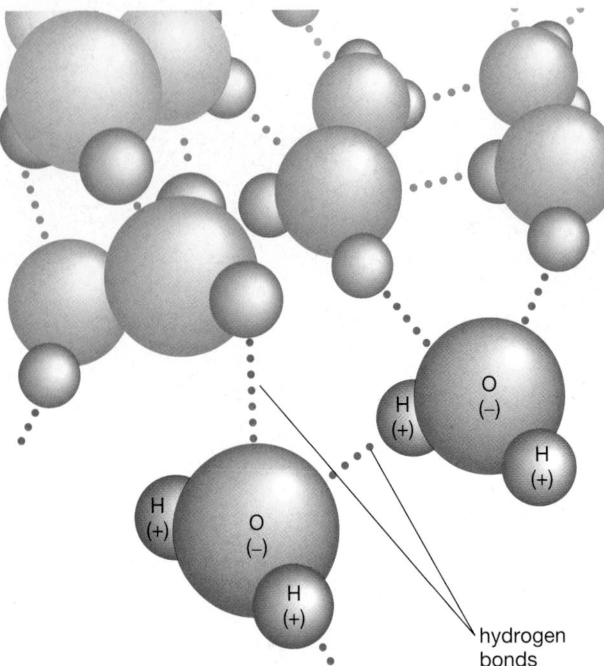

Figure 2-5 Hydrogen bonds The partial charges on different parts of water molecules produce weak hydrogen bonds, with each bond joining the hydrogen of one water molecule to the oxygen of another water molecule.

Figure 2-6 Water as a solvent When a salt crystal is dropped into water, the oppositely charged poles of the water molecules surround the sodium and chloride ions. Notice how the water molecules insulate the ions from the attractiveness of other molecules of salt. The ions disperse, and the whole crystal gradually dissolves. **QUESTION:** *Why is it important to human physiology that sugars dissolve easily in water?*

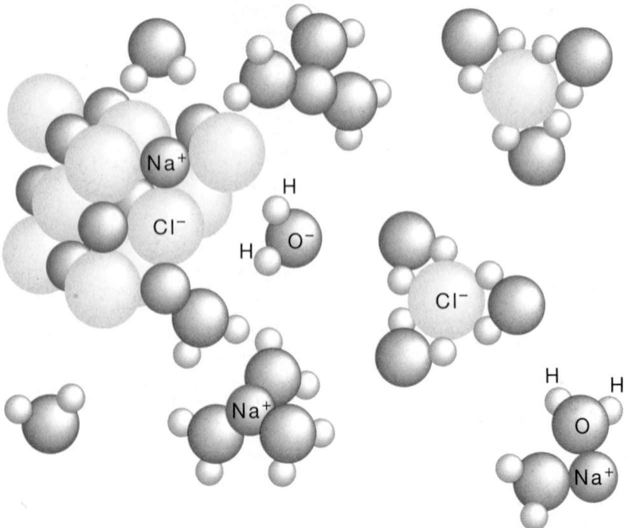

bonds (Fig. 2-5; see Table 2-1). As we shall see shortly, hydrogen bonds give water several unusual properties that are essential to life on Earth. Hydrogen bonds are also important in many biological molecules.

2.3 Why Is Water So Important to Life?

Water is one of the most abundant substances on Earth and plays a crucial role in sustaining life. In fact, most scientists believe that life would be impossible without water. Life on Earth probably originated in water; all living things require water; and 60% to 90% of an organism consists of water. The importance of water stems from some properties of the water molecule that allow it to perform a unique role in support of life. Let's consider some of these properties.

Water Interacts with Many Other Molecules

Water enters into many of the chemical reactions in living cells. For example, water supplies the oxygen that green plants release into the air during photosynthesis. Water is also used when animals digest the molecules in food. Conversely, water is produced in the reactions that produce proteins, fats, and sugars. Whenever a chemical reaction occurs in a living cell, water is very likely to be involved.

Many Molecules Dissolve Easily in Water

Why is water so important in biological chemical reactions? One reason is that water is an extremely good **solvent**; that is, it is capable of dissolving a wide range of substances, including proteins, salts, and sugars. Because water can dissolve so many molecules, the watery environment inside a cell provides an excellent setting for the countless chemical reactions essential to life.

Water is such an excellent solvent because it is a polar molecule, with positive and negative poles. Thus, if a crystal of table salt (sodium chloride) is dropped into water, the positively charged ends of water molecules will be attracted to and will surround the negatively charged chloride ions in the salt crystal. At the same time, the negatively charged poles of water molecules will surround the positively charged sodium ions. As water molecules enclose the sodium and chloride ions and shield them from interacting with each other, the ions separate from the crystal and drift away in the water; the ionic bonds are broken and the salt dissolves (Fig. 2-6).

Water also dissolves molecules that are held together by polar covalent bonds. Its positive and negative poles are attracted to oppositely charged regions of dissolving molecules. Ions and polar molecules that dissolve readily in water are termed *hydrophilic* (Greek for "water-loving") because of their electrical attraction for water molecules. (Molecules that are uncharged and nonpolar and do not dissolve in water are called *hydrophobic*. Fats and oils are examples.) Many biological molecules, including sugars and amino acids, are hydrophilic and dissolve in water. In addition, many gases, such as oxygen and carbon dioxide, also dissolve in water. Thus, fish swimming in a lake use oxygen that is dissolved in the water and release carbon dioxide into the water.

Water Molecules Tend to Stick Together

In addition to interacting with other molecules, water molecules interact with each other. Because hydrogen bonds link individual water molecules, liquid water has high **cohesion**; that is, water molecules have a tendency to stick together. Cohesion among water molecules at the water's surface produces **surface tension**, the tendency for the water surface to resist being broken. If you've ever

(a)

(b)

Figure 2-7 Cohesion among water molecules ***(a)*** With webbed feet bearing specialized scales, the basilisk lizard of South America makes use of surface tension, caused by cohesion, to support its weight as it races across the surface of a pond. ***(b)*** Within a redwood tree, cohesion holds water molecules together in continuous strands from the roots to the highest leaves, which may be more than 300 feet above the ground.

experienced the slap and sting of a belly flop into a swimming pool, you've discovered firsthand the power of surface tension. Surface tension can support fallen leaves, some spiders and water insects, and even a running lizard (**Fig. 2-7a**).

Cohesion plays an important role in land plants. A plant absorbs water through its roots, but is then faced with the big problem of moving the water to its leaves, which can be more than 300 feet up in a tall tree (**Fig. 2-7b**). The problem is solved by cohesion. The tiny tubes that connect the leaves, stem, and roots are filled with water, and when water molecules evaporate from the leaves, water is pulled up the tubes from below to fill the empty space. The system works because the hydrogen bonds between water molecules are stronger than the weight of the water in the tubes. Even in a 300-foot-tall tree, the water "chain" doesn't break.

Another property of water that helps plants get water from their roots to their leaves is adhesion. *Adhesion* describes water's tendency to stick to surfaces that have a slight charge that attracts polar water molecules. Adhesion helps water move within small spaces, such as the thin tubes in plants that carry water from roots to leaves. If you stick the end of a narrow glass tube into water, the water will move a short distance up the tube. Put some water in a narrow glass bud vase or test tube and you will see that the upper surface is curved. The water is pulling itself up the sides of the glass by adhesion to the surface of the glass and cohesion among water molecules.

Water Can Form Ions

Although water is generally regarded as a stable compound, individual water molecules constantly gain, lose, and swap hydrogen atoms. As a result, at any given time about two of every billion water molecules are *ionized*—that is, broken apart into hydrogen ions (H^+) and hydroxide ions (OH^-).

O
H H
O
H
(−)
+
H
(+)
water
(H_2O)
hydroxide ion
(OH^-)
hydrogen ion
(H^+)

A hydroxide ion has gained an electron from the hydrogen atom in a water molecule, and it has a negative charge. The hydrogen ion, which has lost its electron, now has a positive charge. Pure water contains equal concentrations of hydrogen ions and hydroxide ions.

In many solutions, however, the concentrations of H^+ and OH^- are not the same. If the concentration of H^+ is greater than the concentration of OH^-, the solution is *acidic*. An **acid** is a substance that releases hydrogen ions when it is dissolved in water. For example, when hydrochloric acid (HCl) is added to pure

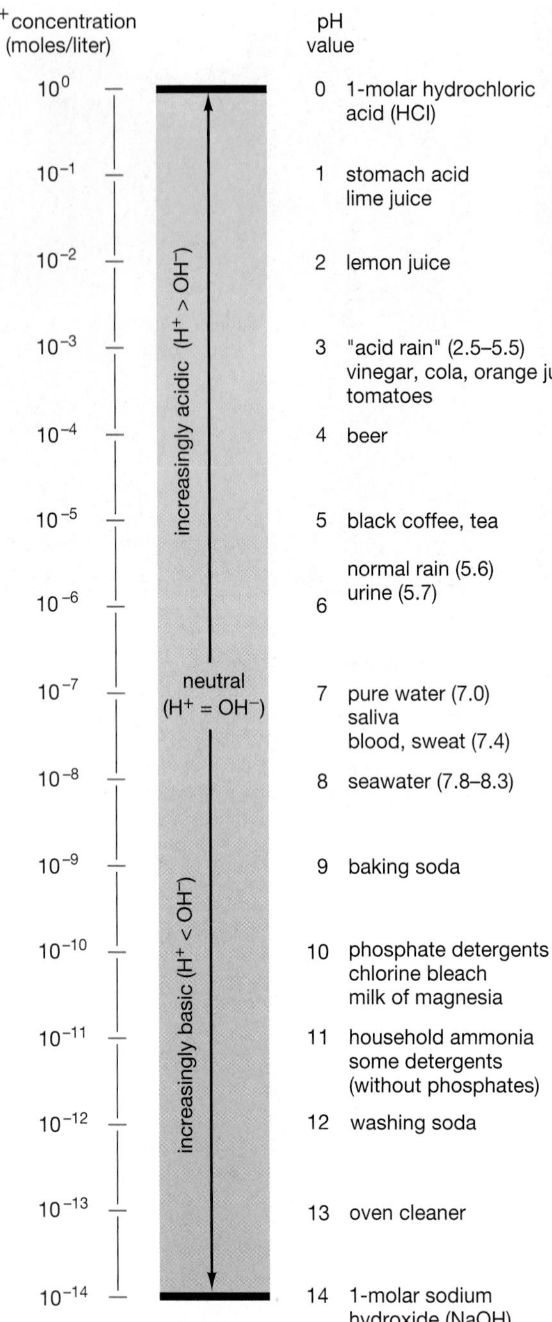

Figure 2-8 The pH scale The pH scale expresses the concentration of hydrogen ions in a solution on a scale of 0 (very acidic) to 14 (very basic). **QUESTION:** *How would the concentration of hydrogen ions in a cup of tea change if you added lemon juice to it?*

water, almost all of the HCl molecules separate into H^+ and Cl^-. Now the concentration of H^+ is much greater than the concentration of OH^-, and the solution is acidic. (Many acidic substances, such as lemon juice and vinegar, have a sour taste because the sour-taste receptors on your tongue are specialized to respond to the excess of H^+.)

If the concentration of OH^- is greater than that of H^+, the solution is *basic*. A **base** is a substance that combines with hydrogen ions, reducing their number. For instance, if sodium hydroxide (NaOH) is added to water, the NaOH molecules separate into Na^+ and OH^-. The OH^- combines with H^+, reducing the number of H^+ ions. The solution then contains an excess of OH^- and is basic.

pH Measures Acidity

The degree of acidity is expressed on the **pH scale**, in which neutrality (equal numbers of H^+ and OH^-) is assigned the number 7. Acids have a pH below 7; bases have a pH above 7 (**Fig. 2-8**). Pure water, with equal concentrations of H^+ and OH^-, has a pH of 7. Each unit on the pH scale represents a tenfold change in the concentration of H^+. Thus, beer (pH 4) has a concentration of H^+ that is 10 times greater than that of coffee (pH 5), and a carbonated soft drink (pH 3) has a concentration of H^+ that is 10,000 times higher than that of water (pH 7).

A Buffer Maintains a Solution at a Constant pH

In most mammals, including humans, both the cell interior and the fluids that bathe the cells are nearly neutral (pH about 7.3 to 7.4). Small increases or decreases in pH may cause drastic changes in both the structure and function of biological molecules, leading to the death of cells or entire organisms. Nevertheless, living cells seethe with chemical reactions that take up or give off H^+. How, then, does the pH remain constant overall?

The answer lies in the many buffers found in organisms. A **buffer** is a substance that tends to maintain a solution at a constant pH by accepting or releasing H^+ in response to small changes in H^+ concentration. If the H^+ concentration rises, buffers combine with them. If the H^+ concentration falls, buffers release H^+. The result in both cases is that the concentration of H^+ is restored to its original level.

2.4 Why Is Carbon So Important to Life?

Organisms contain a tremendous variety of substances, but a great many of these diverse molecules contain carbon. Carbon is so widespread in living things because the carbon atom is so versatile. Carbon can combine with other atoms in many different ways to form a huge number of different molecules.

This vast array of combinations is possible because a carbon atom has four electrons in its outermost shell, leaving room for four more. Therefore, carbon atoms are able to form many bonds. They become stable by sharing four electrons with other atoms, forming up to four covalent bonds. Carbon-containing molecules can contain many carbon atoms and can assume a variety of complex shapes, including chains, branches, and rings. Carbon thus forms the basis for an amazing diversity of molecules and makes it possible for living things to construct the many different substances required to sustain life.

In chemistry, molecules that have a carbon skeleton and also contain some hydrogen atoms are known as **organic molecules**. The carbon skeletons of organic molecules can be quite complex, but carbon alone does not account for the diversity of organic molecules. Instead, groups of atoms, called **functional groups**, attach to the carbon backbone and determine the characteristics and chemical reactivity of the molecules. These functional groups are far less stable than the carbon backbone and are more likely to participate in chemical reactions. The common functional groups found in organic molecules are shown in Table 2-2.

TABLE 2-2 Important Functional Groups in Biological Molecules

Group	Structure	Properties	Types of Molecules
Hydrogen (—H)	—H	Polar or nonpolar, depending on which atom hydrogen is bonded to; involved in dehydration synthesis and hydrolysis	Almost all organic molecules
Hydroxyl (—OH)	—O—H	Polar; involved in dehydration synthesis and hydrolysis	Carbohydrates, nucleic acids, alcohols, some acids, and steroids
Carboxyl (—COOH)	—C(=O)—O—H	Acidic; negatively charged when H^+ separates from it; involved in peptide bonds	Amino acids, fatty acids
Amino (—NH_2)	—N(H)H	Basic; may bond an additional H^+ becoming positively charged; involved in peptide bonds	Amino acids, nucleic acids
Phosphate (—H_2PO_4)	—O—P(=O)(O—H)(O—H)	Acidic; up to two negative charges when H^+ separates from it; links nucleotides in nucleic acids; energy-carrier group in ATP	Nucleic acids, phospholipids
Methyl (—CH_3)	—C(H)(H)H	Nonpolar; tends to make molecules hydrophobic	Many organic molecules; especially common in lipids

2.5 How Are Biological Molecules Joined Together or Broken Apart?

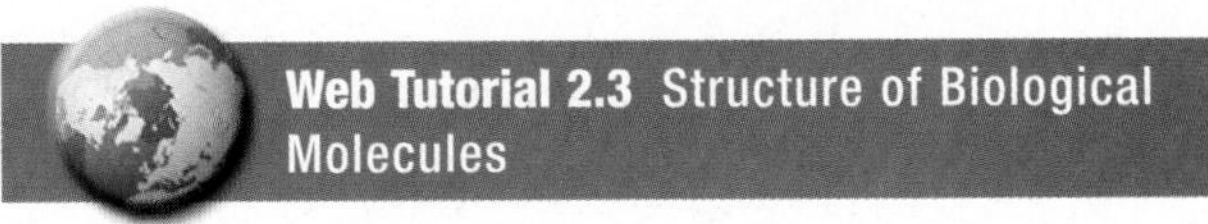

In living organisms, large, complex molecules are constructed by first assembling smaller molecules and then hooking them together. Small organic molecules (for example, *sugars*) are used as subunits to construct longer molecules (for example, *starches*), like cars in a train.

Biological molecules must be both built and broken down. Water plays a central role in both types of reaction. When complex biological molecules are built in an organism, water is often released as a by-product. When molecules, such as those serving as food, are broken down into subunits that an organism can use, water is usually consumed.

The chemical reactions that link subunits together to make large biological molecules are called **dehydration synthesis** (→) reactions. In a dehydration synthesis reaction, a hydrogen atom (—H) is removed from one subunit and a hydroxyl group (—OH) is removed from a second subunit, creating openings in the outer electron shells of the two subunits. These openings are then filled by electrons shared between the subunits, creating a covalent bond that links them. The free hydrogen and hydroxyl ions combine to form a molecule of water (H_2O).

The reverse reaction, **hydrolysis** (→) ("to break apart with water"), can split a large molecule into individual subunits.

TABLE 2-3 The Principal Biological Molecules

Class of Molecule	Principal Subtypes (subunits in parentheses)	Example	Function
Carbohydrate: Normally contains carbon, oxygen, and hydrogen, in the approximate formula $(CH_2O)_n$	(*Monosaccharide*: Simple sugar)	Glucose	Important energy source for cells; subunit of most polysaccharides
	Disaccharide: Two monosaccharides bonded together	Sucrose	Principal sugar transported throughout bodies of land plants
	Polysaccharide: Many monosaccharides (usually glucose) bonded together	Starch Glycogen Cellulose	Energy storage in plants Energy storage in animals Structural material in plants
Lipid: Contains high proportion of carbon and hydrogen; usually nonpolar and insoluble in water	*Oil, fat*: Three fatty acids bonded to glycerol	Peanut oil Beef fat	Energy storage in animals, some plants
	Wax: Variable numbers of fatty acids bonded to long-chain alcohol	Waxes in plant cuticle (surface covering)	Waterproof covering on leaves and stems of land plants
	Phospholipid: Polar phosphate group and two fatty acids bonded to glycerol	Phosphatidylcholine	Part of membranes in cells
	Steroid: Four fused rings of carbon atoms with functional groups attached	Cholesterol	Part of membrane of cells; precursor of other steroids such as testosterone, bile salts
Protein: Chains of amino acids; contain carbon, hydrogen, oxygen, nitrogen, and sulfur	*(Amino acids)*	Keratin	Principal component of hair
		Silk	Principal component of silk moth cocoons and spider webs
		Hemoglobin	Transport of oxygen in vertebrate blood
Nucleic acid: Made of nucleotide subunits; may consist of a single nucleotide or long chain of nucleotides	*Long-chain nucleic acids*	Deoxyribonucleic acid (DNA)	Genetic material of all living cells
		Ribonucleic acid (RNA)	Genetic material of some viruses; in living cells, essential in transfer of genetic information from DNA to protein
	(Single nucleotides)	Adenosine triphosphate (ATP)	Principal short-term energy-carrier
		Cyclic adenosine monophosphate (cyclic AMP)	Molecule in cells; intracellular messenger

Considering how complicated living things are, it might surprise you to learn that nearly all biological molecules fall into one of only four general categories: carbohydrates, lipids, proteins, or nucleic acids (Table 2-3).

2.6 What Are Carbohydrates?

Carbohydrates are molecules composed of carbon, hydrogen, and oxygen in the approximate ratio of 1:2:1. Carbohydrates can be small, water-soluble sugars or long chains that are made by stringing sugar subunits together. A carbohydrate consisting of just one sugar molecule is called a simple sugar, or *monosaccharide*. When two monosaccharides are linked, they form a *disaccharide*. Three or more form a *polysaccharide*.

Carbohydrates are important energy sources for most organisms. Consider a breakfast that includes blueberry pancakes, syrup, and orange juice. The pancakes consist mainly of carbohydrates that were originally stored in the seeds of wheat or other grains. The sugar that sweetens the syrup, blueberries, and orange juice was also stored by plants as an energy source. The carbohydrate molecules that once served the plants that manufactured them now provide energy to the humans who consume them. Other carbohydrates, such as cellulose and similar molecules, provide structural support for individual cells or even for the entire bodies of organisms as diverse as plants, fungi, bacteria, and insects.

Most simple sugars have a backbone of three to seven carbon atoms. Each carbon atom in the backbone generally has both a hydrogen (—H) and a hydroxyl group (—OH) attached to it, so carbohydrates have the general chemical formula $(CH_2O)_n$, where n is the number of carbons in the backbone. This formula explains the origin of the name "carbohydrate," which literally means "carbon plus water." When dissolved in water, the carbon backbone of a simple sugar usually "circles up" to form a ring (Fig. 2-9). Simple sugars also assume the ring form when they link together to form disaccharides and polysaccharides.

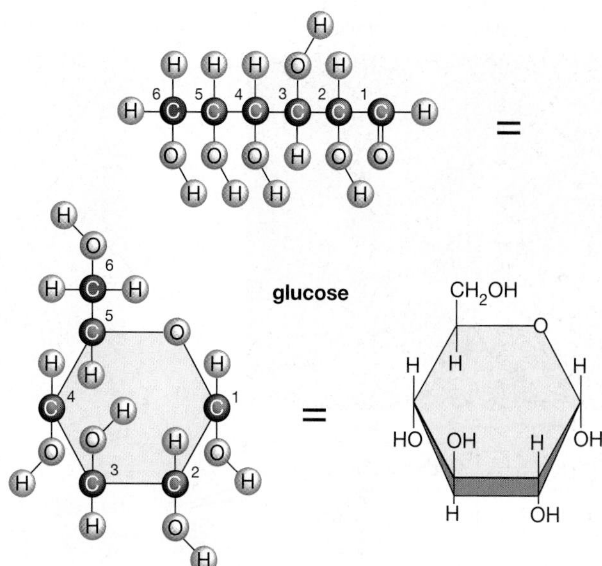

Figure 2-9 A simple sugar Three diagrams of glucose, a six-carbon monosaccharide. The top diagram represents the molecule in its linear (straight) form. The two diagrams at the bottom show two different ways to represent the cyclic (ring) form of glucose. In the ring diagram on the right, the carbon at each corner of the ring is not labeled; this is a convention always used on this type of diagram.

A Variety of Simple Sugars Occurs in Organisms

Glucose is the most common simple sugar in organisms and is the subunit of which most polysaccharides are made. Glucose has six carbons, so its chemical formula is $C_6H_{12}O_6$. Many organisms synthesize other simple sugars that have the same chemical formula as glucose but have slightly different structures. These include fructose (the "corn sugar" found in corn syrup and also the molecule that makes fruits taste sweet). Some other common simple sugars, such as ribose and deoxyribose, have five carbons.

Disaccharides Store Energy and Serve As Building Blocks

Simple sugars, especially glucose and its relatives, have a short life span in a cell. Most are either quickly broken down, so that their chemical energy is freed to fuel various cellular activities, or are linked together to form disaccharides or polysaccharides (Fig. 2-10). Disaccharides are often used for short-term energy storage, especially in plants. Common disaccharides include sucrose (table sugar: glucose plus fructose) and lactose (milk sugar: glucose plus galactose). When cells require energy, disaccharides are broken apart into their monosaccharide subunits to release energy.

Polysaccharides Store Energy and Provide Support

Try chewing a cracker for a long time. Does it taste sweeter the longer you chew? It should, because the cracker contains starch, a polysaccharide that breaks down into its sweet-tasting glucose subunits as you chew. Certain polysaccharides, such as starch (in plants) and glycogen (in animals), are used mainly for long-term energy storage. Starch is commonly formed in roots and seeds, most commonly as huge, branched chains of up to half a million glucose subunits. Glycogen molecules, which are stored as an energy source in the liver and muscles of animals such as humans, are generally much smaller than starch molecules.

Figure 2-10 Manufacture of a disaccharide The disaccharide sucrose is made by a dehydration synthesis reaction in which a hydrogen (—H) is removed from glucose and a hydroxyl group (—OH) is removed from fructose, another simple sugar. The two simple sugars join by a covalent bond, and a water molecule forms in the process.

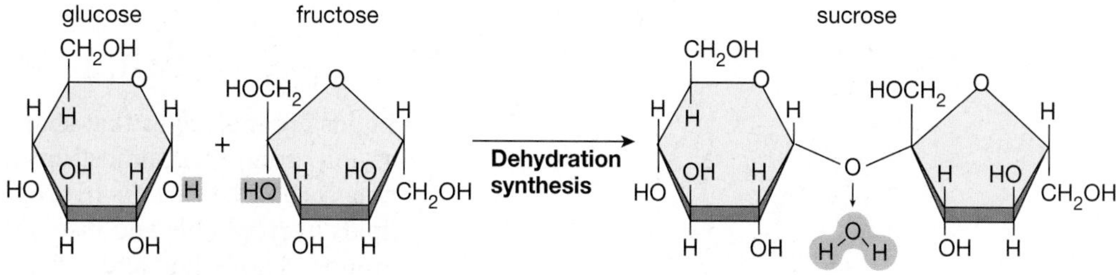

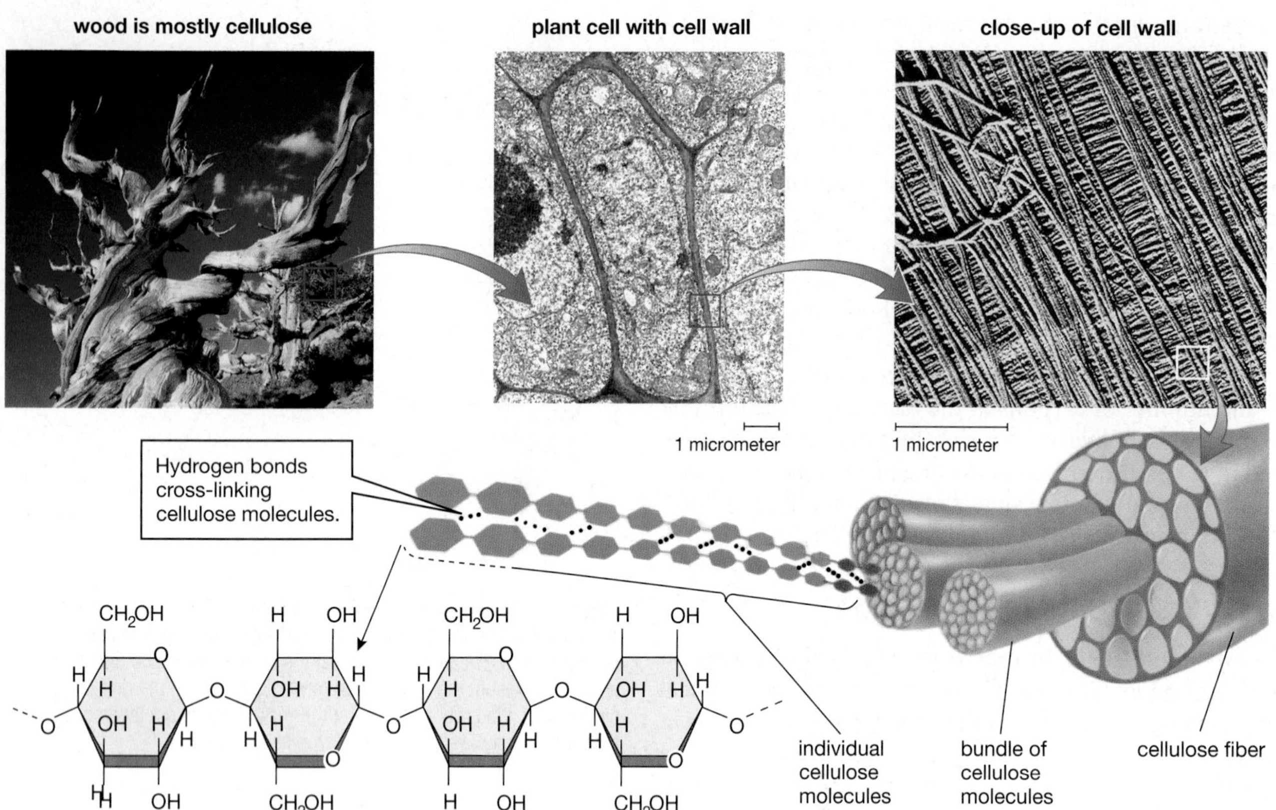

Figure 2-11 Cellulose structure Cellulose, like starch, is composed of glucose subunits. Unlike starch, however, cellulose has great structural strength, due partly to the arrangement of parallel molecules of cellulose into long, cross-linked fibers. **QUESTION:** *Many types of plastic are composed of molecules derived from cellulose, but engineers are working hard to develop plastics based on starch molecules. Why might starch-based plastics be an improvement over existing types of plastic?*

Many organisms also use polysaccharides as structural materials. For example, the polysaccharide *chitin* is the main component of the hard body coverings of insects, spiders, crabs, and lobsters. One of the most important structural polysaccharides is *cellulose*, which provides support for plant cells and makes up about half the bulk of a typical tree trunk (**Fig. 2-11**). Cellulose is by far the most abundant organic molecule on Earth (not that surprising when you consider the vast fields and forests that blanket much of our planet). Ecologists estimate that about a trillion tons of cellulose are made each year.

Like starch, cellulose consists of long chains of glucose subunits. Cellulose's digestibility, however, differs sharply from that of starch. Most animals can readily digest starch, but they cannot digest cellulose at all. For most animals, cellulose is roughage or fiber, material that passes undigested through the digestive tract. The only exceptions are animals such as cows or termites, which have special one-celled organisms living in their digestive tracts. These microscopic organisms are among the few that are able to break down cellulose, so the animals in whose guts they live can eat leaves, wood, and other foods that contain a lot of cellulose. The microbes digest the cellulose, and the animal hosts can absorb the nutrient molecules that are freed.

2.7 What Are Lipids?

Lipids are molecules that share two important features. First, they contain large regions composed almost entirely of hydrogen and carbon, with nonpolar carbon–carbon or carbon–hydrogen bonds. Second, these nonpolar regions make lipids hydrophobic and insoluble in water. Lipids are classified into three major groups: (1) oils, fats, and waxes; (2) phospholipids; and (3) steroids.

(a) Fat

(b) Wax

Figure 2-12 Lipids *(a)* Fat is an efficient way to store energy. If this bear stored the same amount of energy in carbohydrates instead of fat, she probably would be unable to walk! *(b)* Wax is a highly saturated lipid that remains very firm at normal outdoor temperatures. Its rigidity serves well in the strong but thin-walled hexagons of this honeycomb.

Oils, Fats, and Waxes Contain Only Carbon, Hydrogen, and Oxygen

Oils, fats, and waxes have several features in common. First, they contain only carbon, hydrogen, and oxygen. They also include one or more *fatty acid* subunits, which are long chains of carbon and hydrogen with a *carboxyl group* (—COOH) at one end. Finally, they usually do not have ring structures.

Fats and oils form by dehydration synthesis from three fatty acid subunits and one molecule of glycerol, a short, three-carbon molecule with one hydroxyl group (—OH) for each carbon molecule.

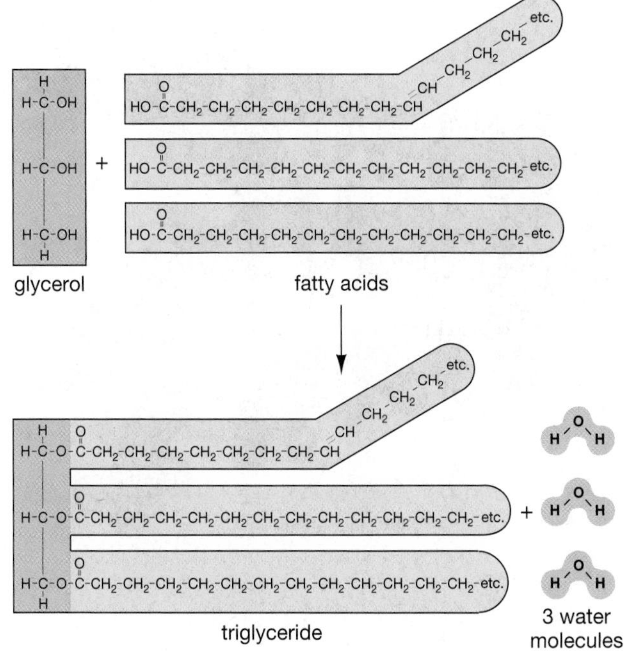

Fats and oils have a high concentration of chemical energy, about 9.3 Calories per gram, compared with 4.1 for sugars and proteins (the Calorie is a unit that measures the energy content of foods). Because fats are so high in calories, fat substitutes such as olestra may be especially appealing to dieters. In an olestra molecule, the position that would be occupied by glycerol in a real fat is instead occupied by a sucrose molecule, which has six, seven, or eight fatty acids attached to it, instead of the three of a normal fat. Apparently, the large number of fatty acid chains prevents digestive enzymes from reaching the digestible sucrose at the center of the olestra molecule. The molecule thus never breaks down into fragments that can be absorbed by the body, and it is excreted unchanged.

Fats and oils are used for long-term energy storage in both plants and animals. For example, during summer and fall, bears consume more energy than they spend and store fat on their bodies, which tides them over during their winter hibernation. Because fats store the same energy with less weight than do carbohydrates, fat is an efficient way for animals to store energy (Fig. 2-12a).

The difference between a fat (such as beef fat), which is a solid at room temperature, and an oil (such as peanut oil), which is liquid at room temperature, lies in their fatty acids. In fats, all carbons in the fatty acid chains are joined with single covalent bonds (one pair of electrons shared between the two atoms). The remaining bond positions on the carbons are occupied by hydrogens. Such fatty acids are said to be saturated, because they are "saturated" with hydrogens; they have as many hydrogens as possible. Saturated fatty acids tend to be very straight, and they nestle closely together, forming solid lumps at room temperature.

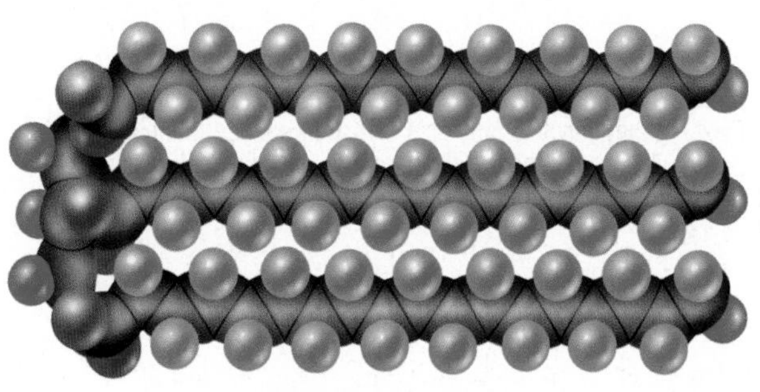

Beef fat (saturated)

HEALTH WATCH

CHOLESTEROL—FRIEND AND FOE

Cholesterol is a steroid with a bad reputation. Why are so many products now advertising themselves as "cholesterol free" or "low in cholesterol"? After all, cholesterol is a crucial component of cell membranes. It is also the raw material for the production of bile (which helps us digest fats), vitamin D, and both male and female sex hormones.

Although cholesterol is crucial to life, medical researchers have found that people with excessively high levels of cholesterol in their blood are at increased risk for heart attacks and strokes. Unfortunately, the cholesterol builds up "silently" and gives no warning signs. A person may not know that anything is wrong until he or she actually suffers a heart attack. Cholesterol contributes to the formation of obstructions in arteries, called *plaques*, which in turn can promote the formation of blood clots. These clots can break loose and block an artery carrying blood to the heart, causing a heart attack, or to the brain, causing a stroke.

Have you heard of "good cholesterol" and "bad cholesterol"? Because cholesterol molecules are nonpolar, they do not dissolve in blood (which is mostly water). Instead, the cholesterol molecules are transported through blood in packets surrounded by special carrier molecules called *lipoproteins* (phospholipids plus proteins). Cholesterol in high-density lipoprotein packets ("HDL cholesterol," which has more protein and less lipid) is the good kind. These packets transport cholesterol to the liver, where it is removed from circulation. Cholesterol in low-density lipoprotein packets ("LDL cholesterol," with less protein and more lipid) is the bad kind. This is the form in which cholesterol circulates to cells throughout the body and can be deposited on artery walls. A high ratio of HDL (good) to LDL (bad) is correlated with reduced risk of heart disease. A complete cholesterol screening test will distinguish between these two forms in your blood.

Where does cholesterol come from? Cholesterol comes from animal-derived foods; it is essentially nonexistent in plants. Egg yolks are a particularly rich source, and meat, whole milk, and butter contain it as well. Another source of cholesterol is your own body, which synthesizes cholesterol from other lipids. Because of genetic differences, some people's bodies manufacture more than others'. People with high cholesterol (about 25% of all adults in the United States) can often reduce their levels by eating a diet low in both cholesterol and saturated fats. For people with excessively high cholesterol who are unable to reduce it adequately by changing their diets, doctors often prescribe cholesterol-reducing drugs.

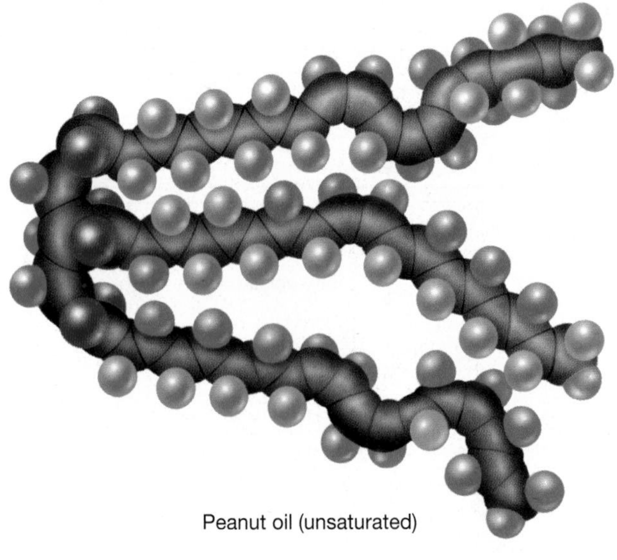
Peanut oil (unsaturated)

In oils, some of the carbons in the fatty acid chains are joined by double covalent bonds (two pairs of electrons shared between the two atoms). Consequently, there are fewer attached hydrogens, and the fatty acid is said to be unsaturated. Unsaturated fatty acids tend to have bends and kinks in the fatty acid chains.

The kinks keep oil molecules apart; as a result, oil is liquid at room temperature. An oil can be converted to a fat by breaking the double bonds between carbons, replacing them with single bonds, and adding hydrogens to the remaining bond positions. This is the "hydrogenated oil" listed in the ingredients on a box of margarine, which makes the margarine solid at room temperature.

Although **waxes** are chemically similar to fats, they are not a food source. We and most other animals cannot digest them. They are highly saturated, making them solid at normal outdoor temperatures. Waxes form a waterproof coating over the leaves and stems of land plants. Some animals synthesize waxes as waterproofing (for mammalian fur, for example), and a few others, such as bees, use waxes to build elaborate structures (Fig. 2-12b).

Phospholipids Have Water-Soluble Heads and Water-Insoluble Tails

The membrane that separates the inside of a cell from the outside world contains several types of *phospholipids*. A phospholipid is similar to an oil, except that one of the three fatty acids is replaced by a smaller, phosphate-containing subunit. This subunit sits at one end of the phospholipid molecule, where it forms a polar

Figure 2-13 Phospholipids Phospholipids are similar to fats and oils, except that only two fatty acid tails are attached to the glycerol backbone. The backbone is bonded to a polar head composed of a phosphate group plus another functional group.

polar head (hydrophilic)
glycerol backbone
fatty acid tails (hydrophobic)

head that is water soluble. The two fatty acids form tails that are not soluble in water. Thus, a phospholipid has two dissimilar ends: a hydrophilic head attached to hydrophobic tails (Fig. 2-13). As you will see in Chapter 3, this dual nature of phospholipids is crucial to the structure and function of cell membranes.

Steroids Consist of Four Carbon Rings Fused Together

Steroids are structurally different from the other lipids. All steroids are composed of four rings of carbon fused together with various functional groups protruding from them (Fig. 2-14; note the basic steroid "skeleton" in color). One type of steroid is cholesterol. Cholesterol is a vital component of the membranes of animal cells and is also used by cells to synthesize other steroids. Other steroids made from cholesterol include male and female sex hormones, the hormones that regulate salt levels, and bile, which assists in fat digestion. Why, then, has cholesterol gotten so much bad publicity? Find out in "Health Watch: Cholesterol—Friend and Foe."

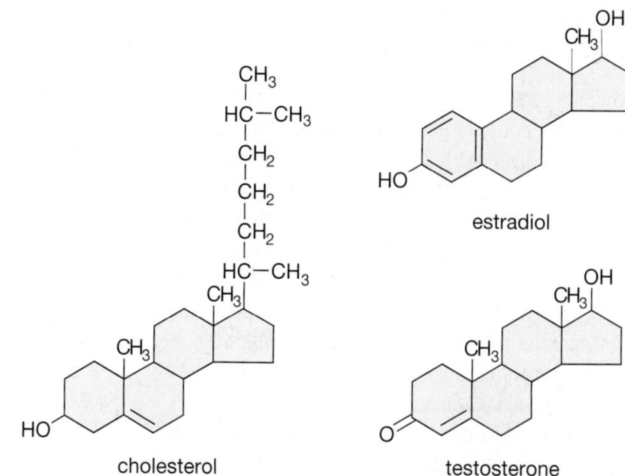

Figure 2-14 Steroids Other steroids are synthesized from cholesterol. All steroids have almost the same molecular structure (colored rings). Great differences in steroid function result from the differences in functional groups attached to the rings. Notice the similarity in structure between the male sex hormone testosterone and the female sex hormone estradiol (a type of estrogen). **QUESTION:** *Why are steroid hormones, after traveling in the bloodstream, able to act by binding with molecules inside the cell nucleus, whereas other types of hormones (i.e., not steroids) act only by interacting with molecules on the outside of the cell membrane?*

2.8 What Are Proteins?

Proteins perform many different functions in organisms. An especially important role is played by *enzymes*, proteins that guide almost all the chemical reactions that occur inside cells. Because each enzyme assists only one or a few specific reactions, most cells contain hundreds of different enzymes. Other types of proteins are used for structure (Fig. 2-15) or energy storage. Proteins may also function in transport or movement (for example, a protein carries oxygen in the blood, and others help muscle cells move).

Proteins Are Formed from Chains of Amino Acids

A molecule of **protein** consists of one or more chains of amino acids. All **amino acids** have the same fundamental structure, in which a central carbon is bonded to four different functional groups: an amino group ($—NH_2$); a carboxyl group ($—COOH$); a hydrogen; and a variable group (represented by the letter R). →

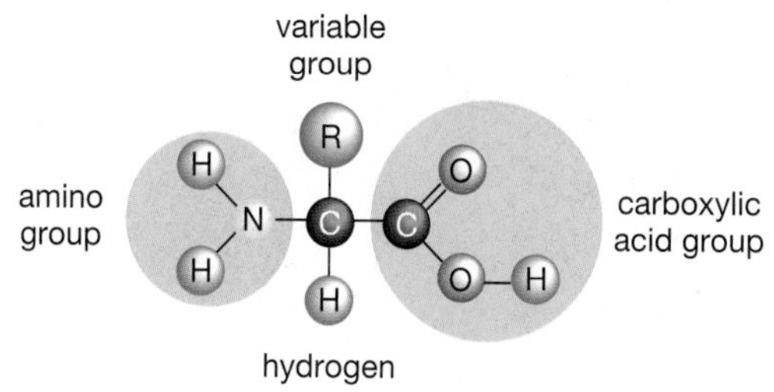

Each of the 20 amino acids that are found in proteins has a different R group, so amino acids differ in their chemical and physical properties—size, water solubility, electrical charge, and so on. Therefore, the sequence of the amino acids in a protein dictates the properties of the protein and determines whether it is an enzyme or a hormone or a structural protein.

Figure 2-15 Structural proteins

(a) Hair

(b) Horn

(c) Silk

EARTH WATCH

DEADLY ELEMENTS

Combinations of elements form our bodies, our surroundings, and everything we eat or drink. Unfortunately, however, certain elements can harm us if we ingest them. Lead and mercury are two well-known examples of elements whose atoms have the capacity to harm us, and recognition of this danger has led to unleaded gasoline, programs to remove lead-based paint from homes, and the gradual disappearance of thermometers filled with mercury. Nonetheless, significant amounts of these dangerous substances remain in our air, water, and food supplies.

Another hazardous element that is widespread in the environment is arsenic. In high doses, arsenic is a strong poison that can kill within hours. It is also quite dangerous in lower doses. In particular, steady consumption of smaller amounts of arsenic can cause cancer in humans, as studies of some unfortunate populations in Chile and Taiwan have shown. The ability of arsenic to cause cancer is bad news indeed, because arsenic is found in the drinking water of a huge number of people around the world. An estimated 35 million people in Bangladesh alone drink water that is heavily contaminated with arsenic. In the United States, levels of arsenic are high in many public water supplies, especially in the western states.

How does arsenic get into drinking water? In many cases, it seems to have dissolved from naturally occurring deposits in rocks. Industrial waste, however, is an important source of arsenic pollution. The waste from mining operations (Fig. E2-1) is especially likely to introduce previously contained arsenic to a water supply.

Ultimately, however, the question of how arsenic got into the water is less important than the question of just how much arsenic it takes to endanger human health. The U.S. government first formally recognized the danger of arsenic in drinking water in 1962, when it agreed to set a standard for safe levels of arsenic in water. Finally, after years of scientific study, the U.S. Environmental Protection Agency (EPA) ordered in January 2001 that the concentration of arsenic in drinking water should not exceed 10 micrograms per liter (sometimes also expressed as 10 parts per billion). Water supplies currently containing higher levels of arsenic are supposed to comply with the new regulation by 2006.

You can find information about arsenic levels in groundwater in different parts of the United States at **http://webserver.cr.usgs.gov/trace/arsenic**.

Figure E2-1 Arsenic pollution from mining waste endangers drinking water Much of the arsenic in our water supply comes from mining and industrial waste, but naturally occurring deposits of arsenic are also major contributors to the problem.

Amino Acids Join to Form Chains by Dehydration Synthesis

Like lipids and polysaccharides, proteins form by dehydration synthesis. The nitrogen of the amino group ($-NH_2$) of one amino acid is joined to the carbon of the carboxyl group ($-COOH$) of another amino acid by a single covalent bond. This bond is called a *peptide bond*, and the resulting chain of two amino acids is called a *peptide* (Fig. 2-16). More amino acids are added, one by one, until the protein is complete. Amino acid chains in living cells vary in length from three to thousands of amino acids. Often, the word "protein," or "polypeptide," is reserved for long chains—say, 50 or more amino acids in length—and "peptide" is used for shorter chains.

Three-Dimensional Shapes Give Proteins Their Functions

The phrase "amino acid chains" may evoke images of proteins as floppy, featureless structures, but they are not. Instead, proteins are highly organized molecules that fold themselves into complex, three-dimensional shapes (Fig. 2-17). Each different type of protein has a different shape, because each one has a different sequence of amino acids, and these sequence differences cause shape differences.

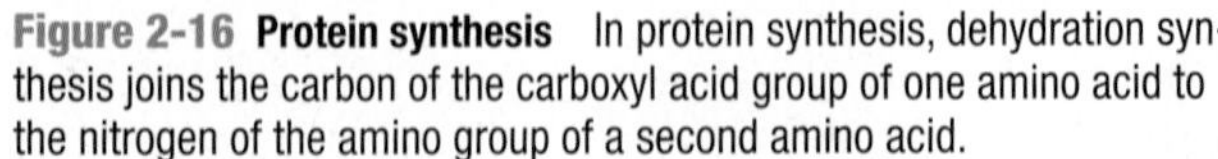

Figure 2-16 Protein synthesis In protein synthesis, dehydration synthesis joins the carbon of the carboxyl acid group of one amino acid to the nitrogen of the amino group of a second amino acid.

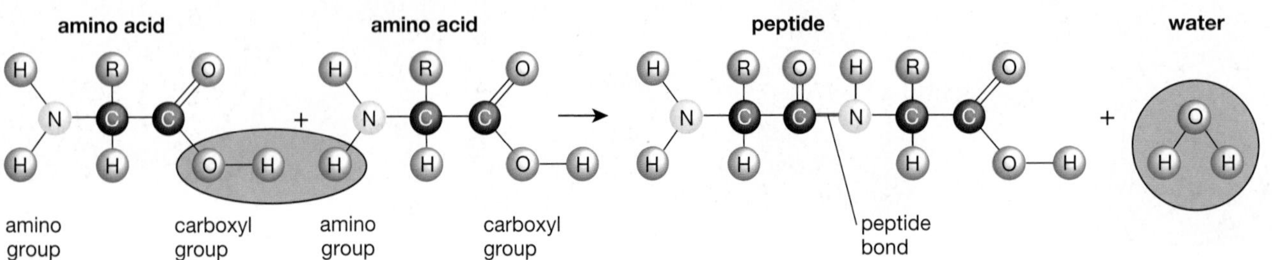

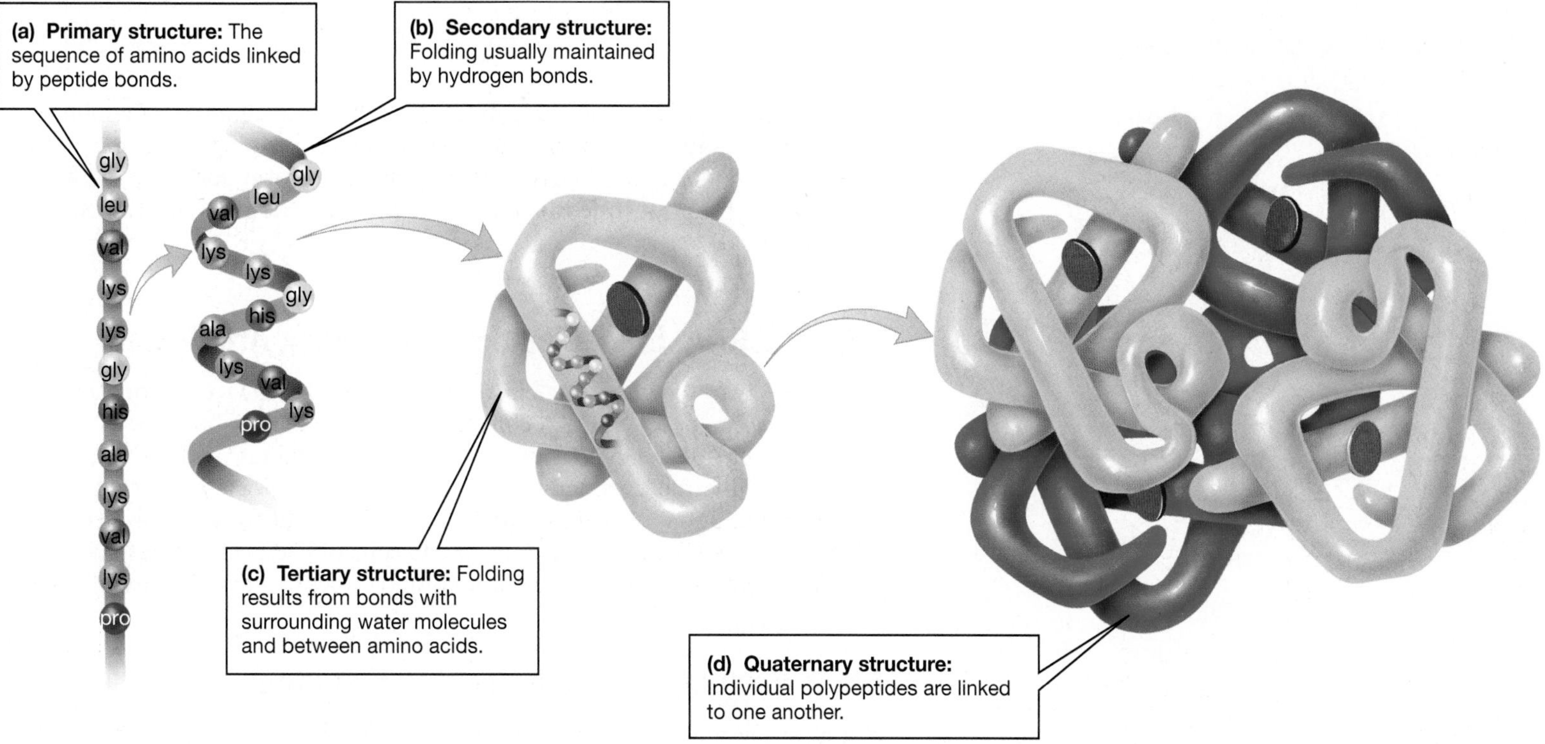

Figure 2-17 Protein chains fold into complex shapes Hemoglobin, the oxygen-carrying protein in red blood cells, is a combination of four peptide subunits (two each of two different types). Even minor alterations to its shape can prevent hemoglobin from doing its job. **QUESTION:** *Why do most proteins, when heated, lose their ability to function?*

A protein's *primary structure* is simply the unique sequence of its long amino acid chain, but most chains also have a *secondary structure*. The secondary structure is caused by hydrogen bonds that bend the chain in a characteristic pattern, which in most proteins is either a spiral shape or a pleated shape. The hydrogen bonds result from attraction between the weakly positive hydrogen atom in a peptide bond and the weakly negative oxygen atom in another, nearby peptide bond. Thus, hydrogen bonds form between the carboxyl groups and the amino groups in peptide bonds all along the length of the chain, causing the chain to fold up.

A protein's secondary structure may be further modified into complex *tertiary stuctures*, caused by interactions among the different R groups along the chain. Because each amino acid has a distinctive R group with distinctive chemical characteristics, the different amino acids in a protein may attract or repel one another, causing further bending and folding into complex, irregular shapes. Finally, protein shape can be still further modifed when two or more folded polypeptides join together to form even more complicated *quaternary structures*.

A protein's shape is what enables it to perform its function. If its shape is disrupted, the protein may no longer be able to function correctly, even if the peptide bonds between amino acids remain intact. Proteins whose shape has been disrupted are said to be **denatured**. There are many ways to denature a protein. For example, when an egg is fried, the heat of the frying pan denatures the albumin protein in the egg white, causing the egg white to change from clear to white and from liquid to solid. Sterilization using heat or ultraviolet rays kills bacteria and viruses by denaturing the proteins they need to live. Salty or acidic solutions also denature proteins—dill pickles are preserved in this way.

2.9 What Are Nucleic Acids?

Nucleic acids are long chains of similar but not identical subunits called *nucleotides*. All nucleotides have a three-part structure: (1) a five-carbon sugar (ribose or deoxyribose), (2) a phosphate group, and (3) a nitrogen-containing molecule called a *base* that differs among nucleotides.→

phosphate
base
sugar

Deoxyribose nucleotide

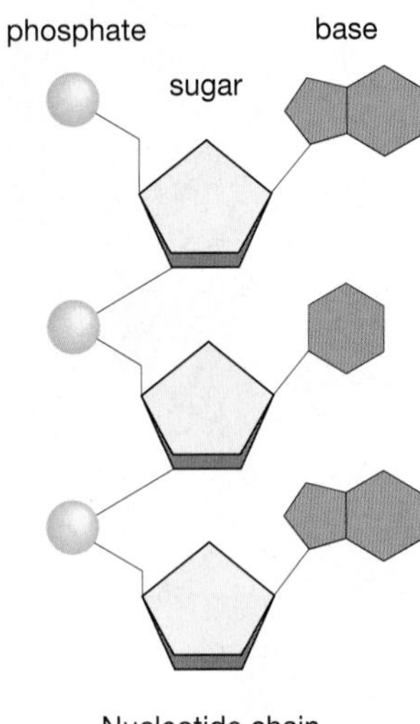

Nucleotide chain

There are two types of nucleotides, the ribose nucleotides (containing the sugar ribose) and the deoxyribose nucleotides (containing the sugar deoxyribose). Nucleotides string together in long chains to form nucleic acids, with the phosphate group of one nucleotide covalently bonded to the sugar of another.

DNA and RNA, the Molecules of Heredity, Are Nucleic Acids

There are two types of nucleic acids: *deoxyribonucleic acid*, or *DNA*, and *ribonucleic acid*, or *RNA*. DNA consists of chains of deoxyribose nucleotides millions of units long. DNA is found in all living things, and its sequence of nucleotides, like the dots and dashes of a biological Morse code, spells out the genetic information needed to construct the proteins of each organism. Chains of RNA are copied from the DNA in the nucleus of each cell. These RNA copies carry the message of DNA's genetic code and direct the manufacture of proteins.

Other Nucleotides Perform Other Functions

Not all nucleotides are part of nucleic acids. Some exist singly in the cell or as parts of other molecules. Cyclic nucleotides, such as *cyclic adenosine monophosphate (cyclic AMP)*, act as messengers within cells, carrying information to other molecules in the cell. Nucleotides with extra phosphate groups, such as *adenosine triphosphate (ATP)*, carry energy from place to place within a cell. They pick up energy where it is produced (during photosynthesis, for example) and deliver it to drive energy-demanding reactions elsewhere (say, to manufacture a protein). Other nucleotides (NAD^+ and FAD) are known as electron carriers and transport energy in the form of high-energy electrons.

IMPROVING ON NATURE? *REVISITED*

Many of those who were most concerned about the safety of the artificial fat substitute olestra were not persuaded by the movie theater experiment described at the beginning of this chapter. The skeptics argued that the experimental subjects consumed only a small amount of olestra and ate it only once. Surely, contended the skeptics, such limited exposure could not establish olestra's safety.

A later study increased subjects' exposure to olestra. In this study, 60 subjects were isolated in a hospital ward for nine days so that researchers could control their diets and directly monitor their digestive health. After three days to establish each subject's baseline health, each patient's diet was supplmented for six days with either wheat germ or olestra (or, in the control group, with nothing at all). The amounts of added wheat germ and olestra were much higher than would be eaten in a normal diet. The result: The digestive health of subjects who consumed wheat germ was significantly worse than that of the control group, but the digestion of subjects who consumed olestra was no different from that of the control group. The researchers concluded that "consumption of olestra in excess of usual snacking conditions did not result in a significant increase ... of common gastrointestinal symptoms."

In 2003, after reviewing the results of all research on the health effects of olestra, the U.S. Food and Drug Administration (FDA) decided that products containing olestra would no longer be required to bear the warning label that had been required since the product was first approved in 1996. In the view of the policy makers at the FDA, the available scientific evidence demonstrates that olestra is safe for human consumtion.

The FDA's decision did not settle the question to everyone's satisfaction. Michael Jacobsen of the Center for Science in the Public Interest reports that consumer complaints continue, and he declared "that this product was ever allowed on the market at all will go down in history as one of the biggest blunders at the FDA."

CONSIDER THIS: Some experts argue that we should encourage people to use artificial sweeteners and fake fats, because obesity will decline if people can eat appealing food while limiting fat and sugar consumption. Critics of fake foods, however, contend that people should avoid the risks associated with artificial food additives and, instead, should consume nutritious foods that are naturally low in sugar and fats. What do you think?

LINKS TO LIFE: Health Food?

Some reactions in cells give rise to molecules that contain atoms with unpaired electrons in their outer shells. This type of molecule, called a *free radical*, is very unstable and reacts readily with nearby molecules, capturing an electron to complete its outer shell. But, by stealing an electron from another molecule, it creates a new free radical and begins a chain reaction that can lead to the destruction of biological molecules crucial to life. Damage caused by free radicals contributes to a variety of human ailments, including heart disease, nervous system disorders such as Alzheimer's disease, and some forms of cancer. Fortunately, other molecules, called *antioxidants*, react with free radicals and render them harmless. Our bodies synthesize several antioxidants, and others can be obtained from a healthy diet. Vitamins E and C, for example, are antioxidants.

Although it is difficult to do controlled studies on the effects of antioxidants in the human diet, there is some indirect evidence that a diet high in antioxidants is beneficial to health. For example, the low incidence of heart disease among the French (many of whom who eat a relatively high-fat diet) may be due in part to antioxidants in wine, which the French consume regularly. The French also eat considerably more fruits and vegetables, which are high in antioxidants, than do Americans.

Now, perhaps amazingly, it appears that chocolate, often a source of guilt for those who indulge in it, might contain antioxidants and thus be a type of health food. Researchers have given us an excuse to eat chocolate and feel good about it. Cocoa powder (the dark, bitter powder made from the seeds inside the cacao pod) contains high concentrations of flavenoids, which are powerful antioxidants, chemically related to those found in wine. So, does this finding mean that consumption of chocolate reduces the risk of cancer and heart disease? No studies have been done yet, but one suspects there will be no shortage of volunteer subjects for the research. Although weight gained from eating too much chocolate candy could certainly counteract any positive effects of cocoa powder, chocoholics now have reason to relax and enjoy some modest indulgence.

Chapter Review

Summary of Key Concepts

2.1 What Are Atoms?

An element is a substance that can neither be broken down nor converted to different substances by ordinary chemical means. The smallest possible particle of an element is the atom, which is itself composed of a central nucleus, containing protons and neutrons, and electrons outside the nucleus. All atoms of a given element have the same number of protons, which is different from the number of protons in the atoms of every other element. Electrons orbit the nucleus in electron shells, at specific distances from the nucleus. Each shell can contain a fixed maximum number of electrons.

2.2 How Do Atoms Form Molecules?

The chemical reactivity of an atom depends on the number of electrons in its outermost electron shell. An atom is most stable, and therefore least reactive, when its outermost shell is either completely full or empty. Atoms combine to form molecules, which are held together by chemical bonds. Oppositely charged ions may be held together by ionic bonds. When two atoms share electrons, covalent bonds form. In a nonpolar covalent bond, the two atoms share electrons equally. In a polar covalent bond, one atom may attract the electron more strongly than the other atom does; in this case, the strongly attracting atom bears a slightly negative charge, and the weakly attracting atom bears a slightly positive charge. Some polar covalent bonds give rise to hydrogen bonding, the attraction between charged regions of individual polar molecules.

2.3 Why Is Water So Important to Life?

The water molecule is important to organisms because of its ability to interact with many other molecules and to dissolve many polar and charged substances, to participate in chemical reactions, and to cohere to itself.

2.4 Why Is Carbon So Important to Life?

Organisms depend on carbon's ability to form a huge variety of different molecules. Carbon-containing molecules are so diverse because the carbon atom is able to form many types of bonds. This ability, in turn, allows organic molecules (molecules with a backbone of carbon and hydrogen atoms) to form many complex shapes, including chains, branches, and rings.

2.5 How Are Biological Molecules Joined Together or Broken Apart?

Most large biological molecules are synthesized by linking together many smaller subunits. Chains of subunits are connected by covalent bonds through dehydration synthesis. These chains may be broken apart by hydrolysis reactions. The most important organic molecules fall into four classes: carbohydrates, lipids, proteins, and nucleic acids. Their major characteristics are summarized in Table 2-3.

2.6 What Are Carbohydrates?

Carbohydrate molecules are generally composed of carbon, hydrogen, and oxygen in the ratio 1 carbon : 2 hydrogens : 1 oxygen. Carbohydrates include sugars, starches, and cellulose. Sugars (monosaccharides and disaccharides) are used for temporary storage of energy and for the construction of other molecules. Starches and glycogen are polysaccharides that serve for longer-term energy storage in plants and animals, respectively. Cellulose and related polysaccharides form the cell walls of bacteria, fungi, plants, and some microorganisms.

2.7 What Are Lipids?

Lipids are nonpolar, water-insoluble molecules with diverse chemical structures. They include oils, fats, waxes, phospholipids, and steroids. Lipids are used for energy storage (oils and fats), as waterproofing for the outside of plants and animals (waxes), as the principal component of cellular membranes (phospholipids), and as hormones (steroids).

2.8 What Are Proteins?

Proteins are chains of amino acids. The sequence of amino acids in the chain determines the structure of a protein. A protein is functional when folded into its characteristic three-dimensional shape. Proteins include enzymes (which guide chemical reactions), structural molecules (hair, horn), hormones (insulin), and transport molecules (hemoglobin).

2.9 What Are Nucleic Acids?

Nucleic acid molecules are chains of nucleotides. Each nucleotide is composed of a phosphate group, a sugar group, and a nitrogen-containing base. The two types of nucleic acids are deoxyribonucleic acid (DNA) and ribonucleic acid (RNA). Nucleotides that function singly include intracellular messengers (cyclic AMP) and energy-carrier molecules (ATP).

Key Terms

acid *p. 23*
amino acid *p. 31*
atom *p. 18*
atomic nucleus *p. 18*
atomic number *p. 18*
base *p. 24*
buffer *p. 24*
carbohydrate *p. 26*
chemical bond *p. 19*
chemical reaction *p. 20*
cohesion *p. 22*
covalent bond *p. 20*
dehydration synthesis *p. 25*
denature *p. 33*
electron *p. 18*
electron shell *p. 18*
element *p. 18*
functional group *p. 24*
hydrogen bond *p. 21*
hydrolysis *p. 25*
ion *p. 20*
ionic bond *p. 20*
isotope *p. 18*
lipid *p. 28*
molecule *p. 19*
neutron *p. 18*
nonpolar molecule *p. 21*
nucleic acid *p. 33*
organic molecule *p. 24*
pH scale *p. 24*
polar molecule *p. 21*
protein *p. 31*
proton *p. 18*
radioactive *p. 18*
solvent *p. 22*
surface tension *p. 22*
wax *p. 30*

Thinking Through the Concepts

Multiple Choice

1. *When an atom ionizes, what happens?*
   a. It shares one or more electrons with another atom.
   b. It emits energy as it loses extra neutrons.
   c. It gives up or takes up one or more electrons.
   d. It shares a hydrogen atom with another atom.
   e. none of the above

2. *A covalent bond forms*
   a. when two ions are attracted to one another
   b. between adjacent water molecules, producing surface tension
   c. when one atom gives up its electron to another atom
   d. when two atoms share electrons
   e. between water molecules and fat globules

3. *What is the defining characteristic of an acid?*
   a. It donates hydrogen ions.
   b. It accepts hydrogen ions.
   c. It will donate or accept hydrogen ions, depending on the pH.
   d. It has an excess of hydroxide ions.
   e. It has a pH greater than 7.

4. *Which of the following is not a function of polysaccharides in organisms?*
   a. energy storage in plants
   b. storage of hereditary information
   c. formation of cell walls
   d. structural support
   e. energy storage in animals

5. *Proteins differ from one another because*
   a. the peptide bonds linking amino acids differ from protein to protein
   b. the sequence of amino acids in the polypeptide chain differs from protein to protein
   c. each protein molecule contains its own unique sequence of sugar molecules
   d. the number of nucleotides in each protein varies from molecule to molecule
   e. the number of nitrogen atoms in each amino acid differs from the number in all others

6. *Which of the following statements about lipids is false?*
   a. A wax is a lipid.
   b. Unsaturated fats are liquid at room temperature.
   c. The body doesn't need any cholesterol.
   d. Both male and female sex hormones are steroids.
   e. Beef fat is highly saturated.

Review Questions

1. Distinguish atoms from molecules; elements from compounds; and protons, neutrons, and electrons from each other.
2. Compare and contrast covalent bonds, ionic bonds, and hydrogen bonds.
3. Describe how water dissolves a salt.
4. Define *acid*, *base*, and *buffer*. How do buffers reduce changes in pH when hydrogen ions or hydroxide ions are added to a solution? Why is this phenomenon important in organisms?
5. Which elements are common components of biological molecules?
6. List the four principal types of biological molecules, and give an example of each.
7. What roles do nucleotides play in organisms?
8. Distinguish among the following: monosaccharide, disaccharide, and polysaccharide. Give two examples of each and their functions.
9. Describe the manufacture of a protein from amino acids.

Applying the Concepts

1. **IS THIS SCIENCE?** Headlines on a magazine cover proclaim: "Turn fat into muscle!" Evaluate this claim from a scientific standpoint.
2. Drugstores sell many different brands of "antacid" remedies, which are intended to bring relief from "acid stomach." Each brand claims to eliminate symptoms faster than its competitors. How do these compounds work? Use your knowledge of acids, bases, and buffers to design an experiment to determine which brand of antacid works best.
3. Many of water's unique properties are the result of its polar covalent bonds, which allow water molecules to form hydrogen bonds with each other. What if water molecules instead had nonpolar covalent bonds? Using information from this chapter, make a list of hypotheses about the ways in which this change might affect the properties of water. Describe how each change would affect living things. Design an experiment to test one of your hypotheses.

For More Information

Atkins, P. W. *Molecules*. New York: Scientific American Library, 1987. A layperson's introduction to atoms and molecules, with superb illustrations.

Burdick, A. "Cement on the Half Shell." *Discover*, February 2003. Mussels produce a protein polymer that is waterproof and incredibly strong.

Goodsell, D. S. *The Machinery of Life*. New York: Springer, 1993. Goodsell depicts the molecules of a cell in all their three-dimensional, interactive glory. A great way to gain a feel for the beauty and intricacy of the organic molecules of life.

Hill, J. W., and Kolb, D. K. *Chemistry for Changing Times*. 10th ed. Upper Saddle River, NJ: Prentice Hall, 2004. A chemistry textbook for nonscience majors that is both clearly readable and thoroughly enjoyable.

King, J., Haase-Pettingell, C., and Gossard, D. "Protein Folding and Misfolding." *American Scientist*, September–October 2002. Protein folding holds the key to proteins' diverse functions.

Kunzig, R. "Arachnomania." *Discover*, September 2001. Researchers work to unravel the mystery of spider silk protein and develop a process to synthesize it.

Morrison, P., and Morrison, P. *Powers of Ten*. New York: W. H. Freeman, 1982. A fascinating journey from the universe to the nucleus of an atom.

WEB TUTORIAL

To access a Web Tutorial visit http://www.prenhall.com/audesirk4. *Log in to the Web site selected by your instructor, navigate to this chapter, and select the appropriate Web Activity number.*

2.1 Interactive Atoms
Estimated Time: 10 minutes

An introduction to the basic chemistry needed to understand biological structures and processes.

2.2 Water and Life
Estimated Time: 10 minutes

Explore the properties of water and why this molecule is essential for life.

2.3 Structure of Biological Molecules
Estimated Time: 10 minutes

Review the chemical structures of the most important biological macromolecules.

2.4 Functions of Macromolecules
Estimated Time: 5 minutes

Explore the function of the major macromolecules.

2.5 Web Investigation: Improving on Nature?
Estimated time: 10 minutes

Whereas most artificial sweeteners, food colors, and fat replacers can be used for any food product, olestra is currently approved by the U.S. Food and Drug Administration (FDA) for use in "savory snacks" (for instance, potato chips) only. Some experts say that it is safe. Others think the side effects outweigh the benefits. In this exercise we will examine the pros and cons of the issue.

Chapter 3

Cell Membrane Structure and Function

Our bodies cannot function without cholesterol, but too much can threaten our health. Excess cholesterol in our blood can contribute to hard deposits (colored gray in this photo) that block arteries and cause heart attacks.

AT A GLANCE

Can Teens Have Heart Attacks?

Did you know that high cholesterol can be a problem for young people? One of every 500 or so people has a genetic defect that causes elevated blood cholesterol. This disorder is known as familial hypercholesterolemia (FH). Chances are good that a young person you know has a copy of the defective FH gene and is unknowingly at increased risk for high cholesterol and heart disease.

Some unfortunate people carry *two* copies of the defective gene and are born with an extremely high level of blood cholesterol. It is not unusual for a person with this severe form of FH to die of a heart attack while still in his or her teens or twenties.

What causes FH? Answering that question is the key to developing treatments for the disorder. Most research on the question has focused on a protein that helps cells pull cholesterol out of the blood. In particular, researchers have tested the hypothesis that lack of this protein causes symptoms similar to those found in FH patients.

One group of researchers worked with mice that were genetically engineered to be unable to produce the cholesterol-removing protein. The researchers predicted that the genetically engineered mice (which lack the protein) would have higher levels of cholesterol than normal mice. An experiment confirmed their prediction: Cholesterol levels were nine times higher in the engineered mice. Then, when the missing protein was injected into the engineered mice, their levels of cholesterol declined to near normal. The researchers concluded that, in mice, the protein was required to maintain normal cholesterol levels, and lack of the protein caused FH-like symptoms. Tests on human FH victims confirmed that they, like the engineered mice, did not produce the cholesterol-removing protein.

How could a single defective protein have such an extreme effect on a person's health? As you read this chapter, see if you can guess how the cholesterol-removing protein performs its function, and why a defective version might cause familial hypercholesterolemia.

Inside our bodies, cell membranes shoulder much of the task of keeping blood cholesterol levels within bounds. This micrograph shows portions of the cell membranes (colored yellow) of two adjacent cells.

3.1 What Does the Plasma Membrane Do?

The cell is the smallest unit of life. Each cell is surrounded by a thin **plasma membrane**, which isolates the cell's contents from the external environment. The membrane also acts as a gatekeeper, controlling which substances are allowed to pass in or out and transferring chemical messages from the external environment to the cell's interior. These are formidable tasks for a structure so thin that 10,000 plasma membranes stacked on one another would scarcely equal the thickness of this page.

At first glance, a plasma membrane might appear to be a simple film surrounding a cell, similar to a soap bubble. Membranes, however, are complex structures that contain many different components, each of which performs a particular function.

3.2 What Is the Structure of the Plasma Membrane?

The ability of the plasma membrane to do its jobs is closely tied to the way membranes are structured. The overall organization of membranes can be described as proteins floating in a double layer of lipids. The lipids are responsible for the isolating function of membranes, and the proteins regulate the exchange of substances across the membrane and communication with the cell's environment.

Membranes Are "Fluid Mosaics"

One way to picture the structure of a membrane is to think of it as a **fluid mosaic**. A fluid is a liquid or a gas—that is, a substance that changes shape without breaking apart and whose molecules move freely relative to one another. The word "mosaic" refers to anything that resembles a mosaic artwork made by setting small colored pieces, such as tile, in mortar. A membrane, then, when viewed at high magnification, looks something like a lumpy, constantly shifting mosaic of tiles. A double layer of phospholipids forms a thick but still liquid "mortar" for the mosaic, and various proteins are the "tiles," which can move about within the phospholipid layers (**Fig. 3-1**). This movement gives a dynamic, ever-changing quality to membranes (even though the actual ingredients of the membrane remain relatively constant).

The Phospholipid Bilayer Is the Fluid Portion of the Membrane

As you learned in Chapter 2, a phospholipid consists of two very different parts: a polar, hydrophilic head and a pair of nonpolar, hydrophobic tails.

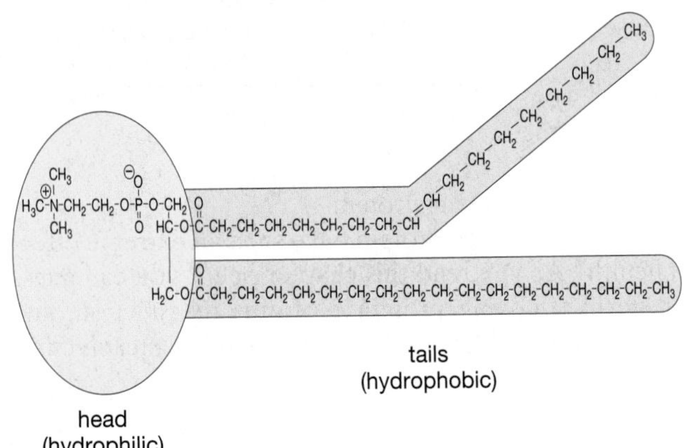

Both the internal and external environments of a cell are watery. Inside the cell, the cytoplasm, which consists of all of a cell's internal contents (except the nucleus), is mostly water. Outside the cell, there is more water. Single-celled organisms may live in fresh water, in the ocean, or in the moisture that clings to soil particles and other surfaces. In multicelled organisms, cells are bathed in a fluid that is mostly water. So, plasma membranes separate a cell's watery cytoplasm from its watery external environment.

Under these wet conditions, phospholipids spontaneously arrange themselves into a double layer called a **phospholipid bilayer**. In this arrangement, the phospholipid heads form the outer surfaces, because they are **hydrophilic**, which

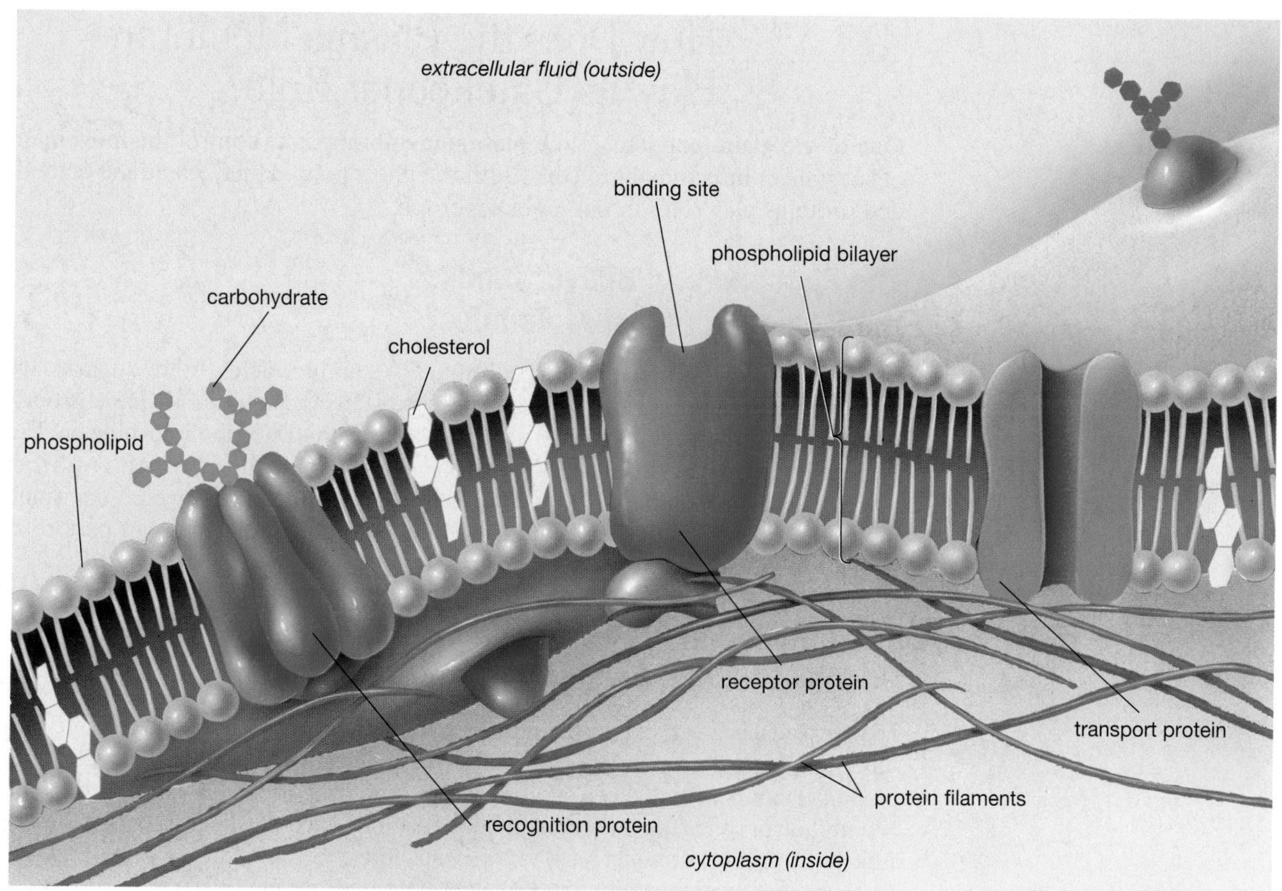

Figure 3-1 The plasma membrane is a fluid mosaic The plasma membrane is a bilayer of phospholipids in which various proteins are embedded. Many proteins have carbohydrates attached to them. The wide variety of membrane proteins fall mostly into three categories: recognition proteins, receptor proteins, and transport proteins.

means that they are attracted to water molecules and tend to dissolve in water. In contrast, the phospholipid tails are **hydrophobic**, so they repel water and "hide" inside the membrane. ➔

Hydrogen bonds can form between water and the phospholipid heads, so each hydrophilic head faces either the cytoplasm or the extracellular fluid. Because individual phospholipid molecules are not bonded to one another, individual phospholipids can move about easily within each layer.

In most cells, the phospholipid bilayer of membranes also contains cholesterol molecules. Cholesterol makes the bilayer stronger, more flexible, and less permeable to water-soluble substances such as ions or simple sugars. (You may want to review Fig. 2-14 and "Health Watch: Cholesterol—Friend and Foe" in Chapter 2.)

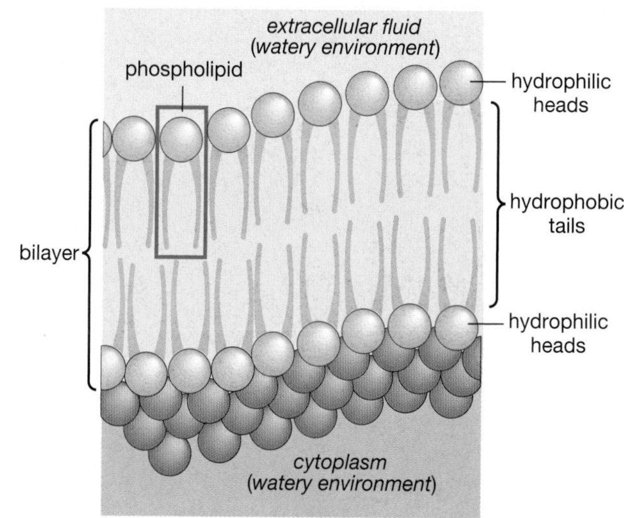

A Mosaic of Proteins Is Embedded in the Membrane

Thousands of proteins are embedded within or attached to the surface of a membrane's phospholipid bilayer. Some of them can move about within the fluid phospholipid bilayer, but others are anchored to a network of protein filaments within the cytoplasm. Many of the proteins have carbohydrate groups attached to them, especially to the parts that stick outside the cell. Together, this collection of proteins is responsible for moving substances across the membrane and for communicating with other cells.

3.3 How Does the Plasma Membrane Play Its Gatekeeper Role?

One of the main functions of the plasma membrane is to control the movement of substances into and out of cells. Both the phospholipid bilayer and the embedded proteins play roles in this gatekeeper job.

The Phospholipid Bilayer Blocks the Passage of Most Molecules

Most biological molecules, including salts, amino acids, and sugars, are hydrophilic: polar and water soluble. These substances cannot easily pass through the nonpolar, hydrophobic fatty acid tails within the phospholipid bilayer. The phospholipid bilayer thus plays the main role in isolating the cell's contents from the external environment. The isolation is not complete, however. Very small molecules, such as water, and uncharged, lipid-soluble molecules can pass relatively freely through the lipid bilayer.

The Embedded Proteins Selectively Transport, Respond to, and Recognize Molecules

Most molecules that cross the membrane do so with the assistance of membrane proteins. In addition to providing transportation services, membrane proteins also play an important role in a cell's responses to the substances in its environment. Membrane proteins fall into three major categories, each of which serves a different function: transport proteins, receptor proteins, and recognition proteins (see Fig. 3-1).

Transport proteins allow the movement of hydrophilic (water-soluble) molecules through the plasma membrane. They do so either by forming channels through which molecules pass or by grabbing onto molecules and carrying them across the membrane.

Receptor proteins deliver chemical messages to the cell. They trigger responses inside the cell when specific molecules outside the cell, such as hormones or nutrients, bind to them. For example, when molecules of the hormone insulin bind to receptor proteins on the surface of one of the cells in your body, the cell responds by activating transport proteins that move glucose molecules across the plasma membrane and into the cell.

Recognition proteins serve as identification tags and cell-surface attachment sites. For example, the cells of your immune system use recognition proteins to distinguish your own cells from those of disease-causing invaders. Thus, a bacterium is recognized as a foreign invader to be destroyed while your blood cells are ignored, because the two types of cells have different recognition proteins on their surfaces.

Each of the three broad categories of membrane proteins contains many different proteins, each of which has a distinctive shape and does a very specific job. Thus, each type of transport protein carries only a particular type of molecule across the plasma membrane, each type of receptor protein binds only to one particular kind of molecule, and different cells bear distinctive collections of recognition proteins.

3.4 What Is Diffusion?

Before continuing our discussion of the transport of substances across the cell's plasma membrane, let's look at some characteristics of molecules in fluids. We begin with some definitions:

- The *concentration* of molecules in a fluid is the number of molecules in a given unit of volume.

- A *gradient* is a physical difference between two regions of space that causes molecules to move from one region to another. Molecules frequently encounter gradients of concentration, pressure, and electrical charge.

Molecules in Fluids Move in Response to Gradients

Concentration gradients are important because they influence the movement of molecules or ions within a fluid. For example, consider perfume molecules moving from an open bottle into the air (remember that both gases and liquids are fluids). The perfume is moving from an area of high perfume concentration (the air inside the bottle) to an area of low perfume concentration (the air outside the bottle). In other words, the perfume molecules move in response to a **concentration gradient**, a difference in concentration between one region and another. The movement of molecules from regions of high concentration to regions of low concentration is called **diffusion**. We refer to such movements as going "down" the concentration gradient (from high to low). If there are no other factors opposing diffusion, it will continue until the substance is evenly dispersed throughout the fluid.

A Drop of Dye in Water Illustrates Diffusion

To watch diffusion in action, place a drop of food coloring in a glass of water, and check its progress every few minutes. With time, the drop will seem to spread out and become paler, until eventually, even without stirring, the entire glass of water will be uniformly faintly colored. Molecules of dye move from the region of high dye concentration into the surrounding water where the dye concentration is low (Fig. 3-2). Simultaneously, some water molecules enter the dye droplet, and the net movement of water is from the high water concentration outside the drop into the lower water concentration inside the drop.

At first, there is a very steep concentration gradient, and the dye diffuses rapidly. As the concentration differences lessen, the dye diffuses more and more slowly. In other words, the greater the concentration gradient, the faster the rate of diffusion. However, as long as the concentration of dye within the expanding drop is greater than the concentration of dye in the rest of the glass, the net movement of dye will continue until the dye becomes uniformly dispersed in the water. At that point, the concentration gradient has been eliminated, and the rate of diffusion has therefore dropped to zero. Water and dye molecules continue to move, of course, but they move at random in all directions, with no net movement in any particular direction.

Figure 3-2 Diffusion of a dye in water

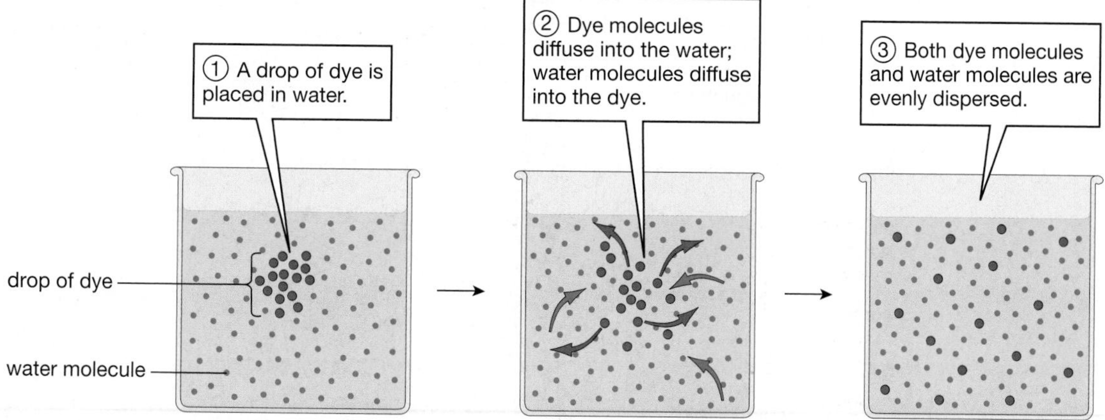

Summing Up: The Principles of Diffusion

1. Diffusion is the movement of molecules down a gradient from high concentration to low concentration.
2. The greater the concentration gradient, the faster the rate of diffusion.
3. If no other processes intervene, diffusion will continue until the concentration gradient is eliminated.

3.5 What Is Osmosis?

Like other molecules, water diffuses from regions of high water concentration to regions of low water concentration. What do we mean when we describe a solution as having a "high water concentration" or a "low water concentration"? The answer is simple: Pure water has the highest water concentration. Any substance added to pure water displaces some of the water molecules, and the resulting solution will have a lower water content than pure water. The higher the concentration of dissolved substances, the lower the concentration of water.

The movement of water across membranes from areas of high water concentration to areas of low water concentration has such dramatic and important effects that we refer to it by a special name: **osmosis**. Osmosis moves water across **selectively permeable** membranes, which are so called because they allow the passage of some molecules but prevent the passage of other molecules. A very simple selectively permeable membrane might have pores just large enough for water to pass through but small enough to prevent the passage of sugar molecules (**Fig. 3-3a**). Consider a bag made of a special plastic that is permeable to water but not to sugar. What will happen if we place a sugar solution in the bag and then immerse the sealed bag in pure water? The principles of osmosis tell us that the bag will swell. If it is weak enough, it will eventually burst (**Fig. 3-3b**).

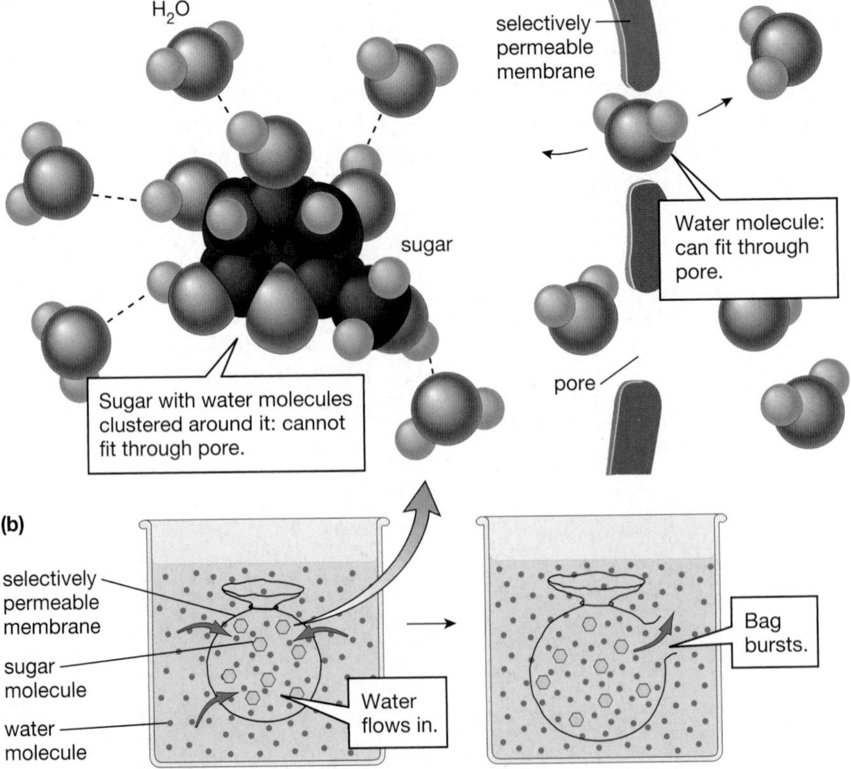

Figure 3-3 Osmosis Water molecules diffuse down their concentration gradient across a selectively permeable membrane, from a region of high water concentration to a region of lower water concentration. **QUESTION:** *Imagine a container of glucose solution, divided into two compartments (A and B) by a membrane that is permeable to water and glucose but not to sucrose. If some sucrose is added to compartment A, how will the contents of compartment B change?*

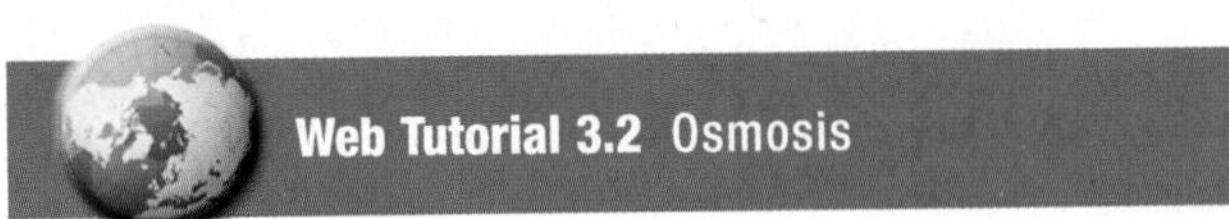

Summing Up: The Principles of Osmosis

1. Osmosis is the diffusion of water across a selectively permeable membrane.
2. Dissolved substances reduce the concentration of water molecules in a solution.
3. Water moves across a membrane down its concentration gradient from a high concentration of water molecules to a low concentration of water molecules.

3.6 How Do Diffusion and Osmosis Affect Transport across the Plasma Membrane?

As shown in Table 3-1, there are two main types of transport across membranes. One of these types is **passive transport**. During passive transport, substances move into or out of cells down concentration gradients. Cells have concentration gradients across their plasma membranes because the composition of the cytoplasm is very different from that of the extracellular fluid. The direction of movement across membranes by passive transport is determined entirely by concentration gradients, but whether a particular substance can cross is controlled by the lipids and proteins of the membrane. Passive transport requires no expenditure of energy. It may be accomplished by simple diffusion, facilitated diffusion, or osmosis.

Plasma Membranes Are Selectively Permeable

Molecules are able to diffuse across membranes, and many molecules do in fact cross plasma membranes by diffusion, driven by differences between their concentration in the cytoplasm and in the external environment. Not all molecules, however, are equally likely to be able to diffuse across a membrane, because plasma membranes are selectively permeable. Some molecules can pass through but other molecules cannot.

Some Molecules Move across Membranes by Simple Diffusion

Molecules that dissolve in lipids, such as ethyl alcohol and vitamin A, easily diffuse across the phospholipid bilayer, as do very small molecules, including water and dissolved gases such as oxygen and carbon dioxide. This type of passive transport

TABLE 3-1 Transport across Membranes

Passive transport	Movement of substances across a membrane by traveling down a concentration, pressure, or electrical charge gradient. Does not require the cell to expend energy.
Simple diffusion	Diffusion of water, dissolved gases, or lipid-soluble molecules through the phospholipid bilayer.
Facilitated diffusion	Diffusion of (normally water-soluble) molecules through a channel or carrier protein.
Osmosis	Diffusion of water across a selectively permeable membrane (a membrane that is more permeable to water than to dissolved molecules).
Energy-requiring transport	Movement across a membrane of substances that travel against a concentration gradient. Requires the cell to expend energy.
Active transport	Movement of individual small molecules or ions through membrane-spanning proteins, using cellular energy, normally ATP.
Endocytosis	Movement into a cell of large particles that are engulfed as the plasma membrane forms sacs that enter the cytoplasm.
Exocytosis	Movement out of a cell of materials that are enclosed in a membranous sac that moves to the cell surface, fuses with the plasma membrane, and opens to the outside, allowing its contents to diffuse away.

(a) Simple diffusion

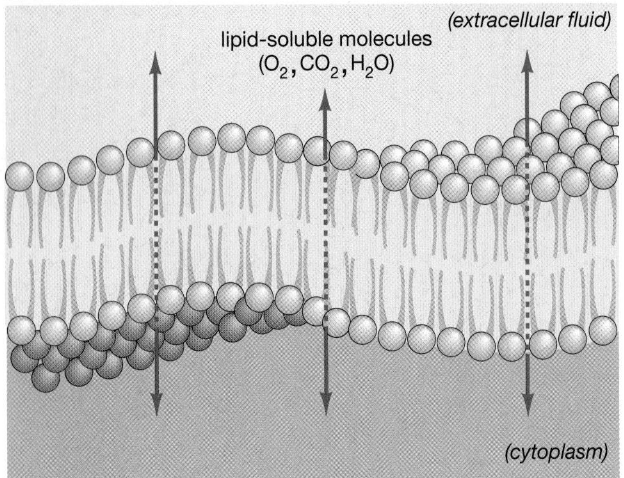

(b) Facilitated diffusion through a channel

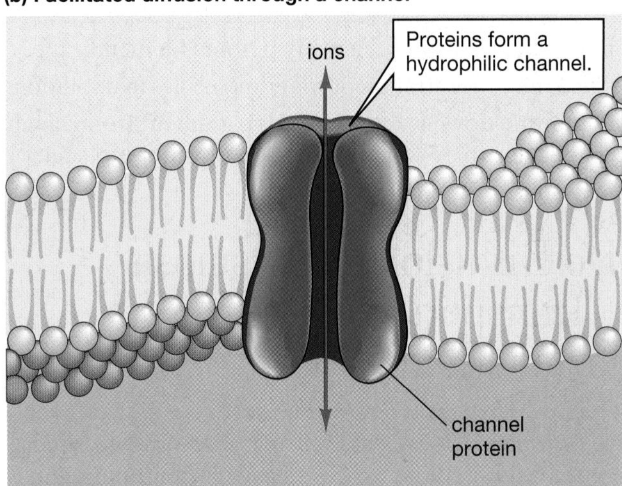

(c) Facilitated diffusion through a carrier

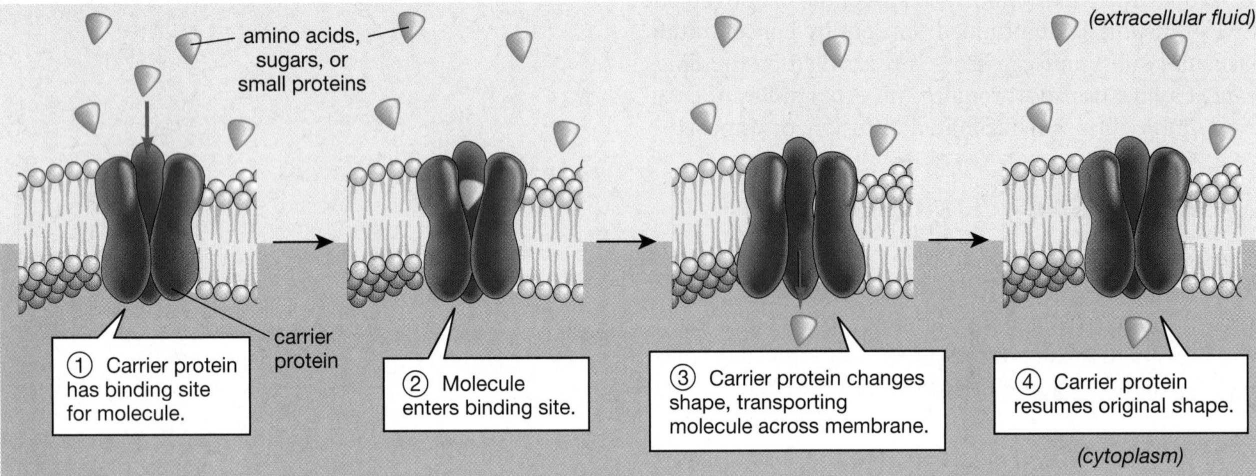

Figure 3-4 Diffusion through the plasma membrane ***(a)*** Lipid-soluble molecules and gases such as oxygen and carbon dioxide can pass by simple diffusion directly through the phospholipids. ***(b)*** Some water-soluble molecules enter or exit the cell by facilitated diffusion through a channel protein. ***(c)*** Certain molecules cross a membrane by facilitated diffusion through a carrier protein that changes shape to allow the passage. **EXERCISE:** *Imagine an experiment that measures the initial rate of diffusion into cells placed in sucrose solutions of various concentrations. Sketch one graph (initial diffusion rate versus solution concentration) that shows the result expected if diffusion is simple and another graph that shows the result expected for facilitated diffusion.*

is called **simple diffusion** (Fig. 3-4a). Generally, the rate of simple diffusion depends on the concentration gradient across the membrane, the size of the molecule, and how easily it dissolves in lipids. Large concentration gradients, small molecule size, and high lipid solubility all increase the rate of simple diffusion.

Other Molecules Cross the Membrane by Facilitated Diffusion

Most water-soluble molecules, such as ions (K^+, Na^+, Ca^{2+}), amino acids, and simple sugars, cannot move through the phospholipid bilayer on their own. These molecules can diffuse across only by **facilitated diffusion**, a type of passive transport in which diffusion is aided by either of two types of transport proteins: channel proteins or carrier proteins.

Channel proteins form pores, or channels, in the lipid bilayer through which certain ions can cross the membrane (Fig. 3-4b). In general, each type of channel protein is specialized and allows only particular ions to pass through. Nerve cells, for example, have separate channels for potassium ions (K^+), sodium ions (Na^+), and calcium ions (Ca^{2+}).

A **carrier protein** grabs onto a specific molecule on one side of a membrane and carries it to the other side. Each different type of carrier protein is able to bind to a specific molecule (typically an amino acid, sugar, or small protein). Binding triggers a change in the shape of the carrier that allows the bound molecule to pass through the protein and across the plasma membrane (Fig. 3-4c).

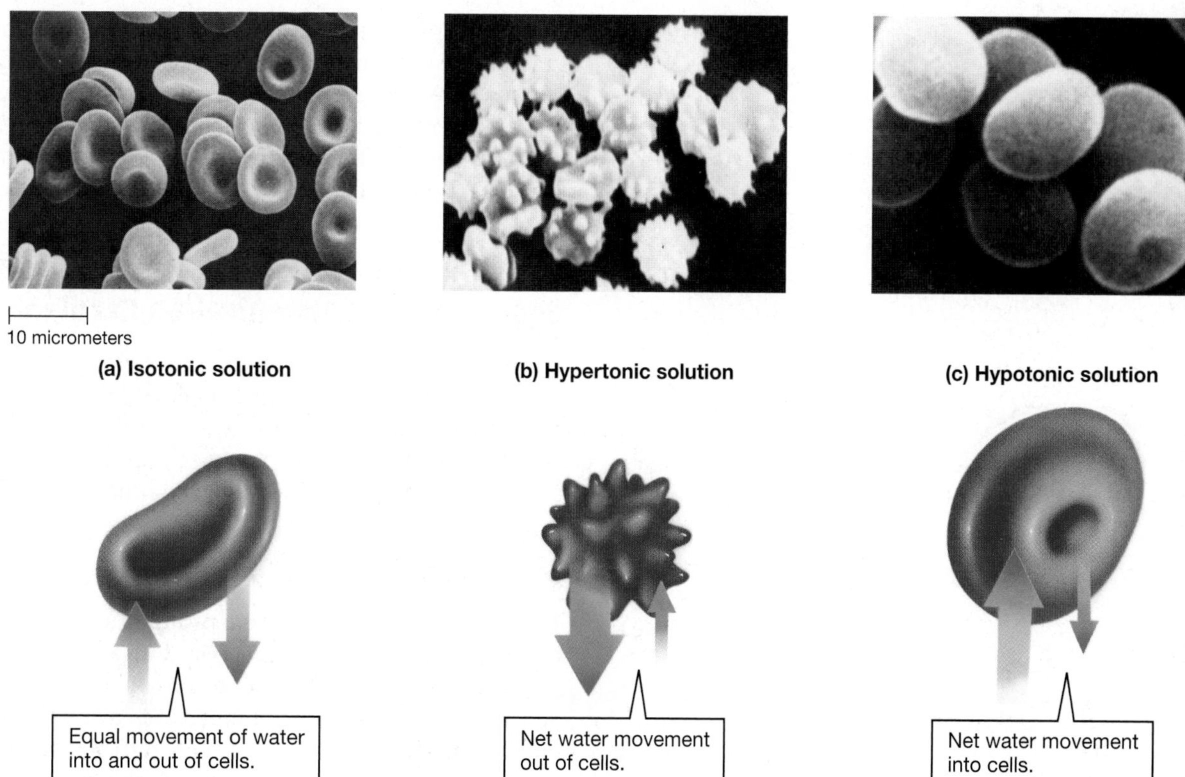

Figure 3-5 The effects of osmosis Red blood cells are normally suspended in the fluid environment of the blood. ***(a)*** If red blood cells are immersed in an isotonic salt solution, which has the same concentration of dissolved substances as the blood cells do, there is no net movement of water across the plasma membrane. The red blood cells keep their characteristic dimpled disk shape. ***(b)*** A hypertonic solution, with too much salt, causes water to leave the cells, shriveling them up. ***(c)*** A hypotonic solution, with less salt than is in the cells, causes water to enter, and the cells swell. **QUESTION:** *A freshwater fish swims in a solution that is hypotonic compared with the fluid inside its body. Why don't freshwater fish swell up and burst?*

Water Can Move across Plasma Membranes by Osmosis

Most plasma membranes are highly permeable to water. The flow of water across them depends on whether water is more highly concentrated inside or outside the cell. Ordinarily, the extracellular fluid outside the cell has the same concentration as the cytoplasm inside the cell. In that case, the extracellular fluid is said to be *isotonic* to the cytoplasm, and there will be no net movement of water into or out of the cell (**Fig. 3-5a**). Note that the *types* of dissolved particles are seldom the same inside and outside the cells, but the *total concentration* of all dissolved particles is equal, with the result that the water concentration inside is equal to that outside the cells.

If red blood cells are taken out of the body and immersed in salt solutions of varying concentrations, the effects of osmosis become dramatically apparent. If the solution has a higher salt concentration than the cytoplasm of the red blood cell (that is, if the solution has a lower water concentration), water will leave the cells by osmosis. The cells in such a *hypertonic* solution will shrivel up until the concentrations of water inside and outside become equal (**Fig. 3-5b**). If, on the other hand, the solution has little salt (is *hypotonic*), water will enter the cells, causing them to swell (**Fig. 3-5c**). This process explains why your fingers wrinkle up after a long bath. It may seem as if your fingers are shrinking, but they're not. Instead, water is moving into the outer skin cells of your fingers, swelling them more rapidly than the cells underneath and causing the wrinkling.

3.7 How Do Molecules Move against a Concentration Gradient?

Passive transport, though undeniably efficient, could never by itself meet all of a cell's needs for moving molecules across membranes. Passive transport can carry substances only down concentration gradients, but all cells also need to move some materials "uphill" across their plasma membranes, against concentration gradients. For example, it is often necessary to move nutrients from the environment, where they are less concentrated, into the cell's cytoplasm, where they are more concentrated. Passive transport would instead move nutrient molecules out of the cell, so a different type of transport—one that requires energy—is necessary (see Table 3-1).

Active Transport Uses Energy to Move Molecules against Their Concentration Gradients

During **active transport**, the cell uses energy to move substances *against* a concentration gradient. A helpful analogy for understanding the difference between passive and active transports is to consider what happens when you ride a bike. If you don't pedal, you can go only downhill, as in passive transport. However, if you put enough energy into pedaling, you can go uphill as well, as in active transport.

Membrane Proteins Regulate Active Transport

In active transport, membrane proteins use energy to move individual molecules across the plasma membrane (Fig. 3-6). Active-transport proteins span the width of the membrane and have two active sites where other molecules can bind. One active site binds a particular molecule, say a calcium ion. The second site binds an energy-carrier molecule, normally adenosine triphosphate (ATP). The ATP donates energy to the active-transport protein, causing it to change shape and move the calcium ion across the membrane. Active-transport proteins are often called *pumps*, in analogy to water pumps, because they use energy to move molecules "uphill" against a concentration gradient.

Cells Engulf Particles or Fluids by Endocytosis

Sometimes cells must acquire particles that are too large to move across a membrane by either passive transport or active transport. One way in which cells can accomplish this task is by an energy-consuming process called **endocytosis**. During endocytosis, the plasma membrane engulfs an extracellular particle or a droplet of fluid and pinches off a membranous sac called a **vesicle**, with the particle or fluid inside, into the cytoplasm.

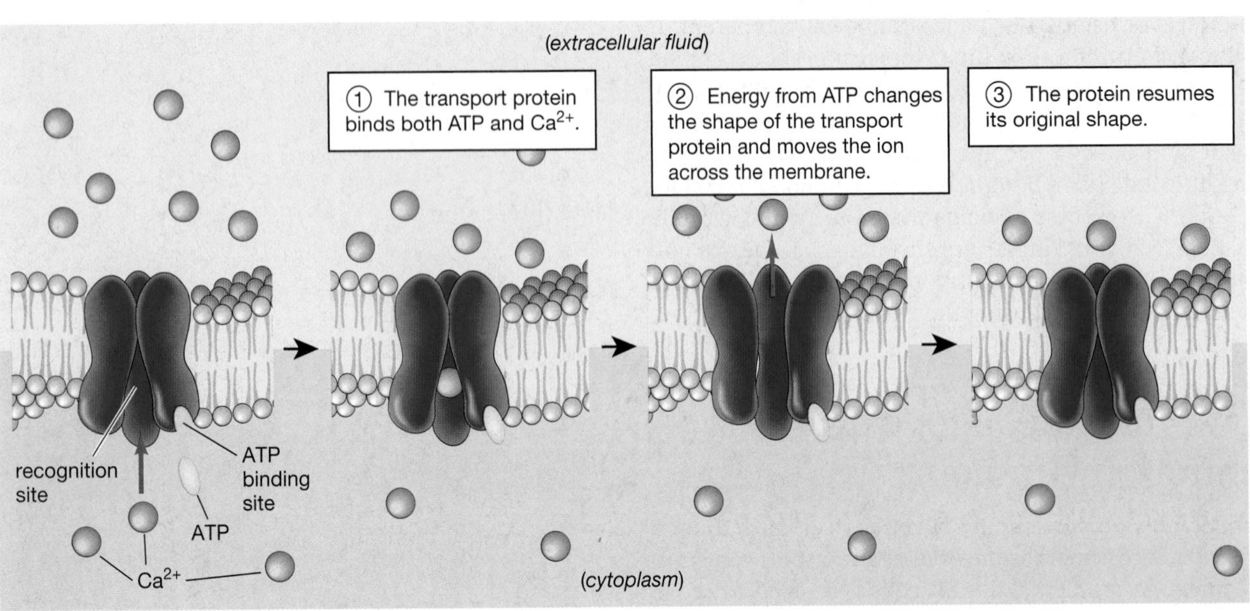

Figure 3-6 Active transport Active transport uses cellular energy to move molecules across the plasma membrane, often against a concentration gradient. An active-transport protein binds ATP and the molecule to be transported, and then changes shape to move the ion across the membrane.

(a) Pinocytosis

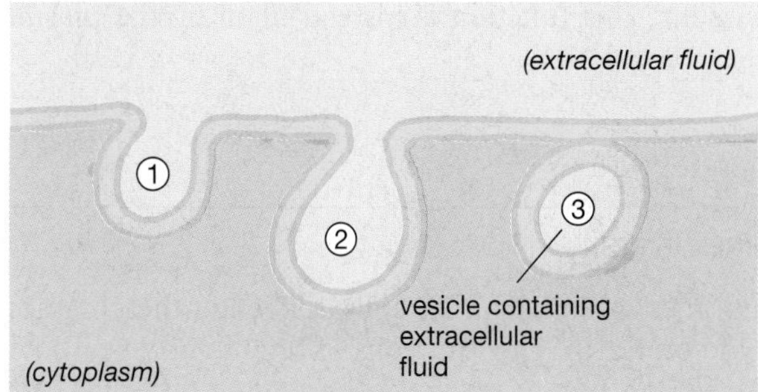

① A dimple forms in the plasma membrane, which ② deepens and surrounds the extracellular fluid. ③ The membrane encloses the extracellular fluid, forming a vesicle.

(b) Phagocytosis

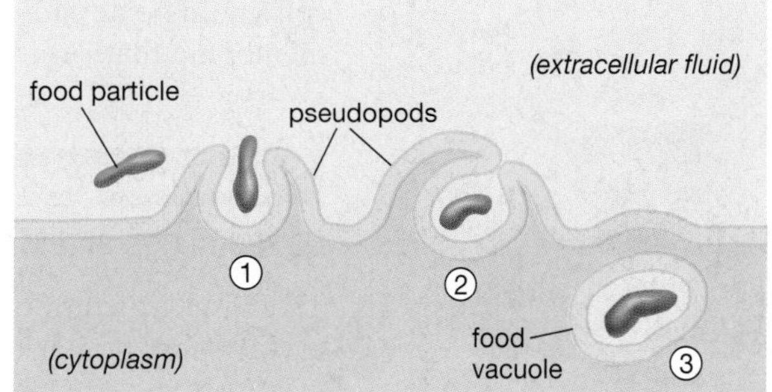

① The plasma membrane extends pseudopods toward an extracellular particle (for example, food). ② The ends of the pseudopods fuse, encircling the particle. ③ A vesicle called a food vacuole is formed containing the engulfed particle.

Figure 3-7 Two types of endocytosis

Endocytosis can proceed in several different ways, depending on the size and nature of the material to be captured. When a drop of liquid must be moved into a cell, a very small patch of plasma membrane dimples inward as it surrounds extracellular fluid and buds off into the cytoplasm as a tiny vesicle (**Fig. 3-7a**), a process called *pinocytosis.*

Another process, *phagocytosis,* is used to pick up larger particles, such as when an amoeba engulfs food or when a white blood cell destroys bacteria that have invaded a human body. In phagocytosis, the cell first extends parts of its surface membrane. These extensions are called *pseudopodia* (singular, pseudopod). The tips of the pseudopodia fuse around the particle and carry it to the interior of the cell inside a vesicle (**Fig. 3-7b**).

Exocytosis Moves Material out of the Cell

Cells often use the reverse of endocytosis, a process called **exocytosis**, to dispose of unwanted materials, such as the waste products of digestion, or to secrete materials, such as hormones (**Fig. 3-8**). During exocytosis, a vesicle carrying material to be expelled moves to the cell surface, where the vesicle's membrane fuses with the cell's plasma membrane. The vesicle then opens to the extracellular fluid, and its contents diffuse out.

Some Plasma Membranes Are Surrounded by Cell Walls

The plasma membranes of plants, fungi, and many types of bacteria lie inside stiff coatings called **cell walls**. Cell walls, which are produced by the cells they surround, support and protect otherwise fragile cells. For example, cell walls allow plants and mushrooms to resist the forces of gravity and blowing winds so that they can stand erect on land. Tree trunks, which can support enormous weight, demonstrate just how great the collective strength of cell walls can be. Despite their strength, cell walls are normally porous, permitting easy passage of small molecules such as minerals, water, oxygen, carbon dioxide, amino acids, and sugars. The cell wall,

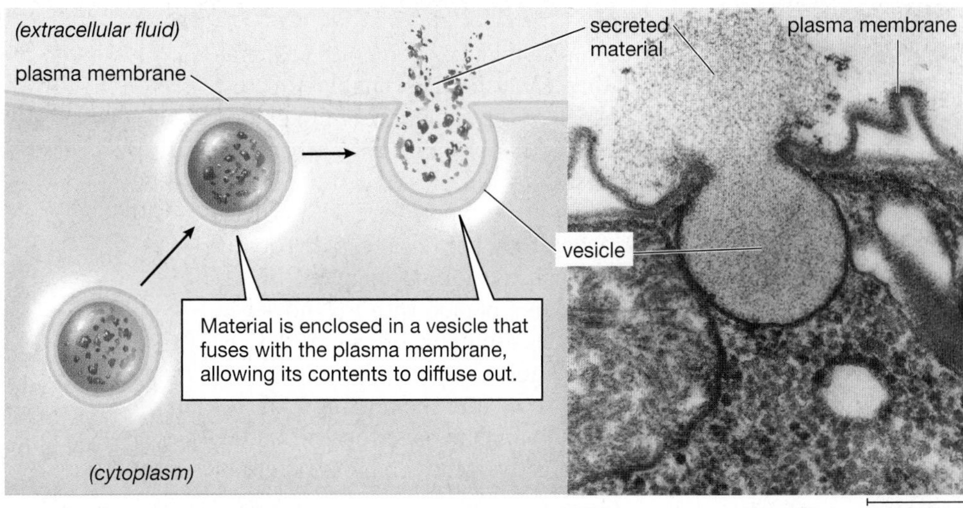

Figure 3-8 Exocytosis Exocytosis is functionally the reverse of endocytosis.

Figure 3-9 Caribou browse on the frozen Alaskan tundra How do they conserve body heat?

however, does not control the interaction between a plant, fungal, or bacterial cell and its external environment. This function is instead filled by the plasma membrane that lies inside the cell wall.

Evolutionary Connections

Caribou Legs and Membrane Diversity

The basic structure of membranes is similar in all cells, reflecting the common evolutionary heritage of all life on Earth. The functions of cell membranes, however, can vary tremendously among organisms and even among different cells within a single organism. These differences in membrane function stem largely from differences in membrane proteins and phospholipids.

To see how environmental conditions can influence the evolution of membrane components, consider the caribou, animals that live in very cold regions of North America (Fig. 3-9). During the long arctic winters of these regions, temperatures plummet far below freezing. These conditions have favored the evolution of anatomical specializations that permit the temperature of the lower legs to drop almost to freezing (thus allowing caribou to conserve body heat and avoid using precious energy to keep their lower legs warm). In contrast, the upper legs and main trunk of the body remain at about 99°F (37°C).

The radically different temperatures in different body regions place different demands on the plasma membranes of cells. Phospholipids in the cell membranes of the upper leg must be very different from those near the hooves, if membranes in both places are to remain fluid. Remember, the membrane of a cell needs to be somewhat fluid to allow the membrane proteins to move to sites where they are needed.

The fluidity of a membrane is a function of the fatty acid tails of its phospholipids: Unsaturated fatty acids remain fluid at lower temperatures than do saturated fatty acids (see Chapter 2). Caribou have a range of fatty acids in the phospholipids of the cells in their legs. The membranes of cells near the chilly hoof have lots of unsaturated fatty acids, whereas the membranes of cells near the warmer trunk have more saturated ones. This arrangement gives the plasma membranes throughout the leg the proper fluidity despite great differences in temperature.

CAN TEENS HAVE HEART ATTACKS? *REVISITED*

As you may have guessed, familial hypercholesterolemia occurs when circulating cholesterol is unable to cross plamsa membranes. Much of the cholesterol in the blood travels in the form of low-density lipoprotein (LDL), particles consisting of thousands of cholesterol molecules (and other lipids) bound to a protein molecule. In a healthy person, LDL particles are removed from the blood when they bind to receptor protein molecules on the surfaces of liver cells and are engulfed by the cells (an example of endocytosis). In a person with FH, however, the LDL receptor protein molecules are defective, so most LDL remains in the blood, where it accumulates to dangerous levels.

One way to treat FH would be to restore the missing LDL receptor proteins. Recently, many researchers have focused on developing methods for replacing defective genes in the liver cells of FH patients. For example, several groups of researchers have successfully inserted working LDL receptor genes into defective human cells that lacked them. Although this result was obtained in cells growing in culture in the lab, researchers are hopeful that they will see a similar outcome in a future clinical trial with FH patients.

CONSIDER THIS: Both being overweight and having high blood cholesterol increase the risk of heart disease. Many controlled experiments have shown that reducing the amount of saturated fat in a person's diet reduces blood cholesterol. Currently popular low-carbohydrate diets for weight loss, however, encourage consumption of high-fat foods. Should physicians recommend low-carb diets to overweight patients?

LINKS TO LIFE: Membrane Killers

Many poisonous snakes have venom that works, at least in part, by destroying the plasma membranes of victims' cells. For example, the Western diamondback rattlesnake, which is responsible for most of the 10 to 15 snakebite deaths in the United States each year, possesses a venom that is rich in enzymes called *phospholipases*. Phospholipases break down the phospholipids of plasma membranes, causing cells to rupture and die. In the victim's bloodstream, the phospholipases in the rattler's venom attack red blood cells, reducing the blood's ability to carry oxygen and causing the victim to become short of breath. Once carried to muscles, phospholipases also attack muscle cell membranes.

Bacterial parasites of humans may also produce toxins that harm plasma membranes. Some disease-causing bacteria, for example, produce toxins that attack membrane proteins, thus disrupting a cell's ability to control the flow of molecules across its plasma membrane. The bacterium that causes cholera, a serious disease that is mostly eradicated in the United States but is still prevalent in less-developed countries, produces a toxin that disrupts membrane transport proteins in the cells that line the human digestive tract. These proteins normally regulate the process that causes water to move out of the large intestine, across the intestinal lining, and into the bloodstream. When the cholera toxin binds to the surfaces of the intestinal cells and prevents the normal movement of molecules, large amounts of water move into, rather than out of, the large intestine. This influx of water causes severe diarrhea, dehydration, and sometimes death.

Chapter Review

Summary of Key Concepts

3.1 What Does the Plasma Membrane Do?

The plasma membrane isolates the cytoplasm from the external environment. It also regulates the flow of materials into and out of the cell and the flow of chemical messages into the cell.

3.2 What Is the Structure of the Plasma Membrane?

The plasma membrane consists of a bilayer of phospholipids with proteins embedded in it.

3.3 How Does the Plasma Membrane Play Its Gatekeeper Role?

The phospholipid bilayer blocks most molecules from passing through (though some small molecules and lipid-soluble molecules do pass through). Larger molecules can cross the membrane only with the help of membrane proteins. Transport proteins regulate the movement of water-soluble substances through the membrane. Receptor proteins bind molecules in the external environment, triggering changes in the cell. Recognition proteins serve as identification tags and attachment sites.

3.4 What Is Diffusion?

Diffusion is the movement of particles from regions of higher concentration to regions of lower concentration.

3.5 What Is Osmosis?

Osmosis is the diffusion of water across a selectively permeable membrane and down its concentration gradient. Dissolved substances decrease the concentration of free water molecules.

3.6 How Do Diffusion and Osmosis Affect Transport across the Plasma Membrane?

In simple diffusion, water, dissolved gases, and lipid-soluble molecules diffuse through the phospholipid bilayer. In facilitated diffusion, water-soluble molecules cross the membrane through protein channels or with the assistance of carrier proteins. In all these types of passive transport, molecules move down their concentration gradients, and cellular energy is not required. Water often moves across the plasma membranes by osmosis. Osmosis does not require cellular energy.

3.7 How Do Molecules Move against a Concentration Gradient?

In active transport, proteins spanning the membrane use cellular energy (ATP) to drive the movement of molecules across the plasma membrane, usually against concentration gradients. Large molecules (for example, proteins), particles of food, microorganisms, and extracellular fluid may be acquired by endocytosis. The secretion of substances such as hormones and the excretion of wastes from a cell are accomplished by exocytosis.

Key Terms

active transport *p. 48*
carrier protein *p. 46*
cell wall *p. 49*
channel protein *p. 46*
concentration gradient *p. 43*
diffusion *p. 43*
endocytosis *p. 48*
exocytosis *p. 49*
facilitated diffusion *p. 46*
fluid mosaic *p. 40*
hydrophilic *p. 40*
hydrophobic *p. 41*
osmosis *p. 44*
passive transport *p. 45*
phospholipid bilayer *p. 40*
plasma membrane *p. 40*
receptor protein *p. 42*
recognition protein *p. 42*
selectively permeable *p. 44*
simple diffusion *p. 46*
transport protein *p. 42*
vesicle *p. 48*

Thinking Through the Concepts

Multiple Choice

1. *Active transport through the plasma membrane occurs through the action of*
   a. diffusion
   b. membrane proteins
   c. DNA
   d. water
   e. osmosis
2. *The following is characteristic of a plasma membrane:*
   a. It separates the cell contents from its environment.
   b. It is permeable to certain substances.
   c. It is a lipid bilayer with embedded proteins.
   d. It contains pumps for moving molecules against their concentration gradient.
   e. all of the above
3. *If an animal cell is placed into a solution whose concentration of dissolved substances is higher than that inside the cell,*
   a. the cell will swell
   b. the cell will shrivel
   c. the cell will remain the same size
   d. the solution is described as hypertonic
   e. both b and d are correct
4. *Small, nonpolar hydrophobic molecules such as fatty acids*
   a. pass readily through a membrane's lipid bilayer
   b. diffuse very slowly through the lipid bilayer
   c. require special channels to enter a cell
   d. are actively transported across cell membranes
   e. must enter the cell via endocytosis
5. *Which of the following would be least likely to diffuse through a lipid bilayer?*
   a. water
   b. oxygen
   c. carbon dioxide
   d. sodium ions
   e. the small, nonpolar molecule butane
6. *Which of the following processes causes substances to move across membranes without the expenditure of cellular energy?*
   a. endocytosis
   b. exocytosis
   c. active transport
   d. diffusion
   e. pseudopodia

Review Questions

1. Describe and diagram the structure of a plasma membrane. What are the two principal types of molecules in plasma membranes?
2. What are the three main categories of proteins commonly found in plasma membranes, and what is the function of each?
3. Define *diffusion,* and compare that process with osmosis. How do these two processes help plant leaves remain firm?
4. Define *hypotonic, hypertonic,* and *isotonic.* What would happen to an animal cell immersed in each of the three types of solution?
5. Describe the following types of transport processes: simple diffusion, facilitated diffusion, active transport, pinocytosis, phagocytosis, and exocytosis.

Applying the Concepts

1. **IS THIS SCIENCE?** The inventors of the drug ezetimibe demonstrated that the drug prevents cholesterol from moving across the membranes of the cells that line the intestine. On the basis of this finding, the inventors claim that taking ezetimibe can reduce blood cholesterol. Do you agree? Provide a scientifically sound explanation of your answer. (It may help to review the material presented in "Can Teens Have Heart Attacks?" in this chapter).
2. Different cells have somewhat different plasma membranes. For example, the plasma membrane of a *Paramecium*, a single-celled organism that lives in ponds, is only about 1% as permeable to water as the plasma membrane of a human red blood cell. Why do you think that *Paramecium* has evolved a membrane with such low water permeability? What molecular differences do you think might account for this low water permeability? On the basis of your answers to these questions, develop a hypothesis about how body size and habitat affect the evolution of plasma membrane structure. On the basis of your hypothesis, what testable predictions can you make about how the plasma membranes of different species will differ?
3. A preview question for Chapter 22: The integrity of the plasma membrane is essential for cellular survival. Could the immune system make use of this fact to destroy foreign cells that have invaded the body? How might cells of the immune system disrupt membranes of foreign cells? (Two hints: Virtually all cells can secrete proteins, and some proteins form pores in membranes.)
4. A preview question for Chapter 17: Plant roots take up minerals (inorganic ions such as potassium) that are dissolved in the water of the soil. The concentration of such ions is usually much lower in the soil water than in the cytoplasm of root cells. Design the plasma membrane of a hypothetical mineral-absorbing cell, with special reference to mineral-permeable channel proteins and mineral-transporting active-transport proteins. Justify your choice of channel proteins and active-transport proteins.

For More Information

Kunzig, R., "They Love the Pressure." *Discover*, August 2001. Living at depths that exert a pressure of 15,000 pounds per square inch necessitates more membranes in deep-sea dwellers.

McNeil, P. L. "Cell Wounding and Healing." *American Scientist*, May–June 1991. The fluidity of plasma membrane phospholipids makes cells able to withstand minor damage.

Rothman, J. E., and Orci, L. "Budding Vesicles in Living Cells." *Scientific American*, March 1996. Membranes within cells form vesicles that transport materials inside the cell. Researchers are discovering the mechanisms by which these vesicles form.

Sharon, N., and Lis, H. "Carbohydrates in Cell Recognition." *Scientific American*, January 1993. Carbohydrates, usually attached to proteins, identify cells, serve as parts of receptors in hormone binding, and regulate the attachment and movement of cells.

WEB TUTORIAL

To access a Web Tutorial visit http://www.prenhall.com/audesirk4. *Log in to the Web site selected by your instructor, navigate to this chapter, and select the appropriate Web Activity number.*

3.1 Membrane Structure and Transport

Estimated time: 10 minutes

Explore how the cell membrane controls what enters and leaves a cell.

3.2 Osmosis

Estimated time: 5 minutes

Explore how osmosis affects cells.

3.3 Web Investigation: Can Teens Have Heart Attacks?

Estimated time: 15 minutes

Cholesterol is an essential component of cell membranes, yet too much cholesterol in the diet can lead to heart disease. In the case of individuals with familial hypercholesterolemia, a mutation in a specific gene makes high cholesterol levels inevitable. In this exercise you can find out about what the future holds for people with familial hypercholesterolemia.

Chapter 4

Cell Structure and Function

Scientists studying animal limb regeneration hope that their findings may one day allow humans to regrow lost limbs.

AT A GLANCE

Can Lost Limbs Grow Back?

When a rocket-propelled grenade exploded inside his Humvee, "Jim Williamson" lost his left leg. As did many U.S. soldiers wounded in Iraq, Jim survived injuries that might have killed him in an earlier war, before recent advances in the treatment of traumatic injuries. These advances have increased the number of people who, like Jim, are living without limbs. The growing number of amputees has increased scientific interest in a longstanding question: Can humans regrow lost body parts?

If a person loses a finger, or a hand, or a whole limb, thick scar tissue forms at the site of the wound, and the missing structure does not regrow. This failure stands in stark contrast to the regenerative powers of many other animals. A starfish that loses an arm quickly regrows the lost part, as does a crab that loses a claw, a lizard that loses its tail, or a salamander that loses a leg.

Biologists have hypothesized that such animals produce substances that prevent scar formation and cause cells at the wound site to reproduce and replace the missing tissues. In one recent test of this hypothesis, researchers made an extract of tissue from a regenerating salamander limb and applied it to cultures of mouse muscle cells. Mice cannot regenerate lost parts, and their mature muscle cells normally form ropy, nongrowing fibers of muscle tissue. Under the influence of the newt extract, however, 18% of mature mouse muscle cells reverted to an earlier state, dividing over and over again as if still part of a young, growing mouse. The researchers concluded that some chemical in the salamander's regenerating tissue converted the mouse cells into a growing state.

If a salamander such as this axolotl loses a limb, it grows back.

Studying the process of regeneration in salamanders might help identify substances that could someday be applied to human wounds to promote regeneration instead of scar formation. How might such substances change the way cells behave?

4.1 What Features Are Shared by All Cells?

Cells are the smallest units that retain the properties of life. All living things, from the tiniest bacterium to the largest whale, are composed of cells. The smallest organisms consist of only a single cell, but larger organisms can contain trillions of cells, each specialized to perform a specific function. Because cells can be specialized for many different tasks, and because there are many different species of single-celled organisms, there are many different types of cells. Despite their diversity, cells share certain features, which are described below.

Cells Are Enclosed by a Plasma Membrane

As we saw in Chapter 3, all cells are surrounded by a **plasma membrane**. The plasma membrane consists of a phospholipid bilayer in which proteins are embedded, and it controls the movement of substances into and out of the cell. A description of plasma membrane structure and function can be found in Chapter 3.

Cells Use DNA As a Hereditary Blueprint

All cells contain **deoxyribonucleic acid (DNA)**, genetic material that stores the instructions for making all the other parts of the cell and for producing new cells. Each cell inherits its genetic material from the cells that give rise to it, so all cells in a single body generally contain identical DNA molecules.

Cells Contain Cytoplasm

The **cytoplasm** consists of all the material inside the plasma membrane and outside the DNA-containing region. The fluid portion of the cytoplasm contains water, salts, and an assortment of organic molecules. It forms a thick soup of proteins, lipids, carbohydrates, salts, sugars, amino acids, and nucleotides (see Chapter 2). Most of the cell's *metabolic activities*—that is, the sum of all the biochemical reactions that underlie life—occur in the cell cytoplasm. For example, the manufacture of proteins takes place in the cytoplasm on special structures called ribosomes.

Cells Obtain Energy and Nutrients from Their Environment

To maintain themselves, all cells must continuously acquire and expend energy. The ultimate source of this energy is sunlight. Cells in some organisms, such as plants, can harness the sun's energy directly and store it in high-energy molecules. Once solar energy has been captured in this way, some of the energy-containing molecules become available to cells that are not capable of capturing solar energy on their own. For example, an animal may consume plant tissue, thereby making the energy captured by the plant cells available to its own cells.

In addition to energy, cells need nutrients. Carbon, nitrogen, oxygen, minerals, and other building blocks of biological molecules are found in the environment—the air, water, rocks, and living things. As with energy, cells must obtain these materials from their environment.

Table 4-1 summarizes some features shared by all cells.

TABLE 4-1 Features Common to All Cells

Molecular components	Proteins, amino acids, lipids, carbohydrates, sugars, nucleotides, DNA, RNA
Structural components	Plasma membrane, cytoplasm, ribosomes
Metabolism	Extracts energy and nutrients from the environment; uses energy and nutrients to build, repair, and replace cellular parts

Cell Function Limits Cell Size

Most cells are small, ranging from about 1 to 100 micrometers (millionths of a meter) in diameter (**Fig. 4-1**). Why are cells so small? The answer lies in the method by which cells move nutrients and wastes across the plasma membrane. As you

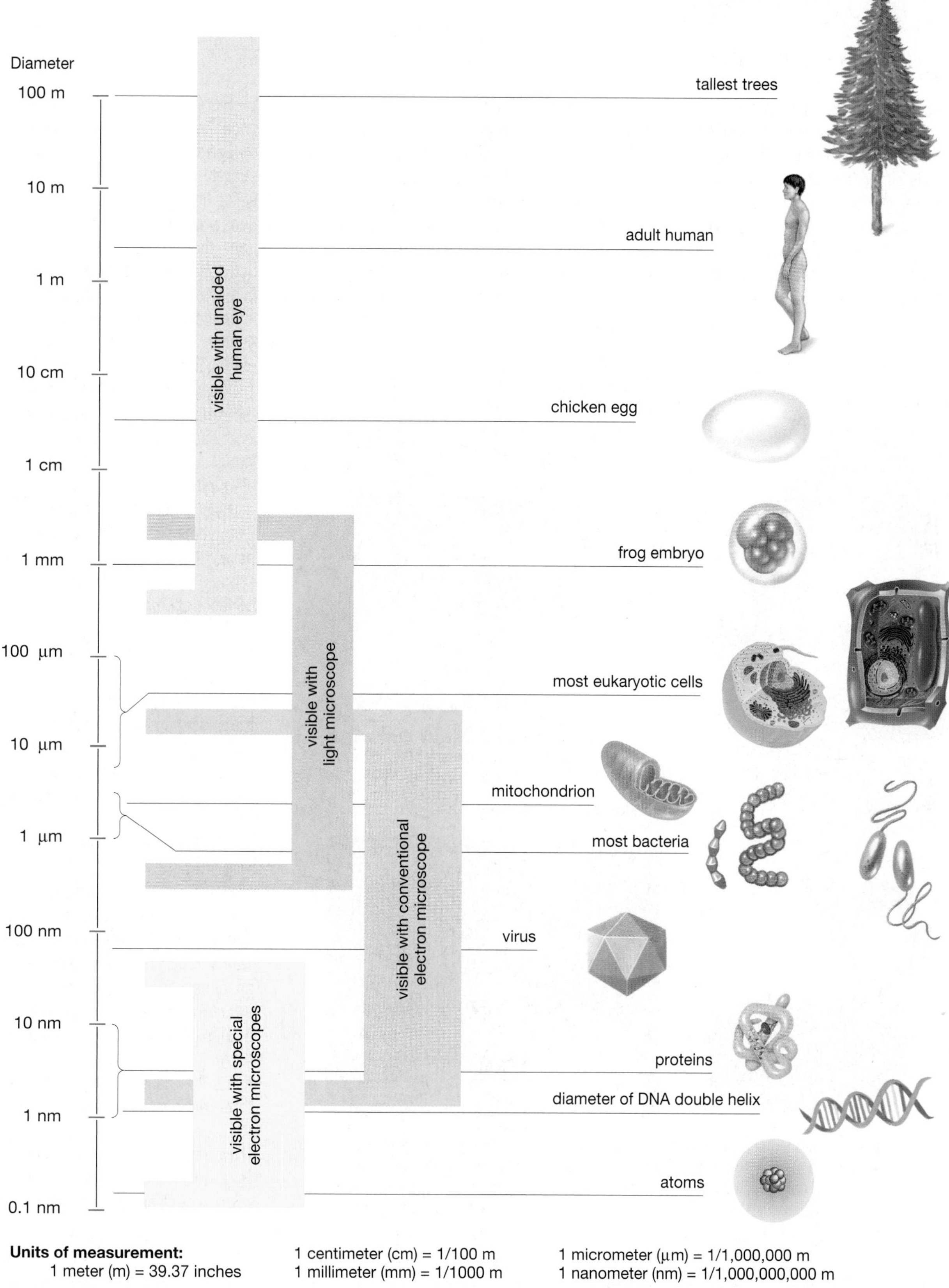

Units of measurement:
1 meter (m) = 39.37 inches
1 centimeter (cm) = 1/100 m
1 millimeter (mm) = 1/1000 m
1 micrometer (μm) = 1/1,000,000 m
1 nanometer (nm) = 1/1,000,000,000 m

Figure 4-1 Relative sizes Dimensions commonly encountered in biology range from about 100 meters (the height of the tallest redwoods) to a few micrometers (the diameter of most cells) to a few nanometers (the diameter of many large molecules).

SCIENTIFIC INQUIRY

THE SEARCH FOR THE CELL

Human understanding of the cellular nature of life came slowly. In 1665, English scientist and inventor Robert Hooke reported observations that he made through a primitive microscope. He aimed his instrument at a thin piece of cork (which comes from the bark of an oak tree) and saw "a great many little Boxes" (Fig. E4-1a). Hooke called the boxes "cells" because he thought they resembled the tiny rooms, or cells, occupied by monks. He wrote that in the living oak and other plants, "These cells [are] fill'd with juices."

By the 1670s, Dutch microscopist Anton van Leeuwenhoek was observing a previously unknown world through simple microscopes of his own construction (Fig. E4-1b). His descriptions of myriad "animalcules" (his term for single-celled organisms) in rain, pond water, and well water caused quite an uproar, because water was consumed without treatment in those days. Leeuwenhoek made careful observations of an enormous range of microscopic specimens, including red blood cells, sperm, and the eggs of small insects.

(a) 17th century microscope

(b) Leeuwenhoek's microscope

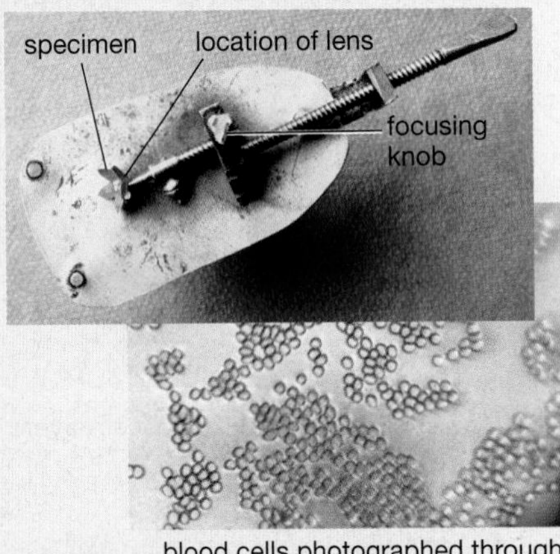

(c) Electron microscope

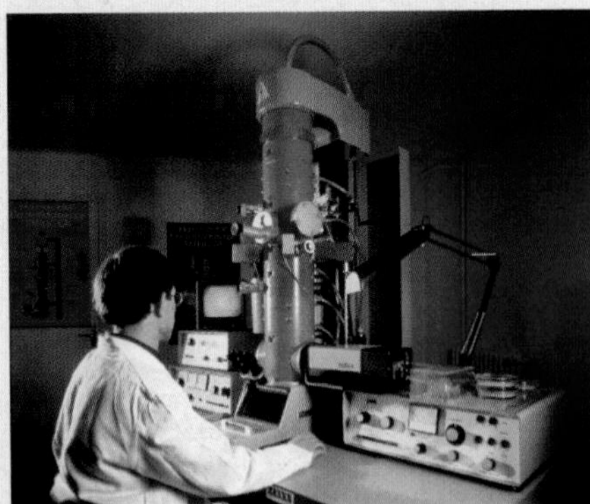

Figure E4-1 Microscopes yesterday and today ***(a)*** Robert Hooke's drawings of the cells of cork, as he viewed them with an early light microscope similar to the one shown here. Only the cell walls remain. ***(b)*** One of Leeuwenhoek's microscopes and a photograph of blood cells taken through a Leeuwenhoek microscope. The specimen is viewed through a tiny hole just underneath the lens. ***(c)*** A scanning electron microscope, which creates a three-dimensional image of a specimen's surface.

More than a century passed before biologists began to understand more about the role of cells in life on Earth. Microscopists first noted that plants consist entirely of cells. The thick wall surrounding all plant cells, first observed by Hooke, made their observations easier. Animal cells, however, escaped notice until the 1830s, when German zoologist Theodor Schwann saw that cartilage contains cells that "exactly resemble [the cells of] plants." In 1839, after studying cells for years, Schwann was confident enough to publish his *cell theory*, calling cells the elementary particles of both plants and animals. By the mid-1800s, German botanist Matthias Schleiden had further refined science's view of cells when he wrote: "It is... easy to perceive that the vital process of the individual cells must form the first, absolutely indispensable fundamental basis" of life.

Ever since the pioneering efforts of Hooke and Leeuwenhoek, biologists, physicists, and engineers have collaborated to improve the capabilities of microscopes. Today's microscopes fall into two basic categories: *light microscopes* and *electron microscopes*.

Light microscopes use lenses, usually made of glass, to focus and magnify light rays that either pass through or bounce off a specimen. Light microscopes provide a wide range of images, depending on how the specimen is illuminated and whether it has been stained (Fig. E4-2a). The resolving power of light microscopes—that is, the smallest structure that can be seen—is about 1 micrometer (1 μm, a millionth of a meter).

Electron microscopes use beams of electrons instead of light. The electrons are focused by magnetic fields rather than by lenses. Some types of electron microscopes can resolve structures as small as a few nanometers (billionths of a meter). *Transmission electron microscopes* (TEMs) pass electrons through a thin specimen and can reveal the details of interior cell structure, including organelles and plasma membranes (Fig. E4-2b). *Scanning electron microscopes* (SEMs; see Fig. E4-1c) bounce electrons off specimens that have been coated with metals and provide three-dimensional images. SEMs can be used to view the surface details of structures that range in size from entire insects down to cells and even organelles (Fig. E4-2c,d). Photographs made through electron microscopes are often called **electron micrographs**.

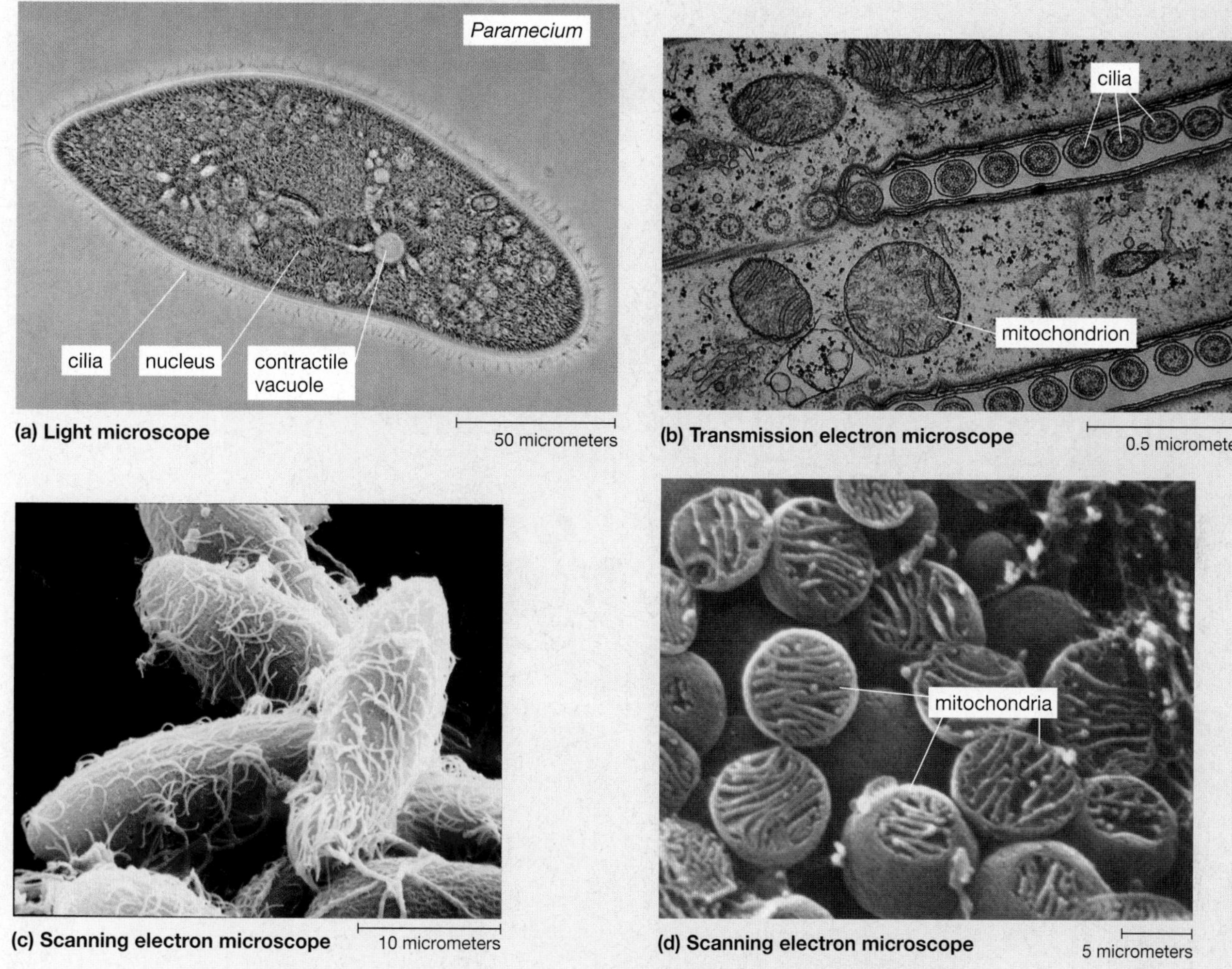

Figure E4-2 A comparison of microscope images ***(a)*** A living *Paramecium* (a single-celled freshwater organism) photographed through a light microscope. ***(b)*** A TEM photo showing the basal bodies at the bases of the cilia that cover *Paramecium*. Mitochondria are also visible. ***(c)*** A false-color SEM photo of *Paramecium*. ***(d)*** An SEM photo at much higher magnification, showing mitochondria (many of which are sliced open) within the cytoplasm.

learned in Chapter 3, many nutrients and wastes move into and out of cells by *diffusion*, the movement of molecules from places of high concentration to places of low concentration. Diffusion, however, is a slow process; oxygen molecules take more than 200 days to diffuse over a distance of 4 inches in cytoplasm. At such speeds, diffusion is effective only over short distances. If cells were any larger than they are, vital materials could not diffuse from the plasma membrane to the innermost portion of the cell (or vice versa) quickly enough to be useful.

Further, as a cell enlarges, its volume increases more rapidly than its surface area does. For example, a cell that doubles in radius becomes eight times greater in volume but only four times greater in surface area. ➔

So, as a cell grows, its volume of cytoplasm (and thus the amount of nutrients and wastes that must be exchanged with the environment) grows much faster than the area of plasma membrane through which the exchanges are made. If cells get too big, metabolic needs will overwhelm the available membrane surface area.

distance to center (r)	1.0	2.0	3.0
surface area ($4\pi r^2$)	12.6	50.3	113.1
volume ($4/3\ \pi r^3$)	4.2	33.5	113.1
area/volume	3.0	1.5	1.0

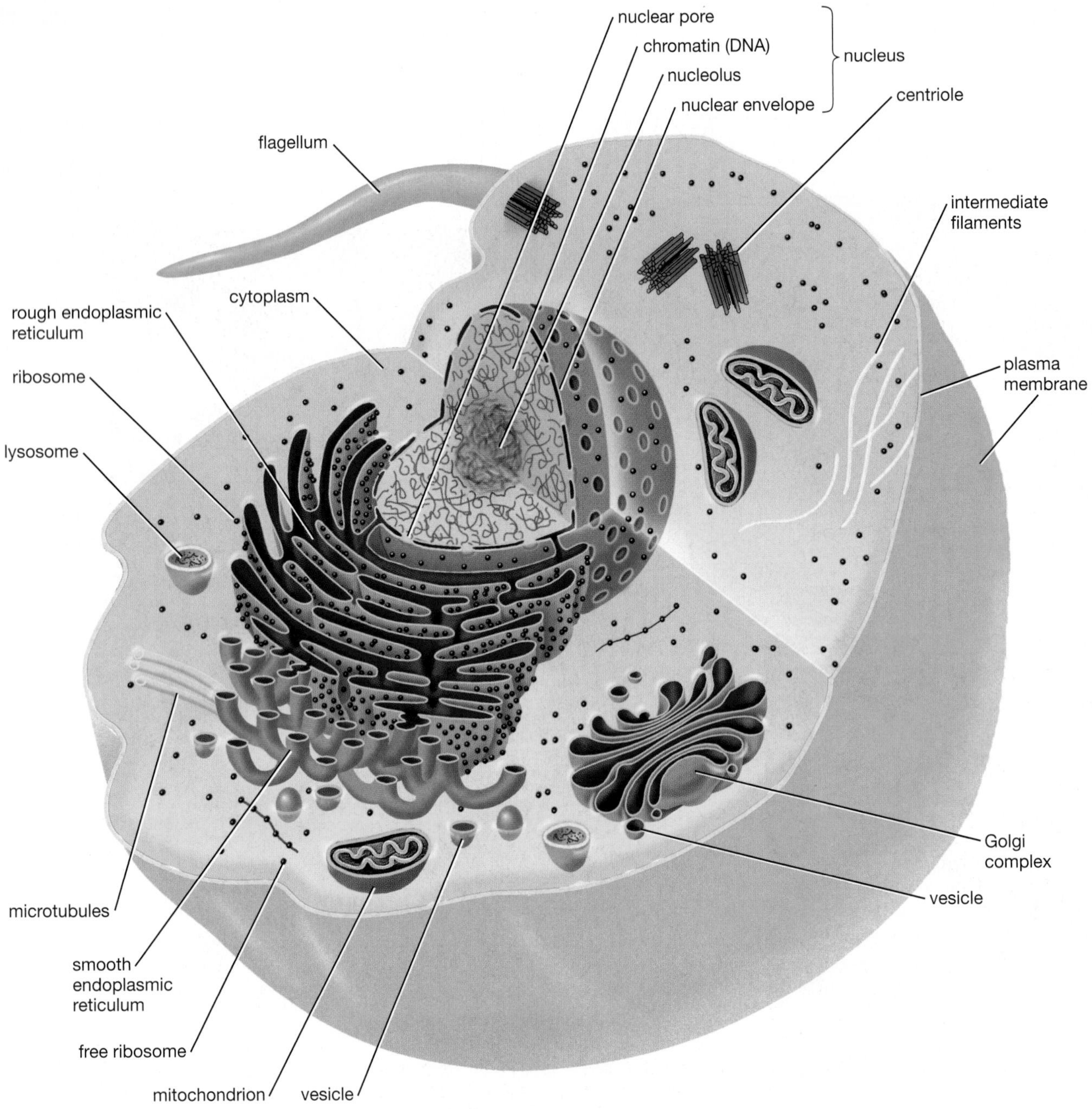

Figure 4-2 **A generalized animal cell**

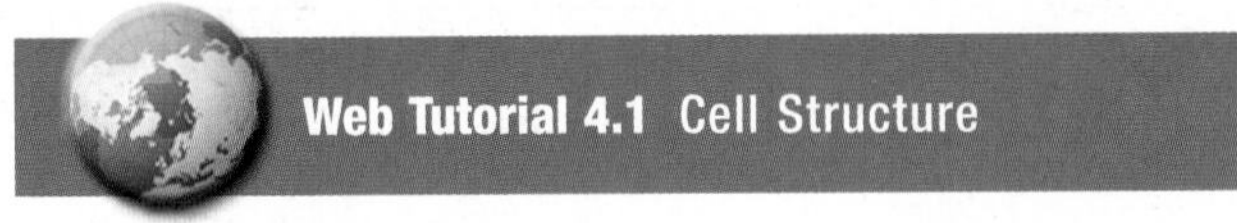

4.2 How Do Prokaryotic and Eukaryotic Cells Differ?

Despite the features that are shared by all cells, cells can be grouped into two fundamentally different types. The first type, the **prokaryotic cell**, is found only in two groups of single-celled organisms, the bacteria and the archaea. Biologists often refer to bacteria and archaea as *prokaryotes*. The second type of cell, which makes up the bodies of all other living things, is called the **eukaryotic cell**, and organisms with these cells are known as *eukaryotes*. One striking difference between the two cell types is that the DNA of eukaryotic cells is contained within a separate, membrane-bound structure called the **nucleus**. In contrast, the genetic material of prokaryotic cells is not enclosed within a membrane.

The earliest cells on Earth were probably prokaryotic. Eukaryotic cells arose much later, perhaps as the result of a partnership among different kinds of prokaryotic cells. This scenario for the origin of eukaryotic cells is described in more detail in Chapter 15.

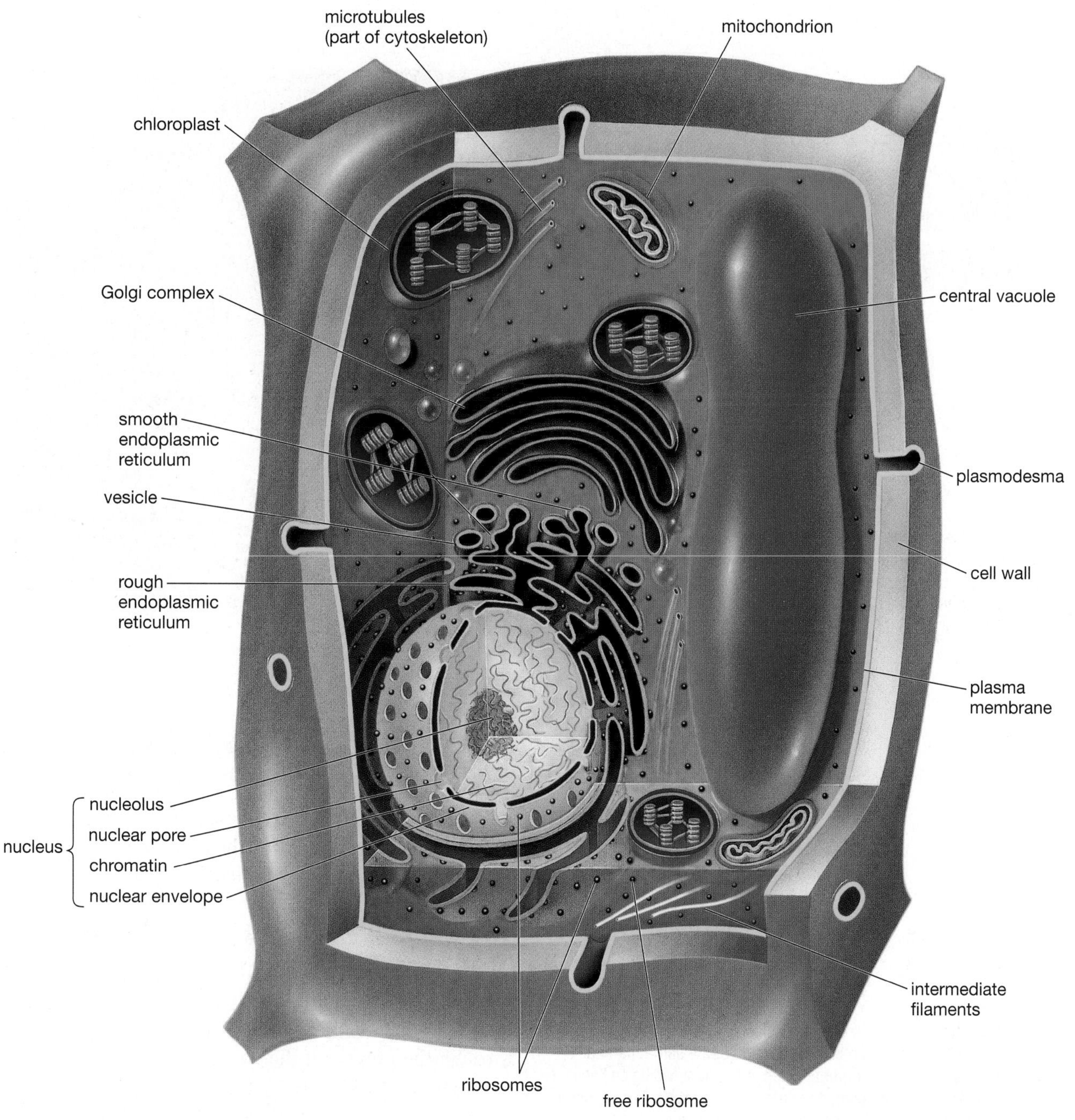

Figure 4-3 A generalized plant cell QUESTION: *Of the nucleus, ribosome, chloroplast, and mitochondrion, which appeared earliest in the history of life?*

4.3 What Are the Main Features of Eukaryotic Cells?

Eukaryotic cells differ from prokaryotic cells in many ways. For one thing, they are usually larger than prokaryotic cells—typically more than 10 micrometers in diameter (see Fig. 4-1). Also, the cytoplasm of eukaryotic cells houses membrane-enclosed structures called **organelles** that perform specific functions within the cell. A network of protein fibers, the cytoskeleton, gives shape and organization to the cytoplasm of eukaryotic cells. Many of the organelles are attached to the cytoskeleton.

Different types of eukaryotic cells may contain different kinds of organelles. For example, animal cells (**Fig. 4-2**) have a few organelles that are not found in plant cells (**Fig. 4-3**), and vice versa. As you read the following sections, which describe the structures of the cell in more detail, you may wish to refer to the illustrations in **Figures 4-2** and **4-3**. Bear in mind that these pictures represent generic, "typical" animal and plant cells. The appearance of real cells varies greatly, depending on a cell's function and the kind of organism in which it occurs.

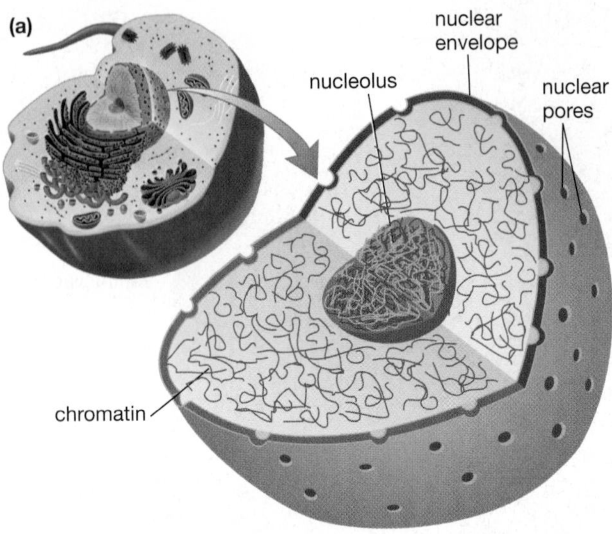

Figure 4-4 The nucleus ***(a)*** The nucleus is bounded by a nuclear envelope. Inside are chromatin (DNA and associated proteins) and a nucleolus. ***(b)*** An electron micrograph of a yeast cell that was frozen and broken open to reveal its internal structures. The large nucleus, with nuclear pores penetrating its nuclear envelope, is clearly visible.

4.4 What Role Does the Nucleus Play?

The nucleus is usually the largest organelle in a cell. It consists of three readily distinguishable parts (Fig. 4-4). The *nuclear envelope* separates the interior of the nucleus from the cytoplasm. Inside the nuclear envelope, the nucleus contains a granular-looking material called *chromatin* and a darker region called the *nucleolus*.

The Nuclear Envelope Controls Passage of Materials

The nucleus is isolated from the rest of the cell by a **nuclear envelope** that consists of a double membrane. The membrane is perforated with tiny membrane-lined channels called *pores*. Water, ions, and small molecules can pass freely through the pores, but the passage of large molecules, such as proteins, pieces of ribosomes, and RNA, is regulated by special "gatekeeper proteins" that line each nuclear pore. These gatekeepers permit the passage of certain molecules and prevent the passage of others. DNA does not cross the nuclear membrane. It remains within the nucleus for the life of the cell.

Figure 4-5 Chromosomes The chromosomes in this light micrograph of a dividing cell in an onion root tip are made of the same material (DNA and proteins) as the chromatin seen in the nondividing cell adjacent to it but are in a more compact state. **QUESTION:** *Why doesn't chromatin remain in its condensed form in nondividing cells?*

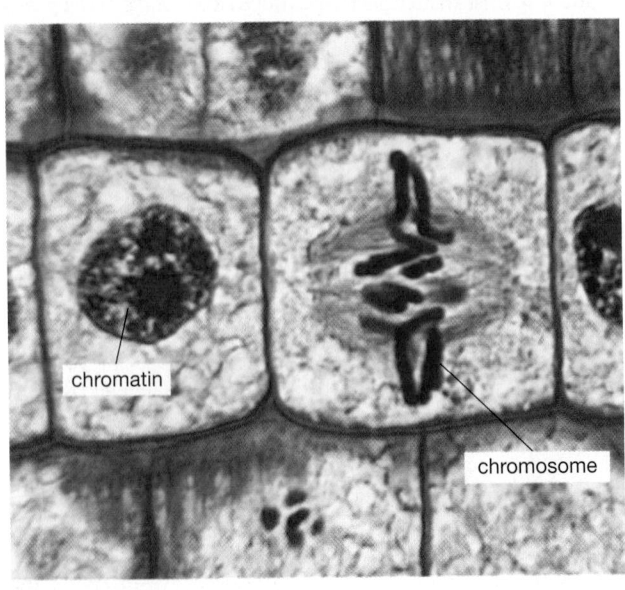

The Nucleus Contains Chromosomes

The DNA molecules contained inside the nucleus are closely associated with certain kinds of protein molecules, and this DNA-protein complex is known as **chromatin**. A eukaryotic cell's chromatin is arranged in a set of long, thread-shaped structures called **chromosomes**. When cells divide, each chromosome coils upon itself, becoming thicker and shorter. The resulting "condensed" chromosomes are easily visible under even relatively low-power microscopes (Fig. 4-5).

Ribosome Components Are Made at the Nucleolus

Most eukaryotic nuclei have one or more darkly staining regions called *nucleoli* (a **nucleolus** is shown in Fig. 4-4a). The nucleolus consists of DNA, RNA, proteins, and ribosomes in various stages of construction.

Nucleoli are the sites where the components of **ribosomes** are manufactured. Finished components leave the nucleus and move to the cytoplasm, where they come together to form complete ribosomes. A ribosome is composed of RNA and proteins and serves as a kind of "workbench" for the manufacture of proteins. Just as a workbench can be used to construct many different objects, a ribosome can be used to synthesize any of the thousands of proteins made by a cell. In electron micrographs, ribosomes appear as dark granules, either distributed in the cytoplasm (**Fig. 4-6**) or clustered along membranes (see Figs. 4-2 and 4-3).

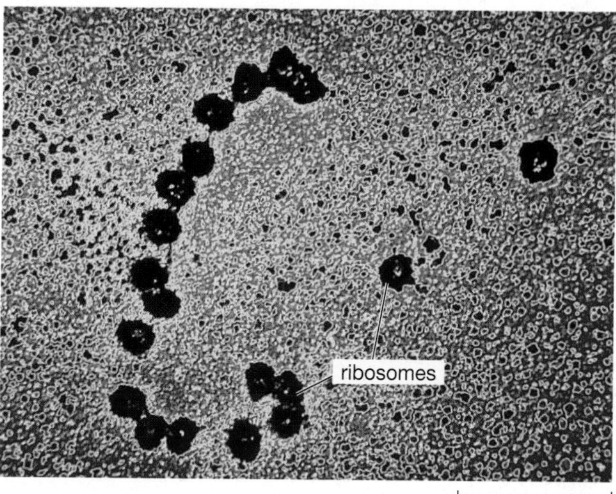

Figure 4-6 Ribosomes

4.5 What Roles Do Membranes Play in Eukaryotic Cells?

Eukaryotic cells have an elaborate system of membranes that not only enclose the cell but also create internal compartments within the cytoplasm. This membrane system is composed of the plasma membrane and several organelles, including the endoplasmic reticulum, nuclear envelope, Golgi complex, and membrane-enclosed sacs such as lysosomes (see Figs. 4-2 and 4-3).

The Plasma Membrane Isolates the Cell and Helps It Interact with Its Environment

The plasma membrane forms the outer boundary of a cell, enclosing the cytoplasm. It is a complex structure that performs the seemingly contradictory functions of separating the cytoplasm of the cell from the outside environment and providing for the transport of selected substances into or out of the cell. In addition to a plasma membrane, the cells of plants, fungi, and some single-celled organisms have a stiff *cell wall* that forms an outer protective coating.

The Endoplasmic Reticulum Forms Channels within the Cytoplasm

The endoplasmic reticulum (ER) is a series of interconnected, membrane-enclosed channels in the cytoplasm (**Fig. 4-7**); the ER membrane is continuous with the nuclear membrane. Eukaryotic cells have two forms of ER: rough and

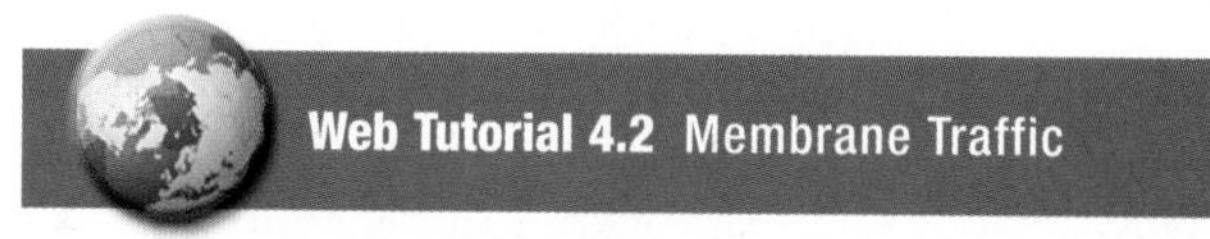

Figure 4-7 Endoplasmic reticulum There are two types of endoplasmic reticulum: rough ER, coated with ribosomes, and smooth ER, without ribosomes. Although in electron micrographs the ER looks like a series of tubes and sacs, it is actually a maze of folded sheets and interlocking channels.

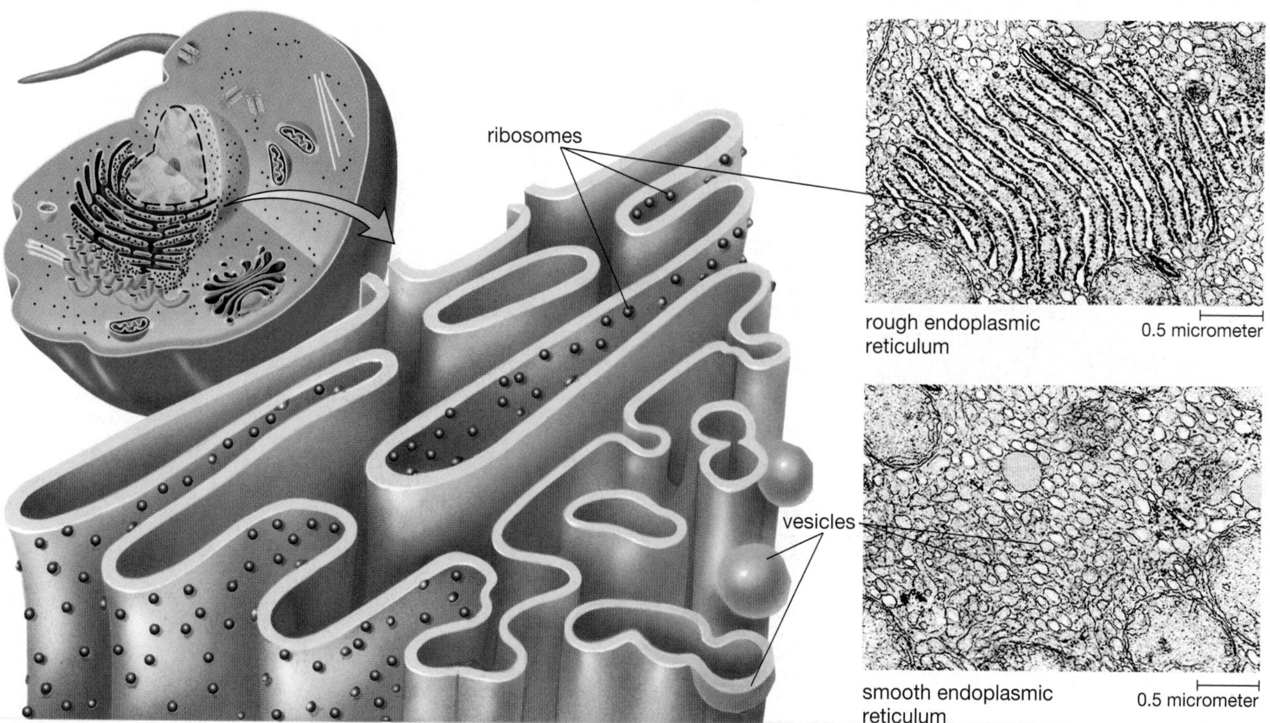

smooth. Numerous ribosomes stud the outside of the **rough endoplasmic reticulum**; in contrast, **smooth endoplasmic reticulum** lacks ribosomes.

The rough ER and smooth ER have different functions. Rough ER, with its embedded ribosomes, provides sites where proteins are manufactured. As they are made, many of the proteins move through the ER membrane into the ER interior. The proteins then move through channels in the ER and accumulate in pockets. The pockets bud off, forming membrane-bound sacs called **vesicles**, that carry their protein cargo to its destination.

Smooth ER provides sites for the manufacture of other molecules that, like the proteins made in the rough ER, can be packaged for shipment elsewhere in the cell. Among the molecules made at the smooth ER are the phospholipids and cholesterol that form membranes. Because smooth ER can manufacture phospholipids and rough ER can make membrane proteins, ER is the cell's site for membrane construction. Most of this new membrane is used for new or replacement ER, but some of it moves toward the nucleus to replace nuclear membrane or into the cytoplasm to maintain other membranes of the cell. Smooth ER also performs other functions in certain types of cells. For example, enzymes bound to the smooth ER in liver cells help detoxify substances such as drugs and alcohol.

The Golgi Complex Sorts, Chemically Alters, and Packages Important Molecules

The **Golgi complex** is a specialized set of membranes derived from the endoplasmic reticulum (Fig. 4-8). It looks very much like a stack of flattened sacs and functions as a kind of transfer station for the substances that are transported around the cell. Vesicles arrive from the ER and fuse with one face of the Golgi complex, adding their membrane to the Golgi complex and emptying their contents into the Golgi sacs. At the same time, other vesicles bud off the Golgi complex on the opposite face of the stack, carrying away proteins, lipids, and other complex molecules. These vesicles then move to other parts of the cell or to the plasma membrane for export.

As molecules pass through the Golgi, they are sorted and sometimes undergo further processing. Proteins and lipids received from the ER are sorted according

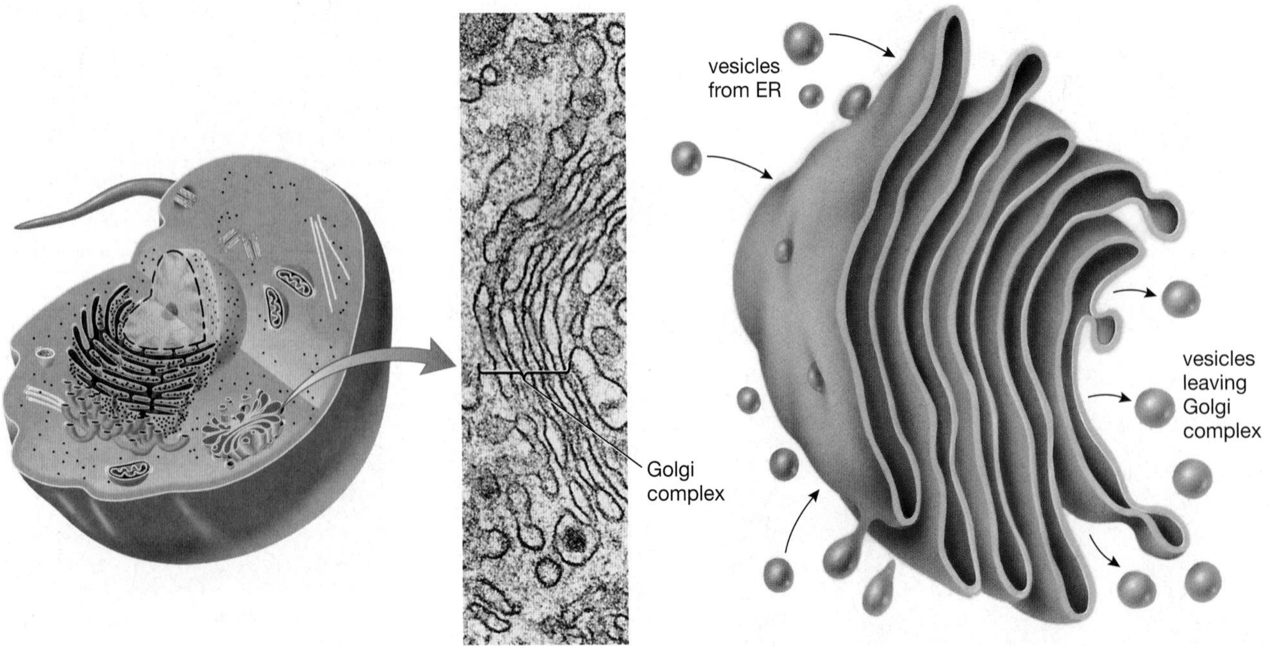

Figure 4-8 The Golgi complex The Golgi complex is a stack of flat membranous sacs. Vesicles transport material from the endoplasmic reticulum (ER) to the Golgi complex (and vice versa) and from the Golgi complex to plasma membrane, lysosomes, and vesicles. Arriving vesicles join the Golgi complex on one face, and departing vesicles bud off from the opposite face.

to their destinations. For example, proteins that will remain inside the cell are separated from those that the cell will secrete. The Golgi complex modifies some molecules. For instance, it adds sugars to proteins to make glycoproteins.

We Can Follow the Travels of a Secreted Protein

To understand how the membranous organelles work together, let's look at the secretion of an antibody within your body. An antibody is a protein, secreted by a type of white blood cell, that binds to foreign invaders (such as bacteria) and helps destroy them. The antibody protein is manufactured on ribosomes of the rough ER and then packaged into vesicles formed from the ER membrane. These vesicles travel to the Golgi complex, where the membranes fuse, releasing the protein into the Golgi complex. Here, carbohydrates are attached to the protein, which is then repackaged into vesicles formed from Golgi membrane. The vesicle containing the completed antibody travels to the plasma membrane and fuses with it, releasing the antibody outside the cell, where it will make its way into the bloodstream to help defend your body against infection.

Lysosomes Serve As the Cell's Digestive System

Some of the proteins manufactured in the ER and sent to the Golgi complex are digestive enzymes that can break down proteins, fats, and carbohydrates into their component subunits. In the Golgi complex, these enzymes are packaged in vesicles called **lysosomes**. One function of lysosomes is to digest excess cellular membranes or defective organelles. After identifying these organelles, the cell encloses them in vesicles made of membrane from the ER. These vesicles fuse with lysosomes, and digestive enzymes within the lysosome enable the cell to recycle valuable materials from the defunct organelles.

4.6 Which Other Structures Play Key Roles in Eukaryotic Cells?

Eukaryotic cells include a number of other important structures. Among them are vacuoles, mitochondria, chloroplasts, and the elements of the cytoskeleton. Some eukaryotic cells also have structures called cilia and flagella that help with locomotion.

Vacuoles Regulate Water and Store Substances

Most cells contain one or more vacuoles, which are fluid-filled sacs surrounded by a single membrane. For example, many plant cells contain a large **central vacuole** (see Fig. 4-3). Filled mostly with water, the central vacuole helps keep the right amount of water in the cell. It also provides a dump site for hazardous wastes, which plant cells often cannot excrete. Some plant cells store extremely poisonous substances, such as sulfuric acid, in their vacuoles. These substances deter animals from feeding on the otherwise nutritious leaves. Vacuoles may also store sugars and amino acids not immediately needed by the cell.

Mitochondria Extract Energy from Food Molecules

The energy that cells need to survive, grow, and reproduce is harvested in the **mitochondria** (singular, mitochondrion). The mitochondrion is sometimes called the powerhouse of the cell because it is the organelle in which energy is extracted from sugar molecules and stored in the high-energy bonds of ATP. Once the energy is stored in ATP molecules, it is available to power the many energy-consuming reactions that occur in every cell. Mitochondria are round, oval, or tubular sacs

(a)

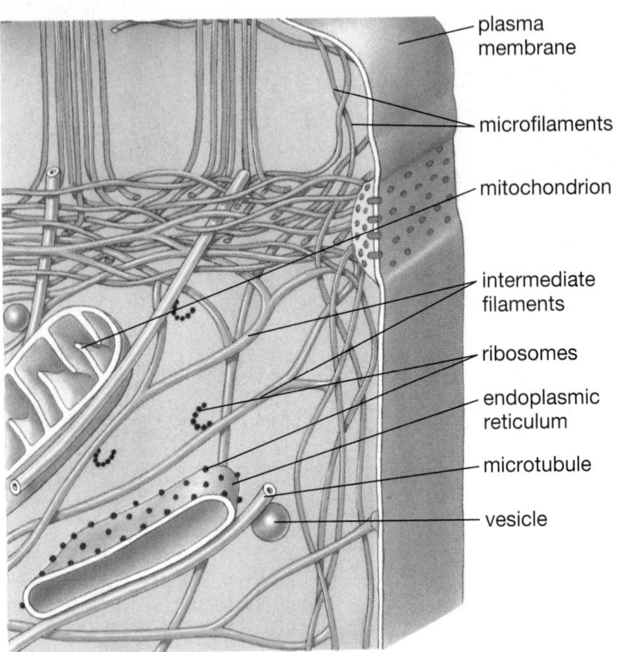

(b)

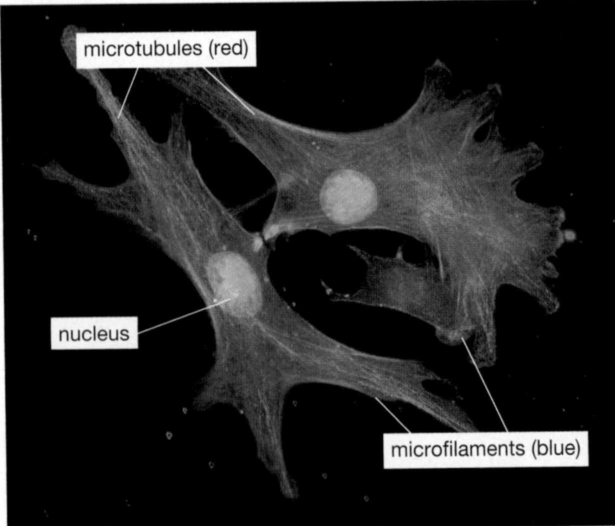

Figure 4-9 The cytoskeleton *(a)* The cytoskeleton gives shape and organization to eukaryotic cells. It consists of three types of proteins: microtubules, intermediate filaments, and microfilaments. *(b)* These cells have been treated with fluorescent stains to reveal microtubules and microfilaments, as well as the nucleus.

made of a pair of membranes. The outer mitochondrial membrane is smooth, but the inner membrane loops back and forth to form deep folds (see Figs. 4-2 and 4-3). The structure of the mitochondrion is described in greater detail in Chapter 7, as is the mitochondrion's role in energy production.

Chloroplasts Capture Solar Energy

Plant cells contain **chloroplasts**, organelles that capture energy directly from sunlight and store it in sugar molecules. Chloroplasts are the site of photosynthesis, the energy-capturing process on which all life ultimately depends. Each chloroplast is surrounded by a double membrane and contains stacks of hollow membranous sacs (see Fig. 4-3). The green pigment chlorophyll gives chloroplasts a green color (plants are green because they contain chloroplasts). The structure of chloroplasts and the reactions of photosynthesis are described in more detail in Chapter 6.

The Cytoskeleton Provides Shape, Support, and Movement

Organelles do not drift haphazardly about the cytoplasm. Most are attached to a network of protein fibers called the **cytoskeleton** (Fig. 4-9). Several types of protein fibers, including thin *microfilaments*, medium-sized *intermediate filaments*, and thick **microtubules**, make up the cytoskeleton.

The cytoskeleton performs the following important functions:

- *Cell shape*. In cells without cell walls, the cytoskeleton determines the shape of the cell.
- *Cell movement*. Many types of cells move about. Much of this movement is generated when microfilaments and microtubules assemble, disassemble, and slide past one another.
- *Organelle movement*. Microtubules and microfilaments move organelles from place to place within a cell. For example, the vesicles that bud off of the ER and Golgi complex are guided to their destinations by the cytoskeleton.
- *Cell division*. Microtubules and microfilaments are essential to cell division in eukaryotic cells. When eukaryotic nuclei divide, microtubules move the chromosomes into the daughter nuclei. When cells of animals divide, the cytoplasm is divided by a ring of microfilaments that pinches the "waist" of the parent cell around its middle.

Cilia and Flagella Move the Cell or Move Fluid Past the Cell

Both **cilia** and **flagella** (singular, cilium and flagellum) are slender, movable extensions of the plasma membrane. They typically contain microtubules that extend along their length. Some single-celled eukaryotic organisms use cilia or flagella to move about. Most animal sperm rely on flagella for movement. Some small animals use cilia for movement, swimming by the coordinated beating of rows of cilia, like the oars on a Roman galley ship. More commonly, however, cilia are used to move fluids and suspended particles past a surface. Cells with cilia line such diverse structures as the gills of oysters (moving food- and oxygen-rich water), the oviducts of female mammals (moving the eggs along from the ovary to the uterus), and the respiratory tracts of most land vertebrates (clearing mucus that carries debris and microorganisms from the windpipe and lungs).

The main differences between cilia and flagella lie in their length, number, and the direction of the force they generate. In general, cilia are shorter (about 10 to 25 micrometers long) and more numerous than flagella. They provide force in a direction parallel to the plasma membrane, like the oars in a canoe. This force is

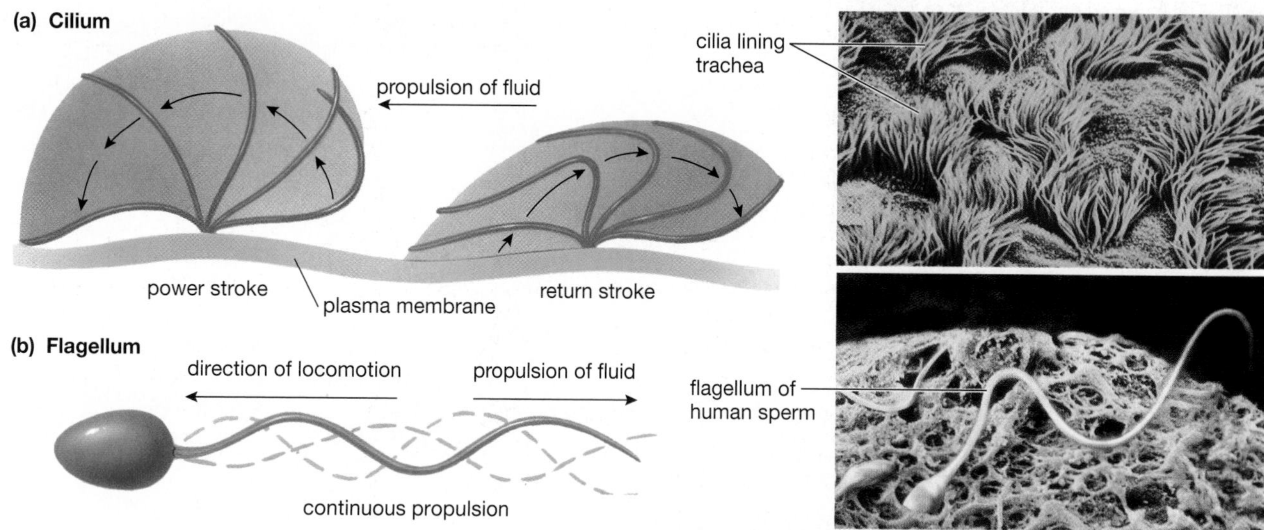

Figure 4-10 How cilia and flagella move *(a)* (Left) Cilia usually "row," providing a force of movement parallel to the plasma membrane. Their movement resembles the arms of a swimmer doing the breast stroke. (Right) Scanning electron micrograph of cilia lining the trachea (which conducts air to the lungs); these cilia sweep out mucus and trapped particles. *(b)* (Left) Flagella move in a wavelike motion, providing continuous propulsion perpendicular to the plasma membrane. In this way, a flagellum attached to a sperm can move the sperm straight ahead. (Right) A human sperm cell on the surface of a human egg cell.

accomplished through a "rowing" motion. Flagella are longer (50 to 75 micrometers), usually are fewer in number, and provide force perpendicular to the plasma membrane, like the engine on a motorboat (**Fig. 4-10**).

4.7 What Are the Features of Prokaryotic Cells?

Although Earth's larger, more conspicuous organisms are made up of eukaryotic cells, prokaryotic cells are extremely abundant. The number of prokaryotes in a single handful of soil is greater than the total number of humans that ever lived. Invisible to us, prokaryotic bacteria and archaea inhabit virtually every nook and cranny of the planet, including on and inside our own bodies. You may be tempted to view the comparatively simple structure of prokaryotic cells as an indication that prokaryotes are primitive and inferior to eukaryotes. That simple structure, however, has formed the basis of enormous biological success.

Prokaryotic cells are usually described as lacking organelles, but recent research has shown that some prokaryotes contain membrane-bound, organelle-like structures that help maintain the correct level of acidity in the cell. Nonetheless, prokaryotic cells lack nuclei, mitochondria, choloroplasts, Golgi complex, ER, and the other complex organelles that characterize eukaryotic cells. Prokaryote cytoplasm does, however, contain ribosomes, the workbenches on which proteins are manufactured in both prokaryotes and eukaryotes. The cytoplasm also may contain food granules that store energy-rich materials, such as glycogen.

Most prokaryotic cells are very small (less than 5 micrometers long) and have a relatively simple internal structure (compare **Fig. 4-11** with Figs. 4-2 and 4-3). In general, they are surrounded by a relatively stiff cell wall, which provides shape and protects the cell. Some prokaryotes can move, propelled by whiplike protrusions that undulate or spin. However, these "flagella" of prokaryotic cells do not contain microtubules and have no evolutionary relationship to eukaryotic flagella or cilia.

The cytoplasm of most prokaryotic cells is relatively uniform in appearance. Each prokaryotic cell has a single, circular strand of DNA. It is usually coiled, attached to the plasma membrane, and concentrated in a region of the cell called the **nucleoid**. The nucleoid is not, however, separated from the rest of the cytoplasm by a membrane.

Prokaryotic cells are compared with the eukaryotic cells of plants and animals in Table 4-2.

Figure 4-11 A generalized prokaryotic cell

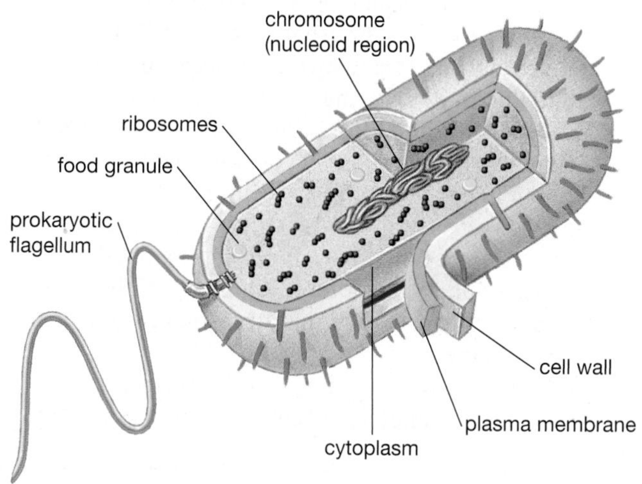

TABLE 4-2 Cell Structures, Their Functions, and Their Distribution in Living Cells

Structure	Function	Prokaryotes	Plants	Animals
Cell surface				
Cell wall	Protects, supports cell	Present	Present	Absent
Plasma membrane	Isolates cell contents from environment; regulates movement of materials into and out of cell; communicates with other cells	Present	Present	Present
Organization of genetic material				
Genetic material	Encodes information needed to construct cell and control cellular activity	DNA	DNA	DNA
Chromosomes	Contain and control use of DNA	Single, circular, no proteins	Many, linear, with proteins	Many, linear, with proteins
Nucleus	Membrane-bound container for chromosomes	Absent	Present	Present
Nuclear envelope	Encloses nucleus; regulates movement of materials into and out of nucleus	Absent	Present	Present
Nucleolus	Synthesizes components of ribosomes	Absent	Present	Present
Cytoplasmic structures				
Mitochondria	Produce energy by aerobic metabolism	Absent	Present	Present
Chloroplasts	Perform photosynthesis	Absent	Present	Absent
Ribosomes	Provide site of protein synthesis	Present	Present	Present
Endoplasmic reticulum	Synthesizes membrane components and lipids	Absent	Present	Present
Golgi complex	Modifies and packages proteins and lipids; synthesizes carbohydrates	Absent	Present	Present
Lysosomes	Contain digestive enzymes	Absent	Present	Present
Central vacuole	Contains water and wastes	Absent	Present	Absent
Cytoskeleton	Gives shape and support to cell; positions and moves cell parts	Absent	Present	Present
Centrioles	Synthesize microtubules of cilia and flagella	Absent	Absent (in most)	Present
Cilia and flagella	Move cell through fluid or move fluid past cell surface	Present[a]	Absent (in most)	Present

[a] Many prokaryotes have structures called flagella, but these are not made of microtubules and move in a fundamentally different way than eukaryotic cilia or flagella do.

CAN LOST LIMBS GROW BACK? *REVISITED*

The ability of some animals to regenerate lost parts depends on their cells' ability to turn back the clock. An animal begins life as an embryo with only a few cells, which give rise to all of the many different types of cells that form the different tissues in an animal body. Cells, such as those in embryos, with the ability to give rise to many different tissues are called stem cells.

Regeneration of lost parts requires stem cells. As a body grows and develops, however, it loses most of its stem cells. Thus, if a human loses a leg, the stem cells needed to replace it don't exist, and new ones are not created. Instead, the body's resources are devoted to healing the wound. If a salamander loses a leg, however, something very different happens. Cells near the wound undergo a transformation that turns them into stem cells. A new leg begins to grow.

Regenerative medicine, which proposes that medical treatment could stimulate regeneration in humans, hypothesizes that human cells can respond to the chemical signals that guide regeneration in other species. In recent years, researchers have discovered a number of the proteins involved in the control of regeneration in salamanders, flatworms, zebrafish, and other regenerators. Experiments such as the one described at the beginning of this chapter have shown that these proteins can affect the growth of mammal cells. Will they prove effective at promoting regeneration of damaged human body parts?

CONSIDER THIS: In addition to the effort to induce the human body to regenerate lost parts, a far larger program of research is aimed at medical use of stem cells grown outside the body. Scientists hypothesize that such stem cells could promote regrowth of damaged tissues. Compared with regenerative medicine, what might be the advantages and disadvantages of the "add new stem cells" approach?

LINKS TO LIFE: Spare Parts

Despite their great potential, regenerative medicine and stem cell research have yet to yield treatments that have actually been used on people. An alternative treatment approach that *is* currently available is *tissue engineering*. Tissue engineers build replacements for damaged body tissues by applying cells to "scaffolds" made of special biodegradable plastics. The cells divide and spread over the scaffold to form an artificial tissue that can be surgically implanted.

Tissue engineers have successfully built artificial skin and cartilage (the stiff tissue that supports your nose, ears, and knees). Artificial skin has been especially useful for treating burn victims. Artificial cartilage has been used to rebuild damaged knees and as a bone substitute in replacements for amputated fingers.

The latest application of tissue engineering is in artificial corneas (the outer layer of the eye). These are built by growing cornea cells on the surface of a very thin plastic membrane. Engineered corneas have been implanted in several patients with severe eye damage, and the results have been promising, with most of the recipients experiencing improved vision. The days of long, often futile, waits for cornea donors may be coming to an end.

The scientists who now shape cells into living cartilage, skin, and corneas might someday be able to sculpt cells into working livers, kidneys, and lungs. Researchers have already managed to grow bone, heart valves, tendons, intestines, blood vessels, and breast tissue on plastic scaffolds and to implant some of these tissues into animals. With each passing day, our ability to control the behavior of cells gets a little better.

Chapter Review

Summary of Key Concepts

4.1 What Features Are Shared by All Cells?

The smallest units of life are single cells. Cells are the functional units of multicellular organisms, and every living organism is made up of one or more cells. Cells are limited in size because they must exchange materials with their surroundings by diffusion. Diffusion is relatively slow, so the interior of the cell must never be too far from the plasma membrane, and the plasma membrane must have a large surface area (relative to the volume of its cytoplasm) for diffusion. Both of these constraints limit the size of cells.

4.2 How Do Prokaryotic and Eukaryotic Cells Differ?

All cells are either prokaryotic or eukaryotic. Prokaryotic cells are small and relatively simple in structure and are found only in bacteria and archaea. More complex eukaryotic cells make up all other forms of life.

4.3 What Are the Main Features of Eukaryotic Cells?

Eukaryotic cells are generally much larger than prokaryotic cells. Eukaryotic cells also have organelles and a cytoskeleton.

4.4 What Role Does the Nucleus Play?

The nucleus is bounded by the double membrane of the nuclear envelope and contains the cell's genetic material (DNA). Pores in the nuclear envelope regulate the movement of molecules between the nucleus and the cytoplasm. The genetic material of eukaryotic cells is organized into chromosomes, which consist of DNA and proteins. Ribosomes are particles that are the sites of protein synthesis. The nucleolus is the site at which the components of ribosomes are manufactured.

4.5 What Roles Do Membranes Play in Eukaryotic Cells?

The membrane system of a cell consists of the plasma membrane, endoplasmic reticulum (ER), Golgi complex, and vesicles derived from these membranes. Rough ER manufactures many cellular proteins. Smooth ER manufactures phospholipids. The ER is the site of all membrane synthesis within the cell. The Golgi complex is a series of membranous sacs derived from the ER. The Golgi complex processes and modifies materials synthesized in the rough and smooth ER. Some substances in the Golgi complex are packaged into vesicles for transport elsewhere in the cell. Lysosomes are vesicles that contain enzymes that digest food particles and defective organelles.

4.6 Which Other Structures Play Key Roles in Eukaryotic Cells?

All eukaryotic cells contain mitochondria, organelles that use oxygen to complete the breakdown of food molecules, capturing much of their energy as ATP. Plant cells contain chloroplasts, which capture the energy of sunlight during photosynthesis, enabling the cells to manufacture organic molecules, particularly sugars, from simple inorganic molecules. Many eukaryotic cells contain vacuoles that are bounded by a single membrane and that store food or wastes.

The cytoskeleton organizes and gives shape to eukaryotic cells and moves and anchors organelles. The cytoskeleton is composed of microfilaments, intermediate filaments, and microtubules. Cilia and flagella are whiplike extensions of the plasma membrane that contain microtubules. These structures move fluids past the cell or move the cell through its fluid environment.

4.7 What Are the Features of Prokaryotic Cells?

Prokaryotic cells are generally very small and have a relatively simple internal structure. Most are surrounded by a relatively stiff cell wall. The cytoplasm of prokaryotic cells lacks membrane-enclosed organelles and contains a single, circular strand of DNA.

Study Note Figures 4-2 and 4-3 illustrate the overall structure of animal and plant cells, respectively. Table 4-2 lists the principal organelles, their functions, and their occurrence in prokaryotic cells, animal cells, and plant cells.

Key Terms

cell *p. 56*
central vacuole *p. 65*
chloroplast *p. 66*
chromatin *p. 62*
chromosome *p. 62*
cilia *p. 66*
cytoplasm *p. 56*
cytoskeleton *p. 66*
deoxyribonucleic acid (DNA) *p. 56*
electron micrograph *p. 58*
eukaryotic cell *p. 60*
flagella *p. 66*
Golgi complex *p. 64*
lysosome *p. 65*
microtubule *p. 66*
mitochondrion *p. 65*
nuclear envelope *p. 62*
nucleoid *p. 67*
nucleolus *p. 62*
nucleus *p. 60*
organelle *p. 61*
plasma membrane *p. 56*
prokaryotic cell *p. 60*
ribosome *p. 62*
rough endoplasmic reticulum *p. 64*
smooth endoplasmic reticulum *p. 64*
vesicle *p. 64*

Thinking Through the Concepts

Multiple Choice

1. *The outermost boundary of an animal cell is the*
a. plasma membrane
b. nucleus
c. cytoplasm
d. cytoskeleton
e. cell wall

2. *Which organelle contains a eukaryotic cell's chromosomes?*
a. Golgi complex
b. ribosome
c. nucleus
d. mitochondrion
e. chloroplast

3. *Energy is converted to usable form (the high-energy bonds of the ATP molecule) in the*
a. Golgi complex
b. ribosomes
c. nucleus
d. mitochondria
e. chloroplast

4. *Which organelle sorts, chemically modifies, and packages newly synthesized protein?*
a. Golgi complex
b. ribosome
c. nucleus
d. mitochondrion
e. chloroplast

5. *Membrane-enclosed organelles that contain digestive enzymes are called*
a. lysosomes
b. smooth endoplasmic reticula
c. cilia
d. Golgi complexes
e. mitochondria

6. *A series of membrane-enclosed channels studded with ribosomes is called*
a. lysosome
b. Golgi complex
c. rough endoplasmic reticulum
d. mitochondrion
e. smooth endoplasmic reticulum

Review Questions

1. Review the section entitled "Can Lost Limbs Grow Back?" at the beginning of this chapter, focusing especially on the investigation described in the third paragraph. Then, using Figure E1-1 (page 4) as a model, describe the observation, question, hypothesis, prediction, experiment, and conclusion that make up the investigation.
2. Diagram "typical" prokaryotic and eukaryotic cells, and describe their important similarities and differences.
3. Which organelles are common to both plant and animal cells, and which are unique to each?
4. Describe the nucleus, including the nuclear envelope, chromatin, chromosomes, DNA, and the nucleolus.
5. What are the functions of mitochondria and chloroplasts?
6. What is the function of ribosomes? Where in the cell are they typically found?
7. Describe the structure and function of the endoplasmic reticulum and Golgi complex.

Applying the Concepts

1. **IS THIS SCIENCE?** Researchers studying a meteorite made of rock from Mars found small, tubular structures that are very similar to fossil bacteria found on Earth. The Martian rock also contained certain organic molecules that on Earth are found in decomposing cells. Some scientists claimed that these findings are evidence of past life on Mars. Do you agree that science has demonstrated the past existence of living cells on Mars? If so, explain why the evidence is scientifically persuasive. If not, describe the kind of scientific evidence that would persuade you.
2. If muscle biopsies (samples of tissue) were taken from the legs of a world-class marathon runner and a typical couch potato, which would you expect to have a higher density of mitochondria? Why? How would the density of mitochondria in a muscle biopsy from the biceps of a weight lifter compare with those of the runner and the couch potato?
3. One of the functions of the cytoskeleton in animal cells is to give shape to the cell. Plant cells have a fairly rigid cell wall surrounding the plasma membrane. Does the rigid cell wall make a cytoskeleton unnecessary for a plant cell? Defend your answer in terms of other functions of the cytoskeleton.
4. Most cells are very small. What physical and metabolic constraints limit cell size? What problems would an enormous cell encounter? What adaptations might help a very large cell survive?

For More Information

de Duve, C. "The Birth of Complex Cells." *Scientific American*, April 1996. Describes the mechanisms by which the first eukaryotic cells were produced from prokaryotic ancestors.

Ford, B. J. "The Earliest Views." *Scientific American*, April 1998. The author used the original microscopes of Anton van Leeuwenhoek to see the microscopic world as Leeuwenhoek saw it. Photographic images taken through these early and very primitive instruments reveal remarkable detail.

Goodsell, D. S. *The Machinery of Life*. New York: Springer, 1993. Wonderful, drawn-to-scale images of the organelles and molecules of the cell.

Hoppert, M., and Mayer, F. "Prokaryotes." *American Scientist*, November/December 1999. These relatively simple cells actually possess a great deal of internal organization.

WEB TUTORIAL

To access a Web Tutorial visit http://www.prenhall.com/audesirk4. *Log in to the Web site selected by your instructor, navigate to this chapter, and select the appropriate Web Activity number.*

4.1 Cell Structure

Estimated time: 5 minutes

Explore the structure of prokaryotic and eukaryotic cells.

4.2 Membrane Traffic

Estimated time: 5 minutes

This animation presents the details of membrane traffic in the cell, including transport of proteins, phagocytosis, and exocytosis.

4.3 Web Investigation: Can Lost Limbs Grow Back?

Estimated time: 10 minutes

Frogs and starfish can regenerate lost limbs. Why can't humans? In this exercise you can learn about regeneration in vertebrates and what this might mean for the future regeneration of human tissues and organs.

Chapter 5

Energy Flow in the Life of a Cell

In a healthy person's body, enzymes help convert energy stored in fats and carbohydrates to the energy of movement.

AT A GLANCE

A Missing Molecule Makes Mischief

Imagine that you could never eat pizza, hamburgers, ice cream, or bagels with cream cheese. For one out of every 10,000 children born in the United States, eating such foods can lead directly to neurological disorders and severe mental retardation. These children, who have a syndrome known as phenylketonuria (PKU), cannot fully digest proteins. In particular, their bodies cannot break down the amino acid phenylalanine. As a result, phenylalanine builds up in their blood. High levels of phenylalanine are toxic and have devastating effects on developing nervous systems.

PKU was not discovered until 1934, when a woman brought her two severely retarded children to Asbjørn Følling, a Norwegian physician. The distraught mother sought an explanation for her children's condition. As part of his examination, Følling performed routine urine tests, including one in which he added the chemical ferric chloride to the urine. Ferric chloride turns purple when it reacts with ketones (molecules whose presence in urine can indicate diabetes), but in samples from the two children, it turned bright green. No substance was known to cause this change, and a puzzled Dr. Følling began a long series of chemical analyses to identify the mysterious substance in the urine. Eventually, he determined that it was phenylpyruvic acid, an indicator of high levels of phenylalanine. Følling hypothesized that the children's retardation was caused by excess phenylalanine and predicted that other mentally disabled individuals would also have phenylpyruvic acid in their urine. To test his prediction, he obtained urine samples from 430 mentally retarded people and found that 8 of the samples did indeed turn bright green when ferric chloride was added.

When enzymes fail to work properly, the simple act of eating a hamburger can have dangerous consequences.

Why can't PKU victims process phenylalanine? It turns out that the only difference between PKU sufferers and everyone else is a defect in a single type of molecule. What kind of molecule could have such a dramatic effect, turning ordinary foods into poisons?

Figure 5-1 From potential energy to kinetic energy The diver on the platform has potential energy, because the heights of the platform and the pool are different. As he dives, the potential energy is converted to the kinetic energy of the motion of his body. Finally, some of this kinetic energy is transferred to the water, which is set in motion.

5.1 What Is Energy?

Energy is the capacity to do work. In this definition, "work" refers to any of a wide range of actions, including manufacturing molecules, moving objects, and generating heat and light. Energy can be either kinetic or potential, depending on whether it is in use or stored. **Kinetic energy**, or energy of movement, includes light (movement of photons), heat (movement of molecules), electricity (movement of electrically charged particles), and movement of large objects. **Potential energy**, or stored energy, includes chemical energy stored in the bonds that hold atoms together in molecules, electrical energy stored in a battery, and positional energy stored in a diver poised to spring.

Under the right conditions, kinetic energy can be transformed into potential energy, and vice versa. For example, a diver climbing up to a platform converts kinetic energy of movement into potential energy. When the diver jumps off, the potential energy is converted back into kinetic energy (Fig. 5-1).

Energy Cannot Be Created or Destroyed

The laws of thermodynamics describe the basic properties and behavior of energy. The **first law of thermodynamics,** often called the law of conservation of energy, states that energy can neither be created nor destroyed, so the total amount of energy remains constant. Energy can, however, change its form. For example, a car converts the potential chemical energy of gasoline into the kinetic energy of movement and heat. Similarly, a runner converts the potential chemical energy of food into the kinetic energy of movement and heat. In both of these cases, potential energy changes form, but the total amount of energy remains unchanged.

Energy Tends to Become Distributed Evenly

The **second law of thermodynamics** states that whenever energy is converted from one form to another, the amount of energy that is in useful forms decreases. All spontaneous changes result in a more even distribution of energy, which reduces the energy differences that are required to do work. In other words, energy tends to be converted from more useful into less useful forms.

To illustrate the second law, let's examine a car engine burning gasoline. In a moving vehicle, the kinetic energy of movement is much smaller than the amount of chemical energy in the gasoline that was burned to cause the movement. The first law tells us that energy cannot be destroyed, so where is the "missing" energy? It has been transferred to the car's materials and surroundings. The burning gas not only moved the car but also heated up the engine, the exhaust system, and the air around the car. In addition, the friction of tires on the pavement slightly heated the road. No energy is missing. However, the energy that was released as heat has been converted to a less useful form. It merely heated the engine, the exhaust system, the air, and the road. The second law tells us that all conversions of energy bring about a loss of usefulness. No process is 100% efficient.

Matter Tends to Become Less Organized

Regions in which energy is concentrated tend to be more organized than regions in which energy is widely dispersed. For example, the eight carbon atoms in a single molecule of gasoline have a much more orderly arrangement than do the eight carbon atoms in the eight separate, randomly moving molecules of carbon dioxide that are formed when the gasoline molecule burns. Therefore, we can also phrase the second law in terms of the organization of matter: Processes that proceed spontaneously lead to increasing randomness and disorder. This randomness and disorder is called **entropy**. Entropy always tends to increase, a tendency that can be overcome only by adding energy.

We all experience the tendency toward entropy in our homes. Frequent inputs of energy are required to keep debris confined to the trash can, newspapers to folded stacks, books to their shelves, and clothes to drawers and closets. Without our energetic cleaning and organizing efforts, these items tend to end up in their lowest-energy state—a state of disorder. When the ecologist G. Evelyn Hutchinson said, "Disorder spreads through the universe, and life alone battles against it," he was making an eloquent reference to entropy and the second law of thermodynamics.

Living Things Use the Energy of Sunlight to Create Low-Entropy Conditions

If you think about the second law of thermodynamics, you may wonder how life can exist at all. All chemical reactions, including those inside living cells, cause the amount of usable energy to decrease. Matter tends toward increasing randomness and disorder. Given these facts, how can organisms accumulate the concentrated energy and precisely ordered molecules that characterize living things?

The answer is that living things use a continuous input of solar energy to construct complex molecules and maintain orderly structures—to "battle against disorder." This solar energy arrives on Earth in the form of sunlight, which is produced by nuclear reactions in the sun. These reactions produce huge increases in entropy, so the highly organized, low-entropy systems of life do not violate the second law. The increased order of living things is achieved at the expense of an enormous loss of order in the sun. The entropy of the solar system as a whole constantly increases.

Web Tutorial 5.1 Energy and Chemical Reactions

5.2 How Does Energy Flow in Chemical Reactions?

A **chemical reaction** converts one set of chemical substances, the **reactants**, into another set, the **products**. Some chemical reactions release energy, and others consume it. A reaction that releases energy is classified as an **exergonic reaction**, and its products contain less energy than the reactants.

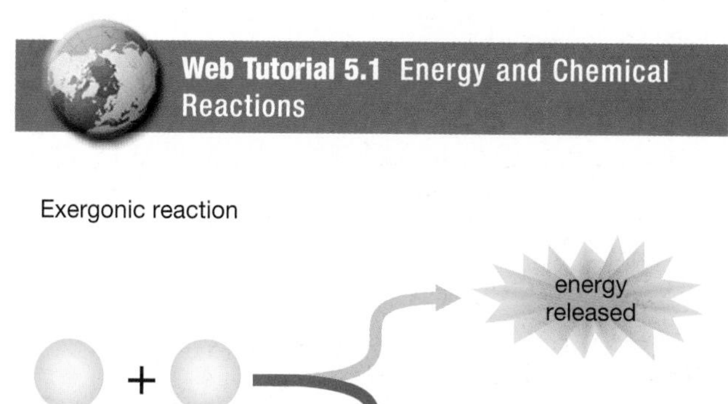

Conversely, an **endergonic reaction** requires an input of energy from some outside source. The products of an endergonic reaction contain more energy than the reactants.

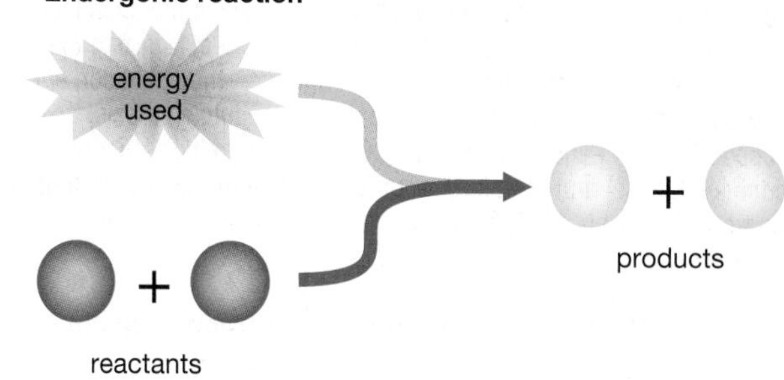

Let's look at two processes that illustrate these types of reactions: burning sugar and photosynthesis.

Exergonic Reactions Release Energy

When sugar is burned by a flame, it reacts with oxygen (O_2) to produce carbon dioxide (CO_2), and water (H_2O), and energy, as described by this equation.

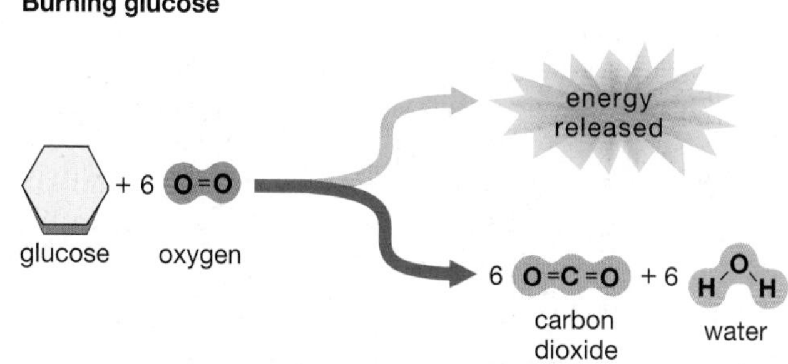

The combustion of sugar is an exergonic reaction, in which the reactants contain more energy than the products. That is, the molecules of sugar and oxygen contain much more energy than the molecules of carbon dioxide and water that are produced when the sugar breaks down. The extra energy is released as heat.

Once started, exergonic reactions proceed without an input of energy. Once ignited, a spoonful of sugar continues to burn spontaneously until all sugar molecules are consumed. It's as if exergonic reactions run "downhill," from high energy to low energy, just as a rock pushed from the top of a hill rolls to the bottom.

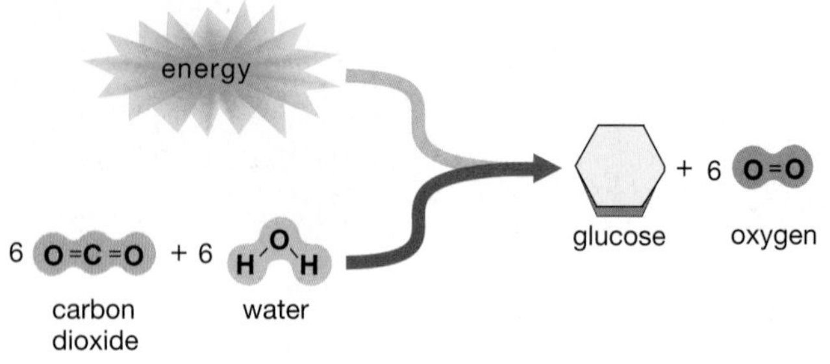

Endergonic Reactions Require an Input of Energy

Unlike the energy-releasing reaction that takes place when sugar burns, many reactions in living things create products that contain *more* energy than the reactants. For example, the sugar produced by photosynthesis contains far more energy than the carbon dioxide and water from which it is formed. (←) Similarly, a protein molecule contains more energy than the individual amino acids that were joined together to build it.

Completing reactions that form complex biological molecules requires an overall input of energy. The energy for photosynthesis, for example, comes from sunlight. We can view endergonic reactions such as photosynthesis as "uphill" reactions, going from low energy to high energy, like pushing a rock to the top of a hill.

All Reactions Require an Initial Input of Energy

All chemical reactions, even exergonic ones that release energy overall, require an initial input of energy to get started. For example, even though burning sugar releases energy, a spoonful of sugar can't burst into flames by itself. The fire doesn't begin until some energy is added. This initial energy input to a chemical reaction is called the **activation energy** and is something like the push you might give a rock poised at the top of a hill to start it rolling down (Fig. 5-2).

Chemical reactions require activation energy to get started because a shell of negatively charged electrons surrounds atoms and molecules. For two molecules to react with each other, their electron shells must be forced together, despite their mutual electrical repulsion. That force requires energy. The usual source of activation energy is the kinetic energy of movement. Molecules moving with sufficient speed collide hard enough to force their electron shells to mingle and react. Because molecules move faster as the temperature increases, most chemical reactions occur more readily at high temperatures than at low temperatures.

Exergonic Reactions May Be Linked with Endergonic Reactions

Endergonic reactions require energy from other sources, and they obtain this energy from exergonic, energy-releasing reactions. In a **coupled reaction**, an exergonic reaction provides the energy needed to drive an endergonic reaction. For example, when you drive a car, the exergonic reaction of burning gasoline provides the energy for the endergonic reaction of starting a stationary car into motion and keeping it moving. Photosynthesis is another coupled reaction. In photosynthesis, the exergonic reaction occurs in the sun, and the endergonic

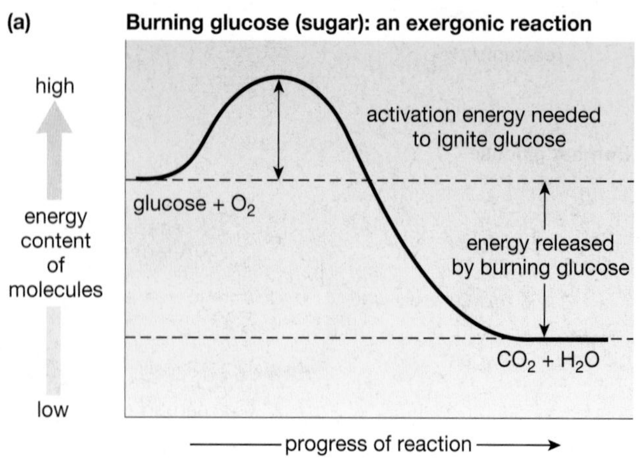

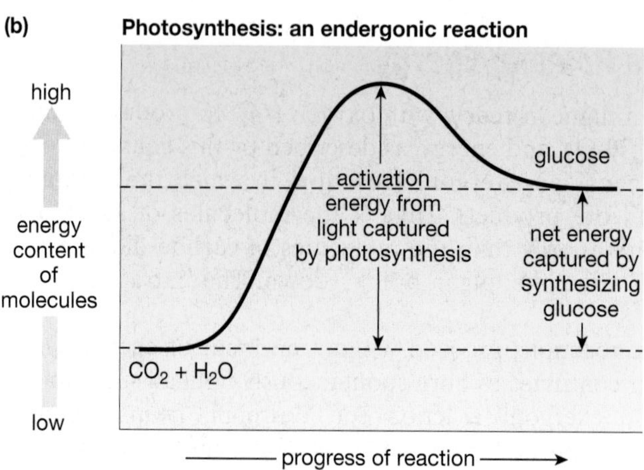

Figure 5-2 Energy relations in exergonic and endergonic reactions ***(a)*** An exergonic ("downhill") reaction, such as the burning of sugar, proceeds from high-energy reactants (here, glucose) to low-energy products (CO_2 and H_2O). The energy difference between the chemical bonds of the reactants and products is released as heat. Starting the reaction, however, requires an initial input of energy—the activation energy. ***(b)*** An endergonic ("uphill") reaction, such as photosynthesis, proceeds from low-energy reactants (CO_2 and H_2O) to high-energy products (glucose) and therefore requires a large input of energy, in this case from sunlight. **QUESTION:** *In addition to heat and sunlight, what are some other sources of activation energy?*

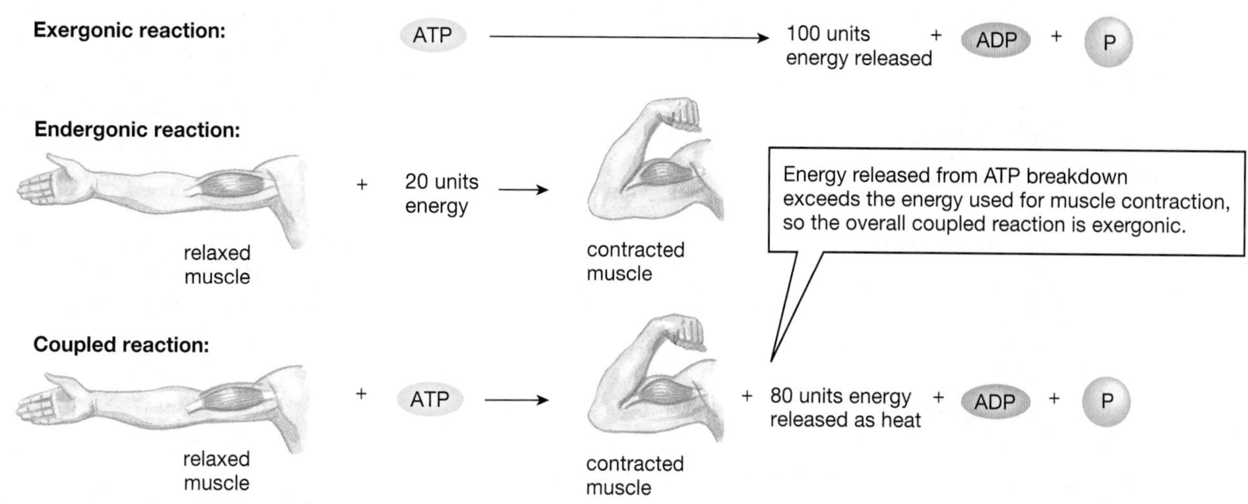

Figure 5-3 A coupled reaction The exergonic reaction of ATP breakdown must precede the endergonic reaction of muscle movement. **QUESTION:** *Why does breakdown of ATP release energy for cellular work?*

reaction occurs in the plant. Because some energy is always lost as heat during the transfer (remember the second law of thermodynamics), the energy provided by an exergonic reaction must be greater than that needed to drive the endergonic reaction.

5.3 How Is Energy Carried between Coupled Reactions?

Cells couple reactions so that the energy released by exergonic reactions is used to drive endergonic, energy-consuming reactions. The exergonic and endergonic parts of coupled reactions often occur in different parts of the cell, so there must be some way to transfer energy from the exergonic reaction that releases energy to the endergonic reaction that consumes it. The job of transferring energy from place to place is done by **energy-carrier molecules**.

Energy carriers work something like rechargeable batteries, picking up an energy charge at an exergonic reaction, moving to another location within the cell, and releasing the energy to drive an endergonic reaction. Because energy-carrier molecules are unstable, they are used only for temporary energy transfer within cells. They are not used to transfer energy from cell to cell, nor are they used for long-term energy storage.

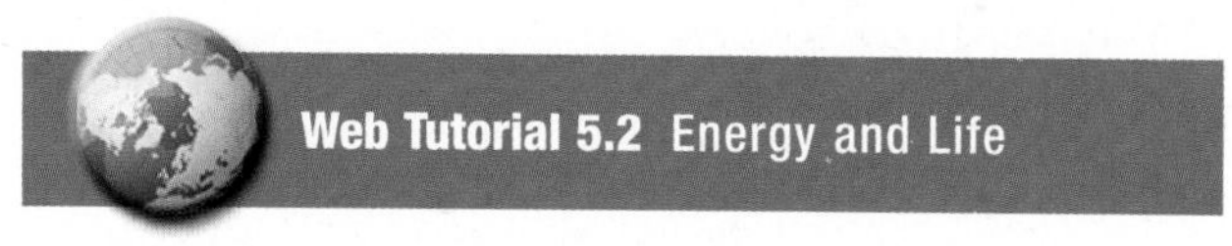

ATP Is the Principal Energy Carrier in Cells

The most common energy-carrier molecule in cells is the nucleotide **adenosine triphosphate**, or **ATP**. ATP provides energy for a wide variety of endergonic reactions, serving as a common currency of energy transfer. ATP is made from **adenosine diphosphate (ADP)** and phosphate in a reaction that uses energy released in cells through glucose breakdown. ➚

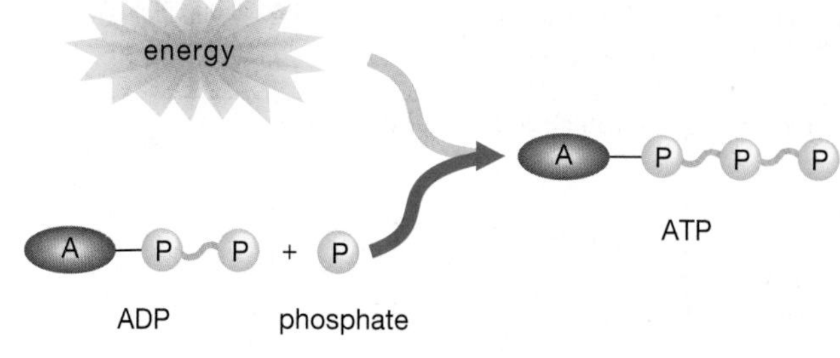

ATP stores energy in its chemical bonds and can carry the energy to sites in the cell that perform energy-requiring reactions. Then, the bond to one of ATP's phosphate groups breaks (yielding ADP and phosphate, which are later recycled to make more ATP). The energy contained in the broken bond is transferred to the energy-requiring reaction. ➔

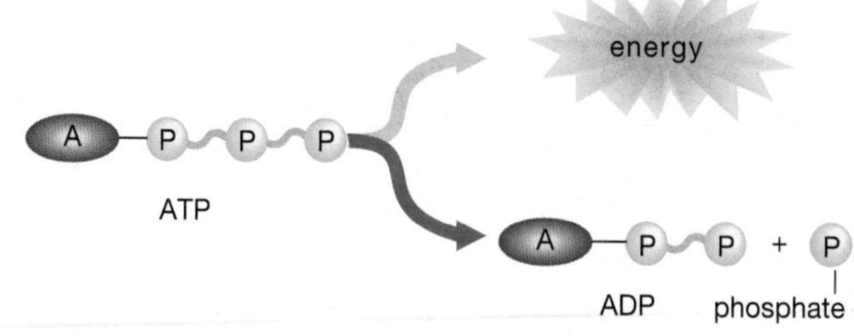

In summary, exergonic reactions (such as glucose breakdown) drive the endergonic reaction that converts ADP to ATP. The ATP molecule moves to a different part of the cell, where the ATP breaks down and liberates some of its energy to drive an endergonic reaction (such as muscle cell contraction, Fig. 5-3). During

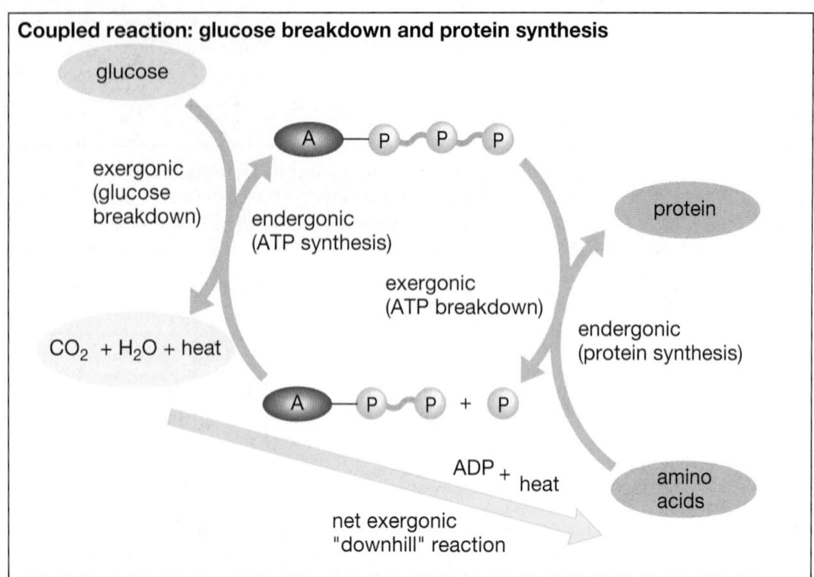

Figure 5-4 Coupled reactions give off heat The overall reaction is "downhill": More energy is produced by the exergonic reaction than is needed to drive the endergonic reaction. The extra energy is released as heat.

these energy transfers, heat is given off and there is an overall loss of usable energy (Fig. 5-4).

The life span of an ATP molecule in a living cell is very short, because this energy carrier is continuously formed, broken down, and remade. If the molecules of ATP that you use just sitting at your desk all day could be captured (instead of recycled), they would weigh nearly 90 pounds. A person running a marathon may use a pound of ATP every minute. (ADP must be quickly converted back to ATP, or it would be a very brief run.) As you can see, ATP is not a long-term energy-storage molecule. Long-term storage is the job of more stable molecules, such as sucrose, glycogen, starch, or fat, that can store energy in your body for hours, days, or months.

Electron Carriers Also Transport Energy within Cells

In addition to ATP, other carrier molecules also transport energy within a cell. **Electron carriers** capture the energetic electrons to which energy is transferred in some exergonic reactions, and the loaded electron carriers then donate the electrons, along with their energy, to molecules participating in endergonic reactions (Fig. 5-5). Common electron carriers include nicotinamide adenine dinucleotide (NAD^+) and its relative flavin adenine dinucleotide (FAD). You will learn more about electron carriers and their role in cells in Chapters 6 and 7.

5.4 How Do Cells Control Their Metabolic Reactions?

A cell is a miniature chemical factory, and the multitude of different chemical reactions that take place in a cell together constitute the cell's **metabolism**. Many of these reactions are linked in sequences called **metabolic pathways** (Fig. 5-6). Photosynthesis (Chapter 6) is one such pathway. Glycolysis, the series of reactions that begin the digestion of glucose (Chapter 7), is another.

A large number of different molecules are present in a typical cell, and this molecular diversity creates a potential problem for the cell's metabolism. With so many potentially reactive molecules present, how does the cell control reactions so that they proceed in orderly metabolic pathways? The most important way in which cells control chemical reactions is through the use of proteins called *enzymes*.

Figure 5-5 Electron carriers An electron-carrier molecule such as NAD^+ picks up an electron generated by an exergonic reaction and holds it in a high-energy outer electron shell. The electron is then deposited, energy and all, with another molecule to drive an endergonic reaction, typically the synthesis of ATP.

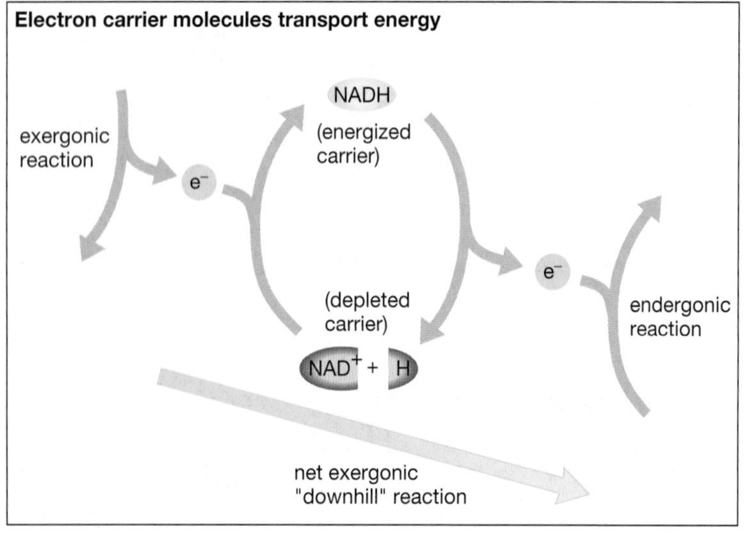

At Body Temperatures, Many Spontaneous Reactions Proceed Too Slowly to Sustain Life

In general, the speed at which a reaction occurs is determined by its activation energy (the amount of energy required to start the reaction; see Fig. 5-2). Reactions with low activation energies can proceed swiftly at body temperature, but reactions with higher activation energies occur only very slowly at such temperatures.

Many metabolic reactions have activation energies high enough that they do not occur spontaneously in the body. For example, at the temperatures found in organisms, sugar molecules almost never spontaneously break down and give up their energy. Nonetheless, the breakdown of sugar is one of the most important energy sources for living cells. Cells are able to take advantage of the energy stored in sugar molecules because of the action of *catalysts*, especially enzymes.

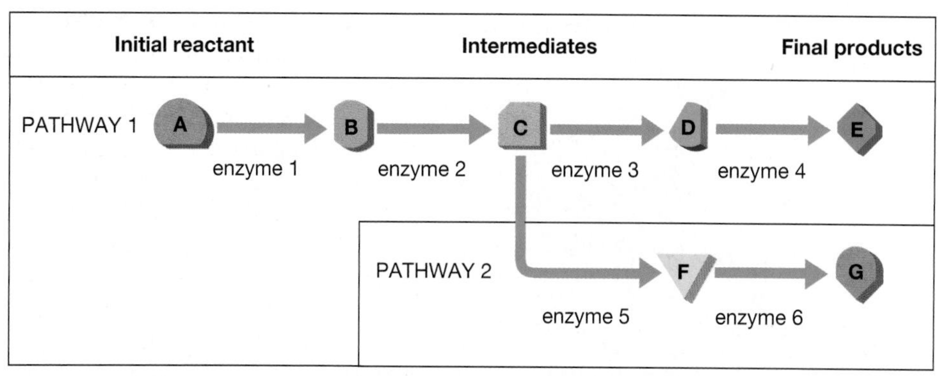

Figure 5-6 Simplified view of metabolic pathways The original reactant molecule, A, undergoes a series of reactions. The product of each reaction serves as the reactant for the next reaction in the pathway or for a reaction in another pathway.

Catalysts Reduce Activation Energy

Catalysts are molecules that speed up a reaction without themselves being used up or permanently altered. They speed up reactions by reducing the activation energy (Fig. 5-7). In the absence of a catalyst, a reaction with high activation energy will proceed only slowly at low temperatures, because few molecules collide hard enough to react. But when the activation energy is lowered by a catalyst, a much higher proportion of molecules move fast enough to react when they collide. Therefore, the reaction proceeds much more rapidly.

As an example of how catalysts work, let's consider the catalytic converters in many automobile engines. When gasoline is burned completely, the final products are carbon dioxide and water. In engines, however, incomplete combustion produces other substances, including poisonous carbon monoxide (CO). In the air, carbon monoxide reacts spontaneously but slowly with oxygen to form less-harmful carbon dioxide (CO_2). This reaction eventually removes CO from the air we breathe, but many vehicles emit so much CO that the spontaneous reaction of CO with O_2 can't keep pace, and unhealthy levels of carbon monoxide accumulate. Enter the catalytic converter. Catalysts in the converter speed up the conversion of CO to CO_2, thereby reducing air pollution.

Note three important principles about all catalysts:

- Catalysts speed up reactions.
- Catalysts can speed up only those reactions that would occur spontaneously anyway (at a much slower rate).
- Catalysts are not consumed in the reactions they promote. No matter how many reactions they participate in, the catalysts themselves are not permanently changed.

Enzymes Are Biological Catalysts

Enzymes are catalysts, usually proteins, that are made by living organisms. In addition to the three characteristics of catalysts that we just described, enzymes have additional attributes that set them apart from nonbiological catalysts. One of the most important of these is that each enzyme is usually very specialized, catalyzing a single reaction or at most only a few different reactions. In metabolic pathways such as those in Figure 5-6, each reaction is catalyzed by a different enzyme.

The Structure of Enzymes Allows Them to Catalyze Specific Reactions

The function of enzymes is closely related to their structure. Each enzyme has a complex three-dimensional shape that includes a "pocket," called the **active site**, into which reactant molecules, called **substrates**, can enter.

Why does each enzyme catalyze only one particular reaction? The active site of each enzyme has a distinctive shape and distribution of electrical charge. Because the enzyme and its substrate must fit together, only certain substrate molecules can enter the active site. As a result, a given enzyme can interact with only one or a small number of substrates. For example, several enzymes are required to completely digest all the proteins we eat, because each enzyme breaks apart only a specific sequence of amino acids.

Figure 5-7 Catalysts reduce activation energy **QUESTION:** *Can a catalyst make a nonspontaneous reaction occur spontaneously?*

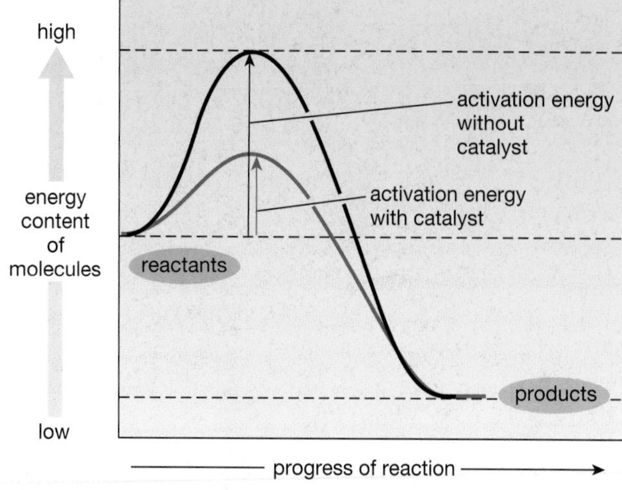

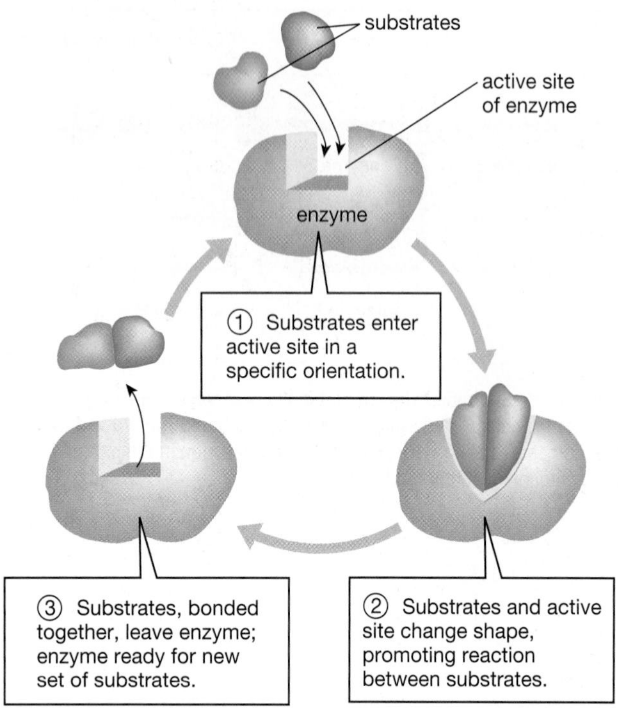

Figure 5-8 The cycle of enzyme-substrate interactions QUESTION: *How would you change reaction conditions if you wanted to increase the rate at which an enzyme-catalyzed reaction produced its product?*

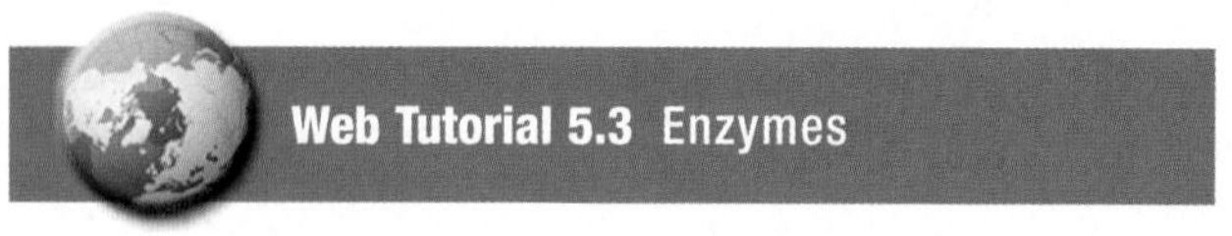

How does an enzyme catalyze a reaction? When substrates enter an enzyme's active site, both the substrates and the active site change shape (Fig. 5-8). Chemical interactions between the active site and the substrate molecules may alter the chemical bonds within the substrates. These changes in the substrates promote the particular chemical reaction catalyzed by the enzyme. When the reaction between the substrates is finished, the products no longer fit properly into the active site and are expelled. The enzyme is ready to accept another set of substrates.

The breakdown or production of a molecule within a cell normally occurs in many separate steps. Just as carving a staircase into a cliff allows the cliff to be climbed one small step at a time, breaking a reaction into small steps (each with a low activation energy), allows the overall reaction to surmount its high total activation energy. Each step is catalyzed by a different enzyme, and each enzyme lowers the activation energy for its particular step (see Fig. 5-7). For example, our cells break down sugar in many small steps, with each step liberating a small amount of energy. This step-by-step procedure is the reason that the reaction can proceed at body temperature, and we are not roasted by the sugar that is "burning" in our bodies.

The Activity of Enzymes Is Influenced by Their Environment

The complex three-dimensional structure of an enzyme is required for proper function, but the shape is sensitive to changes in environmental conditions. The hydrogen bonds that play a key role in determining enzyme structure can be altered by the enzyme's chemical and physical surroundings. Each enzyme functions optimally at a particular pH, salt concentration, and temperature, and deviations from these optimal conditions can change an enzyme's shape and destroy its effectiveness. For example, one reason that dill pickles stay well preserved in a vinegar-salt solution is that that salty, acidic conditions in the solution distort the shape of enzymes used in bacterial metabolism. With their enzymes inoperative, the bacteria that ordinarily cause food to spoil cannot survive.

When temperatures rise too high, the hydrogen bonds that determine enzyme shape may be broken apart and the enzyme rendered useless. Smaller changes in temperature, however, may simply change the rate at which enzyme-catalyzed reactions proceed. At higher temperatures, substrate molecules move more rapidly, and their random movements are more likely to bring them into contact with the active site of an appropriate enzyme. Thus, many reactions are accelerated by moderately higher temperatures and slowed by lower temperatures.

A MISSING MOLECULE MAKES MISCHIEF *REVISITED*

The molecule missing from people with PKU is the enzyme phenylalanine hydroxylase. This enzyme normally catalyzes a key reaction in the metabolic pathway of phenylalanine, namely its conversion to another amino acid, tyrosine. When the hydroxylase enzyme is absent, the reactant phenylalanine is not converted to the product tyrosine. The reactant instead accumulates, reaching toxic levels.

The only known treatment for PKU is to avoid eating proteins (almost all of which contain at least some phenylalanine). So people with PKU cannot eat many foods that the rest of us take for granted, including meat, cheese, and fish. PKU sufferers must also avoid the widespread artificial sweetener aspartame, which contains phenylalanine.

Sticking to the special diet can be difficult. Other treatments would be welcome, and scientists at several labs are working to develop them. One research group is testing a treatment in which young adults with PKU eat a much less restrictive diet but also eat a mixture of amino acids that slows the movement of phenylalanine from the blood into the brain. Because these patients eat protein, the level of phenylalanine in their blood increases, but the researchers predict that the higher levels won't reach and damage the brain.

To test their prediction, the researchers divided 45 patients into two groups. One group ate the usual, restrictive PKU diet, and the other ate a more diverse diet but also took the amino acid supplement. After 12 months, the subjects were given an IQ test, and their scores were compared with those from tests given before the experiment. Neither group's test performance declined. The researchers concluded that their treatment could allow older PKU patients to have a more diverse diet.

CONSIDER THIS: Newborns in all developed nations are screened for PKU, and affected individuals are identified before any neurological damage can occur. Infants are also screened for some other ailments, but there are many potentially detectable conditions for which they are not tested. In your view, what criteria should be used to decide which tests are mandatory for all newborns?

LINKS TO LIFE: Is Milk Just for Babies?

Infants thrive on an all-milk diet, but most adults cannot digest milk. About 75% of people worldwide, including 25% of those in the United States, lose the ability to digest lactose, or "milk sugar," in early childhood, and they become *lactose intolerant*. Roughly 75% of African Americans, Hispanics, and Native Americans, as well as 90% of Asian Americans, are lactose intolerant. Only a relatively small proportion of people, primarily those of northern European descent, retain the ability to digest lactose into adulthood.

Lactose intolerance arises when the body stops producing the enzyme lactase, which catalyses the breakdown of lactose. When someone who lacks this enzyme consumes milk products, undigested lactose draws water into the intestines by osmosis and also feeds intestinal bacteria that produce gas. The excess water and gas lead to abdominal pain, bloating, diarrhea, and flatulence—a rather high price to pay for indulging in ice cream. But compared with the consequences of other enzyme deficiencies, such as PKU, the inability to tolerate milk is a relatively minor inconvenience.

Most people who are lactose intolerant do not need to avoid milk products altogether. Some such people produce enough lactase to tolerate a few servings daily, and most can eat aged cheeses (such as cheddar) and yogurt with live bacteria, which contains relatively little lactose because the bacteria break it down. Those unwilling to forgo ice cream can consume lactase supplements along with dairy products.

Chapter Review

Summary of Key Concepts

5.1 What Is Energy?

Energy is the capacity to do work. Kinetic energy is the energy of movement (light, heat, electricity, movement of large particles). Potential energy is stored energy (chemical energy, positional energy). The first law of thermodynamics states that energy can be neither created nor destroyed, so the total amount of energy remains constant, although it may change in form. The second law of thermodynamics states that any use of energy within a system causes a decrease in the quantity of concentrated, useful energy and an increase in the randomness and disorder of matter (entropy).

5.2 How Does Energy Flow in Chemical Reactions?

Chemical reactions fall into two categories. In exergonic reactions, the product molecules have less energy than do the reactant molecules, so the reaction releases energy. In endergonic reactions, the products have more energy than do the reactants, so the reaction requires an input of energy. Exergonic reactions can occur spontaneously, but all reactions, including exergonic ones, require an input of activation energy to overcome electrical repulsions between reactant molecules. In a coupled reaction, the energy liberated by an exergonic reaction drives the endergonic reaction. Organisms couple exergonic reactions, such as light-energy capture or sugar metabolism, with endergonic reactions, such as the synthesis of organic molecules.

5.3 How Is Energy Carried between Coupled Reactions?

Energy released by chemical reactions within a cell is captured and transported within the cell by energy-carrier molecules, such as ATP and electron carriers. These molecules are the major means by which cells couple exergonic and endergonic reactions that occur at different places in the cell.

5.4 How Do Cells Control Their Metabolic Reactions?

Cellular reactions are linked in metabolic pathways and are regulated in large measure by enzymes. High activation energies cause many reactions, even exergonic ones, to proceed very slowly under normal environmental conditions. Catalysts lower the activation energy and thereby speed up chemical reactions without being permanently changed themselves. Organisms synthesize enzymes that catalyze one or a few reactions. The reactants temporarily bind to the active site of the enzyme, making it easier to form the new chemical bonds of the products. Environmental conditions including pH, salt concentration, and temperature can promote or inhibit the function of enzymes, by altering their three-dimensional structure.

Key Terms

activation energy *p. 76*
active site *p. 79*
adenosine diphosphate (ADP) *p. 77*
adenosine triphosphate (ATP) *p. 77*
catalyst *p. 79*
chemical reaction *p. 75*
coupled reaction *p. 76*
electron carrier *p. 78*
endergonic reaction *p. 75*
energy *p. 74*
energy-carrier molecule *p. 77*
entropy *p. 75*
enzyme *p. 79*
exergonic reaction *p. 75*
first law of thermodynamics *p. 74*
kinetic energy *p. 74*
metabolic pathway *p. 78*
metabolism *p. 78*
potential energy *p. 74*
product *p. 75*
reactant *p. 75*
second law of thermodynamics *p. 74*
substrate *p. 79*

Thinking Through the Concepts

Multiple Choice

1. *According to the first law of thermodynamics, the total amount of energy in the universe*
a. is always increasing
b. is always decreasing
c. varies up and down
d. is constant
e. cannot be determined

2. *What is predicted by the second law of thermodynamics?*
a. Energy is always decreasing.
b. Disorder cannot be created or destroyed.
c. Systems always tend toward greater states of disorder.
d. All potential energy exists as chemical energy.
e. all of the above

3. *Which statement about exergonic reactions is true?*
a. The products have more energy than do the reactants.
b. The reactants have more energy than do the products.
c. They will not proceed spontaneously.
d. Energy input reverses entropy.
e. none of the above

4. *ATP is important in cells because*
a. it transfers energy from exergonic reactions to endergonic reactions
b. it is assembled into long chains that make up cell membranes
c. it acts as an enzyme
d. it accelerates diffusion
e. all of the above

5. *How does an enzyme increase the speed of a reaction?*
a. by changing an endergonic to an exergonic reaction
b. by providing activation energy
c. by lowering the activation energy
d. by decreasing the concentration of reactants
e. by increasing the concentration of products

6. *Which of the following statements about enzymes is (are) true?*
a. They interact with specific reactants (substrates).
b. Their three-dimensional shapes are closely related to their activities.
c. They change the shape of the reactants.
d. They have active sites.
e. all of the above

Review Questions

1. Explain why organisms do not violate the second law of thermodynamics. What is the ultimate energy source for most forms of life on Earth?

2. Define *metabolism*, and explain how reactions can be coupled to one another.

3. What is activation energy? How do catalysts affect activation energy? How does the effect of catalysts change the rate of reactions?

4. Describe some exergonic and endergonic reactions that occur in plants and animals very regularly.

5. Describe the structure and function of enzymes.

Applying the Concepts

1. **IS THIS SCIENCE?** Creationists sometimes critique the theory of evolution with the following argument: "According to evolutionary theory, organisms have increased in complexity through time. However, evolution of increased biological complexity contradicts the second law of thermodynamics. Therefore, evolution is impossible." Develop a scientific response to this argument.
2. **IS THIS SCIENCE?** Some practitioners of alternative medicine believe that people can improve their health by swallowing capsules of various plant and animal enzymes. This "enzyme therapy" is based on the hypothesis that the enzymes will act in the stomach to predigest food, so that the body will have to use less of its own enzymes for digestion and can instead use them to "maintain metabolic harmony" and improve overall health. Evaluate this hypothesis from a scientific viewpoint and suggest an experiment to test it.
3. A preview question for ecology: When a brown bear eats a salmon, does the bear acquire all the energy contained in the body of the fish? Why or why not? What implications do you think this answer would have for the relative abundance (by weight) of predators and their prey? Does the second law of thermodynamics help explain the title of the book *Why Big, Fierce Animals Are Rare*?
4. As you learned in Chapter 2, the subunits of virtually all organic molecules are joined by dehydration synthesis reactions and can be broken apart by hydrolysis reactions. Why, then, does your digestive system produce separate enzymes to digest proteins, fats, and carbohydrates—in fact, several of each type?

For More Information

Baker, J. J. W., and Allen, G. E. *Matter, Energy, and Life*. 4th ed. Reading, MA: Addison-Wesley, 1981. An excellent introduction to chemical-energy principles for those interested in biology.

Farid, R. S. "Enzymes Heat Up." *Science News*, May 9, 1998. Scientists explore new ways to synthesize enzymes that will function at high temperatures.

Fenn, J. *Engines, Energy, and Entropy*. New York: W. H. Freeman, 1982. Elegantly simple introduction to the laws of thermodynamics and their relationship to everyday life.

WEB TUTORIAL

To access a Web Tutorial visit http://www.prenhall.com/audesirk4. *Log in to the Web site selected by your instructor, navigate to this chapter, and select the appropriate Web Activity number.*

5.1 Energy and Chemical Reactions
Estimated time: 10 minutes

Learn about the different types of energy and the laws of thermodynamics.

5.2 Energy and Life
Estimated time: 10 minutes

Explore the importance of ATP and coupled reactions in the cell..

5.3 Enzymes
Estimated time: 5 minutes

Visualize the structure of an enzyme and explore how an enzyme's structure relates to its catalytic function.

5.4 Web Investigation: A Missing Molecule Makes Mischief
Estimated time: 10 minutes

In this exercise we will use online resources to explore some diseases that are caused by enzyme deficiencies, and how newborns are screened for these diseases soon after being born.

Chapter 6

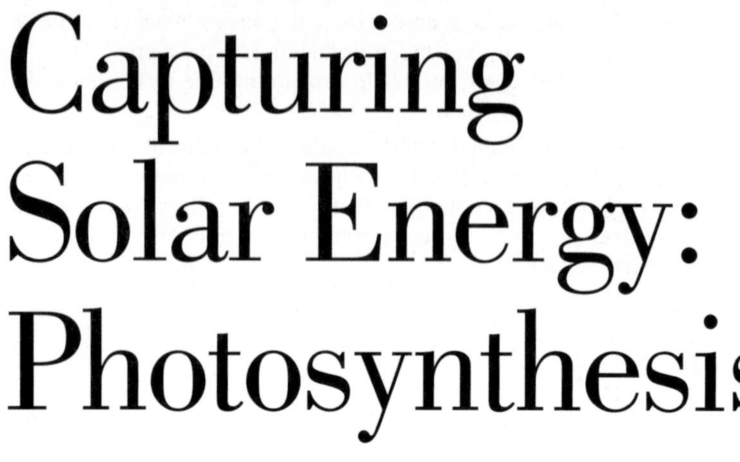

Capturing Solar Energy: Photosynthesis

A giant meteor ended the reign of *Tyrannosaurus* and the other dinosaurs.

AT A GLANCE

Did the Dinosaurs Die from Lack of Sunlight?

About 65 million years ago, a gigantic meteorite plunged out of the sky and crashed into Earth. This dramatic event took place long before there were any humans to observe it, so how do we know it happened? One of the first clues was the observation by geologists that 65-million-year-old rocks at many locations around the world contain large amounts of the element iridium. Iridium is very rare on Earth, and not much is found in rocks older or newer than 65 million years. It is, however, quite common in meteorites.

If the iridium-rich rock layer did indeed result from an impact that hurled meteorite fragments around the globe, then the impact ought to have left a crater. This prediction was confirmed with the discovery of a buried crater, 65 million years old, at the tip of Mexico's Yucatan Peninsula. The crater is a mile deep and 120 miles wide. Calculations suggest that a meteorite large enough to make a crater that size must have been about 6 miles in diameter.

The fossil record reveals that, within a short period after the impact, most of the species on the planet were destroyed. This devastating mass extinction eliminated more than 70% of the species then existing, including dinosaurs. *Triceratops, Tyrannosaurus*, and all the other dinosaur species disappeared forever. Land and sea alike were nearly emptied of life.

That the mass extinction occurred at the same time as the meteorite impact suggests that the two events were connected. But how? Even though the meteorite was huge, its direct killing power must have been limited to the area of impact. Why, then, were thousands of species in all corners of the globe eliminated? Some scientists hypothesize that the most serious damage was done by the impact's disruption of the most important chemical reaction on Earth: photosynthesis.

The meteor's impact darkened the skies and started forest fires worldwide. These disasters drastically reduced photosynthesis, a reduction that had disastrous consequences for the dinosaurs.

6.1 What Is Photosynthesis?

More than 2 billion years ago, some cells evolved the ability to harness the energy in sunlight. Using a process known as **photosynthesis**, these cells captured some of sunlight's energy and stored it as chemical energy. Exploiting this new source of energy with little competition, early photosynthetic cells filled the seas. The type of photosynthesis that eventually became most common (and that is the dominant form today) uses carbon dioxide and water as its raw materials and releases oxygen gas as a by-product.

Photosynthesis is absolutely essential to life on Earth. Before the evolution of photosynthesis, there was very little oxygen gas in the atmosphere. As photosynthetic organisms grew in number, however, oxygen began to accumulate. Today, most living things depend on the oxygen produced by photosynthesis. In addition, virtually all the energy used by living things is captured by photosynthesis. Stored in sugar and other organic molecules, this energy provides nourishment for photosynthetic organisms and for the organisms that eat them.

Figure 6-1 An overview of photosynthetic structures

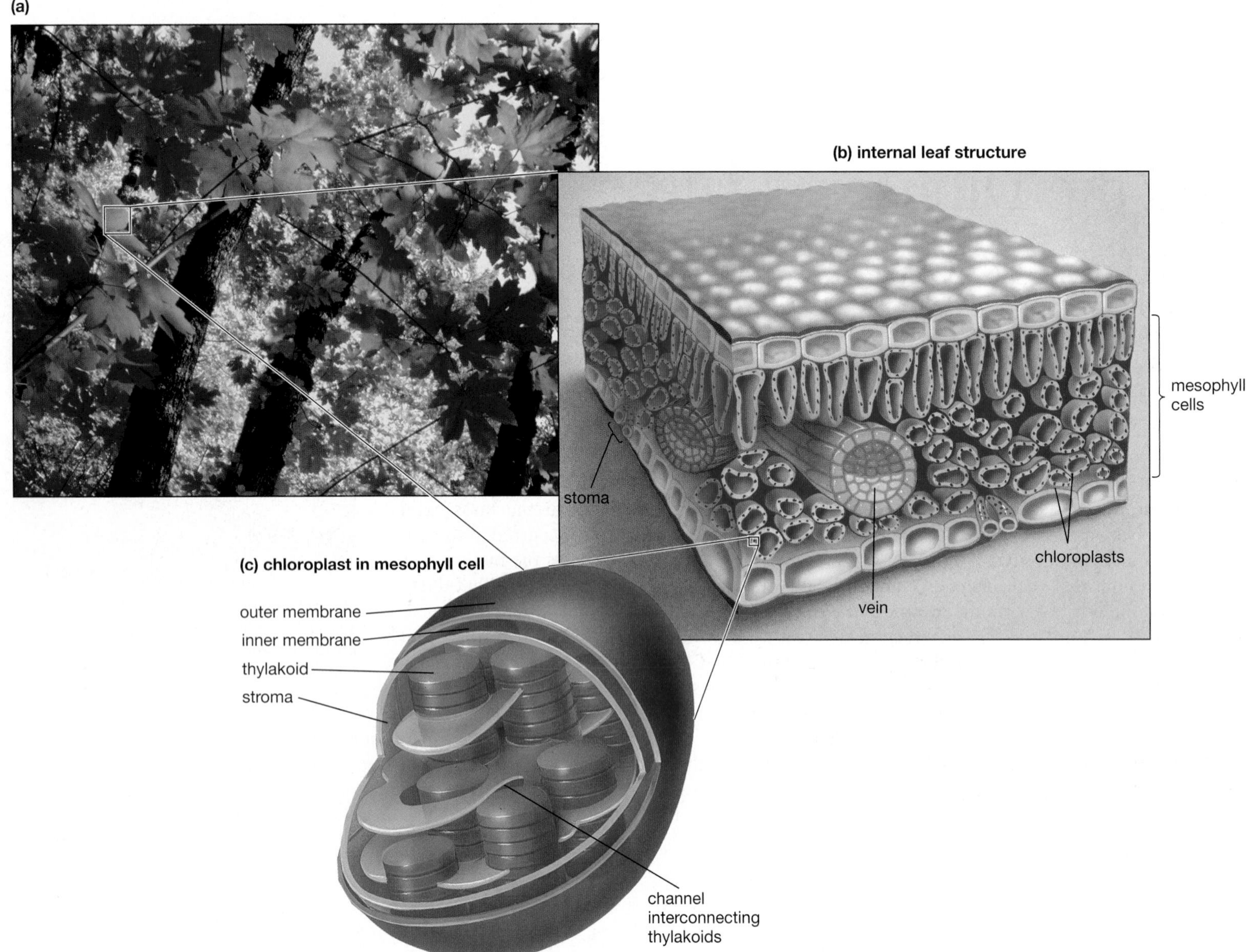

Photosynthesis Converts Carbon Dioxide and Water to Glucose

Starting with the simple molecules of carbon dioxide (CO_2) and water (H_2O), photosynthesis converts the energy of sunlight into chemical energy stored in the bonds of glucose ($C_6H_{12}O_6$). It also releases oxygen. When written in its simplest form, the overall chemical reaction for photosynthesis is:

$$6\ CO_2 + 6\ H_2O + \text{light energy} \longrightarrow C_6H_{12}O_6 + 6\ O_2$$

This basic photosynthetic reaction takes place in almost all photosynthetic organisms, which include plants, seaweeds, and various types of single-celled organisms. Each group of photosynthetic organisms, however, exhibits distinctive variations on the basic process. In this chapter, we will focus on photosynthesis in plants.

Plant Photosynthesis Takes Place in Leaves

Leaves have several features that make them especially well suited for their role as the main location of photosynthesis. For example, the leaves of most land plants are only a few cells thick. Having thin leaves ensures that sunlight can easily penetrate them (**Fig. 6-1a**). In addition, the flattened shape of leaves exposes a large surface area to the sun.

A leaf obtains CO_2 for photosynthesis from the air, through adjustable pores in the leaf surface called **stomata** (singular, stoma). Stomata can open and close, and they open at appropriate times to admit CO_2. Inside the leaf are a few layers of cells collectively called **mesophyll**; photosynthesis occurs mainly in these cells (**Fig. 6-1b**). A system of veins supplies water and minerals to the mesophyll cells and carries the sugars they produce to other parts of the plant.

Leaf Cells Contain Chloroplasts

Leaf cells, especially mesophyll cells, contain large numbers of **chloroplasts**, the organelles in which photosynthesis occurs. Chloroplasts have a double outer membrane that encloses a semifluid medium, the **stroma** (**Fig. 6-1c**). Embedded in the stroma are disk-shaped, interconnected membranous sacs called **thylakoids**. The light-dependent reactions of photosynthesis occur within the membranes of the thylakoids.

Photosynthesis Consists of Light-Dependent and Light-Independent Reactions

The simple equation that summarizes the overall reaction of photosynthesis conceals the fact that photosynthesis actually requires dozens of steps, each catalyzed by a different enzyme. The steps can be divided into two groups, the light-dependent reactions and the light-independent reactions. The two groups of reactions occur at different locations within the chloroplast, but they are linked by energy-carrier molecules.

- In the *light-dependent* reactions, molecules in the membranes of the thylakoids capture sunlight energy and convert some of it into chemical energy stored, for the short term, in two different energy-carrier molecules. The molecules are the familiar energy carrier **adenosine triphosphate (ATP)** and the electron carrier **nicotinamide adenine dinucleotide phosphate (NADPH)**. Oxygen gas is released as a by-product. →
- In the *light-independent* reactions, enzymes in the stroma use the chemical energy of the carrier molecules to power the manufacture of glucose or other organic molecules.

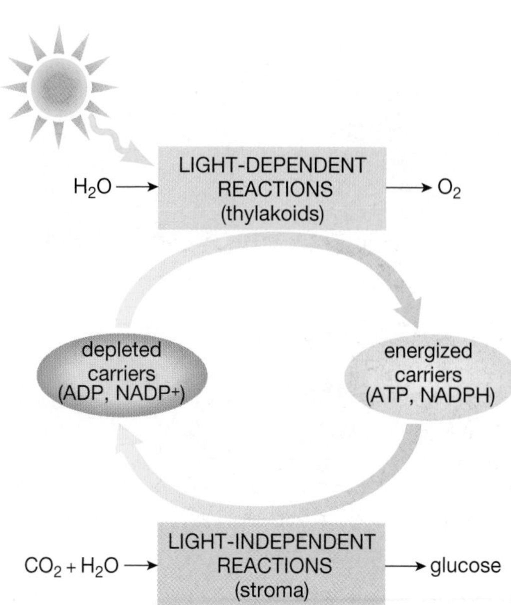

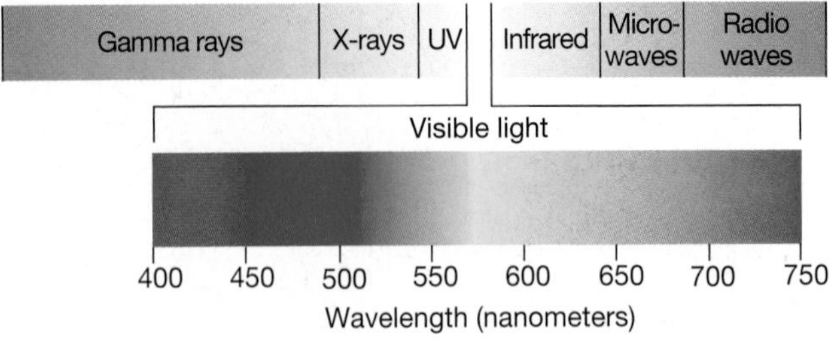

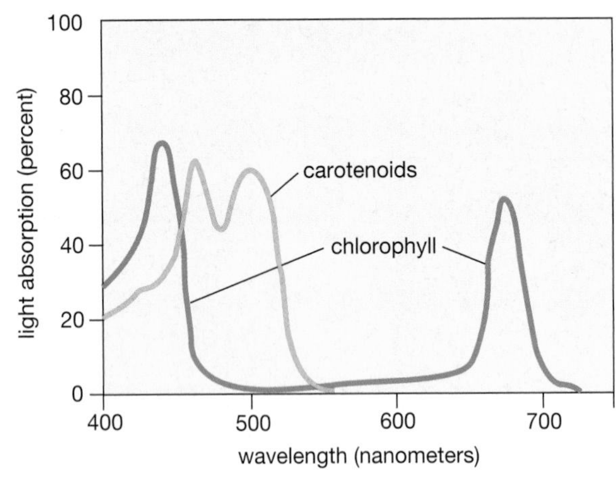

Figure 6-2 Light, chloroplast pigments, and photosynthesis
(a) Visible light, a small part of the electromagnetic spectrum (top line), consists of wavelengths that correspond to the colors of the rainbow. ***(b)*** Chlorophyll (green curve) strongly absorbs violet, blue, and red light. Carotenoids (orange curve) absorb blue and green wavelengths.
QUESTION: *On the basis of the information in this graph, what color are carotenoids?*

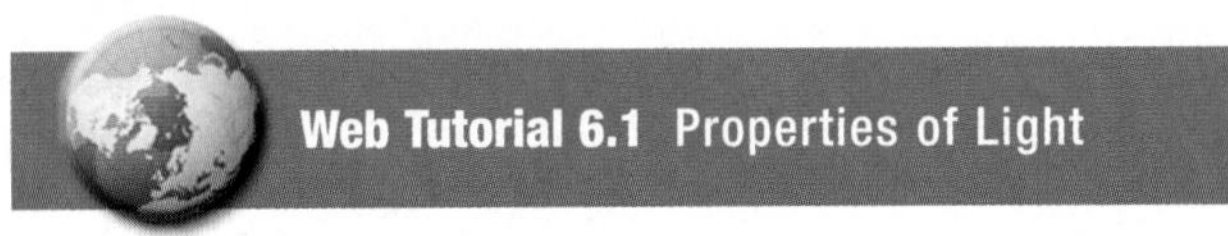

6.2 How Is Light Energy Converted to Chemical Energy?

In the first stage of photosynthesis, the light-dependent reactions convert the energy of sunlight into chemical energy stored in carrier molecules. Sunlight consists of an array of electromagnetic waves of various sizes, but the wavelengths in sunlight are only a portion of the full electromagnetic spectrum. (The size of a wave is measured by its wavelength, and the full electromagnetic spectrum ranges from short-wavelength gamma rays, through ultraviolet, visible, and infrared light, to very long wavelength radio waves). Of the wavelengths emitted by the sun, photosynthesis uses only those that fall into the range of visible light (Fig. 6-2a). Note that, within the range of visible light, we perceive different wavelengths as different colors.

Light Energy Is First Captured by Pigments in Chloroplasts

The job of absorbing light energy for use in photosynthesis falls to certain molecules in the thylakoid membranes of chloroplasts. The membranes contain several types of *pigments* (light-absorbing molecules), each of which absorbs a different range of wavelengths. **Chlorophyll**, the key light-capturing molecule, strongly absorbs violet, blue, and red light but reflects green. The reflected wavelengths reach our eyes, and we see chlorophyll as green (Fig. 6-2b). Most leaves appear green because they are rich in chlorophyll.

Thylakoids also contain other molecules, called accessory pigments, that capture light energy and transfer it to chlorophyll. For example, carotenoids absorb blue and green light, but reflect yellow and orange. In leaves, the colors of carotenoids are usually masked by the presence of chlorophyll. When a leaf dies, however, its chlorophyll usually breaks down before its carotenoids, and a bright yellow or orange color may be revealed.

The Light-Dependent Reactions Generate Energy-Carrier Molecules

The light-dependent reactions take place in special locations called photosystems, which are located in the thylakoid membranes. Each **photosystem** consists of a highly organized assemblage of proteins, chlorophyll, accessory pigments, and electron-carrier molecules. In each thylakoid, there are thousands of photosystems of two types: *photosystem I* (PS I) and *photosystem II* (PS II).

Each photosystem consists of two major parts. The first part, the **light-harvesting complex**, contains about 300 chlorophyll and accessory pigment molecules. These molecules absorb light and pass the energy to a specific chlorophyll molecule called the **reaction center**. The reaction-center chlorophyll is located next to the second part of the photosystem: the **electron transport chain (ETC)**, a series of electron-carrier molecules also embedded in the thylakoid membrane. ←

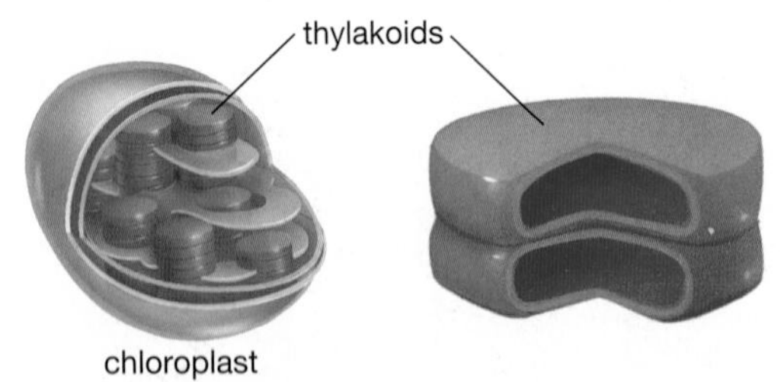

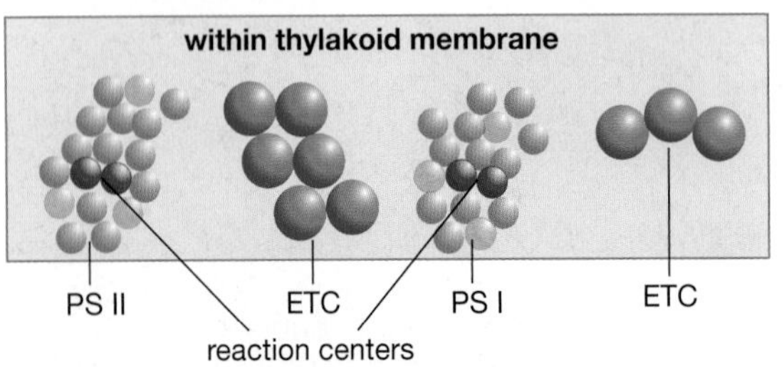

Photosystem II Generates ATP

As you read the following sections about the photosystems, you may wish to consult Figure 6-3 to help you follow the steps. In the first four steps shown, photosystem II captures light energy and produces ATP. ① The light-dependent reactions begin when the light-harvesting complex of photosystem II absorbs energy from light. The energy passes from molecule to molecule until it reaches the reaction center. ② When the reaction-center chlorophyll molecule receives the energy, electrons absorb the energy and jump from the chlorophyll molecule over to ③ the adjacent electron transport chain. After arriving at the electron transport chain, these energetic electrons keep jumping, moving from one carrier molecule to the next. During some of the jumps, the electrons release energy. ④ This released energy powers reactions that synthesize the energy carrier ATP.

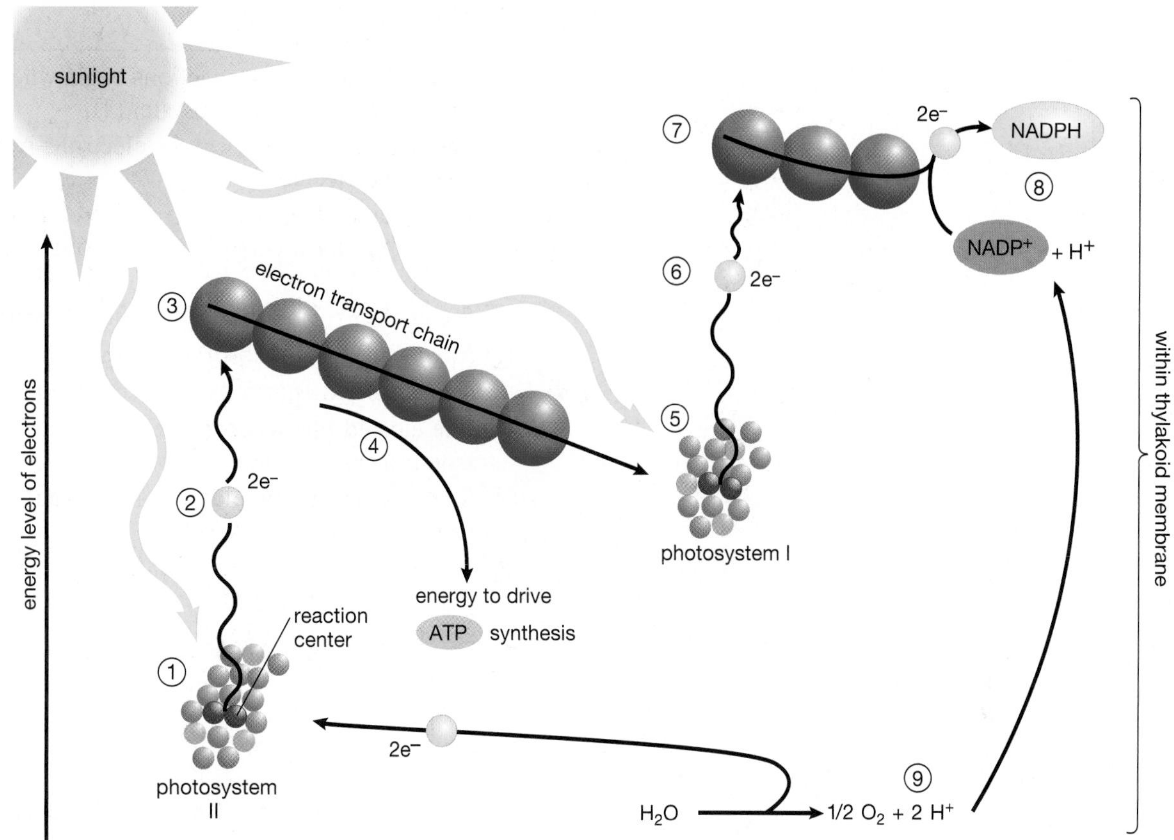

Figure 6-3 The light-dependent reactions of photosynthesis
QUESTION: *If these reactions produce energy-carrier molecules (ATP and NADPH), then why do plant cells need mitochondria?*

Photosystem I Generates NADPH

In the next steps of the light-dependent reactions, photosystem I captures light energy and produces the energy carrier NADPH. ⑤ As photosystem II continues to generate ATP, light rays are also striking the light-harvesting complex of photosystem I. Just as in photosystem II, the energy from the light striking photosystem I is transmitted to the reaction-center chlorophyll, which ⑥ ejects electrons. Almost immediately, these lost electrons are replaced by electrons that have reached the end of photosystem II's electron transport chain and jumped over to photosystem I. ⑦ At the same time, the ejected photosystem I electrons jump to the photosystem I electron transport chain. The electrons move from molecule to molecule through the photosystem I electron transport chain until they reach the electron-carrier molecule $NADP^+$. ⑧ Each $NADP^+$ molecule picks up two electrons and one hydrogen ion, forming NADPH.

Splitting Water Maintains the Flow of Electrons through the Photosystems

Overall, electrons flow from the reaction center of photosystem II, through the photosystem II electron transport chain, to the reaction center of photosystem I, through the photosystem I electron transport chain, and on to form NADPH. To sustain this one-way flow of electrons, photosystem II's reaction center must be continuously supplied with new electrons to replace the ones it gives up. These replacement electrons come from water (step ⑨ in **Fig. 6-3**) by the reaction:

$$H_2O \rightarrow \frac{1}{2} O_2 + 2H^+ + 2\,e^-$$

As the water molecules are split, their oxygen atoms combine to form molecules of oxygen gas, O_2. The oxygen may be used directly by the plant in its own respiration or given off to the atmosphere.

Summing Up: Light-Dependent Reactions

① The light-dependent reactions begin when light is absorbed by the light-harvesting complex of photosystem II.
② The absorbed energy causes electrons from the reaction center of the complex to be ejected.
③ The ejected electrons are transferred to photosystem II's electron transport chain. As the electrons pass through the transport system, they release energy.
④ Some of the energy is used to manufacture ATP.
⑤ Meanwhile, light is absorbed by the light-harvesting complex of photosystem I.
⑥ The absorbed energy ejects electrons from the reaction center.
⑦ The ejected photosystem I electrons are picked up by photosystem I's electron transport chain.
⑧ Some of the energy released as electrons move through the electron transport chain is captured as NADPH.
⑨ A water molecule splits, donating electrons to the photosystem II chlorophyll and generating oxygen as a by-product.

The products of the light-dependent reactions are NADPH, ATP, and O_2.

6.3 How Is Chemical Energy Stored in Glucose Molecules?

The ATP and NADPH molecules generated in the light-dependent reactions are used in the light-independent reactions to manufacture molecules for the long-term storage of energy. The light-independent reactions, so called because they can occur without light as long as ATP and NADPH are available, take place in the fluid stroma that surrounds the thylakoids. There, the reactions make glucose from carbon dioxide and water, using enzymes that are present in the stroma.

The C_3 Cycle Captures Carbon Dioxide

The first step on the path to glucose is to capture carbon dioxide. It is captured in a set of reactions known as the **C_3 cycle** (three-carbon cycle) because some of the important molecules in the cycle have three carbon atoms in them. The reactions are also sometimes called the Calvin-Benson cycle. The C_3 cycle requires the five-carbon sugar **ribulose bisphosphate (RuBP)**, CO_2 (normally from the air), enzymes to catalyze all the reactions, and energy in the form of ATP and NADPH (normally from the light-dependent reactions).

The C_3 cycle begins (and ends) with RuBP. Six molecules of RuBP combine with CO_2 in a series of reactions that yields 12 three-carbon molecules of *phosphoglyceric acid* (PGA). In effect, CO_2 molecules are captured from the atmosphere, a process known as **carbon fixation**, because it "fixes" gaseous CO_2 into a relatively stable organic molecule, PGA (step ① in Fig. 6-4).

As the cycle proceeds, the 12 molecules of PGA are converted to 12 molecules of *glyceraldehyde-3-phosphate* (G3P; step ② in Fig. 6-4), using energy donated by ATP and NADPH (recall that these energy carriers were generated during the light-dependent reactions). Then, through a series of reactions requiring ATP energy, 10 of the G3P molecules are used to regenerate the 6 molecules of RuBP needed to restart the cycle (step ④ in Fig. 6-4).

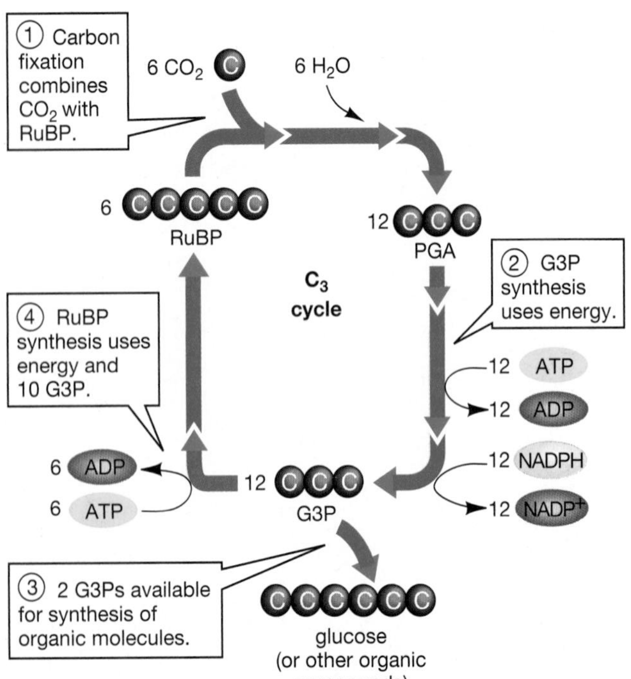

Figure 6-4 The C_3 cycle of carbon fixation Keep track of the carbons as you follow the illustration around. ① Six molecules of RuBP react with 6 molecules of CO_2 and 6 molecules of H_2O to form 12 molecules of PGA. ② The energy of 12 ATPs and the electrons and hydrogens of 12 NADPHs are used to convert the 12 PGA molecules to 12 G3Ps. ③ Two G3P molecules are further processed into glucose. ④ Energy from 6 ATPs is used to rearrange the remaining 10 G3Ps into 6 RuBPs, completing one turn of the C_3 cycle.

Carbon Fixed During the C_3 Cycle Is Used to Synthesize Glucose

Each turn of the C_3 cycle begins and ends with six molecules of RuBP but also captures additional carbon atoms from CO_2. These extra carbons are found in the two "extra" molecules of G3P that are left over after RuBP is regenerated. As indicated in step ③ of **Figure 6-4**, these two G3P molecules (three carbons each) combine to form one molecule of glucose (six carbons).

Summing Up: Light-Independent Reactions

- In the C_3 cycle, 6 molecules of RuBP capture 6 molecules of CO_2 to produce 12 molecules of PGA.
- A series of reactions driven by energy from ATP and NADPH uses the PGA to produce 12 molecules of G3P.
- Ten of these G3P molecules are used to regenerate the 6 RuBP molecules. Thus, 2 molecules of G3P are left over, and they join to form 1 molecule of glucose.
- Each passage through the cycle yields one molecule of glucose and depleted energy carriers (ADP and $NADP^+$) that will be recharged during light-dependent reactions.

6.4 What Is the Relationship between Light-Dependent and Light-Independent Reactions?

As you learn some of the details of the light-dependent and light-independent reactions of photosynthesis, it is important to not lose sight of the close interdependence of the two sets of reactions. The connection between the light-dependent and light-independent reactions is illustrated in **Figure 6-5**, and in the figure on page 87.

Stated simply, the "photo" part of photosynthesis refers to the capture of light energy by the light-dependent reactions. The "synthesis" part refers to the light-independent reactions' use of this captured energy to synthesize glucose. Stated in more detail, the light-dependent reactions in the membranes of the thylakoids use light energy to "charge up" the energy-carrier molecules ADP and $NADP^+$ to form ATP and NADPH. These energized carriers move to the stroma, where their energy is used to drive the light-independent reactions that synthesize glucose. The depleted carriers, ADP and $NADP^+$, then return to the light-dependent reactions for recharging to ATP and NADPH.

Figure 6-5 Two sets of reactions are connected in photosynthesis
QUESTION: *Could a plant survive in an oxygen-free atmosphere?*

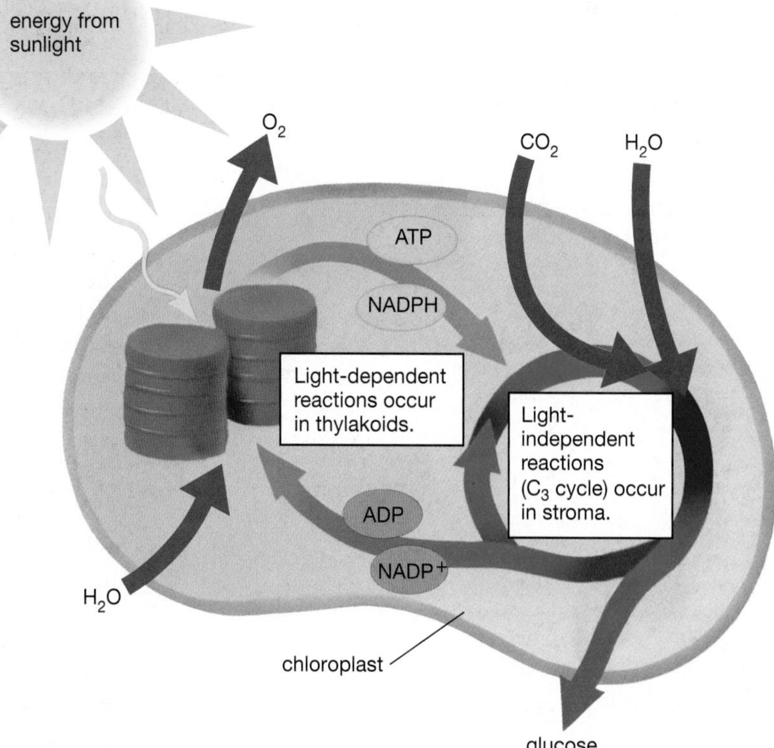

6.5 How Does the Need to Conserve Water Affect Photosynthesis?

Photosynthesis requires carbon dioxide. Therefore, you might think that an ideal leaf would be very porous to allow lots of CO_2 to enter the leaf from the air. For land plants, however, being porous to CO_2 also allows water to evaporate easily from the leaf. Loss of water from leaves can be stressful for land plants and may even be fatal. Leaf structure, therefore, must balance the need to obtain CO_2 with the need to retain water.

EARTH WATCH

CAN FORESTS PREVENT GLOBAL WARMING?

Earth's climate is growing warmer. Although no one can say with certainty what the consequences of this warming trend will be, many biologists and climate scientists fear that a hotter future climate will place extraordinary stress on Earth's inhabitants, including humans. The main cause of global warming appears to be a buildup in the atmosphere of *greenhouse gases*, which cause Earth to retain more heat (a phenomenon described in more detail in Chapter 29). The increase of greenhouse gases stems from the activities of our industrial society, especially from the tremendous amounts of carbon dioxide (a greenhouse gas) that we are adding to the atmosphere from the combustion of wood, gasoline, coal, and other carbon-containing fuels.

In recent years, the looming threat of global warming has prompted governments to seek a worldwide strategy for reducing carbon dioxide emissions. Negotiations have frequently been contentious, as each nation seeks to avoid commitments that would put it at an economic disadvantage. One controversial proposal would have forests counted as offsetting carbon dioxide emissions. Under this idea, a country that agreed to reduce emissions by a certain amount would be able to count its forested area toward that total. After all, the reasoning goes, forests are the site of a great deal of photosynthesis, and photosynthesis consumes carbon dioxide. Consuming carbon dioxide could be just as helpful as reducing the amount produced.

The question of how forests should be counted in carbon dioxide budgets depends on whether forests really do absorb more carbon dioxide than they give off. Although trees undeniably photosynthesize, they also respire; that is, they liberate cellular energy in a process that consumes oxygen and emits carbon dioxide (the subject of Chapter 7). And we must also consider respiration by the animals and other organisms that live in a forest, as well as respiration by the huge number of microorganisms that inhabit forest soils (Fig. E6-1).

Researchers have shown that mature forests do indeed absorb more carbon dioxide than they give off. But, as is so often the case in biology, the situation is not as simple as it first appeared. For example, some research indicates that, as temperature increases, both plants and soil microorganisms respire more rapidly. In a warmer future world, forests might produce more carbon dioxide than they consume. Some scientists have argued, however, that increased respiration might be offset by increased photosynthesis. A recent experiment raises doubts about this scenario. In the experiment, researchers increased the carbon dioxide level in a patch of forest. The trees in the patch experienced a burst of photosynthesis and growth, but only for a short period. The growth spurt quickly used up nutrients in the soil, and with nutrients in short supply, the trees could no longer take advantage of the extra carbon dioxide.

Can forests prevent global warming? Scientists are working hard to uncover the answer to this question. For now, though, the only real answer is that it's too soon to tell.

Figure E6-1 **Forests both consume and emit carbon dioxide** Forest plants photosynthesize, but they also respire. Other forest organisms, such as the mushrooms visible here, also respire.

One solution to the problem of water loss has been the evolution of stomata, pores in leaves that can be opened and closed. When water supplies are adequate, the stomata open, letting in CO_2. If the plant is in danger of drying out, the stomata close. Closing the stomata reduces evaporation, but, unfortunately, it also reduces CO_2 intake and restricts the release of O_2, which is produced during photosynthesis.

When Stomata Are Closed to Conserve Water, Wasteful Photorespiration Occurs

When the stomata close, CO_2 levels inside the leaf drop, O_2 levels rise, and photosynthesis is hampered. The main problem is that, although RuBP combines with CO_2 during the normal C_3 cycle, it can also combine with O_2 (Fig. 6-6a). When RuBP combines with O_2, rather than with CO_2, a wasteful process called **photorespiration** takes place. Photorespiration does not produce any useful cellular energy, and it prevents the C_3 cycle from synthesizing glucose. During hot, dry weather, when the stomata seldom open, photorespiration dominates, and plants may die because they are unable to capture enough energy.

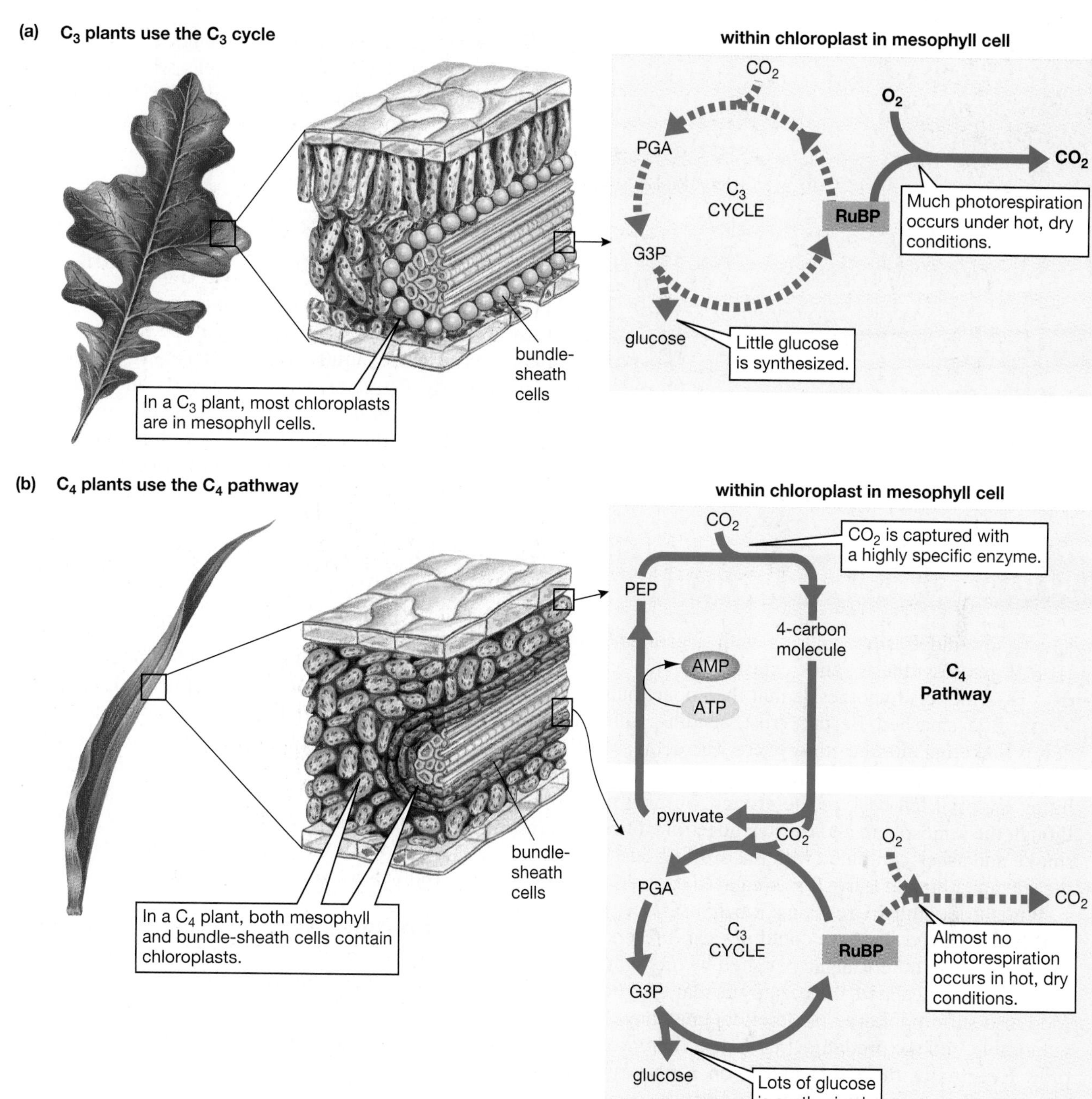

Figure 6-6 Comparison of C_3 and C_4 plants ***(a)*** All carbon fixation in C_3 plants is by the C_3 cycle. ***(b)*** The initial carbon-fixation step in the mesophyll cells of C_4 plants is a reaction between PEP and CO_2, with which O_2 does not compete. The resulting four-carbon molecule moves to the bundle-sheath cells, where it releases CO_2. The bundle-sheath cells thus have higher CO_2 levels, which allow efficient carbon fixation via the C_3 cycle. Note that the regeneration of PEP requires energy: Two phosphates are removed from ATP to produce AMP (adenosine monophosphate). **QUESTION:** *Why do C_3 plants have an advantage over C_4 plants under conditions that are not hot and dry?*

An Alternative Pathway Reduces Photorespiration in C_4 Plants

Some plants, described as *C_4 plants*, have evolved a way to reduce photorespiration and boost photosynthesis during dry weather. These plants, which include species such as corn and crabgrass that thrive in relatively hot and dry conditions, use a two-stage carbon-fixation pathway, the **C_4 pathway**.

In C_4 plants, the mesophyll cells contain a three-carbon molecule called **phosphoenolpyruvate (PEP)** instead of RuBP. Carbon dioxide reacts with PEP to form the four-carbon molecule *oxaloacetate* (for which C_4 plants are named). This reaction is highly specific for CO_2 and is not hindered by high O_2 concentrations.

Oxaloacetate, with its payload of carbon, is then shuttled from mesophyll cells to another group of cells known as **bundle-sheath cells** (Fig. 6-6b). In the bundle-sheath cells, oxaloacetate breaks down, releasing CO_2 again. This release creates a high CO_2 concentration in the bundle-sheath cells, where the regular C_3 cycle can now proceed with less competition from oxygen. The remnant of the shuttle molecule (pyruvic acid) returns to the mesophyll cells, where ATP energy is used to regenerate the PEP molecule.

C_3 and C_4 Plants Are Each Adapted to Different Environmental Conditions

Plants that use the C_4 process to fix carbon are locked into this pathway, which uses up more energy to produce glucose than the C_3 pathway. Nonetheless, C_4 plants have an advantage when light energy is abundant but water is scarce. Consequently, C_4 plants thrive in deserts and in midsummer in temperate climates. However, in places where light levels are low or where water is plentiful and stomata can stay open, the C_3 carbon fixation pathway is more efficient. C_3 plants thus have the advantage in cool, wet, cloudy climates.

DID THE DINOSAURS DIE FROM LACK OF SUNLIGHT? *REVISITED*

How did Earth's collision with a giant meteorite disrupt photosynthesis and exterminate the dinosaurs? Researchers hypothesize that the impact pulverized the meteorite and smashed Earth's crust, sending trillions of tons of debris rocketing into the atmosphere. The debris was flung so high that much of it went into orbit around the planet. Some of the orbiting material fell back to the ground, burning up as it plunged though the atmosphere and causing huge forest fires. The resulting smoke and ashes, combined with the orbiting dust cloud, obscured the sun and plunged Earth into a night that lasted for months.

With little sunlight reaching Earth's surface, the land plants that had survived the fires could not capture enough energy. In the sea, photosynthetic algae perished by the billions. As the photosynthesizers declined, the organisms that depended on them for food also suffered. Large plant-eaters must have been especially vulnerable, and the predators that fed on them were soon without prey. Eventually, the stress of life on a photosynthesis-starved planet spelled extinction for most of Earth's species.

What evidence supports the scenario outlined above? One example is a thin layer of soot found in 65-million-year-old rocks at multiple locations, some quite distant from the impact site. In addition, fossil pollen shows that, 65 million years ago, the vegetation in many places changed from a diverse array of trees, shrubs, and herbs to a community dominated by a few species of fern. Such communities grow today mainly where forests have been recently destroyed by fire. The soot layer and abrupt rise of ferns could be evidence of the global fire that stopped photosynthesis and extinguished the dinosaurs.

CONSIDER THIS: Some paleontologists have concluded that the mass extinction of 65 million years ago was not instantaneous but was spread out over as many as 300,000 years. If this conclusion were correct, could the hypothesis that disrupted photosynthesis caused the extinction also be correct?

LINKS TO LIFE: Have You Thanked a Seaweed Today?

We live in a green, chlorophyll-soaked world, surrounded by photosynthesizers. The world as we know it could not exist without these photosynthetic organisms, because only they have the ability to convert solar energy to body tissue. Photosynthetic organisms capture the energy that the planet's nonphotosynthetic creatures (such as people) need for survival. Without photosynthesizers, the sunlight that bathes Earth would provide heat and light, but not the leaves, stems, fruits, and seeds that we eat. Even meat is a product of photosynthesis. The energy stored in the animals that provide our cheeseburgers, fried chicken, and tuna sandwiches originally came from *their* food—plants and algae, or animals that ate plants and algae. Photosynthesis is the energetic foundation on which ecosystems are constructed.

Our dependence on photosynthesis goes even deeper than its role in making our food. Plants, algae, and photosynthetic bacteria also release into the atmosphere tremendous amounts of oxygen, the main by-product of photosynthetic food production. Thus, the oxygen-rich atmosphere upon which aerobic organisms such as ourselves depend is largely a result of the worldwide presence of photosynthesizers. Photosynthesis gives us both the air we breathe and the food we eat. And, as if food and air were not enough, the oxygen from photosynthesis also forms the ozone layer that sits at the top of Earth's atmosphere. The ozone layer encircles the globe and protects its living inhabitants from the lethal ultraviolet solar radiation that would otherwise make Earth uninhabitable.

Chapter Review

Summary of Key Concepts

6.1 What Is Photosynthesis?

Photosynthesis captures the energy of sunlight to convert the inorganic molecules of carbon dioxide and water into high-energy organic molecules such as glucose. In plants, photosynthesis takes place in the chloroplasts, in two major groups of steps: the light-dependent and the light-independent reactions.

6.2 How Is Light Energy Converted to Chemical Energy?

The light-dependent reactions occur in the thylakoids. Light excites electrons in chlorophyll molecules and transfers the energetic electrons to electron transport chains. The energy of these electrons drives three processes:

- *Photosystem II generates ATP.* Some of the energy from the electrons is used to drive ATP synthesis.
- *Photosystem I generates NADPH.* Some of the energy, in the form of energetic electrons, is added to electron-carrier molecules of $NADP^+$ to make the highly energetic carrier NADPH.
- *Splitting water maintains the flow of electrons through the photosystems.* Some of the energy is used to split water, generating the electrons needed by photosystem II and producing hydrogen ions and oxygen as by-products.

6.3 How Is Chemical Energy Stored in Glucose Molecules?

In the stroma of the chloroplasts, both ATP and NADPH provide the energy that drives the light-independent reactions that synthesize glucose from CO_2 and H_2O. The light-independent reactions constitute the C_3 cycle, and include:

- *Carbon fixation.* Carbon dioxide and water combine with RuBP to form PGA.
- *Synthesis of G3P.* PGA is converted to G3P, using energy from ATP and NADPH.
- *Regeneration of RuBP.* Ten molecules of G3P are used to regenerate six molecules of RuBP, again using ATP energy.

6.4 What Is the Relationship between Light-Dependent and Light-Independent Reactions?

The light-dependent reactions produce the energy carrier ATP and the electron carrier NADPH. Energy from these carriers is used in the synthesis of organic molecules during the light-independent reactions. The depleted carriers, ADP and $NADP^+$, return to the light-dependent reactions for recharging.

6.5 How Does the Need to Conserve Water Affect Photosynthesis?

When stomata close to conserve water, RuBP may combine with O_2 (rather than with CO_2). Such photorespiration prevents carbon fixation and does not generate ATP. Some plants have evolved an additional step for carbon fixation that minimizes photorespiration. In the mesophyll cells of these C_4 plants, CO_2 combines with PEP to form a four-carbon molecule. The four-carbon molecule is transported into adjacent bundle-sheath cells, where it releases CO_2, thereby maintaining a high CO_2 concentration in those cells. This CO_2 is then fixed in the C_3 cycle.

Key Terms

adenosine triphosphate (ATP) *p. 87*
bundle-sheath cell *p. 93*
C_3 cycle *p. 90*
C_4 pathway *p. 93*
carbon fixation *p. 90*
chlorophyll *p. 88*
chloroplast *p. 87*
electron transport chain (ETC) *p. 88*
light-harvesting complex *p. 88*
mesophyll *p. 87*
nicotinamide adenine dinucleotide phosphate (NADPH) *p. 87*
phosphoenolpyruvate (PEP) *p. 93*
photorespiration *p. 92*
photosynthesis *p. 86*
photosystem *p. 88*
reaction center *p. 88*
ribulose bisphosphate (RuBP) *p. 90*
stomata *p. 87*
stroma *p. 87*
thylakoid *p. 87*

Thinking Through the Concepts

Multiple Choice

1. *Photosynthesis is measured in the leaf of a green plant exposed to different wavelengths of light. Photosynthesis is*
   a. highest in green light
   b. highest in red light
   c. highest in blue light
   d. highest in red and blue light
   e. the same at all wavelengths

2. *Where do the light-dependent reactions of photosynthesis occur?*
   a. in the stomata
   b. in the chloroplast stroma
   c. within the thylakoid membranes of the chloroplast
   d. in the leaf cell cytoplasm
   e. in leaf cell mitochondria

3. *The oxygen produced during photosynthesis comes from*
   a. the breakdown of CO_2
   b. the breakdown of H_2O
   c. the breakdown of both CO_2 and H_2O
   d. the breakdown of oxaloacetate
   e. photorespiration

4. *The role of accessory pigments is to*
   a. provide an additional photosystem to generate more ATP
   b. allow photosynthesis to occur in the dark
   c. prevent photorespiration
   d. donate electrons to chlorophyll reaction centers
   e. capture additional light energy and transfer it to the chlorophyll reaction centers
5. *Where do the light-independent, carbon-fixing reactions occur?*
   a. in the guard cell cytoplasm
   b. in the chloroplast stroma
   c. within the thylakoid membranes
   d. at night in the thylakoids
   e. in mitochondria

Review Questions

1. Write the overall equation for photosynthesis. Does the overall equation differ between C_3 and C_4 plants?
2. Draw a diagram of a chloroplast and label it. Explain specifically how chloroplast structure is related to its function.
3. Briefly describe the light-dependent and light-independent reactions. In what part of the chloroplast does each occur?
4. What is the difference between carbon fixation in C_3 and in C_4 plants? Under what conditions does each mechanism of carbon fixation work most effectively?

Applying the Concepts

1. **IS THIS SCIENCE?** Chlorophyll extracted from plants is often sold in capsules. One Web site that offers chlorophyll for sale states that "chlorophyll tablets reduce or eliminate offensive body and breath odors." Do you think this claim is likely to be true? (You might want to research the causes of breath and body odors before answering.) Devise an experiment to test the claim.
2. Suppose an experiment is performed in which plant I is supplied with normal carbon dioxide but with water that contains radioactive oxygen atoms. Plant II is supplied with normal water but with carbon dioxide that contains radioactive oxygen atoms. Each plant is allowed to perform photosynthesis, and the oxygen gas and sugars produced are tested for radioactivity. Which plant would you expect to produce radioactive sugars, and which plant would you expect to produce radioactive oxygen gas? Why?
3. You are called before the Ways and Means Committee of the U.S. House of Representatives to explain why the U.S. Department of Agriculture should continue to fund photosynthesis research. How would you justify the expense of producing, by genetic engineering, an enzyme that catalyzes the reaction of RuBP with CO_2 but prevents RuBP from reacting with oxygen? What are the potential applied benefits of this research?

For More Information

Bazzazz, F. A., and Fajer, E. D. "Plant Life in a CO_2-Rich World." *Scientific American*, January 1992. Burning fossil fuels is increasing CO_2 levels in the atmosphere. This increase could tip the balance between C_3 and C_4 plants.

Falkowski, P. G. "The Ocean's Invisible Forest." *Scientific American*, August 2002. Single-celled photosynthesizers in Earth's oceans consume nearly as much carbon dioxide as land plants do. Can we slow global warming by fertilizing these "forests of the ocean"?

Govindjee and Coleman, W. J. "How Plants Make Oxygen." *Scientific American*, February 1990. The generation of oxygen during photosynthesis is just beginning to be understood.

Hall, D. O., and Rao, K. K. *Photosynthesis*. 6th ed. New York: Cambridge University Press, 1999. An excellent short book recommended to any student interested in finding out more about photosynthesis.

Kring, D. A., and Durda, D. D. "The Day the World Burned." *Scientific American*, December 2003. A nice description of the evidence supporting the hypothesis that wildfires and disrupted photosynthesis played a major role in dinosaur extinction.

Monastersky, R. "Children of the C_4 World." *Science News*, January 3, 1998. What role did a shift in global vegetation toward C_4 photosynthesis play in the evolution of humans?

WEB TUTORIAL

To access a Web Tutorial visit http://www.prenhall.com/audesirk4. *Log in to the Web site selected by your instructor, navigate to this chapter, and select the appropriate Web Activity number.*

6.1 Properties of Light

Estimated time: 5 minutes

Explore the properties of light and how light is captured by green leaves to power photosynthesis.

6.2 Photosynthesis

Estimated time: 10 minutes

Explore the process of photosynthesis, and the flow of electrons through the two photosystems, in animated detail.

6.3 Web Investigation: Did the Dinosaurs Die from Lack of Sunlight?

Estimated time: 15 minutes

This exercise will examine the causes of some recent extinctions in the hope that they might shed some light on the other mass extinctions, including the one that ended the age of dinosaurs.

Chapter 7

Harvesting Energy: Glycolysis and Cellular Respiration

As bicyclists race, the cells of their leg muscles use energy to do work. The reactions that make energy available to the cells require oxygen.

AT A GLANCE

When Athletes Boost Their Blood Counts: Do Cheaters Prosper?

Thousands of spectators cheered wildly as the leaders of the 50-km cross-country ski race entered the home stretch at the 2002 Winter Olympics. As the grueling race drew to its conclusion, the skiers were clearly exhausted, struggling to find the energy for a final burst. One skier, however, came on strong. Johann Mühlegg, competing for Spain, raced to the front of the pack and pulled away, finishing almost 15 seconds ahead of the second-place skier and claiming the gold medal.

Mühlegg's triumph was short lived. Shortly after the race, he was stripped of his medals and expelled from the Games. His offense? Blood doping.

Blood doping increases a person's physical endurance by increasing the blood's ability to carry oxygen. Mühlegg accomplished this by injecting the drug darbepoetin, which mimics the effect of another blood-doping substance, erythropoietin (Epo). Epo is present in normal human bodies, where it stimulates bone marrow to produce more red blood cells. A healthy body produces just enough Epo to ensure that red blood cells are replaced as they age and die. An injection of extra Epo, however, can stimulate the production of a huge number of extra red blood cells. The extra cells greatly increase the oxygen-carrying capacity of the blood.

Do Epo injections really improve endurance? In one study, researchers divided 20 human subjects into two groups, one of which received injections of Epo. After 4 weeks, the subjects were tested for endurance and for oxygen consumption during exercise. Individuals in the Epo group had much better endurance and consumed significantly more oxygen than the control group. The researchers concluded that Epo injections improve endurance and increase the body's capacity to carry oxygen.

Johann Mühlegg is among the elite athletes penalized for artificially boosting the oxygen supply to their cells.

Why is endurance improved by extra oxygen molecules in the bloodstream? Think about this question as we examine the role of oxygen in supplying energy to muscle cells.

7.1 What Is the Source of a Cell's Energy?

Living cells need energy to power the multitude of metabolic reactions needed to sustain life. To power a reaction, however, energy must be in a usable form. To be usable, energy must, in most cases, be stored in the bonds of energy-carrier molecules, especially **adenosine triphosphate (ATP)**. For this reason, some of the most important reactions in cells are the ones that transfer energy from energy-storage molecules to energy-carrier molecules.

Glucose Is a Key Energy-Storage Molecule

Although most cells can harvest energy from a variety of organic molecules, this chapter's description of the harvesting process will focus on glucose. Almost all cells break down glucose for energy at least part of the time. Even when cells draw energy from other types of molecules, those molecules are typically first converted to glucose.

Photosynthesis Is the Ultimate Source of Cellular Energy

The energy released during glucose metabolism was originally acquired by photosynthesis. Photosynthetic cells capture and store the energy of sunlight, and this stored energy is later used by cells. Those cells may be in the photosynthetic organisms themselves, or they may be in other organisms that directly or indirectly consume photosynthesizers. For example, caterpillars eat plants, bluebirds eat caterpillars, and hawks eat bluebirds. The cells that make up the plants, caterpillars, bluebirds, and hawks are all powered by energy stored in glucose that was originally captured by photosynthesis in the plant's leaves.

Glucose Metabolism and Photosynthesis Are Complementary Processes

As photosynthesis captures solar energy for storage in cells, it consumes water and carbon dioxide and produces glucose and oxygen. As glucose metabolism releases stored energy, it consumes glucose and oxygen and produces water and carbon dioxide. Thus, the two processes are almost perfectly complementary. The products of each process provide reactants for the other. This symmetry is apparent in the chemical equations for glucose formation by photosynthesis and for the complete metabolism of glucose back to CO_2 and H_2O:

Photosynthesis:

$$6\,CO_2 + 6\,H_2O + \text{sunlight energy} \longrightarrow C_6H_{12}O_6 + 6\,O_2$$

Complete Glucose Metabolism:

$$C_6H_{12}O_6 + 6\,O_2 \longrightarrow 6\,CO_2 + 6\,H_2O + \text{chemical energy and heat energy}$$

This symmetry might lead you to hypothesize that a cell can convert all of the chemical energy in a glucose molecule to high-energy bonds of ATP. But remember the second law of thermodynamics, which tells us that conversion of energy into different forms always decreases the amount of useful energy. Thus, most of the energy released during the breakdown of glucose is lost as heat rather than converted to the chemical energy of ATP. Nevertheless, a cell can extract a great deal of energy, in the form of ATP, when a glucose molecule is completely broken down.

7.2 How Do Cells Harvest Energy from Glucose?

Glucose metabolism proceeds in stages (Fig. 7-1). The first stage, *glycolysis*, does not require oxygen and proceeds in exactly the same way under both aerobic (with oxygen) and anaerobic (without oxygen) conditions. The products of glycolysis are further processed in the second stage, which is *cellular respiration* if oxygen is available or *fermentation* in the absence of oxygen.

Glycolysis, which occurs in the cytoplasm, splits apart a single glucose molecule (a six-carbon sugar) into two three-carbon molecules of **pyruvate**. The split releases a small amount of chemical energy, which is used to generate two ATP molecules.

The pyruvate produced by glycolysis may enter the **mitochondria**, which are organelles specialized for the aerobic breakdown of pyruvate. There, if oxygen is available, cellular respiration uses oxygen to break pyruvate down completely to carbon dioxide and water, generating an additional 34 or 36 ATP molecules (the number differs from cell to cell). The ATP produced by cellular respiration is so important to most organisms that anything that interferes with its production, such as a sustained lack of oxygen, quickly results in death.

Under anaerobic conditions, the pyruvate produced by glycolysis does not enter the mitochondria to be broken down by cellular respiration. Instead, it remains in the cytoplasm and is usually converted by fermentation into lactate or ethanol. Fermentation does not produce additional ATP energy.

The following sections describe the stages of glucose metabolism in greater detail.

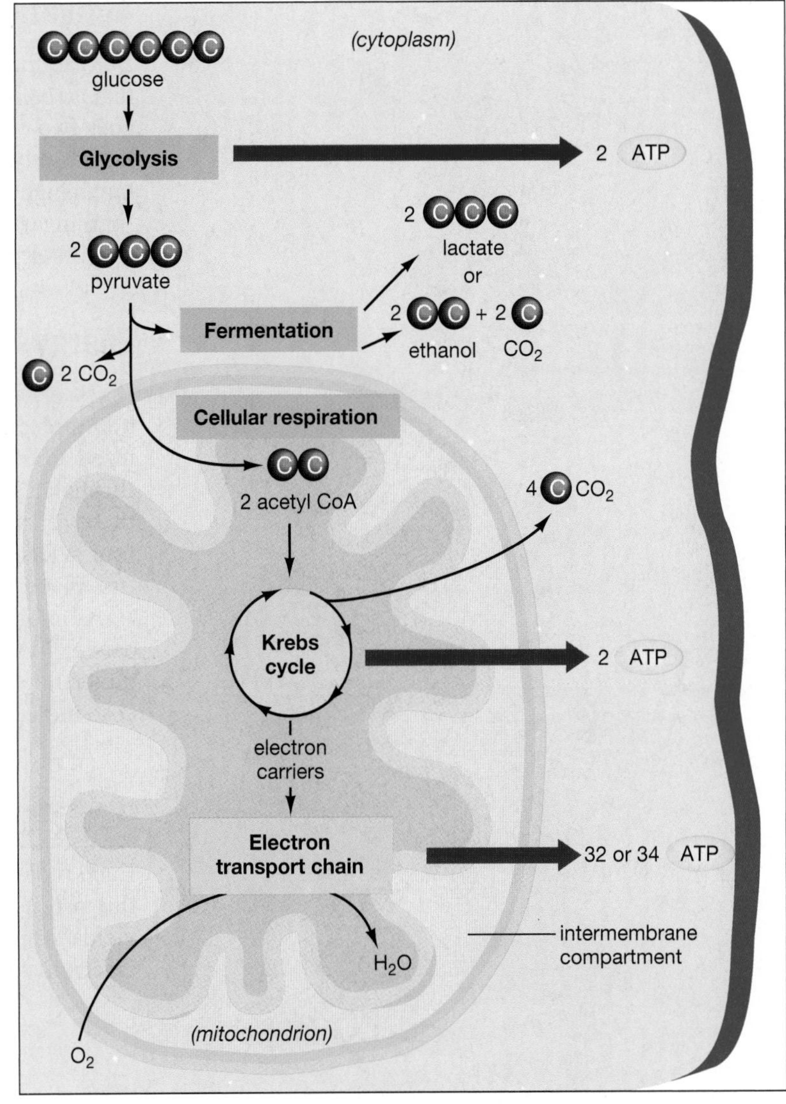

Figure 7-1 An overview of glucose metabolism Refer to this diagram as we progress through the reactions of glycolysis (in the fluid portion of the cytoplasm) and cellular respiration (in the mitochondrion). Glucose is broken down in stages, with energy captured in ATP along the way. Most of the ATP is produced in mitochondria.

7.3 What Happens During Glycolysis?

The reactions that split glucose are collectively called **glycolysis**. Glycolysis, which requires no oxygen, breaks a molecule of glucose into two molecules of pyruvate. The reactions produce relatively few energy carriers, just two molecules of ATP and two molecules of the electron carrier **nicotinamide adenine dinucleotide (NADH)**. Nonetheless, glycolysis is an essential part of glucose metabolism. It consists of two major parts (each with several steps): glucose activation and energy harvest (Fig. 7-2).

Figure 7-2 The essentials of glycolysis

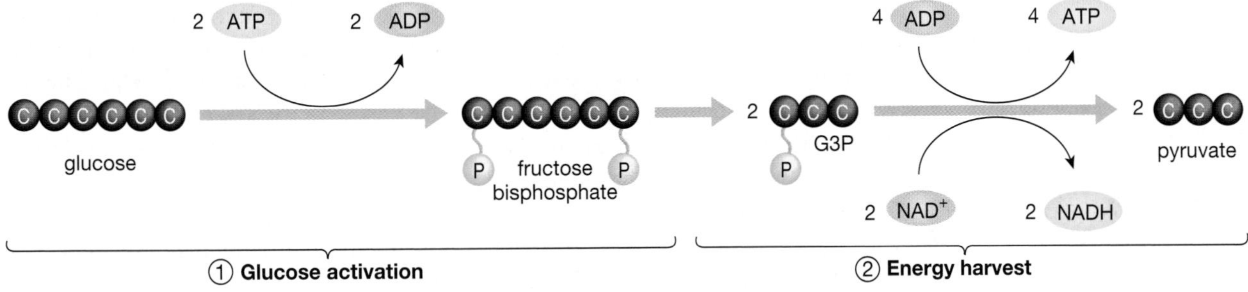

Activation Consumes Energy

Before glucose is broken down to release its energy, it must be activated—a process that actually uses up energy. During glucose activation, a molecule of glucose undergoes two enzyme-catalyzed reactions, each of which uses ATP energy. These reactions use two phosphates from ATP to convert a relatively stable glucose molecule into a highly unstable, activated molecule of fructose biphosphate (Fig. 7-2, step ①). Forming fructose bisphosphate costs the cell two ATP molecules, an investment that is necessary to produce much greater energy returns in the long run.

Energy Harvest Yields Energy-Carrier Molecules

In the energy-harvest steps, fructose bisphosphate splits apart into two three-carbon molecules of glyceraldehyde 3-phosphate (G3P). (In Chapter 6 we encountered G3P in the C_3 cycle of photosynthesis.) Each G3P molecule then goes through a series of reactions that converts it to pyruvate (Fig. 7-2, step ②). During these reactions, two molecules of ATP are generated for each G3P, for a total of four ATPs. Because two ATPs were used to activate the glucose molecule in the first place, there is a net gain of only two ATPs per glucose molecule.

At another step along the way from G3P to pyruvate, the energized electron carrier NADH is produced (Fig. 7-2, step ②). Each molecule of G3P yields one molecule of NADH, so two NADH carrier molecules are formed for each glucose molecule that is converted to pyruvate.

Summing Up: Glycolysis

Each molecule of glucose is broken down to two molecules of pyruvate. During this process, ATP is both consumed and produced, but the process as a whole yields a net increase of two ATP molecules. In addition, two molecules of the electron carrier NADH are formed.

7.4 What Happens During Cellular Respiration?

Cellular respiration is a series of reactions, occurring under aerobic conditions, in which the pyruvate produced by glycolysis is broken down to carbon dioxide and water. Over the course of the reactions, large amounts of ATP are produced. In eukaryotic cells, cellular respiration takes place in the mitochondria.

For an overview of cellular respiration, review Figure 7-3 with the help of the summary below. A more detailed description follows.

① The two molecules of pyruvate produced by glycolysis are transported into a mitochondrion, which is surrounded by two membranes, the inner membrane and the outer membrane. The pyruvate molecules cross both of them and pass into the fluid **matrix** that lies inside the inner membrane.

② Each three-carbon pyruvate is split into CO_2 and a two-carbon molecule, which enters a series of reactions known as the Krebs cycle. The Krebs cycle produces one ATP from each pyruvate and donates energetic electrons to several electron-carrier molecules.

③ The electron carriers donate their energetic electrons to the electron transport chain of the inner membrane.

④ In the electron transport chain, the energy of the electrons is used to transport H^+ from the matrix to the **intermembrane compartment** that lies between the inner and outer membranes.

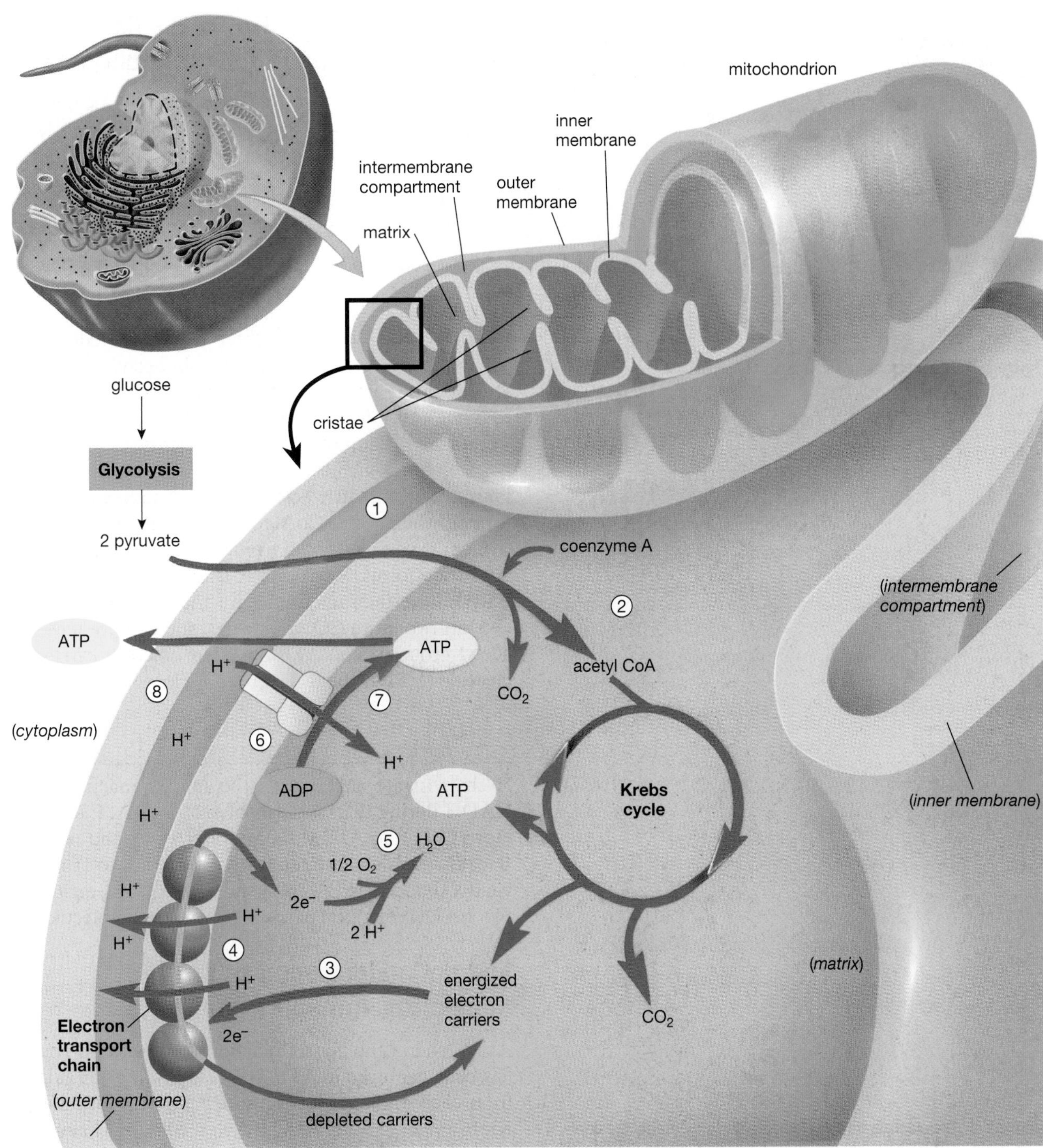

Figure 7-3 Cellular respiration Cellular respiration takes place in mitochondria. A mitochondrion's inner membrane separates the matrix (inner compartment) from the intermembrane compartment (between the inner and outer membranes).

⑤ At the end of the electron transport chain, the electrons combine with O_2 and H^+ to form H_2O.

⑥ The hydrogen ions that were moved to the intermembrane compartment by the electron transport chain move back across the inner membrane, along their concentration gradient.

⑦ As the ions pass back into the matrix, their flow provides the energy to produce ATP.

⑧ ATP moves out of the mitochondrion into the fluid of the cytoplasm, where it provides energy for cellular activities.

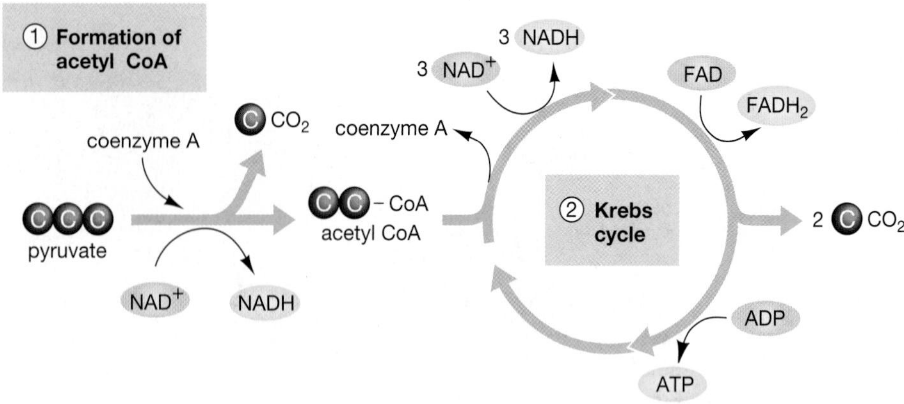

Figure 7-4 The reactions in the mitochondrial matrix

The Krebs Cycle Breaks Down Pyruvate in the Mitochondrial Matrix

The fuel for the reactions of cellular respiration is the pyruvate produced by glycolysis in the cell cytoplasm. The pyruvate moves from the cytoplasm to the mitochondrial matrix, diffusing down its concentration gradient through pores in the two mitochondrial membranes.

After the pyruvate reaches the matrix, it reacts with a molecule called *coenzyme A* (**Fig. 7-4**, step ①). In this reaction, each pyruvate molecule is split into CO_2 and a two-carbon molecule called an acetyl group. The acetyl group immediately attaches to coenzyme A, forming an *acetyl-coenzyme A complex* (acetyl CoA). During this reaction, two energetic electrons and a hydrogen ion are transferred to NAD^+, forming NADH.

The next group of reactions forms the cyclic pathway of the **Krebs cycle** (**Fig. 7-4**, step ②), which is sometimes called the citric-acid cycle because citrate (the ionized form of citric acid) is the first molecule produced in the cycle. Citrate is formed when acetyl CoA combines with the four-carbon molecule oxaloacetate (which, as you may recall from Chapter 6, also plays a role in C_4 photosynthesis). As citrate forms, coenzyme A is released for reuse in further turns of the cycle. Each citrate then undergoes a series of reactions that regenerate oxaloacetate, give off two more CO_2 molecules, and capture most of the energy of the original acetyl group in one ATP and four electron carriers—one **flavin adenine dinucleotide** ($FADH_2$) and three NADH.

Summing Up: The Mitochondrial Matrix Reactions

Each pyruvate that enters the matrix reactions produces one CO_2 and one NADH during the synthesis of acetyl CoA. Each acetyl CoA in turn yields two more CO_2, one ATP, three more NADH, and one $FADH_2$ via the Krebs cycle. Because each glucose feeds two pyruvates to the matrix reactions, the reactions yield a total of six CO_2 molecules, two ATPs, eight NADH electron carriers, and two $FADH_2$ electron carriers per glucose molecule.

Energetic Electrons Are Carried to Electron Transport Chains

At the end of the matrix reactions, only a small portion of the energy of glucose has been captured in ATP. The cell has thus far gained only four ATP molecules from each original glucose molecule: two during glycolysis and two during the Krebs cycle. The cell has, however, captured many energetic electrons in carrier molecules: 2 NADH during glycolysis plus 8 more NADH and 2 $FADH_2$ from the matrix reactions, for a total of 10 NADH and 2 $FADH_2$ (Table 7-1).

The carriers deposit their electrons in **electron transport chains** located in the inner mitochondrial membrane (**Fig. 7-5**, step ①). Within the transport chains, the energetic electrons move systematically from molecule to molecule. During these transfers, energy is released and used to pump hydrogen ions out of the matrix, across the inner mitochondrial membrane, and into the intermembrane compartment (**Fig. 7-5**, step ②).

At the end of the electron transport chain, oxygen finally plays its part. An oxygen atom combines with two hydrogen ions and two energetically depleted electrons to form water (**Fig. 7-5**, step ③). Oxygen thus accepts electrons after they run their course through the electron transport chain. If oxygen were not present to pull spent electrons from the end of the transport chain, the electrons would "pile up" in the chain, and the process of pumping hydrogen ions out of the matrix would come to a halt.

Figure 7-5 The electron transport chain in the inner mitochondrial membrane QUESTION: *If no oxygen is present, how is the rate of ATP production affected?*

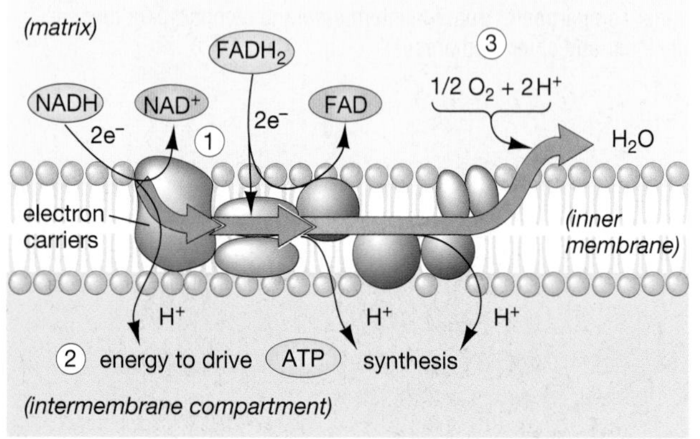

TABLE 7-1 Summary of Glycolysis and Cellular Respiration

Process	Location	Reactions	Electron Carriers Formed	ATP Yield (per glucose molecule)
Glycolysis	Fluid cytoplasm	Glucose breaks down to two pyruvates	2 NADH	2 ATP
Cellular respiration				
Acetyl CoA formation	Matrix of mitochondrion	Each pyruvate combines with coenzyme A to form acetyl CoA and CO_2	2 NADH	
Krebs cycle	Matrix of mitochondrion	Acetyl group of acetyl CoA metabolized to two CO_2	6 NADH, 2 $FADH_2$	2 ATP
Electron transport	Inner membrane, intermembrane compartment	Energy of electrons from NADH and $FADH_2$ used to pump H^+ into intermembrane compartment, H^+ gradient used to synthesize ATP		32–34 ATP

Energy from a Hydrogen-Ion Gradient Is Used to Produce ATP

The hydrogen ions that are pumped out of the matrix become a source of energy for the production of more ATP. The process of pumping hydrogen ions across the inner mitochondrial membrane generates a large H^+ concentration gradient; that is, the concentration of hydrogen ions in the intermembrane compartment becomes high and the concentration in the matrix becomes low. Creating this concentration gradient of hydrogen ions requires an input of energy; it's the chemical equivalent of pumping water upward into an elevated storage tank. Energy is later released when the hydrogen ions are allowed to move down their concentration gradient—like opening the valves of the storage tank and allowing the water to rush out.

In mitochondria, the energy of the hydrogen-ion concentration gradient can be captured because the inner membrane is impermeable to hydrogen ions except at protein channels that are part of ATP-synthesizing enzymes. In the process of **chemiosmosis**, hydrogen ions move down their concentration gradient from the intermembrane compartment to the matrix through these ATP-synthesizing enzymes (see **Fig. 7-3**, steps ⑥ and ⑦). The flow of hydrogen ions provides the energy to synthesize 32 to 34 molecules of ATP for each molecule of glucose that is broken down. This ATP is transported from the matrix to the intermembrane compartment and then diffuses through the outer membrane to the surrounding cytoplasm.

Summing Up: Electron Transport and Chemiosmosis

Electrons from the electron carriers NADH and $FADH_2$ enter the electron transport chain of the inner mitochondrial membrane. Here their energy is used to generate a hydrogen-ion gradient across the inner membrane. The movement of hydrogen ions down the gradient through the pores of ATP-synthesizing enzymes drives the synthesis of 32 to 34 molecules of ATP. At the end of the electron transport chain, two electrons combine with one oxygen atom and two hydrogen ions to form water.

7.5 What Happens During Fermentation?

Many organisms (particularly microorganisms) thrive in places where oxygen is scarce or absent. Such oxygen-free environments can be found in animal intestines, in sediments beneath lakes and oceans, deep beneath the surface of Earth, and in bogs and marshes. Even some of our own body cells must sometimes survive without oxygen for brief periods. The anaerobic conditions in all these places rule out cellular respiration, which requires oxygen as its final electron acceptor. Instead, the products of glycolysis may be metabolized by **fermentation**.

Unlike the reactions of cellular respiration, the reactions of fermentation generate no ATP. Instead, they serve to regenerate the NAD^+ molecules that are needed for glycolysis. This regeneration is necessary because, in the absence of cellular respiration, the electron transport chains of the mitochondria do not help generate NAD^+ by accepting electrons from NADH. If there were no alternative acceptor of electrons from NADH, the cell's supply of NAD^+ would quickly be used up, and glycolysis would halt.

In fermentation, the pyruvate produced by glycolysis acts as an electron acceptor for NADH. The pyruvate is converted to ethanol or lactate, and NADH is converted to NAD^+. Thus, fermentation provides a way to regenerate NAD^+ so that glycolysis can continue.

Some Cells Ferment Pyruvate to Form Alcohol

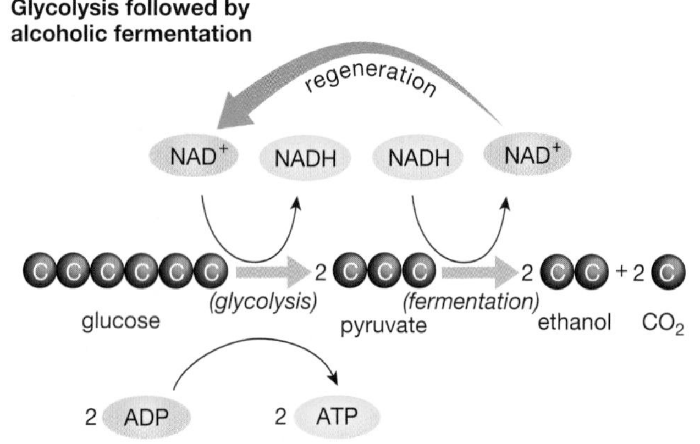

There are two main types of fermentation: one converts pyruvate to ethanol and carbon dioxide and the other converts pyruvate to lactate. The form of fermentation that yields ethanol and CO_2 is known as alcoholic fermentation and is the primary mode of metabolism in many microorganisms. The reactions of alcoholic fermentation use hydrogen ions and electrons from NADH to regenerate NAD^+. ←

Alcoholic fermentation by single-celled fungi called yeasts is responsible for some gastronomic staples: wine, beer, and bread. Wine retains the ethanol product of the yeasts' fermentation of grape sugar, and beer benefits from both the alcohol and carbon dioxide (retained as bubbles) products of fermentation. The yeast in bread dough likewise produces both alcohol and CO_2. The carbon dioxide makes the bread rise, but the alcohol evaporates during baking (**Fig. 7-6a**).

Other Cells Ferment Pyruvate to Lactate

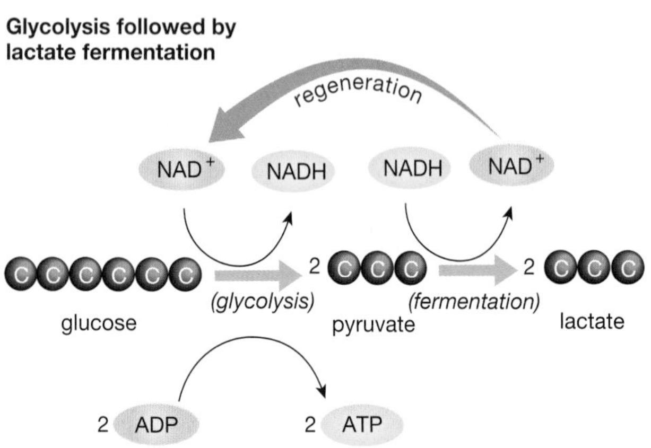

Various microorganisms ferment pyruvate to lactate, including the bacteria that give yogurt, sour cream, and cheese their distinctive flavors. Lactate fermentation also occurs in muscles during vigorous exercise (**Fig. 7-6b**). Working muscles need lots of ATP. They usually get the ATP from cellular respiration, which generates far more ATP than does glycolysis, but cellular respiration is limited by an organism's ability to capture oxygen (by breathing, for example). When you exercise vigorously, you may not be able to supply your muscles with enough oxygen to allow cellular respiration to meet their energy needs.

When deprived of adequate oxygen, your muscles do not immediately stop working. Instead, glycolysis continues for a while, providing its meager two ATP molecules per glucose and generating both pyruvate and NADH. Then, to regenerate NAD^+, muscle cells ferment pyruvate molecules to lactate, using electrons and hydrogen ions from NADH. ←

In high concentrations, however, lactate is toxic to your cells. Soon, it causes intense discomfort and fatigue, forcing you to stop or at least slow down. As you rest, breathing rapidly after your sprint, oxygen once more becomes available and the lactate is converted back to pyruvate.

(a)

(b)

Figure 7-6 Fermentation ***(a)*** Bread dough rises as carbon dioxide is liberated by fermenting yeast. ***(b)*** A runner's lungs and blood vessels may not supply oxygen to her leg muscles fast enough to keep up with the demand for energy, so glycolysis and lactate fermentation must provide the ATP. **QUESTION:** *Some species of bacteria use aerobic respiration and other species use anaerobic (fermenting) respiration. In an oxygen-rich environment, would either type be at a competitive advantage? What about in an oxygen-poor environment?*

Fermentation's Effects Limit Human Muscle Performance

Why is the average speed of the 5000-meter run in the Olympics slower than that of the 100-meter dash? During a sprint, runners' leg muscles use more ATP than cellular respiration can supply, because their bodies cannot deliver enough oxygen to keep up with the demand. With oxygen in short supply, glycolysis and lactate fermentation keep the muscles supplied with ATP. This system can work for a short time, long enough for a sprinter to complete a 100-meter dash. Soon, however, the toxic effects of lactate buildup cause fatigue and cramps. To avoid these effects, distance runners must pace themselves, so that cellular respiration can power their muscles for most of the race. An anaerobic sprint is reserved for the finish.

WHEN ATHLETES BOOST THEIR BLOOD COUNTS: DO CHEATERS PROSPER? *REVISITED*

As you have seen, human cells most efficiently extract energy from glucose when an ample supply of oxygen is available to them. The aim of blood-doping athletes, then, is to extend as long as possible the period in which muscle cells have access to oxygen. During a difficult hill-climb, a skier who has doped his blood with erythropoietin may be able to ski efficiently, his muscle cells using cellular respiration to churn out abundant ATP. At the same time, his "clean" competitor may labor painfully, leg muscles laden with lactate from fermentation.

Because Epo forms naturally in the human body, its abuse is hard to detect. Standard drug-screening procedures cannot distinguish between natural Epo and that injected in blood doping. Sports officials assert that the difficulty of detecting Epo has made it the drug of choice among blood-doping skiers, cyclists, distance runners, and other competitive athletes.

Evidence in support of the hypothesis that Epo abuse is widespread includes a study of blood samples taken from participants in the Nordic Ski World Championships. Researchers predicted that if blood doping with Epo were common among top skiers, contestants' blood would contain abnormally high levels of red blood cells. They found that 36% of the tested skiers had abnormal levels and concluded that many skiers are blood doping.

Contestants at the Olympics are now routinely tested for Epo, but the available tests are not considered to be completely reliable. Meanwhile, researchers continue to explore the chemistry of Epo metabolism in hope of discovering a definitive test for blood doping.

CONSIDER THIS: Advances in gene therapy may one day make it possible to modify athletes' kidney cells so that they have extra copies of the genes that produce Epo. In your view, would such genetically modified individuals be cheating to gain an unfair advantage?

LINKS TO LIFE: A Jug of Wine, a Loaf of Bread, and a Nice Bowl of Sauerkraut

Life would be a little less interesting without fermentation. Some of our most appealing foods and beverages gain their appeal from the waste products of fermentation. We're especially fond of foods dependent on fermentation by yeasts. These single-celled fungi use cellular respiration when oxygen is available, but switch to alcoholic fermentation if they run out of oxygen.

Thus, a mixture of yeast and grape juice eventually accumulates enough alcohol to become wine. As the alcohol level in the fermenting mixture rises, it eventually becomes toxic to the yeasts, which die, bringing fermentation to a halt. Even the most alcohol-tolerant yeast die in a solution that is 15% alcohol, so most wines have an alcohol content lower than that, typically around 10% to 12%.

As you now know, carbon dioxide is also a product of alcoholic fermentation. In most wines, the carbon dioxide is allowed to escape, but sometimes a winemaker wishes to retain the gas. Sparkling wines and champagne are made by adding more yeast and sugar just before the wine is bottled, so that additional fermentation occurs in the sealed bottle, trapping carbon dioxide.

Fermentation also gives bread its airy texture. To make bread, the baker adds yeast to flour, along with some sugar (because yeasts can't ferment the starches that make up flour). The carbon dioxide released during fermentation is trapped within the bread dough, where it forms tiny pockets of gas that give bread its evenly porous structure.

Wine and bread are perhaps the most famous fermentation-assisted foods, but fermentation also contributes to a host of lesser, but nonetheless well-loved, foods. For example, lactate fermentation by bacteria converts sugars in shredded cabbage to lactic acid. The result: sauerkraut, an excellent partner for other fermented foods.

Chapter Review

Summary of Key Concepts

7.1 What Is the Source of a Cell's Energy?

Cells require energy to survive and function. To be usable by a cell, energy must be in the form of chemical energy in the bonds of an energy-carrier molecule, usually ATP. A key component of cellular metabolism consists of reactions that harvest energy from food molecules such as glucose and convert it to ATP energy. The ultimate source of all cellular energy is solar energy that is captured by photosynthesis.

7.2 How Do Cells Harvest Energy from Glucose?

Cells produce usable energy by breaking down glucose to lower-energy compounds and capturing some of the released energy as ATP. In glycolysis, glucose is metabolized in the fluid portion of the cytoplasm to two molecules of pyruvate, generating two ATP molecules. If oxygen is available, the pyruvates are metabolized to CO_2 and H_2O through cellular respiration in the mitochondria, generating much additional ATP.

7.3 What Happens During Glycolysis?

During glycolysis, a molecule of glucose is activated to form fructose bisphosphate. In a series of reactions, the fructose bisphosphate is broken down into two molecules of pyruvate. These reactions produce four ATP molecules and two NADH electron carriers. Because two ATPs were used in the activation steps, the net yield from glycolysis is two ATPs and two NADHs.

7.4 What Happens During Cellular Respiration?

If oxygen is available, cellular respiration can occur. The pyruvates are transported into the matrix of the mitochondria. In the matrix, each pyruvate reacts with coenzyme A to form acetyl CoA plus CO_2. One NADH is also formed at this step. The two-carbon acetyl group of acetyl CoA enters the Krebs cycle, which releases the two carbons as CO_2. One ATP, three NADHs, and one $FADH_2$ are also formed for each acetyl group that goes through the cycle.

The NADHs and $FADH_2$s from glycolysis and the matrix reactions deliver their energetic electrons to the electron transport chain. The energy of the electrons is used to pump hydrogen ions across the inner membrane from the matrix to the intermembrane compartment. At the end of the electron transport chain, the depleted electrons combine with hydrogen ions and oxygen to form water. This is the oxygen-requiring step of cellular respiration. During chemiosmosis, the hydrogen-ion gradient created by the electron transport chain is used to produce ATP, as the hydrogen ions diffuse back across the inner membrane through channels in ATP-synthesizing enzymes. Electron transport and chemiosmosis yield 32 to 34 additional ATPs, for a net yield of 36 to 38 ATPs per glucose molecule.

Figure 7-1 and Table 7-1 summarize the locations, major processes, and overall energy harvest for the complete metabolism of glucose from glycolysis through cellular respiration.

7.5 What Happens During Fermentation?

In the absence of oxygen, the pyruvate produced by glycolysis cannot enter the reactions of cellular respiration. Instead, it is converted by fermentation to lactate or ethanol and CO_2. Fermentation regenerates NAD^+ so glycolysis may continue. However, no additional ATP is gained by fermentation.

Key Terms

adenosine triphosphate (ATP) *p. 100*
cellular respiration *p. 102*
chemiosmosis *p. 105*
electron transport chain *p. 104*
fermentation *p. 106*
flavin adenine dinucleotide ($FADH_2$) *p. 104*
glycolysis *p. 101*
intermembrane compartment *p. 102*
Krebs cycle *p. 104*
matrix *p. 102*
mitochondria *p. 101*
nicotinamide adenine dinucleotide (NADH) *p. 101*
pyruvate *p. 101*

Thinking Through the Concepts

Multiple Choice

1. *Where does glycolysis occur?*
   a. cytoplasm
   b. matrix of mitochondria
   c. inner membrane of mitochondria
   d. outer membrane of mitochondria
   e. stroma of chloroplast

2. *Where does respiratory electron transport occur?*
   a. cytoplasm
   b. matrix of mitochondria
   c. inner membrane of mitochondria
   d. outer membrane of mitochondria
   e. stroma of chloroplast

3. *What is the product of the fermentation of sugar by yeast in bread dough that is essential for the rising of the dough?*
   a. lactate
   b. ATP
   c. ethanol
   d. CO_2
   e. O_2

4. *The majority of ATP produced in aerobic respiration comes from*
   a. glycolysis
   b. the Krebs cycle
   c. chemiosmosis
   d. fermentation
   e. photosynthesis

5. *The process that converts glucose into two molecules of pyruvate is*
   a. glycolysis
   b. fermentation
   c. the Krebs cycle
   d. electron transport
   e. the Calvin-Benson cycle

6. *The process that causes lactate buildup in muscles during strenuous exercise is*
   a. glycolysis
   b. fermentation
   c. the Krebs cycle
   d. electron transport
   e. the Calvin-Benson cycle

Review Questions

1. Starting with glucose ($C_6H_{12}O_6$), write the overall reactions for (a) cellular respiration and (b) fermentation in yeast.
2. Draw a labeled diagram of a mitochondrion, and explain how its structure is related to its function as the powerhouse of the cell.
3. What role do the following play in the complete metabolism of glucose: (a) glycolysis, (b) mitochondrial matrix, (c) inner membrane of mitochondria, (d) fermentation, and (e) NAD^+?
4. Outline the major steps in (a) cellular respiration and (b) fermentation, indicating the sites of ATP production. What is the overall energy harvest (in terms of ATP molecules generated per glucose molecule) for each?
5. Describe the Krebs cycle. In what form is most of the energy harvested?
6. Describe the mitochondrial electron transport chain and the process of chemiosmosis.
7. Why is oxygen necessary for cellular respiration?

Applying the Concepts

1. **IS THIS SCIENCE?** Dinitrophenol (DNP) was a popular diet drug in the 1930s, but its use declined as potential users became aware of its dangerous toxicity. Today, the Food and Drug Administration (FDA) bans prescriptions for DNP, but it is nonetheless readily available for purchase on Web sites that offer it to desperate dieters and obsessive bodybuilders. DNP acts by interfering with an enzyme required for ATP synthesis. On the basis of your understanding of the reactions of cellular respiration, do you think that DNP really causes weight loss? Why or why not? Explain why DNP can kill those who take it.
2. More than a century ago, French biochemist Louis Pasteur described a phenomenon, now called "the Pasteur effect," in the wine-making process. He observed that in a sealed container of grape juice containing yeast, the yeast will consume the sugar very slowly as long as oxygen remains in the container. As soon as the oxygen is gone, however, the rate of sugar consumption by the yeast increases greatly and the alcohol content in the container rises. Discuss the Pasteur effect on the basis of what you know about cellular respiration and fermentation.
3. Imagine that a starving cell reached the stage where every bit of its ATP was depleted and converted to ADP plus phosphate. If that cell were placed in fresh nutrient broth at this point, would it recover and survive? Explain your answer on the basis of what you know of glucose breakdown.

For More Information

Aschwandem, C. "Gene Cheats." *New Scientist,* January 2001. In the near future, athletes may be able to use gene therapy to boost their performance.

Gadsby, P. "The Biology of Sourdough." *Discover,* September 2003. How does fermentation give sourdough bread its distinctive taste and texture?

McCarty, R. E. "H^+-ATPases in Oxidative and Photosynthetic Phosphorylation." *BioScience,* January 1985. A description of the structure and function of the ATP-synthesizing enzymes in mitochondria and chloroplasts.

WEB TUTORIAL

To access a Web Tutorial visit http://www.prenhall.com/audesirk4. *Log in to the Web site selected by your instructor, navigate to this chapter, and select the appropriate Web Activity number.*

7.1 Glucose Metabolism

Estimated time: 15 minutes

Explore the mechanisms cells use to derive energy from the sugar glucose.

7.2 Web Investigation: When Athletes Boost Their Blood Counts: Do Cheaters Prosper?

Estimated time: 15–20 minutes

Explore how blood doping in athletes may improve endurance.

UNIT 2

Inheritance

Inheritance provides for both similarity and difference. All dogs share many similarities because their genes are nearly identical. The enormous variety of size, fur length and color, and bodily proportions results from tiny differences in their genes.

Chapter 8

DNA: The Molecule of Heredity

DNA analysis may reveal whether famed gunslinger Billy the Kid was really killed in New Mexico more than a century ago.

AT A GLANCE

Modern Showdown in the Old West

FORT SUMNER, NEW MEXICO, July 15, 1881 — Sheriff Pat Garrett fired the last shots of the Lincoln County War last night, as he gunned down Billy the Kid in Pete Maxwell's boarding house. The Kid (a.k.a. William Bonney), said to have killed as many as 21 men during his notorious career as a gunfighter and cattle rustler, was wanted for killing two guards during his escape from the Lincoln County jail on April 28. The Kid will be buried in the Fort Sumner cemetery.

And the Kid is still there—or is he? Some say that the Kid escaped to Mexico, eventually moving to Hico, Texas, where he died in 1950 under the name of Brushy Bill Roberts. Others say that he escaped to Arizona, where he lived and died as John Miller. If either of these claims is true, then Garrett must have killed an innocent man and buried the evidence in the Fort Sumner cemetery.

In 2003, the sheriffs of Lincoln County and DeBaca County (where Fort Sumner is located) petitioned to investigate the death of Billy the Kid as a cold case file—really cold. How could anyone hope to decide if the Kid is really buried in Fort Sumner, in Texas, or in Arizona?

From watching *CSI* and dozens of other TV shows, you probably know that DNA analysis can provide convincing proof of identity, if the right samples are available. There aren't any certified samples of the Kid's DNA, so how can DNA be used to decide if the bones in the Kid's grave are really his? For that matter, what's so special about DNA that allows it to be used to identify people, even under the best of circumstances? What, exactly, do forensics scientists compare in two DNA samples, when they say, "We have a match"?

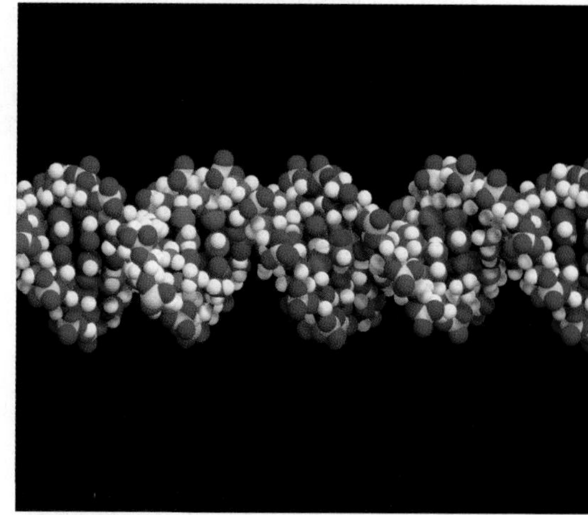

In the structure of DNA lies the secret of inheritance—and perhaps the secret of who is buried in the grave of Billy the Kid.

The Kid's mother, Catherine Antrim, is buried in Silver City, New Mexico. Let's start our investigation with the hypothesis that the DNA from the Kid and his mother are similar. As you study this chapter, think about what it means to say that two DNA samples are similar, why a parent and her child might have similar DNA, and what experiment could be done on DNA samples from the Kid and his mother to find out if Sheriff Garrett really shot the Kid that summer night so long ago.

(a)

(b)

Figure 8-1 Genetic differences Many human traits are inherited, including *(a)* Viggo Mortensen's cleft chin and *(b)* Orlando Bloom's smoothly rounded chin.

8.1 What Are Genes Made of?

For well over a century, biologists have known that inherited characteristics are transmitted from parents to offspring in discrete units called **genes**. For example, genes determine whether a person has a cleft chin, such as the actor Viggo Mortensen, or a rounded chin, like Orlando Bloom (Fig. 8-1). In the early 1900s, biologists discovered that genes are parts of the **chromosomes** that are found in the nucleus of every eukaryotic cell. Then, in the mid-twentieth century, the efforts of many researchers revealed that genes are made of **deoxyribonucleic acid**, or **DNA**, and that a gene is a particular segment of the DNA of a specific chromosome.

The genetic instructions for every organism, from a single-celled bacterium to a blue whale with trillions of cells, are contained in its DNA. These molecular instructions in DNA direct the life of each cell in an organism and confer on each cell its special characteristics. Further, DNA, passed from parent to offspring, transmits inherited characteristics from one generation to the next.

8.2 What Is the Structure of DNA?

Discovering that genes are made of DNA was a great breakthrough but was only the beginning of learning how inheritance works. For example, how is information stored in DNA, how do cells use that information, and how is it transmitted from parent to offspring? To find out, biologists needed to decipher the structure of the DNA molecule.

DNA Is Composed of Four Different Subunits

DNA molecules are composed of four small subunits called **nucleotides**. Each nucleotide in DNA consists of three parts: (1) a phosphate group, (2) a sugar called deoxyribose, and (3) one of four possible nitrogen-containing **bases**: **adenine (A)**, **thymine (T)**, **guanine (G)**, or **cytosine (C)**. ↓

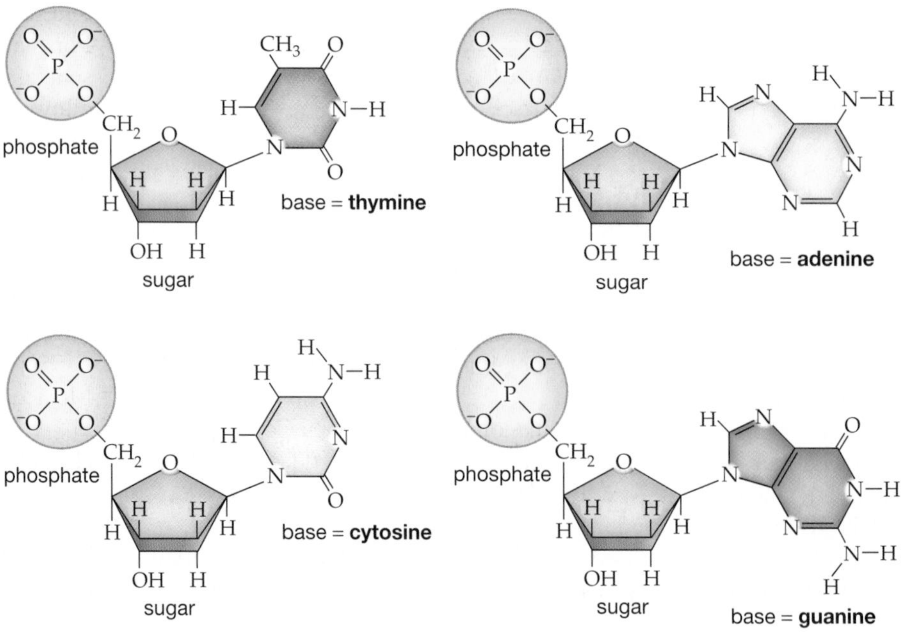

phosphate
sugar
base = **thymine**

phosphate
sugar
base = **adenine**

phosphate
sugar
base = **cytosine**

phosphate
sugar
base = **guanine**

A DNA Molecule Contains Two Nucleotide Strands

Many researchers, working over scores of years, sought to understand the molecular basis of inheritance. Some discovered the nucleotides, the molecular components of DNA. Others showed that DNA is the molecule of inheritance in organisms as diverse as bacteria and people. Still others deciphered the basic structure of DNA: long chains of nucleotides, twisted about like a corkscrew. Finally, in the early 1950s, James Watson and Francis Crick developed a model for the three-dimensional structure of DNA (see "Scientific Inquiry: The Discovery of the Double Helix").

Combining a knowledge of how complex molecules bond together with an intuition that "important biological objects come in pairs," Watson and Crick proposed that a DNA molecule consists of two separate chains of nucleotides, usually called **strands**. Within each strand, the phosphate group of one nucleotide bonds to the sugar of the next nucleotide in the strand. This bonding pattern produces a "backbone" of alternating, covalently bonded sugars and phosphates. The nucleotide bases protrude from this **sugar-phosphate backbone** (Fig. 8-2a).

All the nucleotides in a DNA strand are oriented in the same direction, so the two ends of the strand differ. One end has a "free," or unbonded, sugar, and the other end has a "free," or unbonded, phosphate. (Think of a long line of cars stopped on a crowded one-way street at night: The cars' headlights always point forward, and their taillights always point backward).

Hydrogen Bonds Hold the Two DNA Strands Together in a Double Helix

The two DNA strands are held together by hydrogen bonds that form between the bases that protrude from each strand. These bonds give DNA a ladderlike structure, with the sugar-phosphate backbones forming the sides and the bases forming the rungs (Fig. 8-2a). However, the DNA strands are not straight. Instead, they are twisted about each other to form a **double helix**, much like a ladder twisted into a circular staircase shape (Fig. 8-2b). The two strands in a DNA double helix are oriented in opposite directions. (Again, imagine an evening traffic jam, this time on a two-lane highway: A traffic helicopter pilot overhead sees only the headlights on cars in, say, the northbound lanes and only taillights in the southbound lanes).

Figure 8-2 The Watson-Crick model of DNA structure ***(a)*** Hydrogen bonding between complementary base pairs holds the two strands of DNA together. Three hydrogen bonds (red dotted lines) hold guanine to cytosine, and two hydrogen bonds hold adenine to thymine. Note that each strand has a free phosphate (yellow ball) on one end and a free sugar (blue pentagon) on the opposite end. Further, the two strands run in opposite directions. ***(b)*** Strands of DNA wind about each other in a double helix, like a twisted ladder, with the sugar-phosphate backbone forming the sides and the complementary base pairs forming the rungs. ***(c)*** A space-filling model of the DNA double helix. **QUESTION:** *Which do you think would be more difficult to break apart: an A–T base pair or a C–G base pair? Why?*

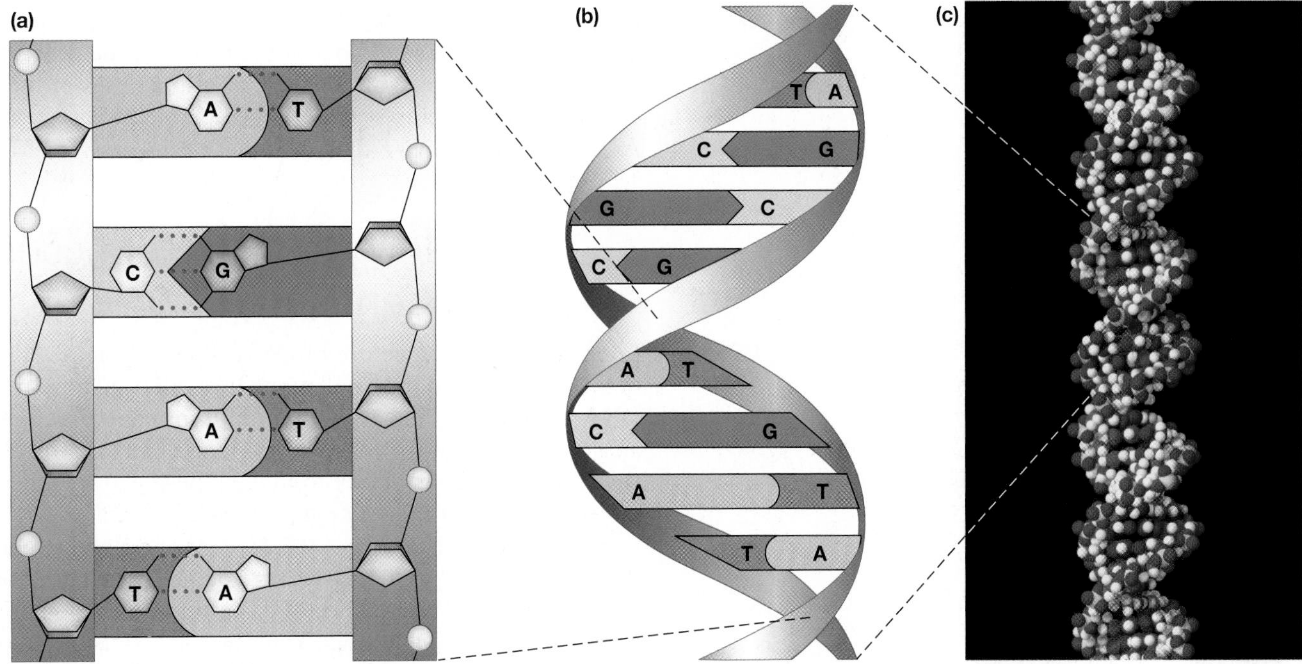

SCIENTIFIC INQUIRY

THE DISCOVERY OF THE DOUBLE HELIX

In the early 1950s, many biologists realized that the key to understanding inheritance lay in the structure of DNA. They also knew that whoever deduced the correct structure of DNA would receive recognition, very possibly the Nobel Prize. Linus Pauling of Caltech was the person most likely to solve DNA structure. He probably knew more about the chemistry of large organic molecules than did any person alive. Pauling, however, had two main handicaps. First, for years he had concentrated on protein research, and therefore he had little data about DNA. Second, he was active in the peace movement. At that time, some government officials, including Senator Joseph McCarthy (remembered today for his vigorous attacks on many individuals as Communists or subversives), considered such activity to be potentially subversive and possibly dangerous to national security. This latter handicap may have proved decisive.

The second most likely competitors were the British scientists Rosalind Franklin and Maurice Wilkins, who were experts in using x-rays to determine the structures of large molecules. In fact, Wilkins and Franklin's x-ray studies made them the only scientists who had very good data about the general shape of the DNA molecule. Unfortunately for them, their methodical approach was slow.

The door was open for the eventual discoverers of the double helix, James Watson and Francis Crick, who had neither Pauling's tremendous understanding of chemical bonds nor Franklin and Wilkins's expertise in x-ray analysis. Watson and Crick did no experiments in the ordinary sense of the word. Instead, they spent their time thinking about DNA, trying to construct a molecular model that made sense and fit the data. Because they were working in England and because Wilkins shared Franklin's data with them (perhaps against her wishes), Watson and Crick were familiar with all the x-ray information relating to DNA.

This x-ray information was just what Pauling lacked. Because of Pauling's presumed subversive tendencies, the U.S. State Department refused to issue him a passport to leave the United States, so he could neither attend meetings at which Wilkins presented the x-ray data nor visit England to talk with Franklin and Wilkins directly. Watson and Crick knew that Pauling was working on DNA structure and were driven by the fear that he would beat them to it. In his book *The Double Helix*, Watson recounts his belief that, if Pauling could have seen the x-ray pictures, "in a week at most, Linus would have the structure."

You might be thinking by now, "But wait just a minute! That's not fair. If the goal of science is to advance knowledge, then everyone should have access to all the data. If Pauling was the best, he should have discovered the double helix first." Perhaps so. But science is an activity of scientists, who, after all, are people. Although virtually all scientists want to see the advancement and benefit of humanity, each individual also wants to be the one responsible for that advancement and to receive the credit and the glory. Linus Pauling remained in the dark about the x-ray pictures of DNA and the clues they gave about its structure.

When Watson and Crick discovered the double helix structure of DNA (**Fig. E8-1**), Watson described it in a letter to Max Delbruck, a friend and adviser at Caltech. He asked Delbruck not to reveal the contents of the letter to Pauling until their structure was formally published. Delbruck, perhaps more of a model scientist, firmly believed that scientific discoveries belong in the public domain and promptly told Pauling all about it. Showing himself to be a great scientist and a great person, Pauling graciously congratulated Watson and Crick on their brilliant solution to the DNA structure. The race was over.

Figure E8-1 The discovery of DNA James Watson and Francis Crick with a model of DNA structure

Take a closer look at the way the bases pair up to form each rung of the double helix ladder (**Fig. 8-2a**). Adenine forms hydrogen bonds only with thymine, and guanine forms hydrogen bonds only with cytosine. These A–T and G–C pairs are called **complementary base pairs**. All of the bases of the two strands of a DNA double helix are complementary to each other. For example, if one strand reads A-T-T-C-C-A-G-G-C-T, then the other strand must read T-A-A-G-G-T-C-C-G-A.

8.3 How Does DNA Encode Information?

Look again at the structure of DNA shown in Figure 8-2. The relative simplicity of the structure and the small number of different bases raise the question of how DNA can encode the vast amount of information required to build and operate an organism. How, for example, can the color of a bird's feathers, the size and shape of its beak, its ability to make a nest, and its song all be determined by a molecule with just four simple parts?

The answer is that it's not the *number* of different subunits but their *sequence* that's important. Within a DNA strand, the four types of bases can be arranged in any order, and each sequence of bases represents a unique set of genetic instructions. An analogy might help: You don't need lots of unique letters to make up a language. English has 26 letters, but Hawaiian has only 12, and computers use only two "letters" (0 and 1, or "on" and "off"). Nevertheless, all three can spell out thousands of different words. A stretch of DNA just 10 nucleotides long can have more than a million possible sequences of the four bases. Because organisms have millions (in bacteria) to billions (in plants or animals) of nucleotides, their DNA molecules can encode a staggering amount of information.

If anyone does try to compare DNA samples from Billy the Kid and his mother, it won't be useful merely to extract DNA from their bones and determine *how many* A, T, C, and G bases each sample contains. Probably any two humans would have virtually identical *amounts* of each base. Instead, forensic experts will determine the base *sequence*. The real Kid's DNA should have nucleotide sequences that are the same as his mother's, but DNA from unrelated imposters should have different sequences.

8.4 How Is DNA Copied?

When cells divide, their genetic information is passed down to their offspring. First, their DNA is copied, with amazing accuracy. Then, the new DNA molecules are parceled out to the offspring (usually called "daughter cells"), so that each cell receives a complete set of genetic instructions. We will explore the processes of cell division in Chapter 10. Here, we will show how the structure of DNA provides the framework for high-precision copying.

Why Does DNA Need to Be Copied?

All of the trillions of cells of your body are the offspring of other cells, going all the way back to when you were a fertilized egg. What's more, nearly every cell of your body contains identical genetic information—the same genetic information that was present in that fertilized egg. Even after you have reached your adult size, millions of your cells continue to divide. For example, your skin cells constantly flake off or are worn away by washing or rubbing on your clothes and must be replaced. Thus, skin cells divide about once a day. To accomplish this, the cells reproduce by a complex process of cell division that produces two daughter cells from a single parental cell. Each daughter cell receives a nearly perfect copy of the parent cell's genetic information. Consequently, the parent cell must copy its DNA before it divides.

DNA Is Copied Before Cell Division

Before a cell divides, it must have two exact copies of its DNA. A process known as **DNA replication** produces two identical DNA double helixes. As you learned in Chapter 4, a chromosome consists of DNA and protein. Each double helix is the DNA component of a single chromosome. A copy of each chromosome will be passed to each of the new daughter cells (see Chapter 10), thus ensuring that each daughter cell receives a copy of each DNA double helix that was originally present in the parental cell.

DNA Replication Produces Two DNA Double Helixes, Each with One Original Strand and One New Strand

Base pairing is the foundation of DNA replication. Remember, bases always pair in the same way: An adenine on one strand pairs with a thymine on the other strand, and a cytosine pairs with a guanine. If one strand reads ATG, for example, then the other strand must read TAC. Therefore, the base sequence of each strand contains all of the information needed to replicate the other strand.

Conceptually, DNA replication is quite simple (Fig. 8-3). Enzymes called **DNA helicases** pull apart the parental DNA double helix, so that the bases of the two parental DNA strands no longer form base pairs with one another. Now DNA

Figure 8-3 Basic features of DNA replication During replication, the two strands of the parental DNA double helix separate. Free nucleotides that are complementary to those in each strand are joined to make new daughter strands. Each parental strand and its new daughter strand then form a new double helix.

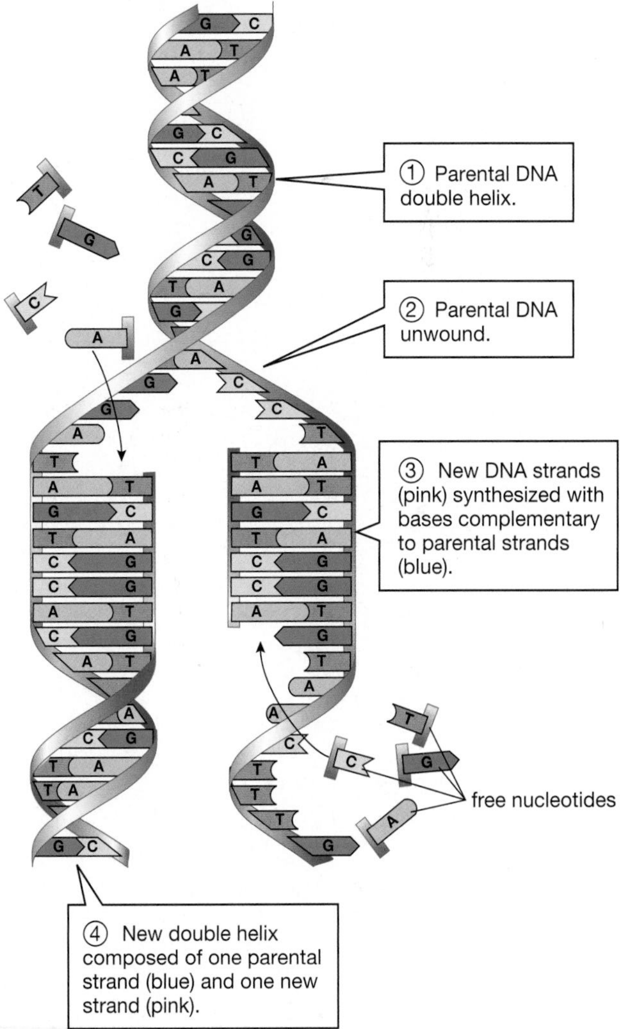

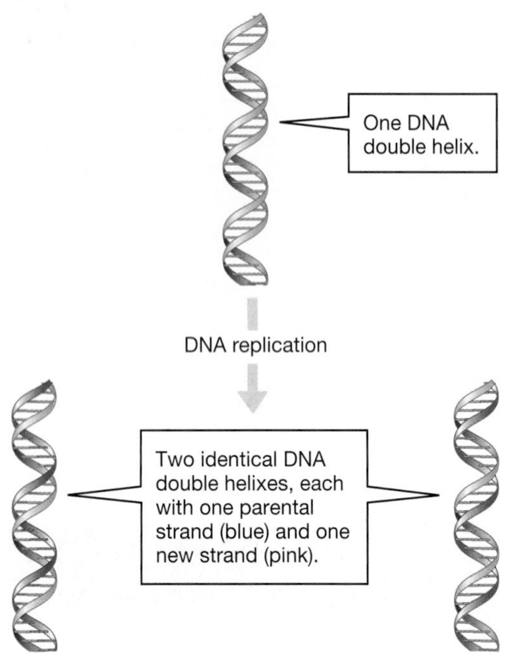

strands complementary to the two parental strands must be synthesized. A second set of enzymes, called **DNA polymerases**, moves along each separated parental DNA strand, matching each base on the strand with free nucleotides that have a complementary base. For example, DNA polymerase pairs an exposed adenine base with a thymine base. DNA polymerase also connects these free nucleotides with one another to form new DNA strands: One new daughter strand is complementary to each of the parental DNA strands. We'll go into more detail on how DNA is replicated in the last section of this chapter, "What Are the Mechanisms of DNA Replication?"

When replication is complete, one parental DNA strand and its complementary daughter DNA strand wind together into one double helix, while the other parental strand and its daughter strand entwine in a second double helix. DNA replication thus keeps, or *conserves*, one parental DNA strand and produces one daughter strand. Hence, the process is called **semiconservative replication**. ←

If no mistakes have been made, the base sequences of both new DNA double helixes are identical to the base sequence of the original, parental DNA double helix, and, of course, to each other.

Proofreading Produces Almost Error-Free Replication of DNA

Complementary base pairing makes DNA replication highly accurate. Nevertheless, DNA replication isn't perfect. DNA polymerase matches bases incorrectly about once in every 10,000 base pairs, partly because replication is so fast (up to about 700 nucleotides per second). However, completed DNA strands contain only about one mistake in every *billion* base pairs. This phenomenal accuracy is ensured by a variety of DNA repair enzymes that "proofread" each daughter strand during and after its synthesis. For example, some forms of DNA polymerase recognize a base-pairing mistake as it is made. This type of DNA polymerase will pause, fix the mistake, and then continue synthesizing more DNA.

Mistakes Do Happen

Despite this amazing accuracy, no organism has perfect, error-free DNA. In addition to mistakes made during DNA replication, the DNA in each cell in your body loses about 10,000 bases every day, simply due to spontaneous chemical breakdown at normal human body temperatures. A variety of environmental conditions can also damage DNA. For example, whenever you go out in the sunshine, DNA in some of your skin cells is damaged by ultraviolet light. Such changes in the base sequence of DNA are called **mutations**. The effects of mutations will be explored in Chapter 9.

8.5 What Are the Mechanisms of DNA Replication?

DNA replication involves three major actions (Fig. 8-4). First, as we have seen, the DNA double helix must be opened up, so that the base sequence can be "read." Then, new DNA strands with base sequences complementary to the two original strands must be synthesized. In eukaryotic cells, these new DNA strands are synthesized in fairly short pieces. Therefore, the third step in DNA replication is to stitch the pieces together to form a continuous strand of DNA. Each step is carried out by a distinct set of enzymes.

DNA Helicase Separates the Parental DNA Strands

DNA helicase ("an enzyme that breaks apart the double helix") breaks the hydrogen bonds between complementary base pairs that hold the two parental DNA strands together. Breaking the bonds separates and unwinds the double

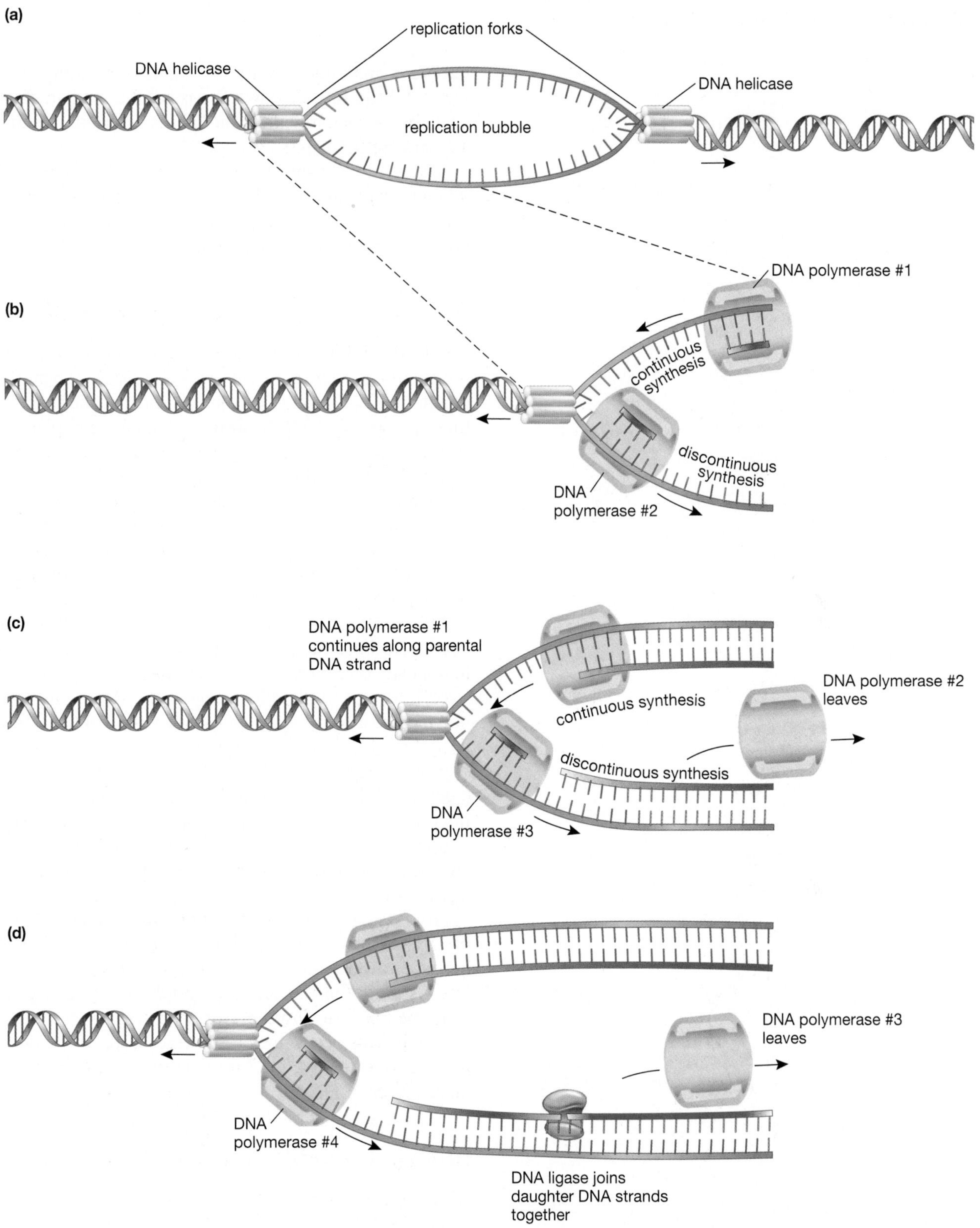

Figure 8-4 Details of DNA replication ***(a)*** DNA helicase separates the parental strands to form a replication bubble. ***(b)*** DNA polymerase synthesizes new pieces of DNA. ***(c)*** DNA helicase and DNA polymerase move along a replication bubble. ***(d)*** DNA ligase joins the small DNA segments into a single daughter strand. **QUESTION:** *During DNA replication, why doesn't DNA polymerase move away from the replication fork on both strands?*

helix, forming a replication "bubble" (Fig. 8-4a). Within the replication bubble, the nucleotide bases of the parental DNA strands are no longer paired with one another. Each replication bubble contains two replication "forks" where the two parental DNA strands have not yet been unwound.

To help visualize this process, imagine that you are driving down a two-lane, undivided road; each lane represents a single strand of a DNA double helix. The two DNA strands of the double helix point in opposite directions (see Fig. 8-2), just as cars in each lane of the road travel in opposite directions. A replication

Web Tutorial 8.2 DNA Replication

bubble is analogous to the two lanes' separating, with a wide median between them. A little farther down the road, the median disappears, and the road once again becomes undivided. The places where the median begins and where it disappears are the forks.

Eukaryotic chromosomes are so long that many DNA helicase enzymes open up many replication bubbles simultaneously, so that all of the DNA can be replicated in a reasonable length of time. The bubbles grow as DNA replication progresses and merge when they meet.

DNA Polymerase Synthesizes New DNA Strands

Replication bubbles are essential because they allow a second enzyme, *DNA polymerase* ("an enzyme that makes a DNA polymer") to get access to the bases of each DNA strand (Fig. 8-4b). At each replication fork, DNA polymerase synthesizes two new DNA strands that are complementary to the two parental strands. During this process, DNA polymerase recognizes an unpaired base in the parental strand and matches it up with a free nucleotide that has the correct complementary base. For example, DNA polymerase pairs up an exposed cytosine base in the parental strand with a guanine base in a free nucleotide. Then, DNA polymerase links the phosphate of the incoming free nucleotide to the sugar of the previously added nucleotide in the growing daughter strand. In this way, DNA polymerase synthesizes the sugar-phosphate backbone of the daughter strand.

DNA polymerase always moves toward the "free sugar" end of a single DNA strand. Because the two strands of the parental DNA double helix are oriented in opposite directions, the new complementary DNA strands will also be synthesized in opposite directions (Fig. 8-4b). Returning to our road analogy, a DNA polymerase enzyme stays on its "own" side of the DNA road, driving in the direction of the free sugar end of the strand.

Now picture how DNA helicase and DNA polymerase work together (Fig. 8-4c). DNA helicase "lands" on the double helix and moves along, unwinding the double helix and separating the strands. Because the two DNA strands run in opposite directions, as a DNA helicase enzyme moves toward the free sugar end of one strand, it is simultaneously moving toward the free phosphate end of the other strand.

Now visualize two DNA polymerases "landing" on the two separated strands of DNA. One DNA polymerase (call it polymerase #1) can follow behind the helicase toward the free sugar end, synthesizing a continuous, complete new DNA strand. On the other strand, however, DNA polymerase #2 moves away from the helicase, and therefore can synthesize only part of a new DNA strand. As the helicase continues to unwind more of the double helix, additional DNA polymerases (#3, #4, etc.) must land on this strand and will in turn synthesize more pieces of DNA.

DNA Ligase Joins Together Segments of DNA

In this way, multiple DNA polymerases synthesize pieces of DNA of varying lengths, as many as 10 million pieces for a single human chromosome. How are all of these pieces sewn together? This is the job of the third major enzyme, **DNA ligase** ("an enzyme that ties DNA together"; Fig. 8-4d). Many DNA ligase enzymes bond together the sugar-phosphate backbones of these fragments of DNA until each daughter strand consists of one long, continuous DNA polymer.

MODERN SHOWDOWN IN THE OLD WEST *REVISITED*

As you learned in this chapter, DNA is the genetic material of all living organisms on Earth. Further, the genetic information in a molecule of DNA is spelled out by its sequence of nucleotides. Finally, when DNA replicates, the nucleotide sequence of the "offspring" DNA is virtually identical to the sequence of the "parent" DNA. How can this information be used to decide if Billy the Kid is really buried in the Fort Sumner cemetery?

Most likely, you've already figured out that DNA from the Kid and from his mother should have similar nucleotide sequences. What you probably don't know is just how similar they should be. The Kid has DNA from both his mother and his father. Therefore, you might think that some of the nucleotide sequences isolated from the Kid's skeleton will have been inherited from his father, and therefore would not match Catherine Antrim's DNA. However, as you will learn in Chapter 15, mitochondria have their own DNA, completely distinct from the DNA in the nucleus of a cell. This DNA is faithfully replicated when mitochondria divide. What's more, *people inherit all of their mitochondria from their mothers*; no mitochondria from your father's sperm entered your mother's egg. Finally, DNA experts can distinguish mitochondrial DNA from nuclear DNA.

Therefore, the sheriffs' hypothesis is that the Kid's mitochondrial DNA will be *identical* to his mother's. Ideally, scientists can isolate mitochondrial DNA from bones in the Kid's grave in Fort Sumner, and from the graves of Brushy Bill Roberts and John Miller, and compare the nucleotide sequences with those from Catherine Antrim's bones. The prediction, of course, is that only the Kid's mitochondrial DNA will match Catherine Antrim's. Various controls will be necessary to be sure that the DNA extracted from the 120-year-old bones is reasonably intact. If the DNA is of good quality, the conclusions should be unambiguous.

Except ... not everyone is convinced that the headstones for the Kid and his mother mark the correct graves. There have been a few floods in the Fort Sumner cemetery, so some graves may have been shuffled a bit, and the Silver City cemetery was moved once, long ago. Can we be certain that the right people are buried in the right places? Undeterred by this potential uncertainty, the sheriffs continue to pursue case number 03-06-136-01. In what may be the oddest twist in the case to date, the governor of New Mexico even appointed an attorney to represent the Kid—with the result that Billy, through his attorney, is petitioning to have himself exhumed!

CONSIDER THIS: Because mammals inherit all of their mitochondria from their mothers, all of the descendants of your great-great-great-grandmother (pick how many greats you want here) should share the same mitochondrial DNA sequences. Not counting mutations, this means that there may be hundreds, or even thousands, of people with the same mitochondrial DNA that you have. How might this fact affect the conclusions drawn in the Billy the Kid case?

LINKS TO LIFE: *Tyrannosaurus rex* Reborn?

Remember *Jurassic Park*? In that blockbuster movie, scientists found a 70 million-year-old mosquito entombed in amber. They delicately inserted a needle into its stomach and sucked out some dinosaur blood cells that the mosquito had dined on just before it died. They extracted the cells' DNA and—*voila!*—made a *Tyrannosaurus rex.* In principle, it seems plausible. After all, DNA contains all the genetic information needed to manufacture an entire tyrannosaurus, doesn't it? So if you could find *T. rex* DNA, why couldn't you make one?

Two reasons, actually. The first is the decay rate of DNA. Remember, DNA spontaneously breaks down and therefore needs constant repair. Well, 70 million years is a *lo-o-ong* time for DNA decay without repair. If you discovered a perfectly preserved, quick-frozen mammoth from 10,000 years ago, the DNA *might* be fairly intact. But *T. rex* DNA inside a 70 million-year-old mosquito? No way.

The second problem is determining which genes to use when. Nearly every cell in your body contains all of the genes that would be needed to replicate you, but your liver doesn't grow hair and your brain doesn't make teeth. During development, each cell of your body uses the genetic information in specific genes, at specific times, to produce a human being. As we will see in Chapter 25, molecules in the fertilized egg begin the process of selecting which genes to turn on and which to turn off. Maybe elephant eggs could regulate mammoth DNA correctly, but it's not likely that lizard or crocodile eggs could properly control the DNA of a *T. rex.*

It's really too bad that we'll never see a live *T. rex,* but you probably wouldn't want to meet one up close and personal, anyway.

Chapter Review

Summary of Key Concepts

8.1 What Are Genes Made of?

Genes, the units of inheritance, are specific sequences of nucleotides in DNA located at specific places on the chromosomes.

8.2 What Is the Structure of DNA?

DNA consists of nucleotides that are linked together into long strands. Each nucleotide consists of a phosphate group, the five-carbon sugar deoxyribose, and a nitrogen-containing base. There are four types of base in DNA: adenine, guanine, thymine, and cytosine. The sugar of one nucleotide is linked to the phosphate of the next nucleotide, forming a sugar-phosphate "backbone" for each strand. The bases stick out from this backbone. Two nucleotide strands wind about one another to form a DNA double helix, which resembles a twisted ladder. The sugar-phosphate backbones form the sides of the ladder. The bases of each strand pair up in the middle of the helix, held together by hydrogen bonds, forming the rungs of the ladder. Only complementary base pairs can bond together in the helix: Adenine bonds with thymine, and guanine bonds with cytosine.

8.3 How Does DNA Encode Information?

Genetic information is encoded as the sequence of nucleotide bases in a DNA molecule, much as the meaning of a word is determined by its sequence of letters. Because DNA molecules are usually millions to billions of nucleotides long, DNA can encode huge amounts of information in its base sequence.

8.4 How Is DNA Copied?

Before a cell can reproduce, it must replicate its DNA so that each daughter cell will receive all of the genetic information contained in the parent cell. During DNA replication, DNA helicase enzymes unwind the two parental DNA strands. Then DNA polymerase enzymes bind to each parental DNA strand. Free nucleotides form hydrogen bonds with complementary bases on the parental strands, and DNA polymerase links the free nucleotides together to form new DNA strands. Therefore, the sequence of bases in each newly formed strand is complementary to the sequence of a parental strand.

Replication is semiconservative because, when DNA replication is complete, both new DNA double helixes consist of one parental DNA strand and one newly synthesized, complementary strand. The two new DNA double helixes are therefore duplicates of the parental DNA double helix.

DNA polymerase and other repair enzymes "proofread" the DNA, minimizing the number of mistakes during replication. Nevertheless, mistakes in DNA replication and damage from environmental agents such as ultraviolet light can change the base sequence of DNA. These sequence changes are called mutations.

8.5 What Are the Mechanisms of DNA Replication?

DNA helicase unwinds the double helix, forming a replication bubble. One DNA polymerase enzyme then binds to each of the unwound strands of DNA. Because the two parental DNA strands are oriented in opposite directions, the DNA polymerase on one parental strand can synthesize a long, continuous daughter strand, but the DNA polymerase on the other parental strand can synthesize only a short daughter strand. As more DNA polymerase molecules bind to this parental strand, they each synthesize a short daughter strand. Finally, DNA ligase connects the sugar-phosphate backbones of these short daughter strands to form a complete, intact strand (see Fig. 8-4).

Key Terms

adenine (A) *p. 114*
base *p. 114*
chromosome *p. 114*
complementary base pair *p. 116*
cytosine (C) *p. 114*
deoxyribonucleic acid (DNA) *p. 114*
DNA helicase *p. 117*
DNA ligase *p. 120*
DNA polymerase *p. 118*
DNA replication *p. 117*
double helix *p. 115*
gene *p. 114*
guanine (G) *p. 114*
mutation *p. 118*
nucleotide *p. 114*
semiconservative replication *p. 118*
strand *p. 115*
sugar-phosphate backbone *p. 115*
thymine (T) *p. 114*

Thinking Through the Concepts

Multiple Choice

1. *How many different possible base sequences are there in a nucleotide chain three nucleotides in length?*
a. 1 **b.** 3
c. 9 **d.** 64
e. more than 64

2. *Because each base pairs with a complementary base, in every DNA molecule the amount of*
a. cytosine equals that of guanine
b. cytosine equals that of thymine
c. cytosine equals that of adenine
d. each nucleotide is unrelated to all others
e. each nucleotide is equal to all others

3. *Semiconservative replication refers to the fact that*
a. each new DNA molecule contains two new single DNA strands
b. DNA polymerase uses free nucleotides to synthesize new DNA molecules
c. certain bases pair with specific bases
d. each parental DNA strand is joined with a new strand containing complementary base pairs
e. mistakes are made during DNA replication

4. *DNA helicase*
a. unwinds the two strands of a DNA double helix
b. converts two single strands of DNA into a double helix
c. is a unique form of DNA
d. adds nucleotides to newly forming DNA molecules
e. proofreads the newly formed DNA strand

5. *DNA polymerase*
a. can advance in either direction along a single strand of DNA
b. cleaves hydrogen bonds that join the two strands of DNA
c. creates a polymer that consists of many molecules of DNA
d. adds appropriate nucleotides to a newly forming DNA strand

6. *Which of the following are incorrectly matched?*
a. complementary base pairs—adenine and cytosine
b. bases—adenine, thymine, cytosine, and guanine
c. nucleotide—phosphate and sugar and base
d. eukaryotic chromosome—DNA and protein
e. enzymes for DNA replication—DNA polymerase, DNA helicase, and DNA ligase

Review Questions

1. If investigators exhume the bodies of Billy the Kid and his mother, will they try to determine the *total amount* of each nucleotide in the DNA samples or the *sequence* of nucleotides? Why?

2. Draw the general structure of a nucleotide. Which parts are identical in all nucleotides, and which can vary?

3. Name the four types of nitrogen-containing bases found in DNA.

4. Which bases are complementary to one another? How are they held together in the double helix of DNA?

5. Describe the structure of DNA. Where are the bases, sugars, and phosphates in the structure?

6. Describe the process of DNA replication.

Applying the Concepts

1. **IS THIS SCIENCE?** Several companies market shampoos that contain DNA, with suggestions that these shampoos might strengthen the hair and nourish the scalp, promote hair growth, or add bounce to permed hair. Assuming that the shampoos can accomplish any or all of those things, do you think that DNA is likely to be the active ingredient? Why or why not?
2. As you learned in "Scientific Inquiry: The Discovery of the Double Helix," scientists in different laboratories often compete with one another to make new discoveries. Do you think this competition helps promote scientific discoveries? Sometimes, researchers in different laboratories collaborate with one another. What advantages does collaboration offer over competition? What factors might provide barriers to collaboration and lead to competition?
3. Genetic information is encoded in the sequence of nucleotides in DNA. Let's suppose that the nucleotide sequence on one strand of a double helix encodes the information needed to synthesize a hemoglobin molecule. Do you think that the sequence of nucleotides on the other strand of the double helix also encodes useful information? Why or why not? (An analogy might help. Suppose that English were a "complementary language," with letters at opposite ends of the alphabet complementary to one another [that is, A is complementary to Z, B to Y, C to X, etc.]. Would a sentence complementary to "To be or not to be" make sense?) Finally, why do you think DNA is double stranded?

For More Information

Crick, F. *What Mad Pursuit: A Personal View of Scientific Discovery*. New York: Basic Books, 1998. A view of the race to determine the structure of DNA, by Francis Crick himself.

Gibbs, W. W. "Peeking and Poking at DNA." *Scientific American (Explorations)*, March 31, 1997. A description of techniques for studying the DNA molecule.

Judson, H. F. *The Eighth Day of Creation.* Cold Spring Harbor, NY: Cold Spring Harbor Laboratory Press, 1993. A very readable historical perspective on the development of genetics.

Radman, M., and Wagner, R. "The High Fidelity of DNA Duplication." *Scientific American,* August 1988. Faithful duplication of chromosomes requires both reasonably accurate initial replication of DNA sequences and final proofreading.

Rennie, J. "DNA's New Twists." *Scientific American,* March 1993. An overview of DNA structure and function, written 40 years after Watson and Crick's initial description.

Watson, J. D. *The Double Helix.* New York, Atheneum, 1968. If you still believe the Hollywood images that scientists are either maniacs or cold-blooded, logical machines, be sure to read this book. Although hardly models for the behavior of future scientists, Watson and Crick are certainly human enough!

Weinberg, R. "How Cancer Arises." *Scientific American,* September 1996. An overview of the molecular basis of cancer: mutations in DNA.

WEB TUTORIAL

To access a Web Tutorial visit http://www.prenhall.com/audesirk4. *Log in to the Web site selected by your instructor, navigate to this chapter, and select the appropriate Web Activity number.*

8.1 DNA Structure
Estimated time: 10 minutes

The structure of the DNA double helix is the key to the ability of this molecule to store and transmit genetic information. In this activity, you will explore the structure of the DNA molecule.

8.2 DNA Replication
Estimated time: 10 minutes

This activity explores the process of DNA replication, which is crucial for the organism growth and reproduction.

8.3 Web Investigation: Modern Showdown in the Old West: Forensic Genetics
Estimated time: 20 minutes

Explore how DNA can be used to convict or exonerate criminals, even "cold" (closed) cases.

Chapter 9

Gene Expression and Regulation

Many of the differences in the body structures of males and females can ultimately be traced to differences in a single gene.

AT A GLANCE

Snakes and Snails and Puppy Dogs' Tails

What are little girls made of? Sugar and spice and everything nice.

What are little boys made of? Snakes and snails and puppy dogs' tails.

—from an old English nursery rhyme

Male and female—so alike, yet so different. The physical differences between men and women are pretty obvious, but for a very long time, biologists had only the vaguest ideas about the genetic bases of those differences. It's been less than a century since Theophilus Painter discovered the Y chromosome, and several more decades passed before it was generally accepted that the Y chromosome usually determines maleness in humans and other mammals. But how?

One hypothesis might be that genes on the Y chromosome encode the male genitalia, so we would predict that anyone with a Y chromosome will have testes and a penis. But males also have all of the other chromosomes that females have (although males have only one X chromosome, rather than the two that females have). Why, then, don't boys develop both male *and* female genitalia? Furthermore, most of the genes needed to produce male sexual characteristics, including the genitalia, are *not* on the Y chromosome. Girls therefore possess these genes, so why don't girls develop both female and male genitalia?

In this chapter, we will examine the flow of information from an organism's genes to its physical characteristics. Just as information in a book remains hidden until someone opens its cover and reads the text, so too the information in genes may, or may not, be used in different organisms, in the various cells of an individual organism, and at different times during the organism's life. Let's begin, then, with the hyphotheses that boys and girls possess all the genes needed to produce both male and female sexual characteristics and that their distinctive physical traits arise because different genes are used during their development.

TABLE 9-1 A Comparison of DNA and RNA

	DNA	RNA	
Strands	2	1	
Sugar	deoxyribose	ribose	
Types of bases	adenine (A), thymine (T) cytosine (C), guanine (G)	adenine (A), uracil (U) cytosine (C), guanine (G)	
Base pairs	DNA:DNA A–T T–A C–G G–C	RNA:DNA A–T U–A C–G G–C	RNA:RNA A–U U–A C–G G–C
Function	Contains genes; sequence of bases in most genes determines the amino acid sequence of a protein	**Messenger RNA (mRNA):** carries the code for a protein-coding gene from DNA to ribosomes **Ribosomal RNA (rRNA):** combines with proteins to form ribosomes, the structures that link amino acids to form a protein **Transfer RNA (tRNA):** carries amino acids to the ribosomes	

9.1 How Is the Information in DNA Used in a Cell?

Information, by itself, doesn't *do* anything. For example, a blueprint may describe the structure of a house in great detail, but unless that information is translated into action, no house will ever be built. Similarly, the base sequence of DNA is the "molecular blueprint" of every cell and contains an incredible amount of information, but DNA cannot carry out any action on its own. So how does DNA determine traits, such as whether you are male or female, or have brown or blue eyes?

Most Genes Contain Information for the Synthesis of a Single Protein

Most chromosomes contain hundreds or thousands of genes, each of which occupies a particular position in the DNA of the chromosome. In most cases, a **gene** is a stretch of DNA encoding the instructions for the manufacture of a single protein, a fact that leads to the general rule "one gene, one protein." Proteins, in turn, are the molecular workhorses of the cell, forming many of its cellular structures and the enzymes that catalyze its chemical reactions.

Technically, the "one gene, one protein" relationship should really be expressed as "one gene, one polypeptide." A polypeptide is a chain of amino acids. Although many proteins consist of a single polypeptide, others are composed of more than one polypeptide subunit. For example, DNA polymerase, a key enzyme in DNA replication (see Chapter 8), is composed of more than a dozen polypeptides, each one encoded by a different gene. Nonetheless, many biochemists informally use the terms *polypeptide* and *protein* interchangeably, and this text uses *protein* to refer to all polypeptides.

RNA Intermediaries Carry the Genetic Information for Protein Synthesis

DNA does not directly guide protein synthesis but instead relies on an intermediary that carries information from the nucleus to the cytoplasm. This intermediary is **ribonucleic acid**, or **RNA**. RNA is similar to DNA but differs structurally in three respects: RNA is normally single stranded, whereas DNA is double stranded; RNA has the sugar ribose instead of deoxyribose in its backbone; and RNA contains the base uracil instead of the base thymine (Table 9-1).

Figure 9-1 Genetic information flows from DNA to RNA to protein

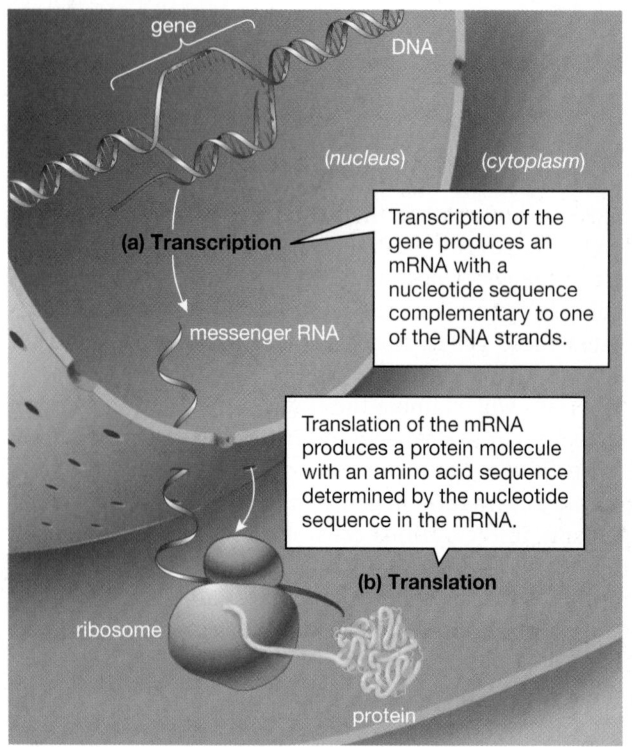

Overview: Genetic Information Is Transcribed into RNA, Then Translated into Protein

Information from DNA is used to direct the synthesis of proteins in a two-step process (Fig. 9-1):

1. During RNA synthesis, or **transcription** (Fig. 9-1a), the information contained in the DNA of a specific gene is copied into one of three types of RNA: **messenger RNA (mRNA)**, **transfer RNA (tRNA)**, or **ribosomal RNA (rRNA)**. Thus, a gene can also be defined as a segment of DNA that can be copied, or transcribed, into RNA. In eukaryotic cells, transcription occurs in the nucleus.
2. During protein synthesis, or **translation** (Fig. 9-1b), the nucleotide sequence in a messenger RNA molecule is used to direct the synthesis of a protein with a specific amino acid sequence. Translation occurs in the cytoplasm.

It is easy to confuse the terms *transcription* and *translation*. Comparing their common English meanings with their biological meanings may help you understand the difference. In everyday English, to *transcribe* means to make a written copy of something, almost always in the same language. In an American courtroom, for example, verbal testimony is transcribed into a written copy, and both the testimony of the witnesses and the transcriptions are in English. In biology, *transcription* is the process of copying information from DNA to RNA using the common "language" of nucleotides. In contrast, the common English meaning of *translation* is to convert words from one language to a different language. Similarly, in biology, *translation* means to convert information from the "nucleotide language" of RNA to the "amino acid language" of proteins.

9.2 What Is the Genetic Code?

To understand the flow of information from DNA to RNA to protein, geneticists first had to overcome the language barrier: How is the language of nucleotide sequences in DNA and messenger RNA translated into the language of amino acid sequences in proteins? This translation relies on a "dictionary" called the **genetic code**, which translates the sequence of bases in nucleic acids into the sequence of amino acids in proteins.

A Sequence of Three Bases Codes for an Amino Acid

Think about how such a genetic code might work. In the simplest possible code, each different base would code for a particular amino acid. Such a simple code is not possible, however, because proteins contain 20 different amino acids while DNA contains only four different bases (A, T, G, and C; see Table 9-1). A system in which a sequence of two bases codes for an amino acid won't work either, because there are only 16 possible combinations of two bases, which is still not enough to encode 20 different amino acids.

A three-base, or *triplet,* sequence, however, gives 64 possible combinations, which is more than enough. Under the assumption that nature operates as economically as possible, biologists hypothesized that the genetic code must be triplet; that is, three bases should specify a single amino acid. In 1961, Francis Crick and three co-workers demonstrated that this hypothesis is correct.

To decipher the "words" of the genetic code, researchers ground up bacteria and isolated the components needed to synthesize proteins. To this mixture, they added artificial mRNA, which allowed them to control what "words" were to be translated. They could then see which amino acids were incorporated into the resulting proteins. For example, an artificial mRNA strand composed entirely of uracil (UUUUUUUU...) directed the mixture to synthesize a protein composed solely of the amino acid phenylalanine. Therefore, the researchers concluded that the triplet UUU must specify phenylalanine. Because the genetic code was deciphered by using these artificial mRNAs, it is usually written in terms of the base triplets in mRNA (rather than in DNA) that code for each amino acid (Table 9-2). These mRNA triplets are called **codons**.

TABLE 9-2 The Genetic Code (Codons of mRNA)

	Second Base								
First Base	U		C		A		G		**Third Base**
U	UUU	Phenylalanine (Phe)	UCU	Serine (Ser)	UAU	Tyrosine (Tyr)	UGU	Cysteine (Cys)	U
	UUC	Phenylalanine	UCC	Serine	UAC	Tyrosine	UGC	Cysteine	C
	UUA	Leucine (Leu)	UCA	Serine	UAA	Stop	UGA	Stop	A
	UUG	Leucine	UCG	Serine	UAG	Stop	UGG	Tryptophan (Trp)	G
C	CUU	Leucine	CCU	Proline (Pro)	CAU	Histidine (His)	CGU	Arginine (Arg)	U
	CUC	Leucine	CCC	Proline	CAC	Histidine	CGC	Arginine	C
	CUA	Leucine	CCA	Proline	CAA	Glutamine (Gln)	CGA	Arginine	A
	CUG	Leucine	CCG	Proline	CAG	Glutamine	CGG	Arginine	G
A	AUU	Isoleucine (Ile)	ACU	Threonine (Thr)	AAU	Asparagine (Asp)	AGU	Serine (Ser)	U
	AUC	Isoleucine	ACC	Threonine	AAC	Asparagine	AGC	Serine	C
	AUA	Isoleucine	ACA	Threonine	AAA	Lysine (Lys)	AGA	Arginine (Arg)	A
	AUG	Methionine (Met) Start	ACG	Threonine	AAG	Lysine	AGG	Arginine	G
G	GUU	Valine (Val)	GCU	Alanine (Ala)	GAU	Aspartic acid (Asp)	GGU	Glycine (Gly)	U
	GUC	Valine	GCC	Alanine	GAC	Aspartic acid	GGC	Glycine	C
	GUA	Valine	GCA	Alanine	GAA	Glutamic acid (Glu)	GGA	Glycine	A
	GUG	Valine	GCG	Alanine	GAG	Glutamic acid	GGG	Glycine	G

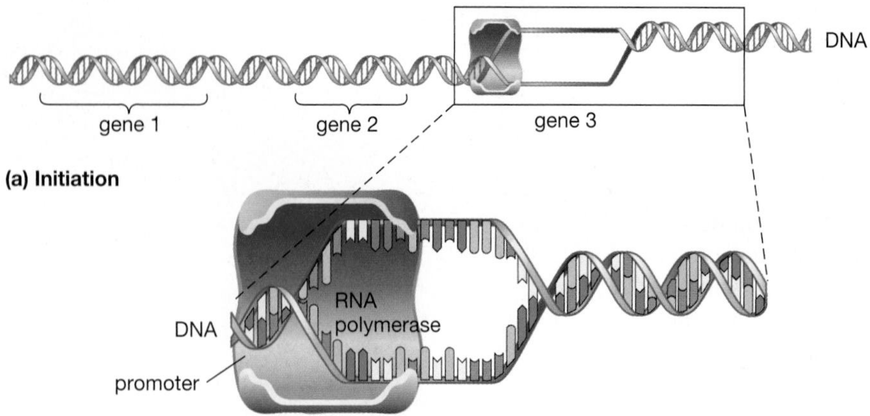

RNA polymerase binds to the promoter region of DNA near the beginning of a gene, separating the double helix near the promoter.

(b) Elongation

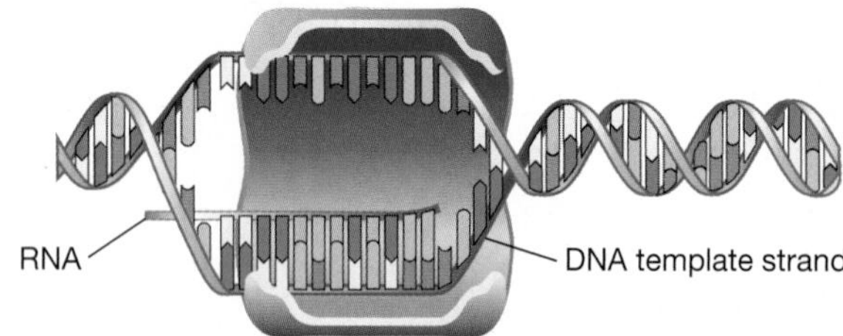

RNA polymerase travels along the DNA template strand, catalyzing the addition of ribose nucleotides into an RNA molecule. The nucleotides in the RNA are complementary to the template strand of the DNA.

(c) Termination

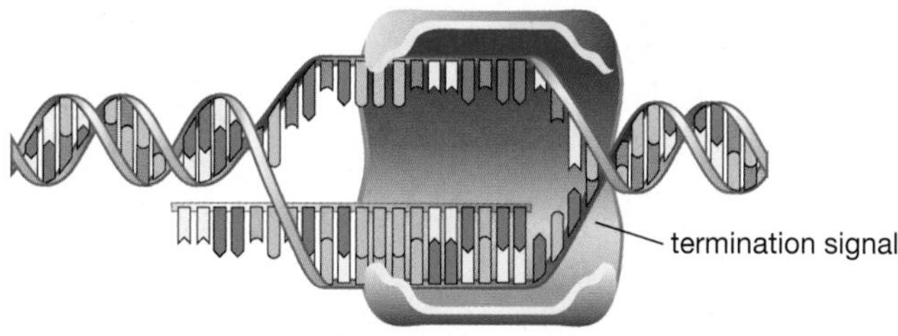

At the end of a gene, RNA polymerase encounters a sequence of DNA called a termination signal. RNA polymerase detaches from the DNA and releases the RNA molecule.

(d) Conclusion of transcription

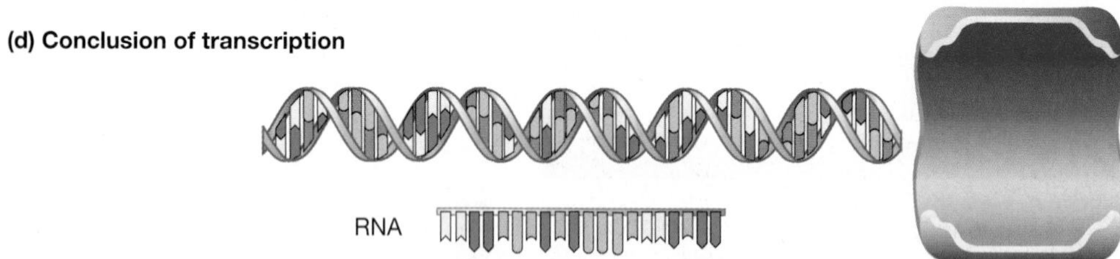

After termination, the DNA completely rewinds into a double helix. The RNA molecule is free to move from the nucleus to the cytoplasm for translation, and RNA polymerase may move to another gene and begin transcription once again.

Figure 9-2 **Transcription is the synthesis of RNA from instructions in DNA** A gene is a segment of a chromosome's DNA. One of the DNA strands will serve as the template for the synthesis of an RNA molecule with bases complementary to the bases in this DNA strand.
QUESTION: *Suppose that you could design an RNA polymerase that would transcribe both strands of a DNA double helix at the same time. Do you think that this would be useful for a cell?*

Just knowing the identities of the codons isn't enough to understand how cells use the genetic code. Why not? For any language to be understood, its users must also know where words begin and end, and where sentences start and stop. How does the cell recognize where codons begin and end, and where the code for an entire protein starts and stops?

All proteins originally begin with the same amino acid, methionine (though it may be removed after the protein is synthesized). Methionine is specified by the codon AUG, which is therefore called the **start codon**. Three codons—UAG,

UAA, and UGA—are **stop codons**, which denote the end of the protein. Because all codons consist of three bases, and the beginning and end of a protein are specified, "spaces" between codons are unnecessary. Why? Consider what would happen if English used only three-letter words: A sentence such as THEDOGSAWTHECAT would be perfectly understandable, even without spaces between the words.

Because the genetic code has three stop codons, 61 nucleotide triplets remain to specify only 20 amino acids. Therefore, most amino acids are specified by several different codons. For example, six different codons specify leucine (see Table 9-2). However, each codon specifies one, and only one, amino acid.

We have now introduced the pathway of information flow from DNA to RNA to protein. We have also described the genetic code by which cells convert the language of nucleotides in DNA and RNA into the language of amino acids in proteins. Now let's examine the two major processes in this pathway—transcription and translation—in more detail.

9.3 How Is the Information in a Gene Transcribed into RNA?

Transcription copies the genetic information of DNA into RNA. We can view transcription as a process consisting of (1) initiation, (2) elongation, and (3) termination. These three steps correspond to the three major parts of most genes: (1) a promoter region at the beginning of the gene, where transcription is started, or initiated; (2) the "body" of the gene, where the RNA strand is elongated; and (3) a termination signal at the end of the gene, where RNA synthesis stops.

Transcription Begins When RNA Polymerase Binds to the Promoter of a Gene

The enzyme **RNA polymerase** synthesizes RNA. To initiate transcription, RNA polymerase must first locate the **promoter** region, an untranscribed sequence of DNA bases that marks the beginning of the gene. When RNA polymerase binds to a promoter, the DNA double helix at the beginning of the gene partially unwinds and transcription begins (Fig. 9-2a).

Elongation Generates a Growing Strand of RNA

RNA polymerase travels down one of the DNA strands, called the **template strand**, unwinding the DNA double helix as it goes. RNA polymerase synthesizes a single strand of RNA with bases that are complementary to those in the DNA (Fig. 9-2b). Base pairing between RNA and DNA is the same as between two strands of DNA, except that uracil in RNA pairs with adenine in DNA (see Table 9-1). After about 10 nucleotides have been added to the growing RNA chain, the first nucleotides in the RNA molecule separate from the DNA template strand, allowing the two DNA strands to rewind into a double helix once again. As transcription continues to elongate the RNA molecule, one end of the RNA drifts away from the DNA while RNA polymerase keeps the other end temporarily attached to the DNA template strand (Fig. 9-3).

Transcription Stops When RNA Polymerase Reaches the Termination Signal

RNA polymerase continues along the template strand until it reaches a sequence of DNA bases known as the termination signal (Fig. 9-2c). At this point, RNA polymerase releases the completed RNA molecule and detaches from the DNA (Fig. 9-2d). The RNA polymerase is then free to bind to another promoter and synthesize another RNA molecule.

Figure 9-3 RNA transcription in action This electron micrograph shows RNA transcription. In the treelike structure in the middle of the micrograph, the central "trunk" (blue) is DNA and the "branches" (red) are RNA molecules. Many RNA polymerase molecules are traveling along the DNA, synthesizing RNA as they go. The beginning of the gene is on the left. The short RNA molecules on the left have just begun to be synthesized, while the long RNA molecules on the right have almost been completed. **QUESTION:** *Why do you think that so many mRNA molecules are being synthesized from the same gene? (Hint: Do organisms always need to have the same amounts of all types of proteins?)*

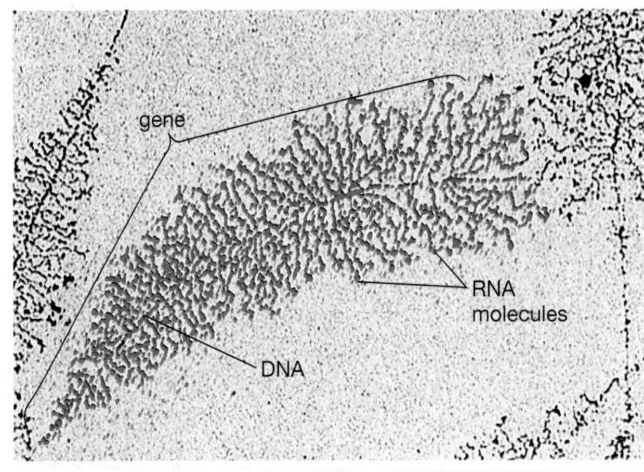

Transcription Is Selective

The transcription of genes into RNA is selective. In any given cell, only a portion of the genes is transcribed into RNA. Some genes are transcribed in all cells because they encode proteins that are essential for the life of any cell. Many genes, however, are transcribed in only certain types of cells. For example, every cell in your body contains the gene for the protein hormone insulin, but that gene is transcribed only in certain cells in your pancreas.

How do cells regulate which genes are transcribed? Proteins that bind to "control regions" of DNA, found near the promoter of a specific gene, block or enhance the binding of RNA polymerase. Different cells contain different proteins that regulate transcription, thereby restricting transcription to the genes needed by a particular type of cell at a particular time. Refer back to "Snakes and Snails and Puppy Dogs' Tails": what causes a boy to develop male sexual characteristics? Let's refine our hypothesis a bit: Perhaps one or more genes on his Y chromosome code for proteins that stimulate transcription of the genes responsible for male sexual development but suppress transcription of the genes that would otherwise cause female sexual development.

9.4 What Are the Functions of RNA?

There are three types of RNA: messenger RNA (mRNA), transfer RNA (tRNA), and ribosomal RNA (rRNA). Each plays a different role in converting the nucleotide sequence of DNA into the amino acid sequence of a protein (Fig. 9-4).

Messenger RNA Carries the Code for a Protein from the Nucleus to the Cytoplasm

All RNA is produced by transcription of DNA, but only mRNA carries the code for the amino acid sequence of a protein (Fig. 9-4a). In eukaryotic cells, mRNA molecules are synthesized in the nucleus and enter the cytoplasm through the

Figure 9-4 Cells synthesize three major types of RNA

(a) Messenger RNA (mRNA)

A U G U G C G A G U U A U G G

The base sequence of mRNA carries the information for the amino acid sequence of a protein.

(b) Ribosome: contains ribosomal RNA (rRNA)

large subunit

catalytic site

1 2

tRNA/amino acid binding sites

small subunit

rRNA combines with proteins to form ribosomes. The small subunit binds mRNA. The large subunit binds tRNA and catalyzes peptide bond formation between amino acids during protein synthesis.

(c) Transfer RNA (tRNA)

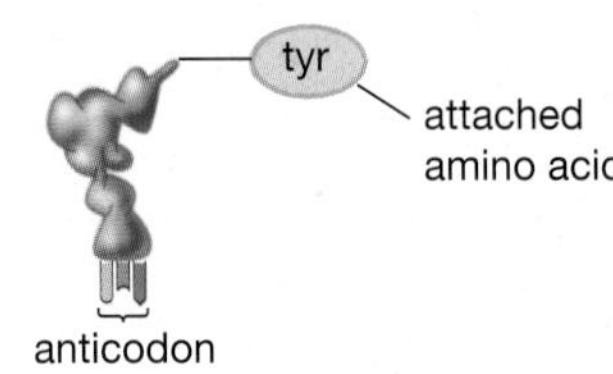

Each tRNA carries a specific amino acid to a ribosome during protein synthesis. The anticodon of tRNA pairs with a codon of mRNA, ensuring that the correct amino acid is incorporated into the protein.

pores in the nuclear envelope. In the cytoplasm, mRNA binds to ribosomes, which synthesize a protein specified by the mRNA base sequence. The gene itself remains safely stored in the nucleus, like a valuable document in a library, while mRNA, like a photocopy, carries the information to the cytoplasm to be used in protein synthesis.

Ribosomal RNA and Proteins Form Ribosomes

Ribosomes carry out translation, the process of synthesizing the proteins encoded in mRNA base sequences. Ribosomal RNA and many different proteins combine to form ribosomes. Each ribosome consists of two subunits—one small and one large (Fig. 9-4b). The small subunit has binding sites for mRNA, a "start" tRNA, and several other proteins that cooperate in reading mRNA and starting protein synthesis. The large subunit has binding sites for two tRNA molecules and a catalytic site where the amino acids are joined together in a growing protein chain. The two subunits remain separate unless they are actively synthesizing proteins. During protein synthesis, the two subunits come together, clasping an mRNA molecule between them.

Transfer RNA Molecules Carry Amino Acids to the Ribosomes

Each cell synthesizes many different types of transfer RNA, one (or sometimes several) for each amino acid. Twenty different enzymes in the cytoplasm, one for each amino acid, recognize the tRNA molecules and attach the correct amino acid (Fig. 9-4c). These "loaded" tRNA molecules deliver the amino acids to the ribosome, where the amino acids are incorporated into the growing protein chain.

The ability of tRNA to deliver the proper amino acid to the growing protein depends on base pairing between tRNA and mRNA. Each tRNA has three exposed bases, called the **anticodon** (see Fig. 9-4c). At the ribosome, the anticodon forms complementary base pairs with an mRNA codon. For example, the mRNA codon AUG base pairs with the anticodon UAC of a tRNA that has methionine attached to its end.

9.5 How Is the Information in Messenger RNA Translated into Protein?

Once transcription has produced a messenger RNA molecule with a particular nucleotide sequence, the mRNA is used during translation to direct the synthesis of a protein with a specific amino acid sequence. Translation occurs in the cytoplasm and involves ribosomes and tRNA molecules. Like transcription, translation has three steps: (1) initiation of protein synthesis, (2) elongation of the protein chain, and (3) termination of translation.

Translation Begins When tRNA and mRNA Bind to a Ribosome

The first AUG codon in mRNA specifies the start of translation. Because AUG also codes for methionine (Table 9-2), all newly synthesized proteins begin with methionine. An *initiation complex* composed of a small ribosomal subunit, a methionine tRNA, and several proteins binds to the end of an mRNA molecule with the AUG start codon (Fig. 9-5a,b). The AUG codon in mRNA forms base pairs with the UAC anticodon of the methionine tRNA. A large ribosomal subunit attaches to the small subunit, completing the assembly of the ribosome (Fig. 9-5c). The ribosome is now ready to begin translation.

Initiation:

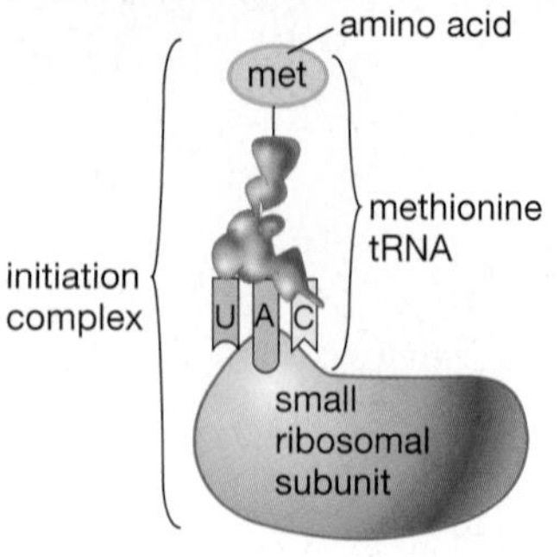

(a) A tRNA with an attached methionine amino acid binds to a small ribosomal subunit, forming an initiation complex.

(b) The initiation complex binds to an mRNA molecule. The methionine (met) tRNA anticodon (UAC) base-pairs with the start codon (AUG) of the mRNA.

(c) The large ribosomal subunit binds to the small subunit. The methionine tRNA binds to the first tRNA site on the large subunit.

Elongation:

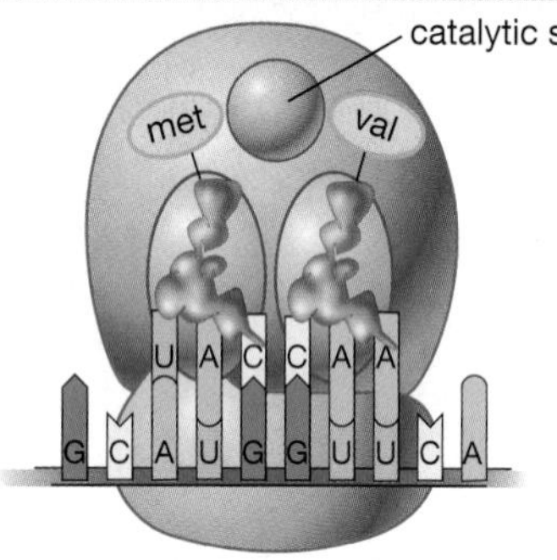

(d) The second codon of mRNA (GUU) base-pairs with the anticodon (CAA) of a second tRNA carrying the amino acid valine (val). This tRNA binds to the second tRNA site on the large subunit.

(e) The catalytic site on the large subunit catalyzes the formation of a peptide bond linking the amino acids methionine and valine. The two amino acids are now attached to the tRNA in the second binding position.

(f) The "empty" tRNA is released and the ribosome moves down the mRNA, one codon to the right. The tRNA that is attached to the two amino acids is now in the first tRNA binding site and the second tRNA binding site is empty.

Termination:

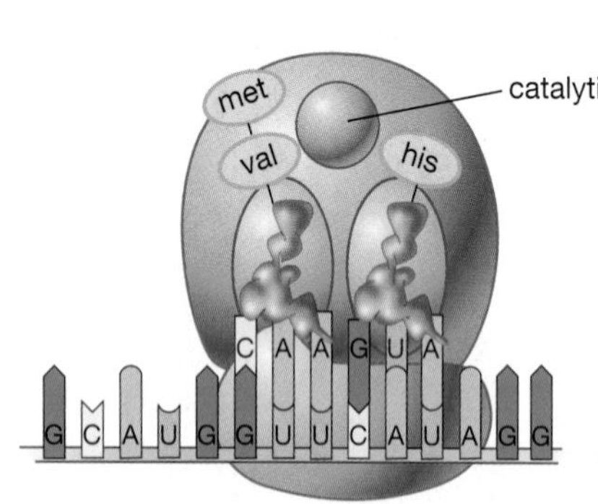

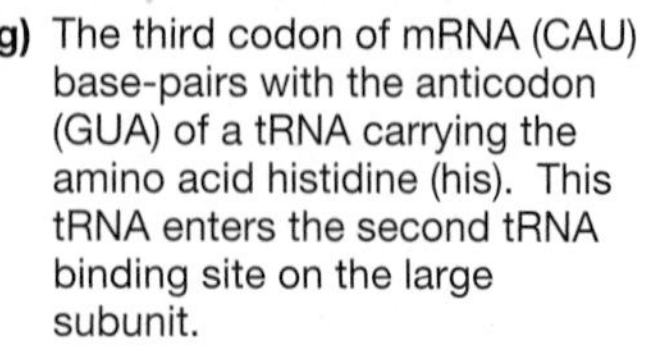

(g) The third codon of mRNA (CAU) base-pairs with the anticodon (GUA) of a tRNA carrying the amino acid histidine (his). This tRNA enters the second tRNA binding site on the large subunit.

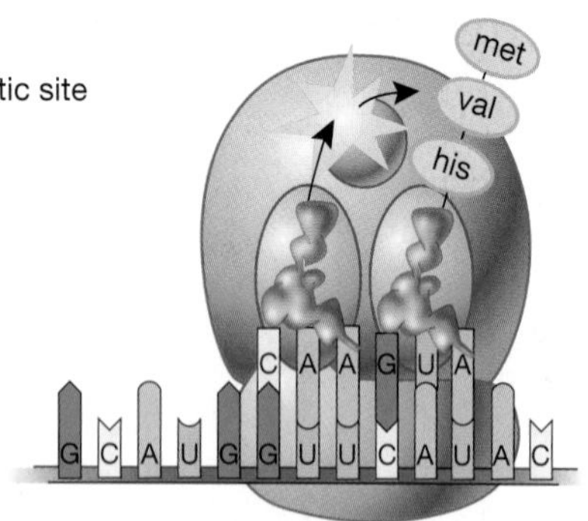

(h) The catalytic site forms a new peptide bond between valine and histidine. A three-amino-acid chain is now attached to the tRNA in the second binding site. The tRNA in the first site leaves, and the ribosome moves one codon over on the mRNA.

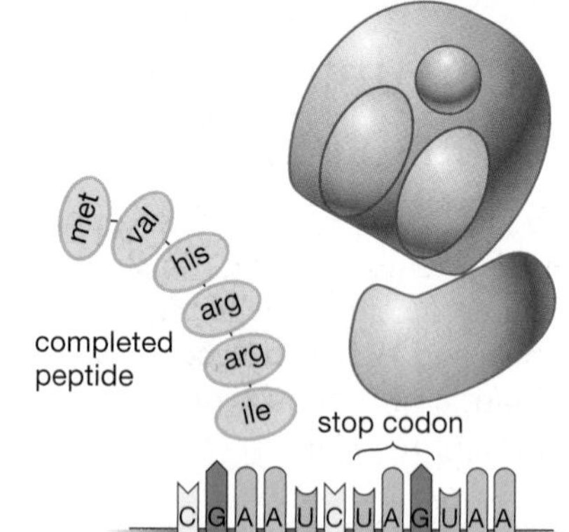

(i) This process repeats until a stop codon is reached; the mRNA and the completed peptide are released from the ribosome, and the subunits separate.

Figure 9-5 Translation is the process of protein synthesis Protein synthesis, or translation, decodes the base sequence of an mRNA into the amino acid sequence of a protein. **QUESTION:** *If all of the guanine molecules (G) visible in this mRNA sequence were changed to uracil (U), how would the translated protein differ from the one shown?*

Elongation Generates a Growing Chain of Amino Acids

The large ribosomal subunit has two tRNA binding sites and a catalytic site. When the first binding site is occupied by a tRNA and its attached amino acid, a second tRNA, with an anticodon complementary to the next mRNA codon, moves into the second binding site. The catalytic site forms a peptide bond between the two amino acids (Fig. 9-5d,e). The ribosome then moves to the next codon, shifting the growing chain from the second to the first binding site, and the pattern repeats, joining a third amino acid to the chain (Fig. 9-5f,g,h). The whole process repeats over and over again as the ribosome moves along the mRNA one codon at a time, elongating the chain.

A Stop Codon Signals Termination

A stop codon in the mRNA molecule signals the ribosome to terminate protein synthesis. Stop codons do not bind to a tRNA. Instead, proteins called release factors bind to the ribosome when it encounters a stop codon, forcing the ribosome to release the finished protein chain and the mRNA (Fig. 9-5i). The ribosome disassembles into large and small subunits, which can then be used to translate another mRNA.

Summing Up: Transcription and Translation

We can now summarize how a cell decodes the genetic information stored in its DNA to synthesize a protein (Fig. 9-6).

a. With few exceptions, such as the genes for tRNA and rRNA, each gene codes for a single protein.

b. Transcription of a protein-coding gene produces a messenger RNA molecule that is complementary to one DNA strand of the gene. Starting from the first AUG, each codon in the mRNA is a sequence of three bases that specifies either an amino acid or a "stop."

c. Enzymes in the cytoplasm attach the appropriate amino acid (based on the tRNA's anticodon) to each tRNA.

d. During translation, tRNAs carry their attached amino acids to the ribosome. There, the bases in tRNA anticodons bind to the complementary bases in mRNA codons, so that the amino acids attached to the tRNAs line up in the sequence specified by the codons. The ribosome links the amino acids together to form a protein.

This chain of events, moving from DNA bases to mRNA codons to tRNA anticodons to amino acids, decodes the information in DNA and results in the synthesis of a protein with a specific amino acid sequence. This amino acid sequence is determined by the base sequence within a single gene.

Figure 9-6 **Complementary base pairing is critical to decoding genetic information** ***(a)*** DNA contains two strands. RNA polymerase uses the template strand to synthesize an RNA molecule. ***(b)*** Bases in the template strand are transcribed into a complementary mRNA. Codons are sequences of three bases that specify an amino acid or a stop during protein synthesis. ***(c)*** Unless it is a stop codon, each mRNA codon forms base pairs with the anticodon of a tRNA molecule that carries a specific amino acid. ***(d)*** The amino acids are linked together to form a protein.

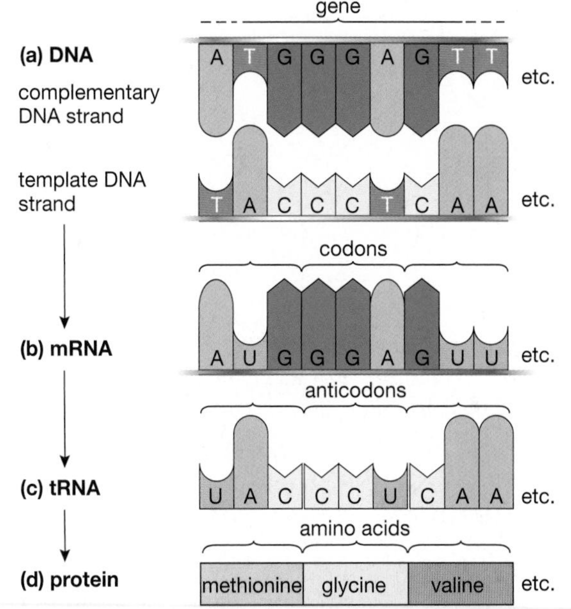

9.6 How Do Mutations Affect Gene Function?

We have all heard of **mutations**: changes in the sequence of bases in a DNA molecule. How do mutations occur, and what are their consequences? Mutations may occur in a variety of ways:

1. There may be a mistake in base pairing during DNA replication as a cell prepares for cell division. Despite the best efforts of proofreading enzymes, a few mistakes occur in each DNA replication.
2. Occasionally, bases change spontaneously because of random movement of atoms in the DNA molecule.

3. Certain chemicals (such as some components of tobacco smoke and aflatoxins synthesized by molds that live on grain and peanuts) and some types of radiation (such as x-rays and ultraviolet rays in sunlight) increase the frequency of base-pairing errors during replication or even induce changes in DNA composition between replications.

Mutations May Be Nucleotide Substitutions, Insertions, or Deletions

There are several types of mutation. Sometimes, during DNA replication, an incorrect pair of nucleotides is incorporated into the growing DNA double helix. These **nucleotide substitutions** are also called **point mutations**, because individual nucleotides in the DNA sequence are changed. An **insertion mutation** occurs when one or more new nucleotide pairs are inserted into a gene. A **deletion mutation** occurs when one or more nucleotide pairs are removed from a gene.

Mutations Affect Proteins in Different Ways

In most cases, mutations are harmful. Just as randomly changing a word in the middle of *Hamlet* would be unlikely to improve Shakespeare's writing, changing the base sequence of a gene is unlikely to improve the performance of the gene's protein product. Instead, a mutation in a gene is likely to disrupt the proper functioning of the gene, and it is also likely to be harmful to the organism in which the mutation occurs. (For an example in which a single mutation has far-reaching consequences, see "Health Watch: Mutations and Gender.") Some mutations, however, have no effect on the organism. In rare cases, a mutation may even be beneficial.

The effects of a mutation on protein structure and function depend on the type of mutation. For example, deletions and insertions of one or two nucleotides usually have catastrophic effects on a gene, because all the codons that follow the deletion or insertion are altered. Think of our sample English sentence, THEDOGSAWTHECAT. Deleting or inserting a letter (deleting the first E, for example) makes all of the following three-letter words nonsensical: THD OGS WAT HEC AT. The protein synthesized from an mRNA containing such a mutation will almost always be nonfunctional.

Point mutations (nucleotide substitutions) within a protein-coding gene, however, will not always render the code useless. Nucleotide substitutions within a protein-coding gene can produce at least four different outcomes. As a concrete example, let's consider mutations that occur in the gene encoding beta-globin, one of the subunits of hemoglobin, the oxygen-carrying protein in red blood cells (Table 9-3). The other type of subunit in hemoglobin is called alpha; a normal hemoglobin molecule consists of two alpha and two beta subunits. In all but the last example, we will consider the results of mutations that occur in the sixth codon (CTC in DNA, transcribed to GAG in mRNA), which specifies glutamic acid, a charged, hydrophilic, water-soluble amino acid.

1. *The protein may be unchanged.* Remember that most amino acids can be encoded by several different codons. If a mutation changes the beta-globin DNA base sequence from CTC to CTT, this sequence still codes for glutamic acid. Therefore, the protein synthesized from the mutated gene remains unchanged (see mutation 1 in Table 9-3).
2. *The new protein may be equivalent to the original one.* Many proteins have regions whose exact amino acid sequence is relatively unimportant. In beta-globin, the amino acids on the outside of the protein must be hydrophilic to keep the protein dissolved in the cytoplasm of red blood cells, but it doesn't usually matter *which* hydrophilic amino acids are on the outside. For example, a family in the Japanese town of Machida was found to contain a mutation from CTC to GTC, replacing glutamic acid (hydrophilic) with glutamine

TABLE 9-3 Effects of Mutations in the Hemoglobin Gene

	DNA (template strand)	mRNA	Amino Acid	Properties of Amino Acid	Functional Effect on Protein	Disease
Original codon 6	CTC	GAG	Glutamic acid	Hydrophilic	Normal protein function	None
Mutation 1	CT**T**	GA**A**	Glutamic acid	Hydrophilic	Neutral, normal protein function	None
Mutation 2	**G**TC	**C**AG	Glutamine	Hydrophilic	Neutral, normal protein function	None
Mutation 3	C**A**C	G**U**G	Valine	Hydrophobic	Loses water solubility; compromises protein function	Sickle-cell anemia
Original codon 17	TTC	AAG	Lysine	Hydrophilic	Normal protein function	None
Mutation 4	**A**TC	**U**AG	Stop codon	Ends translation after amino acid 16	Synthesizes only part of protein; eliminates protein function	Beta-thalassemia

(also hydrophilic; see mutation 2 in Table 9-3). Hemoglobin containing this mutant beta-globin protein functions well. This type of mutation is called a neutral mutation, because it doesn't significantly change the function of the encoded protein.

3. *Protein function may be changed by an altered amino acid sequence.* A mutation from CTC to CAC replaces glutamic acid (hydrophilic) with valine (hydrophobic; see mutation 3 in Table 9-3). This substitution is the genetic defect that causes sickle-cell anemia (see Chapter 11). The valines on the outside of the hemoglobin molecules cause them to clump together, distorting the shape of the red blood cells.

4. *Protein function may be destroyed by a premature stop codon.* A particularly catastrophic mutation occasionally occurs in the seventeenth codon of the beta-globin gene (TTC in DNA, AAG in mRNA). This codon specifies the amino acid lysine. A mutation from TTC to ATC (UAG in mRNA) results in a stop codon, halting translation of beta-globin mRNA before the protein is completed (see mutation 4 in Table 9-3). People who inherit this mutant gene from both their mother and their father do not synthesize any functional beta-globin protein; instead, they manufacture hemoglobin consisting entirely of alpha-globin subunits. This "pure alpha" hemoglobin does not bind oxygen very well. This condition, called beta-thalassemia, can be fatal unless treated with blood transfusions.

Mutations Are the Raw Material for Evolution

Mutations are the ultimate source of all genetic differences among individuals, providing raw material for evolution. Without mutations, all individuals would share the same DNA sequence. Although most mutations are neutral or harmful to the individuals that carry them, a mutation may occasionally improve an individual's ability to survive and reproduce. If such a beneficial mutation is in a cell that gives rise to gametes (sperm or eggs), it may be passed on to the next generation. If organisms possessing the mutant gene produce more offspring than individuals that lack the mutation, the mutant gene and the characteristics it bestows will become more common over time. This process, known as natural selection, is a major cause of evolutionary change, which is described in Unit Three.

9.7 Are All Genes Expressed?

Each cell contains many genes, but only some of them are active at any given time. For example, all of the 25,000 to 30,000 genes in the human genome are present in each body cell, but individual cells *express* (transcribe and translate) only a small fraction of these genes. The particular set of genes that is expressed depends on the type of cell and the needs of the organism. This regulation of gene expression is crucial for proper functioning of individual cells and entire oganisms.

Gene Expression Differs from Cell to Cell and over Time

In organisms with more than one cell, the set of genes that is expressed largely depends on the function of a cell. For example, in humans and other mammals, the cells of hair follicles synthesize the proteins from which hair is made. Muscle cells, in contrast, synthesize large amounts of the proteins actin and myosin, which are necessary for muscle contraction, but they do not synthesize hair proteins.

Gene expression also changes over time, depending on the body's needs from moment to moment. For example, immediately after a baby's birth, the milk-producing cells in a woman's breasts begin to express the gene that encodes casein, the major protein in milk. This change in gene expression allows the mother to produce large amounts of protein-rich milk to feed her baby. Circumstances may even dictate that some genes are never expressed. A human male, for example, does not express the casein gene. However, he will pass a copy of this gene to his daughters, who will express it if they give birth.

Environmental Cues Influence Gene Expression

Changes in an organism's environment also help determine which genes are transcribed. In birds living in temperate climates, for example, the longer days of spring stimulate the sex organs (testes or ovaries) to enlarge and produce sex hormones. The sex hormones in turn cause the birds to produce eggs and sperm, to sing, to mate, and to build nests. The proliferation of cells in the sex organs, the production of hormones by these cells, and the effects of those hormones on other cells throughout the body all result, directly or indirectly, from changes in gene expression.

9.8 How Is Gene Expression Regulated?

A cell may regulate gene expression in many different ways. It may alter the rate of transcription of mRNA, how long a given mRNA molecule lasts before it is broken down, how fast the mRNA is translated into protein, how long the protein lasts, or how fast a protein enzyme catalyzes a reaction. Here, we will describe a few ways in which transcription is regulated.

Regulatory Proteins That Bind to Promoters Alter the Transcription of Genes

As we described earlier in this chapter, transcription is often regulated by molecules that enhance or inhibit the ability of RNA polymerase to bind to the promoter region of a gene. Many steroid hormones accomplish their functions by acting in this manner (see Chapter 23 for a complete description). In birds, for example, the sex hormone estrogen regulates expression of the gene for albumin (the protein in egg whites). The albumin gene is not transcribed in winter, when birds are not breeding and estrogen levels are low. During the breeding season,

HEALTH WATCH

MUTATIONS AND GENDER

Sometime in her early to mid-teens, a girl goes through puberty: Her breasts swell, her hips widen, and she begins to menstruate. In rare instances, however, a girl may develop all of the outward signs of womanhood but not menstruate. Eventually, when it becomes clear that she isn't merely developing a bit late, she reports this symptom to her physician, who may take a blood sample to do a chromosome test. In some cases, the chromosome test gives what might seem to be an impossible result: The girl's sex chromosomes are XY, a combination that would normally give rise to a boy. The reason she has not begun to menstruate is that she lacks ovaries and a uterus, but instead has testes that have remained inside her abdominal cavity. She has about the same concentrations of *androgens* (male sex hormones, such as testosterone) circulating in her blood as would be found in a boy her age. In fact, androgens, produced by the testes, have been present since early in her development. The problem is that her cells cannot respond to them—a rare condition called *androgen insensitivity*.

Androgen insensitivity is caused by a mutation in a gene located on the X chromosome. The affected gene codes for a protein known as an androgen receptor. In normal males, androgen receptor proteins are present in the cytoplasm of many body cells. Male hormones such as testosterone bind to the receptor molecules. The hormone-receptor combination enters the cell nucleus and binds to DNA, stimulating the transcription of genes that help to produce many male features, including the formation of a penis, the descent of the testes into sacs outside the body cavity, and sexual characteristics that develop at puberty, such as a beard and increased muscle mass.

A person with XY chromosomes who inherits a mutant androgen receptor gene will be unable to make a functional androgen receptor protein. Her cells will be unable to respond to the testosterone that the testes produce, and male characteristics will not develop. Thus, a mutation that changes the nucleotide sequence of a single gene, causing a single type of defective protein to be produced, can cause a person who is genetically male (XY) to look like and perceive herself to be a woman (Fig. E9-1).

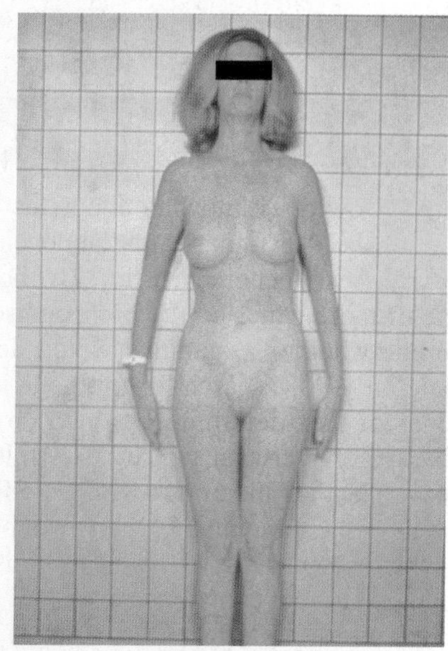

Figure E9-1 Androgen insensitivity leads to female features This individual looks female but is genetically male because she has an X and a Y chromosome. As a result, she has testes (located in her abdomen) that produce testosterone, and she lacks ovaries and a uterus. A mutation in her androgen receptor gene prevents her cells from responding to the testosterone produced by her testes and causes her obviously female appearance.

however, estrogen enters cells in the female reproductive system and binds to a receptor protein in the cytoplasm. The estrogen-receptor combination then enters the nucleus and attaches to the DNA in a region near the promoter of the albumin gene. This attachment makes it easier for RNA polymerase to bind to the promoter. As a result, the cells begin to transcribe large amounts of albumin mRNA, which is translated into the albumin protein needed to make eggs.

Some Regions of Chromosomes Are Condensed and Not Normally Transcribed

Certain parts of eukaryotic chromosomes are in a highly condensed, compact state in which most of the DNA is inaccessible to RNA polymerase. Some of these tightly condensed regions contain genes that are not currently being transcribed. When the product of a gene is needed, the portion of the chromosome containing that gene becomes "decondensed"—loosened so that the nucleotide sequence is accessible to RNA polymerase and transcription can occur.

Entire Chromosomes May Be Inactivated and Not Transcribed

In some cases an entire chromosome may be condensed, making it largely inaccessible to RNA polymerase. An example occurs in the sex chromosomes of female mammals. Male mammals usually have an X and a Y chromosome (XY), and females usually have two X chromosomes (XX). As a consequence, females have the capacity to synthesize twice as much protein from the genes on their two X chromosomes as males, with only one X chromosome. Twice as much is generally unnecessary, and might even be harmful. Consequently, one of the X chromosomes is condensed into a tight mass, in which most of its genes cannot be transcribed.

SNAKES AND SNAILS AND PUPPY DOGS' TAILS *REVISITED*

How does knowing about the control of gene expression help us understand the development of male and female sexual characteristics? By the 1930s, biologists knew that one or more genes on the Y chromosome are essential for determining whether a mammal would develop into a male or a female. In 1990, a search for this gene led to discovery of the SRY gene, for "*s*ex-determining *r*egion on the *Y* chromosome." If the SRY gene is crucial for sex determination, then one might hypothesize that mammals with an SRY gene will be male, and mammals without an SRY gene will be female.

In an elegant test of this hypothesis, researchers took fertilized mouse eggs that contained two X and no Y chromosomes. They injected mouse SRY genes into some of these fertilized eggs and implanted them into surrogate mothers. Sure enough, the mice that developed from these "transgenic" eggs had male sexual characteristics: They developed a penis and testes and behaved like male mice. Mice that lacked a functional SRY gene developed as females, regardless of whether they had two X chromosomes or an X and a Y. The conclusion: Male (XY) mammals have all the genes needed to be female but usually aren't, because they have an SRY gene. Likewise, female (XX) mammals have all the genes needed to be male *except* an SRY gene, so they are usually female.

How does the SRY gene exert such enormous effects? For a few weeks, early in embryonic development, the SRY gene is transcribed in cells that will become the testes. The SRY protein binds to the promoter region and enhances transcription of other genes whose products are essential to testicular development. Once formed, the testes in the embryo secrete testosterone, which stimulates transcription of still other genes, leading to development of the penis and scrotum. Will a child be a boy or a girl? It depends on the carefully regulated expression of many genes, with a single gene, SRY, serving as the initial genetic switch to activate male development.

CONSIDER THIS: We have seen two different ways in which a person with XY sex chromosomes can develop as a female instead of a male: The Y chromosome may have a defective SRY gene, or the X chromosome may have a defective androgen receptor gene. Suppose a 16-year-old girl approaches a physician. The girl is tearful and frightened because she has never menstruated and wants to know what is wrong with her. The doctor orders a chromosome test, and perhaps hormone tests as well. These tests reveal that she has both X and Y chromosomes, but she either has androgen insensitivity or lacks a functional SRY gene.

What should the physician tell her? Clearly, she has to be told that she has no uterus, will never menstruate, and cannot bear children. But beyond that, what? To most people, a person with two X chromosomes is a female, and one with one X and one Y chromosome is a male, and that's that. Should the doctor tell her that she is genetically male although physiologically female? What will this knowledge do to her self-image and consequently her psychological health? What would *you* do? To see how one physician handled this dilemma, see "The Curse of the Garcias," by Robert Marion, in *Discover* magazine, December 2000.

LINKS TO LIFE: Antibiotic Resistance

All life on Earth is related through evolution, sometimes closely (dogs and foxes), sometimes distantly (bacteria and people). Mutations occur constantly, usually at a very low rate. Distantly related organisms may have had a common ancestor millions of years ago. But a lot of mutations may have occurred since then, so that the genes of these organisms now differ by many nucleotides. Medicine takes advantage of these differences to develop antibiotics to treat bacterial infections.

Streptomycin and neomycin, commonly prescribed antibiotics, kill certain bacteria by binding to a specific sequence of RNA in the small subunits of the bacterial ribosomes, inhibiting protein synthesis. Without adequate protein synthesis, the bacteria die. Patients infected by these bacteria don't die when given streptomycin, however, because the small subunits of their eukaryotic ribosomes have a different nucleotide sequence than the bacteria's prokaryotic ribosomes do.

You have probably heard of *antibiotic resistance*, in which bacteria that are frequently exposed to antibiotics evolve defenses against those antibiotics. Bacteria evolve resistance against streptomycin and related antibiotics rather rapidly. Why? It's actually pretty straightforward. If eukaryotic ribosomes are insensitive to streptomycin, then eukaryotic ribosomes must function perfectly well with a different RNA sequence than prokaryotic ribosomes have. Bacteria that are resistant to some of streptomycin's chemical relatives have a mutation that changes just a single nucleotide in their ribosomal RNA from adenine to guanine, which is precisely the nucleotide found at the comparable position in eukaryotic ribosomal RNA.

Genetics, mutations, the mechanisms of protein synthesis, and evolution are important not only to biologists but to physicians, too. In fact, a whole discipline of medicine, called *evolutionary medicine*, has arisen, which uses the evolutionary relationships between people and microbes to help fight disease.

Chapter Review

Summary of Key Concepts

9.1 How Is the Information in DNA Used in a Cell?

Genes are segments of DNA that can be transcribed into RNA and, for most genes, translated into protein. Each gene usually codes for a single protein.

9.2 What Is the Genetic Code?

The genetic code consists of codons, sequences of three bases in messenger RNA, that specify either an amino acid in a protein chain, the beginning of protein synthesis (start codon), or the end of protein synthesis (stop codon).

9.3 How Is the Information in a Gene Transcribed into RNA?

Using a strand of DNA as a template, transcription produces a strand of ribonucleic acid (RNA). Within an individual cell, only certain genes are transcribed. During transcription, RNA polymerase binds to the promoter region of the DNA of a gene and synthesizes a single strand of RNA. This RNA is complementary to the template strand in the gene's DNA double helix.

9.4 What Are the Functions of RNA?

There are three types of RNA: messenger RNA (mRNA), ribosomal RNA (rRNA), and transfer RNA (tRNA). mRNA carries the genetic information of a gene from the nucleus to the cytoplasm, where ribosomes use the information to synthesize a protein. Ribosomes contain rRNA and proteins organized into large and small subunits. Ribosomes are the organelles at which mRNA will be translated into protein. There are many different tRNAs. Each tRNA binds a specific amino acid and carries it to a ribosome for incorporation into a protein.

9.5 How Is the Information in Messenger RNA Translated into Protein?

The start codon of mRNA binds to the small subunit of a ribosome, along with the corresponding (methionine) tRNA. A large subunit binds to the small subunit, forming the complete protein-synthesizing machine. tRNAs deliver the appropriate amino acids to the ribosome for incorporation into the growing protein. Base pairing between the anticodons of the tRNAs and the codons of the mRNA puts the amino acids in the order coded in the mRNA. Two tRNAs, each carrying an amino acid, bind simultaneously to the ribosome's large subunit, which then catalyzes the formation of peptide bonds between the amino acids. As each new amino acid is attached, one tRNA detaches, and the ribosome moves down the mRNA one codon and binds to another tRNA that carries the amino acid specified by that codon. Addition of amino acids to the growing protein continues until a stop codon is reached, signaling the ribosome to disassemble and to release both the mRNA and the newly formed protein.

9.6 How Do Mutations Affect Gene Function?

A mutation is a change in the nucleotide sequence of a gene. Mutations can be caused by mistakes in base pairing during replication, by chemical agents, and by environmental factors such as radiation. Common types of mutations include changes in a nucleotide base pair (point mutations) and insertions or deletions of nucleotide base pairs. Mutations are generally neutral or harmful, but in rare cases a beneficial mutation will encode features that will be favored by natural selection.

9.7 Are All Genes Expressed?

To be expressed, a gene is transcribed and translated, and the resulting protein performs some action within the cell. Which genes are expressed in a cell at any given time depends on the function of the cell, the developmental stage of the organism, and the environment.

9.8 How Is Gene Expression Regulated?

Cells regulate gene expression by altering the rate of transcription of mRNA, how long a given mRNA molecule lasts before it is broken down, how fast the mRNA is translated into protein, how long the protein lasts, and how fast a protein enzyme catalyzes a reaction. The rate of transcription may be regulated by stimulating or repressing transcription of an individual gene, by condensing or exposing large parts of chromosomes, or by condensing entire chromosomes.

Key Terms

anticodon *p. 131*
codon *p. 127*
deletion mutation *p. 134*
gene *p. 126*
genetic code *p. 127*
insertion mutation *p. 134*
messenger RNA (mRNA) *p. 126*
mutation *p. 133*
nucleotide substitution *p. 134*
point mutation *p. 134*
promoter *p. 129*
ribonucleic acid (RNA) *p. 126*
ribosomal RNA (rRNA) *p. 126*
ribosome *p. 131*
RNA polymerase *p. 129*
start codon *p. 128*
stop codon *p. 129*
template strand *p. 129*
transcription *p. 126*
transfer RNA (tRNA) *p. 126*
translation *p. 126*

Thinking Through the Concepts

Multiple Choice

1. *A gene*
   a. is synonymous with a chromosome
   b. is composed of mRNA
   c. is a specific segment of nucleotides in DNA
   d. always encodes information for the synthesis of a protein
   e. specifies the sequence of nutrients required by the body
2. *Which of the following is a single-stranded molecule that contains the information for the assembly of a specific protein?*
   a. transfer RNA
   b. messenger RNA
   c. DNA
   d. promoter
   e. ribosomal RNA
3. *Anticodon is the term applied to*
   a. the list of amino acids that corresponds to the genetic code
   b. the concept that multiple codons sometimes code for a single amino acid
   c. the part of the tRNA that interacts with the codon
   d. the several three-nucleotide stretches that code for "stop"
   e. the part of the tRNA that binds to an amino acid
4. *DNA*
   a. takes part directly in protein synthesis by leaving the nucleus and being translated on the ribosome
   b. takes part indirectly in protein synthesis; the DNA itself stays in the nucleus
   c. has nothing to do with protein synthesis; it is involved only in cell division
   d. is involved in protein synthesis within the nucleus
   e. codes for mRNA but not for tRNA or rRNA
5. *Synthesis of a protein based on the sequence of messenger RNA*
   a. is catalyzed by DNA polymerase
   b. is catalyzed by RNA polymerase
   c. is called translation
   d. is called transcription
   e. occurs in the nucleus
6. *The promoter of a gene*
   a. consists of nucleotides of DNA to which RNA polymerase binds for transcription
   b. is an organelle involved in protein synthesis
   c. is a stop codon
   d. consists of nucleotides of DNA to which DNA polymerase binds for replication
   e. consists of the nucleotides of DNA that code for ribosomal RNA

Review Questions

1. How does RNA differ from DNA?
2. What are the three types of RNA? What is the function of each?
3. Define the following terms: genetic code, codon, anticodon. What is the relationship among the bases in DNA, the codons of mRNA, and the anticodons of tRNA?
4. Diagram and describe protein synthesis.
5. Explain how both transcription and translation require complementary base pairing.
6. Describe some ways in which gene expression is regulated.
7. Define mutation, and give one example of how a gene might mutate. Would you expect most mutations to be beneficial or harmful? Explain your answer.

Applying the Concepts

1. **IS THIS SCIENCE?** About 40 years ago, some researchers reported that they could transfer learning from one animal (*Planaria*, a type of flatworm) to another by feeding trained animals to untrained animals. Further, they claimed that RNA was the active molecule of learning. Given your knowledge of the roles of RNA and protein in cells, do you think that a *specific* memory (for example, remembering the base sequences of codons of the genetic code) could be encoded by a *specific* molecule of RNA and that this RNA could transfer that memory to another person? In other words, someday, could you learn biology by popping an RNA pill? If you could, how would the pill work? If you think the RNA pill wouldn't work, can you propose a reasonable hypothesis for the *Planaria* results? How would you test your hypothesis?
2. As you learned in this chapter, many factors influence gene expression, including hormones. The use of anabolic steroids and growth hormones among athletes has created controversy in recent years. Hormones certainly affect gene expression, but, in the broadest sense, so do vitamins and foods. What do you think are appropriate guidelines for the use of hormones? Should athletes take steroids and growth hormones? Should children at risk of being unusually short be given growth hormones? Should parents be allowed to request growth hormones for their children of normal height in the hope of producing a future basketball player?

For More Information

Gibbs, W. W. "The Unseen Genome: Beyond DNA." *Scientific American*, December 2003. Gene expression can be regulated across generations by modifying the nucleotides of DNA.

Marion, R. "The Curse of the Garcias." *Discover*, December 2000. How one physician diagnosed and counseled a patient with androgen insensitivity.

Nirenberg, M. W. "The Genetic Code: II." *Scientific American*, March 1963. Nirenberg describes some of the experiments in which he deciphered much of the genetic code.

Tjian, R. "Molecular Machines That Control Genes." *Scientific American*, February 1995. Complexes of proteins regulate which genes are transcribed in a cell and therefore help determine the cell's structure and function.

WEB TUTORIAL

To access a Web Tutorial visit http://www.prenhall.com/audesirk4. *Log in to the Web site selected by your instructor, navigate to this chapter, and select the appropriate Web Activity number.*

9.1 Transcription

Estimated time: 10 minutes

The genetic information that is stored in the DNA located in the nucleus of a cell must be conveyed to the cytoplasm, where that information can be translated into a specific sequence of amino acids. The intermediary molecule is RNA, synthesized by the process of transcription. This tutorial investigates the steps involved in transcription.

9.2 Translation

Estimated time: 10 minutes

Transcription produces an RNA transcript of the gene. Now it is the task of the cell to take the information encoded in this messenger RNA and translate it into a specific sequence of amino acids. This occurs on the ribosomes and involves a number of steps illustrated in this tutorial.

9.3 Web Investigation: Snakes and Snails and Puppy Dogs' Tails

Estimated time: 10 minutes

Everyone knows that chromosomes determine gender. Baby girls have two X chromosomes (XX) while baby boys are XY. But what about individuals with XO, XXY, or XYY genomes? Why are there female babies with XY karyotypes (chromosome sets)? Gender biology is much more complicated than one might think.

Chapter 10

The Continuity of Life: How Cells Reproduce

CC the cloned cat (right) and her genetic donor, Rainbow (left), have different fur colors and patterns, even though they share exactly the same genes.

Cloning Controversy

A few years ago, you probably saw the photo of CC, a gray and white kitten peering out of a beaker, on television or in your local newspaper. CC is the first cloned cat. As you probably know, *cloning* is the creation of organisms that are genetically identical to a preexisting organism. Why clone a cat? Well, in 1997, John Sperling (founder of the University of Phoenix) and Joan Hawthorne realized that their beloved spayed mutt, Missy, was getting old. They felt that Missy's genetic makeup was very special, and they didn't want to lose it when Missy died. Hoping to obtain another dog with those genes, Sperling donated $3.7 million to Texas A&M University to develop the technology needed to clone dogs.

The first fruit of the Missyplicity Project was a cat, not a dog, because cats are easier to clone. Even so, CC was the only successful birth out of 87 cloned embryos (though current methods seem to have a higher success rate). There have been many attempts to clone Missy, but as of mid-2004, none have succeeded. Meanwhile, Missy died on July 6, 2002, at the age of 15. Many of her cells remain frozen, awaiting future cloning efforts.

The ultimate goal of the Missyplicity Project is to clone a dog, Missy.

Why can't Sperling and Hawthorne replace Missy the old-fashioned way? Missy was a Border collie–Siberian husky mix. One very simple hypothesis might be that breeding a Siberian husky with a Border collie might produce another Missy. As you probably already know, if you did that experiment, none of the puppies would be a precise duplicate of Missy, nor would they be identical to each other. But how would you explain those results? For that matter, if the Missyplicity Project succeeds, would Missy's clone really be *exactly* the same as the original? As we describe how cells reproduce, see if you can figure out the answers to those questions.

10.1 Why Do Cells Divide?

One of the essential qualities of living things is that they grow and reproduce. Both growth and reproduction depend on **cell division**, the process by which one cell gives rise to two or more cells, usually called *daughter cells*. As you will see in this chapter, there are three distinct types of cell division. Prokaryotic cells such as bacteria divide by a process known as *binary fission* (which means simply "to split in two"). Most eukaryotic cells divide by a process called *mitotic cell division*. In both binary fission and mitotic cell division, the daughter cells are genetic duplicates of the parent cell. A third type of cell division, called *meiotic cell division*, occurs in reproductive structures of multicellular eukaryotic organisms (and in some unicellular eukaryotes). Meiotic cell division produces cells that are *not* genetic duplicates of their parent cells but contain exactly *half* of the DNA of their parents.

We will explore each of these three types of cell division in this chapter, but first let's see how cell division is used in the lives of living organisms.

Cell Division Is Required for Growth and Development

Since your conception as a single fertilized egg, mitotic cell division has produced all the cells in your body and continues every day in many organs, such as your skin and intestines. After cell division, the daughter cells may grow and divide again, or they may become specialized, or *differentiate*. Even though cells may have exactly the same DNA, during differentiation, different cell types use, or *express*, different genes (see Chapter 9). As a result, they may become very different in appearance and function. For example, differences in gene expression cause certain cells of your pancreas to churn out insulin, while other cells in the pancreas produce digestive enzymes.

Cell Division Is Required for Asexual Reproduction

When a single parent produces genetically identical offspring, by either binary fission or mitotic cell division, the process is called *asexual reproduction*. Prokaryotic organisms such as bacteria are unicellular. Therefore, when a bacterium divides by binary fission, it reproduces asexually.

Single-celled eukaryotic organisms, such as *Paramecium* in ponds (**Fig. 10-1a**) and yeast in rising bread (**Fig. 10-1b**), reproduce asexually by mitotic cell division, with two new organisms arising from each preexisting cell. Some multicellular eukaryotic organisms can also reproduce by asexual reproduction. Like its relative the sea anemone, a *Hydra* reproduces by growing a small replica of itself, called a bud, on its body (**Fig. 10-1c**). Eventually, the bud is able to live independently and separates from its parent.

Many plants and fungi are capable of reproducing both asexually and sexually. The beautiful aspen groves of Colorado, Utah, and New Mexico (**Fig. 10-1d**) develop asexually from shoots growing up from the root system of a single parent tree. Although a grove seems to be a population of separate trees, it can be considered a single individual whose multiple trunks are interconnected by a common root system. Aspen can also reproduce by seeds, which are made through sexual reproduction.

Meiotic Cell Division Is Required for Sexual Reproduction

Sexual reproduction in eukaryotic organisms occurs when offspring are produced by the fusion of **gametes** (sperm and eggs) from two parents. Specialized cells in the parent's reproductive system undergo meiotic cell division to produce daughter

Figure 10-1 Asexual reproduction by mitotic cell division ***(a)*** In unicellular microorganisms, such as the protist *Paramecium*, cell division produces two new, independent organisms. ***(b)*** Yeast, a unicellular fungus, reproduces by cell division. ***(c)*** *Hydra*, a freshwater relative of the sea anemone, grows a miniature replica of itself (a bud) on its side. When fully developed, the bud breaks off and assumes independent life. ***(d)*** Trees in an aspen grove are often genetically identical. Each tree grows up from the roots of a single ancestral tree. This photo shows three separate groves near Aspen, Colorado. In fall, the appearance of their leaves shows the genetic identity within a grove and the genetic difference between groves. **QUESTION:** *Are there other possible explanations for when the aspen leaves in these groves change color? How would you test the hypothesis that the timing of when the leaves turn is genetically determined?*

cells with exactly half the DNA of their parent cells (and of the "ordinary" body cells of the rest of the organism). In animals, these cells become gametes. In plants and fungi, these cells first undergo one or more rounds of mitotic cell division, sometimes developing into an entire multicellular organism (see Chapter 18), but eventually some of their daughter cells become gametes.

Because gametes contain exactly half the DNA of the body cells of the parent, when a sperm fertilizes an egg, together they produce an offspring that once again contains the full complement of DNA. This offspring is genetically unique, similar to both parents, but identical to neither. It grows and develops by mitotic cell division and the differentiation of the resulting daughter cells.

Any kind of cell division is just one part of the **cell cycle**, which is the orderly sequence of activities that make up the life of a cell, from one cell division to the next. Both prokaryotic and eukaryotic cells have cell cycles that include growth, DNA replication, and cell division. However, because of the structural and functional differences between these two cell types, prokaryotic and eukaryotic cell cycles differ considerably.

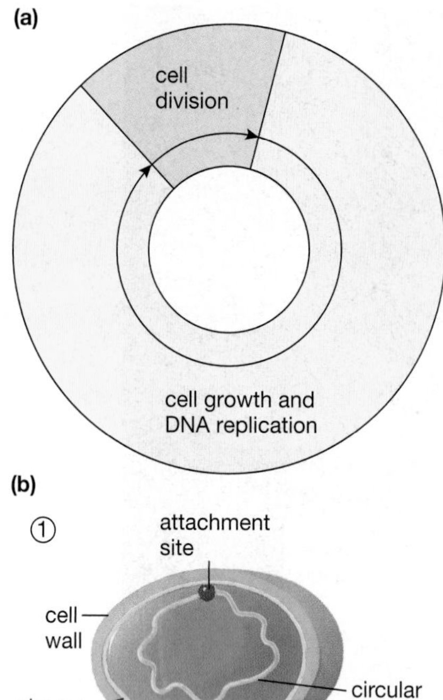

The circular DNA double helix is attached to the plasma membrane at one point.

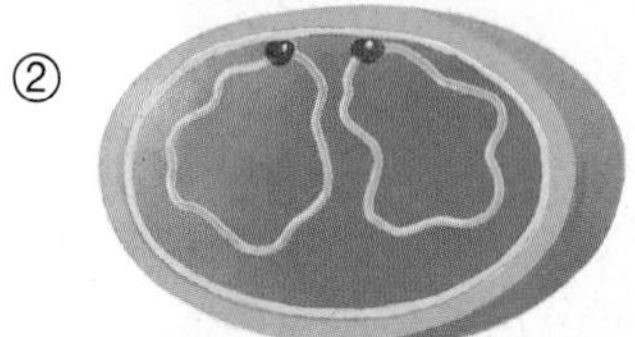

The DNA replicates and the two DNA double helices attach to the plasma membrane at nearby points.

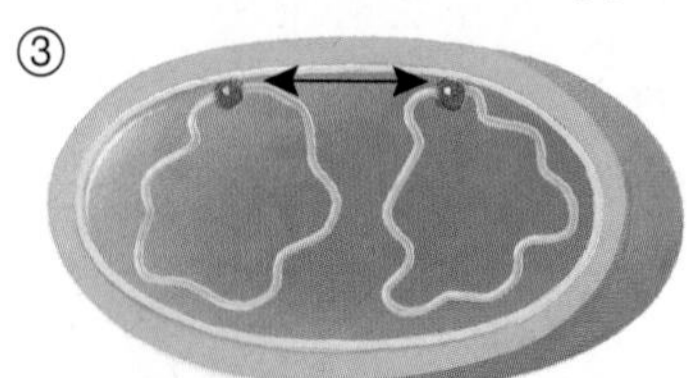

New plasma membrane is added between the attachment points, pushing them further apart.

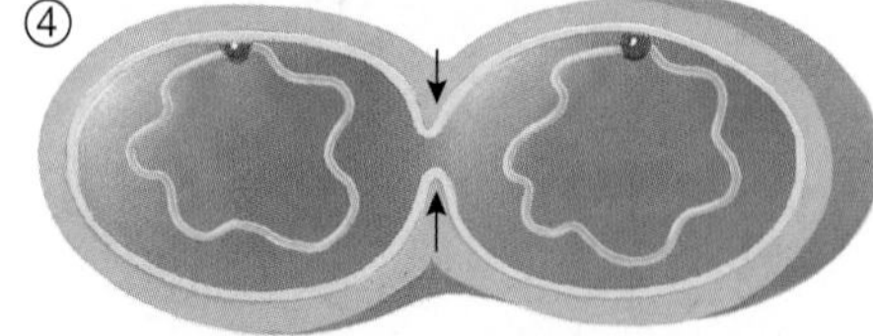

The plasma membrane grows inward at the middle of the cell.

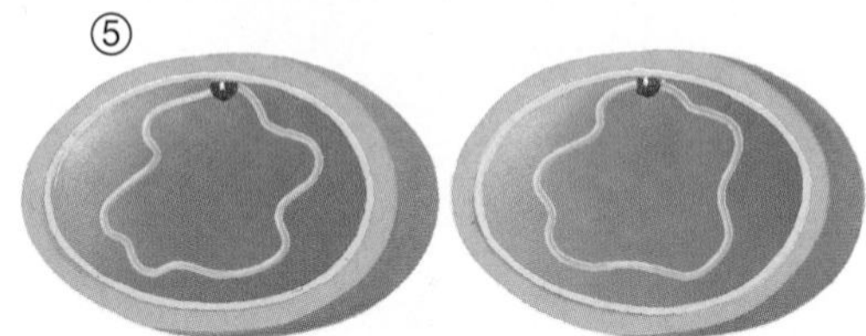

The parent cell divides into two daughter cells.

10.2 What Occurs During the Prokaryotic Cell Cycle?

The prokaryotic cell cycle consists of a relatively long period of growth—during which the cell also replicates its DNA—followed by rapid cell division (Fig. 10-2a). Prokaryotic cells divide by **binary fission**, or "splitting in two." The prokaryotic chromosome is usually a circle of DNA, attached at one point to the plasma membrane (Fig. 10-2b, step ①).

During the "growth phase" of the prokaryotic cell cycle, the DNA is replicated, producing two identical chromosomes that become attached to the plasma membrane at nearby, but separate, points (Fig. 10-2b, step ②). As the cell grows, new plasma membrane is added between the attachment points of the chromosomes, pushing them apart (Fig. 10-2b, step ③). When the cell has approximately doubled in size, the plasma membrane around the middle of the cell rapidly grows inward between the two DNA attachment sites (Fig. 10-2b, step ④). Fusion of the plasma membrane along the equator of the cell completes binary fission, producing two daughter cells, each containing one of the chromosomes (Fig. 10-2b, step ⑤). Because DNA replication produces two identical DNA molecules (except for the occasional mutation), the two daughter cells are genetically identical to one another and to the parent cell.

Under ideal conditions, binary fission in prokaryotes occurs rapidly. For example, the common intestinal bacterium *Escherichia coli* can grow, replicate its DNA, and divide in about 20 minutes. Luckily, the environment in our intestines is usually not ideal for bacterial growth. Otherwise, the bacteria would soon outweigh the rest of our bodies!

10.3 What Occurs During the Eukaryotic Cell Cycle?

The eukaryotic cell cycle is divided into two major phases: interphase and cell division (Fig. 10-3). During **interphase**, the cell acquires nutrients from its environment, grows, and duplicates its chromosomes. During cell division, one copy of each chromosome and usually about half the cytoplasm (including mitochondria, ribosomes, and other organelles) are parceled out into each of the two daughter cells.

Most eukaryotic cells spend the majority of their time in interphase. For example, some cells in human skin, which divide about once a day, spend roughly 22 hours in interphase and only a couple of hours dividing. Immediately after cell division, a newly formed daughter cell enters interphase, during which it grows and may differentiate. During this time, the cell is sensitive to internal and external signals that may cause it to continue through the cell cycle and divide, to differentiate into a particular cell type, or even to commit "cellular suicide."

If these signals trigger continuation of the cell cycle, the cell replicates its DNA. After DNA replication, the cell may grow some more before dividing. Often, if the signals trigger differentiation, the cell exits the cell cycle. It may continue to grow in size, but it does not replicate its DNA or divide. Some differentiated cells, including most of those in your heart muscle, eyes, and brain, never divide again. This is one reason why heart attacks and strokes are so devastating: The dead cells cannot be replaced. Other differentiated cells may return to the cell cycle if they receive the necessary stimuli.

Figure 10-2 The prokaryotic cell cycle ***(a)*** The prokaryotic cell cycle consists of growth and DNA replication (yellow), followed by binary fission (blue). ***(b)*** Binary fission in prokaryotic cells.

Finally, some signals may cause cell death. For example, as an embryo, you had webbed fingers and toes, but the cells of the webbing received "death signals," so the fingers and toes were separated. Far from replicating their DNA, cells undergoing programmed cell death usually produce enzymes that cut up their DNA into little pieces.

There Are Two Types of Division in Eukaryotic Cells: Mitotic Cell Division and Meiotic Cell Division

If a eukaryotic cell continues on through the cell cycle, it may undergo either mitotic cell division or meiotic cell division. These two types of cell division are evolutionarily related, but very different. Mitotic cell division might be thought of as "ordinary" cell division, the kind that occurs during development from a fertilized egg, during asexual reproduction in *Hydra* and aspens, and in your skin, liver, and digestive tract every day. Meiotic cell division, on the other hand, is the specialized type of cell division required for sexual reproduction.

Mitotic Cell Division

Mitotic cell division consists of nuclear division (called **mitosis**) followed by cytoplasmic division (called **cytokinesis**). The word *mitosis* comes from the Greek word for "thread." During mitosis, chromosomes condense and appear as thin, threadlike structures when viewed through a light microscope. Cytokinesis (from the Greek words for "cell movement") is the process by which the cytoplasm is divided between the two daughter cells. As we will see later in this chapter, mitosis gives each daughter nucleus one copy of the parent cell's replicated chromosomes. Cytokinesis usually places one of these daughter nuclei into each daughter cell. Hence, mitotic cell division typically produces two daughter cells that are genetically identical to each other and to the parent cell, and usually contain about equal amounts of cytoplasm.

In addition to its role in asexual reproduction and the growth of multicellular organisms, mitotic cell division is also important in biotechnology. Mitosis produced the nuclei used to create CC the cat (see "Cloning Controversy" at the beginning of this chapter) and Dolly the cloned sheep, which you will read about in "Scientific Inquiry: Carbon Copies: Cloning in Nature and in the Laboratory." Because mitosis produces daughter cells that are genetically identical to the parent cell, CC and Dolly are genetically identical to their respective "nuclear donors" (the animals that provided the nuclei for each cloning procedure).

Finally, mitotic cell division also gives rise to "stem cells." These cells, which are found in both embryos and adults, may produce a wide variety of differentiated cell types, such as nerve cells, immune system cells, or muscle cells. We will discuss some medical applications and ethical implications of stem cells and cloning in Chapter 12.

Meiotic Cell Division

Meiotic cell division is required for sexual reproduction in all eukaryotic organisms. In animals, meiotic cell division occurs only in ovaries and testes. **Meiotic cell division** involves a specialized nuclear division called **meiosis** and two rounds of cytokinesis to produce four daughter cells that can become gametes (eggs or sperm). Gametes carry half of the DNA of the parent cell. Thus, the cells produced by meiotic cell division are not genetically identical to each other *or* to the original cell.

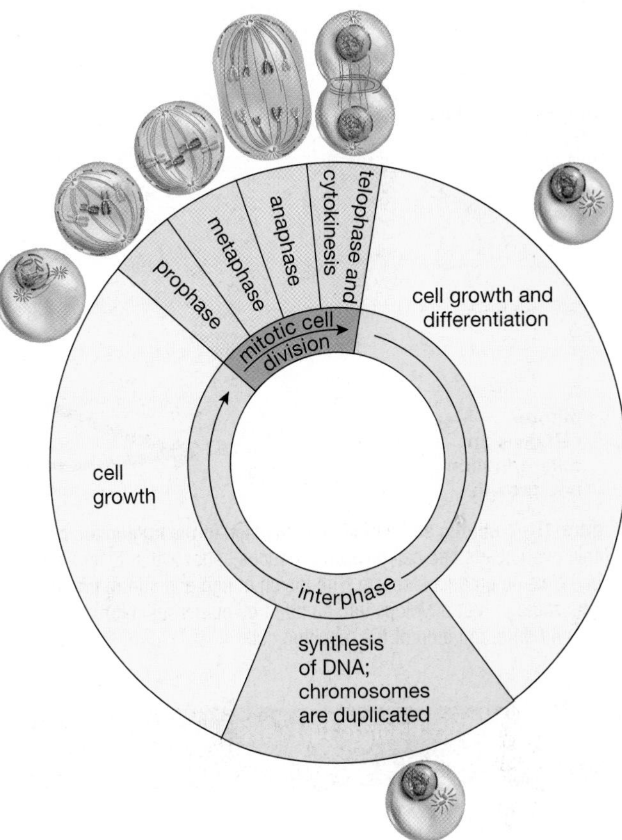

Figure 10-3 The eukaryotic cell cycle The eukaryotic cell cycle consists of interphase (yellow) and mitotic cell division (blue). Some cells remain permanently in interphase and never divide again. As a prelude to sexual reproduction, meiotic cell division replaces mitotic cell division in the cell cycle of certain specialized cells.

The Life Cycles of Eukaryotic Organisms Include Both Mitotic and Meiotic Cell Division

The life cycles of organisms can follow many different patterns, reflecting the great diversity of life on Earth. Generally, however, the life cycles of multicellular eukaryotic organisms include both meiotic and mitotic cell division

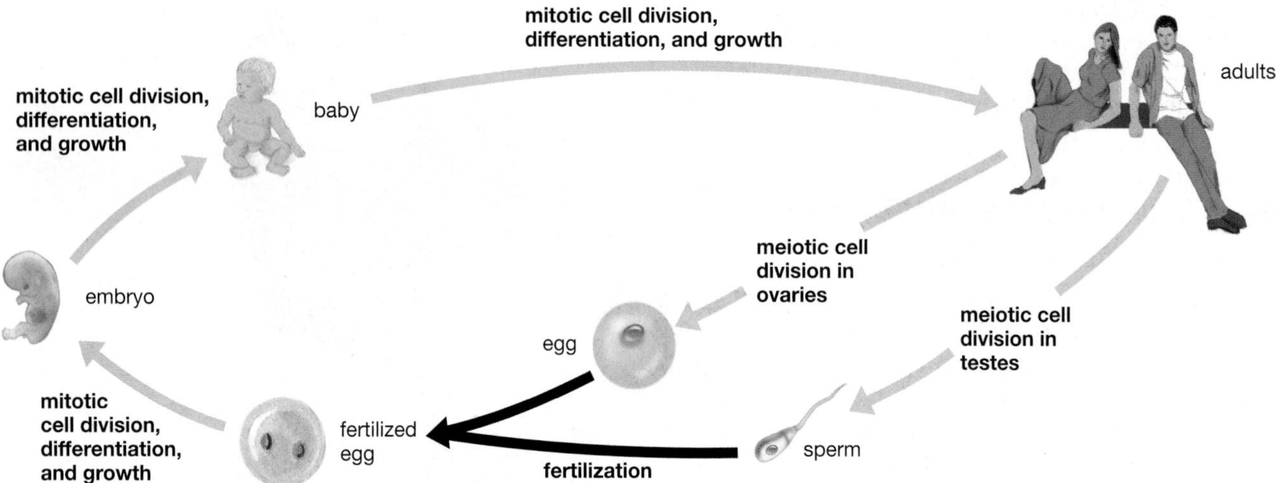

Figure 10-4 Mitotic and meiotic cell division in the human life cycle Within ovaries, meiotic cell division produces eggs; within testes, meiotic cell division produces sperm. The fusion of egg and sperm produces a fertilized egg that develops into an adult by numerous mitotic cell divisions and differentiation of the resulting cells.

(Fig. 10-4). A new generation begins when two gametes fuse, bringing together genes from two parents. Through mitotic cell division and differentiation of the resulting daughter cells, the fertilized egg grows and develops a multicellular body. At some point in the organism's life, meiotic cell division generates new gametes that can unite with other gametes to produce the next generation.

10.4 How Is DNA in Eukaryotic Cells Organized into Chromosomes?

As you know, in eukaryotic cells, DNA is contained in chromosomes in the nucleus. Therefore, to understand how cells can parcel out their DNA very precisely during mitotic and meiotic cell division, we need to look at the structure of chromosomes.

Eukaryotic Chromosomes Consist of DNA Bound to Proteins

Fitting all the DNA of a eukaryotic cell into the nucleus is no trivial task. If it were laid end to end, the DNA in a single cell of your body would be about 6 feet long, and this DNA must fit into a nucleus that is at least a million times smaller! The degree of DNA compaction, or condensation, varies at each stage of the cell cycle. During most of the life of a cell, much of the DNA is extended, making it readily accessible for transcription. In this extended state, individual chromosomes, which consist of a single DNA double helix and many associated proteins, are too thin to be visible in light microscopes.

Chromosomes Condense During Cell Division

During cell division, the chromosomes are sorted out and moved into two daughter nuclei. Just as thread is easier to organize when it is wound onto spools, sorting and transporting chromosomes are easier when they are condensed and shortened. During cell division, proteins fold up the DNA of each chromosome into compact structures that can be seen in a light microscope.

A Chromosome Contains Many Genes

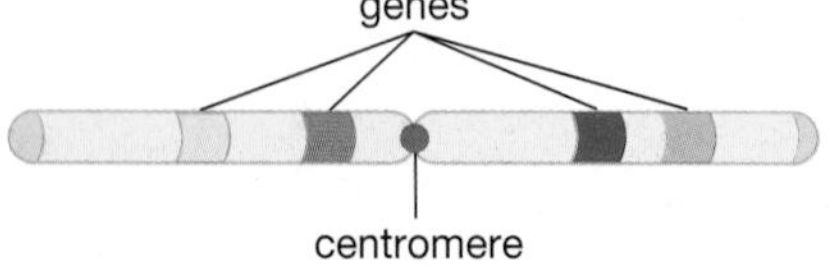

How are chromosomes and genes related? Recall that genes are sequences of DNA from hundreds to thousands of nucleotides long. A single DNA double helix may contain hundreds or even thousands of genes, each occupying a specific place on a chromosome.

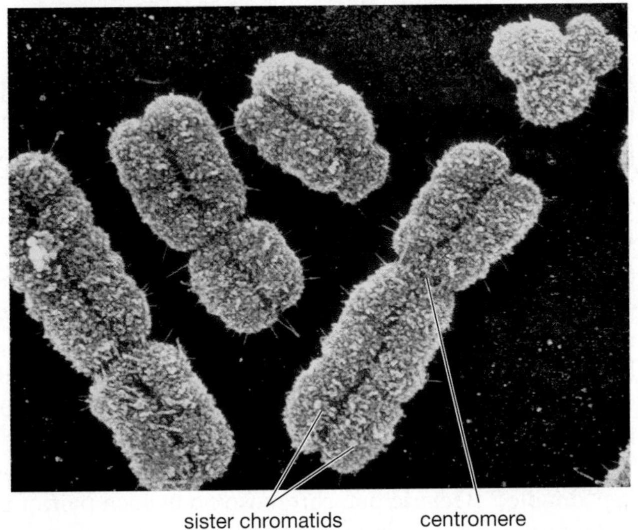

Figure 10-5 Human chromosomes during mitosis The DNA and associated proteins in these duplicated human chromosomes have coiled up into thick, short sister chromatids attached at the centromere. During cell division, the condensed chromosomes are about 5 to 20 micrometers long. At other times, the chromosomes uncoil until they are about 10,000 to 40,000 micrometers long.

Chromosomes vary in length, and therefore in the number of genes they contain. The largest human chromosome, chromosome 1, contains approximately 3000 genes, whereas one of the smallest human chromosomes, chromosome 22, contains only about 600 genes.

Duplicated Chromosomes Separate During Cell Division

When DNA is replicated before cell division (see Chapter 8), the result is a duplicated chromosome consisting of two DNA double helixes attached to each other at a specialized region called the **centromere**. While the two replicated DNA molecules remain attached at the centromere, we refer to each one as a sister **chromatid** (→). Thus, DNA replication produces a duplicated chromosome with two identical sister chromatids (Fig. 10-5).

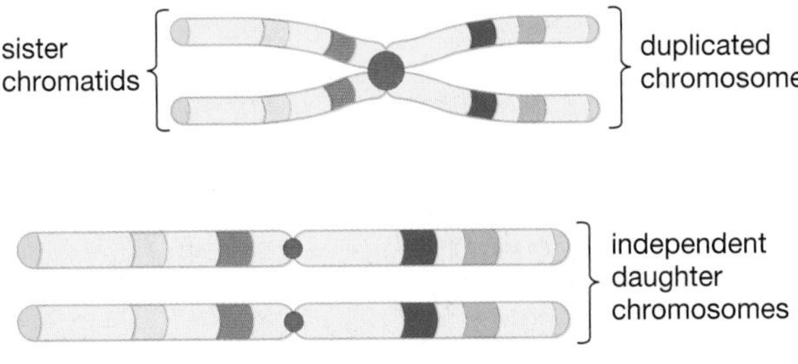

As we will see, during mitotic cell division, the two sister chromatids separate, and each chromatid becomes an independent chromosome that is delivered to one of the two daughter cells. →

Eukaryotic Chromosomes Usually Occur in Pairs

We can view an entire set of chromosomes from a single cell in a preparation called a *karyotype* (Fig. 10-6). The nonreproductive cells of many organisms, including humans, contain pairs of chromosomes. With one exception that we will discuss shortly, the two chromosomes in each pair are the same length and have

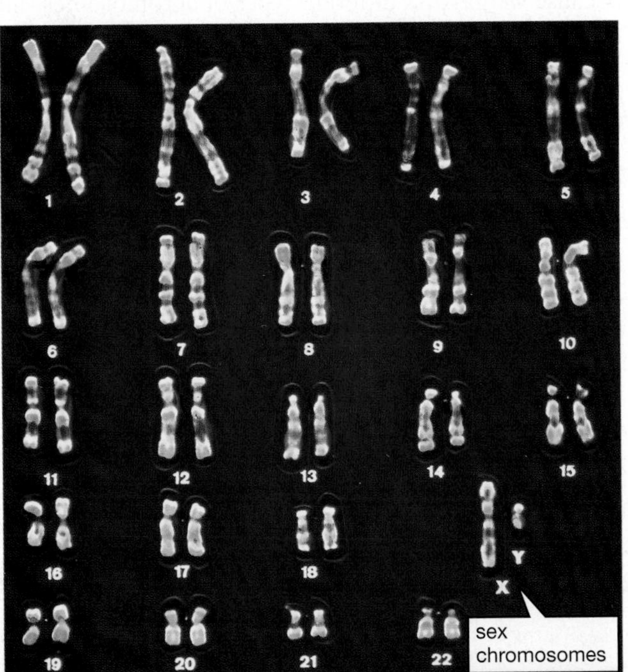

Figure 10-6 The karyotype of a human male This picture was made by staining and photographing one male's chromosomes. Individual chromosomes were then cut out of the photo and arranged in descending order of size. The two chromosomes above each number are a homologous pair. The two chromosomes of a pair are similar in size and staining patterns, and have similar genetic material. Notice that the Y chromosome is much smaller than the X chromosome. If this karyotype had been from a female, there would be two X chromosomes and no Y chromosome.

the same staining pattern. This similarity in size, shape, and staining occurs because each chromosome in a pair carries the same genes arranged in the same order. Chromosomes that contain the same genes are called *homologous chromosomes* or **homologues**, from Greek words that mean "to say the same thing."

Despite their name, homologues usually don't say *exactly* the "same thing." Why not? Mistakes in DNA replication, ultraviolet rays in sunlight, and certain chemicals in food, water, or air can cause mutations. When a mutation occurs in one homologue but not the other, their nucleotide sequences will differ. If mutations occur in the chromosomes of cells that give rise to sperm or eggs, they may be passed from parent to offspring. As a result of mutations, accumulating in a species over millions of years, no pair of homologous chromosomes has exactly the same nucleotide sequence.

Cells containing pairs of homologous chromosomes are called **diploid**, meaning "double." Because one chromosome in each pair of homologues was inherited from the mother (in her egg), these are usually called the *maternal chromosomes*. The chromosomes in each pair inherited from the father (in his sperm) are called *paternal chromosomes*.

Consider an ordinary cell of the human body, such as a skin cell. It has 23 pairs of chromosomes, for a total of 46. There are two copies (the two homologues) of chromosome 1, two copies of chromosome 2, and so on, up through chromosome 22. These chromosomes, which have similar appearance, have similar DNA sequences, and are paired in diploid cells of both sexes, are called **autosomes**. The cell also has two **sex chromosomes**: either two X chromosomes (if you're female) or an X and a Y chromosome (if you're male). The X and Y chromosomes are quite different in size (see Fig. 10-6) and genetic composition. However, as we will see, the X and Y chromosomes behave as a pair during meiotic cell division.

Not All Cells Have Paired Chromosomes

Most body cells are diploid. However, during sexual reproduction, cells in the ovaries or testes undergo meiotic cell division to produce gametes (sperm or eggs). Gametes contain only one chromosome from each pair of autosomes and one of the two sex chromosomes. Cells such as gametes that contain only one of each type of chromosome are called **haploid** (meaning "half"). In humans, a haploid cell contains one each of the 22 autosomes, plus either an X or a Y sex chromosome, for a total of 23 chromosomes. (Think of a haploid cell as one that contains *half* the diploid number of chromosomes, or one of each type of chromosome. A diploid cell contains two of each type of chromosome.)

In biological shorthand, the number of different types of chromosomes in a species is called the *haploid number* and is designated n. For humans, $n = 23$ because we have 23 different types of chromosomes (autosomes 1 to 22 plus one sex chromosome). Diploid cells contain $2n$ chromosomes. As we have seen, the body cells of humans have 46 (2×23) chromosomes. Every species has a specific number of chromosomes in its cells, from just a handful (e.g., 6 in mosquitoes) to hundreds (in shrimp and some plants).

Not all organisms are diploid. The bread mold *Neurospora*, for example, has haploid cells for most of its life cycle. Some plants, on the other hand, have more than two copies of each type of chromosome, with 4, 6, or even more copies of each chromosome in each cell.

Web Tutorial 10.2 Mitosis

10.5 How Does Mitotic Cell Division Produce Genetically Identical Daughter Cells?

As we described earlier, mitotic cell division consists of mitosis (nuclear division) and cytokinesis (cytoplasmic division). Remember, the DNA has already been replicated during interphase, so at the beginning of mitosis each chromosome consists of two sister chromatids attached to one another at the centromere.

For convenience, biologists divide mitosis into four phases, based on the appearance and behavior of the chromosomes: (1) *prophase*, (2) *metaphase*, (3) *anaphase*, and (4) *telophase* (Figs. 10-7 and 10-8). Cytokinesis usually occurs during telophase.

During Prophase, the Chromosomes Condense and Are Captured by the Spindle

During **prophase**, three major events occur: (1) the duplicated chromosomes condense; (2) the **spindle microtubules** form; and (3) the chromosomes are captured by the spindle (Figs. 10-7b and 10-8b,c).

At the beginning of prophase, the duplicated chromosomes coil up and condense. As the chromosomes condense, the nucleolus disappears.

Next, the spindle develops. In all eukaryotic cells, the proper movement of chromosomes during mitosis depends on spindle microtubules. The spindle microtubules assemble, grow outward, and eventually surround the nucleus like a football-shaped basket. The tips of the football, from which the spindle microtubules radiate, are called the *spindle poles*.

Animal cells contain a pair of microtubule-containing organelles called **centrioles** near the nucleus. Centrioles produce cilia and flagella, so ciliated or flagellated eukaryotic cells need centrioles. During interphase, a new pair of centrioles

Figure 10-7 The cell cycle in a plant cell In these light micrographs, the chromosomes are stained bluish purple and the spindle microtubules are stained pink to red. Other parts of the cells are not stained and are not visible in these photographs. Compare these photographs with the photographs and drawings of mitotic cell division in an animal cell, shown in Figure 10-8.

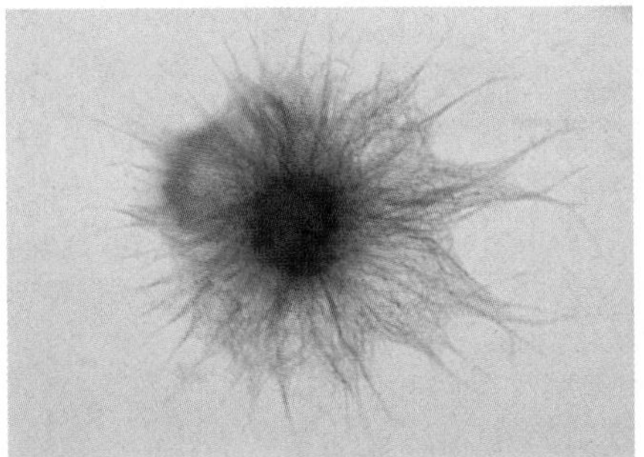

(a) Interphase in a seed cell: The chromosomes (blue) are in the thin, extended state and appear as a mass in the center of the cell. The spindle microtubules (red) extend outward from the nucleus to all parts of the cell.

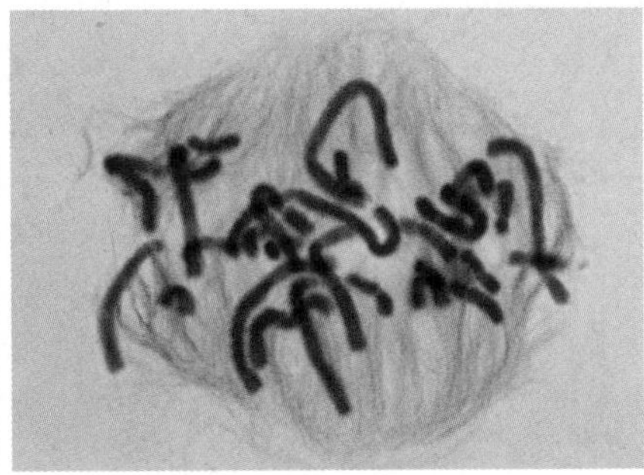

(b) Late prophase: The chromosomes (blue) have condensed and attached to the spindle microtubules (red).

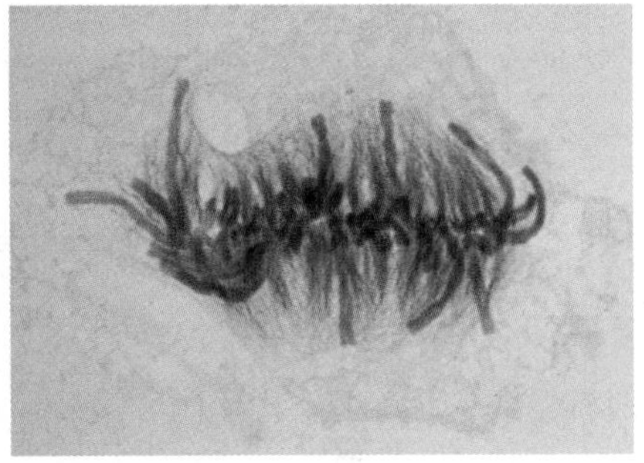

(c) Metaphase: The chromosomes have moved to the equator of the cell.

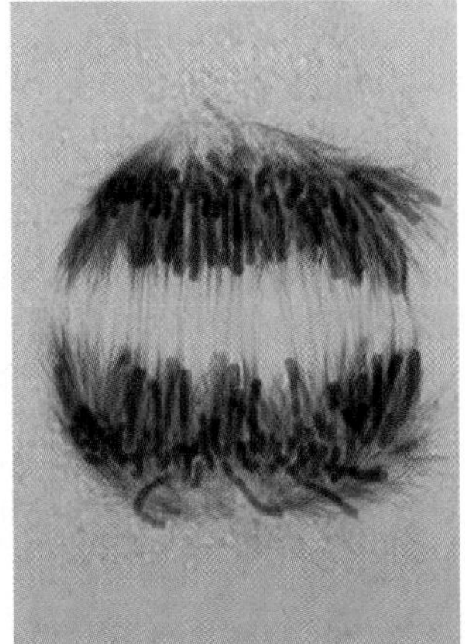

(d) Anaphase: Sister chromatids have separated, and one set has moved toward each pole.

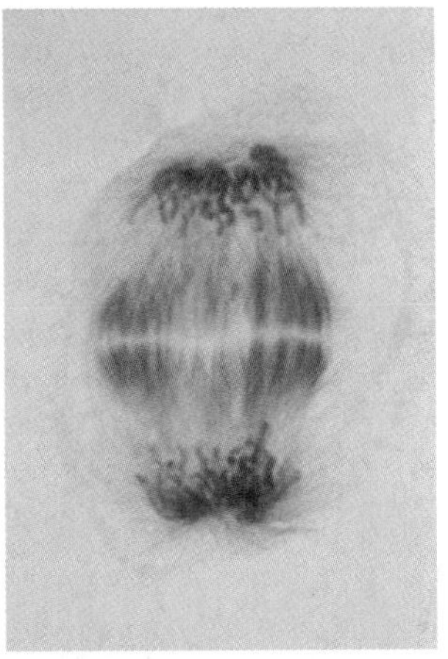

(e) Telophase: The chromosomes have gathered into two clusters, one at the site of each future nucleus.

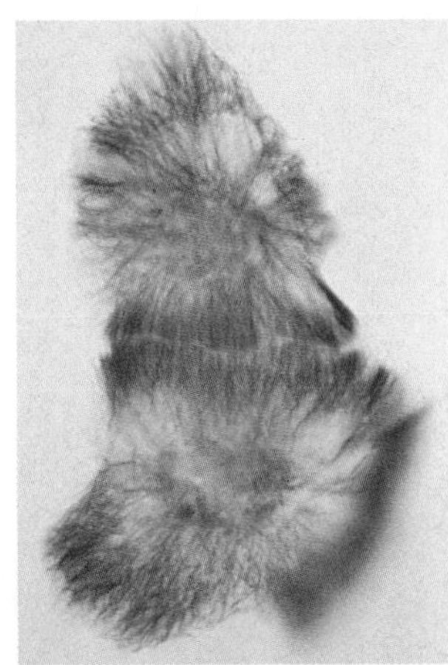

(f) Resumption of interphase: The chromosomes are relaxing again into their extended state. The spindle microtubules are disappearing, and the microtubules of the two daughter cells are rearranging into the interphase pattern.

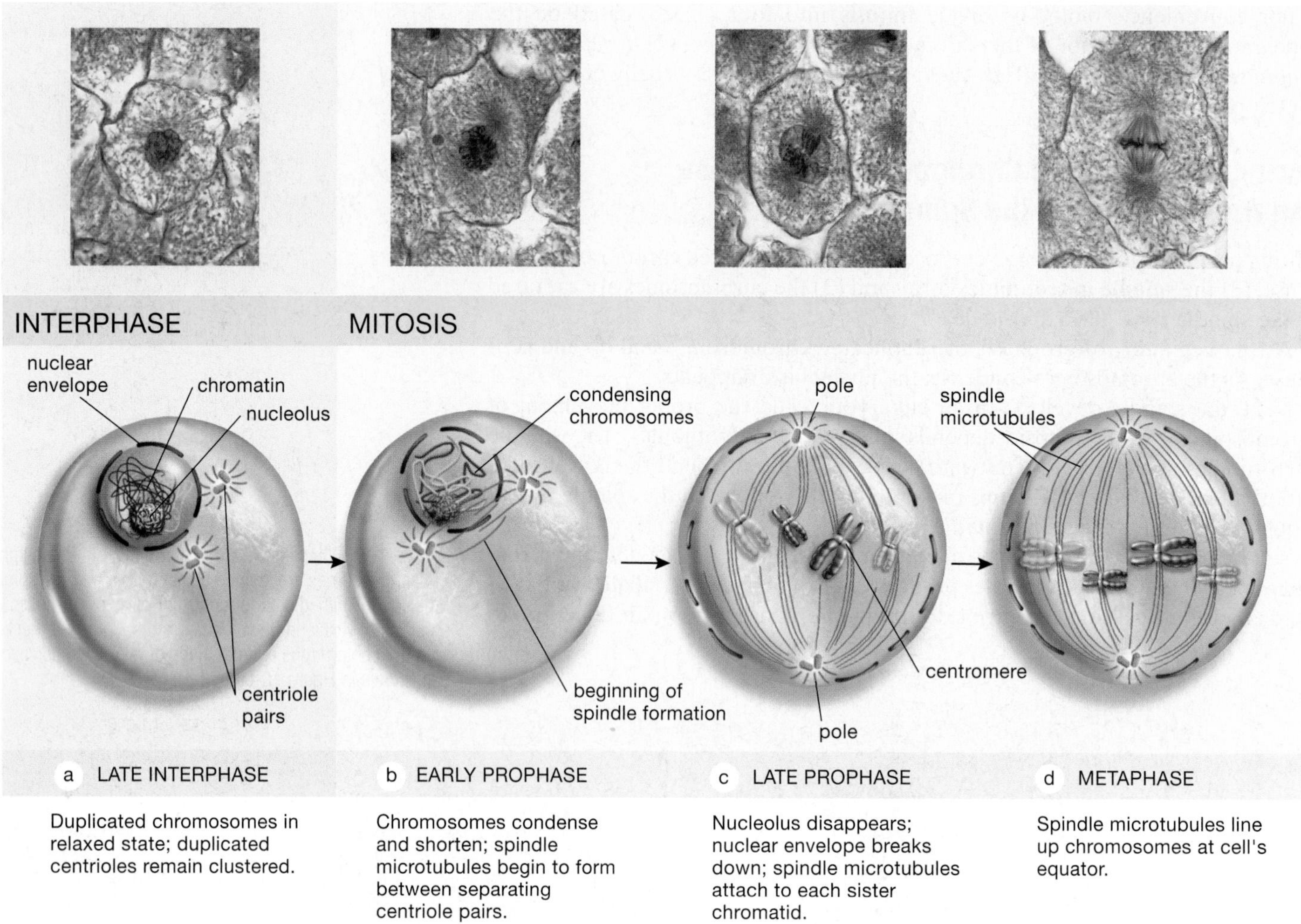

Figure 10-8 Mitotic cell division in an animal cell QUESTION: *What would be the genetic consequences for the daughter cells if one set of sister chromatids failed to separate at anaphase?*

forms near the previously existing pair (**Fig. 10-8a**). During prophase, the centriole pairs migrate with the spindle poles to opposite sides of the nucleus (**Fig. 10-8b, c**), so when the cell divides, each daughter cell receives a centriole. Although the centrioles *move with* the spindle poles, they do not *produce* the spindle microtubules. Therefore, the cells of plants and fungi, which do not contain centrioles, can make fully functional spindles.

As the spindle microtubules form a complete basket around the nucleus, the nuclear envelope disintegrates, releasing the chromosomes. Proteins located at the centromere of each chromosome bind to spindle microtubules. In each duplicated chromosome, one sister chromatid binds to microtubules leading to one spindle pole, while the other sister chromatid binds to microtubules leading to the opposite spindle pole (**Figs. 10-7b** and **10-8c**). Some spindle microtubules do not attach to chromosomes. Instead, they have free ends that overlap along the cell's equator.

During Metaphase, the Chromosomes Line Up along the Equator of the Cell

At the end of prophase, each duplicated chromosome is connected to both spindle poles. During **metaphase**, the microtubules attached to each duplicated chromosome lengthen and shorten, until each chromosome lines up along the equa-

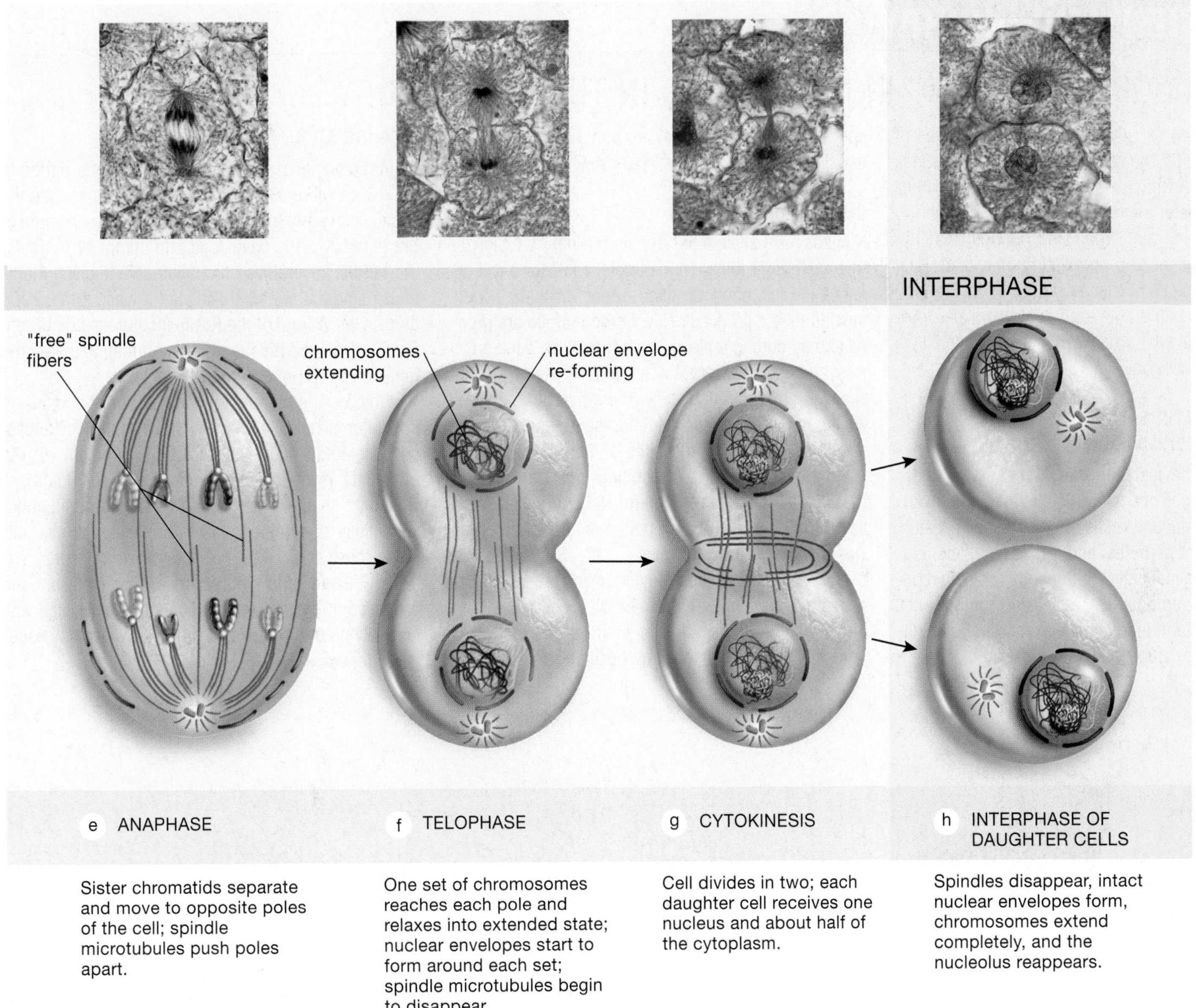

e ANAPHASE

Sister chromatids separate and move to opposite poles of the cell; spindle microtubules push poles apart.

f TELOPHASE

One set of chromosomes reaches each pole and relaxes into extended state; nuclear envelopes start to form around each set; spindle microtubules begin to disappear.

g CYTOKINESIS

Cell divides in two; each daughter cell receives one nucleus and about half of the cytoplasm.

h INTERPHASE OF DAUGHTER CELLS

Spindles disappear, intact nuclear envelopes form, chromosomes extend completely, and the nucleolus reappears.

tor of the cell, with one sister chromatid facing each pole (**Figs. 10-7c** and **10-8d**). The "free" spindle microtubules, not attached to chromosomes, still overlap at the equator.

During Anaphase, Sister Chromatids Separate and Move to Opposite Poles of the Cell

At the beginning of **anaphase** (**Figs. 10-7d** and **10-8e**), the sister chromatids separate, becoming independent daughter chromosomes. One of the two daughter chromosomes derived from each original parental chromosome moves along the spindle microtubules to each pole of the cell. At the same time, the unattached spindle microtubules lengthen, pushing the poles of the cell apart and forcing the cell into an oval shape (**Fig. 10-8e**). The two clusters of chromosomes that form at opposite poles of the cell each contain one copy of every chromosome that was in the parent cell.

SCIENTIFIC INQUIRY

CARBON COPIES: CLONING IN NATURE AND IN THE LABORATORY

The word *cloning* usually brings to mind images of Dolly the sheep, CC the cat, or even *Star Wars: Attack of the Clones*, but nature has been cloning for hundreds of millions of years. **Cloning** is the creation of one or more individual organisms (**clones**) that are genetically identical to a preexisting individual. How are clones produced, either in nature or in the lab? Why is cloning such a hot—and controversial—topic in the news? And why do we discuss cloning in a chapter on cell division?

CLONING IN NATURE: THE ROLE OF MITOTIC CELL DIVISION

Let's address the last question first. As you know, there are two types of cell division: mitotic and meiotic. Sexual reproduction relies on meiotic cell division, the production of gametes, and fertilization, and usually produces genetically unique offspring. In contrast, asexual reproduction relies on mitotic cell division. Because mitotic cell division creates daughter cells that are genetically identical to the parent cell, offspring produced by asexual reproduction are genetically identical to their parents—they're clones.

CLONING ON THE FARM

Humans have been in the cloning business a lot longer than you might think. For example, consider navel oranges, which don't produce seeds. Without seeds, how do they reproduce? Navel orange trees are propagated by cutting a piece of stem from an adult navel tree and grafting it onto the top of the root of a seedling orange tree, usually of a different type. (Why would the seedling usually not be a navel orange seedling?) Therefore, the cells of the above-ground, fruit-bearing parts of the resulting tree are clones of the original navel orange stem. Navel oranges originated from a single mutant bud of an orange tree that was discovered in Brazil in the early 1800s and has been propagated asexually ever since. Three navel orange trees were brought from Brazil to Riverside, California, in the 1870s. (One of them is still there!) All American navels are clones of these three trees.

CLONING ADULT MAMMALS

Animal cloning isn't a recent development, either. In the 1950s, John Gurdon and his colleagues inserted nuclei from early frog embryos into eggs, and some of the resulting cells developed into complete frogs. By the 1990s, several labs had been able to clone mammals using embryonic nuclei, but it wasn't until 1996 that Dr. Ian Wilmut of the Roslin Institute in Edinburgh, Scotland, cloned the first adult mammal, to create the famous sheep Dolly (see Fig. E10-1).

Why is it important to clone an *adult* animal? In agriculture, it is usually only worthwhile to clone adults, because adults show the traits that one might wish to propagate (such as increased quantity or quality of wool on a sheep). If the valuable traits of the adult are genetically determined, all of its clones will also express those traits. Cloning of embryos is usually not useful, because the embryonic cells would have been produced by sexual reproduction in the first place, and normally no one could tell if the embryo had any especially desirable traits.

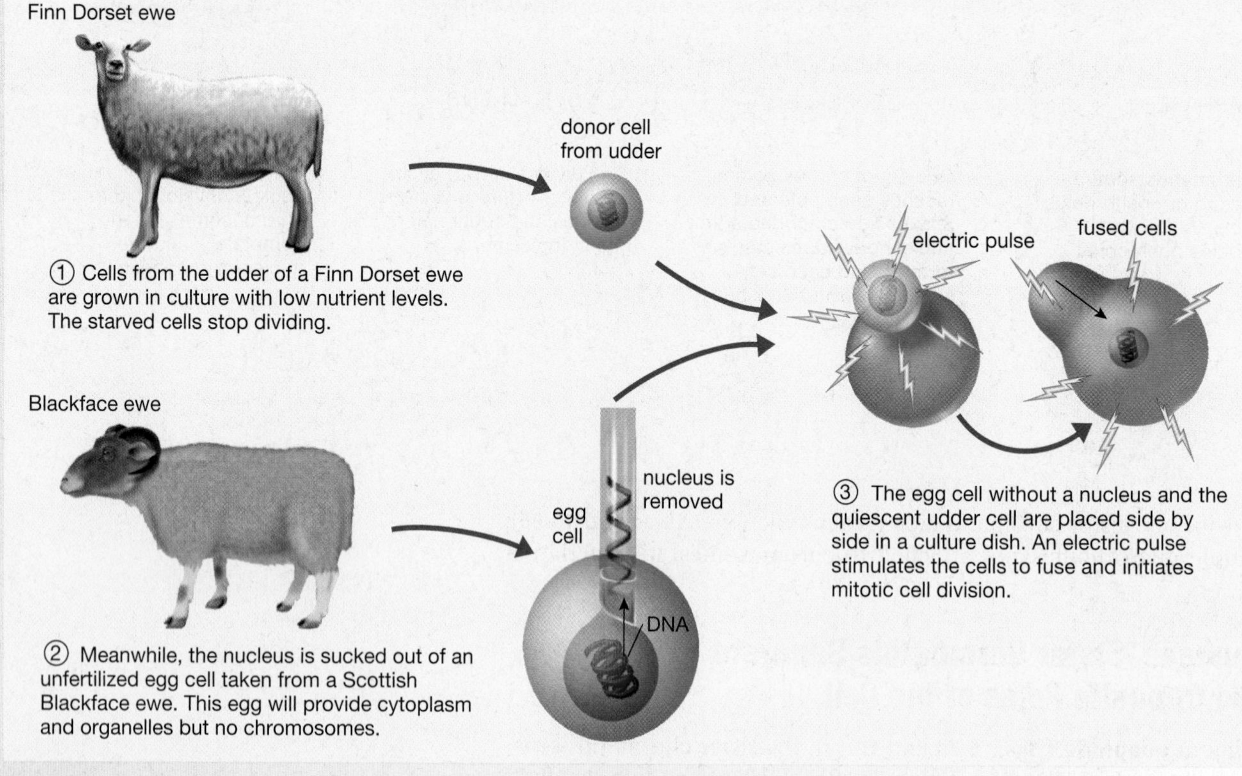

Figure E10-1 The making of Dolly

For some medical applications, too, cloning of adults is essential. Suppose that a pharmaceutical company genetically engineered a cow that secreted a valuable molecule, such as an antibiotic, in its milk (see Chapter 12). These techniques are extremely expensive and somewhat hit-or-miss, so the company may successfully produce only one profitable cow. This cow could then be cloned, creating a whole herd of antibiotic-producing cows. Cloned cows that produce more milk or meat already exist.

Cloning might also help to rescue endangered species, many of which don't reproduce well in zoos. As Richard Adams of Texas A&M, home of the Missyplicity Project, put it, "You could repopulate the world [with an endangered species] in a matter of a couple of years. Cloning is not a trivial pursuit."

CLONING: AN IMPERFECT TECHNOLOGY

Unfortunately, cloning mammals is inefficient and beset with difficulties. An egg is subjected to severe trauma when its nucleus is sucked out or destroyed and a new nucleus is inserted (Fig. E10-1). Often, the egg may simply die. Molecules in the cytoplasm that are needed to control development may be lost or moved to the wrong places, so that even if the egg survives and divides, it may not develop properly. If the eggs develop into viable embryos, the embryos must then be implanted into the uterus of a surrogate mother. Many clones die or are aborted during gestation, sometimes with fatal consequences for the surrogate mother. Even if the clone survives gestation and birth, it may have defects, commonly a deformed head, lungs, or heart. Given the high failure rate—it took 277 tries to produce Dolly, and 87 to make CC—cloning mammals is an expensive proposition.

To make things even more problematic, "successful" clones often have hidden defects. Dolly, for example, seemed to be born with "middle-aged" chromosomes. Nucleotide sequences on the ends of chromosomes, called telomeres, help to keep chromosomes intact during cell division. Every time cells divide, their telomeres shorten. When the telomeres get too short, the cells die, or at least stop dividing. Therefore, many people predicted that Dolly would age prematurely. In fact, Dolly developed arthritis when she was $5\frac{1}{2}$ and was euthanized with a serious lung disease when she was $6\frac{1}{2}$, relatively young ages for a sheep. Of course, no one can say for sure if these health problems occurred because she was a clone.

THE FUTURE OF CLONING

As a technology, cloning shows great promise. As the process becomes more routine, however, it will bring ethical questions. Hardly anyone objects to navel oranges, and few people would refuse antibiotics or other medicinal products from cloned livestock, but many people think that cloning pets is, at best, a frivolous luxury and might even be a form of animal abuse. We will return to the ethics of cloning, especially cloning people, in "Cloning Controversy Revisited" at the end of this chapter.

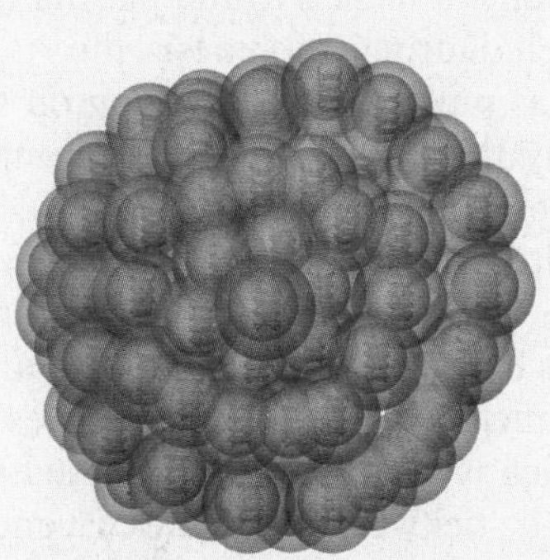

④ The cell divides, forming an embryo that consists of a hollow ball of cells.

⑤ The ball of cells is implanted into the uterus of another Blackface ewe.

⑥ The Blackface ewe gives birth to Dolly, a female Finn Dorset lamb, a genetic twin of the Finn Dorset ewe.

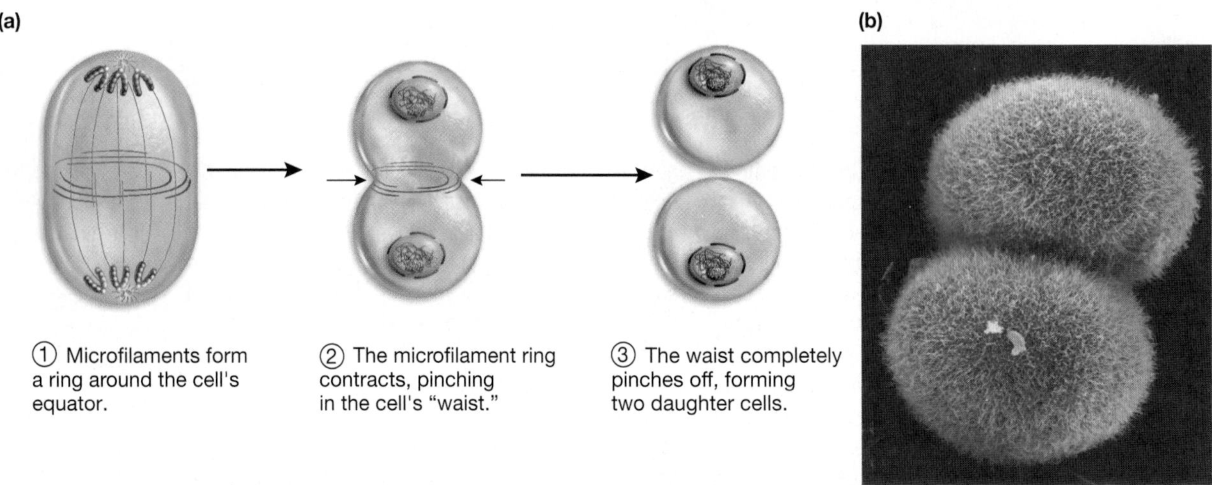

Figure 10-9 Cytokinesis in an animal cell *(a)* A ring of microfilaments just beneath the plasma membrane contracts around the equator of the cell, pinching it in two. *(b)* This scanning electron micrograph shows the two daughter cells nearly separated.

During Telophase, Nuclear Envelopes Form Around Both Groups of Chromosomes

When the chromosomes reach the poles, **telophase** begins (Figs. 10-7e and 10-8f). The spindle microtubules disintegrate, and a nuclear envelope forms around each group of chromosomes. The chromosomes revert to their extended state, and the nucleoli reappear. In most cells, cytokinesis occurs during telophase, placing each daughter nucleus into a separate cell (Fig. 10-8g).

During Cytokinesis, the Cytoplasm Is Divided between Two Daughter Cells

In dividing animal cells, microfilaments attached to the plasma membrane form a ring around the equator of the cell. During cytokinesis, the ring contracts and constricts the cell's equator, much as pulling the drawstring on a pair of sweatpants tightens the waist. Eventually the "waist" constricts completely, dividing the cytoplasm into two new daughter cells (Fig. 10-9).

Cytokinesis in plant cells is quite different, perhaps because their stiff cell walls make it impossible to divide one cell into two by pinching at the waist. Instead, carbohydrate-filled vesicles bud off the Golgi complex and line up along the cell's equator between the two nuclei (Fig. 10-10). The vesicles fuse, producing a structure called the *cell plate*, which is shaped like a flattened sac, surrounded by plasma membrane, and filled with sticky carbohydrates. Eventually, the edges of the cell plate merge with the original plasma membrane of the cell. The carbohydrates contained in the vesicles become the cell wall between the two daughter cells.

After cytokinesis, daughter cells of both plants and animals enter interphase (Figs. 10-7f and 10-8h).

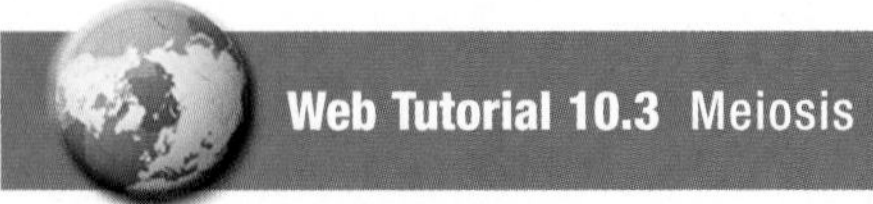

10.6 How Does Meiotic Cell Division Produce Haploid Cells?

The key to sexual reproduction is meiotic cell division, the production of haploid cells with unpaired chromosomes from diploid parent cells with paired chromosomes. Each daughter cell receives one of each pair of homologous chromo-

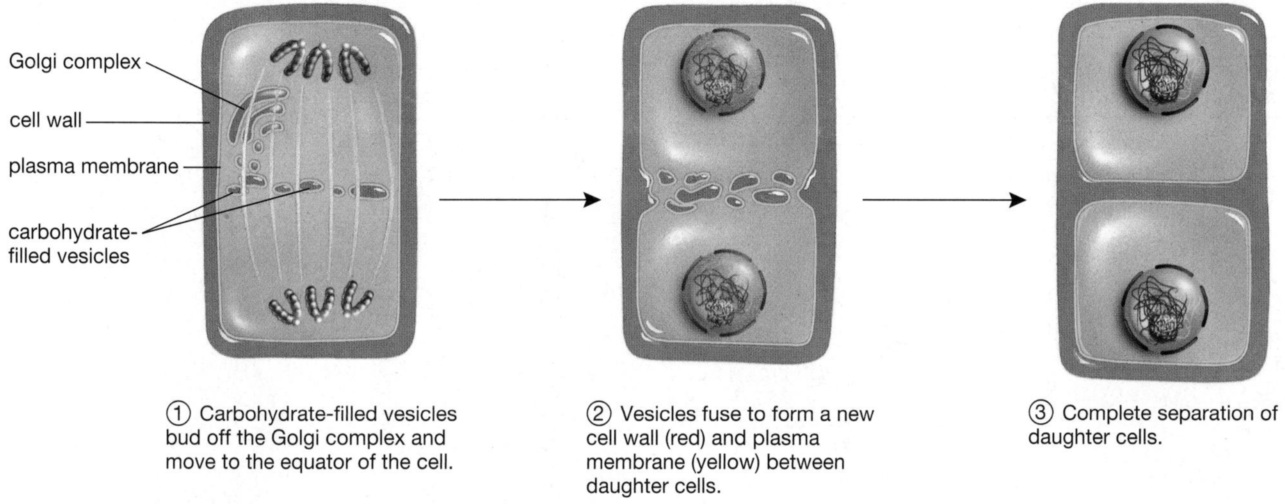

Figure 10-10 Cytokinesis in a plant cell

somes. For example, each diploid cell in a human body contains 23 pairs of chromosomes, so meiotic cell division produces gametes with 23 chromosomes, one from each pair. Cytokinesis in meiotic cell division is similar to cytokinesis in mitotic cell division. Therefore, we will describe only meiosis, the division of the nuclei, in the following sections.

Meiosis Produces Four Haploid Daughter Nuclei

Many of the events in meiosis are similar to those of mitosis. A major difference is that meiosis includes *two* nuclear divisions, so each parent cell yields *four* daughter cells. The two nuclear divisions are known as meiosis I and meiosis II. In meiosis I, homologous chromosomes pair up but sister chromatids remain connected to each other (this step differs from mitosis). In meiosis II, chromosomes behave as they do in mitosis: Sister chromatids separate and are pulled to opposite poles of the cell.

The phases of meiosis have the same names as the roughly equivalent phases in mitosis, followed by a I or II to distinguish the two nuclear divisions.

Meiosis I Separates Homologous Chromosomes into Two Haploid Daughter Nuclei

The chromosomes are replicated in interphase before meiosis I. As we saw in mitosis, the sister chromatids of each duplicated chromosome are attached to one another at the centromere.

During Prophase I, Homologues Pair Up

Meiosis I differs considerably from mitosis. In mitosis, each chromosome moves independently, and homologous chromosomes do not interact. In contrast, during *prophase I* of meiosis, pairs of homologues line up side by side (**Fig. 10-11a**). Proteins bind the homologues together so that they match up exactly along their entire length. The two homologues in a bound pair often intertwine, forming crosses, or

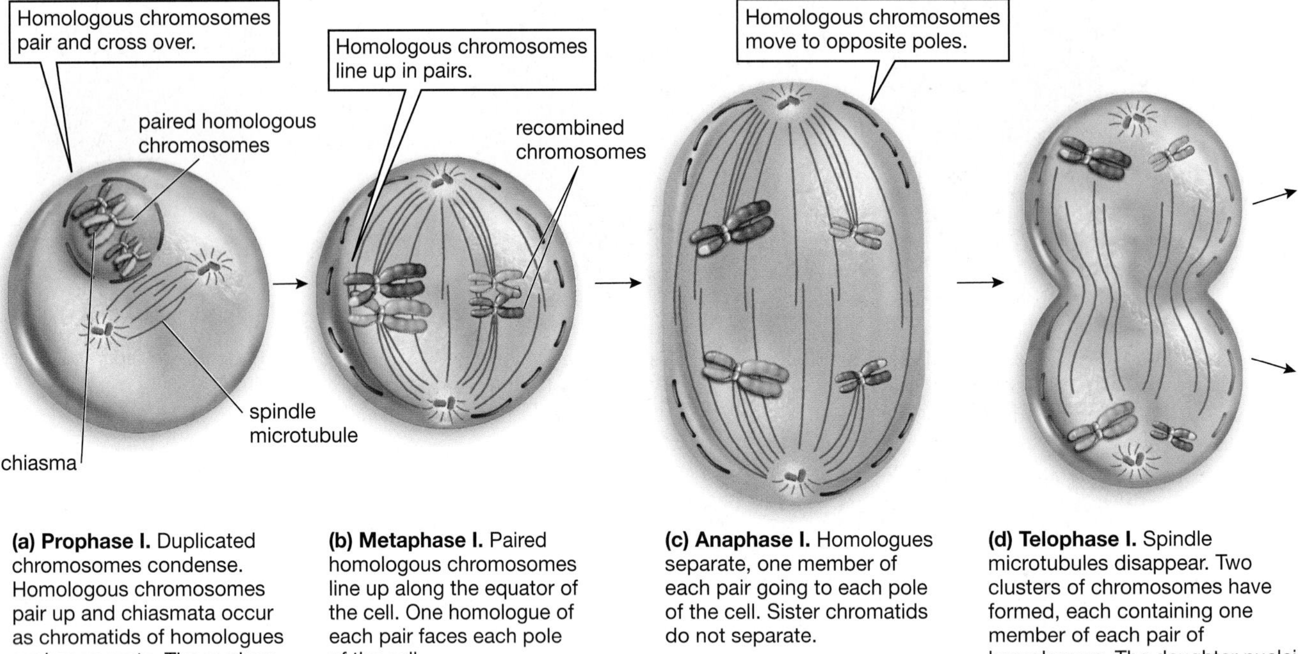

Figure 10-11 Meiotic cell division in an animal cell In these diagrams, two pairs of homologous chromosomes are shown, large and small. The yellow chromosomes are from one parent, and the violet chromosomes are from the other parent.
QUESTION: *What would be the consequences for the gametes if one pair of homologues failed to separate at anaphase I? What would happen if meiosis I was normal but a pair of sister chromatids failed to separate at anaphase II?*

chiasmata (singular, **chiasma**; Fig. 10-12). Eventually, the proteins leave the chromosomes, but the chiasmata continue to hold the homologues together.

As prophase I continues, spindle microtubules begin to assemble outside the nucleus. Near the end of prophase I, the nuclear envelope breaks down and the chromosomes attach to spindle microtubules.

Paired Homologues Exchange DNA

Homologous chromosomes often exchange DNA at the chiasmata. This mutual exchange of DNA is known as **crossing over** (Fig. 10-12). After crossing over, neither chromosome is quite the same as it was before. The result of crossing over is genetic **recombination**: the formation of new combinations of genes on a chromosome.

Figure 10-12 Chiasmata Chiasmata are locations where nonsister chromatids (chromatids from different members of a homologous pair of chromosomes) cross and connect with each other, helping to hold homologues together. Many chiasmata are also the sites of exchange of DNA between chromosomes (crossing over).

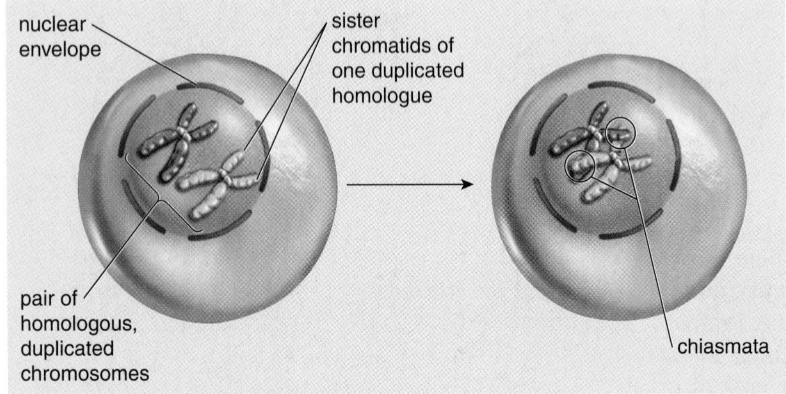

During Metaphase I, Paired Homologues Line Up at the Equator of the Cell

During *metaphase I*, interactions between the chromosomes and the spindle microtubules move the paired homologues to the equator of the cell (see Fig. 10-11b). The pairing of homologues in meiosis I makes the arrangement of chromosomes along the equator very different from the arrangement in mitosis. In mitosis, *individual* duplicated chromosomes line up along the equator, but during metaphase I of meiosis, *homologous pairs* of duplicated chromosomes line up along the equator.

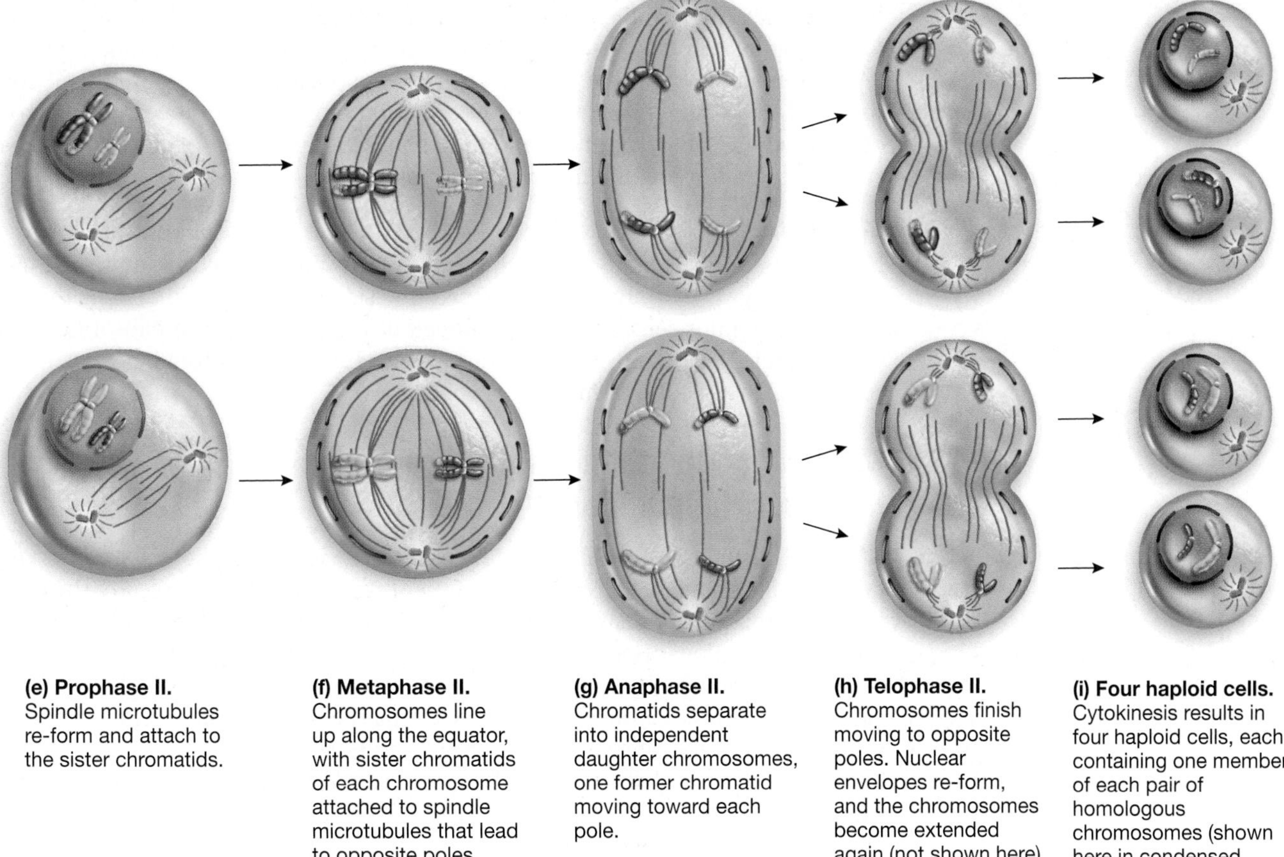

(e) Prophase II. Spindle microtubules re-form and attach to the sister chromatids.

(f) Metaphase II. Chromosomes line up along the equator, with sister chromatids of each chromosome attached to spindle microtubules that lead to opposite poles.

(g) Anaphase II. Chromatids separate into independent daughter chromosomes, one former chromatid moving toward each pole.

(h) Telophase II. Chromosomes finish moving to opposite poles. Nuclear envelopes re-form, and the chromosomes become extended again (not shown here).

(i) Four haploid cells. Cytokinesis results in four haploid cells, each containing one member of each pair of homologous chromosomes (shown here in condensed state).

Attachment of Chromosomes to Spindles Differs between Mitosis and Meiosis I

The behavior of chromosomes during meiosis I is determined by how they attach to the spindle microtubules. It might be helpful, then, to compare chromosome attachment in meiosis I with that in mitosis (Fig. 10-13).

In mitosis, each duplicated chromosome attaches independently to the spindle (Fig. 10-13a). Further, each sister chromatid attaches to spindle microtubules that pull toward opposite poles. Thus, during anaphase, the sister chromatids separate and move to opposite poles.

In meiosis I, however, chiasmata temporarily hold the homologues together, and each *pair of homologues* attaches to the spindle as a unit (Fig. 10-13b). One member of each pair of homologous chromosomes attaches to spindle microtubules leading to one pole, while the other member of the pair attaches to spindle microtubules leading to the opposite pole (Fig. 10-13b). In addition, the sister chromatids do *not* separate into independent chromosomes during meiosis I.

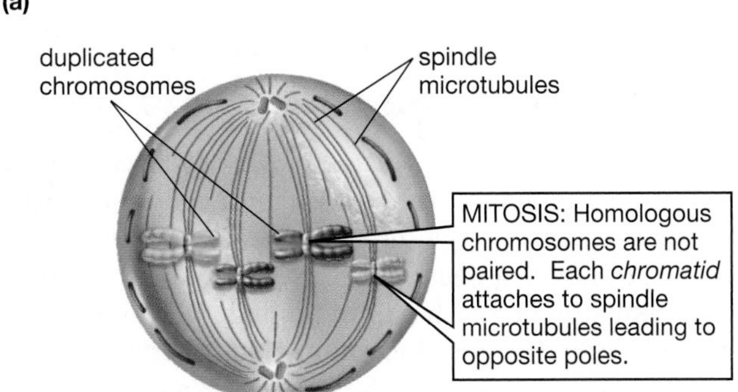

Figure 10-13 A comparison of chromosome attachment to spindles during mitosis and meiosis I ***(a)*** In mitosis, homologous chromosomes are not paired. The sister chromatids are attached to spindle microtubules that lead to opposite poles. When the sister chromatids separate during anaphase, the newly independent daughter chromosomes move to opposite poles of the cell. ***(b)*** In meiosis I, homologous chromosomes are paired. Both sister chromatids of each member of a pair of homologous chromosomes attach to microtubules leading to opposite spindle poles. At anaphase I, sister chromatids of each chromosome remain together, moving to the same pole, but the two members of each pair of homologous chromosomes separate and move to opposite poles.

During Anaphase I, Homologous Chromosomes Separate

In *anaphase I*, the spindle microtubules attached to each pair of homologues pull toward opposite poles. The chiasmata slide apart, allowing the homologues to separate from one another (Fig. 10-11c). One duplicated chromosome (still consisting of two sister chromatids) from each homologous pair moves to each pole of the dividing cell. Thus, at the end of anaphase I, the cluster of chromosomes at each pole contains one member of each pair of homologous chromosomes.

After Telophase I and Cytokinesis, There Are Two Haploid Daughter Cells

In *telophase I*, the spindle microtubules disappear and the nuclear envelope may reappear (Fig. 10-11d). In most cases, cytokinesis takes place and divides the cell into two daughter cells. Each daughter cell has only one of each pair of homologous chromosomes, and is therefore haploid. Each chromosome, however, still consists of two sister chromatids.

Meiosis II Separates Sister Chromatids into Four Haploid Daughter Cells

Meiosis II usually begins immediately after meiosis I ends. There is no intervening interphase, and the chromosomes are not replicated again. Typically, the chromosomes remain condensed. Meiosis II is virtually identical to mitosis, although it occurs in haploid cells.

During *prophase II*, the spindle microtubules re-form (Fig. 10-11e). The duplicated chromosomes attach individually to spindle microtubules as they would in mitosis. Each sister chromatid attaches to spindle microtubules that extend to one or the other pole of the cell. During *metaphase II*, the duplicated chromosomes line up at the cell's equator (Fig. 10-11f). During *anaphase II*, the sister chromatids separate and move to opposite poles (Fig. 10-11g). As *telophase II* and cytokinesis conclude meiosis II, nuclear envelopes form, the chromosomes relax into their extended, noncondensed state, and the cytoplasm divides (Fig. 10-11h,i). Usually, both daughter cells of meiosis I undergo meiosis II, producing a total of four haploid cells from the original parental diploid cell.

Table 10-1 summarizes mitotic and meiotic cell division, showing both the similarities and differences between the two types of cell division.

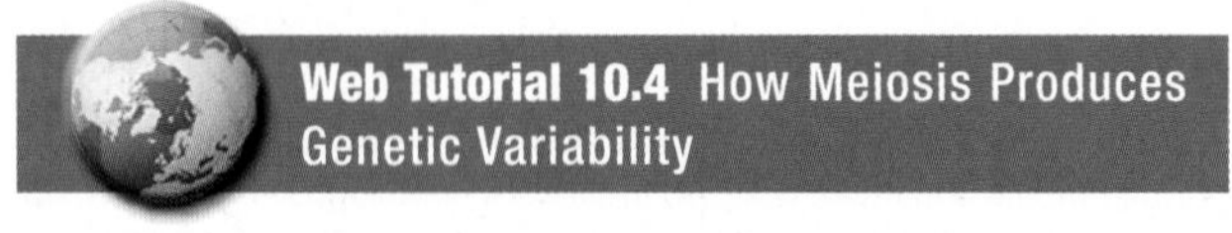

10.7 How Do Meiotic Cell Division and Sexual Reproduction Produce Genetic Variability?

Life has been shaped by evolution, and a crucial prerequisite for evolutionary change is genetic variability among organisms (we will explore this theme at length in Unit Three). Mutations are the ultimate source of new genes, but meiosis and sexual reproduction play key roles in ensuring that individuals carry different combinations of genes. These processes promote genetic variability in three ways.

Shuffling of Homologues Creates Novel Combinations of Chromosomes

One important way in which meiosis produces genetic variability is the random assortment of homologues to daughter cells at meiosis I. Remember that the two chromosomes in each homologous pair have very *similar*, but usually not *identical*, DNA sequences. At metaphase I, paired homologues line up at the cell's equator. For each pair of homologues, one chromosome faces one pole and the other chromosome faces the opposite pole. The determination of which chromosome faces which pole, however, is random.

TABLE 10-1 A Comparison of Mitotic and Meiotic Cell Divisions in Animal Cells

Feature	Mitotic Cell Division	Meiotic Cell Division
Cells in which it occurs	Body cells	Gamete-producing cells
Final chromosome number	Diploid—2*n*; two copies of each type of chromosome (homologous pairs)	Haploid—1*n*; one member of each homologous pair
Number of daughter cells	Two, identical to the parent cell and to each other	Four, containing recombined chromosomes due to crossing over
Number of cell divisions per DNA replication	One	Two
Function in animals	Development, growth, repair and maintenance of tissues, asexual reproduction	Gamete production for sexual reproduction

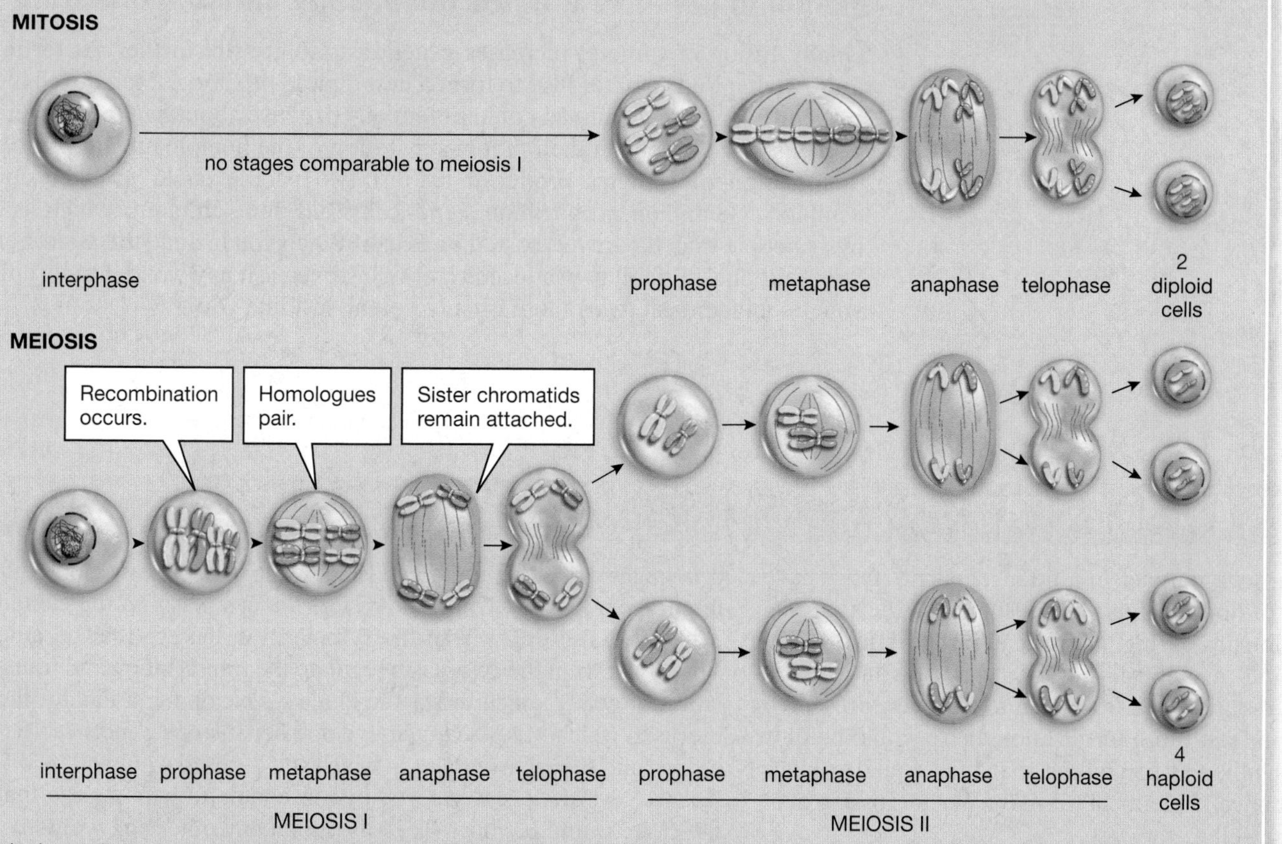

In these diagrams, comparable phases are aligned. In both mitosis and meiosis, chromosomes are replicated during interphase. Meiosis I, with the pairing of homologous chromosomes, formation of chiasmata, exchange of DNA parts, and separation of homologues to form haploid daughter nuclei, has no counterpart in mitosis. Meiosis II, however, is similar to mitosis.

To see the effects of this random arrangement of homologues, let's consider meiosis in mosquitoes, which have three pairs of homologous chromosomes ($n = 3, 2n = 6$). For the sake of simplicity, we'll represent these chromosomes as large, medium, and small. To keep track of the homologues, let's color-code one set of chromosomes yellow and the other set violet. At metaphase I, the chromosomes can align in any one of four configurations. ➔

These four possible metaphase configurations yield eight possible combinations of chromosomes ($2^3 = 8$) when they separate during anaphase I. ➔

Each of these chromosome clusters will then undergo meiosis II to produce two gametes (see Fig. 10-11g–i). A single mosquito, with 3 pairs of homologous chromosomes, therefore produces gametes that each contain one of 8 possible sets of chromosomes. A single human, with 23 pairs of homologous chromosomes, can theoretically produce gametes that contain one of more than 8 million (2^{23}) different combinations of chromosomes.

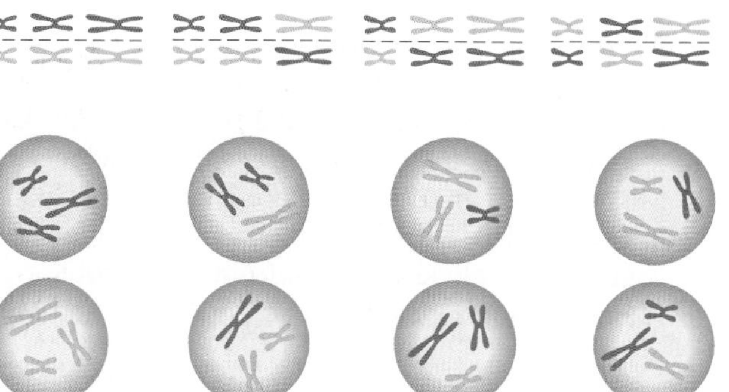

Crossing Over Creates Chromosomes with Novel Combinations of Genetic Material

Crossing over is a second way in which meiosis produces genetic variability. It yields chromosomes with new combinations of genetic material. In fact, these new combinations may have never existed before, because homologous chromosomes cross over in new and different places at each meiotic division. Even though a man produces about 100 million sperm each day, in his whole life he may never produce two that carry exactly the same DNA! In all likelihood, every sperm and every egg is genetically unique.

Fusion of Gametes Creates Genetically Variable Offspring

Finally, fusion of gametes increases genetic variability still further. At fertilization, two haploid gametes fuse to form a new diploid offspring. Even if we ignore crossing over, a single human could potentially produce about 8 million different gametes, based solely on the random separation of the homologues. In principle, then, fusion of gametes produced by just two people could produce about 8 million $\times$ 8 million = 64 trillion genetically different children, which is far more people than have ever existed on Earth! When you include the almost endless additional variability produced by crossing over, is it any wonder that (unless you are an identical twin) there is truly no one just like you?

CLONING CONTROVERSY *REVISITED*

Why are no two puppies in a litter identical? Dogs have 39 pairs of homologous chromosomes, so the shuffling of homologues during meiosis could theoretically produce 2^{39}—550 billion—genetically different gametes in a single dog. Even if gametes from Missy's parents had been preserved and used for *in vitro* (test tube) fertilization, there would be only one chance in 550 billion $\times$ 550 billion that they could produce another puppy genetically identical to Missy. When crossing over is added to the mix, well....

Mitosis, however, *does* produce genetically identical daughter cells. Missy could be cloned by removing the nucleus from an unfertilized egg of a different dog and replacing it with a nucleus from one of Missy's cells. The resulting egg—containing Missy's chromosomes and cytoplasm from a different dog—would be implanted in a surrogate mother's uterus. There, it would divide by mitotic cell division and eventually produce a full-grown dog. Therefore, we might hypothesize that a Missy clone would be exactly like Missy.

On the basis of what you learned in the past few chapters, would it really? First, mutations would cause the DNA in the nucleus taken from Missy's cell to be at least a few nucleotides different from the DNA in the fertilized egg that developed into the original Missy. Second, mitochondria are inherited solely from the mother, in her egg. Therefore, the mitochondria in the egg that would become Missy's clone would be from the egg donor, not from Missy. Metabolic differences between the donor's and Missy's mitochondria might make the energy level or life span of Missy's clone different from Missy's. Third, the rest of the cytoplasm in the cloned egg would also differ from the cytoplasm in the egg that became Missy. Factors in the cytoplasm regulate the transcription and translation of genes, which affect the clone's development. Finally, the uterus in which Missy's clone develops, her surrogate mother's behavior, and the environment in which she is raised would all be different. Nevertheless, it is certainly reasonable to hypothesize that the clone would be more like Missy than any other dog would be. How important would the differences be? How many Missy clones, raised under what types of conditions, would you need to study to find out? Is an appropriately designed experiment even feasible?

CONSIDER THIS: What about human cloning? As of mid-2004, a company called Clonaid claimed that 13 cloned human babies had been born, although there has been no independent confirmation of their claim. Assuming that the technology exists to clone people, is this a good idea? CC the cat was the only survivor of 87 embryos implanted in 8 different surrogate mothers. A cloned Brahma bull, Second Chance, was in intensive care for two weeks with respiratory and cardiovascular disorders. With today's limited technology, it's hard to argue with Rudolf Jaenisch of the Whitehead Institute for Biomedical Research that human cloning is "just criminal." But what about in 2010, or 2050, if the technology is "perfected"? What do you think?

LINKS TO LIFE: Cancer—Cell Division Run Amok

Mitotic cell division is essential for the development of multicellular organisms from fertilized eggs, as well as for routine maintenance of certain body parts, such as the skin and the lining of the digestive tract. Unfortunately, uncontrolled cell division is a menace to life: cancer. Normally, cell division is regulated by a balance between growth-stimulating and growth-inhibiting factors. Mutations and certain types of viruses (which insert some of their DNA into the host cell's chromosomes) may either stimulate cell division inappropriately or reduce a cell's sensitivity to growth-inhibiting factors. These cells are usually eliminated by the immune system, but occasionally a renegade cell survives and reproduces. Because mitotic cell division faithfully transmits genetic information from cell to cell, all daughter cells of the original cancerous cell will themselves be cancerous.

Why does medical science, which has conquered smallpox, measles, and a host of other diseases, have such a difficult time curing cancer? One reason is that both normal and cancerous cells use the same machinery for cell division, so treatments that slow down the multiplication of cancer cells also reduce the maintenance of essential body parts, such as the stomach, intestine, and blood cells. Truly effective and *selective* treatments for cancer must target cell division only in the malignant cells. Although great strides have been made, this daunting problem has yet to be solved.

Chapter Review

Summary of Key Concepts

10.1 Why Do Cells Divide?

There are three types of cell division. Prokaryotic cells divide by binary fission. Eukaryotic cells divide by mitotic or meiotic cell division. Binary fission and mitotic cell division produce daughter cells that are genetically identical to the parent cell. Meiotic cell division produces cells with half of the parent cell's DNA (one of each pair of homologous chromosomes).

Growth of multicellular eukaryotic organisms and replacement of cells that die during an organism's life occur through mitotic cell division and differentiation of the daughter cells. Asexual reproduction of prokaryotic, unicellular organisms occurs through binary fission. Asexual reproduction of eukaryotic organisms (unicellular or multicellular) occurs by mitotic cell division.

Meiotic cell division is required for sexual reproduction. In animals, meiotic cell division produces gametes (sperm or eggs) with half the DNA of an ordinary body cell. Fusion of gametes then creates a fertilized egg that has a genetic makeup different from that of either parent.

10.2 What Occurs During the Prokaryotic Cell Cycle?

The prokaryotic cell cycle consists of growth, DNA replication, and division by binary fission. The two daughter cells are genetically identical to the parent cell.

10.3 What Occurs During the Eukaryotic Cell Cycle?

The eukaryotic cell cycle consists of interphase and cell division. During interphase, the cell grows and duplicates its DNA. It may also differentiate during interphase. Some cells remain permanently in interphase and do not divide again. Most cells that do divide do so by mitotic cell division, which produces two daughter cells that are genetically identical to their parent cell. Specialized reproductive cells undergo meiotic cell division, producing four daughter cells that contain half the DNA of the parent cell.

Multicellular eukaryotic organisms begin life as a fertilized egg. Mitotic cell divisions and differentiation of daughter cells produce the organism's body. At some point in the life of the organism, meiotic cell division produces gametes that can fuse with other gametes to produce a new fertilized egg, starting the life cycle anew.

10.4 How Is DNA in Eukaryotic Cells Organized into Chromosomes?

Each eukaryotic chromosome consists of a single DNA double helix and proteins that organize the DNA. During cell growth, the chromosomes are extended and accessible for transcribing their genes. During cell division, the chromosomes condense into short, thick structures. Eukaryotic cells typically contain pairs of homologous chromosomes. Homologues have virtually identical appearance because they carry the same genes with similar nucleotide sequences. Cells with pairs of homologous chromosomes are diploid. Cells with only a single member of each chromosome pair are haploid.

10.5 How Does Mitotic Cell Division Produce Genetically Identical Daughter Cells?

The chromosomes are duplicated during interphase, before mitosis (nuclear division). A replicated chromosome consists of two identical sister chromatids that remain attached to one another at the centromere during the early stages of mitosis. Mitosis consists of four phases (see Fig. 10-8):

- **Prophase:** The chromosomes condense and attach to the spindle microtubules that form at this time.
- **Metaphase:** The chromosomes move along their attached spindle microtubules to the equator of the cell.
- **Anaphase:** The two chromatids of each duplicated chromosome separate and move along the spindle microtubules to opposite poles of the cell.
- **Telophase:** The chromosomes relax into their extended state, and nuclear envelopes re-form around each new daughter nucleus.

Usually, cytokinesis (division of the cytoplasm) occurs during telophase, producing two new daughter cells, each containing one of the nuclei produced during mitosis. Cytokinesis occurs somewhat differently in animal and plant cells.

10.6 How Does Meiotic Cell Division Produce Haploid Cells?

During interphase before meiotic cell division, chromosomes are duplicated. The cell then undergoes two divisions—meiosis I and meiosis II—to produce four haploid daughter cells (Fig 10-11).

- **Meiosis I:** During prophase I, homologous duplicated chromosomes, each consisting of two chromatids, pair up and exchange parts by crossing over. During metaphase I, homologues move together as a pair to the cell's equator. The two members of each pair face opposite poles of the cell. Homologous chromosomes separate during anaphase I, and two nuclei form during telophase I. Each daughter nucleus receives only one member of each pair of homologues and is therefore haploid. The sister chromatids remain attached to each other throughout meiosis I.
- **Meiosis II:** Both daughter nuclei go through meiosis II, which resembles mitosis in a haploid cell. The duplicated chromosomes move to the cell's equator during metaphase II. The two chromatids of each chromosome separate and move to opposite poles of the cell during anaphase II. During telophase II, four haploid nuclei are produced, two from each of the nuclei formed by meiosis I. Cytokinesis completes the production of four haploid cells, each containing one of the nuclei formed during telophase II.

10.7 How Do Meiotic Cell Division and Sexual Reproduction Produce Genetic Variability?

Genetic variability arises in three ways. First, the random shuffling of homologous maternal and paternal chromosomes creates new chromosome combinations. Second, crossing over creates chromosomes with nucleotide sequences that may never have occurred on a single chromosome before. Because of crossing over, a parent probably never produces any two gametes that are completely identical. Third, the fusion of two genetically unique gametes at fertilization creates genetically unique offspring.

Key Terms

anaphase *p. 153*
autosome *p. 150*
binary fission *p. 146*
cell cycle *p. 145*
cell division *p. 144*
centriole *p. 151*
centromere *p. 149*
chiasma (plural, **chiasmata**) *p. 158*
chromatid *p. 149*
clone *p. 154*
cloning *p. 154*
crossing over *p. 158*
cytokinesis *p. 147*
diploid *p. 150*
gamete *p. 144*
haploid *p. 150*
homologue *p. 150*
interphase *p. 146*
meiosis *p. 147*
meiotic cell division *p. 147*
metaphase *p. 152*
mitosis *p. 147*
mitotic cell division *p. 147*
prophase *p. 151*
recombination *p. 158*
sex chromosome *p. 150*
spindle microtubule *p. 151*
telophase *p. 156*

Thinking Through the Concepts

Multiple Choice

1. *At which stage of mitosis are chromosomes arranged at the equator of the cell?*
   a. anaphase
   b. telophase
   c. metaphase
   d. prophase
   e. interphase
2. *A diploid cell contains in its nucleus*
   a. an even number of chromosomes
   b. an odd number of chromosomes
   c. one member of each pair of homologous chromosomes
   d. either an even or an odd number of chromosomes
   e. four sister chromatids of each chromosome
3. *Synthesis of new DNA occurs during*
   a. prophase
   b. interphase
   c. mitosis
   d. cytokinesis
   e. formation of the cell plate
4. *When do homologous chromosomes pair up?*
   a. only in mitosis
   b. only in meiosis I
   c. only in meiosis II
   d. in both mitosis and meiosis
   e. in neither mitosis nor meiosis
5. *Curiously, there is no crossing over of any chromosome in the male fruit fly* Drosophila, *which has four pairs of chromosomes. How many different combinations of chromosomes are possible in a male fruit fly's sperm?*
   a. 2
   b. 4
   c. 8
   d. 16
   e. many more than the above

Review Questions

1. Define mitosis and cytokinesis. What would a daughter cell look like if cytokinesis did not occur after mitosis?
2. Diagram the stages of mitosis. How does mitosis ensure that each daughter nucleus receives a full set of chromosomes?
3. Define the following terms: homologue, centromere, chromatid, diploid, haploid.
4. Describe and compare cytokinesis in animal cells and in plant cells.
5. Diagram the events of meiosis. At which stage do homologous chromosomes separate?
6. In what ways are mitosis and meiosis similar? In what ways are they different?
7. Describe how meiosis provides for genetic variability. If an animal had a haploid number of 2 (no sex chromosomes), how many genetically different types of gametes could it produce? (Assume no crossing over.) How many could it produce if it had a haploid number of 5?

Applying the Concepts

1. **IS THIS SCIENCE?** Many therapies for cancer have side effects, including nausea, vomiting, and hair loss. There are claims in many magazines and Internet sites that a variety of herbs can cure cancer, with few or no side effects. Given what you learned about cell division and cancer in this chapter, do you think that these claims are reasonable? Why or why not? What types of experiments could you perform to test the claims?

2. Some animal species can reproduce either asexually or sexually, depending on the state of the environment. Asexual reproduction tends to occur in stable, favorable environments; sexual reproduction is more common in unstable or unfavorable circumstances. Discuss the advantages or disadvantages this pattern might have on survival of the species in an evolutionary sense or on survival of individuals.

For More Information

Axtman, K. "Quietly, Animal Cloning Speeds Onward." *Christian Science Monitor*, October 23, 2001. A discussion of the successes and failures of mammalian cloning.

Grant, M. C. "The Trembling Giant." *Discover*, October 1993. Aspen groves are really single individuals: huge, slowly spreading from the roots of the original parent tree, and potentially almost immortal.

Lanza, R. P., Dresser, B. L., and Damiani, P. "Cloning Noah's Ark." *Scientific American*, November 2000. Cloning rare and endangered species may offer hope to prevent extinction.

Travis, J. "A Fantastical Experiment." *Science News*, April 5, 1997. A clear description of the cloning of Dolly the sheep and some of its implications.

Wilmut, I. "Cloning for Medicine." *Scientific American*, December 1998. Explanation of why cloning experiments might have medical applications.

WEB TUTORIAL

To access a Web Tutorial visit http://www.prenhall.com/audesirk4. *Log in to the Web site selected by your instructor, navigate to this chapter, and select the appropriate Web Activity number.*

10.1 Cell Division in Humans
Estimated time: 5 minutes

This animation provides a brief overview of the cell cycle in humans.

10.2 Mitosis
Estimated time: 10 minutes

This activity describes the process of mitosis and will help you understand the steps into which mitosis is divided.

10.3 Meiosis
Estimated time: 10 minutes

In order for the fusion of sperm and egg to produce a cell with the correct number of chromosomes, the sperm and egg must each have one half the number of chromosomes as an adult cell. To accomplish this, certain cells in the body undergo a type of division, called meiosis, which reduces the chromosome number. This activity will lead you through the steps of meiosis.

10.4 How Meiosis Produces Genetic Variability
Estimated time: 5 minutes

Everyone is unique. How does the process of meiosis produce the amazing diversity we see around us? In this activity, review the processes that produce such great genetic diversity.

10.5 Web Investigation: Cloning Controversy
Estimated time: 20 minutes

The Missyplicity Project has raised questions about ethics and science. Explore some of these questions and find possible answers in this web investigation.

Chapter 11

Patterns of Inheritance

Olympic silver medalist Flo Hyman was struck down by Marfan syndrome at the height of her career.

AT A GLANCE

Sudden Death on the Court

Flo Hyman, graceful, athletic, and over 6 feet tall, was one of the best woman volleyball players of all time. A star of the 1984 silver medal American Olympic volleyball team, Hyman later joined a professional Japanese team. In 1986, taken out of a game for a short breather, she died while sitting quietly on the bench. How could this happen to someone so young and fit?

Flo Hyman had a genetic disorder called Marfan syndrome. Marfan syndrome is surprisingly common, affecting about one in 5000 people, including President Abraham Lincoln and the Egyptian pharoah Akhenaten. People with Marfan syndrome are typically tall and slender and have unusually long limbs and large hands and feet. These characteristics helped Flo Hyman to become an outstanding volleyball player. Unfortunately, Marfan syndrome can also be deadly.

An autopsy showed that Hyman died from a ruptured aorta, the massive artery that carries blood from the heart to most of the body. Why did Hyman's aorta break? What does a weak aorta have in common with tallness and large hands? Marfan syndrome is caused by a mutation in the gene that encodes a protein called fibrillin, which forms long fibers that give elasticity and strength to connective tissue. Many parts of the body contain connective tissue, including tendons, ligaments, and artery walls. Defective fibrillin molecules weaken connective tissue, sometimes with tragic consequences. Fibrillin mutations apparently also stimulate growth, causing people with Marfan syndrome to grow tall and lanky.

How did Flo Hyman get this disorder? Did she inherit it from her parents? Or was it a new mutation (most likely in the DNA of either her mother's egg or her father's sperm that fertilized it)? Because new mutations are rare, let's hypothesize that Hyman inherited a defective gene from her parents. Geneticists can't do experiments, in the usual sense, on people, but they can gather other evidence to help them determine modes of inheritance. As you read this chapter, ask yourself a few questions: What evidence would you need to decide if Hyman's case of Marfan syndrome was a new mutation or inherited from her parents? If it was inherited, did it come from both parents, or could she inherit it from just one? If Hyman had had children before she died, would they be likely to have Marfan syndrome?

What did volleyball star Flo Hyman and President Abraham Lincoln have in common?

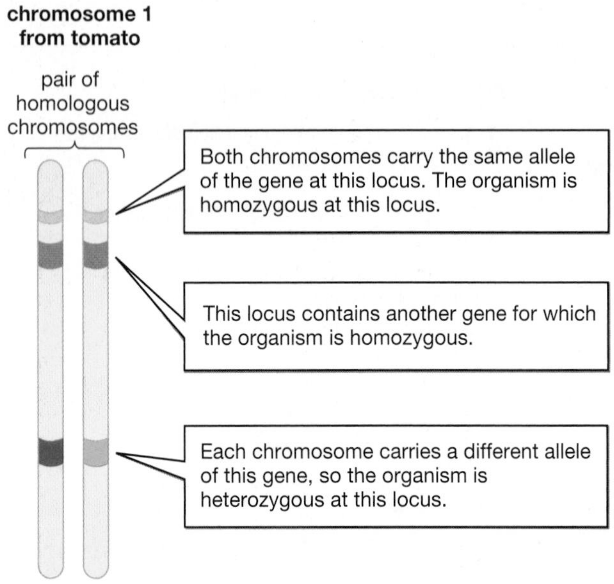

chromosome 1 from tomato

pair of homologous chromosomes

Both chromosomes carry the same allele of the gene at this locus. The organism is homozygous at this locus.

This locus contains another gene for which the organism is homozygous.

Each chromosome carries a different allele of this gene, so the organism is heterozygous at this locus.

Figure 11-1 The relationships among genes, alleles, and chromosomes Homologues carry the same gene loci but may have the same or different alleles at corresponding loci.

11.1 What Is the Physical Basis of Inheritance?

Inheritance is the process by which the characteristics of individuals are passed to their offspring. As you learned in previous chapters, DNA carries genetic information, in the form of sequences of nucleotides. In most cases, segments of DNA ranging from a few hundred to many thousands of nucleotides are the genes that encode the information needed to synthesize a specific protein. DNA together with various proteins make up chromosomes. Genes, therefore, are parts of chromosomes. Finally, chromosomes are passed from cell to cell and from organism to organism during reproduction. Inheritance, then, occurs when genes are transmitted from parent to offspring.

We will begin our exploration with a brief overview of the structures—genes and chromosomes—that form the physical basis of inheritance. In this chapter, we will confine our discussion to diploid organisms, including most plants and animals, that reproduce sexually by the fusion of haploid gametes.

Genes Are Sequences of Nucleotides at Specific Locations on Chromosomes

A gene's physical location on a chromosome is called its **locus** (plural, **loci**). Each member of a pair of homologous chromosomes carries the same genes, located at the same loci. Will the nucleotide sequences at the same locus of a pair of homologues always be identical? Think back to Chapters 8 and 9. Errors in DNA replication, certain chemicals, radiation—all of these can cause mutations that change the nucleotide sequence of DNA. *Different* nucleotide sequences at the *same* locus on two homologous chromosomes are called **alleles**. Different alleles of a gene may produce different forms of a given characteristic, such as brown versus blue eye color.

Figure 11-2 Gregor Mendel

An Organism's Two Alleles May Be the Same or Different

A diploid organism has pairs of homologous chromosomes. The two chromosomes in a pair have the same gene loci. Therefore, the organism has two copies of each gene. If both homologous chromosomes have the *same* allele at a locus, the organism is said to be **homozygous** at that locus. For example, the chromosomes in **Figure 11-1** are homozygous at two loci. If two homologous chromosomes have *different* alleles at a locus, the organism is **heterozygous** at that locus and is sometimes referred to as a *hybrid*. The chromosomes in **Figure 11-1** are heterozygous at one locus.

Whether an individual is homozygous or heterozygous at a locus affects the genetic makeup of the gametes produced by that individual. Because homologous chromosomes separate during meiosis (as we saw in Chapter 10), each gamete receives one chromosome from each pair of homologues. As a result, each gamete receives only one allele for each gene. If an individual is homozygous at a particular locus, all the gametes it produces contain the same allele at that locus. In contrast, an individual that is heterozygous at a locus produces two kinds of gametes: half contain one allele, and half contain the other allele.

11.2 How Were the Principles of Inheritance Discovered?

In the mid-1800s, an Austrian monk, Gregor Mendel (**Fig. 11-2**), deduced the common patterns of inheritance and many essential facts about genes, alleles, and the distribution of alleles in gametes and offpsring during sexual reproduction. This was long before DNA, chromosomes, or meiosis had been discovered. Because his experiments are succinct, elegant examples of science in action, let's follow Mendel's paths of discovery.

intact pea flower

flower dissected to show reproductive structures

carpel (female, produces eggs)

stamen (male, produces pollen)

Figure 11-3 Flowers of the edible pea In the intact pea flower (left), the lower petals form a container enclosing the reproductive structures. Pollen normally cannot enter the flower from outside, so peas usually self-fertilize. If the flower is opened (right), it can be cross-pollinated by hand.

Doing It Right: The Secrets of Mendel's Success

There are three key steps to any successful experiment in biology: (1) choosing the right organism with which to work, (2) designing and performing the experiment correctly, and (3) analyzing the data properly. Mendel was the first geneticist to complete all three steps.

Mendel's choice of the edible pea as an experimental subject was critical to the success of his experiments. The petals of a pea flower enclose all the flower structures, including both male and female parts (Fig. 11-3). Stamens, the male reproductive structures of a flower, produce pollen. Each pollen grain contains sperm. Eggs are produced in ovaries at the base of the carpel, the female reproductive structure of the flower. Pollination allows sperm to fertilize an egg. Pea plants normally reproduce by **self-fertilization**, a process in which the egg cells in a flower are fertilized by sperm from the pollen of the *same* flower. The petals of the pea flower prevent another flower's pollen from entering.

Although peas normally self-fertilize, plant breeders can open a pea flower and remove its stamens, preventing self-fertilization. Then, they can dust the sticky tip of the carpel with pollen from a flower of another plant. Fertilizing a flower's eggs by sperm from a flower of a different plant is called **cross-fertilization**. In this way, two plants can be mated to see what types of offspring they produce.

Mendel's experimental design was brilliant. In contrast to earlier scientists, who usually tried to understand inheritance by comparing *entire organisms* with their offspring, Mendel studied *individual characteristics* (usually called *traits*). Further, he chose traits with unmistakably different forms, such as white versus purple flowers. He also began by studying only one trait at a time. Finally, Mendel followed the inheritance of these traits for several generations, counting the *numbers of offspring* with each type of trait. When he analyzed these numbers, the basic patterns of inheritance became clear. Numerical analysis was an innovation in Mendel's time, but today, statistics is an essential tool in virtually every field of biology.

11.3 How Are Single Traits Inherited?

Mendel started simply. He first raised varieties of pea plants that were true-breeding for different forms of a single trait. The term *true-breeding* refers to organisms with a trait, such as purple flowers, that is always inherited by all offspring that result from self-fertilization. In his earliest experiments, Mendel cross-fertilized plants with different true-breeding traits, saved the resulting hybrid seeds, and then grew the seeds to observe the characteristics of the resulting plants.

In one of these experiments, Mendel cross-fertilized white-flowered plants with purple-flowered ones. These plants are called the parental generation, denoted by the letter *P*. When he grew the resulting seeds, he found that all the first-generation offspring (the "first filial," or F_1, generation) produced purple flowers. →

What had happened to the white color? The flowers of the F_1 hybrids were just as purple as the flowers of their purple parents. The white color seemed to have disappeared in these offspring.

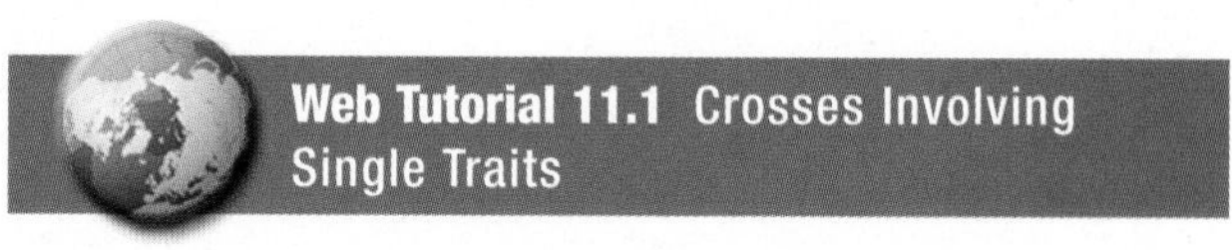

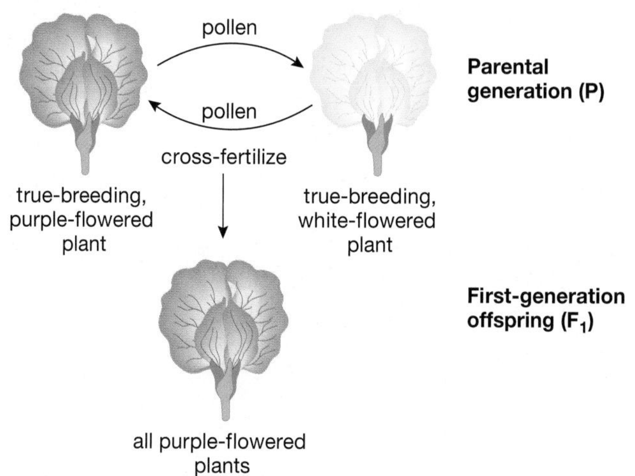

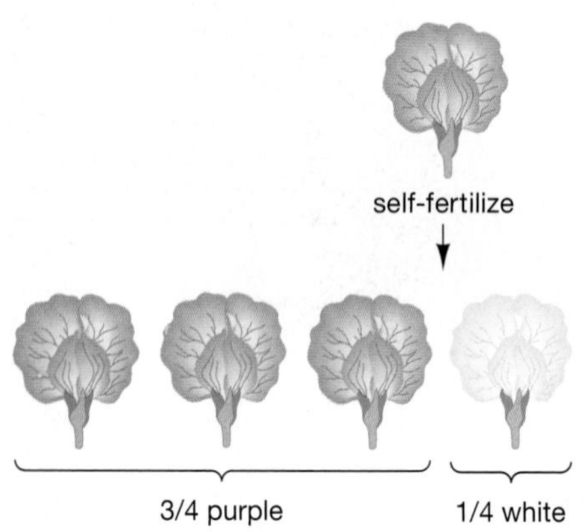

Mendel then allowed the F_1 flowers to self-fertilize, collected the seeds, and planted them the next spring. In the second generation (F_2), about three-fourths of the plants had purple flowers and one-fourth had white flowers.

The exact numbers were 705 purple and 224 white, or a ratio of about 3 purple to 1 white. This result showed that the gene that produced white flowers had not disappeared but had only been "hidden."

Mendel allowed the F_2 plants to self-fertilize and produce a third (F_3) generation. He found that all the white-flowered F_2 plants were true-breeding; that is, they produced only white-flowered offspring. For as many generations as he had time and patience to raise, white-flowered parents always gave rise to white-flowered offspring.

The purple-flowered F_2 plants were a different story. They were of two types: about one-third were true-breeding for purple. The other two-thirds produced both purple- and white-flowered offspring, again in the ratio of 3 to 1. In other words, the F_2 generation included one-fourth true-breeding purple plants, one-half hybrid purple, and one-fourth true-breeding white.

The Pattern of Inheritance of Single Traits Can Be Explained by the Inheritance of Alleles of a Single Gene

Mendel's results, supplemented by modern knowledge of the behavior of genes and chromosomes, allow us to develop a five-part hypothesis to explain the inheritance of single traits.

1. Each trait is determined by pairs of distinct physical units that we now call genes. Each organism has two alleles for each gene, such as the gene that determines flower color. One allele of the gene is present on each of two homologous chromosomes.
2. The pairs of alleles on homologous chromosomes separate from each other during gamete formation, so each gamete receives only one allele from each pair. This conclusion is known as Mendel's **law of segregation**: the two alleles of a gene segregate (separate) from one another during meiosis. When a sperm fertilizes an egg, the resulting offspring receives one allele from the father and one from the mother.
3. Chance determines which allele ends up in any given gamete. Because homologous chromosomes separate at random during meiosis, the distribution of alleles to the gametes is also random.
4. When two different alleles are present in an organism, one—the **dominant** allele—may mask the expression of the other—the **recessive** allele. The recessive allele, however, is still present. Both the dominant and the recessive alleles are passed into the individual's gametes. In Mendel's experiments with flower color, the allele for purple flowers is dominant, and the allele for white flowers is recessive.
5. True-breeding organisms have two copies of the same allele for a given gene. Therefore, they are homozygous at this gene locus. All the gametes from a homozygous individual have the same allele at that locus.

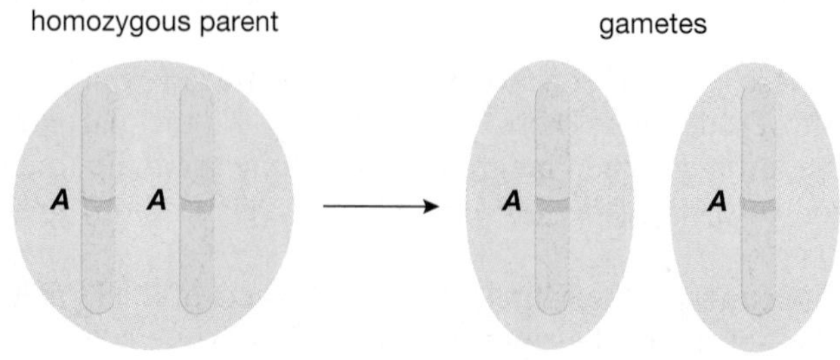

Individuals with two different alleles at a locus are heterozygous. Half of a heterozygote's gametes contain one allele for that gene, and half contain the other allele.

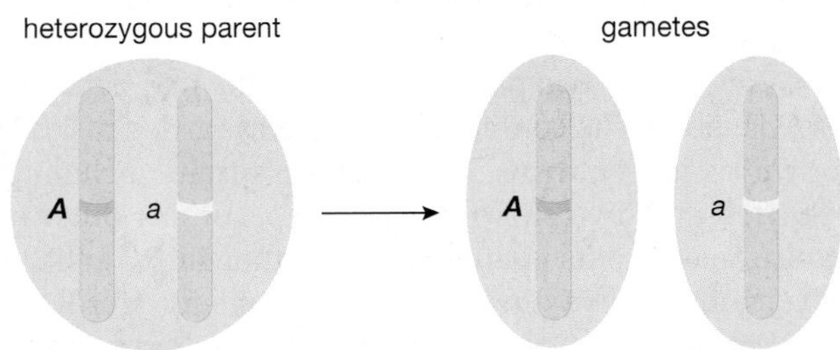

Let's see how this hypothesis explains the results of Mendel's experiments with flower color. Using letters to represent the different alleles, we will assign the uppercase letter *P* to the dominant allele for purple and the lowercase letter *p* to the recessive allele for white. (By Mendel's convention, the dominant allele is represented by a capital letter.) A true-breeding (homozygous) purple-flowered plant has two alleles for purple flowers (*PP*), whereas a white-flowered plant has two alleles for white flowers (*pp*). All the sperm and eggs produced by a *PP* plant carry the *P* allele, and all the sperm and eggs of a *pp* plant carry the *p* allele.

Mendel's F_1 hybrid offspring were produced when *P* sperm fertilized *p* eggs or when *p* sperm fertilized *P* eggs. In both cases, the F_1 offspring were *Pp*. Because *P* is dominant to *p*, all the offspring were purple.

Mendel's F_2 offspring resulted from self-fertilization of the heterozygous F_1 plants. Each gamete produced by a heterozygous *Pp* plant has an equal chance of receiving either the *P* allele or the *p* allele. That is, the heterozygous plant produces equal numbers of *P* and *p* sperm and equal numbers of *P* and *p* eggs. When a *Pp* plant self-fertilizes, each type of sperm has an equal chance of fertilizing each type of egg.

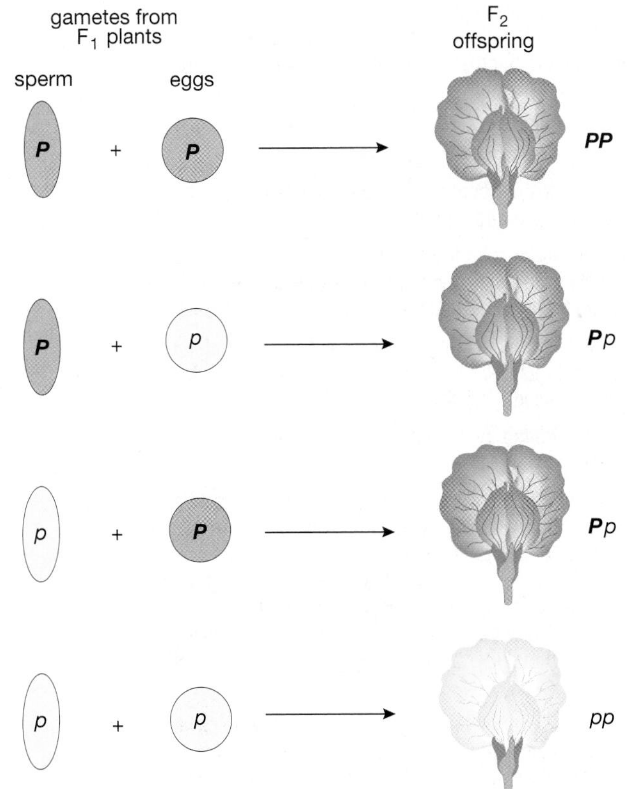

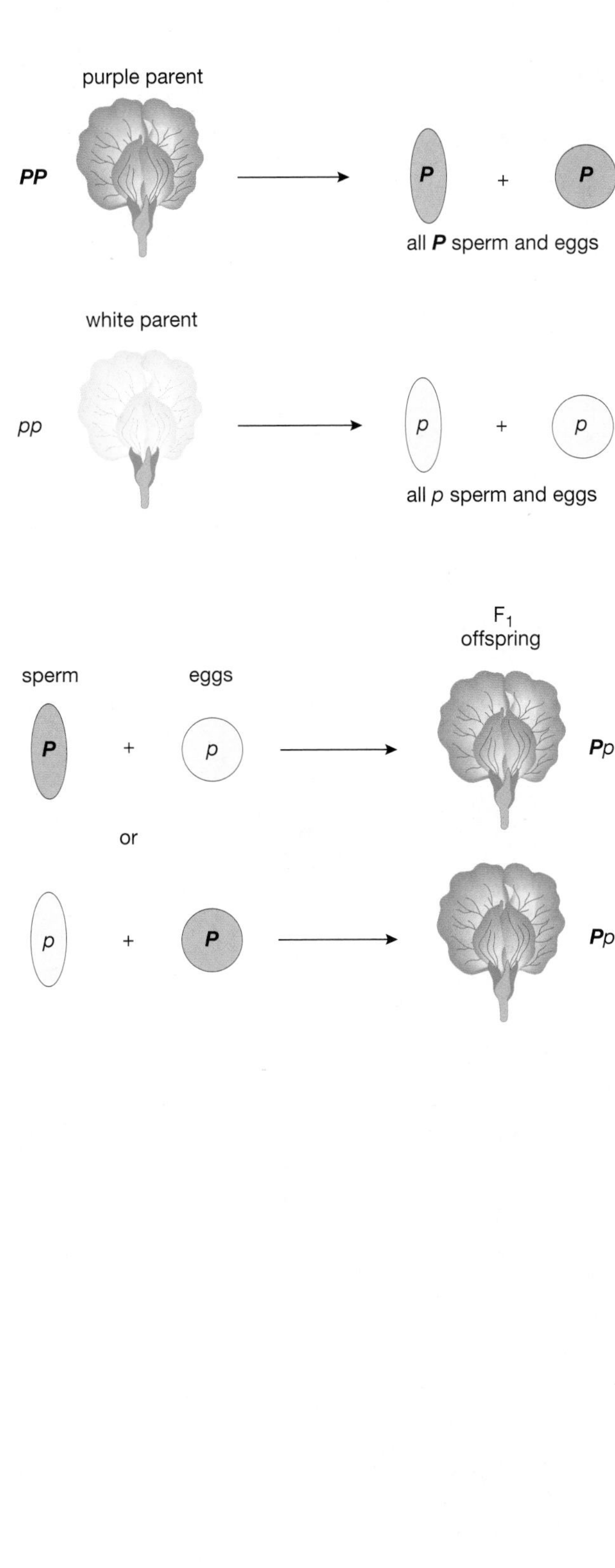

Therefore, Mendel's F_2 generation contained three types of offspring (*PP*, *Pp*, and *pp*), and the three types were in the approximate proportions of $\frac{1}{4}PP$, $\frac{1}{2}Pp$, and $\frac{1}{4}pp$. Because both *PP* and *Pp* plants have purple flowers, three-fourths of the plants in the F_2 generation were purple flowered, and one-fourth were white flowered.

Our hypotheses imply that two plants that look alike may actually carry different combinations of alleles. The combination of alleles carried by an organism (for example, *PP* or *Pp*) is its **genotype**. The organism's traits, including its outward appearance, behavior, digestive enzymes, blood type, or any other observable or measurable feature, make up its **phenotype**. Even though *PP* and *Pp* plants have different genotypes, they have the same phenotype for purple flower color. Therefore, the F_2 generation of Mendel's peas contained three genotypes ($\frac{1}{4}PP$, $\frac{1}{2}Pp$, and $\frac{1}{4}pp$) but only two phenotypes ($\frac{3}{4}$ purple and $\frac{1}{4}$ white).

Simple "Genetic Bookkeeping" Can Predict Genotypes and Phenotypes of Offspring

The **Punnett square** method, named after R. C. Punnett, a famous geneticist of the early 1900s, is a convenient way to predict the genotypes and phenotypes of offspring. Figure 11-4 shows how to use a Punnett square to predict the proportions of offspring that arise from the self-fertilization of an organism that is heterozygous for a single trait. As you use this "genetic bookkeeping" technique, keep in mind that, in a real experiment, the offspring will occur only in *approximately* the predicted proportions. Some sperm do not fertilize an egg, and some eggs are not fertilized by any sperm, so the alleles in those gametes never show up in the next generation. Let's consider an example. Each time a baby is conceived, it has a 50:50 chance of becoming a boy or a girl. However, many families with two children do not have one girl and one boy. The 50:50 ratio of girls to boys only occurs if we average the genders of the children in many families.

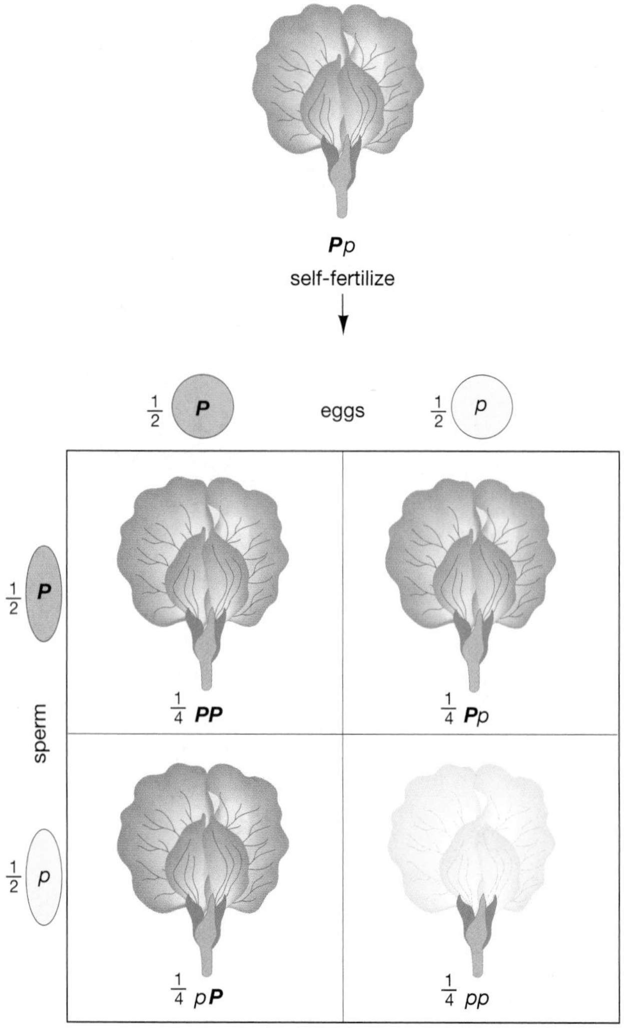

Figure 11-4 The Punnett square method The Punnett square method allows you to predict both genotypes and phenotypes of specific crosses. Here we use it for a cross between plants that are heterozygous for a single trait, flower color.

1. Assign letters to the different alleles. Use uppercase for dominant and lowercase for recessive.
2. Determine all the types of genetically different gametes that can be produced by the male and female parents.
3. Draw the Punnett square, with the columns labeled with the egg genotypes and the rows labeled with the sperm genotypes. (We have included the fractions of these genotypes with each label.)
4. Fill in the genotype of the offspring in each box by combining the genotype of sperm in its row with the genotype of the egg in its column. (We have placed the fractions in each box.)
5. Count the number of offspring with each genotype. (Note that *Pp* is the same as *pP*.)
6. Convert the number of offspring of each genotype to a fraction of the total number of offspring. In this example, out of four fertilizations, only one is predicted to produce the *pp* genotype, so $\frac{1}{4}$ of the total number of offspring produced by this cross is predicted to be white. To determine phenotypic fractions, add the fractions of genotypes that would produce a given phenotype. For example, purple flowers are produced by $\frac{1}{4}PP + \frac{1}{4}Pp + \frac{1}{4}pP$, for a total of $\frac{3}{4}$ of the offspring.

QUESTION: *If you crossed many heterozygous purple-flowered peas, and counted the flower colors of 600 offspring, what numbers of purple and white-flowered offspring would Mendel's hypothesis predict? What would you conclude if you found 440 purple and 160 white-flowered offspring?*

Mendel's Hypothesis Can Predict the Outcome of New Types of Single-Trait Crosses

You have probably recognized that Mendel used the scientific method, observing results and formulating a hypothesis based on them. The scientific method has another step: to see if the hypothesis accurately predicts the results of further experiments. For example, on the basis of his hypothesis that the heterozygous F_1 plants had one allele for purple flowers and one for white (*Pp*), Mendel predicted the outcome of cross-fertilizing these *Pp* plants with homozygous recessive white plants (*pp*): There should be equal numbers of *Pp* (purple) and *pp* (white) offspring. This is indeed what he found. Setting up the cross in a Punnett square can help you to see why.

Referring back to "Sudden Death on the Court," when a person with Marfan syndrome marries a person without the syndrome, their children have a 50% chance of inheriting the condition. Do you think that Marfan syndrome is inherited as a dominant or a recessive allele? Why? Check your reasoning in "Sudden Death Revisited" at the end of the chapter.

11.4 How Are Multiple Traits Inherited?

Having determined how single traits are inherited, Mendel turned to the more complex question of how multiple traits are inherited. He again used edible peas and again studied traits that are expressed as two alternative phenotypes, but this time he crossed plants that differed in two traits, for example, seed color (yellow or green) and seed shape (smooth or wrinkled; Fig. 11-5). From earlier crosses of plants with these traits, he already knew that the "smooth" allele of the gene that controls seed shape (*S*) is dominant to the "wrinkled" allele (*s*) and that the "yellow" allele of the seed color gene (*Y*) is dominant to the "green" allele (*y*).

Mendel crossed a true-breeding plant with smooth, yellow seeds (*SSYY*) with a true-breeding plant with wrinkled, green seeds (*ssyy*). All the F_1 offspring had the genotype *SsYy* and had smooth, yellow seeds. Mendel allowed these F_1 plants to self-fertilize. He found that the F_2 generation consisted of 315 plants with smooth yellow seeds, 101 with wrinkled yellow seeds, 108 with smooth green seeds, and 32 with wrinkled green seeds, a ratio of about 9:3:3:1. Mendel crossed a variety of plants that were heterozygous for two traits and always found that the F_2 generation had phenotypic ratios of about 9:3:3:1.

Mendel Concluded That Multiple Traits Are Inherited Independently

Mendel realized that these results could be explained if the genes for seed color and seed shape were inherited independently. For each trait, he predicted that three-fourths of the offspring would show the dominant phenotype and one-fourth would show the recessive phenotype, producing a 3:1 ratio. This is just what Mendel observed. There were 423 plants with smooth seeds (of either color) and 133 with wrinkled seeds (a ratio of about 3:1), and 416 plants with yellow seeds (of either shape) and 140 with green seeds (also about 3:1). Figure 11-6 shows how a Punnett square can be used to predict the outcome of a cross

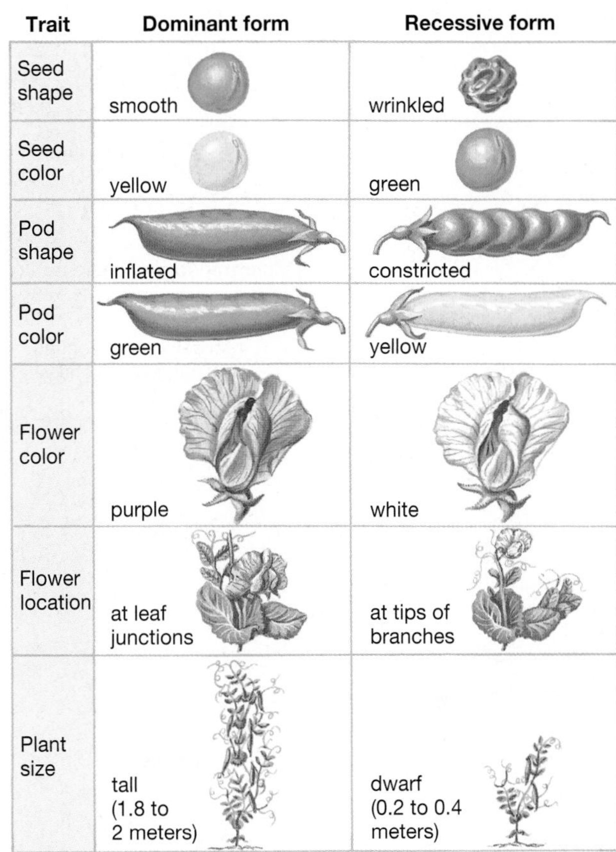

Figure 11-5 Traits of pea plants that Mendel studied

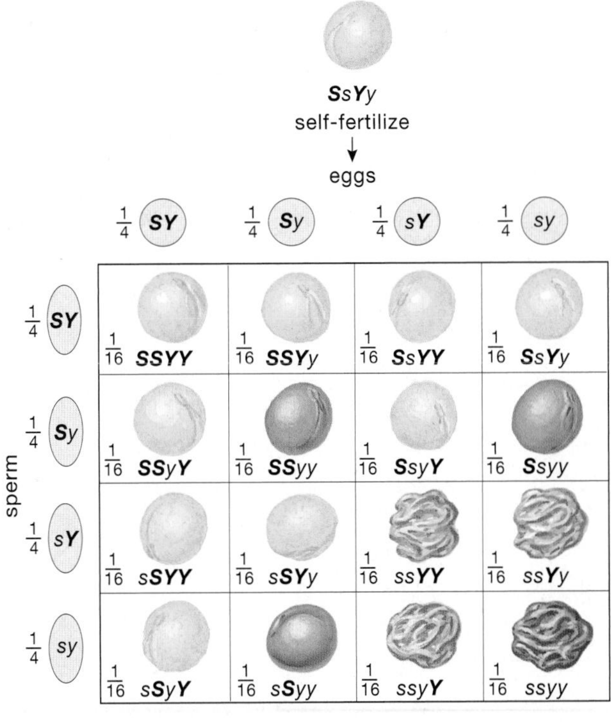

Figure 11-6 Predicting genotypes and phenotypes for a cross between parents that are heterozygous for two traits In pea seeds, yellow color (*Y*) is dominant to green (*y*) and smooth shape (*S*) is dominant to wrinkled (*s*). In this cross, both parents are heterozygous for each trait (or a single individual heterozygous for both traits self-fertilizes). There are now 16 boxes in the Punnett square. In addition to predicting all the genotypic combinations and the 9:3:3:1 overall phenotypic ratio, the Punnett square predicts $\frac{3}{4}$ yellow seeds, $\frac{1}{4}$ green seeds, $\frac{3}{4}$ smooth seeds, and $\frac{1}{4}$ wrinkled seeds, just as we would expect from crosses made of each trait separately.

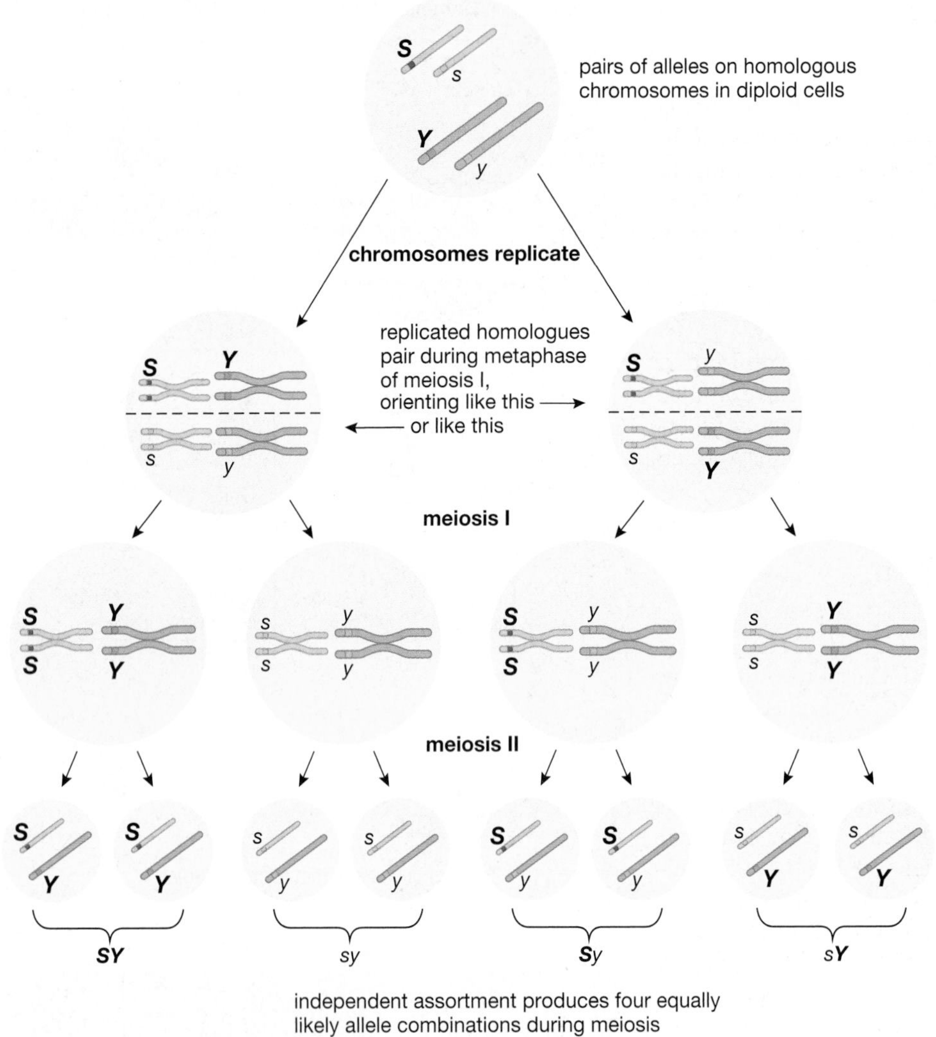

Figure 11-7 Independent assortment of alleles Chromosome movements during meiosis produce independent assortment of alleles, as shown here for two genes. Each possible combination is equally likely, producing gametes in the proportions $\frac{1}{4}$ *SY*, $\frac{1}{4}$ *sy*, $\frac{1}{4}$ *sY*, and $\frac{1}{4}$ *Sy*.

between organisms that are heterozygous for two traits, and how two independent 3:1 ratios combine to produce an overall 9:3:3:1 ratio.

Mendel's conclusion that two or more distinct traits are inherited independently is known as the **law of independent assortment**. Multiple traits are inherited independently because the alleles of one gene are distributed to gametes independently of the alleles of other genes. The physical basis for this independence is found in the events of meiosis (see Chapter 10). When paired homologous chromosomes line up during metaphase I, homologues face the poles of the cell at random, and the orientation of any one homologous pair does not influence the orientation of other pairs. Therefore, when the homologues separate during anaphase I, the alleles of genes on different chromosomes are distributed, or "assorted," independently (Fig. 11-7). Note that only genes that are located on different chromosomes assort independently.

In an Unprepared World, Genius May Go Unrecognized

Gregor Mendel presented his theories of inheritance in 1865, and they were published the following year. His paper did not mark the beginning of genetics. In fact, it didn't make any impression at all during his lifetime. Mendel's experiments, which eventually spawned one of the most important scientific theories in all of biology, simply vanished from the scene. Apparently, very few biologists read his paper, and those who did failed to recognize its significance or discounted it because it contradicted prevailing ideas of inheritance.

It was not until 1900 that three biologists—Carl Correns, Hugo de Vries, and Erich Tschermak—working independently and knowing nothing of Mendel's work, rediscovered the principles of inheritance. No doubt to their intense disappointment, when they searched the scientific literature before publishing their results, they found that Mendel had scooped them more than 30 years earlier. To their credit, they graciously acknowledged the important work of the Austrian monk, who had died in 1884.

11.5 How Are Genes Located on the Same Chromosome Inherited?

As we have seen, each chromosome contains many genes (see Fig. 11-1). *Chromosomes*, not *genes*, are separated during meiosis. For this reason, genes on the same chromosome do not normally assort independently.

Genes on the Same Chromosome Tend to Be Inherited Together

Genetic **linkage** is the inheritance of genes as a group because they are on the same chromosome. One of the first pairs of linked genes to be discovered was found in the sweet pea, a different species from Mendel's garden pea. In sweet peas, the gene for flower color and the gene for pollen grain shape are located on the same chromosome. Thus, alleles of these genes normally *assort together* during meiotic cell division and are *inherited together*.

Consider a sweet pea plant with red flowers and long pollen that has the following pair of homologous chromosomes. ➔

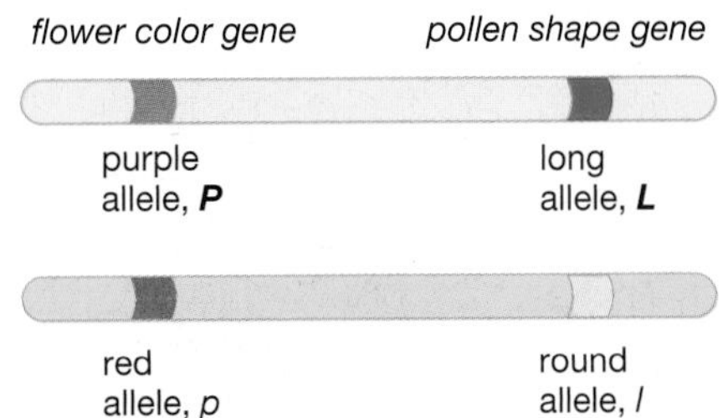

Note that the allele for purple flower color and the allele for long pollen shape are located on one homologue. The allele for red flower color and the one for round pollen shape are located on the other homologue. The gametes produced by this sweet pea plant are likely to have *either* alleles for purple and long *or* alleles for red and round. This pattern of inheritance violates the law of independent assortment, which states that the alleles for flower color and pollen shape should segregate independently. Instead, they tend to stay together during meiotic cell division and to be inherited together.

Crossing Over Can Create New Combinations of Linked Alleles

Genes on the same chromosome do not always stay together. In the sweet pea cross just described, for example, an F_2 generation commonly includes a few plants in which the genes for flower color and pollen shape have been inherited as if they were not linked. What accounts for these exceptions?

Recall from Chapter 10 that homologous chromosomes may exchange DNA during prophase I of meiosis, in a process called **crossing over**. Normally, there is at least one exchange between each pair of homologous chromosomes during each meiotic cell division. Look back at the sweet pea chromosomes in the previous section. The homologous chromosomes started out with either the purple-long or the red-round combination of alleles. If crossing over occurs between the color locus and the pollen shape locus, there will be new combinations of alleles, purple-round or red-long, on the recombined chromosomes. Then, when the homologues separate at anaphase I, the chromosomes that each haploid daughter cell receives will have different combinations of alleles from those of the parent cell.

11.6 How Is Sex Determined?

In animals of many species, an individual's sex is determined by a special pair of chromosomes, the **sex chromosomes**. In mammals, for example, males and females have the same total number of chromosomes, but females have two identical sex chromosomes, called X chromosomes, whereas males have one X chromosome and one Y chromosome (Fig. 11-8). (Recall from Chapter 9 that the Y chromosome contains a gene, known as *SRY*, that acts as a genetic "switch" to activate other genes that cause the development of male characteristics.) The rest of the chromosomes, called **autosomes**, are found in both sexes.

How is the sex of a fertilized egg determined? Even though the Y is much smaller than the X, the X and Y chromosomes act as homologues, pairing up during prophase of meiosis I and separating during anaphase I. Therefore, during gamete formation, the sex chromosomes segregate. Each egg receives one X chromosome (because females are XX), and each sperm receives either an X or

Figure 11-8 Photomicrograph of human sex chromosomes Notice the small size of the Y chromosome, which carries relatively few genes.

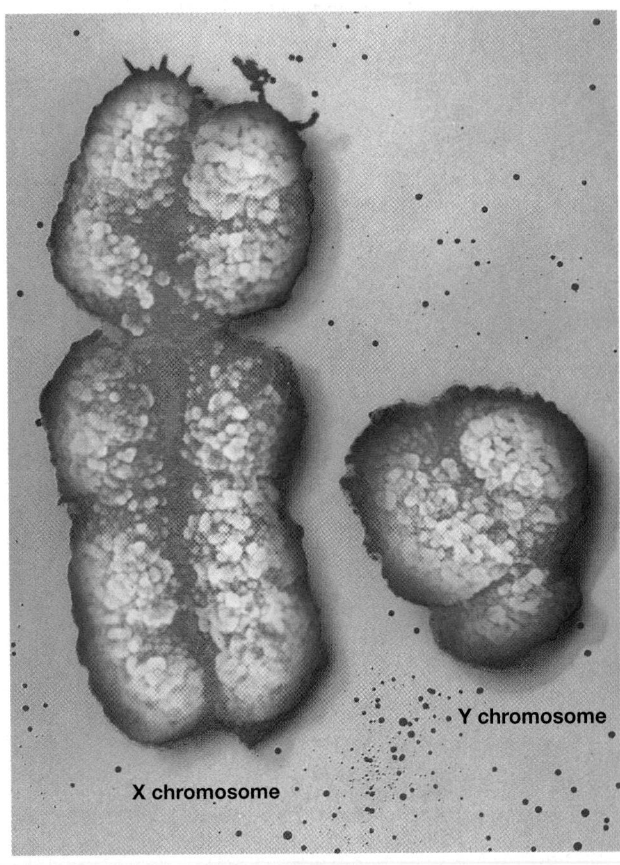

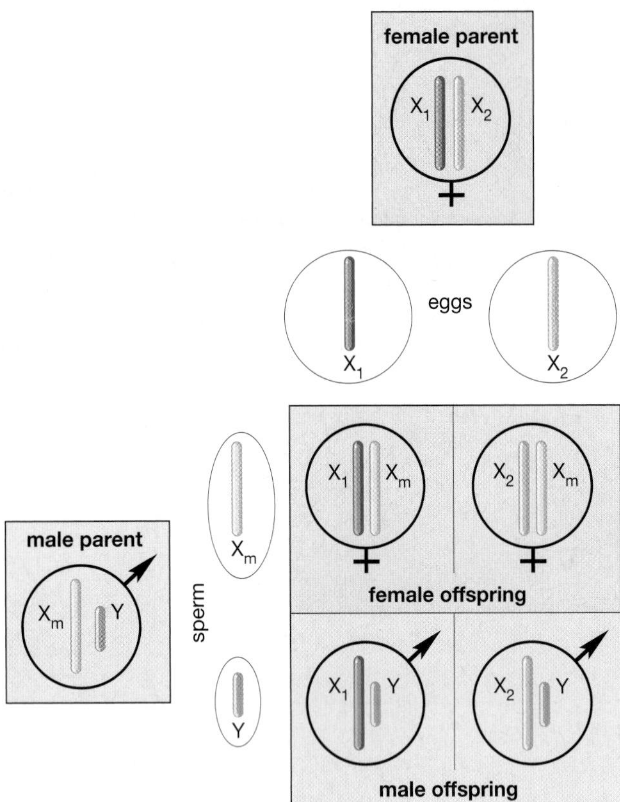

Figure 11-9 Sex determination in mammals Male offspring receive their Y chromosome from the father. Female offspring receive the father's X chromosome (labeled X_m). Both male and female offspring receive an X chromosome (either X_1 or X_2) from the mother.

a Y chromosome (because males are XY). An egg that is fertilized by an X-bearing sperm yields a female offspring. An egg fertilized by a Y-bearing sperm produces a male offspring (Fig. 11-9).

11.7 How Are Sex-Linked Genes Inherited?

Genes that are found on one sex chromosome but not on the other are called **sex-linked**. In many animal species, the Y chromosome carries only a few genes. As of late 2003, only 78 genes had been found on the human Y chromosome. Most of these play a role in male reproduction. In contrast, the human X chromosome contains over 1000 genes, few of which have a specific role in reproduction. Therefore, most of the genes on the human X chromosome have no counterpart on the Y chromosome, including genes for color vision, blood clotting, and certain structural proteins in muscles.

Because they have two X chromosomes, females can be either homozygous or heterozygous for genes on the X chromosome, and normal dominant versus recessive relationships among alleles apply. In contrast, males express *all* the alleles they have on their single X chromosome, whether those alleles are dominant or recessive. For this reason, recessive traits encoded by genes on the X chromosome always show up phenotypically in males. In humans such traits include color blindness, hemophilia, and certain types of muscular dystrophy. We will return to this concept later in the chapter.

11.8 Do the Mendelian Rules of Inheritance Apply to All Traits?

In our discussion of patterns of inheritance thus far, we looked at only the simplest kinds of traits: those that are completely controlled by a single gene, those with only two possible alleles of each gene, and those in which one allele is completely dominant to the other, recessive, allele. Most traits, however, are influenced in more varied and subtle ways, so Mendel's rules of inheritance require some further modifications.

Incomplete Dominance Produces Intermediate Phenotypes

When one allele is completely dominant over a second allele, heterozygotes with one dominant allele have the same phenotype as homozygotes with two dominant alleles. However, relationships between alleles are not always this simple. When the heterozygous phenotype is *intermediate* between the two homozygous phenotypes, the pattern of inheritance is called **incomplete dominance**.

In snapdragons, for example, crossing homozygous red-flowered plants (RR) with homozygous white-flowered ones ($R'R'$) does not produce F_1 hybrids with red flowers. Instead, the F_1 (RR') flowers are pink. (Incompletely dominant alleles are often given uppercase symbols with a superscript prime to denote the different alleles, such as R and R'.) The alleles still remain unchanged from generation to generation. In the F_2 generation, the red and white colors of the homozygotes are as pure as ever (Fig. 11-10). The F_2 offspring include about $\frac{1}{4}$ red (RR), $\frac{1}{2}$ pink (RR'), and $\frac{1}{4}$ white ($R'R'$) flowers.

How can an allele be incompletely dominant? For some traits, *two* copies of a particular allele are required for expression of the dominant phenotype. For example, the R allele in snapdragons codes for an enzyme that catalyzes the formation of red pigment. The R' allele codes for a defective, nonfunctional enzyme. Plants with the RR genotype produce lots of red pigment and have red flowers; those with the $R'R'$ genotype produce no pigment and have white flowers. When only one copy of the R allele is present, an intermediate amount of red pigment is produced, so the flowers are pink. The R allele is incompletely dominant because heterozygotes exhibit a phenotype that is intermediate between the phenotypes of homozygous dominants and homozygous recessives.

A Single Gene May Have Multiple Alleles

A single individual can have only two alleles for any gene, one on each homologous chromosome. However, if we could sample *all* the individuals of a species, we would typically find dozens of alleles for every gene. One eye-color gene in fruit flies, for example, has more than a thousand alleles. Depending on how the alleles are combined, the eyes of fruit flies can be white, yellow, orange, pink, brown, or red. There are hundreds of alleles for both Marfan syndrome and cystic fibrosis (see "Scientific Inquiry: Cystic Fibrosis"), each of which arose as a new mutation.

Human blood types are an example of multiple alleles of a single gene. The blood types A, B, AB, and O arise as a result of three different alleles (for simplicity, we will designate them A, B, and o) of a single gene located on chromosome 9. This gene codes for an enzyme responsible for adding sugar molecules to the ends of glycoproteins that protrude from the surfaces of red blood cells. Alleles A and B code for enzymes that add different sugars to the glycoproteins (we will call the resulting molecules glycoproteins A and B). Allele o codes for a nonfunctional enzyme that doesn't add any sugar. A person may have one of six genotypes: AA, BB, AB, Ao, Bo, or oo.

Alleles A and B are dominant to o. Therefore, people with genotypes AA or Ao have only type A glycoproteins and have type A blood. Those with genotypes BB or Bo synthesize only type B glycoproteins and have type B blood. Homozygous recessive oo individuals lack both types of glycoproteins and have type O blood. In people with type AB blood, however, both enzymes are present, so the plasma membranes of their red blood cells have both A and B glycoproteins. When heterozygotes express phenotypes of *both* of the homozygotes (in this case, both A and B glycoproteins), the pattern of inheritance is called **codominance**, and the alleles are said to be *codominant* to one another.

People make antibodies to the type of glycoproteins that they lack. These antibodies are immune system proteins in blood plasma that bind to foreign glycoproteins by recognizing different sugars. The antibodies cause red blood cells that bear foreign glycoproteins to clump together and to rupture, clogging small blood vessels and damaging vital organs such as the brain, heart, lungs, or kidneys. Therefore, blood types must be matched carefully before a blood transfusion is made.

Type O blood, lacking any sugars, is not attacked by antibodies in A, B, or AB blood, so it can be transfused safely to all other blood types. (The antibodies present in transfused blood become too diluted to cause problems.) People with type O blood are called universal donors. But O blood carries antibodies to both A and B glycoproteins, so type O individuals can receive transfusions of only type O blood. Can you predict the blood type of people called universal recipients? Table 11-1 summarizes blood types and transfusion characteristics.

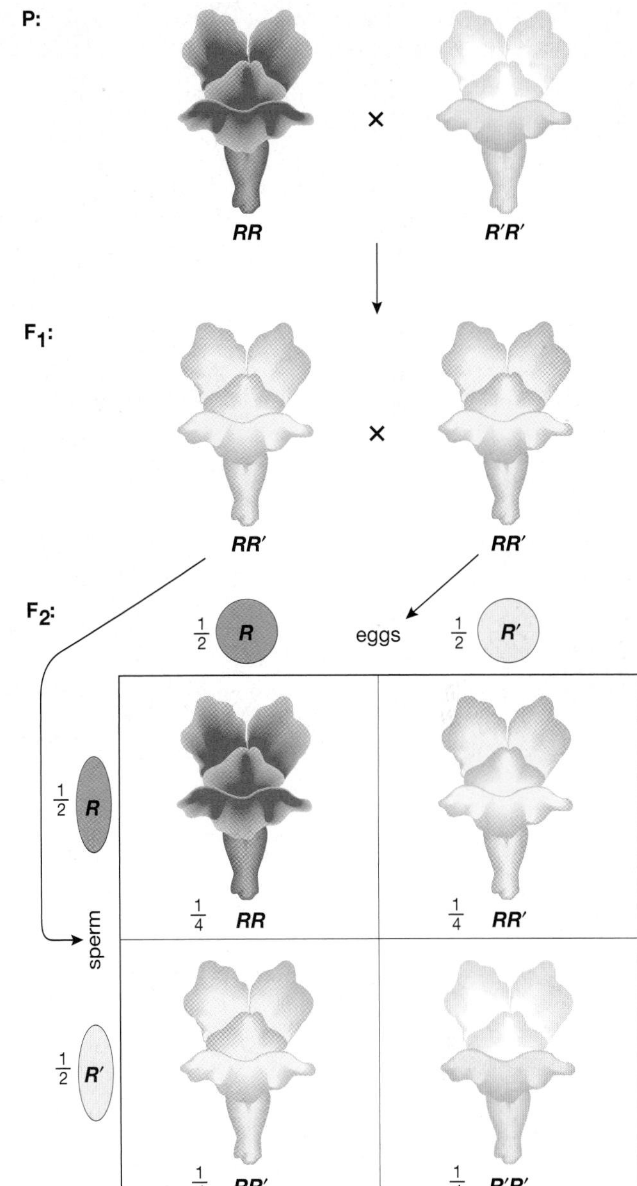

Figure 11-10 Incomplete dominance The inheritance of flower color in snapdragons is an example of incomplete dominance. (In such cases, we use capital letters for both alleles, here R and R'.) Heterozygtes (RR') have pink flowers, whereas the homozygotes are red (RR) or white ($R'R'$). **QUESTION:** *Is it possible for plant breeders to develop a true-breeding pink-flowered snapdragon?*

A Single Trait May Be Influenced by Several Genes

If you look around your class, you are likely to see people of varied heights, skin colors, and body builds, to consider just a few obvious traits. These traits are governed not by single genes but by interactions among two or more genes, a phenomenon called **polygenic inheritance**.

In humans, polygenic inheritance is usually quite complex. As you might imagine, the more genes that contribute to a single trait, the greater the number of phenotypes and the finer the distinctions among them. When more than three genes contribute to a trait, differences among phenotypes are small, and it is extremely difficult to classify the phenotypes reliably. For example, at least three

TABLE 11-1 Human Blood Group Characteristics

Blood Type	Genotype	Red Blood Cells	Has Plasma Antibodies to:	Can Receive Blood from:	Can Donate Blood to:	Frequency in U.S.
A	*AA* or *Ao*	A glycoprotein	B glycoprotein	A or O (no blood with B glycoprotein)	A or AB	40%
B	*BB* or *Bo*	B glycoprotein	A glycoprotein	B or O (no blood with A glycoprotein)	B or AB	10%
AB	*AB*	Both A and B glycoproteins	Neither A nor B glycoprotein	AB, A, B, O (universal recipient)	AB	4%
O	*oo*	Neither A nor B glycoprotein	Both A and B glycoproteins	O (no blood with A or B glycoprotein)	O, AB, A, B (universal donor)	46%

or four genes, each with at least two alleles, control skin pigmentation in people. Exposure to the sun further alters skin color, with the result that humans show almost continuous variation from very dark to very light skin.

A Single Gene May Have Multiple Effects on Phenotype

We have just seen that several different genes may influence a single trait. The reverse is also true: A single gene may influence several different traits, a phenomenon called **pleiotropy**. A good example is the *SRY* gene, discovered in 1990 on the Y chromosome (see Chapter 9). *SRY* codes for a protein that activates other genes, which in turn code for proteins that cause an embryo to develop as a male. Testes form and secrete sex hormones that stimulate the development of male reproductive structures, such as the penis and scrotum, and influence many other characteristics, including facial hair, body size, and muscle development. In this way, a single gene, *SRY*, has wide-ranging effects on many traits.

Figure 11-11 Environmental influence on phenotype Expression of the gene for black fur in the Himalayan rabbit depends on an interaction between genotype and environment. The gene is expressed only on the cooler parts of the rabbit's body.

The Environment Influences the Expression of Genes

An organism's phenotype is not just the sum of its genes. The environment in which an organism lives also plays a large role in determining phenotype. A striking example of environmental effects on gene action is provided by the Himalayan rabbit, which, like the Siamese cat, has pale body fur but black ears, nose, tail, and feet. The Himalayan rabbit actually has the genotype for black fur all over its body. However, the enzyme that produces the black pigment is inactive at temperatures above about 93 °F (34 °C). Because most of a rabbit's body surface is warmer than 93 °F, most of the rabbit's fur is pale. Its extremities, however, are usually cooler than the rest of the body, so its ears, nose, tail, and feet have black fur (Fig. 11-11).

All traits that are influenced by genes are also influenced by environmental factors. Even traits that seem to be under rigid genetic control, such as the color of pea flowers, depend on environmental conditions in the sense that pea plants

SCIENTIFIC INQUIRY

CYSTIC FIBROSIS

Woe to that child which when kissed on the forehead tastes salty. He is bewitched and soon must die.

—17th-century English saying

This adage is based on a remarkably accurate diagnostic tool for the most common recessive genetic disorder in the U.S. and Europe—cystic fibrosis. About 30,000 Americans, 3000 Canadians, and 20,000 Europeans have cystic fibrosis. The story of this disease is a blend of physiology, medicine, and both Mendelian and molecular genetics.

Let's begin with a child's salty forehead. Sweat is mostly water and cools the body by evaporating from the skin. However, sweat also contains a lot of salt (sodium chloride) when it is first secreted by the sweat glands, about as much as in blood. As sweat moves through tubes connecting the secreting cells with the surface of the skin, most of the salt is reclaimed if you are perspiring slowly enough. How? Pumps in the plasma membranes of the cells lining the tubes transport negatively charged chloride ions out of the sweat and return them to the body fluids. Positively charged sodium ions follow along by electrical attraction. Cystic fibrosis is caused by defective chloride pumps: Salt stays in the sweat, so the skin tastes salty.

Salty sweat isn't very harmful, but, unfortunately, the cells lining the lungs have the same chloride pumps. In the lungs, the pumps move chloride onto the surface of the airways. As you recall from Chapter 3, water "follows" ions by osmosis, so the secreted sodium and chloride ions cause water to move to the airway surfaces. Some cells of the airways also secrete mucus. Ideally, the water dilutes the mucus, so that the fluid on the airway surfaces is thin and watery. Why does this matter? The mucus, along with bacteria and debris that it traps, is swept out of the lungs by cilia on the cells. In cystic fibrosis, because of the reduced chloride secretion, not much water reaches the airway surfaces, so the mucus is thick and the cilia can't move it very well. The mucus clogs the airways, and bacteria remain in the lungs, causing frequent infections. Even if a person survives the infections, the lungs usually become permanently damaged. Mucus also builds up in the stomach and intestines, reducing the absorption of nutrients and causing malnutrition. Before modern medical care, most people with cystic fibrosis died by 4 or 5 years of age. Even now, the average life span is only about 30 to 35 years.

Mutations in the gene that encodes the protein of the chloride pump cause cystic fibrosis. Researchers have identified nearly 1000 mutations in this gene. Some mutations introduce a stop codon into the middle of the mRNA, which cuts off translation before the pump protein is completed. Others change the protein in ways that reduce the pumping rate. The most common mutation prevents the pump protein from moving through the endoplasmic reticulum and Golgi apparatus to the plasma membrane. Overall, about 1 American in 30 carries one of these mutations.

Why is cystic fibrosis inherited as a recessive trait? People who are heterozygous, with one normal allele and one copy of any of these mutations, produce enough chloride pump proteins to provide adequate chloride transport. Therefore, they are phenotypically normal; that is, they produce watery secretions in their lungs and do not develop cystic fibrosis. Someone with two defective alleles will have no functioning chloride pumps and will develop the disease.

Can anything be done to prevent, cure, or control the symptoms of cystic fibrosis? Because cystic fibrosis is a genetic disorder, the only way to prevent the disease is to prevent the birth of affected infants. However, people usually don't know if they are carriers and therefore don't know if their children may inherit the disease. Treatments that reduce lung damage in cystic fibrosis include physical manipulation to drain the lungs, medicines that open the airways (similar to those used by people with asthma), and frequent, even continuous, administration of antibiotics (Fig. E11-1). Unfortunately, these treatments postpone, but cannot completely prevent, damage to the lungs, intestine, and other organs.

All of that may change in the next few years. Today, medical labs can identify carriers by a blood test and homozygous recessive embryos by prenatal diagnosis (see "Health Watch: Prenatal Genetic Screening" in Chapter 12). Soon, children with cystic fibrosis may be cured, or close to it, by one of several gene therapies now under development. We will explore both of these applications of biotechnology in Chapter 12.

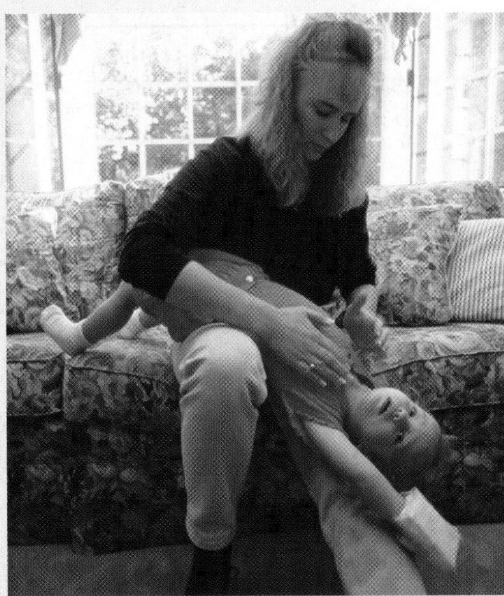

Figure E11-1 Cystic fibrosis A child is treated for cystic fibrosis. Gentle pounding on the chest and back while the child is held upside-down helps to dislodge mucus from the lungs. A device on the child's wrist injects antibiotics into a vein. These treatments combat the numerous lung infections to which cystic fibrosis patients are vulnerable.

will not produce flowers unless the environment provides the correct mix of water, nutrients, sunlight, and so on. More complex traits have correspondingly complex interactions between genetic and environmental influences. For example, the polygenic trait of human skin color is modified by the environmental effects of sun exposure. Height, another polygenic trait, can be reduced by poor nutrition. By the same token, even the most sophisticated human behaviors depend on physical structures (such as the brain) whose construction is influenced by genes. In fact, genetic differences have been shown to influence human traits as diverse as intelligence, sexuality, and susceptibility to cancer. Determining the relative contributions of genes and environment to such complex characteristics is difficult and subject to uncertainty.

(a) A pedigree for a dominant trait

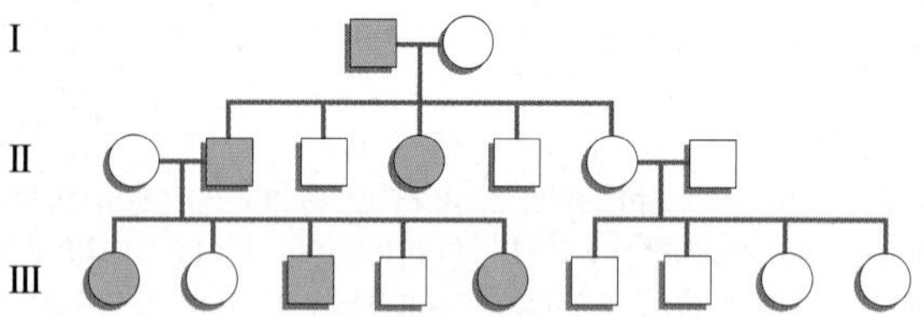

(b) A pedigree for a recessive trait

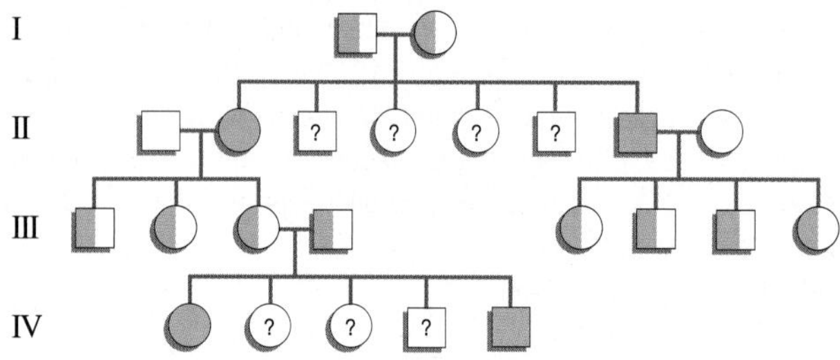

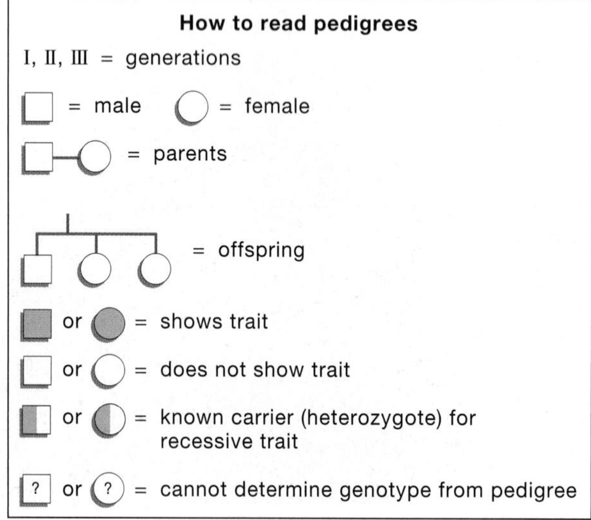

Figure 11-12 Family pedigrees ***(a)*** A pedigree for a dominant trait. Note that any offspring showing a dominant trait *must* have at least one parent with the trait. ***(b)*** A pedigree for a recessive trait. Any individual showing a recessive trait must have been homozygous recessive. If that person's parents did not show the trait, then *both* of the parents must have been heterozygotes (carriers). Note that the genotype cannot be determined for some offspring, who may be either heterozygotes or homozygous dominants.

11.9 How Are Human Genetic Disorders Investigated?

Geneticists are especially interested in genes that influence our susceptibility to disease. Because experimental crosses with humans are out of the question, human geneticists study the inheritance of disease-causing alleles by searching medical, historical, and family records to study past crosses. Records extending across several generations can be arranged in the form of family **pedigrees**, diagrams that show the genetic relationships among a set of related individuals (**Fig. 11-12**).

Careful analysis of pedigrees can reveal whether a particular trait is inherited in a dominant, recessive, or sex-linked pattern. Since the mid-1960s, analysis of human pedigrees, combined with molecular genetic technology, has produced great strides in understanding human genetic diseases. For instance, geneticists now know the genes responsible for dozens of inherited diseases, such as sickle-cell anemia, Marfan syndrome, and cystic fibrosis. Research in molecular genetics promises to increase our ability to predict genetic diseases and perhaps even to cure them, a topic we will explore further in Chapter 12.

11.10 How Are Single-Gene Disorders Inherited?

Many common human traits, including freckles, long eyelashes, cleft chin, and widow's peak hairline, are inherited in a simple Mendelian fashion: Each trait is controlled by a single gene with a dominant and a recessive allele. Here, we will focus on a few examples of medically important disorders that also exhibit simple Mendelian inheritance.

Some Human Genetic Disorders Are Caused by Recessive Alleles

Disorders can be caused by single-gene mutations that damage or destroy key enzymes. In some disorders, however, the effects of such a mutation are masked when one normal allele is present and can generate a sufficient amount of functional enzyme. In such cases, the person who is heterozygous (with one normal and one mutant allele) will remain healthy and have a phenotype that is indistinguishable from that of homozygotes with two copies of the normal allele. In other words, the mutant allele is recessive, and only people who inherit two copies of the mutant allele have the disorder. Cystic fibrosis, which affects about 30,000 Americans, is a recessive disease of this type (see "Scientific Inquiry: Cystic Fibrosis").

Individuals who are heterozygous for a harmful recessive allele have a normal phenotype, but they can pass the defective allele to their offspring. Geneticists estimate that each of us carries 5 to 15 harmful recessive alleles that would cause a serious genetic defect in a homozygote. When a couple has a child, each of the defective alleles in each parent has a 50:50 chance of being passed on, so the probability is very high that the child will inherit some harmful recessive alleles. Fortunately, however, the child is unlikely to be homozygous for a recessive allele for a genetic disease because it is unlikely that both parents carry defective alleles of the *same* gene.

Recessive disorders in offspring are more common if the parents are closely related (especially if they are first cousins or closer). Related couples have inherited some of their genes from a recent common ancestor, so they are much more likely to carry defective recessive alleles of the same genes. Each child born to a couple that shares a harmful recessive allele has a 1 in 4 chance of being affected by the genetic disorder (see Fig. 11-12).

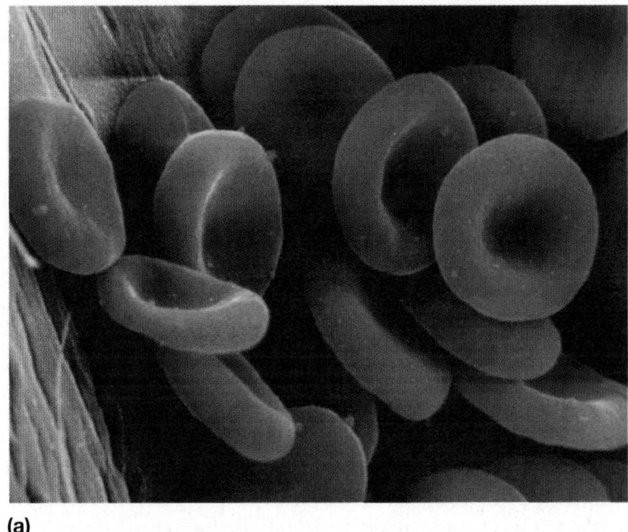

(a)

(b)

Figure 11-13 Sickle-cell anemia ***(a)*** Normal red blood cells are disk shaped and have an indented center. ***(b)*** The red blood cells of a person with sickle-cell anemia become sickle shaped (sickled).

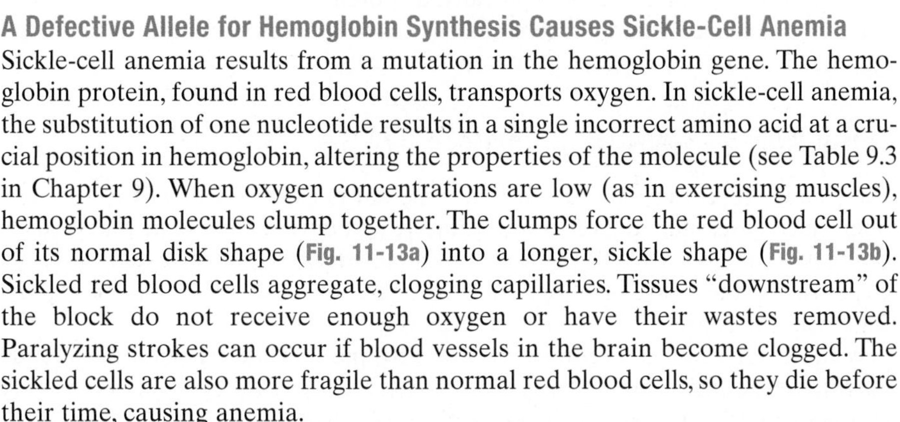

A Defective Allele for Hemoglobin Synthesis Causes Sickle-Cell Anemia

Sickle-cell anemia results from a mutation in the hemoglobin gene. The hemoglobin protein, found in red blood cells, transports oxygen. In sickle-cell anemia, the substitution of one nucleotide results in a single incorrect amino acid at a crucial position in hemoglobin, altering the properties of the molecule (see Table 9.3 in Chapter 9). When oxygen concentrations are low (as in exercising muscles), hemoglobin molecules clump together. The clumps force the red blood cell out of its normal disk shape (**Fig. 11-13a**) into a longer, sickle shape (**Fig. 11-13b**). Sickled red blood cells aggregate, clogging capillaries. Tissues "downstream" of the block do not receive enough oxygen or have their wastes removed. Paralyzing strokes can occur if blood vessels in the brain become clogged. The sickled cells are also more fragile than normal red blood cells, so they die before their time, causing anemia.

People homozygous for the sickle-cell allele synthesize only defective hemoglobin. Therefore, many of their red blood cells become sickled. Although heterozygotes have about half normal and half abnormal hemoglobin, they usually have few sickled cells and are usually not seriously affected; in fact, many world-class athletes are heterozygous for the sickle-cell allele. Because usually only homozygous recessives show symptoms, sickle-cell anemia is a recessive disorder.

About 8% of the African-American population is heterozygous for sickle-cell anemia, but the allele is extremely rare in whites. Why? People who are heterozygous for the sickle-cell allele are less susceptible to malaria than are homozygotes without the allele. Malaria is a potentially fatal disease caused by a parasite transmitted to people by mosquito bites. Before the development of effective medical treatments, malaria was a major killer in the warmer regions of the world, including much of Africa. (Even today, there are hundreds of millions of cases of malaria worldwide.) In Africa, therefore, natural selection favored heterozygotes, who had resistance to malaria, even at the cost of anemia and blood clots in their homozygous children. In colder parts of the world, where malaria is uncommon, the sickle-cell allele has only harmful effects, so it was selected against.

Some Human Genetic Disorders Are Caused by Dominant Alleles

Some genetic diseases are caused by dominant alleles, in which a single defective allele is enough to cause the disorder. Therefore, for diseases caused by dominant alleles to be passed on to offspring, at least one parent must suffer from the disease, which means that this parent must remain healthy enough, long enough, to mature and have children. In some dominant diseases, such as Huntington disease, symptoms do not appear until after the affected person is old enough to have reproduced. Occasionally, a defective dominant allele may result from a new mutation in a normal person's eggs or sperm.

How can a mutant allele be dominant to the normal allele? Some dominant alleles produce an abnormal protein that interferes with the function of the normal one. Other dominant alleles may encode proteins that carry out new, toxic reactions. Finally, dominant alleles may encode a protein that is overactive, performing its function too rapidly or at inappropriate times and places.

One dominant genetic disease in humans is Huntington disease, which causes a slow, progressive deterioration of parts of the brain, resulting in a loss of coordination, flailing movements, personality disturbances, and eventual death. The symptoms of Huntington disease typically do not appear until 30 to 50 years of age. Therefore, many people pass the allele to their children before they suffer the first symptoms. In 1983, molecular genetic technology and painstaking pedigree analysis were used to localize the Huntington gene to a small part of chromosome 4. Geneticists finally isolated the Huntington gene in 1993 and a few years later identified the gene's product, a protein they named huntingtin. The function of normal huntingtin remains unknown. Mutant huntingtin forms large aggregates in nerve cells that ultimately kill the cells.

Some Human Genetic Disorders Are Sex-Linked

If a gene is present on one sex chromosome but not on the other, its pattern of inheritance is sex-linked. For example, the X chromosome contains many genes that have no counterpart on the Y chromosome. Because males have only one X chromosome, they have only one allele for each of these genes. Therefore, this single allele is expressed, even if it is recessive.

A son receives his X chromosome from his mother and passes it only to his daughters (see Fig. 11-9). Thus, sex-linked disorders caused by a recessive allele have a distinctive pattern of inheritance. Such disorders appear far more frequently in males than in females and typically skip generations: An affected male passes the allele to a phenotypically normal daughter, who in turn bears affected sons. Genetic defects caused by recessive alleles of X-chromosome genes include red-green color blindness (**Fig. 11-14**) and hemophilia.

Hemophilia is caused by a recessive allele on the X chromosome that results in a deficiency of one of the proteins needed for blood clotting. People with hemophilia bleed excessively from a wound or from mild damage to internal structures, and they bruise easily. Hemophiliacs often have anemia due to blood loss. Today, modern medical treatment allows many hemophiliacs to live long lives, but even in earlier times some hemophiliac males survived long enough to pass their defective allele to their daughters, who carried the allele and passed it to their sons.

11.11 How Do Errors in Chromosome Number Affect Humans?

In Chapter 10, we examined the intricate processes of meiotic cell division, which ensure that each sperm and each egg receive only one chromosome from each homologous pair. Not surprisingly, this elaborate dance of the chromosomes occasionally misses a step, resulting in gametes that have too many or too few chromosomes. Such errors in meiosis, called **nondisjunction**, can affect the number of either sex chromosomes or autosomes. Most embryos that arise from gametes with abnormal chromosome numbers spontaneously abort, accounting

(a)

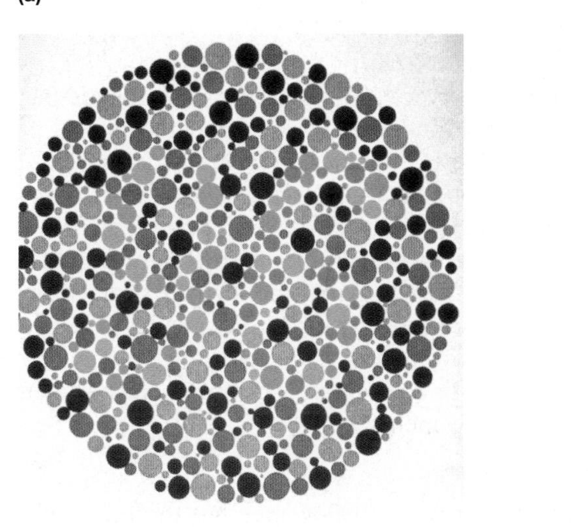

(b)

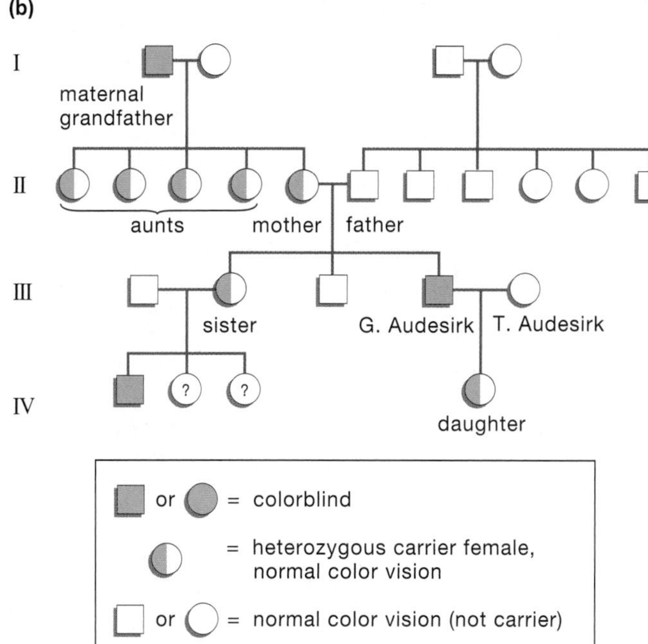

Figure 11-14 Color blindness, a sex-linked recessive trait ***(a)*** This figure, called an *Ishihara chart* after its inventor, distinguishes color-vision defects. People with red-deficient vision see a 6, and those with green-deficient vision see a 9. People with normal color vision see 96. ***(b)*** Pedigree of one of the authors (G. Audesirk), showing sex-linked inheritance of red-green color blindness. Both the author and his maternal grandfather are color deficient. His mother and her four sisters carry the trait but have normal color vision. This pattern of more-common phenotypic expression in males and transmission from affected male to carrier female to affected male is typical of sex-linked recessive traits.

for 20% to 50% of all miscarriages. Some embryos with abnormal chromosome numbers survive to birth or beyond. Abnormal numbers of chromosomes can be diagnosed before birth by examining the chromosomes of fetal cells (see "Health Watch: Prenatal Genetic Screening" in Chapter 12).

Abnormal Numbers of Sex Chromosomes Cause Some Disorders

Because the X and Y chromosomes pair up during meiosis, sperm normally carry either an X or a Y chromosome. Nondisjunction of sex chromosomes in a male, however, produces sperm that have two sex chromosomes (XX, YY, or XY) or that have no sex chromosomes at all (these are designated O). Similarly, nondisjunction in a female can produce XX or O eggs instead of normal eggs with a single X. When these defective sperm or eggs fuse with normal gametes, the resulting offspring have abnormal numbers of sex chromosomes (Table 11-2). The most common abnormalities are XO, XXX, XXY, and XYY. (Genes on the X chromosome are essential to survival, so any embryo without at least one X chromosome spontaneously aborts very early in development.)

TABLE 11-2 Effects of Nondisjunction of the Sex Chromosomes During Meiosis

Nondisjunction in father			
Sex Chromosomes of Defective Sperm	**Sex Chromosomes of Normal Egg**	**Sex Chromosomes of Offspring**	**Phenotype**
O (none)	X	XO	Female—Turner syndrome
XX	X	XXX	Female—Trisomy X
YY	X	XYY	Male—Jacob syndrome
XY	X	XXY	Male—Klinefelter syndrome
Nondisjunction in mother			
Sex Chromosomes of Normal Sperm	**Sex Chromosomes of Defective Egg**	**Sex Chromosomes of Offspring**	**Phenotype**
X	O (none)	XO	Female—Turner syndrome
Y	O (none)	YO	Dies as embryo
X	XX	XXX	Female—Trisomy X
Y	XX	XXY	Male—Klinefelter syndrome

Turner Syndrome (XO)

About one in every 3000 phenotypically female babies has only one X chromosome, a condition known as Turner syndrome. At puberty, hormone deficiencies prevent most XO females from menstruating or developing secondary sexual characteristics, such as enlarged breasts. Treatment with hormones can promote physical development. However, because most women with Turner syndrome lack functional ovaries, hormone treatment cannot reverse their infertility. Other symptoms of Turner syndrome include short stature, folds of skin around the neck, and increased risk of cardiovascular disease, kidney defects, and hearing loss. Because women with Turner syndrome have only one X chromosome, they have more X-linked recessive disorders, such as hemophilia and color blindness, than do XX women.

Trisomy X (XXX)

About one in every 1000 women has three X chromosomes, a condition known as trisomy X or triplo X. Most such women have no detectable defects, except for a tendency to be tall. There is, however, a higher incidence of below-normal intelligence among women with trisomy X. Unlike women with Turner syndrome, most trisomy X women are fertile and, interestingly enough, almost always bear normal XX and XY children. Some unknown process must operate during meiosis to prevent the extra X chromosome from being included in eggs.

Klinefelter Syndrome (XXY)

About one male in every 1000 is born with two X chromosomes and one Y chromosome. Most of these men go through life never realizing that they have an extra X chromosome. However, at puberty, some of these men show mixed secondary sexual characteristics, including partial breast development, broadening of the hips, and small testes. These symptoms are known as Klinefelter syndrome. XXY men are usually infertile, because of low sperm count, but are not impotent. They are usually diagnosed when the affected man and his partner seek medical help for their inability to conceive a baby.

Jacob Syndrome (XYY)

Another common type of sex chromosome abnormality is XYY, present in about one male in every 1000. You might expect that having an extra Y chromosome, which has few genes, would not make very much difference, and this seems to be true in most cases. However, XYY males usually have high levels of testosterone, often have severe acne, and are tall (about two-thirds of XYY males are more than 6 feet tall, compared with the average male height of 5 feet 9 inches). Some appear to have slightly lower scores on IQ tests than their brothers who are XY.

Abnormal Numbers of Autosomes Cause Some Disorders

Nondisjunction of autosomes produces eggs or sperm that are missing an autosome or that have two copies of an autosome. Fusion of one of these abnormal gametes with a normal one (that bears one copy of each autosome) leads to an embryo with either one or three copies of the affected autosome. Embryos that have only one copy of any autosome abort so early in development that the woman never knows she was pregnant. Embryos with three copies of an autosome (trisomy) also usually abort spontaneously. However, a small fraction of embryos with three copies of chromosomes 13, 18, or 21 can develop sufficiently to be born. A baby with trisomy 21 can live into adulthood.

Trisomy 21 (Down Syndrome)

In about one of every 800 to 1000 births, the child inherits an extra copy of chromosome 21, a condition called trisomy 21 or Down syndrome. Children with Down syndrome have several distinctive physical characteristics, including weak muscle tone, a small mouth held partially open because it cannot accommodate the tongue, and distinctively shaped eyelids (**Fig. 11-15**). More serious defects

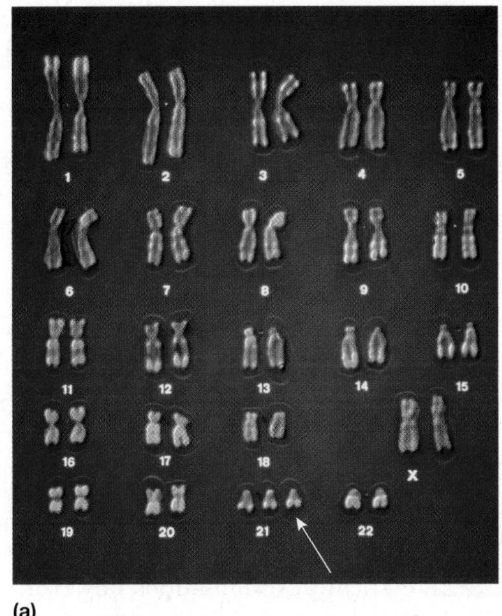

(a)

(b)

Figure 11-15 Trisomy 21, or Down syndrome ***(a)*** A karyotype of a Down syndrome child reveals three copies of chromosome 21. ***(b)*** These girls have the relaxed mouth and distinctively shaped eyes typical of Down syndrome.

include low resistance to infectious diseases, heart malformations, and varying degrees of mental retardation, often severe.

The frequency of nondisjunction in gametes increases with age. There is a small increase in defective sperm with increasing age of the father, and nondisjunction in sperm accounts for about 25% of Down syndrome cases. The mother's age, however, is a more significant factor in the probability of Down syndrome (**Fig. 11-16**). Since the 1970s, it has become more common for couples to delay having children, increasing the probability of trisomy 21. Nevertheless, older women as a group account for far fewer babies than younger women, so the majority of Down syndrome babies are born to young women.

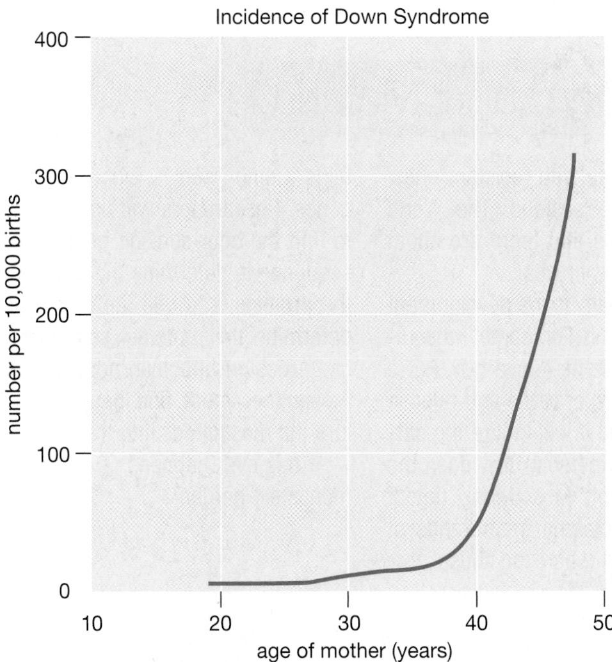

Figure 11-16 Down syndrome frequency increases with maternal age The likelihood of bearing a child with Down syndrome increases rapidly for mothers who are 35 years and older.

SUDDEN DEATH ON THE COURT *REVISITED*

Did Flo Hyman suffer from a new mutation or did she inherit Marfan syndrome? Medical examinations revealed that her father and sister have Marfan syndrome, but her mother and brother do not. Does this finding prove that Hyman inherited the defective allele from her father? As you learned in this chapter, diploid organisms, including people, generally have two alleles of each gene, one on each homologous chromosome. It has been known for many years that one defective fibrillin allele is enough to cause Marfan syndrome. Further, the children of a person with Marfan syndrome have a 50% chance of inheriting the disease. What can we conclude from all of these data?

First, if even one defective fibrillin allele produces Marfan syndrome, then Hyman's mother must carry two normal alleles, because she does not have Marfan syndrome. Second, because new mutations are rare and Hyman's father has Marfan syndrome, it is almost certain that Hyman inherited a defective fibrillin allele from her father.

Third, is Marfan syndrome inherited as a dominant or a recessive condition? Once again, if only one defective allele is enough to cause Marfan syndrome, then this allele must be dominant and the normal allele must be recessive. Finally, if Hyman had borne children, could they have inherited Marfan syndrome from her? For a dominant disorder, any children who inherited her defective allele would develop Marfan syndrome. Therefore, on average, half of her children would have had Marfan syndrome (try working this out with a Punnett square).

At the beginning of this chapter, we suggested that Hyman, Abraham Lincoln, and Akhenaten all had Marfan syndrome. Evaluating the evidence critically, you may wonder how anyone could know if Lincoln or Akhenaten had the disorder. Well, you're right—nobody really knows. The "diagnosis" is based on pictures and descriptions. For example, Akhenaten was the first pharaoh to be depicted with a long head, long arms, and long legs—all typical of Marfan syndrome. As for Lincoln, one visitor to the White House is said to have remarked, "Mr. President, what long legs you have!" To which Lincoln reportedly replied, "Just long enough to reach the ground, Madam." Although other genetic conditions can contribute to tall stature, long limbs, and large hands, the fact that these men were otherwise healthy, successful, and fertile points to Marfan syndrome.

CONSIDER THIS: Marfan syndrome can't be detected in an embryo by a simple biochemical test (yet). Most cystic fibrosis mutations, however, can easily be detected in both heterozygotes and homozygotes, in adults, children, and embryos. A few years ago, some states considered making mandatory testing of couples for cystic fibrosis part of the application for a marriage license. If two heterozygotes marry, each of their children has a 25% chance of having the disorder. Although there is no cure, there will probably be better treatments within a few years. Do you think that carrier screening should be mandatory? If you and your spouse were both heterozygotes, would you seek prenatal diagnosis of an embryo? What would you do if your embryo were destined to be born with cystic fibrosis?

LINKS TO LIFE: Mendel, Mosquitoes, and Malaria

When Gregor Mendel was studying peas almost 150 years ago, he couldn't possibly have foreseen today's applications of genetics. He may have predicted that understanding inheritance would help us to breed more productive crops and farm animals. But what would he have thought of the October 4, 2002, issue of the journal *Science*, devoted almost entirely to the genome of the mosquito that carries malaria? His first reaction might have been, and maybe yours would be too, "Why would anyone waste time and money doing *that*?" In fact, if someone could genetically manipulate mosquitoes so that the malaria parasite couldn't reproduce in them, and could replace wild mosquitoes with resistant ones, billions of people might be ecstatic. The World Health Organization estimates that there are about 300 million cases of malaria each year.

A more immediate possibility is the development of better mosquito repellents. For some mosquitoes, there's nothing like a dame—or a guy. Put a human in the middle of a herd of cattle and release one of these mosquitoes, and it will ignore the cattle and home right in on the person. How does the mosquito do that? By scent. Mosquitoes detect odors with proteins in the plasma membranes of cells on their antennae. And where do these proteins come from? Right—from instructions in genes. Researchers will use the mosquito genome to find the odor-sensing genes, use the nucleotide sequence to determine the amino acid sequence of the proteins, and use the amino acid sequence to determine the proteins' structures. Then they will try to design effective, nontoxic repellents (so the mosquitoes can't find people) and attractants (to lure the mosquitoes into traps).

If this ever happens, it will all have started in a monastery garden.

Chapter Review

Summary of Key Concepts

11.1 What Is the Physical Basis of Inheritance?

Homologous chromosomes carry the same genes but may carry two of the same alleles or two different alleles of a given gene. An organism may thus be homozygous or heterozygous for a particular gene.

11.2 How Were the Principles of Inheritance Discovered?

Gregor Mendel deduced many principles of inheritance in the mid-1800s, before the discovery of DNA, genes, chromosomes, or meiosis. He made these deductions by choosing an appropriate experimental subject, designing his experiments carefully, following offspring for several generations, and analyzing his data statistically.

11.3 How Are Single Traits Inherited?

Traits are inherited in patterns that depend on alleles that parents pass on to their offspring. Each parent provides its offspring with one allele of every gene, so that the offspring inherits two alleles for every gene. The combination of alleles in an offspring determines whether it displays a particular phenotype for a trait. Dominant alleles mask the expression of recessive alleles. Because recessive alleles can be masked, organisms with different genotypes may have the same phenotype. That is, organisms that are homozygous dominant have the same phenotype as do heterozygous organisms. Because each allele segregates randomly during meiosis, we can predict the relative proportions of offspring with a particular trait.

11.4 How Are Multiple Traits Inherited?

If the genes for two traits are located on different chromosome pairs, they assort independently of one another into the egg or sperm. Thus, crossing two organisms that are heterozygous at two loci on different chromosomes produces offspring with 10 different genotypes. If the alleles are typical dominant and recessive alleles, these progeny will display only 4 different phenotypes.

11.5 How Are Genes Located on the Same Chromosome Inherited?

Genes located on the same chromosome are linked and tend to be inherited together. Unless two linked alleles are separated by crossing over during meiosis, they pass together to the offspring.

11.6 How Is Sex Determined?

In many animals, sex is determined by sex chromosomes, often designated X and Y. Female mammals have two X chromosomes, whereas males have one X and one Y chromosome. The Y chromosome has many fewer genes than the X chromosome.

11.7 How Are Sex-Linked Genes Inherited?

Because males have only one copy of most X chromosome genes, recessive alleles on the X chromosome are more likely to be phenotypically expressed in males than in females.

11.8 Do the Mendelian Rules of Inheritance Apply to All Traits?

Not all inheritance follows the simple dominant-recessive pattern:

- In incomplete dominance, heterozygotes have a phenotype that is intermediate between the two homozygous phenotypes.
- Many traits are determined by polygenic inheritance, in which several different genes at different loci contribute to the phenotype.
- Many genes have multiple effects on the phenotype (pleiotropy).
- The environment influences the phenotypic expression of all traits.

11.9 How Are Human Genetic Disorders Investigated?

Studying the genetics of humans is similar to studying the genetics of other animals, except that experimental crosses are not feasible. Analysis of family pedigrees and, more recently, molecular genetic techniques help to determine the mode of inheritance of human traits.

11.10 How Are Single-Gene Disorders Inherited?

Many genetic disorders are inherited as recessive traits, with only homozygous recessive persons showing symptoms of the disease. Heterozygotes carry the recessive allele but do not express the trait. Many other diseases are inherited as simple dominant traits. In such cases, only one copy of the dominant allele is needed to cause the disease symptoms. Some disorders stem from recessive alleles on the X chromosome. These disorders appear more frequently in males, who may pass some of them to their grandsons through their daughters.

11.11 How Do Errors in Chromosome Number Affect Humans?

Errors in meiosis can result in gametes with abnormal numbers of sex chromosomes or autosomes. Many people with abnormal numbers of sex chromosomes have distinguishing physical characteristics. Abnormal numbers of autosomes typically lead to spontaneous abortion early in pregnancy. In rare instances, babies with trisomy for chromosomes 13, 18, or 21 are born. Down syndrome (trisomy 21) is the most common trisomy. Down babies often survive to adulthood but have mental and physical deficiencies.

Key Terms

allele *p. 168*
autosome *p. 175*
codominance *p. 177*
cross-fertilization *p. 169*
crossing over *p. 175*
dominant *p. 170*
genotype *p. 172*
heterozygous *p. 168*
homozygous *p. 168*
incomplete dominance *p. 176*
inheritance *p. 168*
law of independent assortment *p. 174*
law of segregation *p. 170*
linkage *p. 175*
locus (plural, *loci*) *p. 168*
nondisjunction *p. 182*
pedigree *p. 180*
phenotype *p. 172*
pleiotropy *p. 178*
polygenic inheritance *p. 177*
Punnett square *p. 172*
recessive *p. 170*
self-fertilization *p. 169*
sex chromosome *p. 175*
sex-linked *p. 176*

Thinking Through the Concepts

Multiple Choice

1. *An organism is described as* Rr: *red. The* Rr *is the organism's _____; red is the organism's _____; and the organism is _____.*
   a. phenotype; genotype; degenerate
   b. karyotype; hybrid; recessive
   c. genotype; phenotype; heterozygous
   d. gamete; linkage; pleiotropic
   e. zygote; phenotype; homozygous
2. *The 9:3:3:1 ratio is a ratio of*
   a. phenotypes and genotypes in the offspring from a cross of individuals that differ in one trait
   b. phenotypes in the offspring from a cross of individuals that differ in one trait
   c. phenotypes in the offspring from a cross of individuals that differ in two traits
   d. genotypes in the offspring from a cross of individuals that differ in one trait
   e. genotypes in the offspring from a cross of individuals that differ in two traits
3. *A heterozygous pink-flowered snapdragon mated with a white-flowered snapdragon would produce*
   a. all white-flowered plants in the F_1 generation
   b. all pink-flowered plants in the F_1 generation
   c. $\frac{1}{2}$ red-flowered and $\frac{1}{2}$ white-flowered plants in the F_1 generation
   d. $\frac{3}{4}$ pink-flowered and $\frac{1}{4}$ white-flowered plants in the F_1 generation
   e. $\frac{1}{2}$ pink-flowered and $\frac{1}{2}$ white-flowered plants in the F_1 generation
4. *Which is* not *true of sickle-cell anemia?*
   a. It is most common in African Americans.
   b. It involves a one-amino-acid change in hemoglobin.
   c. It involves red blood cells.
   d. It is lethal in heterozygotes because it is dominant.
   e. It affects blood circulation.
5. *Sex-linked disorders such as color blindness and hemophilia are*
   a. caused by genes on the X chromosome
   b. caused by genes on an autosome
   c. caused by genes on the Y chromosome
   d. expressed only in men
   e. expressed only when two chromosomes are homozygous recessive

Review Questions

1. Define the following terms: gene, allele, dominant, recessive, true-breeding, homozygous, heterozygous, cross-fertilization, self-fertilization.
2. Sometimes the term *gene* is used rather casually. Compare and contrast use of the terms *allele* and *locus* as alternatives to *gene*.
3. Explain why genes located on the same chromosome are said to be linked. Why do alleles of linked genes sometimes separate during meiosis? Which genes would you expect to be more tightly linked—genes that are close together on a chromosome or genes that are far apart on a chromosome? Why?
4. Define polygenic inheritance. Why is it possible for a couple to have offspring that are notably different in eye or skin color than either parent?
5. What is sex linkage? In mammals, which sex would be most likely to show recessive sex-linked traits?
6. What is the difference between a phenotype and a genotype? Does knowledge of an organism's phenotype always allow you to determine the genotype? What type of experiment would you perform to determine the genotype of a phenotypically dominant individual?
7. If one (heterozygous) parent of a couple has Huntington disease, calculate the fraction of that couple's children that would be expected to develop the disease. What would be the fraction if both parents were heterozygous?
8. Define nondisjunction, and describe the common syndromes caused by nondisjunction of sex chromosomes and autosomes.

Applying the Concepts

1. **IS THIS SCIENCE?** Many disorders, including heart disease, Alzheimer's disease, and breast cancer, have both genetic and environmental components. For example, most dietary guidelines recommend reducing the amount of salt in the diet, because excessive salt consumption can cause high blood pressure (hypertension), which in turn can cause heart disease. However, high dietary salt causes hypertension principally in genetically "salt-sensitive" people. Unfortunately, most people don't know whether they are salt sensitive. Do you think that dietary recommendations against salt are justified, if only a relatively small proportion of the population is affected? How do you think that scientific data should be used in cases such as this?

2. Mendel's numbers seem almost too perfect to be real, and some believe he may have fudged a bit on his data. Perhaps he continued to collect data until the numbers matched his predicted ratios, then stopped. Recently, there has been much publicity over violations of scientific ethics, including researchers' plagiarizing others' work, using other scientists' methods to develop lucrative patents, or just plain fabricating data. How important an issue is this for society? What are the boundary lines of ethical scientific behavior? How should the scientific community or society "police" scientists? What punishments would be appropriate for violations of scientific ethics?

Genetics Problems

(Note: An additional group of genetics problems, with answers, can be found in the Student Study Companion *and on the* Life on Earth *Web site.)*

1. In certain cattle, hair color can be red (homozygous RR), white (homozygous $R'R'$), or roan (a mixture of red and white hairs, heterozygous RR').
   - **a.** When a red bull is mated to a white cow, what genotypes and phenotypes of offspring could be obtained?
   - **b.** If one of the offspring in (a) were mated to a white cow, what genotypes and phenotypes of offspring could be produced? In what proportion?
2. The palomino horse is golden. Unfortunately for horse fanciers, palominos do not breed true. In a series of matings between palominos, the following offspring were obtained: 65 palominos, 32 cream-colored, 34 chestnut (reddish brown). What is the probable mode of inheritance of palomino coloration?
3. In the edible pea, tall (T) is dominant to short (t), and green pods (G) are dominant to yellow pods (g). List the gametes and offspring phenotypes that would be produced in the following crosses:
   - **a.** $TtGg \times TtGg$
   - **b.** $TtGg \times TTGG$
   - **c.** $TtGg \times Ttgg$
4. In tomatoes, round fruit (R) is dominant to long fruit (r), and smooth skin (S) is dominant to fuzzy skin (s). A true-breeding round, smooth tomato ($RRSS$) is crossed with a true-breeding long, fuzzy tomato ($rrss$). All the F_1 offspring are round and smooth ($RrSs$). When these F_1 plants are crossed with each other, the following F_2 generation is obtained: round and smooth: 43; long and fuzzy: 13. Are the genes for skin texture and fruit shape likely to be on the same chromosome or on different chromosomes? Explain your answer.
5. In the tomatoes of problem 4, an F_1 offspring ($RrSs$) is mated with a homozygous recessive plant ($rrss$). The following offspring are obtained: round and smooth: 583; round and fuzzy: 21; long and fuzzy: 602; long and smooth: 16. What is the most likely explanation for this distribution of phenotypes?
6. In humans, hair color is controlled by two interacting genes. The same pigment, melanin, is present in both brown-haired and blond-haired people, but brown hair has much more of it. The allele for brown hair (B) is dominant to the allele for blond (b). Whether any melanin can be synthesized depends on another gene. The dominant allele (M) allows melanin synthesis; the recessive allele (m) prevents melanin synthesis. Homozygous recessives (mm) are albino (and have white hair). What will be the expected proportions of phenotypes in the children of the following parents?
   - **a.** $BBMM \times BbMm$
   - **b.** $BbMm \times BbMm$
   - **c.** $BbMm \times bbmm$
7. In humans, one of the genes determining color vision is located on the X chromosome. The dominant allele (C) produces normal color vision; red-green color blindness is caused by a recessive allele (c). If a man with normal color vision marries a color-blind woman, what is the probability of their having a color-blind son? A color-blind daughter?
8. In the couple described in problem 7, the woman gives birth to a color-blind but otherwise normal daughter. The husband sues for a divorce on the grounds of adultery. Will his case stand up in court? Explain your answer.

Answers to Genetics Problems

1. a. A red bull (RR) is mated to a white cow ($R'R'$). The bull will produce all R sperm, and the cow will produce all R' eggs. All the offspring will be RR' and will have roan hair.

b. A roan bull (RR') is mated to a white cow ($R'R'$). The bull produces half R and half R' sperm, and the cow produces R' eggs. The Punnett square method:

sperm \ eggs	R'
R	RR'
R'	$R'R'$

Using probabilities:

sperm	egg	offspring
$\frac{1}{2}R$	R'	$\frac{1}{2}RR'$
$\frac{1}{2}R'$	R'	$\frac{1}{2}R'R'$

The predicted offspring will be $\frac{1}{2}RR'$ (roan) and $\frac{1}{2}R'R'$ (white).

2. The offspring are of three types: dark (chestnut), light (cream), and intermediate (palomino). This distribution suggests incomplete dominance, with the allele for chestnut (C) combining with the allele for cream (C') to produce palomino heterozygotes (CC'). We can test this hypothesis by examining the offspring numbers. There are approximately $\frac{1}{4}$ chestnut (CC), $\frac{1}{2}$ palomino (CC'), and $\frac{1}{4}$ cream ($C'C'$). If palominos are heterozygotes, we would expect the cross $CC' \times CC'$ to yield $\frac{1}{4}CC$, $\frac{1}{2}CC'$, and $\frac{1}{4}C'C'$. Our hypothesis is supported.

3. a. $TtGg \times TtGg$. This is a "standard" cross of F_1 heterozygotes whose parents were true breeding for different phenotypes of two traits. Both parents produce TG, Tg, tG, and tg gametes. The expected proportions of offspring are $\frac{9}{16}$ tall green, $\frac{3}{16}$ tall yellow, $\frac{3}{16}$ short green, $\frac{1}{16}$ short yellow.

b. $TtGg \times TTGG$. In this cross, the heterozygous parent produces TG, Tg, tG, and tg gametes. However, the homozygous dominant parent produces only TG gametes. Therefore, all offspring receive at least one T allele for tallness and one G allele for green pods and are tall with green pods.

c. $TtGg \times Ttgg$. The second parent produces two types of gametes, Tg and tg. A Punnett square shows:

sperm \ eggs	Tg	tg
TG	$TTGg$	$TtGg$
Tg	$TTgg$	$Ttgg$
tG	$TtGg$	$ttGg$
tg	$Ttgg$	$ttgg$

The expected proportions of offspring are $\frac{3}{8}$ tall green, $\frac{3}{8}$ tall yellow, $\frac{1}{8}$ short green, $\frac{1}{8}$ short yellow.

4. If the genes were on separate chromosomes, then this would be a cross of individuals heterozygous for two independently assorting traits with expected offspring in the proportions $\frac{9}{16}$ round smooth, $\frac{3}{16}$ round fuzzy, $\frac{3}{16}$ long smooth, and $\frac{1}{16}$ long fuzzy. However, only the parental combinations show up in the offspring, indicating that the genes are on the same chromosome.

5. The genes are on the same chromosome and are quite close together. On rare occasions, crossing over occurs between the two genes, recombining the alleles.

6. a. $BBMM$(brown) $\times$ $BbMm$(brown). The first parent can produce only BM gametes, so all offspring will receive at least one dominant allele for each gene. Therefore, all offspring will have brown hair.

b. $BbMm$(brown) $\times$ $BbMm$(brown). Both parents can produce four types of gametes: BM, Bm, bM, and bm. The Punnett square shows:

sperm \ eggs	BM	Bm	bM	bm
BM	$BBMM$	$BBMm$	$BbMM$	$BbMm$
Bm	$BBMm$	$BBmm$	$BbMm$	$Bbmm$
bM	$BbMM$	$BbMm$	$bbMM$	$bbMm$
bm	$BbMm$	$Bbmm$	$bbMm$	$bbmm$

All mm offspring are albino, so we get the expected proportions $\frac{9}{16}$ brown-haired, $\frac{3}{16}$ blond-haired, $\frac{4}{16}$ albino (white hair).

c. $BbMm$(brown) $\times$ $bbmm$(albino):

sperm \ eggs	bm
BM	$BbMm$
Bm	$Bbmm$
bM	$bbMm$
bm	$bbmm$

The expected proportions of offspring are $\frac{1}{4}$ brown-haired, $\frac{1}{4}$ blond-haired, $\frac{1}{2}$ albino.

7. A man with normal color vision is CY (remember, the Y chromosome does not have the gene for color vision). His color-blind wife is cc. Their expected offspring will be:

sperm \ eggs	c
C	Cc
Y	cY

We therefore expect that all the sons will be color blind and all the daughters will have normal color vision.

8. The husband should win his case. All his daughters must receive one X chromosome, with the C allele, from him and therefore should have normal color vision. If his wife gives birth to a color-blind daughter, her husband cannot be the father (unless there was a new mutation for color blindness in his sperm line, which is very unlikely).

For More Information

Kahn, P. "Gene Hunters Close in on Elusive Prey." *Science*, March 8, 1996. Using pedigrees and biotechnology, researchers are uncovering the causes of diseases that involve complex interactions between genes and the environment.

Mendel Museum of Genetics. Mendel's monastery is located in what is now the Czech Republic. The Mendel Museum in Brno sponsors this Web site, describing Mendel's contributions to the discovery of the principles of inheritance. On the World Wide Web at http://www.mendel-museum.org/eng/1online.

National Institutes of Health. "Genetics Home Reference: Cystic Fibrosis." Last updated, July 2004. On the World Wide Web at http://ghr.nlm.nih.gov/condition=cysticfibrosis.

National Institutes of Health. "Genetics Home Reference: Marfan Syndrome." Last updated, July 2004. On the World Wide Web at http://ghr.nlm.nih.gov/condition=marfansyndrome.

Sapienza, C. "Parental Imprinting of Genes." *Scientific American*, October 1990. It is not quite true that all genes are equal, regardless of whether they have been inherited from mother or father. In some cases, which parent a gene comes from greatly alters its expression in the offspring.

Stern, C., and Sherwood, E. R. *The Origin of Genetics: A Mendel Source Book*. San Francisco: W. H. Freeman, 1966. There is no substitute for the real thing, in this case a translation of Mendel's original paper to the Brünn Society.

WEB TUTORIAL

To access a Web Tutorial visit http://www.prenhall.com/audesirk4. *Log in to the Web site selected by your instructor, navigate to this chapter, and select the appropriate Web Activity number.*

11.1 Crosses Involving Single Traits

Estimated time: 10 minutes

This tutorial will recreate one of Mendel's early experiments and show that a Punnett square can be used to determine how probability figures into genetic analysis.

11.2 Crosses Involving Multiple Traits

Estimated time: 10 minutes

In this tutorial, examine a set of crosses, focusing on two separate genetic traits. See how probability can help explain this situation also.

11.3 Web Investigation: Sudden Death on the Court: Marfan Syndrome

Estimated time: 10 minutes

Marfan syndrome, a genetic disorder, afflicts one in every 3000 to 5000 people of all races and ethnicity. Learn the characteristics of Marfan syndrome, and explore the cellular and biochemical mechanisms underlying its many effects.

Chapter 12

Biotechnology

Eddie Lloyd pumps his fist in joy after his conviction for murder was overturned on the basis of DNA evidence.

AT A GLANCE

Guilty or Innocent?

In 1984, a 16-year-old girl was brutally murdered in Detroit. Later, Eddie Lloyd, by that time a patient in a mental hospital, wrote to the police, offering suggestions about how to solve the murder. Under police interrogation, he confessed to the crime. Because Lloyd would not permit his attorney to plead insanity, the only thing that saved him from execution was that Michigan had repealed the death penalty.

In 1987, an 8-year-old girl was repeatedly raped by an unknown intruder in her own bedroom in Montana. A composite sketch made by a forensic artist led police to Jimmy Ray Bromgard. The little girl said that Bromgard looked like her assailant, but she wasn't sure. Hairs from the crime scene were similar to Bromgard's, and an expert witness testified that there was less than one chance in 10,000 that the hairs were not Bromgard's. Jimmy Ray Bromgard was convicted and sentenced to 40 years in prison.

These cases may appear to be examples of justice in action, except for one flaw: Both Bromgard and Lloyd were innocent, the victims of shoddy investigation, inadequate defense attorneys, expert witnesses who were anything but, and, in Lloyd's case, possible police misconduct. After 15 and 17 years in prison, respectively, Bromgard and Lloyd are now free, thanks to the professors and students of the Innocence Project at the Benjamin Cardozo School of Law at Yeshiva University, and to the science of biotechnology.

You have probably already guessed how the Innocence Project was able to prove Bromgard and Lloyd's innocence: DNA evidence. In the preceding chapters, you learned that genes are made of DNA, that most genes have many different versions called alleles, and that alleles differ in their nucleotide sequences. How can these facts be used in criminal justice?

Jimmy Ray Bromgard leaves the court, a free man after 15 years in prison.

Let's start with the observations that cells contain DNA and that cells are often left behind at a crime scene. We will hypothesize that, with the exception of identical twins, no two people have exactly the same alleles of all of their genes. If this hypothesis is correct, and if we could identify enough alleles in a DNA sample, then we could distinguish a specific person's DNA from the DNA of all the other people in the world.

As you study this chapter, ask yourself a few questions. How can alleles be identified in really small DNA samples? How many alleles would you need to be certain that you correctly identified the person the DNA came from? Might some alleles be more useful than others is obtaining correct identification?

12.1 What Is Biotechnology?

In its broadest sense, **biotechnology** is any use or alteration of organisms, cells, or biological molecules to achieve specific practical goals. Therefore, some aspects of biotechnology really aren't new. For example, people have been using yeast cells to produce bread, beer, and wine for the past 10,000 years. Selective breeding of plants and animals has an equally long history: 8000- to 10,000-year-old fragments of squash, found in a dry cave in Mexico, have larger seeds and thicker rinds than wild squash, suggesting selective breeding for higher nutritional content. Similarly, prehistoric art and animal remains indicate that dogs, sheep, goats, pigs, and camels were domesticated and selectively bred beginning at least 10,000 years ago.

Even today, selective breeding remains an important tool of biotechnology. Modern biotechnology, however, also frequently uses **genetic engineering**, a term that refers to more direct methods for modifying genetic material. Genetically engineered cells or organisms may have had genes deleted, added, or changed. Genetic engineering can be used to learn more about how cells and genes work, to develop better treatments for diseases, to produce valuable biological molecules, and to improve plants and animals for agriculture.

A key tool in genetic engineering is **recombinant DNA**, which is DNA that has been altered to contain genes or portions of genes from different organisms. Large amounts of recombinant DNA can be grown in bacteria, viruses, or yeast and then transferred into other species. Plants and animals that express DNA that has been modified or derived from other species are called **transgenic**, or **genetically modified organisms (GMOs)**.

Since its development in the 1970s, recombinant DNA technology has grown explosively. Today, researchers in almost every area of biology routinely use recombinant DNA technology in their experiments. In the pharmaceutical industy, genetic engineering has become the preferred way to manufacture many products, including human insulin.

Modern biotechnology also includes many methods of manipulating DNA, whether or not the DNA is subsequently put into a cell or an organism. For example, determining the nucleotide sequence of specific pieces of DNA is crucial to forensic science and the diagnosis of inherited disorders.

This chapter provides an overview of modern biotechnology. Our principal emphasis will be on applications of biotechnology and their impacts on society, but we will also briefly describe some of the important methods used in those applications. We will organize our discussion around five major themes: (1) recombinant DNA mechanisms found in nature, mostly in bacteria and viruses; (2) biotechnology in criminal forensics, principally DNA matching; (3) biotechnology in agriculture, specifically the production of transgenic plants and animals; (4) the Human Genome Project; and (5) biotechnology in medicine, focusing on the diagnosis and treatment of inherited disorders.

12.2 How Does DNA Recombine in Nature?

Most people think that a species' genetic makeup is constant, except for the occasional mutation, but genetic reality is far more fluid. Many natural processes can transfer DNA from one organism to another, sometimes even to organisms of different species. Recombinant DNA technologies used in the laboratory are often based on these naturally occurring processes.

Sexual Reproduction Recombines DNA

Sexual reproduction recombines DNA from two different parental organisms. Further, as we saw in Chapter 10, homologous chromosomes exchange DNA by crossing over during meiosis I. Thus, each chromosome in a gamete usually contains a mixture of alleles from the two parental chromosomes. In this sense, every egg and every sperm contain recombinant DNA, derived from the organism's two parents. When a sperm fertilizes an egg, the resulting offspring also contains recombinant DNA.

Transformation May Combine DNA from Different Bacterial Species

Some people think that recombination of DNA within a species during sexual reproduction is "natural" and therefore good, but recombination in the laboratory between different species is "unnatural" and therefore bad. Recombination of DNA between species, however, is not confined to the lab; it also occurs in nature.

Bacteria can undergo several types of recombination that allow gene transfer between species. In a process called **transformation**, for example, bacteria pick up pieces of DNA from the environment (Fig. 12-1). The DNA may be part of the

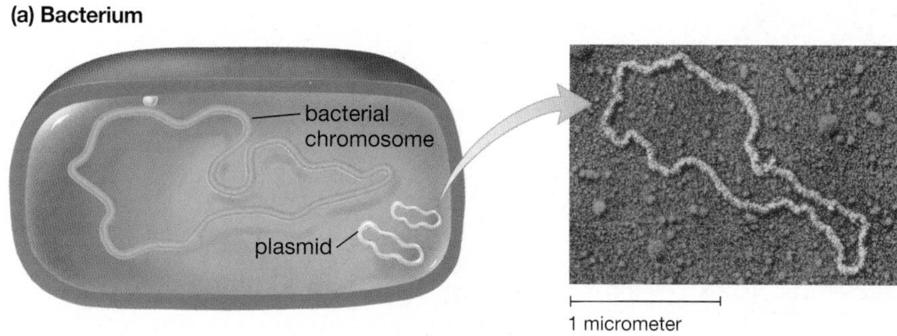

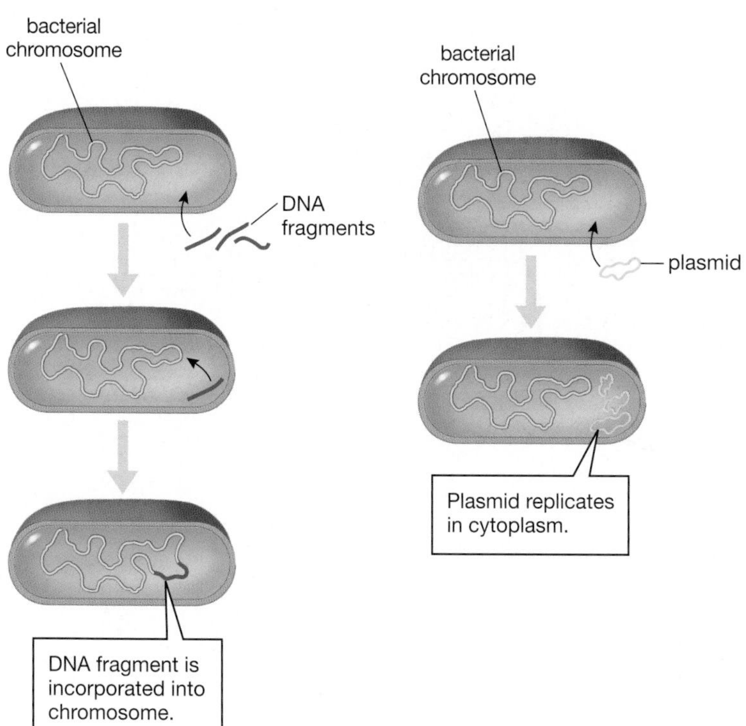

Figure 12-1 **Recombination in bacteria** ***(a)*** In addition to their large circular chromosome, bacteria commonly possess small rings of DNA called plasmids, which often carry additional useful genes. Bacterial transformation occurs when living bacteria take up ***(b)*** fragments of chromosomes or ***(c)*** plasmids.

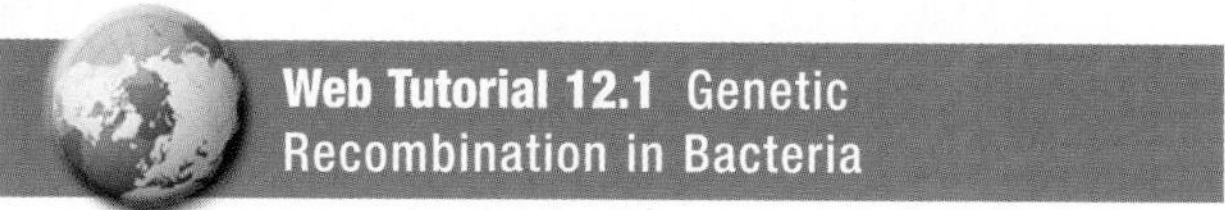

chromosome of another bacterium (**Fig. 12-1b**), even from another species, or may be in the form of tiny circular DNA molecules called **plasmids** (**Fig. 12-1c**). A single bacterium may contain dozens or even hundreds of copies of a plasmid. When the bacterium dies, it releases its plasmids into the environment, where they may be taken up by bacteria of the same or different species. In addition, living bacteria can often pass their plasmids directly to other living bacteria. Sometimes, plasmids may pass from a bacterium to a yeast cell, thereby moving genes from a prokaryotic cell to a eukaryotic cell.

Viruses May Transfer DNA between Species

Viruses, which are often little more than genetic material encased in a protein coat, transfer their genetic material to cells that they infect (called *host cells*). Once inside, the viral genes replicate and use the host cell's enzymes and ribosomes to synthesize viral proteins. The replicated genes and new viral proteins assemble inside the host cell, forming new viruses that are released to infect other cells (**Fig. 12-2**).

Figure 12-2 Viruses may transfer genes between cells

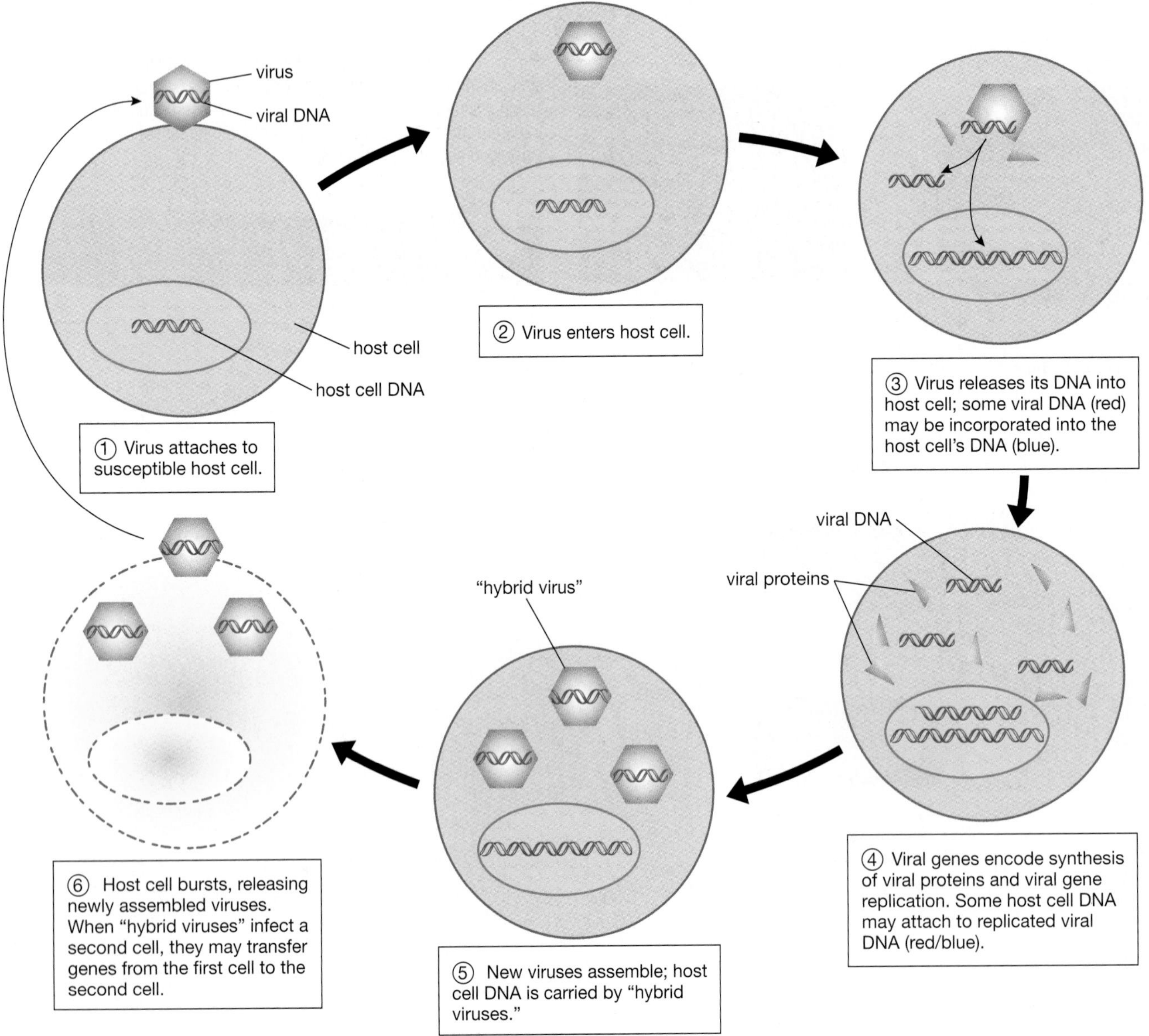

Some viruses can transfer genes from one organism to another. In these instances, the virus inserts its DNA into a host cell's chromosome. The viral DNA may remain there for days, months, or even years. Every time the cell divides, it replicates the viral DNA along with its own DNA. When new viruses are finally produced, some of the host's genes may be incorporated into the viral DNA. If such recombinant viruses infect other cells and insert their DNA into the new host cells' chromosomes, pieces of the previous host cell's DNA will also be inserted. Most of the time, viruses move host DNA between different individuals of a single species. However, some, such as the influenza virus, infect more than one host species and may transfer genes from one species to another.

12.3 How Is Biotechnology Used in Forensics?

The applications of biotechnology vary, depending on the goals of those who use it. Forensic scientists need to identify victims and criminals; biotechnology firms need to identify specific genes and insert them into organisms such as bacteria, cattle, or crop plants; and biomedical firms and physicians need to detect defective alleles and, ideally, devise ways to fix them or to insert normally functioning alleles into patients. We will begin by describing a few common methods of manipulating DNA, using their application to forensics as a specific example.

In the Jimmy Bromgard case, the Innocence Project needed to find out if semen samples collected from the rape victim in 1987 came from Bromgard. Lab technicians used two techniques that have become commonplace in virtually all DNA labs. First, they *amplified* (made many copies of) the DNA so that they had enough material to analyze. Then they determined whether the DNA from the semen samples matched Bromgard's DNA.

The Polymerase Chain Reaction Amplifies DNA

Developed by Kary B. Mullis of the Cetus Corporation in 1986, the **polymerase chain reaction (PCR)** produces virtually unlimited amounts of selected pieces of DNA. PCR is so crucial to molecular biology that it earned Mullis a share in the Nobel Prize for Chemistry in 1993. Let's see how PCR amplifies a specific piece of DNA.

When we described DNA replication in Chapter 10, we omitted some of its real-life complexity. One of the things we did not discuss is crucial to PCR: DNA polymerase, by itself, doesn't know where to start copying a strand of DNA. Normally, when a DNA double helix is unwound, enzymes put a little piece of complementary RNA, called a *primer*, on each strand. DNA polymerase then recognizes this "primed" region of DNA as the place to start replicating the rest of the DNA strand. For PCR, a DNA synthesizer is used to make two artificial DNA primers. These primers "tell" DNA polymerase where to start copying.

In a small test tube, DNA is mixed with primers, free nucleotides, and a special heat-resistant DNA polymerase, isolated from microbes that live in hot springs (see "Scientific Inquiry: Hot Springs and Hot Science"). PCR consists of a three-step cycle of heating and cooling (Fig. 12-3a), which is repeated as many times as necessary to generate enough copies of the DNA segment (Fig. 12-3b).

PCR synthesizes DNA in a geometric progression ($1 \rightarrow 2 \rightarrow 4 \rightarrow 8$, etc.), so 20 PCR cycles make about a million copies, and a little over 30 cycles make a billion copies. Each cycle takes only a few minutes, so PCR can produce billions of copies of a DNA segment in a single afternoon, starting, if necessary, from a single molecule of DNA. The DNA is then available for forensics, sequencing, cloning, making transgenic organisms, or many other purposes.

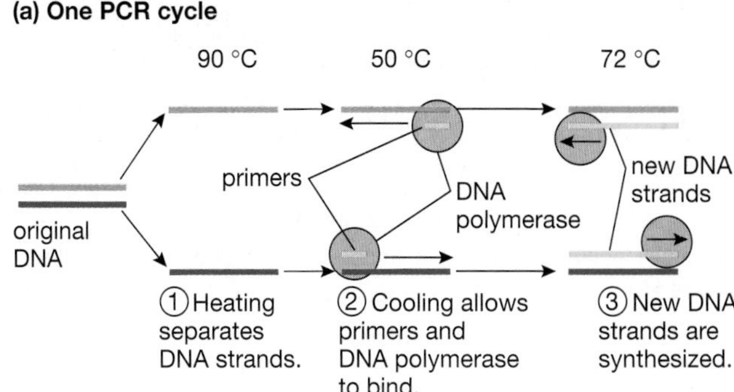

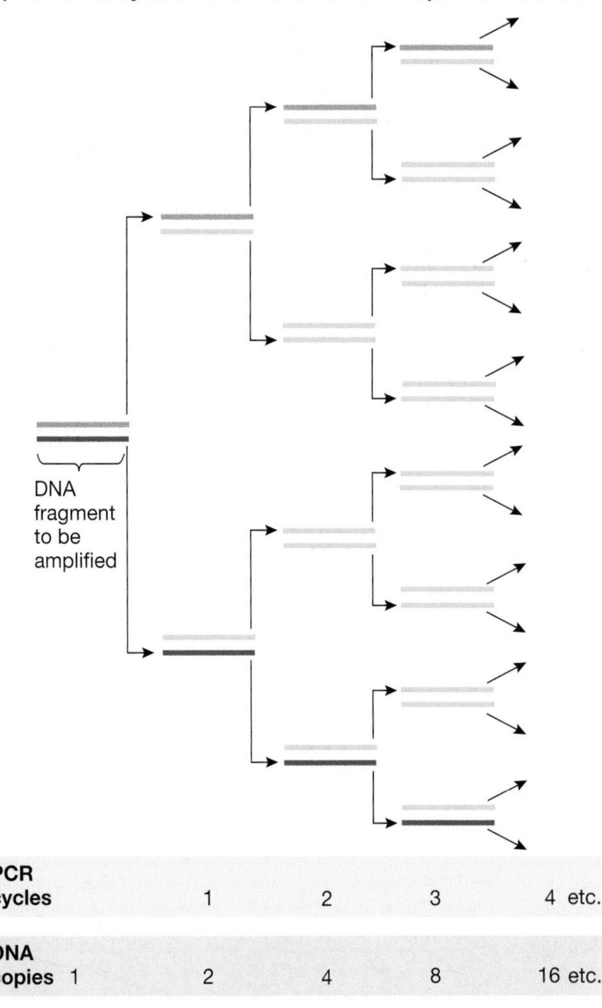

PCR cycles		1	2	3	4 etc.
DNA copies	1	2	4	8	16 etc.

Figure 12-3 **PCR copies a specific DNA sequence** ***(a)*** The polymerase chain reaction consists of a cycle of heating, cooling, and warming that is repeated 20 to 30 times. ***(b)*** Each cycle doubles the amount of target DNA. After just 20 cycles, a million copies of the target DNA have been synthesized. **QUESTION:** *Why do you think that the reaction is warmed up to 72 °C for DNA synthesis [part (a) of the figure]? Hint: Reread "Scientific Inquiry: Hot Springs and Hot Science." Given the normal living conditions of* Thermus aquaticus, *do you think that its DNA polymerase would work most rapidly at 50 °C or 72 °C?*

SCIENTIFIC INQUIRY

HOT SPRINGS AND HOT SCIENCE

At a hot spring, such as those found in Yellowstone National Park, water literally boils out of the ground, gradually cooling as it flows to the nearest stream (Fig. E12-1). You might think that such springs, scalding hot and often containing poisonous metals and sulfur compounds, must be lifeless. However, closer examination often reveals a diversity of microorganisms, each adapted to a different temperature zone in the spring. Back in 1966, in a Yellowstone hot spring, Thomas Brock of the University of Wisconsin discovered *Thermus aquaticus*, a bacterium that lives in water as hot as 176 °F (about 80 °C).

When Kary Mullis first developed the polymerase chain reaction, he encountered a major technical difficulty. The DNA solution must be heated almost to boiling to separate the double helix into single strands, then cooled so DNA polymerase can synthesize new DNA, and this process must be repeated over and over again. "Ordinary" DNA polymerase, like most proteins, is ruined, or *denatured*, by high temperatures. Therefore, new DNA polymerase had to be added after every heat cycle, a process that was expensive and labor intensive. Enter *Thermus aquaticus*. Like other organisms, it replicates its DNA when it reproduces. But because it lives in hot springs, it has a particularly heat-resistant DNA polymerase. When DNA polymerase from *T. aquaticus* is used in PCR, it needs to be added to the DNA solution only once, at the start of the reaction.

Figure E12-1 Thomas Brock surveys Mushroom Spring The c in hot springs arise from minerals dissolved in the water and fron ous types of microbes that live at different temperatures.

Differences in Short DNA Segments Can Identify Individuals

After years of painstaking work, forensics experts have found that small segments of DNA, called *short tandem repeats (STRs)*, can be used to identify people with astonishing accuracy (Fig. 12-4). Think of STRs as very short, stuttering genes. Each STR is *short* (consisting of 2 to 5 nucleotides), *repeat*ed (about 5 to 15 times), and *tandem* (having all the repetitions right alongside one another). Just like any gene, different people may have different alleles of the STRs. In the case of an STR, each allele is simply a different number of repeats of the same few nucleotides.

In 1999, British and American law enforcement agencies agreed to use a standard set of 10 to 13 STRs that vary greatly among individuals. A perfect match of 10 STRs between a suspect's DNA and the DNA found at a crime scene means that there is less than one chance in a trillion that the two DNA samples did not come from the same person. What's more, the DNA around STRs doesn't degrade very rapidly, so even old DNA samples usually have intact STRs.

Forensics labs use PCR primers that amplify only the DNA that includes and immediately surrounds the STRs. Because STR alleles vary in how many times they repeat, they vary in size: An STR with more repeats has more nucleotides and is larger. Therefore, a forensic lab needs to identify each STR in a DNA sample and to determine its size.

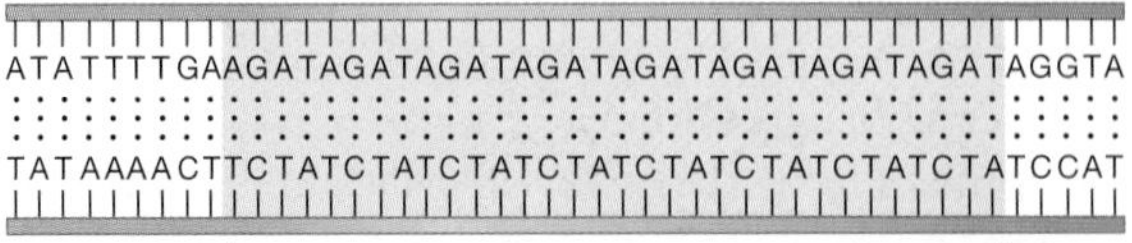

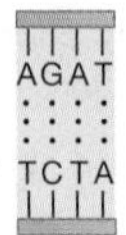

8 side-by-side (tandem) repeats of the same 4-nucleotide sequence,

Figure 12-4 Short tandem repeats This STR consists of the sequence AGAT, repeated from 7 to 13 times in different individuals.

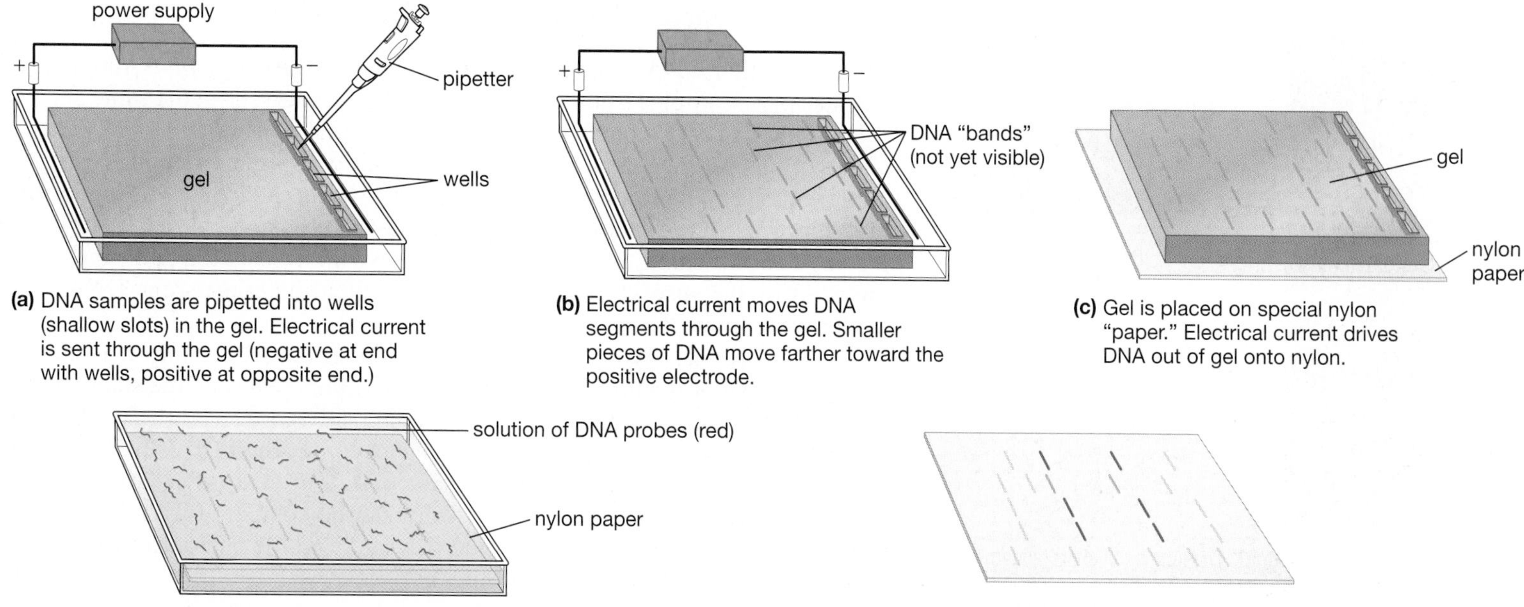

(a) DNA samples are pipetted into wells (shallow slots) in the gel. Electrical current is sent through the gel (negative at end with wells, positive at opposite end.)

(b) Electrical current moves DNA segments through the gel. Smaller pieces of DNA move farther toward the positive electrode.

(c) Gel is placed on special nylon "paper." Electrical current drives DNA out of gel onto nylon.

(d) Nylon paper with DNA is bathed in a solution of labeled DNA probes (red) that are complementary to specific DNA segments in the original DNA sample.

(e) Complementary DNA segments are labeled by probes (red bands).

Figure 12-5 **Gel electrophoresis is used to separate and identify segments of DNA**

Gel Electrophoresis Separates DNA Segments

Modern forensics labs use sophisticated, expensive machines to determine the number of STR repeats. Most of these machines are based on two methods that are used in molecular biology labs around the world: first, separating the DNA by size, and second, labeling specific DNA segments.

The mixture of DNA pieces is separated by a technique called **gel electrophoresis** (Fig. 12-5). First, the mixture of DNA segments is loaded into shallow grooves, or wells, in a *gel*, which is simply a porous solid riddled with tiny holes. The gel is put into a chamber with electrodes connected to each end. One electrode is made positive and the other negative, so a current flows between the electrodes through the gel (Fig. 12-5a).

The phosphate groups in the backbones of DNA molecules are negatively charged. When electrical current flows through the gel, the negatively charged DNA fragments move toward the positively charged electrode. Because smaller bits of DNA slip through the holes in the gel more easily than larger pieces do, they move more rapidly toward the positively charged electrode. Eventually the DNA segments are separated by size, forming distinct bands on the gel (Fig. 12-5b).

DNA Probes Are Used to Label Specific Nucleotide Sequences

Unfortunately, the DNA bands are invisible. How can a technician identify which band contains a specific STR? Well, how does nature identify sequences of DNA? Right, by base pairing! The two strands of the DNA double helix are usually separated during gel electrophoresis, allowing pieces of synthetic DNA, called *DNA probes*, to base pair with specific DNA segments in the sample. DNA probes are short pieces of single-stranded DNA that are complementary to the nucleotide sequence of given STRs (or any other DNA of interest). The DNA probes themselves are labeled, either by radioactivity or by attaching colored molecules to them.

When the gel is finished running, the technician transfers the single-stranded DNA segments out of the gel and onto a piece of paper made of nylon (Fig. 12-5c). Then the paper is bathed in a solution containing a specific DNA probe (Fig. 12-5d), which will base pair with, and therefore bind to, only a specific STR, making this STR visible (Fig. 12-5e).

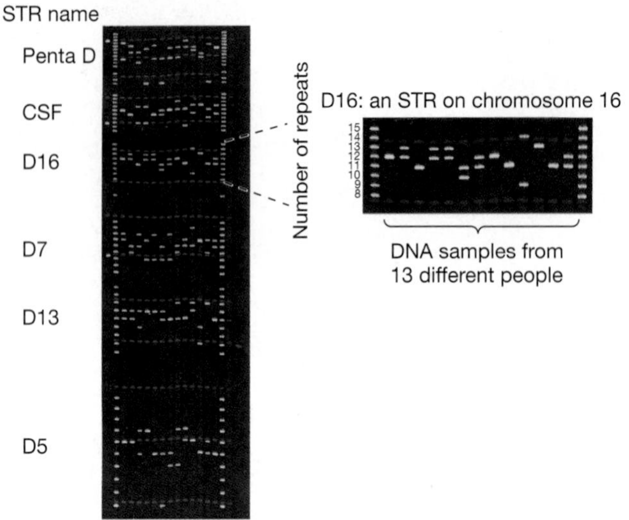

Figure 12-6 DNA fingerprinting The lengths of short tandem repeats of DNA form characteristic patterns on a gel. This gel displays six different STRs (Penta D, CSF, etc.). The evenly spaced yellow-green bands on the far left and far right sides of the gel show the number of repeats of the individual STRs. DNA samples from 13 people were run between these standards, resulting in one or two bands in each vertical lane. In the enlargement of the D16 STR, reading from right to left, the first person's DNA has 12 repeats, the second person's has 13 and 12, the third has 11, and so on. Although some people have the same number of repeats in *some* STRs, none has the same number of repeats of *all* the STRs. *(Photo courtesy of Dr. Margaret Kline, National Institute of Standards and Technology)* **QUESTION:** *For any single person, a given STR always has either one or two bands. Why? Further, single bands are always about twice as bright as each band of a pair. For example, in the D16 STR, the single bands of the first and third DNA samples are twice as bright as the pairs of bands of the second, fourth, and fifth samples. Why?*

A DNA Fingerprint Is Unique to Each Person

To see whether DNA from a crime scene matches a suspect's DNA, forensic technicians place samples from each source into different wells in the same gel. After electrophoresis and labeling the DNA with probes have been completed, the lanes of the gel are compared. By running multiple gels and staining each gel for different STRs, the forensics lab produces a pattern, called a **DNA fingerprint** or DNA profile, of the samples of DNA (Fig. 12-6). If the number of repeats of all of the STRs is the same, then the odds are overwhelming that the crime scene DNA was left by the suspect. If they don't match, then the crime scene DNA cannot belong to the suspect.

In many states, anyone convicted of certain crimes (for example, assault, burglary, or attempted murder) must give a blood sample. Using the standard array of STRs, technicians determine the criminal's DNA fingerprint. This DNA fingerprint is kept on file, either in a state agency or at the FBI, or both. DNA left behind at another crime scene, even years before or years later (after the criminal may have been released from prison), can be compared with the DNA fingerprint file (see "Guilty or Innocent? Revisited" at the end of this chapter for a successful use of DNA fingerprint files).

12.4 How Is Biotechnology Used in Agriculture?

The main goals of agriculture are to grow as much food as possible, as cheaply as possible, with minimal loss from pests such as insects and weeds. Many commercial farmers and seed suppliers have turned to biotechnology to achieve these goals.

Many Crops Are Genetically Modified

Currently, almost all of the genetically modified organisms used in agriculture are plants. In 2002, about 34% of the corn, 71% of the cotton, and 75% of the soybeans grown in the United States were *transgenic*; that is, they contained genes from other species. Crops are most commonly modified to improve their resistance to herbicides or insects or both.

Many herbicides kill plants by inhibiting an enzyme that is used by plants, fungi, and some bacteria—but not animals—to synthesize essential amino acids. Without these amino acids, the plants cannot synthesize proteins, and they die. Many herbicide-resistant transgenic crops have been given a bacterial gene encoding an enzyme that functions even in the presence of these herbicides, so the plants continue to synthesize normal amounts of amino acids and proteins. Herbicide-resistant crops allow farmers to kill weeds without harming their crops. Less competition from weeds means more water, nutrients, and light for the crops, hence larger harvests.

The insect resistance of many crops has been enhanced by giving them a gene, called Bt, from the bacterium *Bacillus thuringiensis*. The protein encoded by the Bt gene damages the digestive tract of insects (but not mammals). Transgenic Bt crops therefore suffer far less damage from insects, and farmers can apply much less pesticide to their fields.

How would a seed company go about making a transgenic plant? Let's examine the process, using insect-resistant Bt plants as an example.

The Desired Gene Is Cloned

The first step is to *clone* (produce identical copies of) the desired gene. Cloning a gene usually involves two tasks: (1) obtaining the gene and (2) inserting it into a plasmid so that huge numbers of copies of the gene can be made.

There are two common ways of obtaining a gene. For a long time, the only practical method was to isolate the gene from the organism that has it. More recently, biotechnologists can often synthesize the gene—or a modified version of it—in the lab, using PCR or DNA synthesizers.

Once the Bt gene has been obtained, why insert it into a plasmid? Plasmids, small circles of DNA in bacteria (see Fig. 12-1), are replicated when the bacteria multiply. Therefore, once the Bt gene has been inserted into a plasmid, producing huge numbers of copies of the gene is as simple as raising lots of bacteria.

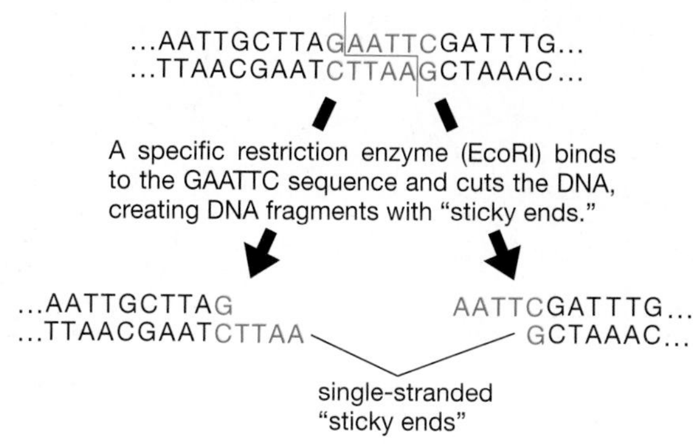

Restriction Enzymes Cut DNA at Specific Nucleotide Sequences

Genes are inserted into plasmids through the use of special enzymes called **restriction enzymes**. Each restriction enzyme cuts DNA at a specific nucleotide sequence. Many restriction enzymes make a "staggered" cut, snipping the DNA in a different location on each of the two strands, so that single-stranded sections hang off the ends of the DNA. These single-stranded regions are commonly called "sticky ends" because they can base pair with, and thus stick to, other single-stranded pieces of DNA with complementary bases.

Cutting Two Pieces of DNA with the Same Restriction Enzyme Allows the Pieces to Be Joined Together

The Bt gene is inserted into a plasmid by using the same restriction enzyme to cut the DNA on either side of the Bt gene and to split open the circle of the plasmid (**Fig. 12-7a**). As a result, the ends of the Bt gene and the opened-up plasmid both have complementary nucleotides in their "sticky ends" and can base pair with each other. When the cut Bt genes and plasmids are mixed together, some of the Bt genes will be temporarily inserted between the ends of the plasmids. DNA ligase (see Chapter 8) is used to bond the Bt genes into the plasmids permanently (**Fig. 12-7b**).

Specialized bacteria that infect only certain types of plants are then transformed with the plasmids (**Fig. 12-7c**). When these bacteria infect a plant cell, the plasmids insert their DNA into the plant cell's chromosomes (**Fig. 12-7d**). Thereafter, whenever that plant cell divides, all of its daughter cells inherit the Bt gene. Hormones stimulate the transgenic plant cells to divide and differentiate into entire plants. These plants are bred to one another, or to other plants, to create commercially valuable crop plants that resist insect attack (**Fig. 12-8**).

Similar techniques can be used to insert medically useful genes into plants, producing medicines down on the "pharm." This possibility will be explored in "Biotechnology Watch: Edible Vaccines," in Chapter 18.

Figure 12-7 Using plasmids to insert DNA into a plant cell

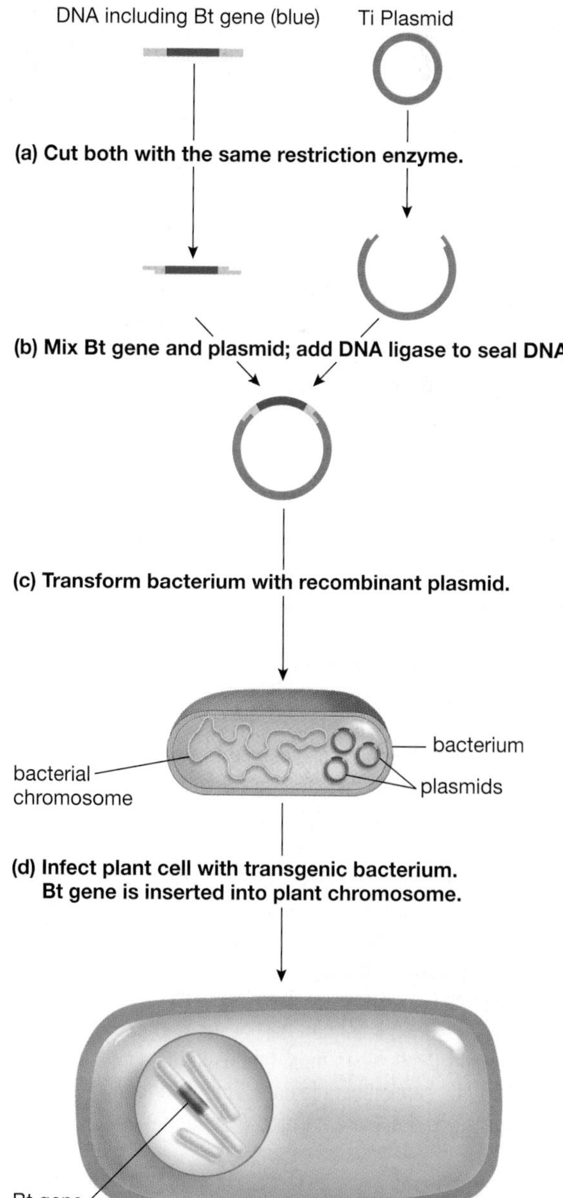

Genetically Modified Animals May Be Useful in Agriculture and Medicine

Creating transgenic animals usually involves injecting the desired DNA into a fertilized egg. The egg is allowed to divide a few times in culture and then is implanted into a surrogate mother. If the offspring are healthy and express the foreign gene, they are bred together to produce homozygous transgenic organisms.

So far, it has proven difficult to produce commercially valuable transgenic livestock. For example, pigs with extra growth-hormone genes indeed grow faster, but they also suffer from arthritis, ulcers, and sterility. Several types of fish with added growth-hormone genes grow much faster than wild-type fish and do not display any obvious ill effects. However, whether "fish farms" should be allowed to grow these fish remains controversial, mainly because of concerns about what would happen if transgenic fish escaped into the wild (see "Biotechnology Watch: Bonanza from the Sea or Frankenfish?").

Because medicines are generally much more valuable than meat, many researchers are developing animals that will produce medicines, such as human

Figure 12-8 Bt plants resist insect attack NewLeaf® potatoes (middle row) have been genetically engineered to express the Bt gene. These potatoes resist attack by Colorado potato beetles, so they are much healthier than nontransgenic plants (right and left rows).

antibodies or other essential proteins. For example, there are sheep whose milk contains a protein, alpha-1-antitrypsin, that may prove valuable in treating cystic fibrosis.

12.5 How Is Biotechnology Used to Learn About the Human Genome?

Virtually all the traits of human beings are influenced by our genes, including gender, size, hair color, intelligence, and susceptibility to disease organisms and toxic substances in the environment. To begin to understand how our genes influence our lives, the Human Genome Project was launched in 1990, with the goal of determining the nucleotide sequence of all the DNA in our entire set of genes, called the human genome.

In 2003, this joint project of molecular biologists in several countries sequenced the human genome with an accuracy of about 99.99%. To many people's surprise, the human genome contains only about 25,000 genes, comprising

BIOTECHNOLOGY WATCH

BONANZA FROM THE SEA OR FRANKENFISH?

Fish are a good source of protein, and the fatty acids found in some species may protect against heart disease. With burgeoning human populations, the United Nations' Food and Agriculture Organization predicts that the demand for fish may double by 2040. However, many fish populations are overfished. How can supply keep up with demand?.

Aquaculture—raising fish in freshwater ponds or enclosures in the oceans—has the potential to produce huge amounts of "domestic" fish. Just as ranching and farming have replaced hunting and gathering on land, aquaculture might do the same in the sea. Farmers raise cattle, pigs, and chickens, rather than deer, wart hogs, or pheasants, because the domestic animals have been bred for centuries for docility, fast weight gain, and high reproductive rate. Aquaculturists are just starting to develop more efficient strains of fish, using genetic engineering rather than centuries of artificial selection.

Perhaps the most famous engineered fish is a strain of Atlantic salmon produced by Aqua Bounty Farms. Aqua Bounty constructed a new "growth gene" for its Atlantic salmon, combining the growth hormone gene from Chinook salmon (a Pacific Ocean species) with promoter sequences that cause the gene to be transcribed all year round. Aqua Bounty salmon, therefore, grow six times faster, reaching marketable size a full year earlier than wild Atlantic salmon (Fig. E12-2).

Before Aqua Bounty salmon can be farmed commercially, several questions must be answered. First, are Aqua Bounty salmon safe to eat? Almost certainly, the answer is yes. People eat Chinook salmon—and, therefore, its growth hormone—with no ill effects. Second, are Aqua Bounty salmon safe to raise? The answer to this question is probably also yes. Salmon, engineered or not, require food to grow. The faster the Aqua Bounty salmon grow, the more they eat. The more they eat, the more uneaten food scraps and feces float away in the water. Large numbers of fast-growing salmon have the potential to produce massive water pollution, but this can almost certainly be solved with ingenuity and money.

The third question is the stickler: What would happen if Aqua Bounty salmon escaped into the wild? Would bioengineered salmon outcompete their wild relatives and replace them in the ocean? Would it matter?

Escape is really only a matter of time. For example, in December 2000, a huge storm destroyed the steel pens of a fish farming operation in Machias Bay in Maine. At least 100,000 fish swam away into the ocean. Vandals, terrorists, neglect—sooner or later, engineered fish would enter the wild. Is that bad? They're still Atlantic salmon, aren't they? Maybe, and maybe not. Wild populations of most species vary greatly. If engineered salmon replaced wild fish, the gene pool of the salmon might shrink to a puddle, leaving them susceptible to changing environmental conditions. Even if the salmon were not adversely affected, what about other species? Voracious engineered salmon might eat a lot more than wild salmon would eat, leaving less for other sea life. On the other hand, most "domesticated" animals are less fit than their wild relatives, so perhaps Aqua Bounty salmon would never replace wild salmon.

Finally, Aqua Bounty plans to breed the fish inland, sterilize the offspring, and ship the sterilized fish to pens along the coast to be grown for market, so even if they did escape, they couldn't breed in the wild. Aqua Bounty's sterilization procedure is 100% effective in small batches in the lab. Will it be 100% effective in commercial operation? Should we take the risk?

In 2002, the U.S. National Research Council (NRC) investigated the safety of genetically modified animals. The NRC decided that the risk to people from eating transgenic animals is minimal, but that the risk to the environment is unknown but potentially large. A lost transgenic cow could be easily caught, but what about fish? In response to such concerns, the state of Washington banned transgenic fish in 2002. How do you think that governments should balance the probable benefits of increasing a valuable food supply with possible damage to the environment?

Figure E12-2 Transgenic salmon The transgenic salmon (bottom) grow much faster than their wild relatives (top).

approximately 2% of the DNA. Some of the other 98% consists of promoters and regions that regulate how often individual genes are transcribed, but it's not really known what most of the DNA does.

What good is it to sequence the human genome? First, many genes were discovered whose functions are completely unknown. Now that these genes have been identified and sequenced, the genetic code allows biologists to predict the amino acid sequences of the proteins they encode. Comparing these proteins with familiar proteins whose functions are already known will enable us to find out what many of these genes do.

Second, knowing the nucleotide sequences of human genes will have an enormous impact on medical practice. In 1990, fewer than 100 genes known to be associated with human diseases had been discovered. By 2003, this number had jumped to over 1400, mostly as a result of the Human Genome Project.

Third, there is no single "human genome" (or else all of us would be identical twins). *Most* of the DNA of everyone on the planet is the same, but each of us also carries our own unique set of alleles. Some of those alleles can cause or predispose people to develop various medical conditions, including Marfan syndrome, sickle-cell anemia, cystic fibrosis (described in earlier chapters), breast cancer, alcoholism, schizophrenia, heart disease, Huntington disease, Alzheimer's disease, and many others. A major impact of the Human Genome Project will be to help diagnose genetic disorders or predispositions, and hopefully to devise treatments or even cures in the future, as we describe in the following sections.

12.6 How Is Biotechnology Used for Medical Diagnosis and Treatment?

For over a decade, biotechnology has been routinely used to diagnose some inherited disorders. Potential parents can learn if they are carriers of a genetic disorder, and often an embryo can be diagnosed early in a pregnancy (see "Health Watch: Prenatal Genetic Screening"). More recently, medical researchers have begun to use biotechnology to attempt to cure, or at least treat, genetic diseases.

DNA Technology Can Be Used to Diagnose Inherited Disorders

A person inherits a genetic disease because he or she inherits one or more dysfunctional alleles. Defective alleles have different nucleotide sequences than functional alleles. Two methods are currently used to find out if a person carries a normal allele or a malfunctioning allele. One relies on the specificity of restriction enzymes to cut certain nucleotide sequences, but not others. In the second, different nucleotide sequences are recognized when DNA probes bind to them through base pairing.

Restriction Enzymes Cut Different Alleles at Different Locations

To understand the first method of finding alleles associated with genetic disease, remember that any given restriction enzyme cuts DNA only at a specific nucleotide sequence. Therefore, a medically important allele may be cut by a particular restriction enzyme, while other alleles are left intact, or vice versa. To illustrate this application of biotechnology, let's consider the diagnosis of sickle-cell anemia.

You may recall that sickle-cell anemia is caused by a point mutation in which thymine replaces adenine near the beginning of the globin gene (see Table 9-3 in Chapter 9). This substitution causes the resulting hemoglobin molecules to clump together, distorting and weakening the red blood cells.

(a) Mst II cuts a normal globin allele (red) in 2 places, but cuts the sickle-cell allele in 1 place.

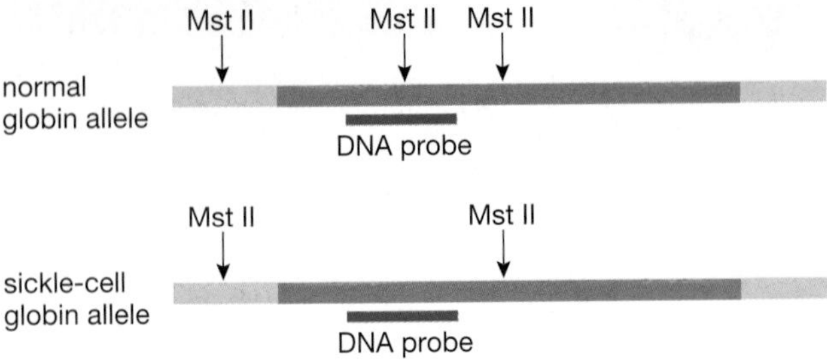

(b) Gel electrophoresis of globin alleles

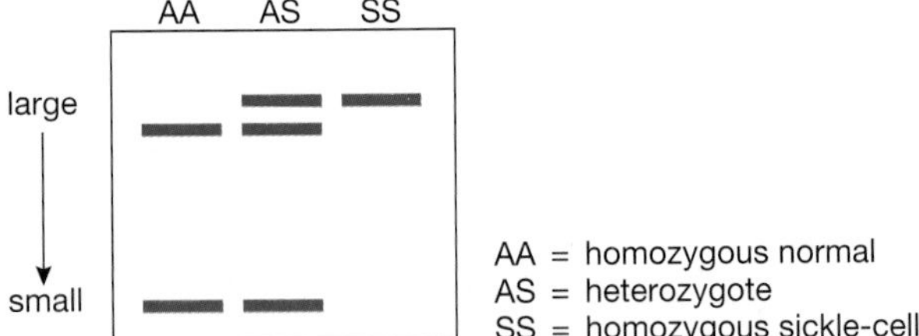

Figure 12-9 Diagnosing sickle-cell anemia with restriction enzymes ***(a)*** The restriction enzyme Mst II cuts both the normal globin allele and the sickle-cell allele (both red) somewhat "ahead" of the globin gene (far left arrow) and near the middle (far right arrow). Mst II also cuts the normal allele, but *not* the sickle-cell allele, in a unique third location (middle arrow). A DNA probe (blue) that is complementary to the DNA on both sides of the unique cut site will label *two* pieces of DNA from the normal allele but only a *single* piece from the sickle-cell allele. ***(b)*** When the cut DNA is run on a gel and labeled with this DNA probe, the large single piece of DNA from the sickle-cell allele is close to the beginning (top) of the gel, while the smaller pieces from the normal allele run farther into the gel (lower down). The pattern of DNA bands shows the genotype of the person from whom the DNA sample was obtained.

One restriction enzyme, called Mst II, cuts DNA near the middle of both the normal and the sickle-cell alleles. It also cuts DNA just outside of both alleles. However, it cuts the normal globin allele, but *not* the sickle-cell allele, in a third location (Fig. 12-9a). How can this unique cut be identified? A DNA probe is synthesized that is complementary to the part of the globin allele spanning the unique cut site. When sickle-cell DNA is cut with Mst II and run on a gel, this probe labels a single large band (Fig. 12-9b). When normal DNA is cut with Mst II, the probe labels two bands, one small and one not quite as large as the sickle-cell band.

The genotypes of parents, children, and fetuses can be determined by this simple test (Fig. 12-9b). Someone who is homozygous for the normal globin allele will have two bands. Someone who is homozygous for the sickle-cell allele will have one band. A heterozygote will have three bands.

Different Alleles Bind to Different DNA Probes

The second method of finding alleles that cause disease is useful in diagnosing cystic fibrosis, a disease caused by a defect in a protein that normally pumps chloride ions across cell membranes (see "Scientific Inquiry: Cystic Fibrosis" in Chapter 11). There are over 1000 different cystic fibrosis alleles, all at the same gene locus, each encoding a slightly different defective chloride pump.

How can you diagnose a disorder with a thousand different alleles? Most of these alleles are extremely rare; only 32 alleles account for about 90% of the cases of cystic fibrosis. Still, 32 alleles is a lot. However, each allele has a unique nucleotide sequence. Therefore, one strand of each allele will form perfect base pairs only with its own complementary strand. Several companies now produce cystic fibrosis "arrays," which are pieces of specialized paper to which DNA probes are bound. Each probe is complementary to a different cystic fibrosis allele (Fig. 12-10).

A person's DNA is tested for cystic fibrosis alleles by cutting it into small pieces, separating them into single strands, and labeling the strands. An array is then bathed in the resulting solution of labeled DNA fragments. Only a perfect complementary strand of the person's DNA will bind to any given probe on the array. Depending on the number of different alleles represented on the array, up to 95% of all cases of cystic fibrosis can be diagnosed by this method.

Figure 12-10 Diagnosing cystic fibrosis with a DNA array DNA from a patient is cut into small pieces, separated into single strands, and labeled (blue, in this diagram). A cystic fibrosis screening array is bathed in this solution of labeled DNA. Each cystic fibrosis allele can bind to only one DNA probe on the array. In this simplified diagram, the patient has one normal allele (upper left) and one defective allele (middle of bottom row).

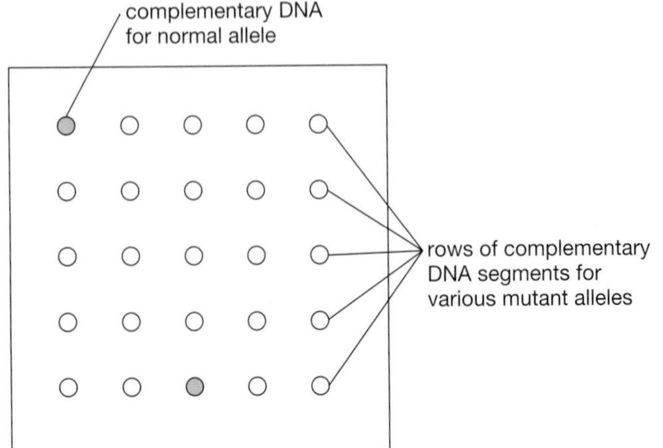

DNA Technology Can Be Used to Treat Disease

DNA technology can be used to treat disease as well as to diagnose it. Thanks to recombinant DNA technology, several medically important proteins are now routinely made in bacteria. Restriction enzymes are used to splice appropriate genes into plasmids, and bacteria are then transformed with these plasmids.

The first human protein made by recombinant DNA technology was insulin. Before 1982, when recombinant human insulin was first licensed for use, the insulin needed by diabetics was extracted from the pancreases of cattle or pigs slaughtered for meat. Although the insulin from these animals is very similar to human insulin, the slight differences cause an allergic reaction in about 5% of diabetics. Recombinant human insulin does not cause allergic reactions.

Some of the types of human proteins produced by recombinant DNA technology are listed in Table 12-1. These proteins, although tremendously helpful and often life saving, do not *cure* inherited disorders. They merely treat the symptoms. They are also easy to use, because they only need to be injected into the bloodstream. Then they circulate throughout the body, causing appropriate responses in the appropriate cells.

What about diseases, such as cystic fibrosis, in which the defective protein must be delivered *inside* cells? And what about actually *curing* inherited disorders? Biotechnology offers great promise here too, although progress has been painfully slow so far. Let's look at two specific examples of how biotechnology may treat, or even cure, devastating illnesses.

TABLE 12-1 Examples of Products Produced by Recombinant DNA Methods

Type of Product	Purpose	Typical Method of Genetic Engineering
Human hormones	Used in treatment of diabetes, growth deficiency, infertility	Human gene inserted into bacteria or cultured mouse cells
Human cytokines (regulators of immune system function)	Used in bone marrow transplants and to treat cancers and viral infections, including hepatitis and genital warts	Human gene inserted into yeast
Antibodies (immune system proteins)	Used to fight infections, cancers, diabetes, organ rejection, and multiple sclerosis	Recombinant antibody genes inserted into cultured hamster cells
Viral proteins	Used to generate vaccines against viral diseases such as Hepatitis B and rabies and for diagnosing viral infections	Viral gene inserted into yeast or harmless virus
Enzymes	Used in treatment of heart attacks, cystic fibrosis, and other diseases and in production of cheese, detergents, dyes, antibiotics, and vitamins	Human, plant, or animal genes inserted into cultured hamster cells, yeast, or bacteria

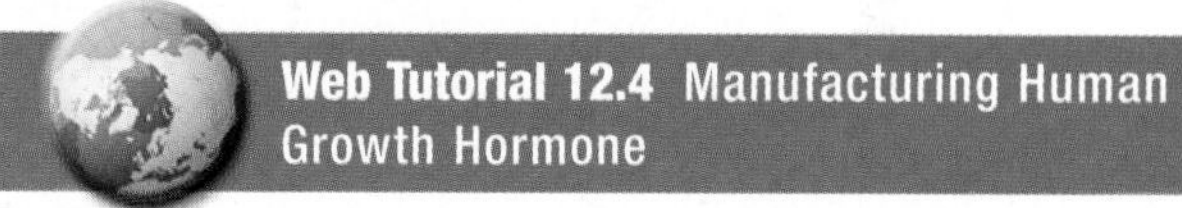

Using Biotechnology to Treat Cystic Fibrosis

Cystic fibrosis causes devastating effects in the lungs, where the lack of chloride transport causes the usually thin, watery fluid lining the airways to become thick and clogged with mucus. Several research groups are developing methods to deliver the allele for normal chloride pumps to the cells of the lungs, get them to make functioning chloride pumps, and insert these pump proteins into their plasma membranes.

The researchers first disable a suitable virus, so that the treatment itself doesn't cause a disease. Cold viruses are often used, because they normally infect cells of the respiratory tract. The DNA of the normal chloride pump allele is then inserted into the DNA of the virus. The recombinant viruses are sprayed into the patient's nose. The viruses enter cells of the lungs and release the normal chloride pump allele. The cells then manufacture pump proteins under the instructions of the normal allele, insert them into their plasma membranes, and transport chloride into the fluid lining the lungs.

The clinical trials under way for such treatments have been reasonably successful, but for only a few weeks. The patients' immune systems probably attack the viruses and eliminate them—and the helpful genes they carry. Because lung cells are continually replaced over time, a single dose "wears off" as the modified cells die. Long-term success will probably require either inhibiting the patients' immune system, which introduces its own set of problems, or designing delivery systems that can "hide" from the immune system.

Using Biotechnology to Cure Severe Combined Immune Deficiency

Like the cells of the lungs, the vast majority of cells in the human body eventually die and are replaced by new cells. In many cases, the new cells come from special populations of cells called **stem cells**. When stem cells divide, they produce daughter cells that can differentiate into several different types of mature cells. It is possible that some stem cells, under the right conditions, might be able to give rise to *any* cell type of the entire body! We will discuss this potential of stem cells in Chapter 25. Here, we will look at a more limited role for stem cells: producing just one or two types of cells.

All the cells of the immune system (mostly white blood cells) originate in the bone marrow. As mature cells die, they are replaced by new cells that arise from the division of stem cells in the bone marrow. Severe combined immune deficiency (SCID) is a rare disorder in which a child fails to develop an immune system. About 1 in 80,000 children is born with some form of SCID. Infections that

HEALTH WATCH

PRENATAL GENETIC SCREENING

Prenatal (before-birth) diagnosis of many genetic disorders, including cystic fibrosis, sickle-cell anemia, Tay-Sachs disease, and Down syndrome, requires samples of fetal cells (cells from the fetus) or chemicals produced by the developing fetus. Presently, there are two main techniques used to obtain these samples: *amniocentesis* and *chorionic villus sampling*. In addition, new techniques for isolating fetal cells from the mother's blood are currently under development. Once samples are collected, they can be used for tests that allow prenatal diagnosis of genetic disorders.

AMNIOCENTESIS A human fetus is surrounded by a watery liquid, which is enclosed in a waterproof membrane called the amnion. As the fetus grows, it sheds some of its own cells into the fluid. When a fetus is 16 weeks or older, this amniotic fluid can be collected safely by a procedure called amniocentesis. A physician determines the position of the fetus by ultrasound scanning, inserts a sterilized needle through the mother's abdominal wall, the uterus, and the amnion, and withdraws 10 to 20 milliliters of fluid (Fig. E12-3). Biochemical analysis may be performed on the fluid immediately, but other types of analysis must wait, because there are very few cells in the fluid sample. For any analyses requiring cells, such as testing for Down syndrome, the cells must first be allowed to multiply in a laboratory culture for a week or two.

CHORIONIC VILLUS SAMPLING The chorion, a membrane produced by the fetus, has many small projections called villi. In chorionic villus sampling (CVS), a physician inserts a small tube into the uterus through the mother's vagina and suctions off a few villi for analysis (see Fig. E12-3). CVS has two major advantages over amniocentesis. First, it can be done much earlier in pregnancy—as early as the eighth week. This factor is especially important if the woman is contemplating a therapeutic abortion in the event that the fetus has a major defect. Second, the sample contains enough fetal cells that analyses can be performed immediately rather than a week or two later. However, chorionic cells are more likely to have abnormal numbers of chromosomes (even when the fetus is normal). CVS also appear to have slightly greater risks than amniocentesis. Finally, CVS cannot detect certain disorders, such as spina bifida. For these reasons, CVS is much less commonly performed than amniocentesis.

FETAL CELLS FROM MATERNAL BLOOD A tiny number of fetal cells cross the placenta and enter the mother's bloodstream as early as the sixth week of pregnancy. Separating fetal cells (perhaps as few as one per milliliter of blood) from the huge numbers of maternal cells is challenging, but it can be done. Several companies now offer paternity testing from maternal blood, but practical genetic screening for inherited disorders seems to be several years in the future.

ANALYZING THE SAMPLES Several types of analyses can be performed on the amniotic fluid or fetal cells. Biochemical analyses may determine the concentration of chemicals in the amniotic fluid. For example, many metabolic disorders can be detected by abnormally low concentrations of key enzymes or by abnormally high concentrations of substances that those enzymes are supposed to break down. Other tests require whole cells. For example, analysis of the chromosomes of the fetal cells can show if the correct number of chromosomes is present or if any chromosomes show structural abnormalities.

Defective alleles, such as those that cause cystic fibrosis or sickle-cell anemia, can be detected by recombinant DNA techniques. In some cases, these techniques have eliminated the need to grow cells in culture before testing. Now, for example, one can extract DNA from just a few cells, use PCR to amplify the hemoglobin gene, and then use restriction enzyme analysis to detect the allele that causes sickle-cell anemia (see Fig. 12-9). If prenatal screening reveals that a fetus is homozygous for the sickle-cell allele, some therapeutic measures can be taken. In particular, regular doses of penicillin after the infant is born will greatly reduce the bacterial infections that otherwise kill about 15% of homozygous children.

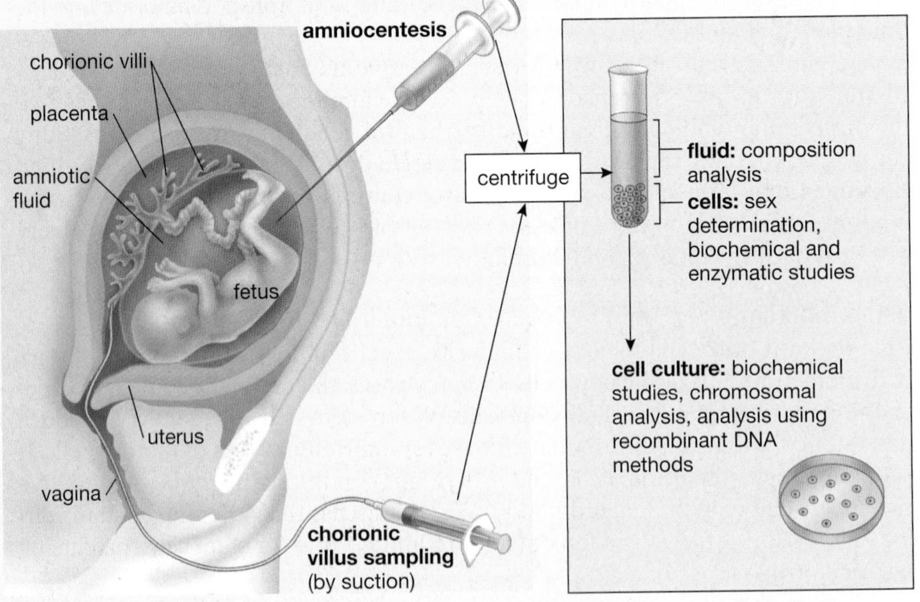

Figure E12-3 Prenatal cell sampling techniques Two methods of obtaining fetal cell samples—amniocentesis and chorionic villus sampling—and some of the tests performed on the fetal cells.

would be trivial in a normal child become life threatening. If the child has an unaffected relative with a similar genetic makeup, a bone marrow transplant from the healthy relative can give the child normal stem cells, so that he or she can develop a functioning immune system. Most SCID victims, however, die before their first birthday.

About half of the cases (over 1000 per year) are what is known as X-linked SCID. As you will see in Chapter 22, white blood cells communicate with one another via hormonelike molecules. These are released by certain white blood cells and bind to receptors on other white blood cells. These "recipient" cells respond by dividing, churning out antibodies, or attacking invading microbes. X-linked SCID is caused by a recessive allele that encodes a defective receptor. Without functioning receptors, white blood cells cannot communicate, and the entire immune response fails. Because the gene is carried on the X chromosome, the affected children are boys (see Chapter 11).

Gene therapy for X-linked SCID begins by removing stem cells from the bone marrow of an affected child. The cells are then infected with a virus carrying the allele for a normal receptor. The virus inserts the normal allele into the stem cells' DNA, and the "cured" stem cells are then injected back into the child's marrow. As these stem cells divide, some of their daughter cells mature into functioning cells of the immune system, while others remain in the bone marrow and continue dividing throughout the patient's lifetime. As many as nine children may have been permanently cured by this procedure.

Unfortunately, the treatment is not without danger. In 2002, two children given this treatment developed leukemia. This devastating complication temporarily shut down not only the SCID therapy but also dozens of other gene therapy trials employing similar viral techniques. Researchers have now identified what caused leukemia in these children and are working to develop safer methods of delivering genes to SCID children.

SCID gene therapy illustrates both the promise and peril of current techniques. Most children with SCID cannot live a normal life, and usually do not live at all, without gene therapy. Is the current gene therapy worth the risk of leukemia, until better methods are found? Gene therapies, like most other biotechnologies, are still in their infancy. Difficult decisions must be made by governments, scientists, physicians, patients, and, in the case of affected children, parents. Science can, perhaps, quantify the risks and benefits of such therapies, but who should decide how much risk is too much?

12.7 What Are the Major Ethical Issues of Biotechnology?

Modern biotechnology offers the promise—some would say the threat—of greatly changing our lives, and the lives of many other organisms on Earth. As Spiderman noted, "With great power comes great responsibility." Is humanity capable of handling the responsibility of biotechnology? Here we will explore two controversies: the use of genetically modified organisms in agriculture and prospects for genetically modifying human beings.

Should Genetically Modified Organisms Be Permitted in Agriculture?

Transgenic crops have clear advantages for farmers. Herbicide-resistant crops allow farmers to rid their fields of weeds, which may reduce harvests by 10% or more, through the use of powerful, nonselective herbicides, at virtually any stage of crop growth. Insect-resistant crops decrease the need to apply synthetic pesticides, saving the cost of the pesticides, tractor fuel, and labor. Because transgenic crops produce larger harvests at less cost, these savings may be passed along to the consumer.

Other potentially useful genetically modified crops, not yet commercially available, include rice with high amounts of beta-carotene (which the body can convert into vitamin A), soybeans or other plants containing more healthful vegetable oils, and drought-resistant corn.

Regardless of potential monetary or health benefits, many people strenuously object to transgenic crops or livestock. They fear that genetically modified organisms (GMOs) may be hazardous to human health or dangerous to the environment.

Are Foods from GMOs Dangerous to Eat?

In most cases, the safety of food derived from GMOs is not a major concern. For example, tests have shown that the Bt protein is not toxic to mammals and should not prove a danger to human health. If growth-enhanced livestock are ever developed, they will simply have more meat, composed of exactly the same proteins that exist in ordinary animals, so they shouldn't be dangerous either.

Another possible danger is that people might be allergic to genetically modified plants. StarLink™ corn, which contained a genetically modified Bt protein, was not approved for human consumption because its engineered Bt protein is digested less readily than normal Bt protein and might cause allergic reactions. The U.S. Food and Drug Administration now monitors all new transgenic crop plants for their potential to cause allergies. Interestingly, several researchers are using genetic engineering to try to make nonallergenic peanuts for people who are allergic to normal peanuts by modifying or removing the genes for the peanut proteins that cause trouble.

In 2003, the U.S. Society of Toxicology studied the risks of genetically modified plants and concluded that the current transgenic plants pose no significant dangers to human health. The society also recognized that past safety does not guarantee future safety and recommended continued testing and evaluation of all new genetically modified plants.

Are GMOs Hazardous to the Environment?

The environmental effects of GMOs are more problematic. Because the genes for herbicide resistance or pest resistance are incorporated into the genome of the transgenic crop, these genes will be in its pollen, too. Wind might carry this pollen miles from a farmer's field. Many crops, including corn, sunflowers, and, in Eastern Europe and the Middle East, wheat, barley, and oats, have wild relatives living nearby. Suppose these wild plants interbred with transgenic crops and became resistant to herbicides or pests? Would they become significant weed problems? Would they displace other plants in the wild, because they would be less likely to be eaten by insects?

Even if transgenic crops have no close relatives in the wild, bacteria and viruses sometimes carry genes between unrelated species. Could viruses spread genes into wild plant populations? No one really knows the answers to these questions.

In 2002, a committee of the U.S. National Academy of Sciences pointed out that crops modified by both traditional methods and recombinant DNA technologies may cause major changes in the environment. The committee also found that the U.S. does not have an adequate system for monitoring changes in ecosystems that might be caused by transgenic crops. It recommended more thorough screening of transgenic plants before they are used commercially, and sustained ecological monitoring after commercialization.

What about transgenic animals? Unlike pollen, most domesticated animals, such as cattle or sheep, are relatively immobile. Further, most have few wild relatives with which they might exchange genes, so the dangers to natural ecosystems appear minimal. However, some transgenic animals, especially fish, may pose significant potential threats (see "Biotechnology Watch: Bonanza from the Sea or Frankenfish?").

Should the Human Genome Be Changed by Biotechnology?

Many of the ethical implications of human applications of biotechnology are fundamentally the same as those connected with other medical procedures. For example, long before biotechnology enabled prenatal testing for cystic fibrosis or sickle-cell anemia, trisomy 21 (Down syndrome) could be diagnosed in embryos by counting the chromosomes in cells taken from the amniotic fluid (see "Health Watch: Prenatal Genetic Screening"). Whether parents should use such information as a basis for therapeutic abortion or to prepare to care for the affected child is an ethical issue that generates considerable debate. Other ethical concerns have arisen purely as a result of advances in biotechnology. For instance, should people be allowed to select, or even *change*, the genomes of their offspring?

Several years ago, a girl in Colorado was born with Franconi anemia, a genetic disorder that is fatal without a bone marrow transplant. Her parents wanted another child—a very special child. They wanted one without Franconi anemia, of course, but they also wanted a child who could serve as a donor for their daughter. They went to Yury Verlinsky of the Reproductive Genetics Institute for help. Verlinsky used the parents' sperm and eggs to create dozens of embryos in culture. The embryos were then tested both for the genetic defect and for tissue compatibility with the couple's daughter. Verlinsky chose an embryo with the desired genotype and implanted it into the mother's uterus. Nine months later, a son was born. Blood from his umbilical cord provided stem cells to transplant into his sister's bone marrow. Today, both children are healthy.

Was this an appropriate use of genetic screening? Should dozens of embryos be created, knowing that the vast majority will be discarded? Is this ethical if it is the only way to save the life of another child? Assuming that it's possible someday, would it be ethical to select embryos that would grow up to be bigger or stronger football players?

Today's technology allows physicians only to select among existing embryos, not to change their genomes. But technologies do exist to alter the genomes of bone marrow stem cells in attempts to cure SCID. Soon, it may become possible to change the genes of fertilized eggs (Fig. 12-11). If such techniques were used to fix SCID or cystic fibrosis, would they be ethical? What about improving prospective athletes? If the technology is developed to cure diseases, it will be difficult to prevent it from being used for nonmedical purposes. Who will determine what is an appropriate use and what is a trivial vanity?

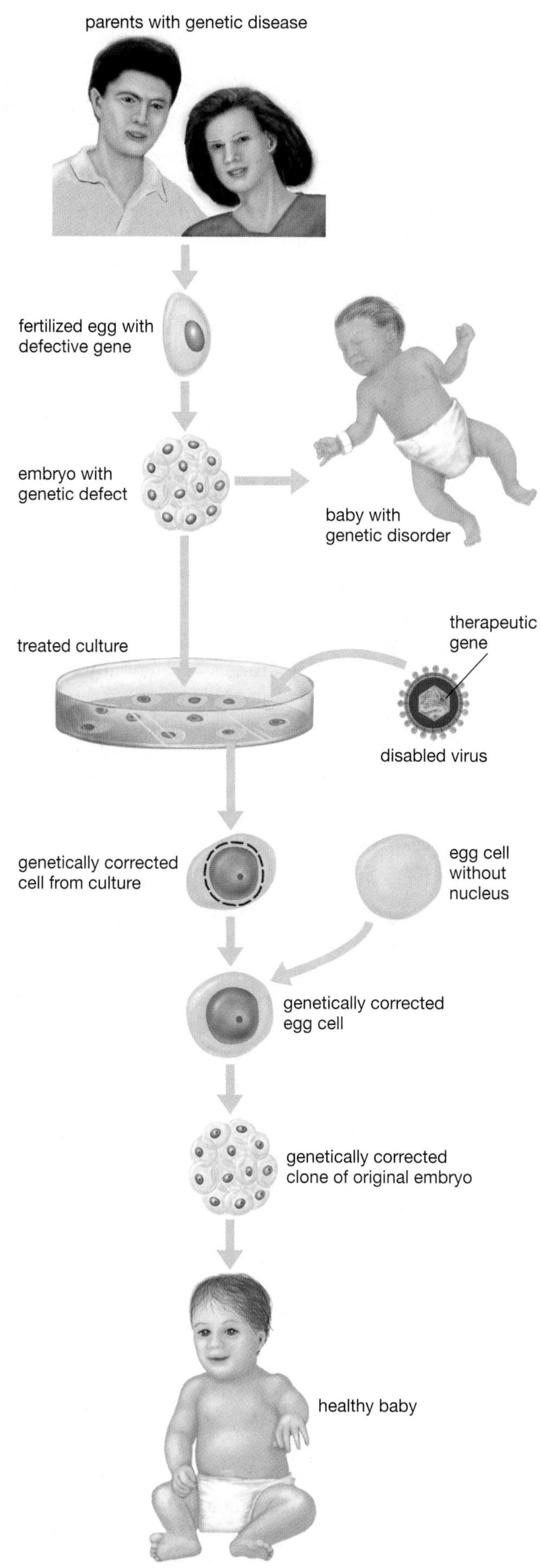

Figure 12-11 Human cloning technology might allow permanent correction of genetic defects In this process, human embryos are derived from eggs fertilized in culture dishes using sperm and eggs from a man and woman, one or both of whom have a genetic disorder. When an embryo containing a defective allele grows into a small cluster of cells, a single cell would be removed from the embryo and a normal allele inserted into its DNA. The repaired nucleus would be injected into another egg (taken from the same woman) whose nucleus had been removed. The repaired egg cell would then be implanted in the woman's uterus for normal development.

GUILTY OR INNOCENT? *REVISITED*

The Innocence Project is strictly a science-based effort—no courtroom theatrics, no picking away at fine points of law and procedure. The Project takes cases only when molecular evidence exists that may exonerate the accused. Without the rapid advances in DNA technology in the past decade, Jimmy Bromgard, Eddie Lloyd, and dozens of other people convicted of serious crimes would still be in prison.

As we have seen in this chapter, the Innocence Project uses data on short tandem repeats collected by forensic scientists, the FBI, and other government agencies. Forensic scientists hypothesized that STR allele differences could be used to identify people. They performed hundreds of experiments to determine which STRs were most variable among individuals and were most likely to remain intact in crime scene samples. Compiling these data, they concluded that a small number of STRs can determine whether crime scene and defendant DNA match, with phenomenal accuracy.

Police and district attorneys also use DNA technology as an investigative tool. In 1990, three elderly women in Goldsboro, North Carolina, were raped; two were murdered. DNA evidence indicated that all three crimes were committed by the same assailant, known only as the "Night Stalker." Over the years, the FBI and many states have slowly built up databases of criminals, each identified by their DNA "fingerprints" of short tandem repeats. In 2001, the Goldsboro police created a DNA fingerprint profile of the Night Stalker from the evidence they had stored for over a decade. They sent the profile to the North Carolina DNA database and discovered a match to a known criminal, who was brought in for questioning. Faced with indisputable DNA evidence, the Stalker confessed. He is now in prison.

CONSIDER THIS: Who are the "heroes" in these stories? The obvious ones, of course, are the professors and law students of the Innocence Project and the members of the Goldsboro Police Department. But what about Thomas Brock, who discovered *Thermus aquaticus* and its unusual lifestyle in Yellowstone hot springs (see "Scientific Inquiry: Hot Springs and Hot Science")? Or Kary Mullis, who discovered PCR? Or the hundreds of biologists, chemists, and mathematicians who, over the last century, developed procedures for gel electrophoresis, labeling DNA, and statistical analysis of sample matching?

Scientists often say that science is worthwhile for its own sake and that it is difficult or impossible to predict which discoveries will lead to the greatest benefits for humanity. Nonscientists are sometimes skeptical of such claims, especially when asked to pay the costs of scientific projects. How do you think public support of science should be allocated? Forty years ago, would *you* have voted to give Thomas Brock public funds to see what types of organisms lived in hot springs?

LINKS TO LIFE: Biotechnology, Privacy, and You

What was your reaction to the methods used by the Goldsboro Police Department to find the Night Stalker? Most people feel that matching DNA from the crime scenes to the Stalker's DNA fingerprint files is a superb example of science and law enforcement working together to protect the public. Some, however, worry that the ever-increasing amount of personal information in government—and possibly corporate—files might be abused if placed in the wrong hands. Of course, the Stalker's DNA fingerprints were on file only because he had committed other serious crimes. How severe a crime do you think warrants DNA fingerprinting? Murder? Assault? Breaking and entering? Should *everyone's* DNA be on file? Every criminal has a first offense, so perhaps more would be caught if there were a universal DNA file.

Others worry about the possible exploitation of biotechnology by health providers or employers. Medical care for a patient with cystic fibrosis, for example, costs about $40,000 per year. Should health insurance rates reflect known genetic diseases? If our ability to understand and predict genetic disorders improves enough, should employers be able to discriminate against people who are genetically predisposed to potentially costly conditions or behaviors, such as schizophrenia, breast cancer, or even risk taking? Several states have passed laws forbidding such discrimination. How do you think society should handle the information revealed by human biotechnology?

Chapter Review

Summary of Key Concepts

12.1 What Is Biotechnology?

Biotechnology is any use or alteration of organisms, cells, or biological molecules to achieve specific practical goals. Modern biotechnology alters genetic material by means of genetic engineering. In many applications of genetic engineering, recombinant DNA is produced by combining DNA from different organisms. Organisms containing modified DNA or DNA from other species are called transgenic, or genetically modified, organisms. Some major goals of genetic engineering are to increase our understanding of gene function, to treat disease, and to improve agriculture.

12.2 How Does DNA Recombine in Nature?

DNA recombination occurs naturally through processes such as bacterial transformation, viral infection, and crossing over during sexual reproduction.

12.3 How Is Biotechnology Used in Forensics?

Specific regions of DNA can be amplified by the polymerase chain reaction (PCR). The most common regions used in forensics are short tandem repeats (STRs). The STRs are separated by gel electrophoresis and made visible with DNA probes. The pattern of STRs is unique to each individual and can be used to match DNA found at a crime scene with DNA from suspects.

12.4 How Is Biotechnology Used in Agriculture?

Many crop plants have been modified by the addition of genes that promote herbicide resistance or pest resistance. Plants may also be modified to produce human proteins, vaccines, or antibodies. Transgenic animals may be produced as well, with properties such as faster growth, increased production of valuable products such as milk, or the ability to produce human proteins, vaccines, or antibodies.

12.5 How Is Biotechnology Used to Learn About the Human Genome?

Techniques of biotechnology were used to discover the complete nucleotide sequence of the human genome. This knowledge will be used to learn the identities and functions of new genes, to discover medically important genes, and to explore the genetic variability among individuals.

12.6 How is Biotechnology Used for Medical Diagnosis and Treatment?

Biotechnology may be used to diagnose genetic disorders such as sickle-cell anemia or cystic fibrosis. For example, in the diagnosis of sickle-cell anemia, restriction enzymes cut normal and defective globin alleles in different locations. The resulting DNA fragments of different lengths may then be separated and identified by gel electrophoresis. In the diagnosis of cystic fibrosis, DNA probes complementary to various cystic fibrosis alleles are placed on a DNA array. Base pairing of a patient's DNA to specific probes on the array identifies which alleles are present in the patient.

Inherited diseases are caused by defective alleles of crucial genes. Genetic engineering may be used to insert functional alleles of these genes into normal cells, into stem cells, or even into eggs to correct the genetic disorder.

12.7 What Are the Major Ethical Issues of Biotechnology?

The use of genetically modified organisms in agriculture is controversial for two major reasons: consumer safety and environmental protection. In general, GMOs contain proteins that are harmless to mammals, are readily digested, or are already found in other foods. The transfer of potentially allergenic proteins to normally nonallergenic foods can be avoided by thorough testing. Environmental effects of GMOs are more difficult to predict. Foreign genes, such as those for pest resistance or herbicide resistance, might be transferred to wild plants, with resulting damage to agriculture or disruption of ecosystems, or both. If they escape, highly mobile transgenic animals might displace their wild relatives.

Key Terms

biotechnology *p. 194*
DNA fingerprint *p. 200*
gel electrophoresis *p. 199*
genetically modified organism (GMO) *p. 194*
genetic engineering *p. 194*
plasmid *p. 196*
polymerase chain reaction (PCR) *p. 197*
recombinant DNA *p. 194*
restriction enzyme *p. 201*
stem cell *p. 205*
transformation *p. 195*
transgenic *p. 194*

Thinking Through the Concepts

Multiple Choice

1. *Restriction enzymes*
- **a.** cut DNA at specific nucleotide sequences
- **b.** may produce DNA fragments with "sticky ends"
- **c.** are used to create recombinant plasmids
- **d.** are used in the diagnosis of some genetic disorders
- **e.** all of the above

2. *Plasmids*
- **a.** are parts of viruses
- **b.** are circles of DNA often found in bacteria
- **c.** are composed chiefly of protein
- **d.** are collected during amniocentesis
- **e.** all of the above

3. *The polymerase chain reaction*
- **a.** is a method of synthesizing human protein from human DNA
- **b.** takes place naturally in bacteria
- **c.** can produce billions of copies of a DNA fragment in several hours
- **d.** uses restriction enzymes
- **e.** is relatively slow and expensive compared with other types of DNA purification

4. *The different alleles of a gene*
- **a.** have different nucleotide sequences
- **b.** may be identified by restriction enzymes
- **c.** base pair with different complementary sequences of DNA probes
- **d.** may produce normal or defective proteins
- **e.** all of the above

5. *Genetic engineers often use _____ to insert foreign DNA into a genome.*
- **a.** bacteria
- **b.** unfertilized egg cells
- **c.** *Bacillus thuringiensis*
- **d.** viruses
- **e.** all of the above

Review Questions

1. Describe three natural forms of genetic recombination and discuss the similarities and differences between recombinant DNA technology and these natural forms of genetic recombination.

2. What is a plasmid? How are plasmids involved in bacterial transformation?

3. What is a restriction enzyme? How can restriction enzymes be used to splice a piece of DNA within a plasmid?

4. Describe the polymerase chain reaction.
5. Describe several uses of genetic engineering in agriculture.
6. Describe several uses of genetic engineering in human medicine.
7. Describe amniocentesis and chorionic villus sampling, including the advantages and disadvantages of each. What are their medical uses?
8. How does gel electrophoresis separate pieces of DNA?
9. How are DNA probes used to identify specific nucleotide sequences of DNA? How are they used in the diagnosis of genetic disorders?

Applying the Concepts

1. **IS THIS SCIENCE?** In a 2004 Web survey conducted by the Canadian Museum of Nature, 84% of the people polled said they would eat a genetically modified banana that contained a vaccine against an infectious disease, while only 47% said they would eat a GM banana with extra vitamin C produced by the action of a rat gene. Do you think that this difference in acceptance of GMOs is scientifically valid? Why or why not?
2. As you may know, many insects have evolved resistance to common pesticides. Do you think that insects might evolve resistance to Bt crops? If this is a risk, do you think that Bt crops should be planted anyway? Why?
3. If you were contemplating having a child, would you want yourself and your spouse tested for the cystic fibrosis allele or other potentially harmful alleles? If both of you were found to carry a harmful recessive allele, how would you deal with this decision?

For More Information

Gibbs, N. "Baby, It's You! And You, and You . . . " *Time*, February 19, 2001. How close are we to cloning a human?

Harder, B. "Born to Heal." *Science News*, March 2004. Should bioengineering and *in vitro* fertilization be used to produce custom-made babies? What, if any, limits should be placed on this technique?

Langridge, W. H. R. "Edible Vaccines." *Scientific American*, September 2000. Current state of attempts to develop plants that produce vaccines or treatments for diseases.

Lanza, R. P., Dresser, B. L., and Damiani, P. "Cloning Noah's Ark." *Scientific American*, November 2000. Using domestic animals as surrogate mothers for endangered species provides some hope for staving off extinction.

Marvier, M. "Ecology of Transgenic Crops." *American Scientist,* March/April 2001. A thoughtful article that weighs the benefits, risks, and uncertainties of bioengineered crops.

Ridley, M. "The Year of the Genome." *Discover*, January 2001. Now that the Human Genome Project has finished sequencing the entire human genome, the real biology begins—figuring out what all of those genes, and especially their myriad alleles, actually do.

Wheelwright, J. "Bad Genes, Good Drugs." *Discover*, April 2002. Researchers are using the results of the Human Genome Project to identify genes that predispose people to diseases, such as Alzheimer's disease, and to develop new drugs to combat those diseases.

Wheelwright, J. "Body, Cure Thyself." *Discover*, March 2002. The promise of gene therapy is vast, but so far, the results have been decidedly mixed.

WEB TUTORIAL

To access a Web Tutorial visit http://www.prenhall.com/audesirk4. *Log in to the Web site selected by your instructor, navigate to this chapter, and select the appropriate Web Activity number.*

12.1 Genetic Recombination in Bacteria
Estimated time: 5 minutes

We often think of DNA and the genes it contains as being static and unchanging. In fact, it is anything but static. This activity will demonstrate some of the different ways that genetic recombination can occur in cells, both prokaryotic and eukaryotic.

12.2 Polymerase Chain Reaction (PCR)
Estimated time: 10 minutes

In this activity, you'll see the process of polymerase chain reaction (PCR), which allows minute amounts of DNA to be amplified into large amounts.

12.3 Human Genome Sequencing
Estimated time: 10 minutes

One strategy for sequencing the genome, called map-based sequencing, is the primary subject of this activity.

12.4 Manufacturing Human Growth Hormone
Estimated time: 10 minutes

In this activity, you'll see how human growth hormone can be produced inexpensively and safely using recombinant DNA techniques.

12.5 Web Investigation: Guilty or Innocent?
Estimated time: 15 minutes

Sometimes, people are wrongly convicted of serious crimes. Later, DNA evidence can clear them by demonstrating that samples of cells left at the scene of the crime do not come from the person convicted of the crime. Take a look at the techniques being used to reexamine cold cases.

UNIT 3

Evolution

The ghostly grandeur of ancient bones evokes images of a lost world. Fossil remnants of extinct creatures, such as this *Triceratops* dinosaur skeleton, provide clues for the biologists who attempt to reconstruct the history of life.

Chapter 13

Principles of Evolution

This sleek, streamlined humpback whale has tiny hip and leg bones embedded in the flesh of its abdomen, a legacy of its evolutionary heritage.

AT A GLANCE

What Good Are Wisdom Teeth?

Have you had your wisdom teeth removed yet? If not, it's probably only a matter of time. Almost all of us will visit an oral surgeon to have our wisdom teeth extracted. There's just no room in our jaws for these rearmost molars, and removing them is the best way to prevent dental disasters. And removal is harmless, because we don't really need wisdom teeth. They're pretty much useless.

If you've already suffered through a wisdom tooth extraction, you may have found yourself wondering why we even *have* these useless molars. Biologists hypothesize that we have them because our apelike ancestors had them and we inherited them, even though we don't really need them. If this hypothesis is correct, then the presence in a living species of structures that have no current function, but that *are* useful in other species, demonstrates the shared ancestry of different species.

Some excellent evidence of the connection between useless traits and evolutionary ancestry is provided by whales. Modern whales have only two limbs: a pair of flippers located toward the front of the body. They have no hind limbs at all. Nonetheless, if you examine the skeleton of a modern whale, you will find some very small pelvis and leg bones embedded in the abdomen, toward the rear of the animal's body.

If these tiny, useless bones really are, as hypothesized, vestiges of actual rear limbs that were present in the whales' evolutionary ancestors, then we would predict the discovery of fossil whales with rear limbs. And, in fact, several species of extinct, limbed whales have been found. One of the more recent discoveries is *Ambulocetus*, a 49-million-year-old whale ancestor with long, webbed feet on its hind legs.

***Ambulocetus*, an ancient relative of today's whales, had well-developed hind legs. Modern whales retain small vestiges of these limbs.**

When you began studying biology, you may not have seen a connection between your wisdom teeth and a whale's missing legs. But the connection is there, provided by the concept that unites all of biology: **evolution**, the process by which the characteristics of a population change over time.

13.1 How Did Evolutionary Thought Evolve?

Modern biology is based on our understanding that life has evolved, but early scientists did not recognize this fundamental principle. The main ideas of evolutionary biology were widely accepted only after the publication of Charles Darwin's work in the late nineteenth century. Nonetheless, the intellectual foundation on which these ideas rest developed gradually over the centuries before Darwin's time.

Early Biological Thought Did Not Include the Concept of Evolution

Pre-Darwinian science, heavily influenced by theology, held that all organisms were created simultaneously by God and that each distinct life-form remained fixed and unchanging from the moment of its creation. This explanation of how life's diversity arose was elegantly expressed by the ancient Greek philosophers, especially Plato and Aristotle. Plato (427–347 B.C.) proposed that each object on Earth was merely a temporary reflection of its divinely inspired "ideal form." Plato's student Aristotle (384–322 B.C.) categorized all organisms into a linear hierarchy that he called the "ladder of Nature."

These ideas formed the basis of the view that the form of each type of organism is permanently fixed. This view reigned unchallenged for nearly 2000 years. By the eighteenth century, however, several lines of newly emerging evidence began to erode the dominance of this static view of creation.

Exploration of New Lands Revealed a Staggering Diversity of Life

The Europeans who explored and colonized Africa, Asia, and the Americas were often accompanied by naturalists who observed and collected the plants and animals of these previously unknown (to Europeans) lands. By the 1700s, the accumulated observations and collections of the naturalists had begun to reveal the true scope of life's variety. The number of species, or different types of organisms, was much greater than anyone had suspected.

Stimulated by the new evidence of life's incredible diversity, some eighteenth-century naturalists began to take note of some fascinating patterns. They noticed, for example, that the species found in one place were different from those found in other places, so that each area had its own distinctive set of species. In addition, the naturalists saw that some of the species in a given location closely resembled one another, yet differed in some characteristics. To some scientists of the day, the differences between the species of different geographical areas and the existence of clusters of similar species within areas seemed inconsistent with the idea that species were fixed and unchanging.

A Few Scientists Speculated That Life Had Evolved

A few eighteenth-century scientists went so far as to speculate that species had, in fact, changed over time. For example, the French naturalist Georges Louis LeClerc (1707–1788), known by the title Comte de Buffon, suggested that perhaps the original creation provided a relatively small number of founding species and that some modern species had been "conceived by Nature and produced by Time"—that is, they had evolved through natural processes.

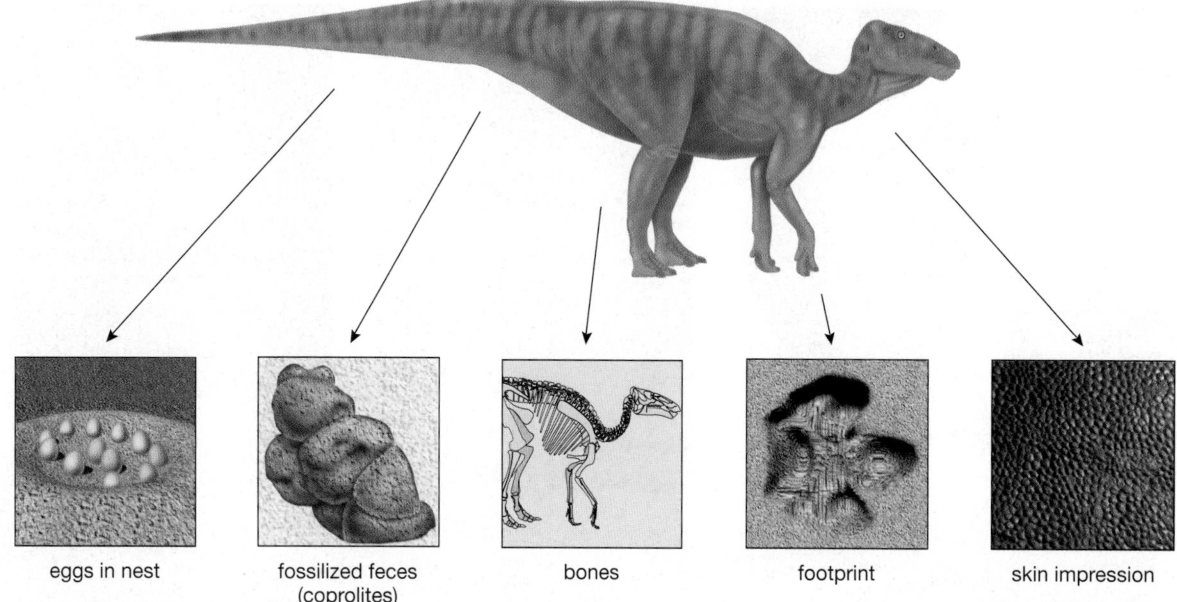

Figure 13-1 Types of fossils Any preserved part or trace of an organism is a fossil.

Fossil Discoveries Showed That Life Has Changed over Time

As Buffon and his contemporaries pondered the implications of new biological discoveries, developments in geology cast further doubt on the idea of permanently fixed species. Especially important was the discovery, during excavations for roads, mines, and canals, of rock fragments that resembled parts of living organisms. People had known of such **fossils** since the fifteenth century, but most thought they were ordinary rocks that wind, water, or people had worked into lifelike forms. As more and more fossils were discovered, however, it became obvious that they were the remains or impressions of plants or animals that had died long ago and had been changed into or in some way preserved in rock (Fig. 13-1).

By the beginning of the nineteenth century, some pioneering investigators realized that how fossils were distributed in rock was also significant. Many rocks occur in layers, with newer layers positioned over older layers (Fig. 13-2). The British surveyor William Smith (1769–1839), who studied rock layers and the fossils embedded in them, recognized that certain fossils were always found in the same layers of rock. Further, the organization of fossils and rock layers was consistent: Fossil type A could always be found in a rock layer resting on top of an older layer containing fossil type B, which in turn rested on top of a still-older layer containing fossil type C, and so on.

Figure 13-2 The Grand Canyon of the Colorado River Layer upon layer of sedimentary rock forms the walls of the Grand Canyon. The layers formed from the accumulation of sediments over more than a billion years.

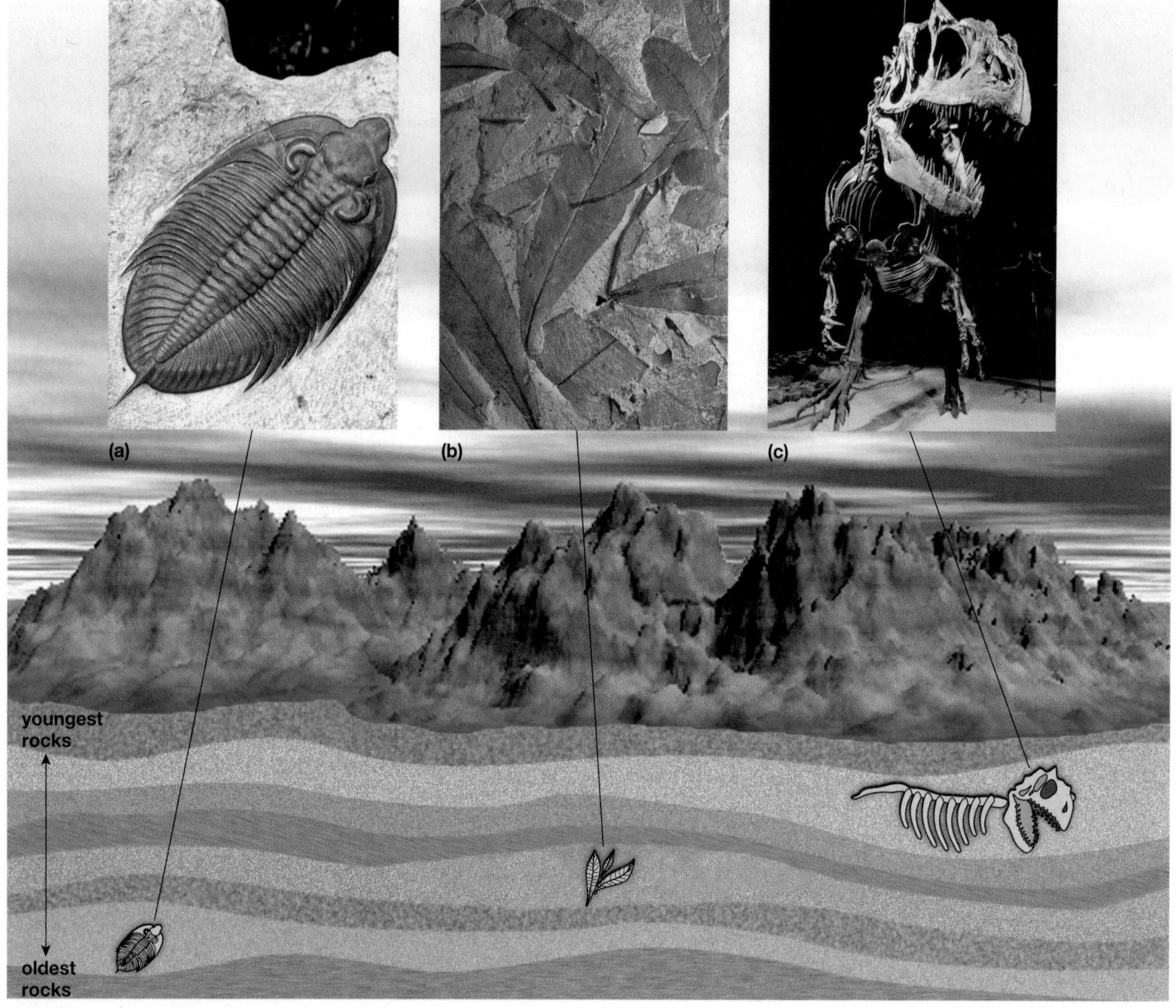

Figure 13-3 Fossils of extinct organisms Fossils provide strong support for the idea that today's organisms were not created all at once but arose over time by the process of evolution. If all species had been created simultaneously, we would not expect ***(a)*** trilobites to be found in older rock layers than ***(b)*** seed ferns, which in turn would not be expected in older layers than ***(c)*** dinosaurs, such as *Allosaurus*. Trilobites became extinct about 230 million years ago, seed ferns about 150 million years ago, and dinosaurs 65 million years ago.

Scientists of the period also discovered that fossil remains showed a remarkable progression. Most fossils found in the oldest layers were very different from modern organisms, and the resemblance to modern organisms gradually increased in progressively younger rocks. Many of the fossils were from plant or animal species that had gone *extinct;* that is, no members of the species still lived on Earth (Fig. 13-3).

Putting all of these facts together, some scientists came to an inescapable conclusion. Different types of organisms had lived at different times in the past.

Some Scientists Devised Nonevolutionary Explanations for Fossils

Despite the growing fossil evidence, many scientists of the period did not accept the proposition that species changed and new ones had arisen over time. To account for extinct species while preserving the notion of a single creation by God, Georges Cuvier (1769–1832) proposed the theory of **catastrophism**. Cuvier, a French paleontologist, hypothesized that a vast supply of species was created initially. Successive catastrophes (such as the Great Flood described in the Bible) produced layers of rock and destroyed many species, fossilizing some of their remains in the process. The organisms of the modern world, he theorized, are the species that survived the catastrophes.

Geology Provided Evidence That Earth Is Exceedingly Old

Cuvier's hypothesis of a world shaped by successive catastrophes was challenged by the work of the geologist Charles Lyell (1797–1875). Lyell, building on the earlier thinking of James Hutton (1726–1797), considered the forces of wind, water, and volcanoes and concluded that there was no need to invoke catastrophes to explain the findings of geology. Don't flooding rivers lay down layers of sediment? Don't lava flows produce layers of basalt? Shouldn't we conclude, then, that layers of rock are evidence of ordinary natural processes, occurring repeatedly over long periods of time? This concept, called **uniformitarianism**, had profound implications, because it implies that Earth is very old.

Before the 1830 publication of Lyell's evidence in support of uniformitarianism, few scientists suspected that Earth could be more than a few thousand years old. Counting generations in the Old Testament, for example, yields a maximum age of 4000 to 6000 years. An Earth this young poses problems for the idea that life has evolved. For example, ancient writers such as Aristotle described wolves, deer, lions, and other organisms that were identical to those present in Europe more than 2000 years later. If organisms had changed so little over that time, how could whole new species possibly have arisen if Earth was created only a couple of thousand years before Aristotle's time?

But if, as Lyell suggested, rock layers thousands of feet thick were produced by slow, natural processes, then Earth must be old indeed, many millions of years old. Lyell, in fact, concluded that Earth was eternal. (Modern geologists estimate that Earth is about 4.5 billion years old; see "Scientific Inquiry: How Do We Know How Old a Fossil Is?" in Chapter 15.)

Lyell (and his intellectual predecessor Hutton) showed that there was enough time for evolution to occur. But what was the mechanism? What process could cause evolution?

Some Pre-Darwin Biologists Proposed Mechanisms for Evolution

One of the first scientists to propose a mechanism for evolution was the French biologist Jean Baptiste Lamarck (1744–1829). Lamarck was impressed by the sequences of organisms in the rock layers. He observed that older fossils tend to be simple, whereas younger fossils tend to be more complex and more like existing organisms. In 1801, Lamarck hypothesized that organisms evolved through the inheritance of acquired characteristics, a process in which the bodies of living organisms are modified through the use or disuse of parts, and these modifications are inherited by offspring. Why would bodies be modified? Lamarck proposed that all organisms possess an innate drive for perfection. For example, if ancestral giraffes tried to improve their lot by stretching upward to feed on leaves that grow high up in trees, their necks became slightly longer as a result. Their offspring would inherit these longer necks and then stretch even farther to reach still higher leaves. Eventually, this process would produce modern giraffes with very long necks indeed.

Today, we understand how inheritance works and can see that Lamarck's proposed evolutionary process could not work as he described it. Acquired characteristics are not inherited. The fact that a prospective father pumps iron doesn't mean that his children will look like Arnold Schwarzenegger. Remember, though, that in Lamarck's time the principles of inheritance had not yet been discovered (Mendel was born a few years before Lamarck's death). In any case, Lamarck's insight that inheritance plays an important role in evolution was an important influence on the later biologists who discovered the key mechanism of evolution.

Darwin and Wallace Proposed a Mechanism of Evolution

By the mid-1800s, a growing number of biologists had concluded that present-day species had evolved from earlier ones. But how? In 1858 Charles Darwin and Alfred Russel Wallace, working separately, provided convincing evidence that evolution was driven by a simple yet powerful process.

Although their social and educational backgrounds were very different, Darwin and Wallace were quite similar in some respects. Both had traveled extensively in the tropics and had studied the plants and animals living there. Both found that some species differed in only a few features (Fig. 13-4). Both were familiar with the fossils that had been discovered, which showed a trend of increasing complexity through time. Finally, both were aware of the studies of Hutton and Lyell, who had proposed that Earth is extremely ancient. These facts suggested to both Darwin and Wallace that species change over time. Both men sought a mechanism that might cause such evolutionary change.

In 1858 Darwin and Wallace each described the same mechanism for evolution in remarkably similar papers that were presented to the Linnaean Society in London. Like Gregor Mendel's manuscript on the principles of genetics, their papers had little impact. The secretary of the society, in fact, wrote in his annual report that nothing very interesting happened that year. Fortunately, the next year Darwin published his monumental *On the Origin of Species by Means of Natural Selection,* which attracted a great deal of attention to the new theory.

13.2 How Does Natural Selection Work?

Darwin and Wallace concluded that life's huge variety of excellent designs arose by a process of descent with modification, in which the members of each generation differ slightly from the members of the preceding generation, and these small changes accumulate over long stretches of time to produce major transformations.

Figure 13-4 Darwin's finches, residents of the Galapagos Islands Darwin studied a group of closely related species of finches on the Galapagos Islands. Each species specializes in eating a different type of food and has a beak of characteristic size and shape, because natural selection has favored the individuals best suited to exploit each food source efficiently. Aside from the differences in their beaks, the finches are quite similar.

(a) Large ground finch, beak suited to large seeds

(b) Small ground finch, beak suited to small seeds

(c) Warbler finch, beak suited to insects

(d) Vegetarian tree finch, beak suited to leaves

They called the process natural selection. The chain of logic leading to their powerful conclusion turns out to be surprisingly simple and straightforward. Darwin and Wallace based their theory of evolution by natural selection on four observations and the conclusions that followed naturally from those observations.

Observation 1: A natural **population**, or all the individuals of one species in a particular area, has the potential to grow rapidly, because organisms can produce far more offspring than are required merely to replace the parents.

Observation 2: Nevertheless, the number of individuals in a natural population tends to remain relatively constant over time.

> **Conclusion 1:** Therefore, more organisms must be born than survive long enough to reproduce. If some individuals fail to survive, it must also be true that organisms compete to survive and reproduce. In each generation, many individuals must die young. Even among those that survive, many must fail to reproduce, produce few offspring, or produce less-fit offspring that, in turn, fail to survive and reproduce.

Observation 3: Individual members of a population differ from one another in many respects, including their ability to obtain resources, withstand environmental extremes, and escape predators.

> **Conclusion 2:** These differences among individuals help determine which individuals survive and reproduce most successfully, thereby leaving the most offspring. This process, in which those individuals whose traits are most advantageous leave a larger number of offspring, is known as **natural selection**.

Observation 4: At least some of the variation in traits that affect survival and reproduction is due to differences that may be passed from parent to offspring.

> **Conclusion 3:** Because the individuals that are best suited to their environment leave more offspring, the traits (and underlying genes) of these individuals are passed to a larger proportion of the individuals in subsequent generations. Over many generations, this differential, or unequal, reproduction among individuals with different genetic makeup changes the overall genetic composition of the population. This process is evolution by natural selection.

Figure 13-5 charts these observations and conclusions, and "Scientific Inquiry: Charles Darwin—Nature Was His Laboratory" tells how Darwin was able to make the observations that led to his understanding of how organisms evolve.

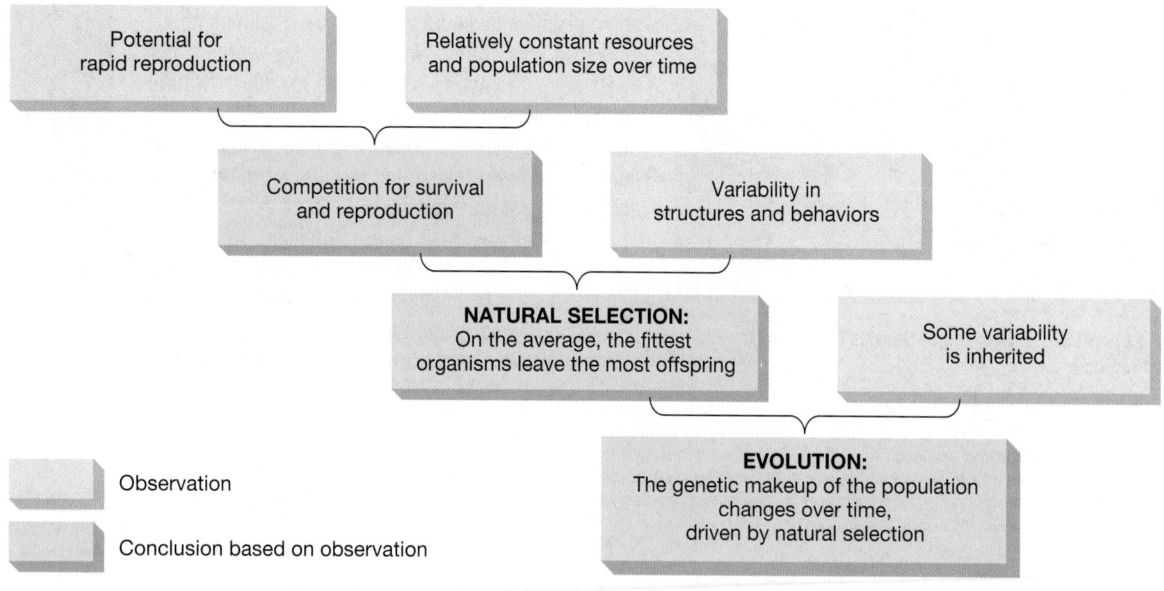

Figure 13-5 A flowchart of evolutionary reasoning This chart is based on the hypotheses of Darwin and Wallace but incorporates ideas from modern genetics. **QUESTION:** *Is sexual reproduction required to generate the variability in structures and behaviors that is necessary for natural selection?*

SCIENTIFIC INQUIRY

CHARLES DARWIN—NATURE WAS HIS LABORATORY

As do many students, Charles Darwin excelled only in subjects that intrigued him. Although his father was a physician, Darwin was uninterested in medicine and unable to stand the sight of surgery. He eventually obtained a degree in theology from Cambridge University, although theology too was of minor interest to him. What he really liked to do was to tramp over the hills, observing plants and animals, collecting new specimens, scrutinizing their structures, and categorizing them.

In 1831, when Darwin was 22 years old (Fig. E13-1), he secured a position as "gentleman companion" to Captain Robert Fitzroy of the HMS *Beagle.* The *Beagle* soon embarked on a 5-year surveying expedition along the coastline of South America and then around the world.

Darwin's voyage on the *Beagle* sowed the seeds for his theory of evolution. In addition to his duties as companion to the captain, Darwin served as the expedition's official naturalist, whose task was to observe and collect geological and biological specimens. The *Beagle* sailed to South America and made many stops along its coast. There Darwin observed the plants and animals of the Tropics and was stunned by the greater diversity of species compared with that of Europe.

Although he had boarded the *Beagle* convinced of the permanence of species, Darwin's experiences soon led him to doubt it. He discovered a snake with rudimentary hindlimbs, calling it "the passage by which Nature joins the lizards to the snakes." Another snake he encountered vibrated its tail like a rattlesnake but had no rattles and therefore made no noise. Similarly, Darwin noticed that penguins used their wings to paddle through the water rather than fly through the air. If a creator had individually created each animal in its present form, to suit its present environment, what could be the purpose behind these makeshift arrangements?

Figure E13-1 A painting of Charles Darwin as a young man

Perhaps the most significant stopover of the voyage was the month spent on the Galapagos Islands off the northwestern coast of South America. There, Darwin found huge tortoises. Different islands were home to distinctively different types of tortoises (Fig. E13-2). Darwin also found several types of finches, and, as with the tortoises, different islands had slightly different finches. Could the differences in these organisms have arisen after they became isolated from one another on separate islands? The diversity of tortoises and finches haunted him for years afterward.

In 1836, Darwin returned to England after 5 years on the *Beagle* and became established as one of the foremost naturalists of his time. But the problem of how isolated populations come to differ from each other gnawed constantly at his mind. Part of the solution came to him from an unlikely source: the writings of an English economist and clergyman, Thomas Malthus. In his *Essay on Population,* Malthus wrote, "It may safely be pronounced, therefore, that [human] population, when unchecked, goes on doubling itself every 25 years, or increases in a geometrical ratio."

Darwin realized that a similar principle holds true for plant and animal populations. In fact, most organisms can reproduce much more rapidly than can humans (consider rabbits, dandelions, and houseflies) and consequently could produce overwhelming populations in short order. Nonetheless, the world is not chest-deep in rabbits, dandelions, or flies: Natural populations do not grow "unchecked" but tend to remain approximately constant in size. Clearly, vast numbers of individuals must die in each generation, and most must not reproduce.

From his experience as a naturalist, Darwin realized that members of a species typically differ from one another. Further, which individuals die without reproducing in each generation is not arbitrary, but depends to some extent on the structures and abilities of the organisms. This observation was the source of the theory of evolution by natural selection. As Darwin's colleague Alfred Wallace put it, "Those which, year by year, survived this terrible destruction must be, on the whole, those which have some little superiority enabling them to escape each special form of death to which the great majority succumbed." Here you see the origin of the expression "survival of the fittest." That "little superiority" that confers greater fitness might be better resistance to cold, more efficient digestion, or any of hundreds of other advantages, some very subtle.

Everything now fell into place. Darwin wrote, "It at once struck me that under these circumstances favorable variations would tend to be preserved, and unfavorable ones to be destroyed." If a favorable variation were inheritable, then the entire species would eventually consist of individuals possessing the favorable trait. With the continual appearance of new variations (due, as we now know, to mutations), which in turn are subject to further natural selection, "the result... would be the formation of new species. Here, then, I had at last got a theory by which to work."

When Darwin finally published *On the Origin of Species* in 1859, his evidence had become truly overwhelming. Although its full implications would not be realized for decades, Darwin's theory of evolution by natural selection has become a unifying concept for virtually all of biology.

Figure E13-2 A Galapagos tortoise

Modern Genetics Confirmed Darwin's Assumption of Inheritance

As you know, the principles of genetics had not yet been discovered when Darwin published *On the Origin of Species.* Observation 4, therefore, was an untested assumption for Darwin and Wallace—and thus a weakness in their theory at the time of its publication. Mendel's later work showed that Darwin was correct in assuming that particular traits can be passed to offspring. In addition, we now also know that the variations in natural populations arise purely by chance, as a result of random mutations in DNA. These new variations can be good, bad, or indifferent; there is no way to ensure that favorable variations will arise.

Natural Selection Modifies Populations over Time

How might natural selection among chance variations change the makeup of a species? In *On the Origin of Species,* Darwin proposed the following example. "Let us take the case of a wolf, which preys on various animals, securing [them] by . . . fleetness. . . . The swiftest and slimmest wolves would have the best chance of surviving, and so be preserved or selected. . . . Now if any slight innate change of habit or structure benefited an individual wolf, it would have the best chance of surviving and of leaving offspring. Some of its young would probably inherit the same habits or structure, and by the repetition of this process, a new variety might be formed." The same argument would apply to the wolf's prey, of which the fastest or most alert would be the most likely to avoid predation and would pass on these traits to their offspring.

Notice that natural selection acts on individuals within a population. Over generations, the population changes as the percentage of its individuals inheriting favorable traits increases. An individual cannot evolve, but a population can.

Although it is easiest to understand how natural selection would cause changes *within* a species, under the right circumstances, the same principles might produce entirely *new* species. In Chapter 14 we will discuss the circumstances that give rise to new species.

13.3 How Do We Know That Evolution Has Occurred?

Virtually all biologists consider evolution to be a fact. Why? Because an overwhelming body of evidence permits no other conclusion. The key lines of evidence come from fossils, comparative anatomy (the study of how body structures differ among species), embryology, and biochemistry and genetics.

Fossils Provide Evidence of Evolutionary Change over Time

If it is true that many fossils are the remains of species ancestral to modern species, we might expect to find progressive series of fossils that start with an ancient, primitive organism, progress through several intermediate stages, and culminate in a modern species. Such series have indeed been found. The best

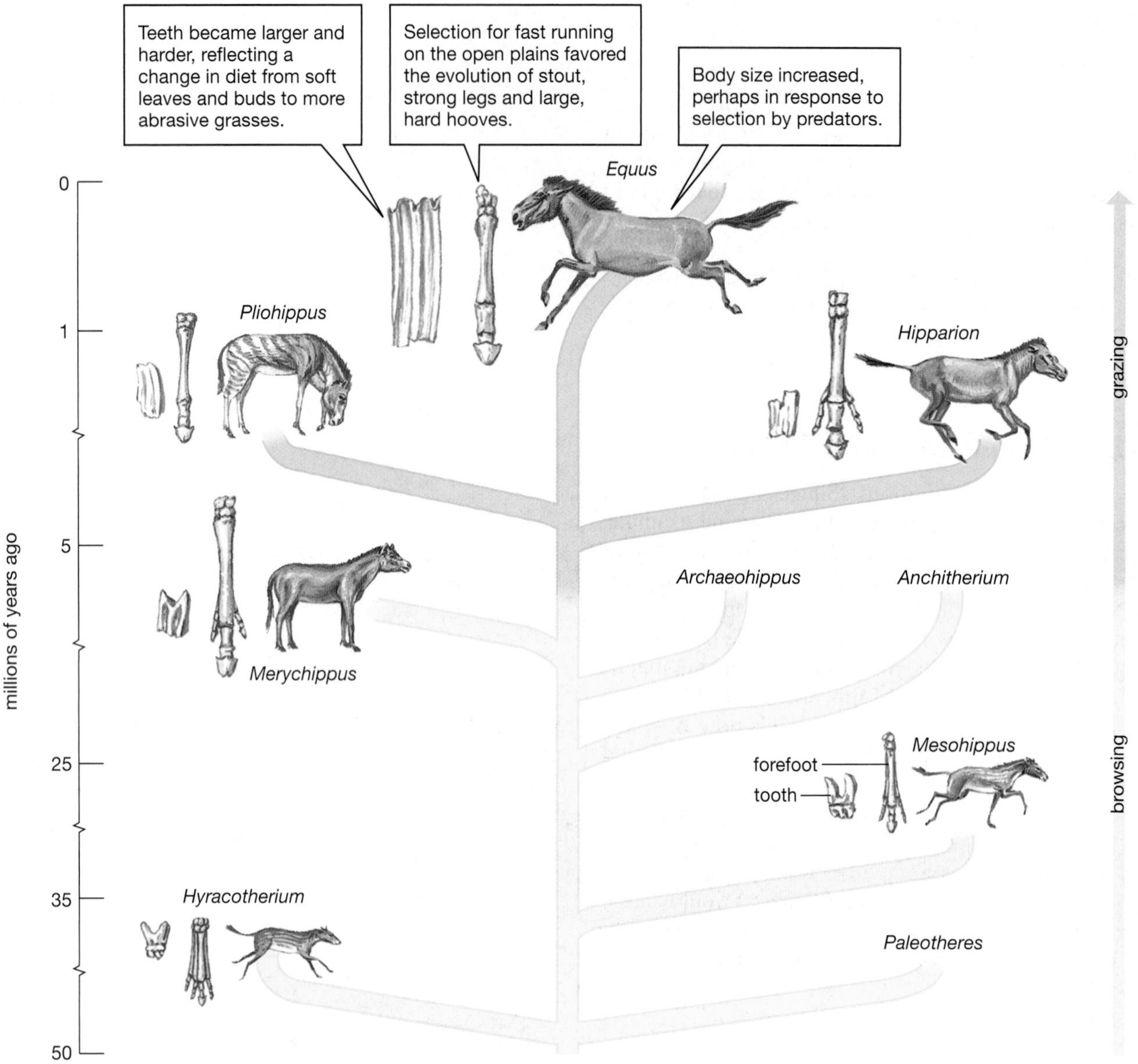

Teeth became larger and harder, reflecting a change in diet from soft leaves and buds to more abrasive grasses.

Selection for fast running on the open plains favored the evolution of stout, strong legs and large, hard hooves.

Body size increased, perhaps in response to selection by predators.

Equus

Pliohippus

Hipparion

Merychippus

Archaeohippus

Anchitherium

Mesohippus

forefoot

tooth

Hyracotherium

Paleotheres

millions of years ago: 0, 1, 5, 25, 35, 50

browsing → grazing

Figure 13-6 The evolution of the horse Over the past 50 million years, horses evolved from small woodland browsers to large plains-dwelling grazers. Three major changes are size, leg anatomy, and tooth anatomy. **QUESTION:** *The fossil history of some kinds of modern organisms, such as sharks and crocodiles, shows that their structure and appearance have changed very little over hundreds of millions of years. Is this lack of change evidence that such organisms have not evolved over that time?*

known one is probably the fossil horse series (**Fig. 13-6**), though series of fossil giraffes, elephants, and mollusks also show the evolution of body structures over time. These fossil series suggest that new species evolved from, and replaced, previous species. Certain sequences of fossil snails have such slight gradations in body structures between successive rock layers that paleontologists cannot easily decide where one species leaves off and the next one begins.

Comparative Anatomy Gives Evidence of Descent with Modification

Fossils provide snapshots of the past that allow biologists to trace evolutionary changes, but careful examination of today's organisms can also uncover evolution's story. Comparing the bodies of organisms of different species can reveal similarities that can be explained only by shared ancestry and differences that could result only from evolutionary change during descent from a common ancestor. In this way, the study of comparative anatomy has supplied strong evidence that different species are linked by a common evolutionary heritage.

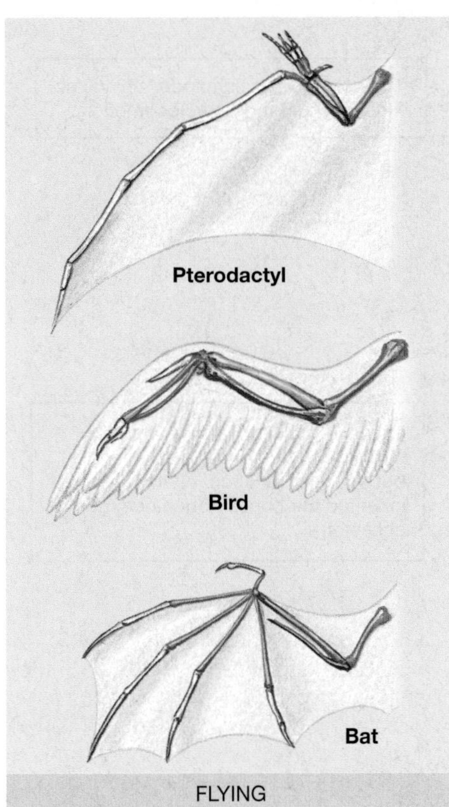

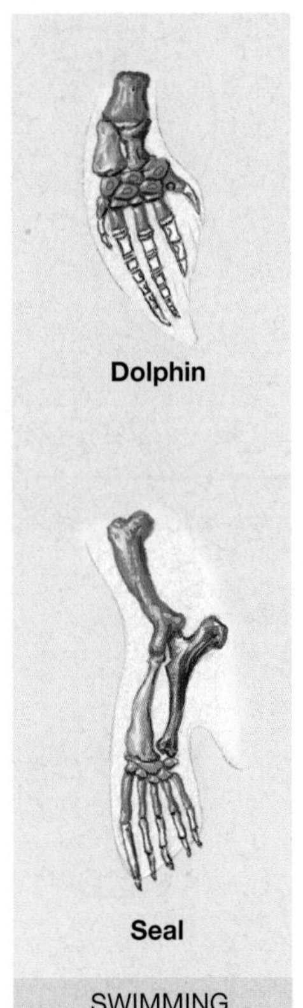

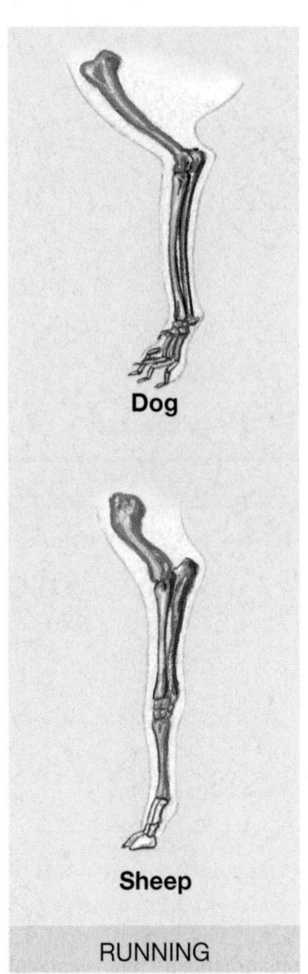

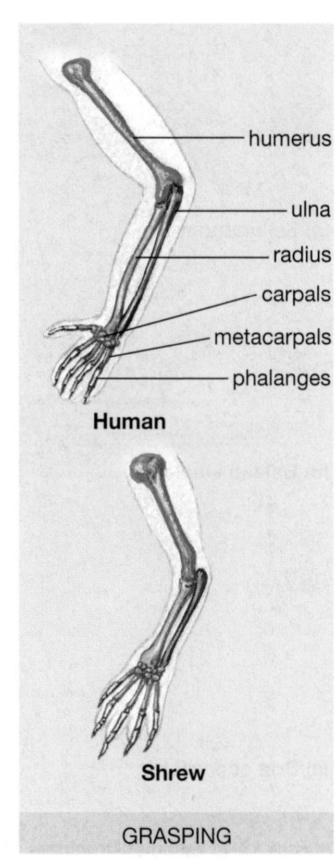

Figure 13-7 Homologous structures Despite wide differences in function, the forelimbs of all of these animals contain the same set of bones, inherited from a common ancestor. The different colors of the bones highlight the correspondences among the various species.

Homologous Structures Provide Evidence of Common Ancestry

The same body structure may be modified by evolution to serve different functions in different species. The forelimbs of birds and mammals, for example, are variously used for flying, swimming, running over several types of terrain, and grasping objects, such as branches and tools. Despite this enormous diversity of function, the internal anatomy of all bird and mammal forelimbs is remarkably similar (Fig. 13-7). It seems inconceivable that the same bone arrangements would be used to serve such diverse functions if each animal had been created separately. Such similarity is exactly what we would expect, however, if bird and mammal forelimbs were derived from a common ancestor. Through natural selection, each has been modified and now performs a particular function. Such internally similar structures are called **homologous structures**, meaning that they have the same evolutionary origin despite any differences in current function or appearance.

Functionless Structures Are Inherited from Ancestors

Evolution by natural selection also helps explain the curious circumstance of **vestigial structures** that serve no apparent purpose. Examples include such things as molar teeth in vampire bats (which live on a diet of blood and therefore don't chew their food) and pelvic bones in whales and certain snakes (Fig. 13-8). Both of these vestigial structures are clearly homologous to structures that are found in—and used by—other vertebrates (animals with a backbone). Their continued existence in animals that have no use for them is best explained as a sort of "evolutionary baggage." For example, the ancestral mammals from which whales evolved had four legs and a well-developed set of pelvic bones. Whales do not have hind legs, yet they have small pelvic and leg bones embedded in their sides. During whale evolution, losing the hind legs provided an advantage, better streamlining the body for movement through water. The result is the modern whale with small, useless, and unused pelvic bones.

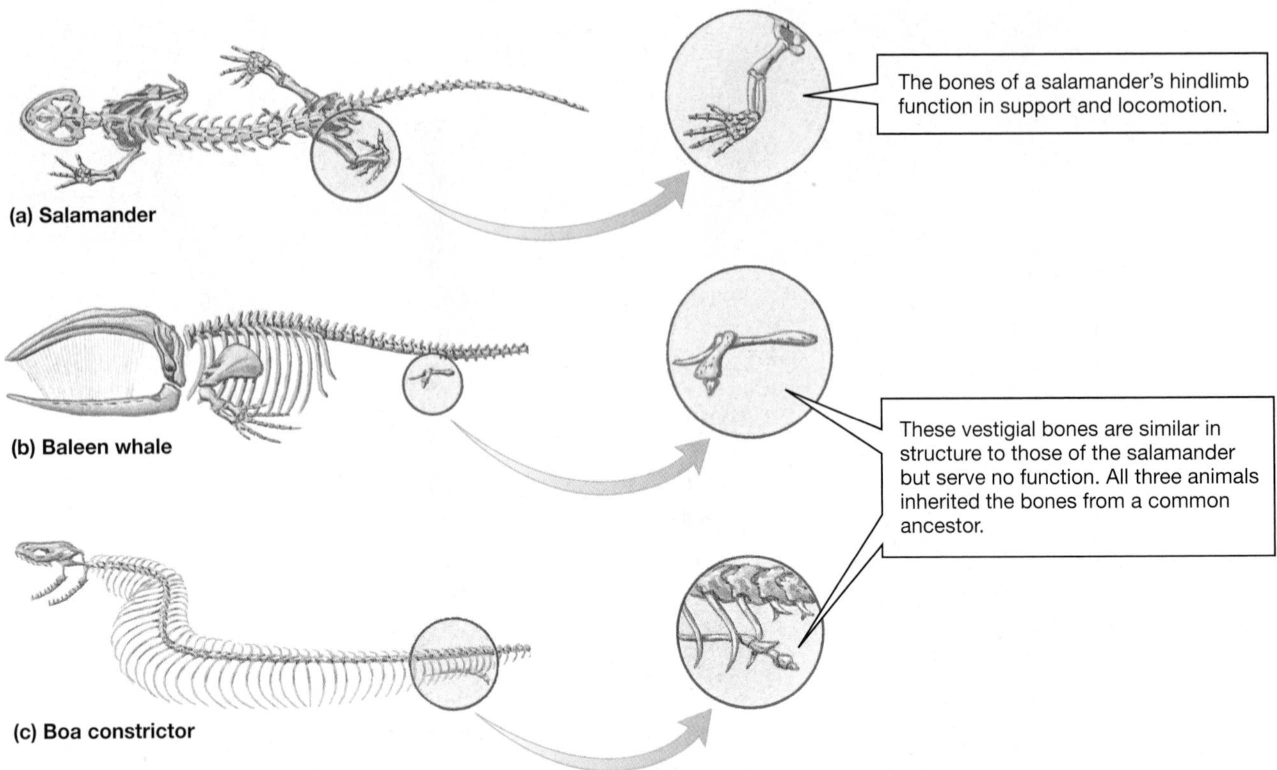

Figure 13-8 Vestigial structures Many organisms have vestigial structures that serve no apparent function. The ***(a)*** salamander, ***(b)*** whale, and ***(c)*** snake all inherited hindlimb bones from a common ancestor. These bones remain functional in the salamander but are vestigial in the whale and snake.

Some Anatomical Similarities Result from Evolution in Similar Environments

The study of comparative anatomy has demonstrated the shared ancestry of life by identifying a host of homologous structures that different species have inherited from common ancestors, but comparative anatomists have also identified many anatomical similarities that do not stem from common ancestry. Instead, these similarities stem from **convergent evolution**, in which natural selection causes nonhomologous structures that serve similar functions to resemble one another. For example, both birds and insects have wings, but this similarity did not arise from evolutionary modification of a structure that both birds and insects inherited from a common ancestor. Instead, the similarity arose from modification of two different, nonhomologous structures that eventually gave rise to superficially similar structures. Because natural selection favored flight in both birds and insects, the two groups evolved superficially similar structures—wings—that are useful for flight. Such outwardly similar but nonhomologous structures are called **analogous structures** (Fig. 13-9). Analogous structures are typically very different in internal anatomy, because the parts are not derived from common ancestral structures.

Embryological Similarity Suggests Common Ancestry

In the early 1800s, German embryologist Karl von Baer noted that all vertebrate embryos (developing organisms in the period from fertilization to birth or hatching) look quite similar to one another early in their development (Fig. 13-10). In their early embryonic stages, fish, turtles, chickens, mice, and humans all develop tails and gill slits. Only fish go on to develop gills, and only fish, turtles, and mice retain substantial tails.

Why do vertebrates that are so different have similar developmental stages? The only plausible explanation is that ancestral vertebrates possessed genes that directed the development of gills and tails. All of their descendants still have those genes. In fish, these genes are active throughout development, resulting in

(a)

(b)

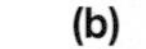

(c)

(d)

Figure 13-9 Analogous structures Convergent evolution can produce outwardly similar structures that differ anatomically. The wings of ***(a)*** insects and ***(b)*** birds and the sleek, streamlined shapes of ***(c)*** seals and ***(d)*** penguins are examples of such analogous structures. **QUESTION:** *Are a peacock's tail and a dog's tail homologous structures or analogous structures?*

adults with fully developed tails and gills. In humans and chickens, these genes are active only during early developmental stages, and the structures are lost or inconspicuous in adults.

Modern Biochemical and Genetic Analyses Reveal Relatedness among Diverse Organisms

Biochemistry and molecular biology provide striking evidence of the evolutionary relatedness of all organisms. At the most fundamental biochemical levels, all living cells are very similar. For example, all cells have DNA as the carrier of

Figure 13-10 Embryological stages reveal evolutionary relationships Early embryonic stages of a ***(a)*** lemur, ***(b)*** pig, and ***(c)*** human, showing strikingly similar anatomical features.

(a)

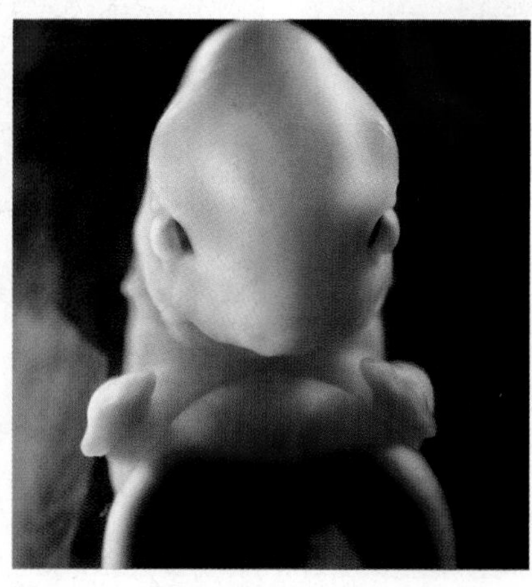
(b)

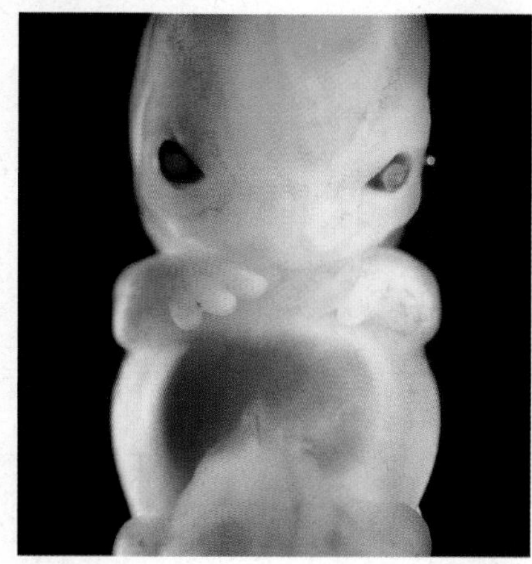
(c)

genetic information. All use RNA, ribosomes, and approximately the same genetic code to translate that genetic information into proteins. All use roughly the same set of 20 amino acids to build proteins, and all use ATP as an intracellular energy carrier. The most plausible explanation for such widespread sharing of such complex and specific biochemical traits is that the traits arose in the common ancestor of all living things, from which all of today's organisms inherited them.

13.4 What Is the Evidence That Populations Evolve by Natural Selection?

We have seen that evidence of evolution comes from several sources. But what is the evidence that evolution occurs by the process of natural selection?

Controlled Breeding Modifies Organisms

One line of evidence supporting evolution by natural selection is **artificial selection**, the breeding of domestic plants and animals to produce specific desirable features. The various breeds of dog provide a striking example of artificial selection (Fig. 13-11). Dogs descended from wolves, and even today, the two will readily cross-breed. With few exceptions, however, modern dogs do not resemble wolves. Some breeds are so different from one another that they would be considered separate species if they were found in the wild. Humans produced these radically different dogs in a few thousand years by doing nothing more than repeatedly selecting individuals with desirable traits for breeding. Therefore, it is quite plausible that natural selection could, by an analogous process acting over hundreds of millions of years, produce the spectrum of living organisms. Darwin was so impressed by the connection between artificial selection and natural selection he devoted a chapter of *Origin of Species* to the topic.

Evolution by Natural Selection Occurs Today

The logic of natural selection gives us no reason to believe that evolutionary change is limited to the past. After all, inherited variation and competition for access to resources are certainly not limited to the past. If Darwin and Wallace were correct that those conditions lead inevitably to evolution by natural selec-

(a)

(b)

Figure 13-11 Dog diversity illustrates artificial selection A comparison of ***(a)*** the ancestral dog (the gray wolf, *Canis lupus*) and ***(b)*** various breeds of dog. Artificial selection by humans has caused great divergence in the size and shape of dogs in only a few thousand years.

tion, then scientific observers and experimenters ought to be able to detect evolutionary change as it occurs. And they have. We consider next some examples that give us a glimpse of natural selection at work.

Brighter Coloration Can Evolve When Fewer Predators Are Present

On the island of Trinidad, guppies live in streams that are also inhabited by several species of larger, predatory fish that frequently dine on guppies. In upstream portions of these streams, however, the water is too shallow for the predators, and guppies are free of danger from predators. When scientists compared male guppies in an upstream area with ones in a downstream area, they found that the upstream guppies were much more brightly colored than the downstream guppies. The scientists knew that the source of the upstream population was guppies that had found their way up into the shallower waters many generations earlier.

The explanation for the difference in coloration between the two populations stems from the sexual preferences of female guppies. The females prefer to mate with the most brightly colored males, so the brightest males have a large advantage when it comes to reproduction. In predator-free areas, male guppies with the bright colors that females prefer have more offspring than duller males. Bright color, however, makes guppies more conspicuous to predators, and therefore more likely to be eaten. Thus, where predators are common, they act as agents of natural selection by eliminating the bright-colored males before they can reproduce. In these areas, the duller males have the advantage and produce more offspring. The color difference between the upstream and downstream guppy populations is a direct result of natural selection.

Natural Selection Can Lead to Pesticide Resistance

Natural selection is also evident in numerous instances of insect pests evolving resistance to the pesticides with which we try to control them. For example, a few decades ago, Florida homeowners were dismayed to realize that roaches were ignoring a formerly effective poison bait called Combat®. Researchers discovered that the bait had acted as an agent of natural selection. Roaches that liked it were consistently killed; those that survived inherited a rare mutation that caused them to dislike glucose, a type of sugar found in the corn syrup used as bait in Combat. By the time researchers identified the problem in the early 1990s, the formerly rare mutation had become common in Florida's urban roach population.

Unfortunately, the evolution of pesticide resistance in insects is a common example of natural selection in action. Such resistance has been documented in more than 500 species of crop-damaging insects, and virtually every pesticide has fostered the evolution of resistance in at least one insect species. We pay a heavy price for this evolutionary phenomenon. The additional pesticides that farmers apply in their attempts to control resistant insects cost almost $2 billion each year in the United States alone and add millions of tons of poisons to Earth's soil and water.

Experiments Can Demonstrate Natural Selection

In addition to observing natural selection in the wild, scientists have also devised numerous experiments that confirm the action of natural selection. For example, one group of evolutionary biologists released small groups of *Anolis sagrei* lizards onto 14 small Bahamian islands that were previously uninhabited by lizards. The original lizards came from a population on Staniel Cay, an island with tall vegetation, including plenty of trees. In contrast, the islands to which the small colonial groups were introduced had few or no trees and were covered mainly with small shrubs and other low-growing plants.

The biologists returned to those islands 14 years after releasing the colonists and found that the original small groups of lizards had given rise to thriving populations of hundreds of individuals. On all 14 of the experimental islands, lizards had legs that were shorter and thinner than lizards from the original source population on Staniel Cay. In just over a decade, it appeared, the lizard populations had changed in response to new environments.

Why had the new lizard populations evolved shorter, thinner legs? Long legs allow greater speed for escaping predators, but shorter legs allow for more agility and maneuverability on narrow surfaces. So, natural selection favors legs that are as long and thick as possible while still allowing sufficient maneuverability. When the lizards were moved from an environment with thick-branched trees to an environment with only thin-branched bushes, the individuals with formerly favorable long legs were at a disadvantage. In the new environment, more agile, shorter-legged individuals were better able to escape predators and survive to produce a greater number of offspring. Thus, members of subsequent generations had shorter legs on average.

Selection Acts on Random Variation to Favor the Phenotypes That Work Best in Particular Environments

Two important points underlie the evolutionary changes just described:

- *The variations on which natural selection works are produced by chance mutations.* The bright coloration in Trinidadian guppies, distaste for glucose in Florida cockroaches, and shorter legs in Bahamian lizards were not *produced* by the female mating preferences, poisoned corn syrup, or thinner branches. The mutations that produced each of these beneficial traits had arisen spontaneously.
- *Natural selection selects for organisms that are best adapted to a particular environment.* Natural selection is not a process for producing ever-greater degrees of perfection. Natural selection does not select for the "best" in any absolute sense, but only for what is best in the context of a particular environment, which varies from place to place and which may change over time. A trait that is advantageous under one set of conditions may become disadvantageous if conditions change. For example, in the presence of poisoned corn syrup, a distaste for glucose yields an advantage to a cockroach, but under natural conditions avoiding glucose would cause the insect to bypass good sources of food.

13.5 A Postscript by Charles Darwin

"It is interesting to contemplate an entangled bank, clothed with many plants of many kinds, with birds singing on the bushes, with various insects flitting about, and with worms crawling through the damp earth, and to reflect that these elaborately constructed forms ... have all been produced by laws acting around us. These laws, taken in the highest sense, being Growth with Reproduction; Inheritance [and] Variability; a Ratio of Increase so high as to lead to a Struggle for Life, and as a consequence to Natural Selection, entailing Divergence of Character and Extinction of less-improved forms. ... There is grandeur in this view of life, with its several powers, having been originally breathed into a few forms or into one; and that, whilst this planet has gone cycling on according to the fixed law of gravity, from so simple a beginning endless forms most beautiful and most wonderful have been, and are being, evolved."

These are the concluding sentences of Darwin's *On the Origin of Species.*

WHAT GOOD ARE WISDOM TEETH? *REVISITED*

Wisdom teeth are but one of dozens of human anatomical structures that appear to serve no function. Darwin himself noted many of these "useless, or nearly useless" traits in the very first chapter of *Origin* and declared them to be prime evidence that humans had evolved from earlier species.

One vestigial structure is the appendix, a narrow tube attached to the large intestine. Although the appendix produces some white blood cells, a person clearly does not need one. Each year, about 300,000 diseased appendixes are surgically removed from Americans, who get along just fine without them. The appendix is probably homologous with the cecum, an extension of the large intestine used for food storage in many plant-eating mammals.

Body hair is another functionless human trait. It seems to be an evolutionary relic of the fur that kept our distant ancestors warm (and that still warms our closest evolutionary relatives, the great apes). Not only do we retain useless body hair, we also still have erector pili, the muscle fibers that allow other mammals to puff up their fur for better insulation. In humans, these vestigial structures just give us goose bumps.

Though humans don't have and don't need a tail, we nonetheless have a tailbone. The tailbone consists of a few tiny vertebrae fused into a small structure at the base of the backbone, where a tail would be if we had one. Some small muscles are attached to the tailbone, but people who are born without one or have theirs surgically removed suffer no ill effects.

CONSIDER THIS: Advocates of creationism argue that if a structure can do anything, it cannot be considered functionless, even if its removal has no effect. Thus, according to this view, wisdom teeth are not evidence of evolution, because they *can* be used to chew if not removed. Do you find this argument persuasive?

LINKS TO LIFE: People Promote High-Speed Evolution

You probably don't think of yourself as a major engine of evolution. Nonetheless, as you go about the routines of your daily life, you are contributing to what is perhaps today's most significant cause of rapid evolutionary change. Human activity has changed Earth's environments tremendously, and when environments change, populations adapt. The biological logic of natural selection, spelled out so clearly by Darwin, tells us that environmental change leads inevitably to evolutionary change. Humans have thus become a major agent of natural selection.

Unfortunately, many of the evolutionary changes we have caused have turned out to be bad news for us. Our liberal use of pesticides has selected for resistant pests that frustrate efforts to protect our food supply. By overmedicating ourselves with antibiotics and other drugs, we have selected for resistant "supergerms" and diseases that are ever more difficult to treat. Heavy fishing in the world's oceans has selected for ever smaller fish (that can slip through nets more easily) and has reduced our ability to extract food from the sea. The rapid evolutionary changes caused by our technology now threaten us. Only strategies based on a sound understanding of the principles of evolution will be capable of solving them.

Chapter Review

Summary of Key Concepts

13.1 How Did Evolutionary Thought Evolve?

Historically, the most common explanation for the origin of species was the divine creation of each species in its present form, and species were believed to remain unchanged after their creation. This view was challenged by evidence from fossils, geology, and biological exploration of the tropics. Since the middle of the nineteenth century, scientists have realized that species originate and evolve by the operation of natural processes that change the genetic makeup of populations.

13.2 How Does Natural Selection Work?

Charles Darwin and Alfred Russel Wallace independently proposed the theory of evolution by natural selection. Their theory can be concisely expressed as three conclusions based on four observations. These are summarized in modern biological terms in Figure 13-5.

13.3 How Do We Know That Evolution Has Occurred?

Many lines of evidence indicate that evolution has occurred, including the following:

- Fossils of ancient species tend to be simpler in form than modern species. Sequences of fossils have been discovered that show a graded series of changes in form. Both of these observations would be expected if modern species evolved from older species.
- Species thought to be related through evolution from a common ancestor show many similar anatomical structures. An example is the forelimbs of amphibians, reptiles, birds, and mammals.

- Stages in early embryological development are quite similar among very different types of vertebrates.
- Similarities in such biochemical traits as the use of DNA as the carrier of genetic information support the notion of descent of related species through evolution from common ancestors.

13.4 What Is the Evidence That Populations Evolve by Natural Selection?

Similarly, many lines of evidence indicate that natural selection is the chief mechanism driving changes in the characteristics of species over time, including the following:

- Inheritable traits have been changed rapidly in populations of domestic animals and plants by selectively breeding organisms with desired features (artificial selection). The immense variations in species produced in a few thousand years of artificial selection by humans makes it seem likely that much larger changes could be wrought by hundreds of millions of years of natural selection.
- Evolution can be observed today. Both natural and human activities drastically change the environment over short periods of time. Characteristics of species have been observed to change significantly in response to such environmental changes.

Key Terms

analogous structure *p. 226*
artificial selection *p. 228*
catastrophism *p. 218*
convergent evolution *p. 226*
evolution *p. 215*
fossil *p. 217*
homologous structure *p. 225*
natural selection *p. 221*
population *p. 221*
uniformitarianism *p. 219*
vestigial structure *p. 225*

Thinking Through the Concepts

Multiple Choice

1. *Your arm is homologous with*
a. a seal flipper
b. an octopus tentacle
c. a bird wing
d. a sea star arm
e. both a and c

2. *All organisms share the same genetic code. This commonality is evidence that*
a. evolution is occurring now
b. convergent evolution has occurred
c. evolution occurs gradually
d. all organisms are descended from a common ancestor
e. life began a long time ago

3. *Which of the following are fossils?*
a. pollen grains buried in the bottom of a peat bog
b. the petrified cast of a clam's burrow
c. the impression a clam shell made in mud, preserved in mudstone
d. an insect leg sealed in plant resin
e. all of the above

4. *In Africa, there is a species of bird called the yellow-throated longclaw. It looks almost exactly like the meadowlark found in North America, but they are not closely related. This is an example of*
a. uniformitarianism
b. artificial selection
c. gradualism
d. vestigial structures
e. convergent evolution

5. *Which of the following are examples of vestigial structures?*
a. your tailbone
b. your nostrils
c. sixth fingers found on some humans
d. your kneecap
e. none of the above

6. *Which of the following would stop evolution by natural selection from occurring?*
a. if humans became extinct because of a disease epidemic
b. if a thermonuclear war killed most living organisms and changed the environment drastically
c. if ozone depletion led to increased ultraviolet radiation, which caused many new mutations
d. if all individuals in a population were genetically identical, and there were no genetic recombination, sexual reproduction, or mutation
e. all of the above

Review Questions

1. Selection acts on individuals, but only populations evolve. Explain why this statement is true.
2. Distinguish between catastrophism and uniformitarianism. How did these hypotheses contribute to the development of evolutionary theory?
3. Describe Lamarck's theory of inheritance of acquired characteristics. Why is it invalid?
4. What is natural selection? Describe how natural selection might have caused differential reproduction among the ancestors of a fast-swimming predatory fish, such as the barracuda.
5. Describe how evolution occurs through the interactions among the reproductive potential of a species, the normally constant size of natural populations, variation among individuals of a species, natural selection, and inheritance.
6. What is convergent evolution? Give an example.
7. How do biochemistry and molecular genetics contribute to the evidence that evolution occurred?

Applying the Concepts

1. **IS THIS SCIENCE?** In discussions of untapped human potential, it is commonly said that the average person uses only 10% of his or her brain. Is this conclusion likely to be correct? Explain your answer in terms of natural selection.
2. **IS THIS SCIENCE?** Both the theory of evolution by natural selection and the theory of special creation (which states that all species were simultaneously created by God) have had an impact on evolutionary thought. Discuss why one is considered to be a scientific theory and the other is not.
3. Does evolution through natural selection produce "better" organisms in an absolute sense? Are we climbing the "ladder of Nature"? Defend your answer.

For More Information

Darwin, C. *On the Origin of Species by Means of Natural Selection.* Garden City, NY: Doubleday, 1960 (originally published in 1859). An impressive array of evidence amassed to convince a skeptical world.

Dennet, D. *Darwin's Dangerous Idea.* New York: Simon & Schuster, 1995. A philosopher's view of Darwinian ideas and their application to the world outside biology. A thought-provoking book that seems to have inspired admiration and condemnation in roughly equal proportions.

Eiseley, L. C. "Charles Darwin." *Scientific American,* February 1956. An essay on the life of Darwin by one of his foremost American biographers. Even if you need no introduction to Darwin, read this as an introduction to Eiseley, author of many marvelous essays.

Gould, S. J. *Ever Since Darwin,* 1977; *The Panda's Thumb,* 1980; and *The Flamingo's Smile,* 1985. New York: W. W. Norton. A series of witty, imaginative, and informative essays, mostly from *Natural History* magazine. Many deal with various aspects of evolution.

Selim, J. "Useless Body Parts." *Discover,* June 2004. A descriptive listing of vestigial human anatomy.

Weiner, J. "Evolution Made Visible." *Science,* January 6, 1995. A clear summary of modern evidence for evolution in action.

Wong, K. "The Mammals That Conquered the Seas." *Scientific American,* May 2002. The story of whale evolutionary history.

WEB TUTORIAL

To access a Web Tutorial visit http://www.prenhall.com/audesirk4. *Log in to the Web site selected by your instructor, navigate to this chapter, and select the appropriate Web Activity number.*

13.1 Analogous and Homologous Structures

Estimated time: 5 minutes

This exercise will help you to learn the difference between structures that are similar because of common origin or because of similar evolutionary pressures.

13.2 Natural Selection for Antibiotic Resistance

Estimated time: 5 minutes

This activity demonstrates the mechanism of natural selection using the example of evolution of an antibiotic resistance trait in a population of the bacteria that cause the human disease tuberculosis.

13.3 Web Investigation: What Good Are Wisdom Teeth?

Estimated time: 10 minutes

Human wisdom teeth, or third molars, are considered to be vestigial structures; they served human ancestors, but now often never develop or erupt. Explore the origins and problems of these troublesome structures.

Chapter 14

How Populations Evolve

We think of hospitals as places in which to seek protection from disease, but they also foster the evolution of drug-resistant supergerms.

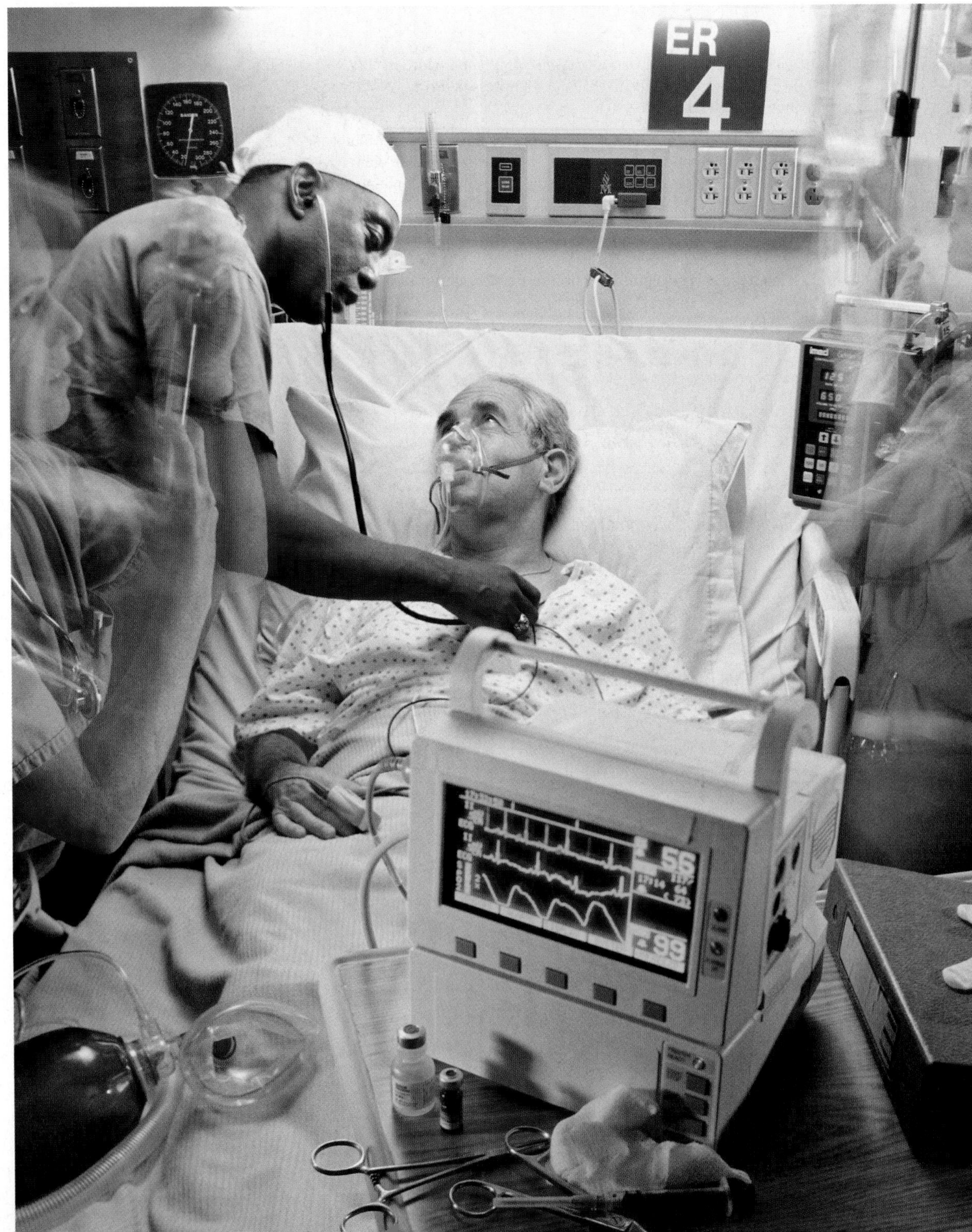

AT A GLANCE

Evolution of a Menace

When you come down with a bad cold, do you drag yourself down to the health center to ask for some antibiotics? If you do, you have probably noticed that physicians are increasingly unlikely to grant a request for antibiotics, especially for illnesses, such as colds and flu, that are not readily cured by antibiotics. This increased caution stems from fears that overuse of antibiotics has contributed to the rise of drug-resistant bacteria.

Increasingly, the bacteria that cause disease are becoming less susceptible to antibiotic drugs. Drug resistance is now common in the bacteria that cause tuberculosis, food poisoning, blood poisoning, dysentery, pneumonia, gonorrhea, meningitis, urinary tract infections, and other ailments. We are experiencing an onslaught of resistant "supergerms."

Why do doctors think they can combat resistant diseases by reducing antibiotic use? Scientists have hypothesized that the rise of antibiotic resistance is a consequence of evolutionary change in populations of bacteria and that the agent of change is natural selection applied by antibiotic drugs. What kind of evidence would help test this hypothesis? One approach is to examine bacterial evolution in a single human body. For example, one group of researchers took a sample of tuberculosis-causing bacteria from a patient who had recently been diagnosed with the disease. Tests showed that the bacteria were susceptible to a number of antibiotics, including rifampin. The patient was treated with rifampin, and his condition improved, but he ultimately suffered a relapse and died. Just after his death, the researchers again sampled bacteria from his lungs.

Although the bacteria from the deceased patient were still sensitive to many antibiotics, they were highly resistant to rifampin. Further analysis of the bacteria showed that they were descendants of the bacteria originally present and thus did not represent a new bacterial population that had invaded the patient during his illness. Does this result demonstrate that bacteria evolve antibiotic resistance through natural selection by antibiotics?

***Staphylococcus aureus*, a common source of human infections, is among the many bacterial species that have evolved resistance to antibiotics.**

14.1 How Are Populations, Genes, and Evolution Related?

The changes that we see in an individual organism as it grows and develops are not evolutionary changes. Instead, evolutionary changes occur from generation to generation, causing descendants to be different from their ancestors. Furthermore, we can't detect evolutionary change across generations by looking at a single set of parents and offspring. For example, if you observed that a 6-foot-tall man had an adult son who stood 5 feet tall, could you conclude that humans were evolving to become shorter? Obviously not. Rather, if you wanted to learn about evolutionary change in human height, you would begin by measuring many humans of many generations to see if the average height is changing over time. Clearly, evolution is a property not of individuals but of populations (a **population** is a group that includes all the members of a species living in a given area).

The recognition that evolution is a population-level phenomenon was one of Darwin's key insights. But populations are composed of individuals, and the actions and fates of individuals determine which characteristics will be passed to descendant populations. In this fashion, inheritance provides the link between the lives of individual organisms and the evolution of populations. We will therefore begin our discussion of the processes of evolution by reviewing some principles of genetics as they apply to individuals. We will then extend those principles to the genetics of populations.

Genes and the Environment Interact to Determine Traits

Each cell of every organism contains genetic information encoded in the DNA of its chromosomes. Recall that a *gene* is a segment of DNA located at a particular place on a chromosome. The sequence of nucleotides in a gene encodes the sequence of amino acids in a protein, usually an enzyme that catalyzes a particular reaction in the cell. At a given gene's location, different members of a species may have slightly different nucleotide sequences, called *alleles*. Different alleles generate different forms of the same enzyme. In this way, various alleles of the gene that influences eye color in humans, for example, help produce eyes that are brown, or blue, or green, and so on.

In any population of organisms, there are usually two or more alleles of each gene. An individual of a diploid species whose alleles of a particular gene are both the same is *homozygous* for that gene, and an individual with different alleles for that gene is *heterozygous*. The specific alleles borne on an organism's chromosomes (its *genotype*) interact with the environment to influence the development of its physical and behavioral traits (its *phenotype*).

Let's illustrate these principles with an example familiar to you from Unit Two. A pea flower is colored purple because a chemical reaction in its petals converts a colorless molecule to a purple pigment. When we say that a pea plant has the allele for purple flowers, we mean that a particular stretch of DNA on one of its chromosomes contains a sequence of nucleotides that codes for the enzyme that catalyzes this reaction. A pea with the allele for white flowers has a different sequence of nucleotides at the corresponding chromosomal position. The enzyme for which that different sequence codes cannot produce purple pigment. If a pea is homozygous for the white allele, its flowers produce no pigment and thus are white.

The Gene Pool Is the Sum of the Genes in a Population

We can often deepen our understanding of a subject by looking at it from more than one perspective. In studying evolution, looking at the process from the point of view of a gene has proven to be an enormously effective tool. In particular, evolutionary biologists have made excellent use of the tools of a branch of genetics called population genetics, which deals with the frequency, distribution, and inheritance of alleles in populations. To take advantage of this powerful aid to understanding evolution, you will need to learn a few of the basic concepts of population genetics.

Population genetics defines the **gene pool** as the sum of all the genes in a population. In other words, the gene pool consists of all the alleles of all the genes in all the individuals of a population. Each particular gene can also be considered to have its own gene pool, which consists of all the alleles of that specific gene in a population. If we added up all the copies of each allele of that gene in all the individuals in a population, we could determine the relative proportion of each allele, a number called the **allele frequency**. For example, a population of 100 pea plants would contain 200 alleles of the gene that controls flower color (because pea plants are diploid, and each plant thus has two copies of each gene). If 50 of those 200 alleles were of the type that codes for white flowers, then we would say that the frequency of that allele in the population is 0.25 (or 25%), because $50/200 = 0.25$.

Evolution Is the Change of Allele Frequencies within a Population

A casual observer might define evolution on the basis of changes in the outward appearance or behaviors of the members of a population. A population geneticist, however, looks at a population and sees a gene pool that just happens to be divided into the packages that we call individual organisms. So any outward changes that we observe in the individuals that make up the population can also be viewed as the visible expression of underlying changes to the gene pool. A population geneticist therefore defines evolution as the changes in allele frequencies that occur in a gene pool over time. Evolution is a change in the genetic makeup of populations over generations.

The Equilibrium Population Is a Hypothetical Population in Which Evolution Does Not Occur

It is easier to understand what causes populations to evolve if the characteristics of a population that would *not* evolve are considered first. In 1908, English mathematician Godfrey H. Hardy and German physician Wilhelm Weinberg independently developed a simple mathematical model now known as the **Hardy-Weinberg principle**. This model showed that, under certain conditions, allele frequencies and genotype frequencies in a population will remain constant no matter how many generations pass. In other words, this population will not evolve. Population geneticists use the term **equilibrium population** for this idealized, nonevolving population in which allele frequencies do not change, as long as the following conditions are met:

- There must be no mutation.
- There must be no **gene flow** between populations. That is, there must be no movement of alleles into or out of the population (as would be caused, for example, by the movement of organisms into or out of the population).
- The population must be very large.
- All mating must be random, with no tendency for certain genotypes to mate with specific other genotypes.
- There must be no natural selection. That is, all genotypes must reproduce with equal success.

Under these conditions, allele frequencies within a population will remain the same indefinitely. If one or more of these conditions is violated, then allele frequencies may change: The population will evolve.

As you might expect, few if any natural populations are truly in equilibrium. What, then, is the importance of the Hardy-Weinberg principle? The Hardy-Weinberg conditions are useful starting points for studying the mechanisms of evolution. In the following sections, we will examine some of the conditions, show that natural populations often fail to meet them, and illustrate the consequences of such failures. In this way, we can better understand both the inevitability of evolution and the processes that drive evolutionary change.

14.2 What Causes Evolution?

Population genetics theory predicts that the Hardy-Weinberg equilibrium can be disturbed by deviations from any of its five conditions. Most evolutionary biologists agree, however, that the most important sources of evolutionary change are mutation, small population size, and natural selection.

Mutations Are the Original Source of Genetic Variability

A population remains in genetic equilibrium only if there are no **mutations** (changes in DNA sequence). Mutations, however, are inevitable. Although cells have efficient methods for protecting the integrity of their genes, some changes in nucleotide sequence slip past the checking and repair systems. An unrepaired mutation in a cell that produces gametes may be passed to an offspring and enter the gene pool of a population.

Mutations Are Rare But Important

How significant is mutation in changing the gene pool of a population? Mutations are rare. For example, only one out of every 100,000 to 1,000,000 human gametes carries a mutation of a given gene. Therefore, mutation by itself is not a major cause of evolution. However, mutations are the source of new alleles, new variations on which other evolutionary processes can work. As such, they are the foundation of evolutionary change. Without mutations there would be no evolution.

Mutations Are Not Goal Directed

A mutation does not arise as a result of, or in anticipation of, environmental necessities. A mutation simply happens and may in turn produce a change in a structure or function of an organism. Whether that change is helpful or harmful or neutral, now or in the future, depends on environmental conditions over which the organism has little or no control (Fig. 14-1). The mutation provides a potential for evolutionary change. Other processes, especially natural selection, may act to spread the mutation through the population or to eliminate it from the population.

Allele Frequencies May Drift in Small Populations

To remain in equilibrium, a population must be so large that chance events have no impact on its overall genetic makeup. Disaster may befall even the fittest organism. Maple seeds that fall into a pond never sprout, and the deer and elk that were killed by the eruption of Mount St. Helens left no descendants. If a population is sufficiently large, chance events are unlikely to alter its genetic composition, because random removal from the gene pool of a few individuals' alleles won't have a big impact on allele frequencies in the population as a whole. In a small population, however, certain alleles may be carried by only a few organisms. Chance events could eliminate some or all such alleles from the population, altering its genetic makeup. This process by which chance events change allele frequencies in a small population is called **genetic drift**.

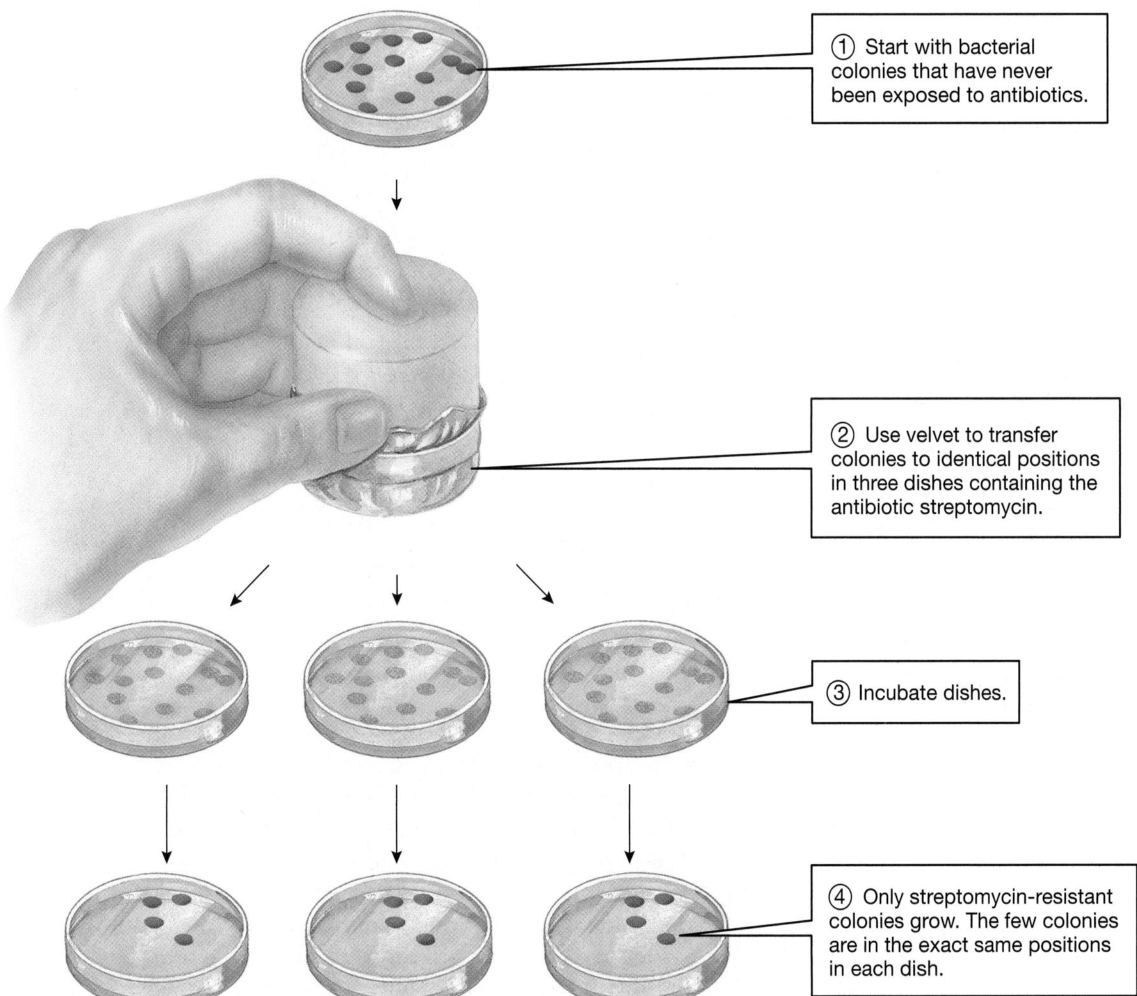

Figure 14-1 Mutations occur spontaneously This experiment demonstrates that mutations occur spontaneously and not in response to environmental pressures. When bacterial colonies that have never been exposed to antibiotics are exposed to the antibiotic streptomycin, only a few colonies grow. The observation that these surviving colonies grow in the exact same positions in all dishes shows that the mutations for resistance to streptomycin were present in the original dish before exposure to the environmental pressure, streptomycin. **QUESTION:** *If it were true that mutations do occur in response to the presence of an antibiotic, how would the result of this experiment have differed from the actual result?*

Population Size Matters

To see how genetic drift works, imagine two populations of amoebas in which each amoeba is either red or blue, and color is controlled by two alleles (*A* and *a*) of a gene. Half of the amoebas in each of our two populations are red and half are blue. One population, however, has only four individuals in it, whereas the other has 10,000.

Now let's picture reproduction in our imaginary populations. Let's select at random half of the individuals in each population and allow them to reproduce by binary fission. To do so, each reproducing amoeba splits in half to yield two amoebas, each of which is the same color as the parent. In the large population, 5000 amoebas reproduce, yielding a new generation of 10,000. What are the chances that all 10,000 members of the new generation will be red? Just about nil. In fact, it would be extremely unlikely for even 3000 amoebas to be red or for 7000 to be red. The most likely outcome is that about half will be red and half blue, just as in the original population. In this large population, then, we would not expect a major change in allele frequencies from generation to generation.

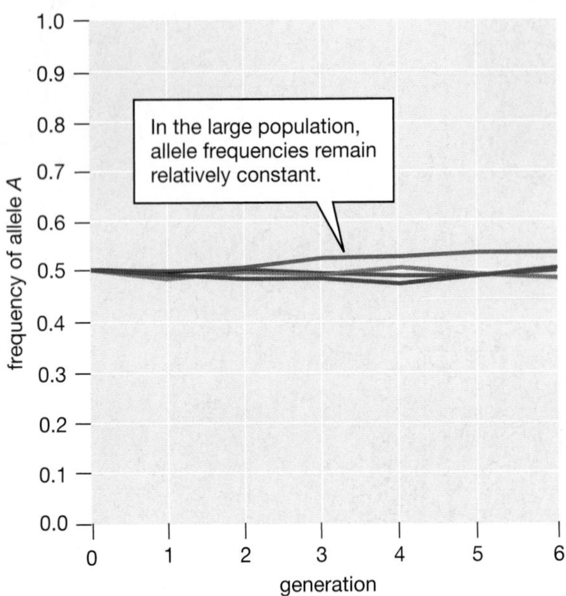

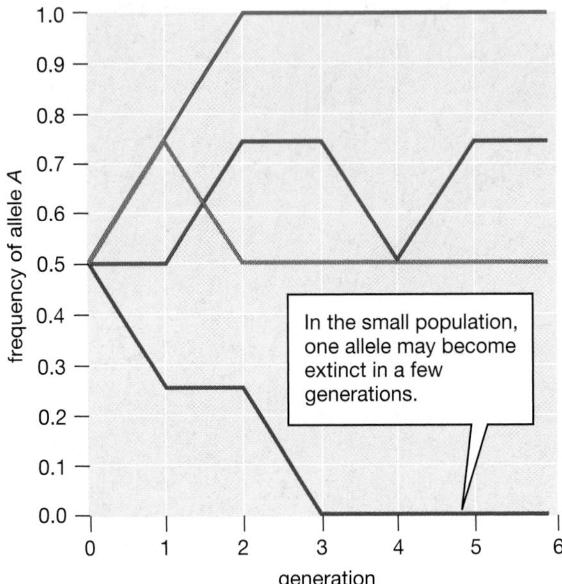

Figure 14-2 The effect of population size on genetic drift Each colored line represents one computer simulation of the change over time in the frequency of allele *A* in a ***(a)*** large and a ***(b)*** small population. In each populations half the alleles were *A* (0.5), and randomly chosen individuals reproduced. **EXERCISE:** *Sketch a graph that shows the result you would predict if the simulation were run four times with a population size of 20.*

One way to test this prediction is to write a computer program that simulates how the allele frequencies of the alleles could change over generations. Figure 14-2a shows the results from four runs of such a simulation. Notice that the frequency of allele *A*, encoding red color, remains close to 0.5, consistent with the expectation that half of the amoebas would be red.

In the small population, the situation is different. Only two amoebas reproduce, and there is a 25% chance that both reproducers will be red. (This outcome is as likely as flipping two coins and having both come up heads.) If only red amoebas reproduce, then the next generation will consist entirely of red amoebas—a relatively likely outcome. It is thus possible, within a single generation, for the allele for blue color to disappear from the population.

Figure 14-2b shows the fate of allele *A* in four runs of a simulation of our small population. In one of the four runs (red line), allele *A* reaches a frequency of 1.0 (100%) in the second generation, meaning that all the amoebas in the third and following generations are red. In another run, the frequency of *A* drifts to 0.0 in the fourth generation (blue line), and the population subsequently is all blue. Thus, one of the two amoeba phenotypes disappeared in half of the simulations.

A Population Bottleneck Is an Example of Genetic Drift

Two causes of genetic drift, called the population bottleneck and the founder effect, further illustrate the effect that small population size may have on the allele frequencies of a species. In a **population bottleneck**, a population is drastically reduced—as a result of a natural catastrophe or overhunting, for example. Then, only a few individuals are available to contribute genes to the next generation. As our amoeba example showed, population bottlenecks can both change allele frequencies and reduce genetic variability (Fig. 14-3a). Even if the population later increases, the genetic effects of the bottleneck may remain for hundreds or thousands of generations (Fig. 14-3b,c).

A special case of a population bottleneck is the **founder effect**, which occurs when isolated colonies are founded by a small number of organisms. A flock of birds, for instance, that becomes lost during migration or is blown off course by a storm may settle on an isolated island. The small founder group may, by chance, have allele frequencies that are very different from the frequencies of the parent population. If they do, the gene pool of the future population in the new location will be quite unlike that of the larger population from which it sprang.

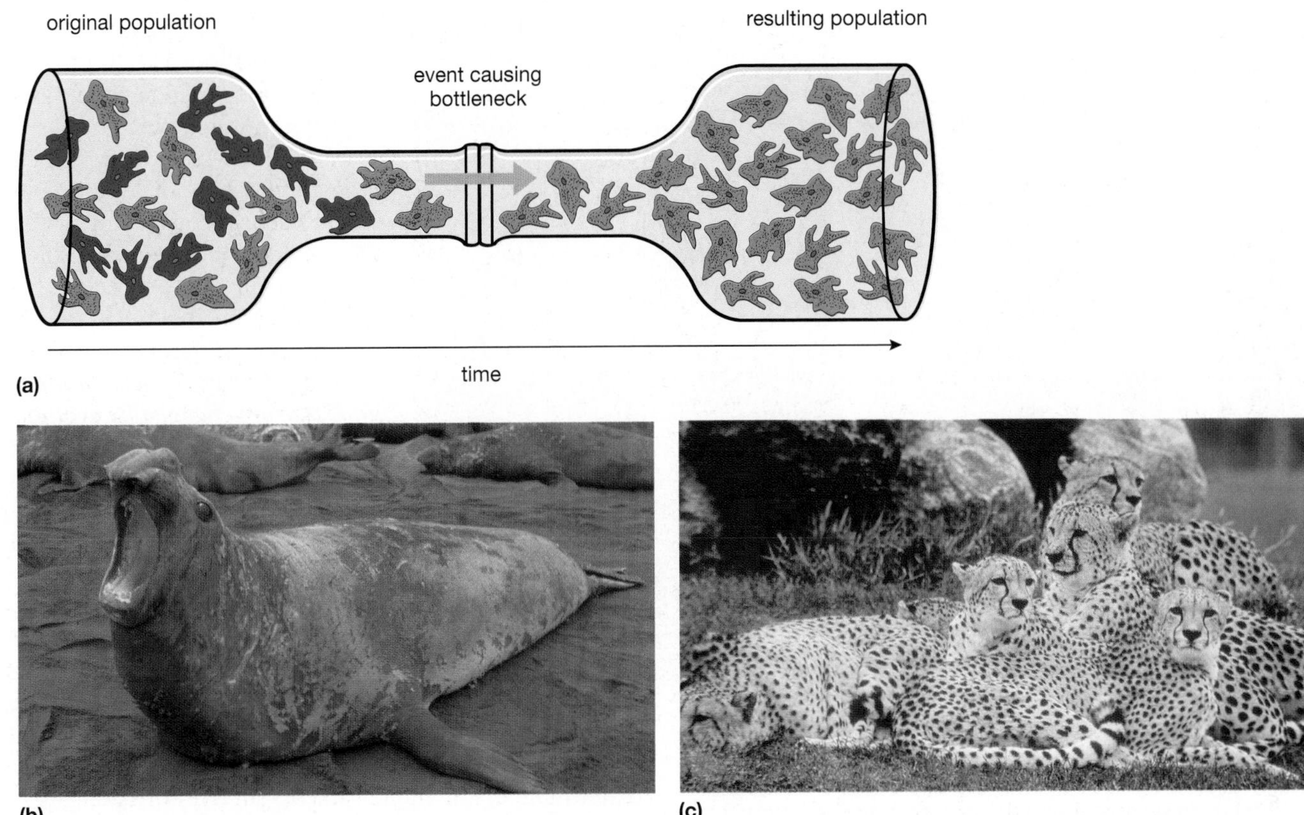

Figure 14-3 **Population bottlenecks reduce variation** ***(a)*** A population bottleneck may drastically reduce genotypic and phenotypic variation because the few organisms that survive may all carry similar sets of alleles. Both ***(b)*** the northern elephant seal, which was hunted almost to extinction in the 1800s, and ***(c)*** the cheetah passed through a population bottleneck in the recent past, resulting in an almost total loss of genetic diversity. **QUESTION:** *If a population grows large again after a bottleneck, genetic diversity will ultimately increase. Why?*

Genotypes Are Not All Equally Beneficial

In a hypothetical equilibrium population, individuals of all genotypes survive and reproduce equally well; that is, no genotype has any advantage over the others. This condition, however, is probably met only rarely, if ever, in real populations. Even though some alleles are neutral, in the sense that organisms possessing any of several alleles are equally likely to survive and reproduce, clearly not all alleles are neutral in all environments. Any time an allele provides, in Alfred Russel Wallace's words, "some little superiority," natural selection favors the individuals who possess it. That is, those individuals have higher reproductive success. This phenomenon is well illustrated by an example concerning an antibiotic drug.

Antibiotic Resistance Evolves by Natural Selection

The antibiotic penicillin first came into widespread use during World War II, when it was used to combat infections in wounded soldiers. Suppose that an infantryman, brought to a field hospital after suffering a gunshot wound in his arm, develops a bacterial infection in that arm. A medic sizes up the situation and resolves to treat the wounded soldier with an intravenous drip of penicillin. As the antibiotic courses through the soldier's blood vessels, millions of bacteria die before they can reproduce. A few bacteria, however, carry a rare allele that codes for an enzyme that destroys any penicillin that comes into contact with the bacterial cell. (This allele is a variant of a gene that normally codes for an enzyme that breaks down the bacterium's waste products.) The bacteria carrying this rare allele are able to survive and reproduce, and their offspring inherit the penicillin-destroying allele. After a few generations, the frequency of the penicillin-destroying allele has soared to nearly 100%, and the frequency of the normal, waste-processing allele has declined to near zero. As a result of natural selection imposed by the antibiotic's killing power, the population of bacteria within the soldier's body has evolved. The gene pool of the population has changed, and natural selection, in the form of bacterial destruction by penicillin, has caused the change.

(a)

(b)

Figure 14-4 A compromise between opposing environmental pressures *(a)* A male giraffe with a long neck is at a definite advantage in combat to establish dominance. *(b)* But a giraffe's long neck forces it to assume an extremely awkward and vulnerable position when drinking. Thus, drinking and male-male contests place opposing evolutionary pressures on neck length.

Penicillin Resistance Illustrates Key Points about Evolution

The example of penicillin resistance highlights some important features of natural selection and evolution.

Natural selection does not cause genetic changes in individuals. The allele for penicillin resistance arose spontaneously, long before penicillin was dripped into the soldier's vein. Penicillin did not cause resistance to appear; its presence merely favored the survival of bacteria with penicillin-destroying alleles over that of bacteria with waste-processing alleles.

Natural selection acts on individuals, but it is populations that are changed by evolution. The agent of natural selection, penicillin, acted on individual bacteria. As a result, some individuals reproduced and some did not. However, it was the population as a whole that evolved as its allele frequencies changed.

Evolution is change in allele frequencies of a population, owing to unequal success at reproduction among organisms bearing different alleles. In evolutionary terminology, the **fitness** of an organism is measured by its reproductive success. In our example, the penicillin-resistant bacteria had greater fitness than the normal bacteria did, because the resistant bacteria produced greater numbers of viable (able to survive) offspring.

Evolutionary changes are not "good" or "progressive" in any absolute sense. The resistant bacteria were favored only because of the presence of penicillin in the soldier's body. At a later time, when the environment of the soldier's body no longer contained penicillin, the resistant bacteria may have been at a disadvantage relative to other bacteria that could process waste more effectively. Similarly, the long necks of male giraffes are helpful when the animals battle to establish dominance, but are a hindrance to drinking (Fig. 14-4). The length of male giraffe necks represents an evolutionary compromise between the advantage of being able to win contests with other males and the disadvantage of vulnerability while drinking water. (The necks of female giraffes are long—though not as long as male necks—because successful males pass the alleles for a long neck to daughters as well as to sons.)

14.3 How Does Natural Selection Work?

Natural selection is not the *only* evolutionary force. As we have seen, mutation provides variability in heritable traits, and the chance effects of genetic drift may change allele frequencies. Further, evolutionary biologists are now beginning to appreciate the power of random catastrophes in shaping the history of life on Earth; massively destructive events may exterminate thriving and failing species alike. Nevertheless, it is natural selection that shapes the evolution of populations as they adapt to their changing environment. For this reason, we will examine natural selection in more detail.

Natural Selection Stems from Unequal Reproduction

To most people, the words **natural selection** are synonymous with the phrase "survival of the fittest." Natural selection evokes images of wolves chasing caribou, of lions snarling angrily in competition over a zebra carcass. Natural selection, however, is not about survival alone. It is also about reproduction. It is certainly true that if an organism is to reproduce, it must survive long enough to do so. In some cases, it is also true that a longer-lived organism has more chances to reproduce. But no organism lives forever, and the only way that its genes can continue into the future is through successful reproduction. When an organism dies without reproducing, its genes die with it. An organism that reproduces lives on, in a sense, through the genes that it has passed to its offspring. Therefore, although evolutionary biologists often discuss survival, partly because survival is usually easier to observe than reproduction, the main issue of natural selection is *differences in reproduction*: Individuals bearing certain alleles leave more offspring (who inherit those alleles) than do other individuals with different alleles.

Natural Selection Acts on Phenotypes

Although we have defined evolution as changes in the genetic composition of a population, it is important to recognize that natural selection cannot act directly on the genotypes of individual organisms. Rather, natural selection acts on phenotypes, the structures and behaviors displayed by the members of a population. This selection of phenotypes, however, inevitably affects the genotypes present in a population, because phenotypes and genotypes are closely tied. For example, we know that a pea plant's height is strongly influenced by the plant's alleles of certain genes. If a population of pea plants were to encounter environmental conditions that favored taller plants, then taller plants would leave more offspring. These offspring would carry the alleles that contributed to their parents' height. Thus, if natural selection favors a particular phenotype, it will necessarily also favor the underlying genotype.

Some Phenotypes Reproduce More Successfully Than Others

As we have seen, natural selection simply means that some phenotypes reproduce more successfully than others do. This simple process is such a powerful agent of change because only the "best" phenotypes pass traits to subsequent generations. But what makes a phenotype the best? Successful phenotypes are those that have the best adaptations to their particular environment. **Adaptations** are characteristics that help an individual survive and reproduce.

An Environment Has Nonliving and Living Components

Individual organisms must cope with an environment that includes not only physical factors but also the other organisms with which the individual interacts. The nonliving (abiotic) component of the environment includes such factors as climate, availability of water, and minerals in the soil. The abiotic environment establishes the "bottom line" requirements that an organism must meet to survive and reproduce. However, many of the adaptations that we see in modern organisms have arisen because of interactions with other organisms—the living (biotic) component of the environment. As Darwin wrote, "The structure of every organic being is related . . . to that of all other organic beings, with which it comes into competition for food or residence, or from which it has to escape, or on which it preys." A simple example illustrates this concept.

A buffalo grass plant sprouts in a small patch of soil in the eastern Wyoming plains. Its roots must be able to take up enough water and minerals for growth and reproduction, and to that extent it must be adapted to its abiotic environment. But even in the dry prairies of Wyoming, this requirement is relatively trivial, provided that the plant is alone and protected in its square yard of soil. In reality, however, many other plants—other buffalo grass plants as well as other grasses, sagebrush bushes, and annual wildflowers—also sprout in that same patch of soil. If our buffalo grass is to survive, it must compete with the other plants for resources. Its long, deep roots and efficient methods of mineral uptake have evolved not so much because the plains are dry as because the buffalo grass must share the dry prairies with other plants. Further, buffalo grass must also coexist with animals that wish to eat it, such as the cattle that graze the prairie (and the bison that grazed it in the past). As a result, buffalo grass is extremely tough. Silica compounds reinforce its leaves, an adaptation that discourages grazing. Over time, tougher, hard-to-eat plants survived better and reproduced more than did less-tough plants—another adaptation to the biotic environment.

Competition Acts As an Agent of Selection

As the buffalo grass example shows, one of the major agents of natural selection in the biotic environment is **competition** with other organisms for scarce resources. Competition for resources is most intense among members of the same species. As Darwin wrote in *On the Origin of Species*, "The struggle almost invariably will be most severe between the individuals of the same species, for they frequent the same districts, require the same food, and are exposed to the same dangers." In other words, no competing organism has such similar requirements for survival as does another member of the same species. Different species may also compete for the same resources, although generally to a lesser extent than do individuals within a species.

Both Predator and Prey Act As Agents of Selection

When two species interact extensively, each exerts strong selection on the other. When one evolves a new feature or modifies an old one, the other typically evolves new adaptations in response. This constant, mutual feedback between two species is called **coevolution**. Perhaps the most familiar form of coevolution is found in predator–prey relationships.

Predation includes any situation in which one organism eats another. In some instances, coevolution between predators (those who do the eating) and prey (those who are eaten) is a sort of "biological arms race," with each side evolving new adaptations in response to "escalations" by the other. Darwin used the example of wolves and deer: Wolf predation selects against slow or careless deer, thus leaving faster, more-alert deer to reproduce and continue the species. In their turn, alert, swift deer select against slow, clumsy wolves, because such predators cannot acquire enough food.

Sexual Selection Favors Traits That Help an Organism Mate

In many animal species, males have conspicuous features such as bright colors, long feathers or fins, or elaborate antlers. Males may also exhibit bizarre courtship behaviors or sing loud, complex songs. Although these extravagant features typically play a role in mating, they also seem to be at odds with efficient survival and reproduction. Exaggerated ornaments and displays may help males gain access to females, but they also make the males more vulnerable to predators. Darwin was intrigued by this apparent contradiction. He coined the term **sexual selection** to describe the special kind of selection that acts on traits that help an animal acquire a mate.

Darwin recognized that sexual selection could be driven either by sexual contests among males or by female preference for particular male phenotypes. Male-male competition for access to females can favor the evolution of features that provide an advantage in fights or ritual displays of aggression (**Fig. 14-5**). Female mate choice provides a second source of sexual selection. In animal species in which females actively choose their mates from among males, females often seem to prefer males with the most elaborate ornaments or most extravagant displays (**Fig. 14-6**). Why?

A popular hypothesis is that male structures, colors, and displays that do not enhance survival might instead provide a female with an outward sign of a male's condition. Only a vigorous, energetic male can survive when burdened with conspicuous coloration or a large tail that might make him more vulnerable to predators. Conversely, males that are sick or under parasitic attack are dull and

Figure 14-5 Competition between males favors the evolution, through sexual selection, of structures for ritual combat Two male bighorn sheep spar during the fall mating season. In many species, the losers of such contests are unlikely to mate, while winners enjoy tremendous reproductive success. **QUESTION:** *If we studied a population of bighorn sheep and were able to identify the father and mother of each lamb born, would you predict that the difference in number of offspring between the most reproductively successful adult and the least successful adult would be greater for males or for females?*

frumpy compared with healthy males. A female that chooses the brightest, most ornamented male is also choosing the healthiest, most vigorous male. By doing so, she gains fitness if, for example, the most vigorous male provides superior parental care to offspring or if he carries alleles for disease resistance that will be inherited by offspring and help ensure their survival. Females thus gain a reproductive advantage by choosing the most highly ornamented males, and the traits (including the exaggerated male ornament) of these flashy males will be passed to subsequent generations.

Figure 14-6 The peacock's showy tail has evolved through sexual selection The ancestors of today's peahens were apparently picky when deciding on a male with which to mate, favoring males with longer and more colorful tails.

14.4 What Is a Species?

Although Darwin brilliantly explained how evolution shapes complex, amazingly well-designed organisms, his ideas did not fully explain life's diversity. In particular, the process of natural selection cannot by itself explain how living things came to be divided into groups, with each group distinctly different from all other groups. When we look at big cats, we don't see a continuous array of different tiger phenotypes that gradually grades into a lion phenotype. We see lions and tigers as separate, distinct types with no overlap. Each distinct type is known as a species.

Biologists Need a Clear Definition of Species

Before we can study the origin of species, we must first clarify our definition of the term. Throughout most of human history, "species" was a poorly defined concept. In pre-Darwinian Europe, the word "species" simply referred to one of the "kinds" produced by the biblical creation. In this view, humans could not possibly know the criteria of the creator, but could only attempt to distinguish among species on the basis of visible differences in structure. In fact, species is Latin for "appearance."

On a coarse scale, it is easy to use quick visual comparisons to distinguish species. For example, warblers are clearly different from eagles, which are obviously different from ducks. But it is far more difficult to distinguish among different species of warblers, eagles, or ducks. How do scientists make these finer distinctions?

Web Tutorial 14.3 What Is Speciation?

Species Are Groups of Interbreeding Populations

Today, biologists define a **species** as a group of populations that evolves independently. Each species follows a separate evolutionary path because alleles do not move between the gene pools of different species. This definition, however, does not clearly state the standard by which such evolutionary independence is judged. The most widely used standard defines species as "groups of actually or potentially interbreeding natural populations, which are reproductively isolated from other such groups." This definition, known as the *biological species concept*, is based on the observation that reproductive isolation (no successful breeding outside the group) ensures evolutionary independence.

The biological species concept has at least two major limitations. First, because the definition is based on patterns of sexual reproduction, it does not help us determine species boundaries among asexually reproducing organisms. Second, it is not always practical or even possible to directly observe whether members of two different groups interbreed. Thus, a biologist who wishes to determine if a group of organisms is a separate species must often make the determination without knowing for sure if group members breed with organisms outside the group.

Despite the limitations of the biological species concept, most biologists accept it for identifying species of sexually reproducing organisms. Nonetheless, scientists who study bacteria and other organisms that mainly reproduce asexually must use alternative definitions of species.

(a)

(b)

Figure 14-7 **Members of a species may differ in appearance** ***(a)*** The myrtle warbler and ***(b)*** Audubon's warbler are members of the same species.

Appearance Can Be Misleading

Biologists have found that some organisms with very similar appearances belong to different species. Conversely, differences in appearance do not always mean that two populations belong to different species. For example, bird field guides published in the 1970s list the myrtle warbler and Audubon's warbler (Fig. 14-7) as distinct species. These birds differ in geographical range and in the color of their throat feathers. More recently, scientists decided that these birds are local varieties of the same species. The reason: Where their ranges overlap, these warblers interbreed, and the offspring are just as vigorous and fertile as the parents.

14.5 How Do New Species Form?

Despite his exhaustive exploration of the process of natural selection, Charles Darwin never proposed a complete mechanism of **speciation**, the process by which new species form. One scientist who did play a large role in describing the process of speciation was Ernst Mayr of Harvard University, an ornithologist (expert on birds) and a pivotal figure in the history of evolutionary biology. Mayr developed the biological species concept discussed above. He was also among the first to recognize that speciation depends on two factors acting on a pair of populations: isolation and genetic divergence.

- *Isolation of populations.* If individuals move freely between two populations, interbreeding and the resulting gene flow will cause changes in one population to soon become widespread in the other as well. Thus, two populations cannot grow increasingly different unless something happens to block interbreeding between them. Speciation depends on isolation.
- *Genetic divergence of populations.* It is not sufficient for two populations simply to be isolated. They will become separate species only if, during the period of isolation, they evolve sufficiently large genetic differences. The differences must be large enough that, if the isolated populations are reunited, they can no longer interbreed and produce vigorous, fertile offspring. Such differences can arise by chance (genetic drift), especially if at least one of the isolated populations is small. Large genetic differences can also arise through natural selection, if the isolated populations experience different environmental conditions.

Geographical Separation of a Population Can Lead to Speciation

New species can arise when an impassible barrier physically separates different parts of a population. Physical separation could occur if, for example, some members of a population of land-dwelling organisms drifted, swam, or flew to a remote oceanic island. Populations of water-dwelling organisms might be split when geological processes such as volcanism or continental drift create new land barriers that subdivide previously continuous seas or lakes. Geological change can also divide terrestrial populations (Fig. 14-8). Portions of populations can become stranded in patches of suitable habitat that become isolated by climate shifts. You can probably imagine many other scenarios that could lead to the geographical subdivision of a population.

Figure 14-8 **Geographical isolation** To determine if these two squirrels are members of different species, we must know if they're "actually or potentially interbreeding." Unfortunately, it is hard to tell, because ***(a)*** the Kaibab squirrel lives only on the north rim of the Grand Canyon and ***(b)*** the Abert squirrel lives only on the south rim. The two populations are geographically isolated but still quite similar. Have they diverged enough since their separation to be considered separate species? On the basis of our current knowledge, it is impossible to say.

(a)

(b)

If two populations become geographically isolated for any reason, there will be no gene flow between them. If the pressures of natural selection differ in the separate locations, then the populations may accumulate genetic differences. Alternatively, genetic differences may arise if one or more of the separated populations is small enough for genetic drift to occur, which may be especially likely in the aftermath of a founder event (in which a few individuals become isolated from the main body of the species). In either case, genetic differences between the separated populations may eventually become large enough to make interbreeding impossible. At that point, the two populations will have become separate species. Most evolutionary biologists believe that geographical isolation followed by speciation has been the most common source of new species, especially among animals.

Figure 14-9 Adaptive radiation More than 300 species of cichlid fish live in Lake Malawi in East Africa. These species are found nowhere else, and all of them descended from a single ancestral population within a million years. This dramatic adaptive radiation has led to a collection of closely related species with an array of adaptations for exploiting the many different food sources in the lake.

Under Some Conditions, Many New Species May Arise

In some cases, many new species arise in a relatively short time. This process, called **adaptive radiation**, can occur when populations of one species invade a variety of new habitats and evolve in response to the differing environmental pressures in those habitats. Adaptive radiation has occurred many times and in many groups of organisms, typically when species encounter a wide variety of unoccupied habitats. For example, episodes of adaptive radiation took place when the ancestors of Darwin's finches colonized the Galapagos Islands, when marsupial mammals first invaded Australia, and when an ancestral cichlid fish species arrived at Lake Malawi (Fig. 14-9). In these examples, the invading species faced no competitors except other members of their own species, and all the available habitats and food sources were rapidly exploited by new species that evolved from the original invaders.

14.6 How Is Reproductive Isolation between Species Maintained?

The process of speciation depends on the evolution of traits that prevent interbreeding. Genetic divergence during the period of isolation is necessary for the origin of a new species. It is not, however, sufficient unless it includes changes that would block reproduction between the two separated groups if they were to come into contact again. If members of a population are unable to interbreed with members of other populations, the population is said to be in **reproductive isolation**.

The traits that prevent interbreeding and maintain reproductive isolation are called **isolating mechanisms**. Isolating mechanisms give a clear benefit to individuals. Any individual that mates with a member of another species will probably produce unfit or sterile offspring, thereby wasting its reproductive effort and contributing nothing to future generations. Thus, natural selection favors traits that prevent mating across species boundaries. Mechanisms that prevent mating between species are called **premating isolating mechanisms**.

When premating isolating mechanisms fail or have not yet evolved, members of different species may mate. If, however, all resulting hybrid offspring die during development, then the two species are still reproductively isolated from one another. In some cases, however, viable hybrid offspring are produced. Even so, if these hybrids are less fit than their parents or are themselves infertile, the two species may still remain separate, with little or no gene flow between them. Mechanisms that prevent the formation of vigorous, fertile hybrids between species are called **postmating isolating mechanisms**.

Figure 14-10 Ecological isolation This female fig wasp is carrying fertilized eggs from a mating that took place within a fig. She will find another fig of the same species, enter it through a pore, lay eggs, and die. Her offspring will hatch, develop, and mate within the fig. Because each species of fig wasp reproduces only in its own particular fig species, each wasp species is reproductively isolated.

Premating Isolating Mechanisms Prevent Mating between Species

Reproductive isolation can be maintained by a variety of mechanisms, but those that prevent mating attempts are especially effective. We next describe the most important types of such premating isolating mechanisms.

Members of Different Species May Be Prevented from Meeting

Members of different species cannot mate if they never get near one another. As we have already seen, *geographical isolation* typically provides the conditions for speciation in the first place. However, we cannot determine if geographically separated populations are actually distinct species. Should the physical barrier separating the two populations disappear (an intervening river might change course, for example), the reunited populations might interbreed freely and not be separate species after all. If they cannot interbreed, then other mechanisms, such as those considered below, must have developed during their isolation. Geographical isolation, therefore, is usually considered to be a mechanism that allows new species to form rather than a mechanism that maintains reproductive isolation between species.

Different Species May Occupy Different Habitats

Two populations that use different resources may spend time in different habitats within the same general area and thus exhibit *ecological isolation*. White-crowned and white-throated sparrows, for example, have extensively overlapping ranges. The white-throated sparrow, however, frequents dense thickets, whereas the white-crowned sparrow inhabits fields and meadows, seldom penetrating far into dense growth. The two species may coexist within a few hundred yards of one another and yet seldom meet during the breeding season. A more dramatic example is provided by the more than 750 species of fig wasp (Fig. 14-10). Each species of fig wasp breeds in (and pollinates) the fruits of a particular species of fig, and each fig species hosts one and only one species of pollinating wasp.

Although ecological isolation may slow down interbreeding, it seems unlikely that it could prevent gene flow entirely. Other mechanisms normally also contribute to reproductive isolation.

Different Species May Breed at Different Times

Even if two species occupy similar habitats, they cannot mate if they have different breeding seasons, a phenomenon called *temporal (time-related) isolation*. Bishop pines and Monterey pines grow together near Monterey on the California coast (Fig. 14-11). Viable hybrids have been produced between these two species in the laboratory. In the wild, however, the two species release their sperm-containing pollen and have eggs ready to receive the pollen at different times: The Monterey pine releases pollen in early spring, the bishop pine in summer. For this reason, the two species never interbreed under natural conditions.

Figure 14-11 Temporal isolation Bishop pines, such as these, and Monterey pines live together in nature. In the laboratory, they produce fertile hybrids. In the wild, however, they do not interbreed, because they release pollen at different times of the year.

Different Species May Have Different Courtship Rituals

Among animals, the elaborate courtship colors and behaviors that so enthrall human observers not only serve as recognition and evaluation signals between male and female, but also prevent mating with members of other species. Signals and behaviors that differ from species to species create *behavioral isolation*. The striking colors and calls of male songbirds, for example, may attract females of their own species, but females of other species treat them with the utmost indifference. Among frogs, males are often impressively indiscriminate, jumping on every female in sight, regardless of the species, when the spirit moves them. Females, however, approach only male frogs that croak the "ribbet" appropriate to their species. If they do find themselves in an unwanted embrace, they utter the "release call," which causes the male to let go. As a result, few hybrids are produced.

BIOTECHNOLOGY WATCH

CLONING ENDANGERED SPECIES

Environmentalists tend to be skeptical of claims that biotechnology can help solve environmental problems, but one group of scientists is determined to use bioengineering to rescue endangered species. Researchers at the biotech company Advanced Cell Technologies (ACT) have embarked on an ambitious plan to clone species that are in danger of extinction. They had their first success a few years ago with the birth of a gaur, a wild ox native to India and southeast Asia (Fig. E14-1). Gaurs are very rare and in danger of extinction, because their habitat is disappearing rapidly.

The gaur was cloned using a preserved skin cell from an individual that had died 8 years earlier. The genetic material from this cell was inserted into a cow egg from which the nucleus had been removed (see "Scientific Inquiry: Carbon Copies, Cloning in Nature and the Lab" in Chapter 10). The resulting cloned embryo was implanted in the uterus of a surrogate mother cow. (Cows were chosen to be the egg donor and surrogate mother because of easy availability and to avoid posing undue risk to any gaur.) The embryo developed properly and the cow gave birth to the gaur calf. Unfortunately, the calf died 2 days later of dysentery, an infection that the researchers say was unrelated to the cloning procedure.

Buoyed by their first success, the cloning team subsequently cloned a banteng, a highly endangered relative of the gaur, also native to southeast Asia. Two banteng embryos were cloned from a cell taken from the frozen tissue of an animal that died in 1980 and were carried to term by cow surrogate mothers. One of the clones survived and now lives with the banteng herd at the San Diego zoo. Zookeepers hope that, when the cloned animal matures, it will breed and add much-needed genetic diversity to the captive herd.

The scientists at ACT plan to continue with the cloning project. They recently received permission from the Spanish government to clone the bucardo, a mountain goat species of Europe that is already extinct (the last captive individual recently died). The scientists have spoken of being especially inspired by the idea of resurrecting an extinct species (they will use preserved bucardo tissues that are in storage in Spain). The researchers have also focused on the endangered and charismatic giant panda, as well as endangered big cats such as tigers, cheetahs, and leopards.

The idea of cloning endangered species is highly controversial among conservationists. Many feel that producing new animals one at a time can do little to truly help endangered species and that, in any case, the ability to artificially reproduce members of a species does little good if the species' habitat has been destroyed. Advocates reply that cloning may be a way to keep a species alive until efforts to restore its habitat are successful and to preserve genetic diversity by saving the genotypes of the most-threatened populations of a species. Many who wish to continue efforts to clone endangered species argue that it doesn't make sense to abandon any approach to conserving species, as we will need more than one strategy to solve such a difficult problem.

Figure E14-1 The gaur, an endangered Asian ox, was recently cloned

Species' Differing Sexual Organs May Foil Mating Attempts

In rare instances, a male and a female of different species attempt to mate. Their attempt is likely to fail. Among animal species with internal fertilization (in which the sperm is deposited inside the female's reproductive tract), the male's and female's sexual organs simply may not fit together. Among plants, differences in flower size or structure may prevent pollen transfer between species because the differing flowers may attract different pollinators. Isolating mechanisms of this type are called *mechanical incompatibilities*.

Postmating Isolating Mechanisms Limit Hybrid Offspring

Premating isolation sometimes fails. When it does, members of different species may mate, and the sperm of one species may reach the egg of another species. Such matings, however, often fail to produce vigorous, fertile hybrid offspring, owing to postmating isolating mechanisms.

One Species' Sperm May Fail to Fertilize Another Species' Eggs

Even if a male inseminates a female of a different species, his sperm may not be able to fertilize her eggs, an isolating mechanism called *gametic incompatibility*. For example, the fluids of the female reproductive tract may weaken or kill sperm of other species. Among plants, chemical incompatibility may prevent the germination of pollen from one species that lands on the stigma (pollen-catching structure) of the flower of another species.

TABLE 14-1 Mechanisms of Reproductive Isolation

Premating isolating mechanisms: factors that prevent organisms of two populations from mating

- **Geographical isolation:** The populations cannot interbreed because a physical barrier separates them.
- **Ecological isolation:** The populations do not interbreed even if they are within the same area because they occupy different habitats.
- **Temporal isolation:** The populations cannot interbreed because they breed at different times.
- **Behavioral isolation:** The populations do not interbreed because they have different courtship and mating rituals.
- **Mechanical incompatibility:** The populations cannot interbreed because their reproductive structures are incompatible.

Postmating isolating mechanisms: factors that prevent organisms of two populations from producing vigorous, fertile offspring after mating

- **Gametic incompatibility:** Sperm from one population cannot fertilize eggs of another population.
- **Hybrid inviability:** Hybrid offspring fail to survive to maturity.
- **Hybrid infertility:** Hybrid offspring are sterile or have low fertility.

Hybrid Offspring May Survive Poorly

If cross-species fertilization does occur, the resulting hybrid may be unable to survive, a situation called *hybrid inviability*. The genetic instructions directing development of the two species may be so different that hybrids abort early in development. If a hybrid animal does survive, it may display behaviors that are mixtures of the two parental types. In attempting to do some things the way species *A* does them and other things the way species *B* does them, the hybrid may be hopelessly uncoordinated and therefore unable to reproduce. Hybrids between certain species of lovebirds, for example, have great difficulty learning to carry nest materials during flight and probably could not reproduce in the wild.

Hybrid Offspring May Be Infertile

Most animal hybrids, such as the mule (a cross between a horse and a donkey) and the liger (a zoo-based cross between a male lion and a female tiger), are sterile. *Hybrid infertility* prevents hybrids from passing on their genetic material to offspring. A common reason for such infertility is the failure of chromosomes to pair properly during meiosis, so eggs and sperm never develop.

Table 14-1 summarizes the different types of isolating mechanisms.

14.7 What Causes Extinction?

Every living organism must eventually die, and the same is true of species. Just like individuals, species are "born" (through the process of speciation), persist for some period of time, and then perish. The ultimate fate of any species is **extinction**, the death of the last of its members. In fact, at least 99.9% of all the species that have ever existed are now extinct. The natural course of evolution, as revealed by fossils, is continual turnover of species as new ones arise and old ones go extinct.

The immediate cause of extinction is probably always environmental change, in either the living or the nonliving parts of the environment. Two major environmental factors that may drive a species to extinction are competition among species and habitat destruction.

Interactions with Other Species May Drive a Species to Extinction

As described earlier, interactions such as competition and predation serve as agents of natural selection. In some cases, these same interactions can lead to extinction rather than to adaptation.

Organisms compete for limited resources in all environments. If a species' competitors evolve superior adaptations and the species doesn't evolve fast enough to keep up, it may become extinct. A particularly striking example of extinction through competition occurred in South America, beginning about 2.5 million years ago. At that time, the isthmus of Panama rose above sea level and formed a land bridge between North America and South America. After the previously separated continents were connected, the mammal species that had evolved in isolation on each continent were able to mix. Many species did indeed expand their ranges, as North American mammals moved southward and South American mammals moved northward. As they moved, each species encountered resident species that occupied the same kinds of habitats and exploited the same kinds of resources. The ultimate result of the ensuing competition was that the North American species diversified and underwent an adaptive radiation that displaced the vast majority of the South American species, many of which went

EARTH WATCH

ENDANGERED SPECIES: FROM GENE POOLS TO GENE PUDDLES

Since the Endangered Species Act was passed in 1973, the United States has had an official policy of protecting rare species (Fig. E14-2). The ultimate goal of the act, however, is not just to protect endangered species, but to restore them. As one U.S. Fish and Wildlife Service official said, "The goal is to get species *off* the list." A species that has been officially listed as endangered can be removed from the list if government biologists determine that its population has grown large enough that it is no longer in danger of extinction from unpredictable events, such as a drought or a disease epidemic. If a species' population reaches this critical size, it is no longer legally "endangered" and it loses the legal protections provided by the act.

Unfortunately, a population that has become small enough to warrant endangered status is likely to undergo evolutionary changes that increase its chances of going extinct. One problem is that, in small populations, mating choices are limited and a high proportion of matings may be between close relatives. This inbreeding increases the odds that offspring will be homozygous for harmful recessive alleles, and these less-fit individuals may die before reproducing, further reducing the size of the population.

The greatest threat to small populations, however, stems from their inevitable loss of genetic diversity. From our discussion of population bottlenecks, it is apparent that, when populations shrink to very small sizes, many of the alleles that were present in the original population will not be represented in the gene pool of the remnant population. Furthermore, we have seen that genetic drift in small populations will cause many of the surviving alleles to subsequently disappear permanently from the population (see Fig. 14-2b). Because genetic drift is a random process, many of the lost alleles will be advantageous ones that were previously favored by selection. Inevitably, the number of different alleles in the population grows ever smaller. As ecologist Thomas Foose aptly put it, "Gene pools are being converted into gene puddles." Even if the size of an endangered population eventually begins to grow, the damage has already been done. Lost genetic diversity is regained only very slowly.

Why does it matter if a population's genetic diversity is low? There are two main risks. First, the fitness of the population as a whole is reduced by the loss of advantageous alleles that underlie adaptive traits. A less-fit population is unlikely to thrive. Second, a genetically impoverished population lacks the variation it will need to adapt when environmental conditions change. When the environment changes, as it inevitably will, a genetically uniform species is less likely to contain individuals well suited to survive and reproduce under the new conditions. A species unable to adapt to changing conditions is at very high risk of extinction.

What can be done to preserve the genetic diversity of species? The best solution, of course, is to preserve plenty of diverse types of habitat so that species never become endangered. The human population, however, has grown so large and appropriated so large a share of Earth's resources that this solution is impossible in many places. For many species, the only solution is to ensure that areas of preserved habitat are large enough to hold populations of sufficient size to contain most of a threatened species' total genetic diversity. If, however, circumstances dictate that preserved areas be small, it is important that the small areas be linked by corridors of the appropriate habitat, so that gene flow among populations in the small preserves can increase the spread of new and beneficial alleles.

Figure E14-2 **Endangered by habitat destruction** The bighorn sheep is among the species designated "endangered" by the U.S. government.

extinct. Clearly, evolution had bestowed on the North American species some (as yet unknown) set of adaptations that enabled their descendants to exploit resources more efficiently and effectively than their South American counterparts could.

Habitat Change and Destruction Are the Leading Causes of Extinction

Habitat change, both contemporary and prehistoric, is the single greatest cause of extinctions. Present-day habitat destruction due to human activities is proceeding at a rapid pace. Many biologists believe that we are presently in the midst of the fastest-paced and most widespread episode of species extinction in the history of life. Loss of tropical forests is especially devastating to species diversity. As many as half the species presently on Earth may be lost over the next 50 years as the tropical forests that contain them are cut for timber and to clear land for cattle and crops (see "Earth Watch: Endangered Species: From Gene Pools to Gene Puddles."). We will discuss extinctions due to prehistoric habitat change in Chapter 15.

Evolutionary Connections

Scientists Don't Doubt Evolution

In the popular press, disagreements among evolutionary biologists are sometimes seen as conflicts about evolution itself. We occasionally read statements implying that new theories are overthrowing Darwin's and casting doubt on the reality of evolution. Nothing could be further from the truth. Despite some disagreements about the details of the evolutionary process, biologists unanimously agree that evolution occurred in the past and is still occurring today. The only argument is about the relative importance of the various mechanisms of evolutionary change in the history of life on Earth, their pace, and which forces were most important in shaping the evolution of particular species. Meanwhile, wolves still tend to catch the slowest caribou, small populations still undergo genetic drift, and habitats still change or disappear. Evolution continues, still generating, in Darwin's words, "endless forms most beautiful."

EVOLUTION OF A MENACE *REVISITED*

Studies of bacterial evolution in individual patients, such as the one discussed at the beginning of this chapter, convincingly demonstrate that antibiotic resistance can evolve in response to the presence of antibiotics. But research on a larger scale is required to demonstrate that this kind of evolution is truly widespread.

Many researchers have examined antibiotic resistance on a larger scale. For example, one study compared bacteria from almost 3000 new, previously untreated cases of tuberculosis with bacteria from about 200 patients who had undergone treatment with antibiotics but suffered a relapse. Among the new, antibiotic-free cases, only 8% of patients were infected with antibiotic-resistant bacteria, but among the relapsed, antibiotic-treated patients, almost 22% carried resistant bacteria. If the patients in this study are representative of patients in general, we can conclude that widespread use of antibiotics increases the prevalence of resistant populations of disease-causing bacteria.

Medical use of antibiotics is the primary source of natural selection for resistance, but antibiotics also select for resistance in environments outside of our bodies. For example, about 20 million pounds of antibiotics are fed to farm animals each year, with the result that the meat we consume can contain resistant bacteria. In addition, recent research has detected widespread presence of antibiotics in soil and water, presumably because antibiotics enter the environment through human and animal wastes. Although researchers are only now beginning to investigate drug resistance in free-living bacteria, our understanding of natural selection strongly suggests that resistance will evolve wherever antibiotics are present. In our fight against disease, we can no longer afford to overlook the principles of evolutionary biology.

CONSIDER THIS: Natural selection acts only on existing variation, so bacteria in natural populations must already carry alleles that help them resist attack by antibiotics. Why are such alleles present in bacterial populations? (Hint: Almost all medically useful antibiotics were originally derived from fungi or bacteria.)

LINKS TO LIFE: Confining the Contagious

A great way to prevent natural selection for drug resistance is to eliminate variation in bacterial phenotypes, preferably by killing every disease-causing bacterium in a victim's body. But if a tuberculosis patient fails to take antibiotics until all of the bacteria in his or her body are completely wiped out, the goal of eliminating variation is not achieved—and the surviving bacteria are especially likely to be resistant. A noncompliant tuberculosis patient thus becomes a factory that produces drug-resistant bacteria and does additional damage by helping spread those bacteria to other people.

How can we contain the threat to public health posed by uncooperative tuberculosis victims? One solution is to round them up and detain them in a location where their daily treatment can be closely supervised and monitored. In many states, tuberculosis patients may be detained and held against their will if they fail to follow a treatment regimen for their disease. In essence, the freedom of such patients is sacrificed to protect society from the negative effects of their behavior. The laws that permit such detention reflect sound evolutionary thinking. Do you believe that detention of uncooperative tuberculosis victims is justified?

Chapter Review

Summary of Key Concepts

14.1 How Are Populations, Genes, and Evolution Related?

Evolution is change in frequencies of alleles in a population's gene pool. Allele frequencies in a population will remain constant over generations only if the following conditions are met: (1) There is no mutation; (2) there is no gene flow; (3) the population is very large; (4) all mating is random; and (5) all genotypes reproduce equally well (that is, there is no natural selection). These conditions are rarely, if ever, met in nature. Understanding what happens when they are not met helps reveal the mechanisms of evolution.

14.2 What Causes Evolution?

- Mutations are random, undirected changes in DNA composition. Although most mutations are neutral or harmful to the organism, some prove advantageous in certain environments. Mutations are rare and do not change allele frequencies very much, but they provide the raw material for evolution.
- In any population, chance events kill or prevent reproduction by some of the individuals. If the population is small, chance events may eliminate a disproportionate number of individuals who bear a particular allele, thereby greatly changing the allele frequency in the population: This is genetic drift.
- The survival and reproduction of organisms are influenced by their phenotypes. Because phenotype depends at least partly on genotype, natural selection tends to favor the reproduction of certain alleles at the expense of others.

14.3 How Does Natural Selection Work?

Natural selection is driven by differences in reproductive success among different genotypes. Natural selection proceeds from the interactions of organisms with both the biotic and abiotic parts of their environments. When two or more species exert mutual environmental pressures on each other for long periods of time, both of them evolve in response. Such coevolution can result from any type of relationship between organisms, including competition and predation. Phenotypes that help organisms mate can evolve by sexual selection.

14.4 What Is a Species?

According to the biological species concept, a species consists of all the populations of organisms that are potentially capable of interbreeding under natural conditions and that are reproductively isolated from other populations.

14.5 How Do New Species Form?

Speciation, the formation of new species, takes place when gene flow between two populations is reduced or eliminated and the populations diverge genetically. Most commonly, speciation follows geographical isolation and subsequent genetic divergence of the separated populations through genetic drift or natural selection.

14.6 How Is Reproductive Isolation between Species Maintained?

Reproductive isolation between species may be maintained by one or more of several mechanisms, collectively known as premating isolating mechanisms and postmating isolating mechanisms. Premating isolating mechanisms include geographical isolation, ecological isolation, temporal isolation, behavioral isolation, and mechanical incompatibility. Postmating isolating mechanisms include gametic incompatibility, hybrid inviability, and hybrid infertility.

14.7 What Causes Extinction?

Factors that cause extinctions include competition among species and habitat destruction.

Key Terms

adaptation *p. 243*
adaptive radiation *p. 247*
allele frequency *p. 237*
coevolution *p. 244*
competition *p. 244*
equilibrium population *p. 237*
extinction *p. 250*
fitness *p. 242*
founder effect *p. 240*
gene flow *p. 237*
gene pool *p. 237*
genetic drift *p. 238*
Hardy-Weinberg principle *p. 237*
isolating mechanism *p. 247*
mutation *p. 238*
natural selection *p. 242*
population *p. 236*
population bottleneck *p. 240*
postmating isolating mechanism *p. 247*
predation *p. 244*
premating isolating mechanism *p. 247*
reproductive isolation *p. 247*
sexual selection *p. 244*
speciation *p. 246*
species *p. 245*

Thinking Through the Concepts

Multiple Choice

1. *Genetic drift is a _____ process.*
   a. random
   b. directed
   c. selection-driven
   d. coevolutionary
   e. uniformitarian
2. *Of the following possibilities, the best way to estimate an organism's evolutionary fitness is to measure the*
   a. size of its offspring
   b. number of eggs it produces
   c. number of eggs it produces over its lifetime
   d. number of offspring it produces over its lifetime
   e. number of offspring it produces over its lifetime that survive to breed
3. *In many species of fireflies, males flash to attract females. Each species has a different flashing pattern. This is probably an example of*
   a. ecological isolation
   b. temporal isolation
   c. geographical isolation
   d. premating isolation
   e. postmating isolation
4. *Under the biological species concept, the main criterion for identifying a species is*
   a. anatomical distinctiveness
   b. behavioral distinctiveness
   c. geographical isolation
   d. reproductive isolation
   e. gametic incompatibility
5. *In terms of changes in allele frequencies, founder events result in*
   a. gradual accumulation of many small changes
   b. large, rapid changes
   c. coevolution
   d. hybridization
   e. mechanical incompatibility
6. *After the demise of the dinosaurs, mammals evolved rapidly into many new forms because of*
   a. the founder effect
   b. a genetic bottleneck
   c. adaptive radiation
   d. geological time
   e. genetic drift

Review Questions

1. What is a gene pool? How would you determine the allele frequencies in a gene pool?
2. Define *equilibrium population*. Outline the conditions that must be met for a population to stay in genetic equilibrium.
3. How does population size affect the likelihood of changes in allele frequencies by chance alone? Can significant changes in allele frequencies (that is, evolution) occur as a result of genetic drift?
4. If you measured the allele frequencies of a gene and found large differences from those predicted by the Hardy-Weinberg principle, would that prove that natural selection is occurring in the population you are studying? Review the conditions that lead to an equilibrium population, and explain your answer.
5. People like to say that "you can't prove a negative." Study the experiment in Figure 14-1 again, and comment on what it demonstrates.
6. What is sexual selection? How is sexual selection similar to and different from other forms of natural selection?
7. Many of the oak tree species in central and eastern North America hybridize (interbreed). Are they "true species"?
8. What are the two major types of reproductive isolating mechanisms? Give examples of each type and describe how they work.

Applying the Concepts

1. **IS THIS SCIENCE?** A question commonly asked by new biology students is: "If humans descended from apes, why are there still apes?" On the basis of your understanding of the process of evolutionary change, provide a scientifically informed answer to this question.
2. **IS THIS SCIENCE?** Southern Wisconsin is home to several populations of gray squirrels (*Sciurus carolinensis*) with black fur. Some observers claim that the black squirrels are a separate species. Design a study to test the hypothesis that the gray squirrels and black squirrels are separate species.
3. In many countries, conservationists are trying to design national park systems so that "islands" of natural area (the big parks) are connected by thin "corridors" of undisturbed habitat. The idea is that this arrangement will allow animals and plants to migrate between refuges. Why would such migration be important?

For More Information

Dawkins, R. *Climbing Mount Improbable.* New York: Norton, 1996. An eloquent book-length tribute to the power of natural selection to design intricate adaptations. The chapter on the evolution of the eye is a classic.

Levy, S. B. "The Challenge of Antibiotic Resistance." *Scientific American*, March 1998. An excellent summary of the public health implications of antibiotic resistance. Also discusses some strategies for ameliorating the problem.

Nicolaou, K. C., and Boddy, C. N. B. "Behind Enemy Lines." *Scientific American*, May 2001. An overview of the effort to devise pharmaceutical solutions to the problem of antibiotic-resistant bacteria.

Quammen, D. *The Song of the Dodo*. New York: Scribner, 1996. Beautifully written exposition of the biology of islands. Read this book to understand why islands are known as "natural laboratories of speciation."

Palumbi, S. R. *The Evolution Explosion*. New York: Norton, 2001. An evolutionary biologist explores cases of rapid evolution caused by humans, including antibiotic resistance, pesticide resistance, and the evolution of the virus that causes AIDS.

Rennie, J. "Fifteen Answers to Creationist Nonsense." *Scientific American*, July 2002. A summary of some common misconceptions espoused by creationists, and the scientific response to them.

Schilthuizen, M. *Frogs, Flies, and Dandelions: Speciation—The Evolution of New Species.* Oxford: Oxford University Press, 2001. A readable and entertaining summary of the latest biological thought on species and speciation.

WEB TUTORIAL

To access a Web Tutorial visit http://www.prenhall.com/audesirk4. *Log in to the Web site selected by your instructor, navigate to this chapter, and select the appropriate Web Activity number.*

14.1 Agents of Change
Estimated time: 10 minutes

Natural selection is just one of the agents of microevolution. In this activity, explore three other agents that can change allele frequencies in a population.

14.2 The Bottleneck Effect
Estimated time: 5 minutes

This activity demonstrates the mechanism of natural selection, using the example of evolution of an antibiotic resistance trait in a population of the bacteria that cause the human disease tuberculosis.

14.3 What is Speciation?
Estimated time: 2 minutes

The animation provides a brief overview of the concept of speciation through reproductive isolation.

14.4 Web Investigation: Evolution of a Menace
Estimated time: 15 minutes

Antibiotic resistance is reaching crisis proportions. Diseases that were all but eradicated only a few years ago are back and resistant to all known treatments. This exercise will take a brief look at the basic biology, medical repercussions, and social implications of bacterial antibiotic resistance.

Chapter 15

The History of Life on Earth

Fossils of feathered dinosaurs such as *Caudipteryx* (shown here in an artist's reconstruction) provide powerful evidence that today's birds descended from dinosaur ancestors.

AT A GLANCE

Dinosaurs Singing in the Backyard?

Dinosaurs no longer walk the Earth, but their descendants are lurking outside your window. The sparrows singing in your backyard, the pigeons perched atop your apartment building, and the crows soaring overhead are all descended from dinosaurs. Although a small, delicate, flying bird may not seem to have much in common with a large, heavy-limbed, earthbound dinosaur, birds are the closest living relatives of the extinct dinosaurs. Still, given the dissimilarity of the two groups, how can we know that dinosaurs are the ancestors of birds? The key evidence comes from fossils, which provide a record of life's history.

Paleontologists (the scientists who study fossils) have long known that birds share ancestry with reptiles. This shared ancestry was revealed by fossils of *Archaeopteryx*, the earliest known bird. *Archaeopteryx* had feathers very much like those of modern birds, but also had some reptilian traits. Unlike modern birds, *Archaeopteryx* had teeth, clawed fingers, and a long, bony tail. These traits suggest that *Archaeopteryx* represents an early moment in the history of birds, when birds still retained many structures that were later lost as birds evolved to become the efficient flying machines that we see today.

Beautifully preserved fossils of *Archaeopteryx*, the first of which was discovered in 1861, provide a window on the past and reveal the link between birds and other reptiles. However, evidence of the link between birds and the particular group of reptiles known as dinosaurs was much longer in coming. Although the hypothesis that birds descended from dinosaurs is not new, convincing fossil evidence in support of the hypothesis was not found until recently.

This 150-million-year-old fossil impression is the oldest evidence of feathers.

Can you guess the main prediction of the birds-from-dinosaurs hypothesis of bird evolution? What fossil evidence would support the prediction? Think about these questions as we review a broad outline of the history of life on Earth.

15.1 How Did Life Begin?

Pre-Darwinian thought held that all species were simultaneously created by God a few thousand years ago. Further, until the nineteenth century most people thought that new members of species sprang up all the time, through **spontaneous generation** from both nonliving matter and other, unrelated forms of life. In 1609 a French botanist wrote, "There is a tree ... frequently observed in Scotland. From this tree leaves are falling; upon one side they strike the water and slowly turn into fishes, upon the other they strike the land and turn into birds." Medieval writings abound with similar observations. Microorganisms were thought to arise spontaneously from broth, maggots from meat, and mice from mixtures of sweaty shirts and wheat.

Experiments Refuted Spontaneous Generation

In 1668, the Italian physician Francesco Redi disproved the maggots-from-meat hypothesis simply by keeping flies (whose eggs hatch into maggots) away from uncontaminated meat, as we saw in Chapter 1. In the mid-1800s, Louis Pasteur in France and John Tyndall in England disproved the broth-to-microorganism idea (Fig. 15-1). Although their work effectively demolished the notion of spontaneous generation, it did not address the question of how life on Earth originated in the first place. Or, as the biochemist Stanley Miller put it, "Pasteur never proved it didn't happen once, he only showed that it doesn't happen all the time."

The First Living Things Arose from Nonliving Ones

For almost half a century, the subject lay dormant. Eventually, biologists returned to the question of the origin of life. In the 1920s and 1930s, Alexander Oparin in Russia and John B. S. Haldane in England noted that today's oxygen-rich atmosphere would not have permitted the spontaneous formation of the complex organic molecules necessary for life. Oxygen reacts readily with other molecules, disrupting chemical bonds. Thus, an oxygen-rich environment tends to keep molecules simple.

Oparin and Haldane speculated that the atmosphere of the young Earth must have contained very little oxygen and that, under such atmospheric conditions, complex organic molecules could have arisen through ordinary chemical reactions. Some kinds of molecules could persist in the lifeless environment of early Earth better than others, and would therefore become more common over time. This chemical version of the "survival of the fittest" is called *prebiotic* (meaning

Figure 15-1 Spontaneous generation refuted Louis Pasteur's experiment disproving the spontaneous generation of microorganisms in broth

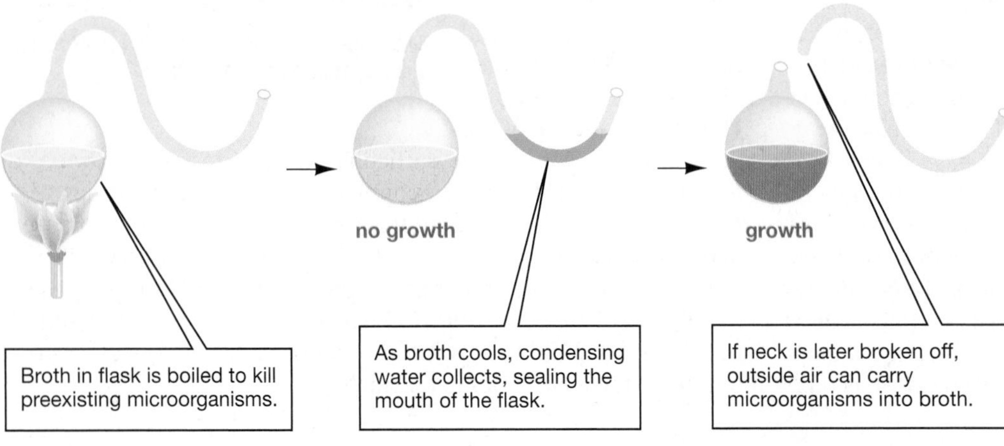

"before life") evolution. In the scenario envisioned by Oparin and Haldane, prebiotic chemical evolution gave rise to progressively more complex molecules and eventually to living organisms.

Organic Molecules Can Form Spontaneously under Prebiotic Conditions

Inspired by the ideas of Oparin and Haldane, Stanley Miller and Harold Urey set out in 1953 to simulate prebiotic evolution in the laboratory. They knew that, on the basis of the chemical composition of the rocks that formed early in Earth's history, geochemists had concluded that the early atmosphere probably contained virtually no oxygen gas, but did contain other substances, including methane, ammonia, hydrogen, and water vapor. Miller and Urey simulated the oxygen-free atmosphere of early Earth by mixing these components in a flask. An electrical discharge mimicked the intense energy of early Earth's lightning storms. In this experimental microcosm, the researchers found that simple organic molecules appeared after just a few days (**Fig. 15-2**).

Similar experiments by Miller and others have produced amino acids, short proteins, nucleotides, adenosine triphosphate (ATP), and other molecules characteristic of living things. Interestingly, the exact composition of the "atmosphere" used in these experiments is unimportant, provided that hydrogen, carbon, and nitrogen are available and that oxygen gas is excluded. Similarly, a variety of energy sources, including ultraviolet light, electrical discharge, and heat, are all about equally effective. Even though we may never know exactly what the earliest atmosphere was like, we can be confident that organic molecules formed spontaneously on early Earth. Additional organic molecules probably arrived from space when meteorites and comets crashed into Earth's surface. (Analysis of present-day meteorites recovered from impact craters on Earth has revealed that some meteorites contain relatively high concentrations of amino acids and other simple organic molecules similar to those generated in the Miller-Urey experiment.)

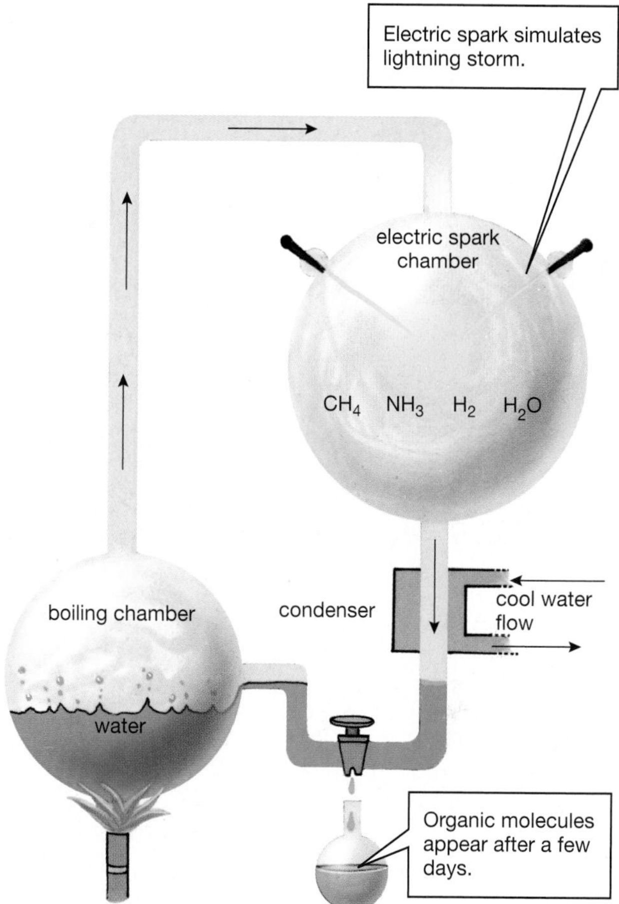

Figure 15-2 The experimental apparatus of Stanley Miller and Harold Urey Life's very earliest stages left no fossils, so evolutionary historians have pursued a strategy of re-creating in the laboratory the conditions that may have prevailed on early Earth. The mixture of gases in the spark chamber simulates Earth's early atmosphere. **QUESTION:** *How would the experiment's result change if oxygen (O_2) were included in the spark chamber?*

Organic Molecules Can Accumulate under Prebiotic Conditions

Prebiotic synthesis was not very efficient or very fast. Nonetheless, in a few hundred million years, large quantities of organic molecules accumulated in the early Earth's oceans. Today, most organic molecules have a short life because they are either digested by living organisms or they react with atmospheric oxygen. Early Earth, however, lacked both life and free oxygen, so molecules would not have been exposed to these threats.

Still, the prebiotic molecules must have been threatened by the sun's high-energy ultraviolet (UV) radiation, because early Earth lacked an ozone layer. The ozone layer is a region high in today's atmosphere that is enriched with ozone (O_3) molecules, which absorb some of the sun's UV light before it reaches Earth's surface. Before the ozone layer formed, UV bombardment, which can break apart organic molecules, must have been fierce. Some places, however, such as those beneath rock ledges or at the bottoms of even fairly shallow seas, would have been protected from UV radiation. In these locations, organic molecules may have accumulated.

Organic Molecules May Have Become Concentrated in Tidal Pools

In the next stage of prebiotic evolution, simple molecules combined to form larger molecules. The chemical reactions that formed the larger molecules required that the reacting molecules be packed closely together. Scientists have proposed several processes by which the required high concentrations might have been achieved on early Earth. One possibility is that shallow pools at the ocean's edge were filled with water by waves crashing onto the shore during high tides. Afterward, during low tides, some of the water in the pools might have evaporated, concentrating the dissolved substances. Given enough cycles of refilling and evaporation, the molecules in these pools could have become a concentrated "primordial soup" in which spontaneous chemical reactions between simple molecules could generate complex organic molecules. These molecules could then have become the building blocks of the first living organisms.

RNA May Have Been the First Self-Reproducing Molecule

Although all living organisms use DNA to encode and store genetic information, it is unlikely that DNA was the earliest informational molecule. DNA can reproduce itself only with the help of large, complex protein enzymes, but the instructions for building these enzymes are encoded in DNA itself. For this reason, the origin of DNA's role as life's information storage molecule poses a "chicken and egg" puzzle: DNA requires proteins, but those proteins require DNA. It is thus difficult to construct a plausible scenario for the origin of self-replicating DNA from prebiotic molecules. It is therefore likely that the current DNA-based system of information storage evolved from an earlier system.

A prime candidate for the first self-replicating informational molecule is RNA. In the 1980s, Thomas Cech and Sidney Altman, working with the single-celled organism *Tetrahymena*, discovered a cellular reaction that was catalyzed not by a protein, but by a small RNA molecule. Because this special RNA molecule performed a function previously thought to be performed only by protein enzymes, Cech and Altman decided to give their catalytic RNA molecule the name **ribozyme**.

In the years since their discovery, researchers have found dozens of naturally occurring ribozymes that catalyze a variety of different reactions, including cutting other RNA molecules and splicing together different RNA fragments. Ribozymes have also been found in the protein manufacturing machinery of cells, where they help catalyze the attachment of amino acid molecules to growing proteins. In addition, researchers have been able to synthesize different ribozymes in the laboratory, including some that can catalyze the replication of small RNA molecules.

The discovery that RNA molecules can act as catalysts for diverse reactions, including RNA replication, provides support for the hypothesis that life arose in an "RNA world." According to this view, the current era of DNA-based life was preceded by one in which RNA served as both the information-carrying genetic molecule and the enzyme catalyst for its own replication. This RNA world may have emerged after hundreds of millions of years of prebiotic chemical synthesis, during which RNA nucleotides would have been among the molecules synthesized. After reaching a sufficiently high concentration, the nucleotides probably bonded together to form short RNA chains.

Let's suppose that, purely by chance, one of these RNA chains was a ribozyme that could catalyze the production of copies of itself. This first self-reproducing ribozyme probably wasn't very good at its job and produced copies with lots of errors. These mistakes were the first mutations. As do modern mutations, most undoubtedly ruined the catalytic abilities of the "daughter molecules," but a few may have been improvements. Such improvements set the stage for the evolution of RNA molecules, as variant ribozymes with increased speed and accuracy of replication reproduced faster, making more copies of themselves and displacing less efficient molecules. Molecular evolution in the RNA world proceeded until, by some still unknown chain of events, RNA gradually receded into its present role as an intermediary between DNA and protein enzymes.

Membrane-Like Microspheres May Have Enclosed Ribozymes

Self-replicating molecules alone do not constitute life; these molecules must be contained within some kind of enclosing membrane. The precursors of the earliest biological membranes may have been simple structures that formed spontaneously from purely physical, mechanical processes. For example, chemists have shown that if water containing proteins and lipids is agitated to simulate waves beating against ancient shores, the proteins and lipids combine to form hollow structures called **microspheres** (Fig. 15-3). These hollow balls resemble living cells in several respects. They have a well-defined outer boundary that separates their internal contents from the external solution. If the composition of the microsphere is right, a "membrane" forms that is remarkably similar in appearance to a real cell membrane. Under certain conditions, microspheres can absorb material from the external solution, grow, and even divide.

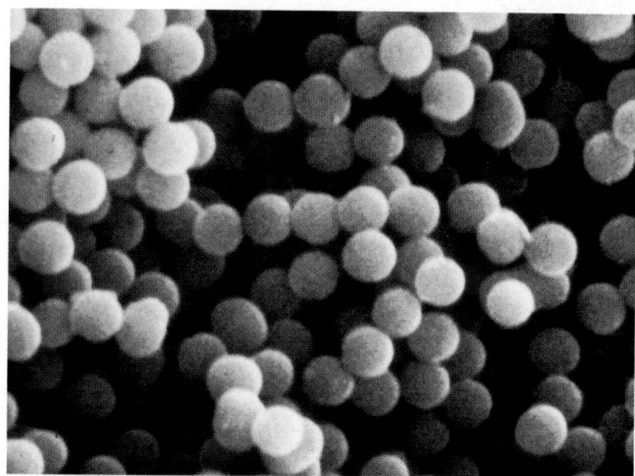

Figure 15-3 Did microspheres enclose the earliest cells? Cell-like microspheres can be formed by agitating proteins and lipids in a liquid medium. Each microsphere in this photo is about 5 μm in diameter.

If a microsphere happened to surround the right ribozymes, it would form something resembling a living cell. We could call it a **protocell**, structurally similar to a cell but not a living thing. In the protocell, ribozymes and any other enclosed molecules would have been protected from free-roaming ribozymes in the primordial soup. Nucleotides and other small molecules might have diffused across the membrane and been used to synthesize new ribozymes and other complex molecules. After sufficient growth, the microsphere may have divided, with a few copies of the ribozymes becoming incorporated into each daughter microsphere. If this process occurred, the path to the evolution of the first cells would be nearly at its end.

Was there a particular moment when a nonliving protocell gave rise to something alive? Probably not. Like most evolutionary transitions, the change from protocell to living cell was a continuous process, with no sharp boundary between one state and the next.

But Did All This Happen?

The above scenario, although plausible and consistent with many research findings, is by no means certain. One of the most striking aspects of origin-of-life research is a great diversity of assumptions, experiments, and contradictory hypotheses. (Iris Fry's *The Emergence of Life on Earth*, cited in the "For More Information" section at the end of this chapter, offers a taste of these controversies.) Researchers disagree as to whether life arose in quiet pools, in the sea, in moist films on the surfaces of clay or iron pyrite (fool's gold), or in furiously hot deep-sea vents. A few researchers even argue that life arrived on Earth from space. Can we draw any conclusions from the research conducted so far? No one knows for sure, but we can make a few observations.

First, the experiments of Miller and others show that amino acids, nucleotides, and other organic molecules, along with simple membrane-like structures, would have formed in abundance on the early Earth. Second, chemical evolution had long periods of time and huge areas of the Earth available to it. Given sufficient time and a sufficiently large pool of reactant molecules, even extremely rare events can occur many times. So, even if prebiotic evolution yielded only simple molecules, the earliest catalysts were not very efficient, and the earliest membranes were simple, the vast expanses of available time and space would have increased the likelihood of each small step on the path from primordial soup to living cell.

Most biologists accept that the origin of life was probably an inevitable consequence of the working of natural laws. We should emphasize, however, that this proposition cannot be definitively tested. The origin of life left no record, and researchers exploring this mystery can proceed only by developing a hypothetical scenario and then conducting laboratory investigations to determine if the scenario's steps are chemically and biologically possible and plausible.

15.2 What Were the Earliest Organisms Like?

When Earth first formed about 4.5 billion years ago, it was quite hot. A multitude of meteorites smashed into the forming planet, and the kinetic energy of these extraterrestrial rocks was converted into heat on impact. Still more heat was released by the decay of radioactive atoms. The rock composing Earth melted, and heavier elements such as iron and nickel sank to the center of the planet, where they remain molten even today. It must have taken hundreds of millions of years for Earth to cool enough to allow water to exist as a liquid. Nonetheless, it appears that life arose in fairly short order once liquid water was available.

The oldest fossil organisms found so far are in rocks that are about 3.5 billion years old. (Their age was determined using radiometric dating techniques; see "Scientific Inquiry: How Do We Know How Old a Fossil Is?".) Chemical traces in older rocks have led some paleontologists to believe that life is even older, perhaps as old as 3.9 billion years. The period in which life began is known as the Precambrian era. This interval was designated by geologists and paleontologists, who have devised a hierarchical naming system of eras, periods, and epochs to delineate the immense span of geological time (Table 15-1).

The First Organisms Were Anaerobic Prokaryotes

The first cells to arise in Earth's oceans were **prokaryotes**, cells whose genetic material was not contained within a nucleus, separate from the rest of the cell. These cells probably obtained nutrients and energy by absorbing organic molecules from their environment. There was no oxygen gas in the atmosphere, so the cells must have metabolized the organic molecules anaerobically. You will recall from Chapter 7 that anaerobic metabolism yields only small amounts of energy.

Thus, the earliest cells were primitive anaerobic bacteria. As these bacteria multiplied, they must have eventually used up the organic molecules produced by prebiotic chemical reactions. Simpler molecules, such as carbon dioxide and water, would still have been very abundant, as was energy in the form of sunlight. What was lacking, then, was not materials or energy itself, but energetic molecules—molecules in which energy is stored in chemical bonds.

Some Organisms Evolved the Ability to Capture the Sun's Energy

Eventually, some cells evolved the ability to use the energy of sunlight to drive the synthesis of complex, high-energy molecules from simpler molecules: In other words, photosynthesis appeared. Photosynthesis requires a source of hydrogen, and the very earliest photosynthetic bacteria probably used hydrogen sulfide gas dissolved in water for this purpose (much as today's purple photosynthetic bacteria do). Eventually, however, Earth's supply of hydrogen sulfide (which is produced mainly by volcanoes) must have run low. The shortage of hydrogen sulfide set the stage for the evolution of photosynthetic bacteria that were able to use the planet's most abundant source of hydrogen: water (H_2O).

Photosynthesis Increased the Amount of Oxygen in the Atmosphere

Water-based photosynthesis converts water and carbon dioxide to energetic molecules of sugar, releasing oxygen as a by-product. The emergence of this new method for capturing energy introduced significant amounts of free oxygen to the atmosphere for the first time. At first, the newly liberated oxygen

TABLE 15-1 The History of Life on Earth

Era	Period	Epoch	Millions of Years Ago	Major Events
Precambrian			4600	Origin of solar system and Earth
			4000–3900	Appearance of first rocks on Earth
			3900–3500	First living cells (prokaryotes)
			3500	Origin of photosynthesis (in cyanobacteria)
			2200	Accumulation of free oxygen in atmosphere
			2000–1700	First eukaryotes
			By 1000	First multicellular organisms
			About 1000	First animals (soft-bodied marine invertebrates)
Paleozoic	Cambrian		544–505	Primitive marine algae flourish; origin of most marine invertebrate types; first fishes
	Ordovician		505–440	Invertebrates, especially arthropods and mollusks, dominant in sea; first fungi
	Silurian		440–410	Many fishes, trilobites, mollusks in sea; first vascular plants; invasion of land by plants; invasion of land by arthropods
	Devonian		410–360	Fishes and trilobites flourish in sea; first amphibians and insects; first seeds and pollen
	Carboniferous		360–286	Swamp forests of tree ferns and club mosses; first conifers; dominance of amphibians; numerous insects; first reptiles
	Permian		286–245	Massive marine extinctions, including last of trilobites; flourishing of reptiles and decline of amphibians; continents aggregated into one landmass, Pangaea
Mesozoic	Triassic		245–208	First mammals and dinosaurs; forests of gymnosperms and tree ferns; breakup of Pangaea begins
	Jurassic		208–146	Dominance of dinosaurs and conifers; first birds; continents partially separated
	Cretaceous		146–65	Flowering plants appear and become dominant; mass extinctions of marine life and some terrestrial life, including last dinosaurs; modern continents well separated
Cenozoic	Tertiary	Paleocene Eocene Oligocene Miocene Pliocene	65–54 54–38 38–23 23–5 5–1.8	Widespread flourishing of birds, mammals, insects, and flowering plants; shifting of continents into modern positions; mild climate at beginning of period, with extensive mountain building and cooling toward end
	Quaternary	Pleistocene Recent	1.8–0.01 0.01–present	Evolution of genus *Homo*; repeated glaciations in Northern Hemisphere; extinction of many giant mammals

Source: University of California Museum of Paleontology, April 2000.

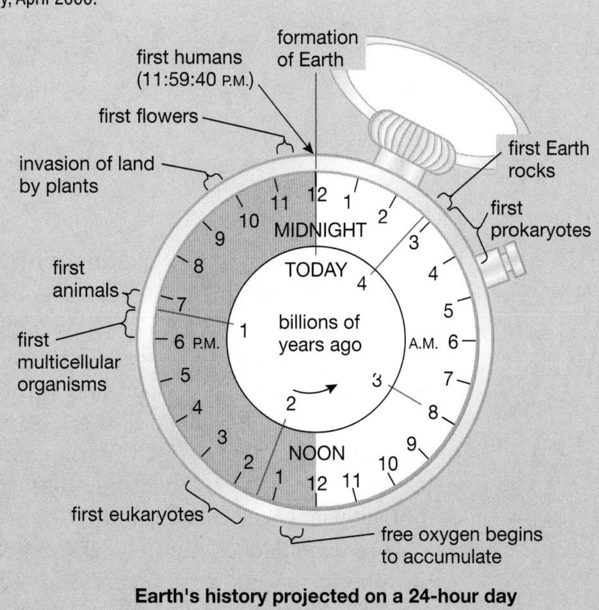

Earth's history projected on a 24-hour day

SCIENTIFIC INQUIRY

HOW DO WE KNOW HOW OLD A FOSSIL IS?

Early geologists could date rock layers and their accompanying fossils only in a *relative* way: Fossils found in deeper layers of rock were generally older than those found in shallower layers. With the discovery of radioactivity, it became possible to determine *absolute* dates, within certain limits.

To determine absolute dates, scientists take advantage of their knowledge that the nuclei of radioactive elements spontaneously break down, or decay, into other elements. For example, carbon-14 (usually written ^{14}C) decays into nitrogen-14 (^{14}N). Each radioactive element decays at a characteristic rate that is independent of temperature, pressure, or the chemical compound of which the element is a part. The time it takes for half the radioactive nuclei to decay is called the *half-life*. The half-life of ^{14}C, for example, is 5730 years. Thus, it takes 5730 years for half of the ^{14}C in a sample to decay into ^{14}N.

How are radioactive elements used to determine the age of rocks? If we know the rate of decay and measure the proportion of decayed nuclei to undecayed nuclei, we can estimate how much time has passed since these radioactive elements were trapped in rock. This process is called *radiometric dating*.

A particularly straightforward dating technique uses the decay of potassium-40 (^{40}K), which has a half-life of about 1.25 billion years, into argon-40 (^{40}Ar). Potassium is a very reactive element commonly found in volcanic rocks such as granite and basalt. Argon, however, is an unreactive gas. Let us suppose that a volcano erupts with a massive lava flow, covering the countryside. All the ^{40}Ar, being a gas, will bubble out of the molten lava, so when the lava cools and solidifies into rock, it will start out with no ^{40}Ar. Any ^{40}K present in the hardened lava will decay to ^{40}Ar, with half of the ^{40}K decaying every 1.25 billion years. The ^{40}Ar gas will be trapped in the rock. A geologist could take a sample of the rock and determine the proportion of ^{40}K to ^{40}Ar (Fig. E15-1). If the analysis finds equal amounts of the two elements, the geologist will conclude that the lava hardened 1.25 billion years ago. With appropriate care, such age estimates are quite reliable. If a fossil is found beneath a lava flow dated at, say, 500 million years, then we know that the fossil is at least that old.

Some radioactive elements, as they decay, can even give an estimate of the age of the solar system. Analysis of uranium, which decays to lead, has shown that the oldest meteorites and moon rocks collected by astronauts are about 4.6 billion years old.

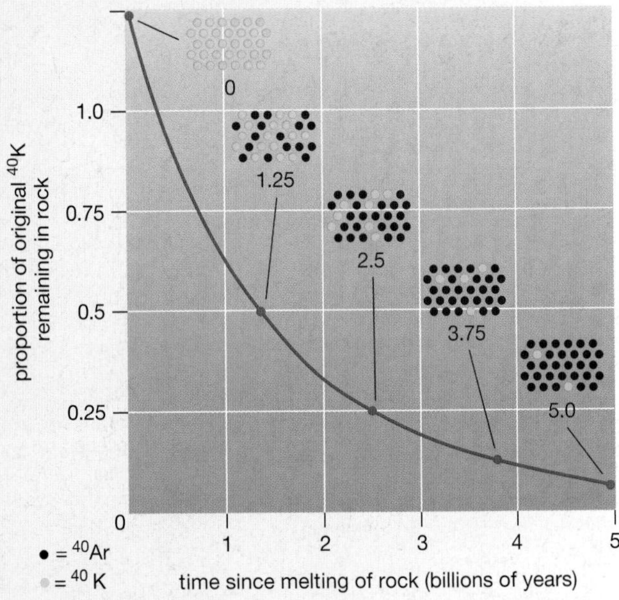

Figure E15-1 The relationship between time and the decay of radioactive ^{40}K to ^{40}Ar EXERCISE: *Uranium-235, with a half-life of 713 million years, decays to lead-207. If you analyze a rock and find that it contains uranium-235 and lead-207 in a ratio of 3:1, how old is the rock?*

was quickly consumed by reactions with other molecules in the atmosphere and in Earth's crust, or surface layer. One especially common reactive atom in the crust was iron, and much of the new oxygen combined with iron atoms to form huge deposits of iron oxide (also known as rust).

After all the accessible iron had turned to rust, the concentration of oxygen gas in the atmosphere began to increase. Chemical analysis of rocks suggests that significant amounts of oxygen first appeared in the atmosphere about 2.2 billion years ago, produced by bacteria that were probably very similar to modern cyanobacteria. (You will undoubtedly breathe in some oxygen molecules today that were expelled 2 billion years ago by one of these early cyanobacteria.) Atmospheric oxygen levels increased steadily until they reached a stable level about 1.5 billion years ago. Since that time, the proportion of oxygen in the atmosphere has been nearly constant, as the amount of oxygen released by photosynthesis worldwide is neatly balanced by the amount that is consumed by aerobic respiration.

Aerobic Metabolism Arose in Response to the Oxygen Crisis

Oxygen is potentially very dangerous to living things, because it reacts with organic molecules, destroying them. Many of today's anaerobic bacteria perish when exposed to what is for them a deadly poison, oxygen. The accumulation of oxygen in the atmosphere of early Earth probably exterminated many organisms and fostered the evolution of cellular mechanisms for detoxifying oxygen. This crisis for evolving life also provided the environmental pressure for the next great advance in the Age of Microbes: the ability to use oxygen in metabolism. This ability not only provides a defense against the chemical action of oxygen, but actually channels oxygen's destructive power through aerobic respiration to generate useful energy for the cell. Because the amount of energy available to a cell is vastly increased when oxygen is used to metabolize food molecules, aerobic cells had a significant selective advantage.

Some Organisms Acquired Membrane-Enclosed Organelles

Hordes of bacteria would offer a rich food supply to any organism that could eat them. There are no fossils of the first predatory cells to roam the seas, but paleobiologists speculate that once a suitable prey population (such as these bacteria) appeared, predation would have evolved quickly. These early predators would have been specialized prokaryotic cells that were able to engulf whole bacteria as prey. According to the most widely accepted hypothesis, these predators were otherwise quite primitive, being capable of neither photosynthesis nor aerobic metabolism. Although they could capture large food particles, namely bacteria, they metabolized them inefficiently. By about 1.7 billion years ago, however, one predator probably gave rise to the first eukaryotic cell.

As you know, the cells of **eukaryotes** differ from prokaryotic cells not only in having a nucleus that contains the cell's genetic material but in other ways as well. Other key eukaryotic structures are the organelles used for energy metabolism: mitochondria and (in plants only) chloroplasts. How did these organelles evolve?

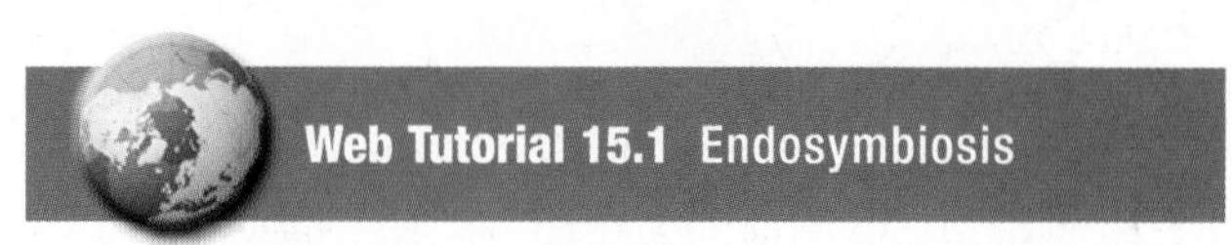

Mitochondria and Chloroplasts May Have Arisen from Engulfed Bacteria

The **endosymbiont hypothesis** proposes that primitive cells acquired the precursors of mitochondria and chloroplasts by engulfing certain types of bacteria. These cells and the bacteria trapped inside them (*endo* means "within") gradually entered into a *symbiotic* relationship, a close association between different types of organisms over an extended time. How might this have happened?

Let's suppose that an anaerobic predatory cell captured an aerobic bacterium for food, as it often did, but for some reason failed to digest this particular prey. The aerobic bacterium remained alive and well. In fact, it was better off than ever, because the cytoplasm of its predator-host was chock-full of half-digested food molecules, the remnants of anaerobic metabolism. The aerobe absorbed these molecules and used oxygen to metabolize them, thereby gaining enormous amounts of energy. So abundant were the aerobe's food resources, and so bountiful its energy production, that the aerobe must have leaked energy, probably as ATP or similar molecules, back into its host's cytoplasm. The anaerobic predatory cell with its symbiotic bacteria could now metabolize food aerobically, gaining a great advantage over other anaerobic cells and leaving a greater number of offspring. Eventually, the endosymbiotic bacterium lost its ability to live independently of its host, and the mitochondrion was born (Fig. 15-4, ① and ②).

One of these successful new cellular partnerships must have managed a second feat: It captured a photosynthetic cyanobacterium and similarly failed to digest its prey. The cyanobacterium flourished in its new host and gradually evolved into the first chloroplast (Fig. 15-4, ③ and ④). Other eukaryotic organelles may have also originated through endosymbiosis. Many biologists believe that cilia, flagella, centrioles, and microtubules may all have evolved from a symbiosis between a spirilla-like bacterium (a form of bacterium with an elongated corkscrew shape) and a primitive eukaryotic cell.

Figure 15-4 The probable origin of mitochondria and chloroplasts in eukaryotic cells **QUESTION:** *Scientists have identified a living bacterium believed to be descended from the endosymbiont that gave rise to mitochondria. Would you expect the DNA sequence of this modern bacterium to be most similar to the sequence of DNA from a plant chloroplast, an animal cell nucleus, or a plant mitochondrion?*

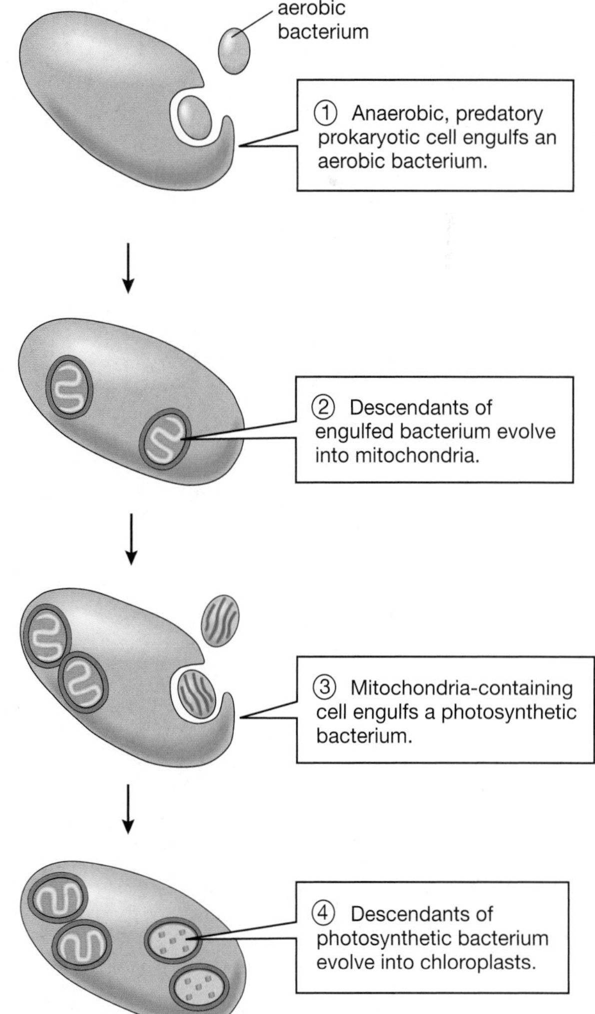

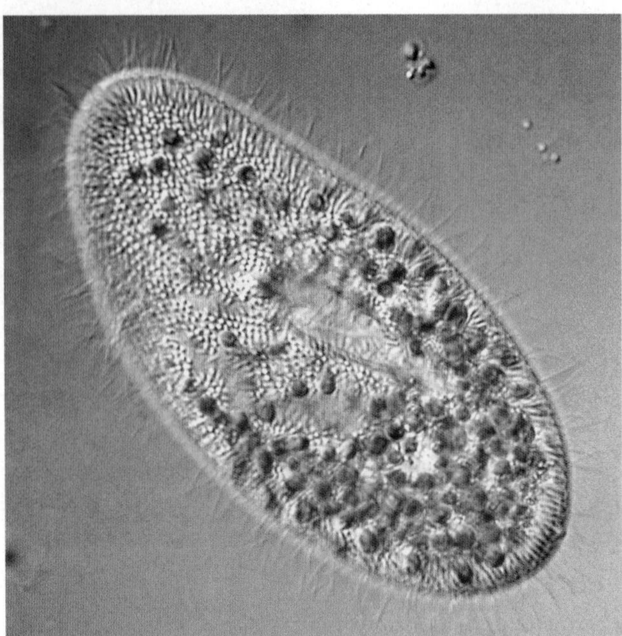

Figure 15-5 Symbiosis within a modern cell The ancestors of the chloroplasts in today's plant cells may have resembled *Chlorella*, the green, photosynthetic, single-celled algae living symbiotically within the cytoplasm of the *Paramecium* pictured here.

Evidence for the Endosymbiont Hypothesis Is Strong

Several types of evidence support the endosymbiont hypothesis. A particularly compelling line of evidence is the many distinctive biochemical features shared by eukaryotic organelles and living bacteria. In addition, mitochondria, chloroplasts, and centrioles each contain their own minute supply of DNA, which many researchers interpret as remnants of the DNA originally contained within the engulfed bacteria.

Another kind of support comes from *living intermediates*, organisms alive today that are similar to hypothetical ancestors and thus help show that a proposed evolutionary pathway is plausible. For example, the amoeba *Pelomyxa palustris* lacks mitochondria but hosts a permanent population of aerobic bacteria that carry out much the same role. Similarly, a variety of corals, some clams, a few snails, and at least one species of *Paramecium* harbor a permanent collection of photosynthetic algae in their cells (Fig. 15-5). These examples of modern cells that host bacterial endosymbionts suggest that we have no reason to doubt that similar symbiotic associations could have occurred almost 2 billion years ago and led to the first eukaryotic cells.

15.3 What Were the Earliest Multicellular Organisms Like?

Once predation had evolved, increased size became an advantage. In the marine environments to which life was restricted, a larger cell could easily engulf a smaller cell and would also be difficult for other predatory cells to ingest. Most larger organisms can also move faster than small ones, making successful predation and escape more likely. But enormous single cells have problems. Oxygen and nutrients going into the cell and waste products going out must diffuse through the plasma membrane. The larger a cell becomes, the less surface membrane is available per unit volume of cytoplasm.

There are only two ways that an organism larger than a millimeter or so in diameter can survive. First, it can have a low metabolic rate so that it doesn't need much oxygen or produce much carbon dioxide. This strategy seems to work for certain very large single-celled algae. Alternatively, an organism can be multicellular; that is, it may consist of many small cells packaged into a larger, unified body.

Some Algae Became Multicellular

The oldest fossils of multicellular organisms are about 1 billion years old and include impressions of the first multicellular algae, which arose from single-celled eukaryotic cells containing chloroplasts. Multicellularity would have provided at least two advantages for these seaweeds. First, large, many-celled algae would have been difficult for single-celled predators to engulf. Second, specialization of cells would have provided the potential for staying in one place in the brightly lit waters of the shoreline, as rootlike structures burrowed in sand or clutched onto rocks, while leaflike structures floated above in the sunlight. The green, brown, and red algae lining our shores today—some, such as the brown kelp, more than 200 feet long—are the descendants of these early multicellular algae.

Animal Diversity Arose in the Precambrian Era

In addition to fossil algae, billion-year-old rocks have yielded fossil traces of animal tracks and burrows. This evidence of early animal life notwithstanding, fossils of animal bodies first appear in Precambrian rocks laid down between 610 million and 544 million years ago. Some of these ancient invertebrate animals

Figure 15-6 Diversity of ocean life during the Silurian period
(a) Life characteristic of the oceans during the Silurian period, 440 million to 410 million years ago. Among the most common fossils from that time are ***(b)*** the trilobites and their predators, the nautiloids, and ***(c)*** the ammonites. ***(d)*** This living *Nautilus* is very similar in structure to the Silurian nautiloids, showing that a successful body plan may exist virtually unchanged for hundreds of millions of years.

(animals lacking a backbone) are quite different in appearance from any animals that appear in later fossil layers and may represent types of animals that left no descendants. Others fossils in these rock layers, however, appear to be ancestors of today's animals. Ancestral sponges and jellyfish appear in the oldest layers, followed later by ancestors of worms, mollusks, and arthropods.

The full range of modern invertebrate animals, however, does not appear in the fossil record until the Cambrian period, marking the beginning of the Paleozoic era, about 544 million years ago. (The phrase "fossil record" is a shorthand reference to the entire collection of all fossil evidence that has been found to date.) These Cambrian fossils reveal an adaptive radiation that had already yielded a diverse array of complex body plans. Almost all of the major groups of animals on Earth today were already present in the early Cambrian. The apparently sudden appearance of so many different kinds of animals suggests that the earlier evolutionary history that produced such an impressive range of different animal forms is not preserved in the fossil record.

The early diversification of animals was probably driven in part by the emergence of predatory lifestyles. Coevolution of predator and prey led to the evolution of new features in many kinds of animals. By the Silurian period (440 million to 410 million years ago), mud-skimming, armored trilobites were preyed on by ammonites and the chambered nautilus, which still survives almost in almost unchanged form in deep Pacific waters (Fig. 15-6).

Many animals of the Paleozoic era were more mobile than their evolutionary predecessors. Predators gain an advantage by being able to travel over wide areas in search of suitable prey, and the ability to make a speedy escape is an advantage for prey. The evolution of efficient movement was often associated with the evolution of greater sensory capabilities and more complex nervous systems. Senses for detecting touch, chemicals, and light became highly developed, along with nervous systems capable of handling the sensory information and directing appropriate behaviors.

About 530 million years ago, one group of animals—the fishes—developed a new form of support for the body: an internal skeleton. These early fishes were inconspicuous members of the ocean community, but by 400 million years ago fishes were a diverse and prominent group. By and large, the fishes proved to be faster than the invertebrates, with more-acute senses and larger brains. Eventually, they became the dominant predators of the open seas.

15.4 How Did Life Invade the Land?

A compelling subplot in the long tale of life's history is the story of life's invasion of land after more than 3 billion years of a strictly watery existence. In moving to solid ground, organisms had many obstacles to overcome. Life in the sea provides buoyant support against gravity, but on land an organism must bear its weight against the crushing force of gravity. The sea provides ready access to life-sustaining water, but a terrestrial organism must find adequate water. Sea-dwelling plants and animals can reproduce by means of mobile sperm or eggs, or both, which swim to each other through the water, but the gametes of land-dwellers must be protected from drying out.

Despite the obstacles to life on land, the vast empty spaces of the Paleozoic landmass represented a tremendous evolutionary opportunity. The potential rewards of terrestrial life were especially great for plants. Water strongly absorbs light, so even in the clearest water, photosynthesis is limited to the upper few hundred meters of depth, and usually much less. Out of the water, the dazzling brightness of the sun permits rapid photosynthesis. Furthermore, terrestrial soils are rich storehouses of nutrients, whereas seawater tends to be low in certain nutrients, particularly nitrogen and phosphorus. Finally, the Paleozoic sea swarmed with plant-eating animals, but the land was devoid of animal life. The plants that first colonized the land would have had ample sunlight, untouched nutrient sources, and no predators.

Some Plants Became Adapted to Life on Dry Land

In moist soils at the water's edge, a few small green algae began to grow, taking advantage of the sunlight and nutrients. They didn't have large bodies to support against the force of gravity, and, living right in the film of water on the soil, they could easily obtain water. About 400 million years ago, some of these algae gave rise to the first multicellular land plants. Initially simple, low-growing forms, land plants rapidly evolved solutions to two of the main difficulties of plant life on land: obtaining and conserving water and staying upright despite gravity and winds. Waterproof coatings on aboveground parts reduced water loss by evaporation, and rootlike structures delved into the soil, mining water and minerals. Specialized cells formed tubes called vascular tissues to conduct water from roots to leaves. Extra-thick walls surrounding certain cells enabled stems to stand erect.

Primitive Land Plants Retained Swimming Sperm and Required Water to Reproduce

Reproduction out of water presented challenges. As do animals, plants produce sperm and eggs, which must be able to meet to produce the next generation. The first land plants had swimming sperm, presumably much like those of some of today's marine algae (some of which have swimming eggs as well). Consequently, the earliest plants were restricted to swamps and marshes, where the sperm and eggs could be released into the water, or to areas with abundant rainfall, where the ground would occasionally be covered with water. Later, plants with swimming sperm prospered during periods in which the climate was warm and moist. For example, the Carboniferous period (360 million to 286 million years ago) was characterized by vast forests of giant tree ferns and club mosses (Fig. 15-7). The coal we mine today is derived from the fossilized remains of those forests.

Figure 15-7 The swamp forest of the Carboniferous period The treelike plants in this artist's reconstruction are tree ferns and giant club mosses, most species of which are now extinct.

Seed Plants Encased Sperm in Pollen Grains

Meanwhile, some plants inhabiting drier regions had evolved a means of reproduction that no longer depended on water. The eggs of these plants were retained on the parent plant, and the sperm were encased in drought-resistant pollen grains that blew on the wind from plant to plant. When the pollen grains landed near an egg, they released sperm cells directly into living tissue, eliminating the need for a surface film of water. The fertilized egg remained on the parent plant, where it developed inside a seed, which provided protection and nutrients for the developing embryo within.

The earliest seed-bearing plants appeared in the late Devonian period (375 million years ago) and produced their seeds along branches, without any specialized structures to hold them. By the middle of the Carboniferous period, however, a new kind of seed-bearing plant had arisen. These plants, called **conifers**, protected their developing seeds inside cones. Conifers, which did not depend on water for reproduction, flourished and spread during the Permian period (286 to 245 million years ago), when mountains rose, swamps drained, and the climate became much drier. The good fortune of the conifers, however, was not shared by the tree ferns and giant club mosses, which, with their swimming sperm, largely went extinct.

Flowering Plants Enticed Animals to Carry Pollen

About 140 million years ago, during the Cretaceous period, the flowering plants appeared, having evolved from a group of conifer-like plants. Many flowering plants are pollinated by insects and other animals, and this mode of pollination seems to have conferred an evolutionary advantage. Flower pollination by animals can be far more efficient than pollination by wind. Wind-pollinated plants must produce an enormous amount of pollen because the vast majority of pollen grains fail to reach their target. Flowering plants also evolved other advantages, including more-rapid reproduction and, in some cases, much more rapid growth. Today, flowering plants dominate the land, except in cold northern regions, where conifers still prevail.

Some Animals Became Adapted to Life on Dry Land

Soon after land plants evolved, providing potential food sources for other organisms, animals emerged from the sea. The first animals to move onto land were **arthropods** (the group that today includes insects, spiders, scorpions, centipedes, and crabs). Why arthropods? The answer seems to be that they already possessed certain structures that, purely by chance, were suited to life on land. Foremost among these structures was the external skeleton, or **exoskeleton**, a hard covering surrounding the body, such as the shell of a lobster or crab. Exoskeletons are both waterproof and strong enough to support a small animal against the force of gravity.

Figure 15-8 A fish that walks on land Some modern fishes, such as this mudskipper, walk on land. As did the ancient lobefin fishes that gave rise to amphibians, mudskippers use their strong pectoral fins to move across dry areas in their swampy habitats. **QUESTION:** *Does the mudskipper's ability to walk on land constitute evidence that lobefin fishes were the ancestors of amphibians?*

For millions of years, arthropods had the land and its plants to themselves, and for tens of millions of years more, they were the dominant land animals. Dragonflies with a wingspan of 28 inches (70 centimeters) flew among the Carboniferous tree ferns, while millipedes 6.5 feet (2 meters) long munched their way across the swampy forest floor. Eventually, however, the arthropods' splendid isolation came to an end.

Amphibians Evolved from Lobefin Fishes

About 400 million years ago, a group of Silurian fishes called the lobefins appeared, probably in freshwater. **Lobefins** had two important features that would later enable their descendants to colonize land: (1) stout, fleshy fins with which they crawled about on the bottoms of shallow, quiet waters, and (2) an outpouching of the digestive tract that could be filled with air, like a primitive lung. One group of lobefins colonized very shallow ponds and streams, which shrank during droughts and often became oxygen poor. By taking air into their lungs, these lobefins could obtain oxygen anyway. Some began to use their fins to crawl from pond to pond in search of prey or water, as some modern fish do today (Fig. 15-8).

The benefits of feeding on land and moving from pool to pool favored the evolution of a group of animals that could stay out of water for longer periods and that could move about more effectively on land. With improvements in lungs and legs, **amphibians** evolved from lobefins, first appearing in the fossil record about 350 million years ago. To an amphibian, the Carboniferous swamp forests were a kind of paradise: no predators to speak of, abundant prey, and a warm, moist climate. As had the insects and millipedes, some amphibians evolved gigantic size, including salamanders more than 10 feet (3 meters) long.

Despite their success, the early amphibians were not fully adapted to life on land. Their lungs were simple sacs without very much surface area, so they had to obtain some of their oxygen through their skin. Therefore, their skin had to be kept moist, a requirement that restricted them to swampy habitats where they wouldn't dry out. Further, amphibian sperm and eggs could not survive in dry surroundings and had to be deposited in watery environments. So, although amphibians could move about on land, they could not stray too far from the water's edge. Along with the tree ferns and club mosses, amphibians declined when the climate turned dry at the beginning of the Permian period about 286 million years ago.

Figure 15-9 A reconstruction of a Cretaceous forest By the Cretaceous period, flowering plants dominated terrestrial vegetation. Dinosaurs, such as the predatory pack of 6-foot-long *Velociraptors* shown here, were the preeminent land animals. Although small by dinosaur standards, *Velociraptors* were formidable predators with great running speed, sharp teeth, and deadly, sickle-like claws on their hind feet.

Reptiles Evolved from Amphibians

As the conifers were evolving on the fringes of the swamp forests, a group of amphibians was also evolving adaptations to drier conditions. These amphibians ultimately gave rise to the **reptiles**, which had three major adaptations to life on land. First, reptiles evolved shelled, waterproof eggs that enclosed a supply of water for the developing embryo. Thus, eggs could be laid on land without the reptiles' having to venture back to the dangerous swamps full of fish and amphibian predators. Second, ancestral reptiles evolved scaly, waterproof skin that helped prevent the loss of body water to the dry air. Finally, reptiles evolved improved lungs that were able to provide the entire oxygen supply for an active animal. As the climate dried during the Permian period, reptiles became the dominant land vertebrates, relegating amphibians to the swampy backwaters where most remain today.

A few tens of millions of years later, the climate returned to more moist and stable conditions. This period saw the evolution of some very large reptiles, in particular the dinosaurs. The variety of dinosaur forms was enormous—from predators (Fig. 15-9) to plant eaters, from those that dominated the land to others that took to the air, to still others that returned to the sea. Dinosaurs were among the most successful animals ever, if we consider persistence as a measure of success. They flourished for more than a hundred million years, until about 65 million years ago, when the last dinosaurs went extinct. No one is certain why they died out, but the aftereffects of a gigantic meteorite's impact with Earth seems to have been the final blow (as discussed in the following section).

Even during the age of dinosaurs, many reptiles remained quite small. One major difficulty faced by small reptiles is maintaining a high body temperature. Being active on land is helped by a warm body that maximizes the efficiency of the nervous system and muscles. But a warm body loses heat to the environment unless the air is also warm. Heat loss is a big problem for small animals, which have a larger surface area per unit of weight than do larger animals. Many species of small reptiles have retained slow metabolisms and have coped with the heat-loss problem by developing lifestyles in which they remain active only when the air is sufficiently warm. Two groups of small reptiles, however, independently followed a different evolutionary pathway: they developed insulation. One group evolved feathers, and another evolved hair.

Reptiles Gave Rise to Both Birds and Mammals

In ancestral birds, insulating feathers helped retain body heat. Consequently, these animals could be active in cool habitats and during the night, when their scaly relatives became sluggish. Later, some ancestral birds evolved longer, stronger feathers on their forelimbs, perhaps under selection for better ability to glide from trees or to jump after insect prey. Ultimately, feathers evolved into structures capable of supporting powered flight. Fully developed, flight-capable feathers are present in 150-million-year-old fossils, so the earlier insulating structures that eventually developed into flight feathers must have been present well before that time.

The earliest mammals coexisted with the dinosaurs but were small creatures, probably living in trees and being active mostly at night. When the dinosaurs went extinct, mammals colonized the habitats left empty by the extinctions. Mammal species prospered, diversifying into the array of modern forms.

Unlike birds, which retained the reptilian habit of laying eggs, mammals evolved live birth and the ability to feed their young with secretions of the mammary (milk-producing) glands. Ancestral mammals also developed hair, which provided insulation. Because the uterus, mammary glands, and hair do not fossilize, we may never know when these structures first appeared, or what their intermediate forms looked like. Recently, however, a team of paleontologists found bits of fossil hair preserved in coprolites, which are fossilized animal feces. These coprolites, found in the Gobi Desert of China, were deposited by an anonymous predator 55 million years ago, so mammals have presumably had hair at least that long.

15.5 What Role Has Extinction Played in the History of Life?

If there is a moral to the great tale of life's history, it is that nothing lasts forever. The story of life can be read as a long series of evolutionary dynasties, with each new dominant group rising, ruling the land or the seas for a time and, inevitably, falling into decline and extinction. Dinosaurs are the most famous of these fallen dynasties, but the list of extinct groups known only from fossils is impressively long. Despite the inevitability of extinction, however, the overall trend has been for species to arise at a faster rate than they disappear, so the number of species on Earth has tended to increase over time.

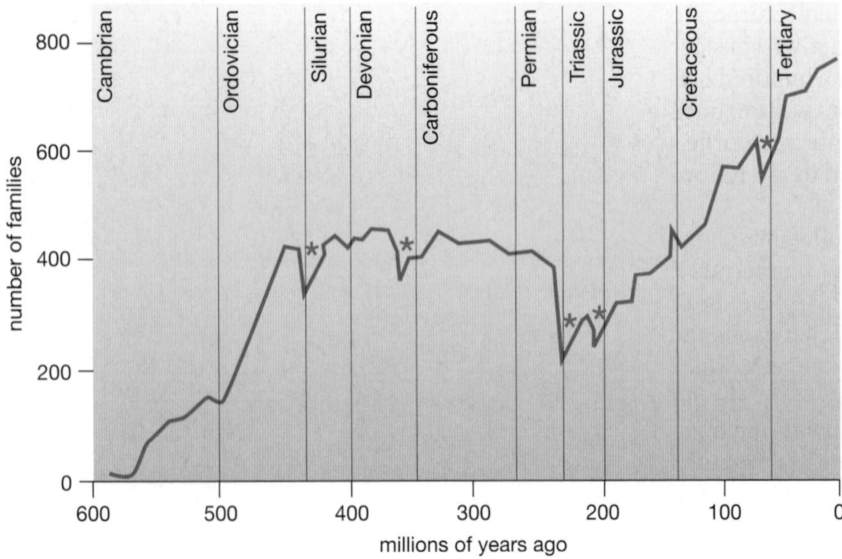

Figure 15-10 Mass extinctions This graph plots number of marine-animal groups against time, as reconstructed from the fossil record. Note the general trend toward an increasing number of groups, punctuated by periods of sometimes rapid extinction. Five of these declines, marked by blue asterisks, are so steep that they qualify as catastrophic mass extinctions. **QUESTION:** *If extinction is the ultimate fate of all species, how can the total number of species have increased over time?*

Evolutionary History Has Been Marked by Periodic Mass Extinctions

Over much of life's history, dynastic succession has proceeded in a steady, relentless manner. This slow and steady turnover of species, however, has been interrupted by episodes of **mass extinction** (Fig. 15-10). These mass extinctions are characterized by the relatively sudden disappearance of a wide variety of species over a large part of Earth. In the most catastrophic episodes of extinction, more than half of the planet's species disappeared. The worst episode of all, which occurred 245 million years ago at the end of the Permian period, wiped out more than 90% of the world's species, and life came perilously close to disappearing altogether.

Climate Change Contributed to Mass Extinctions

Mass extinctions have had a profound impact on the course of life's history, repeatedly redrawing the picture of life's diversity. What could have caused such dramatic changes in the fortunes of so many species? Many evolutionary biologists believe that changes in climate must have played an important role. When the climate changes, as it has done many times over the course of Earth's history, organisms that are adapted for survival in one climate may be unable to survive under in a drastically different climate. In particular, at times when warm climates gave way to drier, colder climates with more variable temperatures, species may have gone extinct after failing to adapt to the harsh new conditions.

One cause of climate change is the changing positions of continents. Earth's surface is divided into portions called plates, which include the continents and the seafloor. The solid plates slowly move above a viscous but fluid layer. This movement is called **plate tectonics**. As the plates wander, their positions may change in latitude (Fig. 15-11). For example, 350 million years ago much of North America was located at or near the equator, an area characterized by consistently warm and wet tropical weather. But plate tectonics carried the continent up into temperate and arctic regions. As a result, the tropical climate was replaced by a regime of seasonal changes, cooler temperatures, and less rainfall. Plate tectonics continues today; the Atlantic Ocean, for example, widens by a few centimeters each year.

Catastrophic Events May Have Caused the Worst Mass Extinctions

Geological data indicate that most mass extinction events coincided with periods of climatic change. To many scientists, however, the rapidity of mass extinctions suggests that the slow process of climate change could not, by itself, be responsible for such large-scale disappearances of species. Perhaps more sudden events also play a role. For example, catastrophic geological events, such as massive volcanic eruptions, could have had devastating effects. Geologists have found evidence of past volcanic eruptions so huge that they make the 1980 Mount St. Helens explosion look like a firecracker by comparison. Even such gigantic eruptions, however, would directly affect only a relatively small portion of Earth's surface.

The search for the causes of mass extinctions took a fascinating turn in the early 1980s when Luis and Walter Alvarez proposed that the extinction event of 65 million years ago, which wiped out the dinosaurs and many other species, was caused by the impact of a huge meteorite. The Alvarezes' idea was met with great skepticism when it was first introduced, but geological research since that time has generated a great deal of evidence that a massive impact did indeed occur 65 million years ago. In fact, researchers have identified the Chicxulub crater, a 100-mile-wide crater buried beneath the Yucatan Peninsula of Mexico, as the impact site of a giant meteorite, 10 miles in diameter, that collided with Earth just at the time that dinosaurs disappeared.

Could this immense meteorite strike have caused the mass extinction that coincided with it? No one knows for sure, but scientists suggest that such a massive impact would have thrown so much debris into the atmosphere that the entire planet would have been plunged into darkness for a period of years. With little light reaching the planet, temperatures would have dropped precipitously and the photosynthetic capture of energy (upon which all life ultimately depends) would have declined drastically. The worldwide "impact winter" would have spelled doom for the dinosaurs and a host of other species.

15.6 How Did Humans Evolve?

Scientists are intensely interested in the origin and evolution of humans, especially in the evolution of the gigantic human brain. The outline of human evolution that we present in this section is a synthesis of current thought on the subject, but we note that it is speculative, because the fossil evidence of human evolution is comparatively scarce. Paleontologists disagree about the interpretation of the fossil evidence, and many ideas may have to be revised as new fossils are found.

Humans Inherited Some Early Primate Adaptations for Life in Trees

Humans are members of a mammal group known as **primates**, which also includes lemurs, monkeys, and apes. The oldest primate fossils are 55 million years old, but because primate fossils are relatively rare compared with those of many other animals, the first primates probably arose considerably earlier but left no fossil record. Early primates probably fed on fruits and leaves and were adapted for life in the trees. Many modern primates retain the tree-dwelling lifestyle of their ancestors (Fig. 15-12). The common heritage of humans and other primates is reflected in a set of physical characteristics that was present in the earliest primates and that persists in many modern primates, including humans.

Binocular Vision Provided Early Primates with Accurate Depth Perception

One of the earliest primate adaptations seems to have been large, forward-facing eyes (see Fig. 15-12). Jumping from branch to branch is risky business unless an animal can accurately judge where the next branch is located. Accurate depth perception was made possible by binocular vision, provided by forward-facing eyes with overlapping fields of view. Another key adaptation was color vision. We cannot, of course, tell if a fossil animal had color vision, but since modern primates

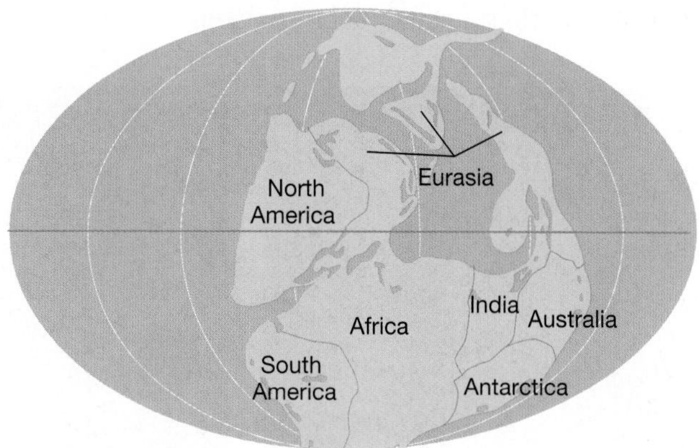

(a) 340 million years ago

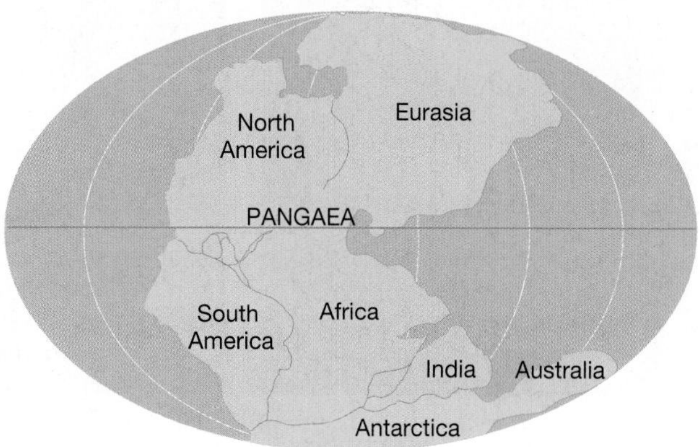

(b) 225 million years ago

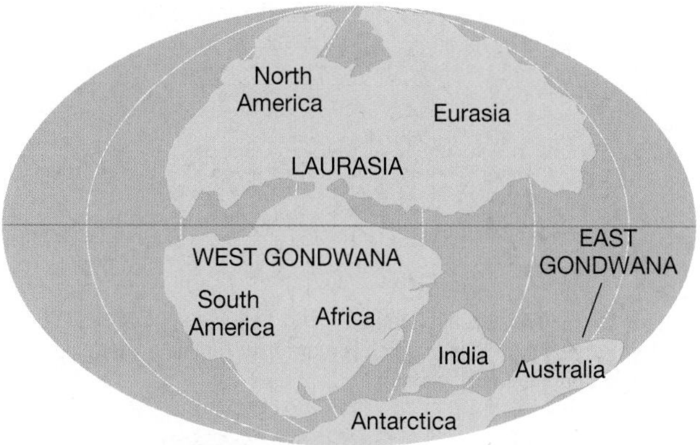

(c) 135 million years ago

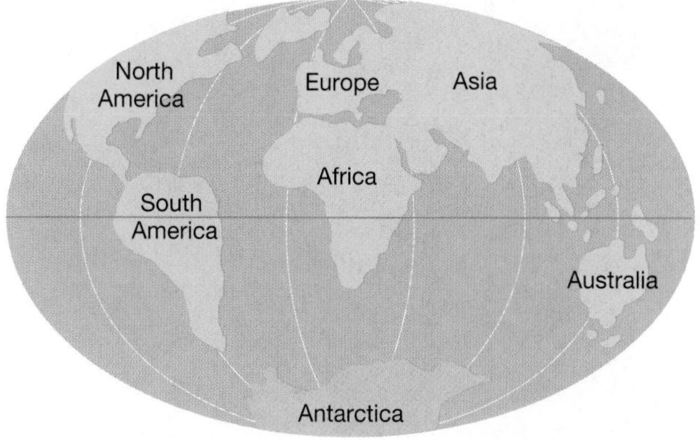

(d) Present

Figure 15-11 Continental drift from plate tectonics The continents are passengers on plates moving on Earth's surface as a result of plate tectonics. ***(a)*** About 340 million years ago, much of what is now North America was positioned at the equator. ***(b)*** All the plates eventually fused together into one gigantic landmass, which geologists call Pangaea. ***(c)*** Gradually Pangaea broke up into Laurasia and Gondwanaland, which itself eventually broke up into West and East Gondwana. ***(d)*** Further plate motion eventually resulted in the modern positions of the continents.

Figure 15-12 Representative primates The *(a)* tarsier, *(b)* lemur, and *(c)* lion-tail macaque monkey all have a relatively flat face, with forward-looking eyes providing binocular vision. All also have color vision and grasping hands. These features, retained from the earliest primates, are shared by humans.

have excellent color vision, it seems reasonable to assume that earlier primates did too. Many primates feed on fruit, and color vision helps to detect ripe fruit among a bounty of green leaves.

Early Primates Had Grasping Hands

Early primates had long, grasping fingers that could wrap around and hold onto tree limbs. This adaptation to tree dwelling was the basis for later evolution of human hands that could perform both a precision grip (used by modern humans for delicate maneuvers such as manipulating small objects, writing, and sewing) and a power grip (used for powerful actions, such as swinging a club or thrusting with a spear).

A Large Brain Facilitated Hand-Eye Coordination and Complex Social Interactions

Primates have brains that are larger, relative to their body size, than the brains of almost all other animals. No one really knows for certain which environmental forces favored the evolution of large brains. It seems reasonable, however, that controlling and coordinating rapid locomotion through trees, dexterous movements of the hands, and binocular, color vision would be facilitated by increased brain power. Most primates also have fairly complex social systems, which probably require relatively high intelligence. If sociality promoted increased survival and reproduction, then there would have been environmental pressures for the evolution of larger brains.

The Oldest Hominid Fossils Are from Africa

On the basis of comparisons of DNA from modern chimps, gorillas, and humans, researchers estimate that the **hominid** line (humans and their fossil relatives) diverged from the ape lineage sometime between 5 million and 8 million years ago. The fossil record, however, suggests that the split must have occurred at the early end of that range. Paleontologists working in the African country of Chad in 2002 discovered fossils of a hominid, *Sahelanthropus tchadensis*, that lived more than 6 million years ago (Fig. 15-13). *Sahelanthropus* is clearly a hominid, as it shares several anatomical features with later members of the group. But because this oldest known member of our family also exhibits other features that are more characteristic of apes, it may represent a point on our family tree that is close to the split between apes and hominids.

In addition to *Sahelanthropus,* two other hominid species, *Ardipithecus ramidus* and *Orrorin tugenensis*, are known from fossils appearing in rocks that are between 4 million and 6 million years old. Our knowledge of these hominids is limited, because only a few specimens have been found so far, mostly in recent discoveries that typically include only small portions of skeletons. A more extensive record of early hominid evolution does not begin until about 4 million years ago. That date marks the beginning of the fossil record of the genus *Australopithecus* (Fig. 15-14), a group of African hominid species with brains larger than those of their prehominid forebears but still much smaller than those of modern humans.

Figure 15-13 The earliest hominid This nearly complete skull of *Sahelanthropus tchadensis*, which is more than 6 million years old, is the oldest hominid fossil yet found.

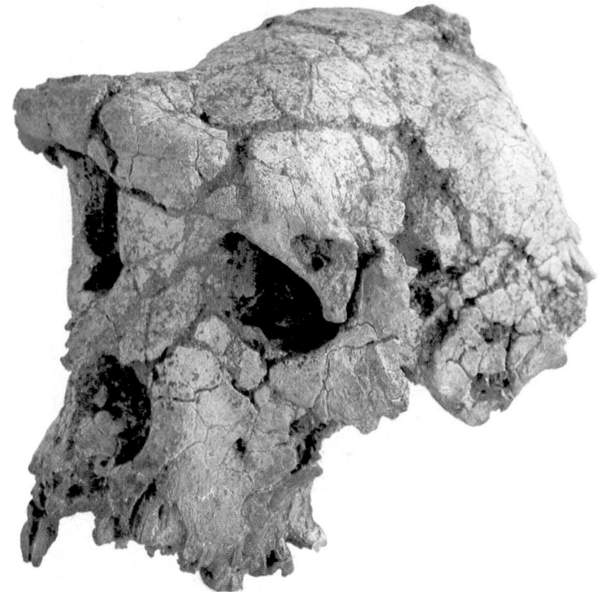

The Earliest Hominids Could Stand and Walk Upright

The earliest australopithecines (as the various species of *Australopithecus* are collectively known) had legs that were shorter, relative to their height, than those of modern humans, but their knee joints allowed them to straighten their legs fully, permitting efficient bipedal (upright, two-legged) locomotion. Footprints

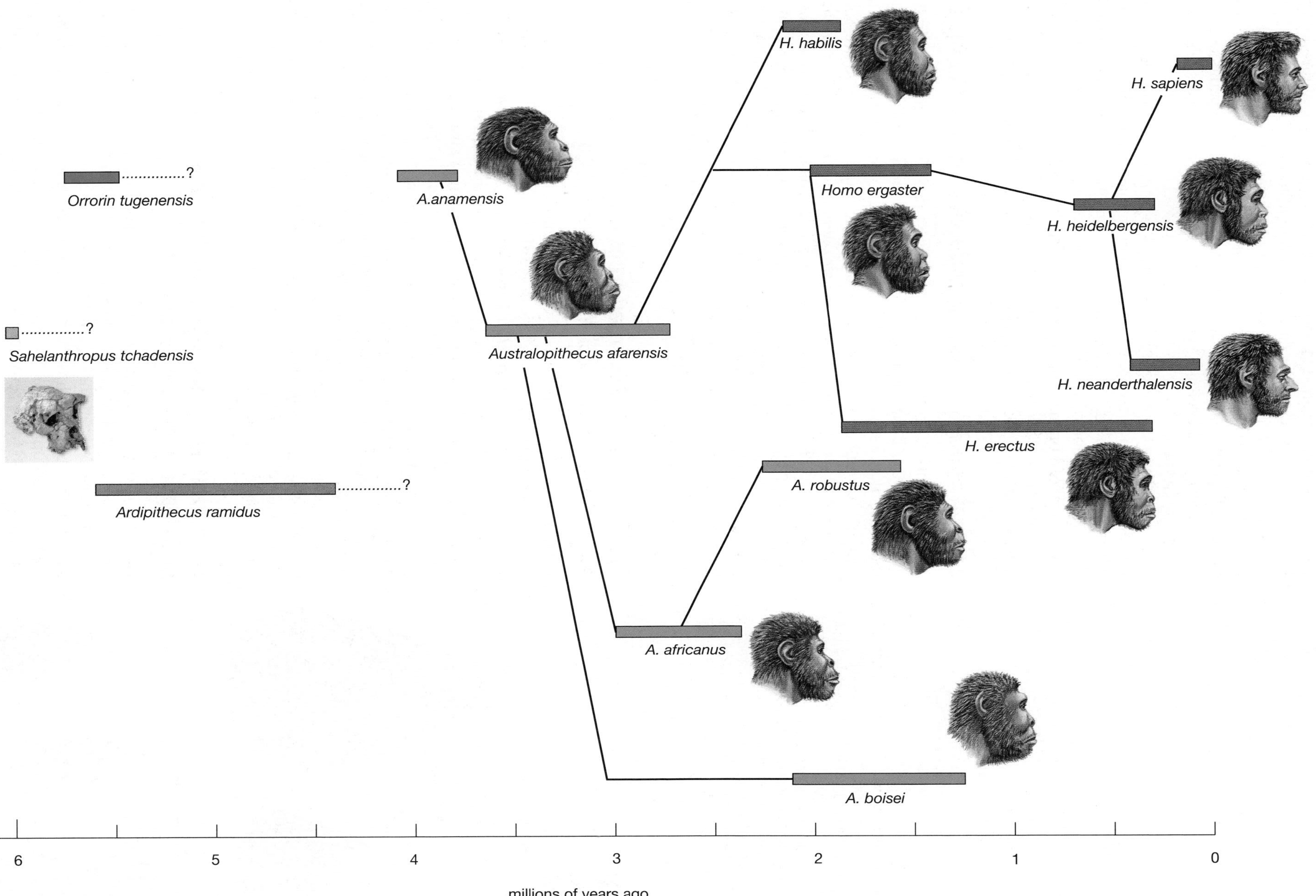

Figure 15-14 A possible evolutionary tree for humans This hypothetical family tree shows facial reconstructions of representative specimens. Although many paleontologists consider this to be the most likely human family tree, there are several alternative interpretations of the known hominid fossils. Fossils of the earliest hominids are scarce and fragmentary, so the relationship of these species to later hominids remains unknown.

almost 4 million years old, discovered in Tanzania by anthropologist Mary Leakey, show that even the earliest australopithecines could, and at least sometimes did, walk upright. Upright posture may have evolved even earlier. The discoverers of *Sahelanthropus* and *Orrorin* argue that the leg and foot bones of these earliest hominids have characteristics that indicate bipedal locomotion, but this conclusion will remain speculative until more complete skeletons of these species are found.

The reasons for the evolution of bipedal locomotion among the early hominids remain poorly understood. Perhaps hominids that could stand upright gained an advantage in gathering or carrying food in their forest habitat. Whatever its cause, the early evolution of upright posture was extremely important in the evolutionary history of hominids, because it freed their hands from use in walking. Later hominids were thus able to carry weapons, manipulate tools, and eventually achieve the cultural revolutions produced by modern *Homo sapiens*.

Several Species of *Australopithecus* Emerged in Africa

The oldest australopithecine species, represented by fossilized teeth, skull fragments, and arm bones, was unearthed near an ancient lake bed in Kenya from sediments that were dated, by means of radioactive isotopes, as between 3.9 million and 4.1 million years old (see "Scientific Inquiry: How Do We Know How Old a Fossil Is?"). It was named *Australopithecus anamensis* by its discoverers (*anam* means "lake" in the local Ethiopian language). The second most ancient australopithecine, called *Australopithecus afarensis*, was discovered in the Afar region of Ethiopia. Fossil remains of this species as old as 3.9 million years have been unearthed. The *A. afarensis* line apparently gave rise to other species. All of the australopithecine species had apparently gone extinct by 1.2 million years ago, but one of them (*A. afarensis* in the interpretation shown in Fig. 15-14) first gave rise to a new branch of the hominid family tree, the genus *Homo*.

Figure 15-15 Representative hominid tools ***(a)*** *Homo habilis* produced only fairly crude chopping tools called *hand axes*, usually unchipped on one end to hold in the hand. ***(b)*** *Homo ergaster* manufactured much finer tools. The tools were typically sharp all the way around the stone, so at least some of these blades were probably tied to spears rather than held in the hand. ***(c)*** Neanderthal tools were works of art, with extremely sharp edges made by flaking off tiny bits of stone. In comparing these weapons, note the progressive increase in the number of flakes taken off the blades and the corresponding decrease in flake size. Smaller, more numerous flakes produce a sharper blade and suggest more insight into tool making, more patience, finer control of hand movements, or perhaps all three.

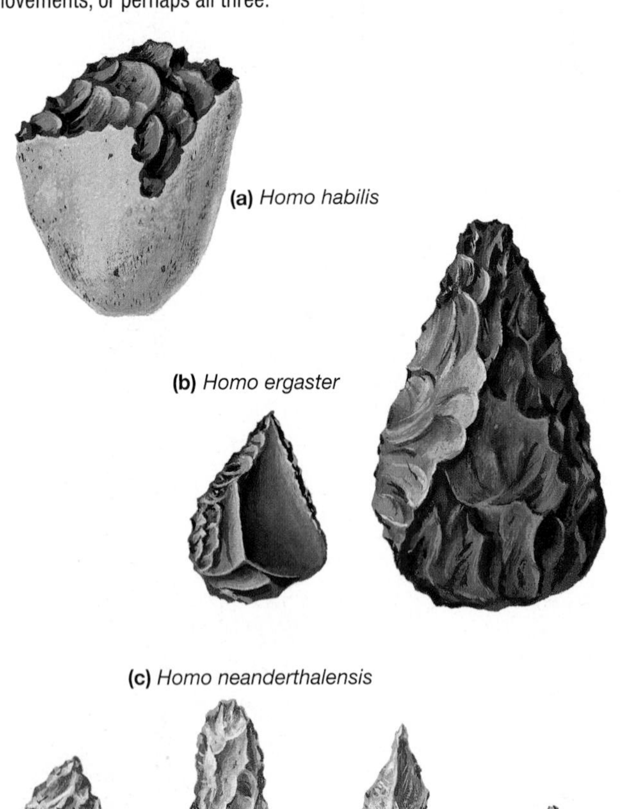

The Genus *Homo* Diverged from the Australopithecines 2.5 Million Years Ago

Hominids that are sufficiently similar to modern humans to be placed in the genus *Homo* first appear in African fossils that are about 2.5 million years old. Among the earliest African *Homo* fossils are *H. habilis* (see Fig. 15-14), a species whose body and brain were larger than those of the australopithecines but which retained the apelike long arms and short legs of their australopithecine ancestors. In contrast, the skeletal anatomy of *H. ergaster*, a species whose fossils first appear 2 million years ago, has limb proportions more like those of modern humans. This species is believed by many paleoanthropologists (scientists who study human origins) to be on the evolutionary branch that led ultimately to our own species, *H. sapiens*. In this view, *H. ergaster* was the common ancestor of two distinct branches of hominids. The first branch led to *H. erectus*, which was the first hominid species to leave Africa. The second branch from *H. ergaster* ultimately led both to the species commonly known as the Neanderthals and to *H. sapiens*, modern humans.

The Evolution of *Homo* Was Accompanied by Advances in Tool Technology

Hominid evolution is closely tied to the development of tools, a hallmark of hominid behavior. The oldest tools discovered so far were found in 2.5-million-year-old East African rocks, concurrent with the early emergence of the genus *Homo*. Early *Homo*, whose cheek teeth were much smaller than those of the genus's australopithecine ancestors, might have first used stone tools to break and crush tough foods that were hard to chew. Hominids constructed their earliest tools by striking one rock with another to chip off fragments and leave a sharp edge behind. Over the next several hundred thousand years, tool-making techniques in Africa gradually became more advanced. By 1.7 million years ago, tools were more sophisticated, with flakes chipped symmetrically from both sides of a rock to form double-edged tools ranging from hand axes, used for cutting and chopping, to points, probably used on spears (**Fig. 15-15a,b**). *Homo ergaster* and other bearers of these weapons presumably ate meat, probably acquired from both hunting and scavenging for the remains of prey killed by other predators. Double-edged tools were carried to Europe at least 600,000 years ago by migrating hominid populations, and the Neanderthal descendants of these emigrants took stone-tool construction to new heights of skill and delicacy (**Fig. 15-15c**).

Neanderthals Had Large Brains and Excellent Tools

Neanderthals first appeared in the European fossil record about 150,000 years ago and, by about 70,000 years ago, had spread throughout Europe and western Asia. By 30,000 years ago, however, Neanderthals were extinct. Contrary to the popular image of a hulking, stoop-shouldered "caveman," Neanderthals were quite similar to modern humans in many ways. Although more heavily muscled, Neanderthals walked fully erect, were dexterous enough to manufacture finely crafted stone tools, and had brains that, on average, were slightly larger than those of modern humans. Many European Neanderthal fossils show heavy brow ridges and a broad, flat skull, but others, particularly from areas around the eastern shores of the Mediterranean Sea, are somewhat more physically similar to *H. sapiens*.

Despite the physical and technological similarities between *H. neanderthalensis* and *H. sapiens*, there is no solid archeological evidence that Neanderthals ever developed an advanced culture that included such characteristically human endeavors as art, music, and rituals. Some anthropologists argue that, because their skeletal anatomy shows that they were physically capable of making the sounds required for speech, Neanderthals might have acquired language. This interpretation of Neanderthal anatomy, however, is not unanimously accepted. In general, the available evidence of the Neanderthal way of life is limited and open to different interpretations, and anthropologists are engaged in a sometimes heated debate about how advanced Neanderthal culture became.

Though some anthropologists argue that Neanderthals were simply a variety of *H. sapiens*, most agree that Neanderthals were a separate species. Dramatic evidence in support of this hypothesis has come from two groups of researchers who were able to isolate and analyze DNA from two Neanderthal skeletons that were more than 30,000 years old. The researchers determined the nucleotide sequence of a Neanderthal gene and compared it with the same gene from a large number of modern humans from various parts of the world. The Neanderthal gene sequence was very different from that of modern humans, indicating that the evolutionary branch leading to Neanderthals diverged from the ancestral human line hundreds of thousands of years before the emergence of modern *H. sapiens*.

Modern Humans Emerged Only 150,000 Years Ago

The fossil record shows that anatomically modern humans appeared in Africa about 150,000 years ago. The location of these fossils suggests that *Homo sapiens* originated in Africa, but most of our knowledge about our own early history comes from European and Middle Eastern fossil *H. sapiens* collectively known as Cro-Magnons (after the district in France in which their remains were first discovered). Cro-Magnons appeared about 90,000 years ago. They had domed heads, smooth brows, and prominent chins (just like us). Their tools were precision instruments similar to the stone tools used until recently in many parts of the world.

Behaviorally, Cro-Magnons seem to have been similar to, but more sophisticated than, Neanderthals. Artifacts from 30,000-year-old Cro-Magnon archeological sites include elegant bone flutes, graceful carved ivory sculptures, and evidence of elaborate burial ceremonies (**Fig. 15-16**). Perhaps the most remarkable

Figure 15-16 Paleolithic burial This 24,000-year-old grave shows evidence that Cro-Magnon people ritualistically buried their dead. The body was covered with a dye known as red ocher, then buried with a headdress made of snail shells and a flint tool in its hand.

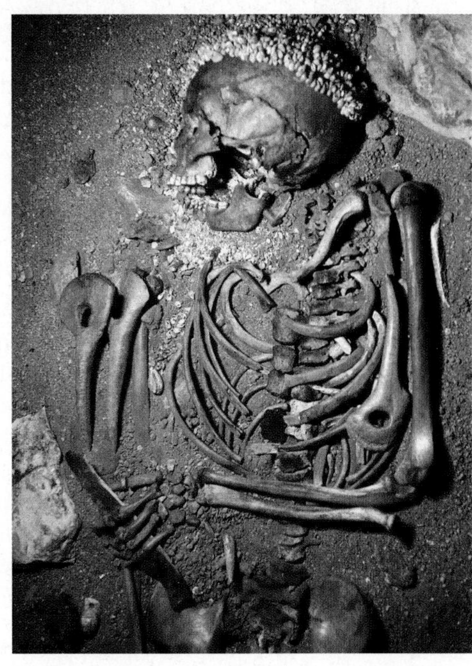

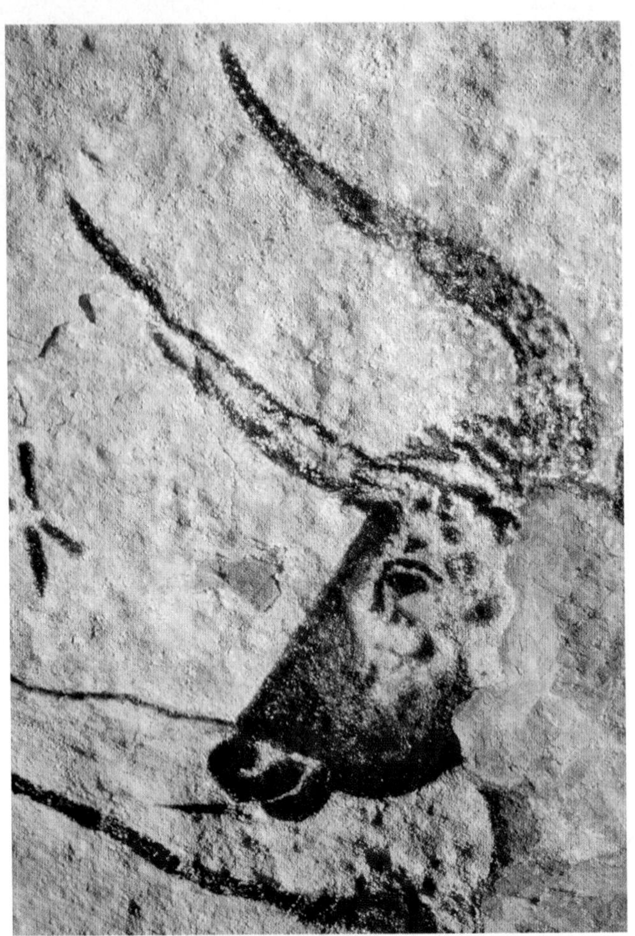

Figure 15-17 The sophistication of Cro-Magnon people Cave paintings by Cro-Magnons have been remarkably preserved by the relatively constant underground conditions of a cave in Lascaux, France.

accomplishment of Cro-Magnons is the magnificent art left in caves in places such as Altamira in Spain and Lascaux and Chauvet in France (Fig. 15-17). The oldest cave paintings so far found are more than 30,000 years old, and even the oldest ones make use of sophisticated artistic techniques. No one knows exactly why these paintings were made, but they attest to minds as fully human as our own.

Cro-Magnons and Neanderthals Lived Side by Side

Cro-Magnons coexisted with Neanderthals in Europe and the Middle East for perhaps as many as 50,000 years before the Neanderthals disappeared. Some researchers believe that Cro-Magnons interbred extensively with Neanderthals, so Neanderthals were essentially absorbed into the human genetic mainstream. Other scientists disagree, citing mounting evidence such as the fossil DNA described earlier, and suggest that later-arriving Cro-Magnons simply overran and displaced the less-well-adapted Neanderthals.

Neither hypothesis does a good job of explaining how the two kinds of hominids managed to occupy the same geographical areas for such a long time. The persistence in one area of two similar but distinct groups for tens of thousands of years seems inconsistent with both interbreeding and direct competition. Perhaps the competition between *H. neanderthalensis* and *H. sapiens* was indirect, so that the two species were able to coexist for a time in the same habitat, until the superior the ability of *H. sapiens* to exploit the available resources slowly drove Neanderthals to extinction.

DINOSAURS SINGING IN THE BACKYARD? *REVISITED*

Biologists who hypothesized that birds descended from dinosaurs turned to the fossil record to test their hypothesis. They reasoned that, if the hypothesis were correct, feathers could well have been present in dinosaurs before birds evolved. Thus, they predicted that paleontologists would eventually find fossils of dinosaurs with feathers.

In 1998, the predicted fossils were finally uncovered. In rock deposits near the little village of Sihetun in northeastern China, fossil hunters extracted the exquisitely preserved remains of some previously undiscovered types of dinosaurs. Along the margin of some of these clearly dinosaurian fossil skeletons, impressions of what appeared to be feathers were plainly visible. For the first time, scientists had solid evidence that some dinosaurs had feathers.

In the years since the first discoveries in China, many additional feathered dinosaurs have been unearthed, including some with especially well preserved fossil feathers. The fossils include more than a dozen species of feathered dinosaurs. To most paleontologists, these fossils collectively demonstrate that feathers did not originate with birds, but were present in a group of dinosaurs, known as theropods, that gave rise to birds. Some paleontologists; argue that probably all theropods had feathers, including *Velociraptor* (famously portrayed in *Jurassic Park;* see Fig. 15-9) and the fearsome predator *Tyranusaurus*.

Eventually all dinosaurs, feathered and unfeathered, went extinct. Their evolutionary cousins the birds, however, lived on and evolved into the diverse group that brightens our backyards today.

CONSIDER THIS: An alternative hypothesis to explain the existence of feathered dinosaurs is convergent evolution. Is it possible that dinosaurs and birds are unrelated, and that each group separately evolved feathers? What is the likelihood of convergent evolution of feathers? What evidence might, if found, cause you to reject the conclusion that birds are descended from dinosaurs and accept the hypothesis of convergent evolution?

LINKS TO LIFE: Amateurs on the Cutting Edge

Do you want to contribute to the advancement of science? Hunt for fossils. Unlike almost every other area of science, paleontology offers opportunities for nonprofessionals to make important discoveries. You will never hear of a high school student characterizing a genetic sequence for the Human Genome Project or designing a new anticancer drug, but a group of high school students did find a rare complete fossil skeleton of an *Allosaurus* dinosaur. Students in a college geology course found a set of dinosaur skeletons that is now on display in a museum. One of the most famous of all fossil discoveries, the complete *Tyrannosaurus* skeleton known as Sue, was found by an amateur fossil hunter.

The discovery of Sue was a spectacular example of a contribution to science made by an amateur, but the find also called attention to the sometimes seamy financial aspects of fossil hunting. Sue's discovery sparked a fierce legal battle to determine who owned the dinosaur skeleton. Ultimately, Sue's bones were auctioned, and the Field Museum in Chicago paid almost $13 million for the skeleton. (The proceeds went to the owner of the land on which Sue was found, not to the fossil hunter who found the skeleton.)

One unfortunate consequence of fossils' monetary value is that national parks and other public and private lands around the world are plagued by fossil poachers. So, if you decide to try fossil hunting, make sure you have permission to explore your collecting area. A course offered by a college or natural history museum can help you get started.

Chapter Review

Summary of Key Concepts

15.1 How Did Life Begin?

Before life arose, lightning, ultraviolet light, and heat formed organic molecules from water and the components of primordial Earth's atmosphere. These molecules probably included nucleic acids, amino acids, short proteins, and lipids. By chance, some molecules of RNA may have had enzymatic properties, catalyzing the assembly of copies of themselves from nucleotides in Earth's waters. These may have been the forerunners of life. Protein-lipid microspheres enclosing these ribozymes may have formed the first protocells.

15.2 What Were the Earliest Organisms Like?

The oldest fossils, about 3.5 billion years old, are of prokaryotic cells that fed by absorbing organic molecules that had been synthesized in the environment. Because there was no free oxygen in the atmosphere, their energy metabolism must have been anaerobic. As the cells multiplied, they depleted the organic molecules that had been formed by prebiotic synthesis. Some cells developed the ability to synthesize their own food molecules by using simple inorganic molecules and the energy of sunlight. These earliest photosynthetic cells were probably ancestors of today's cyanobacteria.

Photosynthesis releases oxygen as a by-product, and by about 2.2 billion years ago significant amounts of free oxygen were accumulating in the atmosphere. Aerobic metabolism, which generates more cellular energy than does anaerobic metabolism, probably arose about this time.

Eukaryotic cells had evolved by about 1.7 billion years ago. The first eukaryotic cells probably arose as symbiotic associations between predatory prokaryotic cells and other bacteria. Mitochondria may have evolved from aerobic bacteria engulfed by predatory cells. Similarly, chloroplasts may have evolved from photosynthetic cyanobacteria.

15.3 What Were the Earliest Multicellular Organisms Like?

Multicellular organisms evolved from eukaryotic cells and first appeared, in the seas, about 1 billion years ago. Multicellularity offers several advantages, including greater size. In plants, increased size offered some protection from predation. Specialization of cells allowed plants to anchor themselves in the nutrient-rich, well-lit waters of the shore. For animals, multicellularity allowed more-efficient predation and more-effective escape from predators. These in turn provided environmental pressures for faster locomotion, improved senses, and greater intelligence.

15.4 How Did Life Invade the Land?

The first land organisms were probably algae. The first multicellular land plants appeared about 400 million years ago. Life on land required special adaptations for support of the body, reproduction, and the acquisition, distribution, and retention of water, but the land also offered abundant sunlight and protection from aquatic herbivores. Soon after land plants evolved, arthropods invaded the land. Absence of predators and abundant land plants for food probably facilitated the invasion of the land by animals.

The earliest land vertebrates evolved from lobefin fishes, which had leglike fins and a primitive lung. A group of lobefins evolved into the amphibians about 350 million years ago. Reptiles evolved from amphibians, with several further adaptations for life on land: internal fertilization, waterproof eggs that could be laid on land, waterproof skin, and better lungs. Birds and mammals evolved independently from separate groups of reptiles. A major advance in the evolution of both birds and mammals was insulation over the body surface in the form of feathers and hair.

15.5 What Role Has Extinction Played in the History of Life?

The history of life has been characterized by constant turnover of species as species go extinct and are replaced by new ones. Mass extinctions, in which large numbers of species disappear within a relatively short time, have occurred periodically. Mass extinctions were probably caused by some combination of climate change and catastrophic events, such as volcanic eruptions and meteorite impacts.

15.6 How Did Humans Evolve?

One group of mammals evolved into the tree-dwelling primates. Between 20 million and 30 million years ago, some primates descended from the trees, and these were the ancestors of apes and humans. The australopithecines arose in Africa about 4 million years ago. These hominids walked erect, had larger brains than did their forebears, and fashioned primitive tools. One group of australopithecines gave rise to a line of hominids in the genus *Homo*, which in turn gave rise to modern humans.

Key Terms

amphibian *p. 270*
arthropod *p. 269*
conifer *p. 269*
endosymbiont hypothesis *p. 265*
eukaryotes *p. 265*
exoskeleton *p. 269*
hominid *p. 274*
lobefin *p. 270*
mass extinction *p. 272*
microsphere *p. 261*
plate tectonics *p. 272*
primate *p. 273*
prokaryote *p. 262*
protocell *p. 261*
reptile *p. 270*
ribozyme *p. 260*
spontaneous generation *p. 258*

Thinking Through the Concepts

Multiple Choice

1. *There was no free oxygen in the early atmosphere because most of it was tied up in*
   - **a.** water
   - **b.** ammonia
   - **c.** methane
   - **d.** rock
   - **e.** radioactive isotopes
2. *RNA became a candidate for the first information-carrying molecule when Tom Cech and Sidney Altman discovered that some RNA molecules can act as enzymes that*
   - **a.** degrade proteins
   - **b.** turn light into chemical energy
   - **c.** split water and release oxygen gas
   - **d.** synthesize copies of themselves
   - **e.** synthesize DNA
3. *The earliest living organisms were*
   - **a.** multicellular
   - **b.** eukaryotes
   - **c.** prokaryotes
   - **d.** photosynthesizers
   - **e.** aerobic
4. *Which two of the following observations support the endosymbiont hypothesis for the origin of chloroplasts and mitochondria from ingested bacteria?*
   - **a.** Aerobic respiration takes place in mitochondria.
   - **b.** Mitochondria have their own DNA.
   - **c.** Photosynthesis takes place in chloroplasts.
   - **d.** Chloroplasts have their own DNA.
   - **e.** Some bacteria use anaerobic respiration.
5. *Early marine-dwelling arthropods were well suited for life on land because an exoskeleton*
   - **a.** can support an animal's weight against the pull of gravity
   - **b.** allows a wide diversity of body types
   - **c.** resists drying
   - **d.** absorbs light
   - **e.** both a and c
6. *The evolution of the shelled, waterproof egg was an important event in vertebrate evolution because it*
   - **a.** led to the Cambrian explosion
   - **b.** was the first example of parents' caring for their young
   - **c.** allowed the colonization of freshwater environments
   - **d.** freed organisms from having to lay their eggs in water
   - **e.** allowed internal fertilization of eggs

Review Questions

1. What is the evidence that life might have originated from nonliving matter on early Earth? What kind of evidence would you like to see before you would accept this hypothesis?
2. If the first cells with aerobic metabolism were so much more efficient at producing energy, why didn't they drive to extinction cells with only anaerobic metabolism?
3. Explain the endosymbiont hypothesis for the origin of chloroplasts and mitochondria.
4. Name two advantages of multicellularity for plants and two for animals.
5. What advantages and disadvantages would terrestrial existence have had for the first plants to invade the land? For the first land animals?
6. Outline the major adaptations that emerged during the evolution of vertebrates, from fish to amphibians to reptiles to birds and mammals. Explain how these adaptations increased the fitness of the various groups for life on land.
7. Outline the evolution of humans from early primates. Include in your discussion such features as binocular vision, grasping hands, bipedal locomotion, tool making, and brain expansion.

Applying the Concepts

1. **IS THIS SCIENCE?** The "panspermia" hypothesis proposes that life did not originate on Earth, but instead arose elsewhere in the galaxy. According to the hypothesis, cells arrived on Earth via meteorites or pieces of comets that fell to Earth from space. Is panspermia a plausible alternative to other hypotheses for the origin of life on Earth? What kinds of evidence would provide support for the panspermia hypothesis?
2. Do you think that studying our ancestors can shed light on the behavior of modern humans? Why or why not?
3. What do you think was the most significant event in the history of life? Explain your answer.

For More Information

de Duve, C. "The Birth of Complex Cells." *Scientific American*, April 1996. Narrative describing the origin of complex eukaryotic cells by repeated instances of endosymbiosis.

Fenchel, T., and Finlay, B. J. "The Evolution of Life without Oxygen." *American Scientist*, January–February 1994. Clues to the origin of the first eukaryotic cells are provided by symbiotic relationships of organisms in oxygen-free environments.

Fry, I. *The Emergence of Life on Earth: A Historical and Scientific Overview*. Brunswick, NJ: Rutgers University Press, 2000. A thorough review of research and hypotheses concerning the origin of life.

Hay, R. L., and Leakey, M. D. "The Fossil Footprints of Laetoli." *Scientific American*, February 1982. The actual footprints of a hominid family were discovered by Hay and Leakey in volcanic ash 3.5 million years old.

Maynard Smith, J., and Szathmary, E. *The Origins of Life: From the Birth of Life to the Origin of Language*. New York: Oxford University Press, 1999. A thought-provoking review of the major shifts that have occurred over the 3.5-billion-year history of life.

Monastersky, R. "The Rise of Life on Earth." *National Geographic*, March 1998. A beautifully illustrated and engaging description of current ideas and evidence of how life arose.

Tattersall, I. "How We Came to Be Human." *Scientific American*, December 2001. A discussion of the evolutionary origin of language, art, symbolic representation, and other distinctictive traits of modern humans.

Tattersall, I. "Once We Were Not Alone" *Scientific American*, January 2000. A overview of the evolutionary history that led to modern *Homo sapiens*, with illustrations of some of the hominids that preceded us.

Ward, P. D. *The End of Evolution: On Mass Extinctions and the Preservation of Biodiversity.* New York: Bantam Books, 1994. An engaging first-person account of a paleontologist's investigation of the causes of mass extinction.

WEB TUTORIAL

To access a Web Tutorial visit http://www.prenhall.com/audesirk4. *Log in to the Web site selected by your instructor, navigate to this chapter, and select the appropriate Web Activity number.*

15.1 Endosymbiosis

Estimated time: 5 minutes

This activity demonstrates how cellular symbiosis may have resulted in the complex organelles of eukaryotes: mitochondria and chloroplasts.

15.2 Evolutionary Timescales

Estimated time: 5 minutes

In this activity, you will explore the evolutionary history of three important types of organisms: land plants, terrestrial vertebrate animals, and primates.

15.3 Web Investigation: Dinosaurs Singing in the Backyard

Estimated time: 15 minutes

Birds evolved from dinosaurs? It's not an obvious connection, yet fossil evidence supports conclusions of an evolutionary relationship between these dissimilar organisms. This exercise will give you the opportunity to consider the changes that led to dinosaurs singing in your backyard.

Chapter 16

The Diversity of Life

More than half of all known species are insects. Beetles, such as the harlequin beetle shown here, are by far the largest group of insects.

AT A GLANCE

An Unsolved Mystery: How Many Species *Are* There?

Do you know how many species are on Earth? How might you go about finding out? One possibility might be to simply count them. You could comb the scientific literature to find all the species that scientists have discovered and named, and then tally up the total number. If you did that, you'd end up with a count of roughly 1.4 million species. But you still wouldn't know how many species are on Earth.

Why doesn't counting work? Because most of the planet's species remain undiscovered. Relatively few scientists are engaged in the search for new species, and many undiscovered species are small and inconspicuous or live in poorly explored habitats such as the floor of the ocean or the topmost branches of tropical rain forests.

The gap between our knowledge of species and their actual number was recently illustrated by scientists who sequenced the DNA in a sample of water from the Sargasso Sea. The sample contained 1.2 million different genes, suggesting the presence of around 1800 microscopic species. If 1800 mostly unknown species are present in one test tube of seawater, the current total of about 5000 known species of bacteria worldwide must represent only a tiny fraction of true bacterial diversity. The same is true for virtually every other group of organisms.

If we are unable to directly count all the species on Earth, how can we estimate their number? Scientists' general approach has been to extrapolate to a worldwide scale from data for a smaller area. For example, researcher Terry Erwin calculated that the world's tropical rain forests must hold about 30 million species of insects, based on the number of beetle species he found in the canopy of a particular species of tree. How do you think Erwin extrapolated from his data to all of Earth's tropical rain forests? We will return to that question after a tour of life's diversity.

Biologists sampling life in the sea have found that Earth's oceans contain a multitude of undiscovered species.

16.1 How Are Organisms Named and Classified?

Organisms are placed into categories on the basis of their evolutionary relationships. There are eight major categories: domain, kingdom, phylum, class, order, family, genus, and species. These categories form a nested hierarchy in which each level includes all of the other levels below it. Each domain contains a number of kingdoms; each kingdom contains a number of phyla; each phylum includes a number of classes; each class includes a number of orders; and so on. As we move down the hierarchy, smaller and smaller groups are included. Each category is increasingly narrow and specifies groups whose common ancestor is increasingly recent. Table 16-1 describes some examples of classifications of specific organisms.

Each Species Has a Unique, Two-Part Name

The scientific name of an organism is formed from the two smallest categories, the genus and the species. Each genus includes a group of very closely related species, and each species within a genus includes populations of organisms that can potentially interbreed under natural conditions. Thus, the genus *Sialia* (bluebirds) includes the eastern bluebird (*Sialia sialis*), the western bluebird (*Sialia mexicana*), and the mountain bluebird (*Sialia currucoides*)—very similar birds that normally do not interbreed (**Fig. 16-1**).

Each two-part scientific name is unique, so referring to an organism by its scientific name rules out any chance of ambiguity or confusion. For example, the bird *Gavia immer* is commonly known in North America as the common loon, in Great Britain as the northern diver, and by still other names in non-English-speaking countries. But the Latin scientific name *Gavia immer* is recognized by biologists worldwide, overcoming language barriers and allowing precise communication.

Note that, by convention, scientific names are always underlined or *italicized*. The first letter of the genus name is always capitalized, and the first letter of the species name is always lowercase. The species name is never used alone but is always paired with its genus name.

Classification Originated As a Hierarchy of Categories

Aristotle (384–322 B.C.) was among the first to attempt to formulate a logical, standardized language for naming living things. Based on characteristics such as structural complexity, behavior, and degree of development at birth, he classified about 500 organisms into 11 categories. Aristotle's categories formed a hierarchical structure, with each category more inclusive than the one beneath it, a concept that is still used today.

TABLE 16-1 Classification of Selected Organisms, Reflecting Their Degree of Relatedness*

	Human	Chimpanzee	Wolf	Fruit Fly	Sequoia Tree	Sunflower
Domain	**Eukarya**	**Eukarya**	**Eukarya**	**Eukarya**	**Eukarya**	**Eukarya**
Kingdom	**Animalia**	**Animalia**	**Animalia**	**Animalia**	**Plantae**	**Plantae**
Phylum	**Chordata**	**Chordata**	**Chordata**	Arthropoda	Coniferophyta	Anthophyta
Class	**Mammalia**	**Mammalia**	**Mammalia**	Insecta	Coniferosida	Dicotyledoneae
Order	**Primates**	**Primates**	Carnivora	Diptera	Coniferales	Asterales
Family	Hominidae	Pongidae	Canidae	Drosophilidae	Taxodiaceae	Asteraceae
Genus	*Homo*	*Pan*	*Canis*	*Drosophila*	*Sequoiadendron*	*Helianthus*
Species	*sapiens*	*troglodytes*	*lupus*	*melanogaster*	*giganteum*	*annuus*

* Boldface categories are those that are shared by more than one of the organisms classified. Genus and species names are always italicized or underlined.

Building on this foundation more than 2000 years later, Swedish naturalist Carl von Linné (1707–1778)—who called himself Carolus Linnaeus, a latinized version of his given name—laid the groundwork for the modern classification system. He placed each organism into a series of hierarchically arranged categories on the basis of its resemblance to other life-forms, and he also introduced the scientific name composed of genus and species.

Nearly 100 years later, Charles Darwin (1809–1882) published *On the Origin of Species*, which demonstrated that all organisms are connected by common ancestry. Biologists then began to recognize that the categories ought to reflect the pattern of evolutionary relatedness among organisms. The more categories two organisms share, the closer their evolutionary relationship.

Figure 16-1 **Three species of bluebird** Despite their obvious similarity, these three species of bluebird remain distinct because they do not interbreed. The three species shown are (from left to right) the eastern bluebird (*Sialia sialis*), the western bluebird (*Sialia mexicana*), and the mountain bluebird (*Sialia currucoides*).

Biologists Identify Features That Reveal Evolutionary Relationships

Biologists seek to reconstruct the tree of life, but they must do so without much direct knowledge of evolutionary history. Because they can't see into the past, they must infer the past as best they can, on the basis of similarities among living organisms. Not just any similarity will do, however. Some observed similarities stem from convergent evolution in organisms that are not closely related, and such similarities are not useful for classification. Instead, biologists value the similarities that exist because two kinds of organisms both inherited the characteristic from a common ancestor. Therefore, an important task is to distinguish informative similarities caused by common ancestry from similarities that result from convergent evolution. In the search for informative similarities, biologists look at many kinds of characteristics.

Anatomy Plays a Key Role in Classification

Historically, the most important and useful distinguishing characteristics have been anatomical. Biologists look carefully at similarities in external body structure (see Fig. 16-1) and in internal structures, such as skeletons and muscles. For example, homologous structures, such as the finger bones of dolphins, bats, seals, and humans (see Fig. 13-7), provide evidence of a common ancestor. To detect relationships between more closely related species, biologists may use microscopes to discern finer details—the number and shape of the "teeth" on the tonguelike radula of a snail, the shape and position of the bristles on a marine worm, or the external structure of pollen grains of a flowering plant (Fig. 16-2).

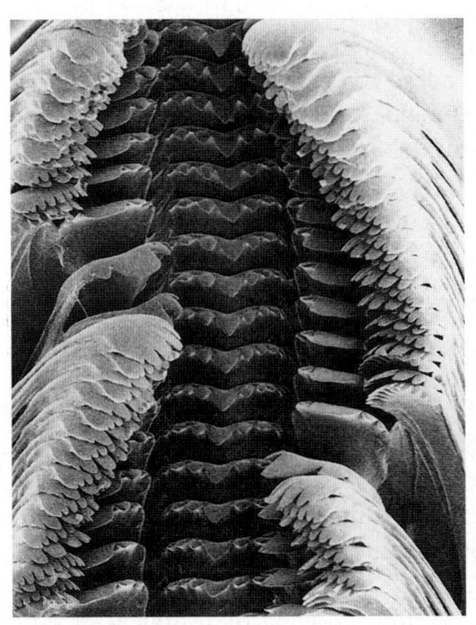

(a)

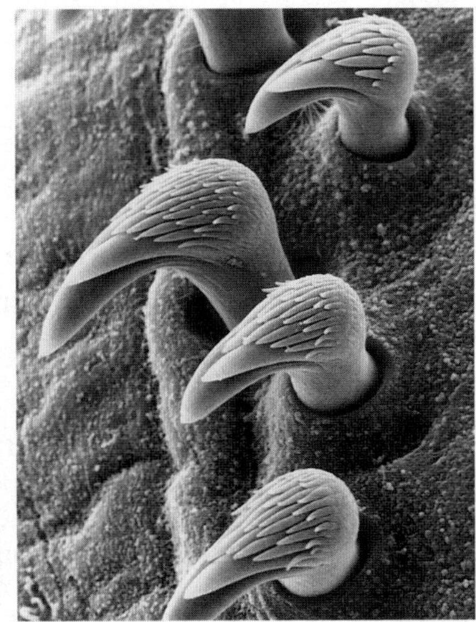

(b)

(c)

Figure 16-2 **Microscopic structures may be used to classify organisms** ***(a)*** The "teeth" on a snail's tonguelike radula (a structure used in feeding), ***(b)*** the bristles on a marine worm, and ***(c)*** the shape and surface features of pollen grains are characteristics potentially useful in classification. Such finely detailed structures can reveal similarities between species that are not apparent in larger and more obvious structures.

Molecular Similarities Are Also Useful for Classification

The anatomical characteristics shared by related organisms are expressions of underlying genetic similarities, so it stands to reason that evolutionary relationships among species must also be reflected in genetic similarities. Of course, direct genetic comparisons were not possible for most of the history of biology. Since the 1980s, however, advances in the techniques of molecular genetics have revolutionized studies of evolutionary relationships.

For the first time, the nucleotide sequence of DNA (that is, genotype), rather than phenotypic features such as appearance or behavior, can be used to investigate relatedness among different types of organisms. Genetic relatedness among organisms can also be evaluated by examining the structure of their chromosomes. Among the findings derived from this technique is that the chromosomes of chimpanzees and humans are extremely similar, showing that these two species are very closely related.

Web Tutorial 16.2 Tree of Life

16.2 What Are the Domains of Life?

Before 1970, all forms of life were classified into two kingdoms: Animalia and Plantae. All bacteria, fungi, and single-celled photosynthetic eukaryotes were considered to be plants, and the single-celled eukaryotes were classified as animals. As scientists learned more about fungi and microorganisms, however, it became apparent that the two-kingdom system oversimplified evolutionary history. To help rectify this problem, Robert H. Whittaker in 1969 proposed a five-kingdom classification that was eventually adopted by most biologists.

The Five-Kingdom System Improved Classification

Whittaker's five-kingdom system placed all prokaryotic organisms into a single kingdom and divided the eukaryotes into four kingdoms. The designation of a separate kingdom (called Monera) for the prokaryotes reflected growing recognition that the evolutionary pathway of these tiny, single-celled organisms had diverged from that of the eukaryotes early in the history of life. Among the eukaryotes, the five-kingdom system recognized three kingdoms of multicellular organisms (Plantae, Fungi, and Animalia) and placed all of the remaining, mostly single-celled eukaryotes in a single kingdom (Protista).

Because it more accurately reflected current understanding of evolutionary history, the five-kingdom system was an improvement over the old two-kingdom system. As understanding continued to grow, however, our view of life's most fundamental categories needed yet another revision. The pioneering work of microbiologist Carl Woese has shown that biologists had overlooked a fundamental event in the early history of life, one that demands a new and more-accurate classification of life.

Figure 16-3 Two domains of prokaryotic organisms Although superficially similar in appearance, ***(a)*** *Vibrio cholerae* and ***(b)*** *Methanococcus jannaschi* are less closely related than a mushroom and an elephant. *Vibrio* is in the domain Bacteria, and *Methanococcus* is in the Archaea.

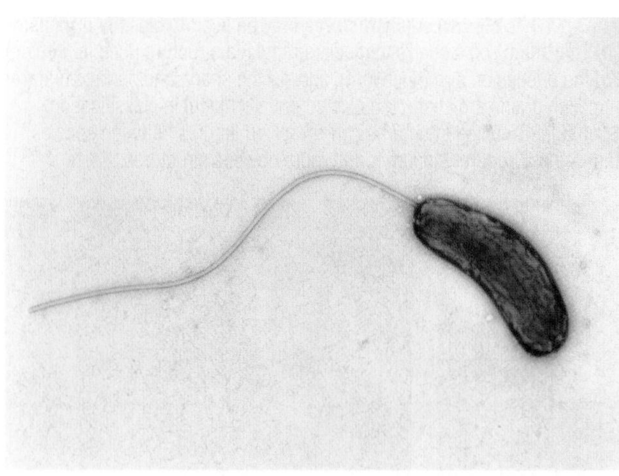

(a)

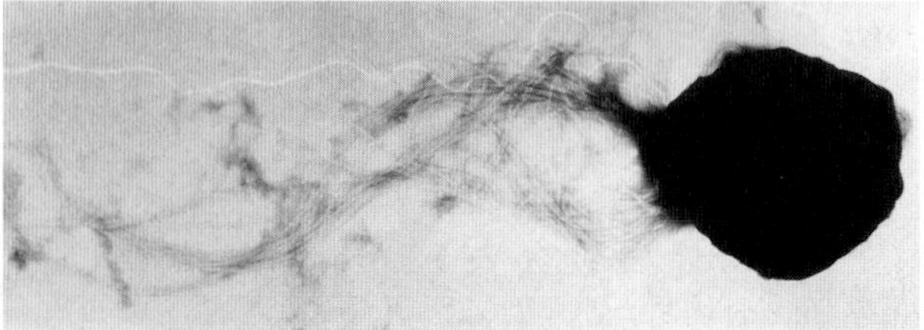

(b)

A Three-Domain System More Accurately Reflects Life's History

Woese and other biologists interested in the evolutionary history of microorganisms have studied the biochemistry of prokaryotic organisms. These researchers, focusing on nucleotide sequences of the RNA that is found in the organisms' ribosomes, discovered that the supposed kingdom Monera actually included two very different kinds of organisms. Woese has dubbed these two groups the Bacteria and the Archaea (Fig. 16-3).

Despite superficial similarities in their appearance under the microscope, the Bacteria and the Archaea are radically different. These two groups are no more closely related to one another than either one is to any eukaryote. The tree of life split into three parts very early in the history of life, long before the appearance of plants, animals, and fungi. As a result of this new understanding, the five-kingdom system has been replaced by a classification that divides life into three domains: **Bacteria**, **Archaea**, and **Eukarya** (Fig. 16-4).

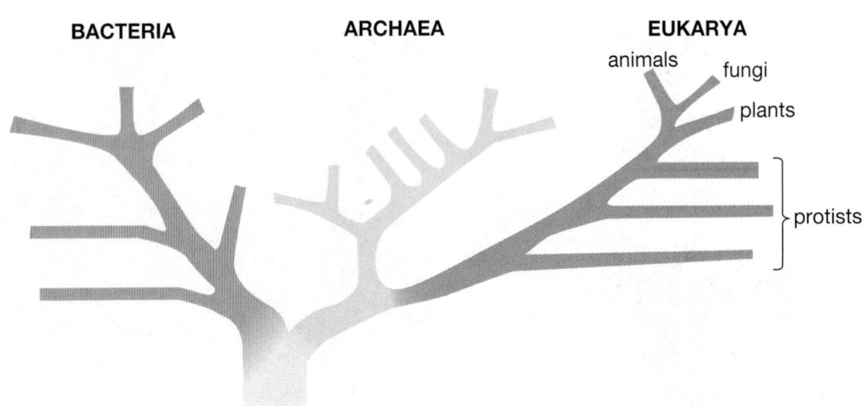

Figure 16-4 The tree of life The three domains of life represent the earliest branches in evolutionary history.

Kingdom-Level Classification Remains Unsettled

The move to a three-domain classification system has required systematists to reexamine the kingdoms within each domain, and the process of establishing kingdoms is still under way. If we accept that the striking differences between plants, animals, and fungi demand that each of these evolutionary lineages retains its kingdom status, then the logic of classification requires that we also assign kingdom status to groups that branch off of the tree of life earlier than these three groups of multicellular eukaryotes. Following this logic, biologists recognize about 15 kingdoms among the Bacteria and three or so kingdoms among the Archaea. Biologists also recognize additional kingdoms within the Eukarya, reflecting a number of very early evolutionary splits within the diverse array of single-celled eukaryotes formerly lumped together in the kingdom Protista. Biologists, however, have yet to reach a consensus about the precise definitions of new prokaryotic and eukaryotic kingdoms, though new information about the evolutionary history of single-celled organisms is emerging rapidly. Thus, kingdom-level classification is in a state of transition as biologists strive to incorporate the latest information.

This text's descriptions of the diversity of life sidestep the unsettled state of life's kingdoms. The prokaryotic domains Archaea and Bacteria are discussed without reference to kingdom-level relationships. Among the eukaryotes, fungi, plants, and animals are treated as distinct evolutionary units, and the generic term "protist" designates the diverse collection of eukaryotes that are not members of these three kingdoms.

16.3 Bacteria and Archaea

Two of life's three domains, Bacteria and Archaea, consist entirely of prokaryotes, single-celled microbes that lack organelles such as the nucleus, chloroplasts, and mitochondria. Both bacteria and archaea are normally very small, ranging from about 0.2 to 10 micrometers in diameter; in comparison, the diameters of eukaryotic cells range from about 10 to 100 micrometers. About 250,000 average-sized bacteria or archaea could congregate on the period at the end of this sentence, though a few species of bacteria are larger. The largest known bacterium is as much as 700 micrometers in diameter, making it visible to the naked eye.

Bacteria and Archaea Are Fundamentally Different

Bacteria and archaea are superficially similar in appearance under the microscope, but have striking differences in their structural and biochemical features that reveal the ancient evolutionary separation between them. For example, the cell walls of bacterial cells contain *peptidoglycan*, but the cell walls of archaea do not. Bacteria and archaea also differ in the structure and composition of plasma membranes, ribosomes, and RNA polymerases and in basic processes such as transcription and translation.

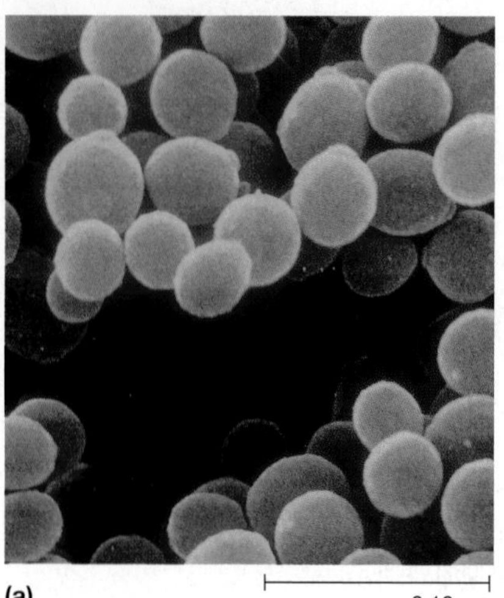

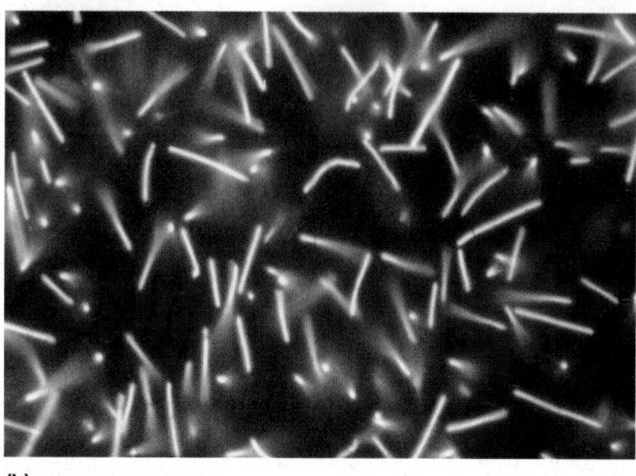

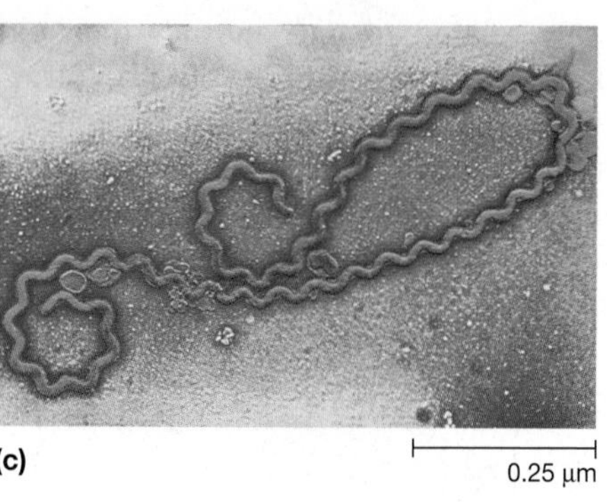

Figure 16-5 Three common prokaryote shapes ***(a)*** Spherical bacteria of the genus *Micrococcus,* ***(b)*** rod-shaped archaea of the genus *Methanopyrus,* and ***(c)*** corkscrew-shaped bacteria of the species *Leptospirosis interrogans.*

Classification of Prokaryotes within Each Domain Is Difficult

The sharp biochemical differences between archaea and bacteria make distinguishing the two domains a straightforward matter, but classification within each domain poses challenges. Prokaryotes are tiny and structurally simple and simply do not exhibit the huge array of anatomical and developmental differences that are used to infer the evolutionary history of plants, animals, and other eukaryotes. Consequently, prokaryotes have been classified on the basis of such features as shape, means of locomotion, pigments, nutrient requirements, the appearance of colonies (groups of individuals that descended from a single cell), and staining properties.

In recent years our understanding of the evolutionary history of the prokaryotic domains has been greatly expanded by comparisons of DNA and RNA nucleotide sequences. Prokaryote classification, however, is a rapidly changing field, and consensus on kingdom-level classification has thus far proved elusive. With new DNA sequence data being generated at a furious pace, and with new and distinctive types of bacteria and archaea being discovered and described on a regular basis, the revision of prokaryote classification schemes will likely continue for some time to come.

Prokaryotes Differ in Shape and Structure

The cell walls that surround prokaryotic cells give characteristic shapes to different types of bacteria and archaea. The most common shapes are rodlike, spherical, and corkscrew-shaped (Fig. 16-5). Some bacteria and archaea have **flagella** (singular, flagellum). These are simpler in structure than eukaryotic flagella. Prokaryote flagella may appear singly at one end of a cell, in pairs (one at each end of the cell), as a tuft at one end of the cell (Fig. 16-6), or scattered over the entire cell surface. Flagella can rotate rapidly, propelling the organism through its liquid environment.

Figure 16-6 The prokaryote flagellum An archaean of the genus *Aquifex* uses its flagella to move toward favorable environments.

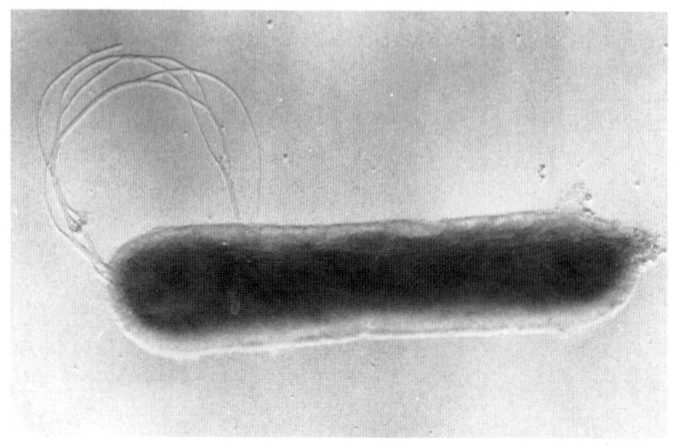

Many Bacteria Form Films on Surfaces

The cell walls of some bacterial species are surrounded by sticky layers of protective slime, composed of polysaccharide or protein, which protects the bacteria and helps them adhere to surfaces. In many cases, slime-secreting bacteria of one or more species aggregate in colonies to form communities known as *biofilms.* One familiar biofilm is dental plaque, which is formed by the bacteria that inhabit the mouth. The protection afforded by biofilms helps defend the embedded bacteria against a variety of attacks, including those launched by antibiotics and disinfectants. As a result, biofilms formed by bacteria harmful to humans can be very difficult to eradicate. The persistence of biofilms is unfortunate, because the surfaces on which biofilms form include contact lenses, surgical sutures, and medical equipment such as catheters. In addition, many infections of the human body take the form of biofilms, including those responsible for tooth decay, gum disease, and ear infections.

Protective Endospores Allow Some Bacteria to Withstand Adverse Conditions

When environmental conditions become inhospitable, many rod-shaped bacteria form protective structures called **endospores**. An endospore, which forms inside a bacterium, contains genetic material and a few enzymes encased within a thick protective coat (Fig. 16-7). Metabolic activity ceases until the spore encounters favorable conditions, at which time metabolism resumes and the spore develops into an active bacterium.

Endospores are resistant to even extreme environmental conditions. Some can withstand boiling for an hour or more. Endospores are also able to survive for extraordinarily long periods. In the most extreme example of such longevity, scientists recently discovered bacterial spores that had been sealed inside rock for 250 million years. After being carefully extracted from their rocky "tomb," the spores were incubated in test tubes. Amazingly, live bacteria developed from the ancient spores, which were older than the oldest dinosaur fossils.

Endospores are one of the main reasons that the bacterial disease anthrax has become an agent of biological terrorism. The bacterium that causes anthrax forms endospores, which provide the means by which terrorists (or governments) can disperse the bacteria. The spores can be stored indefinitely and can survive the harsh conditions they might encounter while traveling to their destination, including the stress of a missile launch and high-altitude travel. When they reach their target, the spores can survive dispersal into the atmosphere, remaining viable until inhaled by a potential victim.

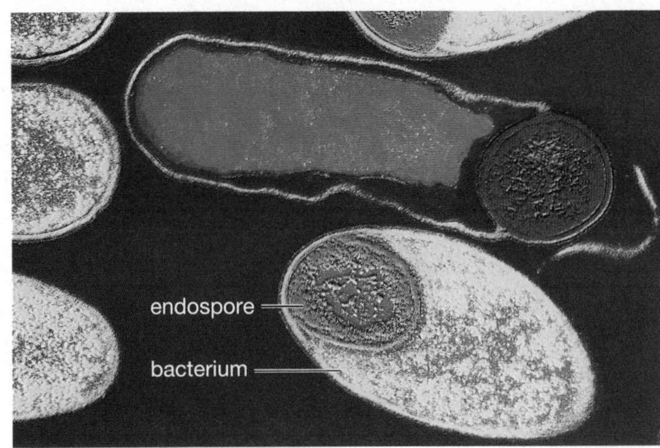

Figure 16-7 **Spores protect some bacteria** Resistant endospores have formed inside bacteria of the genus *Clostridium*, which causes the potentially fatal food poisoning called botulism. **QUESTION:** *What might explain the observation that most endospore-forming bacteria are species that live in soil?*

Prokaryotes Reproduce by Binary Fission

Most prokaryotes reproduce asexually by a simple form of cell division called binary fission, which produces genetically identical copies of the original cell (Fig. 16-8). Under ideal conditions, a prokaryotic cell can divide about once every 20 minutes, potentially giving rise to sextillions (10^{21}) of offspring in a single day. This rapid reproduction allows prokaryotes to exploit temporary habitats, such as a mud puddle or warm pudding. Rapid reproduction also allows bacterial populations to evolve quickly. Recall that many mutations, the source of genetic variability, are the result of mistakes in DNA replication during cell division. Thus, the rapid, repeated cell division of prokaryotes provides ample opportunity for new mutations to arise and also allows mutations that enhance survival to spread quickly.

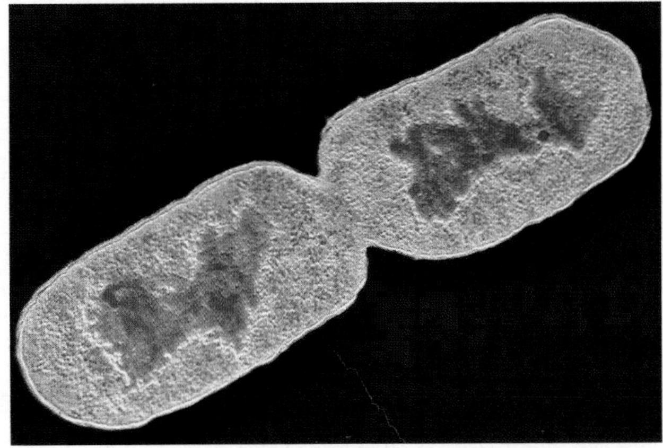

Figure 16-8 **Reproduction in prokaryotes** Prokaryotic cells reproduce by binary fission. In this color-enhanced electron micrograph, *Escherichia coli*, a normal resident of the human intestine, is dividing. Red areas are genetic material. **QUESTION:** *What is the main advantage of binary fission, compared with sexual reproduction?*

Prokaryotes Are Specialized for Specific Habitats

Prokaryotes occupy virtually every habitat, including those where extreme conditions keep out other forms of life. For example, some bacteria thrive in near-boiling environments, such as the hot springs of Yellowstone National Park (Fig. 16-9). Many archaea live in even hotter environments, including springs where the water actually boils or deep-ocean vents, where superheated water is spewed through cracks in Earth's crust at temperatures of up to 230 °F (110 °C). Prokaryotes can also survive at extremely high pressures, such as are found 1.7 miles (2.8 kilometers) below Earth's surface, where scientists recently discovered a new bacterial species. Bacteria and archaea are also found in very cold environments, such as Antarctic sea ice.

Even extreme chemical conditions fail to impede invasion by prokaryotes. Thriving colonies of bacteria and archaea live in the Dead Sea, where a salt concentration seven times that of the oceans precludes all other life, and in waters

Figure 16-9 **Some prokaryotes thrive in extreme conditions** Hot springs harbor bacteria and archaea that are both heat tolerant and mineral tolerant. Several species of bacteria paint these hot springs in Yellowstone National Park with vivid colors, and each is confined to a specific area determined by temperature range. The bacterial pigments function in photosynthesis.

that are as acidic as vinegar or as alkaline as household ammonia. Of course, rich bacterial communities also reside in a full range of more moderate habitats, including in and on the healthy human body. But an animal need not be healthy to harbor bacteria. Recently, a colony of bacteria was found dormant within the intestinal contents of a mammoth that had lain in a peat bog for 11,000 years.

No single species of prokaryote, however, is as versatile as these examples may suggest. In fact, most prokaryotes are specialists. One species of archaea that inhabits deep-sea vents, for example, grows optimally at 223 °F (106 °C) and stops growing altogether at temperatures below 194 °F (90 °C). Clearly, this species could not survive in a less extreme habitat. Bacteria that live on the human body are also specialized; different species colonize the skin, the mouth, the respiratory tract, the large intestine, and the urogenital tract.

Prokaryotes Exhibit Diverse Metabolisms

Prokaryotes are able to colonize diverse habitats partly because they have evolved diverse methods of acquiring energy and nutrients from the environment. For example, unlike eukaryotes, many prokaryotes are **anaerobes**; their metabolisms do not require oxygen. Their ability to inhabit oxygen-free environments allows prokaryotes to exploit habitats that are off-limits to eukaryotes. Some anaerobes, such as many of the archaea found in hot springs and the bacterium that causes tetanus, are actually poisoned by oxygen. Others are opportunists, engaging in anaerobic respiration when oxygen is lacking and switching to aerobic respiration (a more efficient process) when oxygen becomes available. Many prokaryotes, of course, are strictly aerobic, and require oxygen at all times.

Whether aerobic and anaerobic, different prokaryote species can extract energy from an amazing array of substances. Prokaryotes subsist not only on the sugars, carbohydrates, fats, and proteins that we normally think of as foods, but also on compounds that are inedible or even poisonous for humans, including petroleum, methane (the main component of natural gas), and solvents such as benzene and toluene. Prokaryotes can even metabolize inorganic molecules, including hydrogen, sulfur, ammonia, iron, and nitrite. The process of metabolizing inorganic molecules sometimes yields by-products that are useful to other organisms. For example, certain bacteria release sulfates or nitrates, crucial plant nutrients, into the soil.

Some species of bacteria use photosynthesis to capture energy directly from sunlight. Like green plants, photosynthetic bacteria possess chlorophyll. Most species produce oxygen as a by-product of photosynthesis, but some, known as the sulfur bacteria, use hydrogen sulfide (H_2S) instead of water (H_2O) in photosynthesis, releasing sulfur instead of oxygen. No photosynthetic archaea are known.

Prokaryotes Perform Functions Important to Other Organisms

Many eukaryotic organisms depend on close associations with prokaryotes. For example, most animals that eat leaves, including cattle, rabbits, koalas, and deer, can't actually digest plant material themselves. Instead, they depend on certain bacteria that have the unusual ability to break down cellulose, the principal component of plant cell walls. Some of these bacteria live in the animals' digestive tracts, where they help liberate nutrients from plants that the animals are unable to break down themselves. Without the bacteria, leaf-eating animals could not survive.

Prokaryotes Capture the Nitrogen Needed by Plants

Humans could not live without plants, and plants are entirely dependent on bacteria. In particular, plants are unable to capture nitrogen from that element's most abundant reservoir, the atmosphere. Plants need nitrogen to grow. To acquire it, they depend on **nitrogen-fixing bacteria**, which live both in soil and in specialized *nodules*, small, rounded lumps on the roots of certain plants (legumes, which include alfalfa, soybeans, lupines, and clover). The nitrogen-fixing bacteria capture nitrogen gas (N_2) from air trapped in the soil and combine it with hydrogen to produce ammonium (NH_4^+), a nitrogen-containing nutrient that plants can use directly.

Prokaryotes Are Nature's Recyclers

Prokaryotes play a crucial role in recycling waste. Many substances that we consider to be waste can serve as food for archaea and bacteria, most species of which obtain energy by breaking down complex organic (carbon-containing) molecules. The range of compounds attacked by prokaryotes is staggering. Nearly anything that human beings can synthesize, including detergents and the poisonous solvent benzene, can be destroyed by some prokaryote.

The term "biodegradable" (meaning "broken down by living things") refers largely to the work of prokaryotes. Even oil is biodegradable. Soon after the tanker *Exxon Valdez* dumped 11 million gallons of crude oil into Prince William Sound, Alaska, researchers from Exxon sprayed oil-soaked beaches with a fertilizer that encouraged the growth of natural populations of oil-eating bacteria. Within 15 days, the oil deposits on these beaches were noticeably reduced in comparison with unsprayed areas. Prokaryotes are much less successful, however, at degrading most kinds of plastic. Therefore, the search for useful, economical plastics that are also biodegradable is a major focus of industrial research.

The appetite of some prokaryotes for nearly any organic compound is the key to their important role as decomposers in ecosystems. While feeding themselves, bacteria break down the waste products and dead bodies of plants and animals, freeing nutrients for reuse. The recycling of nutrients provides the basis for continued life on Earth.

Some Bacteria Pose a Threat to Human Health

Despite the benefits some bacteria provide, the feeding habits of certain bacteria threaten our health and well-being. These **pathogenic** (disease-producing) bacteria synthesize toxic substances that cause disease symptoms. (So far, no pathogenic archaea have been identified.)

Some Anaerobic Bacteria Produce Dangerous Poisons

Some bacteria produce toxins that attack the nervous system. Examples of such pathogens include *Clostridium tetani*, which causes tetanus, and *Clostridium botulinum*, which causes botulism (a sometimes lethal food poisoning). Both of these bacterial species are anaerobes that survive as spores until introduced into a favorable, oxygen-free environment. For example, a deep puncture wound may allow tetanus bacteria to penetrate a human body and reach a place where they will be protected from contact with oxygen. As they multiply, the bacteria release their paralyzing poison into the bloodstream. For botulism bacteria, a sealed container of canned food that has been improperly sterilized may provide a haven. Thriving on the nutrients in the can, these anaerobes produce a toxin so potent that a single gram could kill 15 million people. Perhaps inevitably, this potent poison has caught the attention of biological weapon designers, who are presumed to have added it to their arsenals.

Humans Battle Bacterial Diseases Old and New

Bacterial diseases have had a significant impact on human history. Perhaps the most infamous example is bubonic plague, or "Black Death," which killed 100 million people during the mid-fourteenth century. In many parts of the world, one-third or more of the population died. Plague is caused by the highly infectious bacteria *Yersinia pestis*, which is spread by fleas that feed on infected rats and then move to human hosts. Although bubonic plague has not reemerged as a large-scale epidemic, about 2000 to 3000 people worldwide are still diagnosed with the disease each year.

Some bacterial pathogens seem to emerge suddenly. Lyme disease, for example, was unknown until 1975. This disease, named after the town of Old Lyme,

Connecticut, where it was first described, is caused by the spiral-shaped bacterium *Borrelia burgdorferi*. The bacterium is carried by deer ticks, which transmit it to the humans they bite. At first, the symptoms resemble flu, with chills, fever, and body aches. If untreated, weeks or months later the victim may experience rashes, bouts of arthritis, and in some cases abnormalities of the heart and nervous system. Both physicians and the general public are becoming more familiar with the disease, so more victims are receiving treatment before serious symptoms develop.

Perhaps the most frustrating pathogens are those that come back to haunt us long after we believed that we had them under control. Tuberculosis, a bacterial disease once almost vanquished in developed countries, is again on the rise in the United States and elsewhere. Two sexually transmitted bacterial diseases, gonorrhea and syphilis, have reached epidemic proportions around the globe. Cholera, a water-transmitted bacterial disease that flourishes when raw sewage contaminates drinking water or fishing areas, is under control in developed countries but remains a major killer in poorer parts of the world.

Some Common Bacterial Species Can Be Harmful

Some pathogenic bacteria are so widespread and ubiquitous that we cannot expect to ever be totally free of their damaging effects. For example, different species of the abundant streptococcus bacterium produce several diseases. One streptococcus causes strep throat. Another, *Streptococcus pneumoniae*, causes pneumonia by stimulating an allergic reaction that clogs the lungs with fluid. Yet another streptococcus has gained fame as the "flesh-eating bacterium." A small percentage of people who become infected with this one experience severe symptoms, described luridly in tabloid newspapers with such headlines as "Killer Bug Ate My Face." About 800 Americans each year are victims of necrotizing fasciitis (as the "flesh-eating" infection is more properly known), and about 15% of these victims die. The streptococci enter through broken skin and produce toxins that either destroy flesh directly or stimulate an overwhelming and misdirected attack by the immune system against the body's own cells. A limb can be destroyed in hours, and in some cases only amputation can halt the rapid tissue destruction. In other cases, these rare strep infections sweep through the body, causing death within a matter of days.

One of the most common bacterial inhabitants of the human digestive system, *Escherichia coli*, is also capable of doing harm. Different populations of *E. coli* may differ genetically, and some genetic differences can transform this normally benign species into a pathogen. One particularly notorious strain, known as O157:H7, infects about 70,000 Americans each year, about 60 of whom die from its effects. Most O157:H7 infections result from consumption of contaminated beef. About a third of the cattle in the United States carry O157:H7 in their intestinal tracts, and the bacteria can be transmitted to humans when a slaughterhouse inadvertently grinds some gut contents into hamburger. Once in a human digestive system, O157:H7 bacteria attach firmly to the wall of the intestine and begin to release a toxin that causes intestinal bleeding and that spreads to and damages other organs as well. The best defense against O157:H7 is to cook all meat thoroughly.

Most Bacteria Are Harmless

Although some bacteria assault the human body, most of the bacteria with which we share our bodies are harmless, and many are beneficial. For example, the normal bacterial community in the vagina creates an environment that is hostile to infections by parasites such as yeasts. The bacteria that harmlessly inhabit our intestines are an important source of vitamin K. As the late physician, researcher, and author Lewis Thomas so aptly put it, "Pathogenicity is, in a sense, a highly skilled trade, and only a tiny minority of all the numberless tons of microbes on the Earth has ever been involved in it; most bacteria are busy with their own business, browsing and recycling the rest of life."

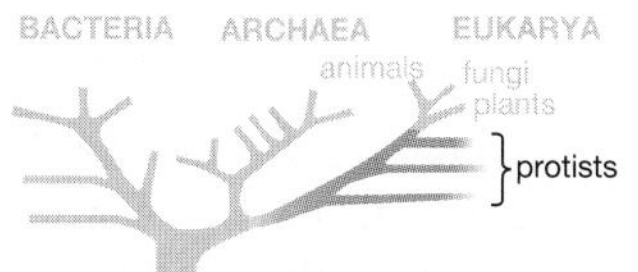

16.4 Protists

The third domain, Eukarya, includes all eukaryotic organisms. The most conspicuous Eukarya are members of the kingdoms Fungi, Plantae, and Animalia, which we discuss later in this chapter. The remaining eukaryotes constitute a diverse collection of organisms collectively known as **protists**. The term "protist" does not describe a true evolutionary unit united by shared features, but is a term of convenience that means "any eukaryote that is not a plant, animal, or fungus."

Most Protists Are Single Celled

Although some protists are large, multicellular organisms, most are single-celled and invisible to us as we go about our daily lives. If we could somehow shrink to their microscopic scale, we might be more impressed with their spectacular and beautiful forms, their varied and active lifestyles, their astonishingly diverse modes of reproduction, and the structural and physiological innovations that are possible within the limits of a single cell. In reality, however, their small size makes them challenging to observe. A microscope and a good supply of patience are required to appreciate the majesty of protists.

Past classifications of protists grouped species according to their mode of nutrition, but improved understanding of protists has revealed that the old categories did not accurately reflect evolutionary history. Nonetheless, biologists still use terminology that refers to groups of protists that share particular characteristics but are not necessarily related. For example, photosynthetic protists are collectively known as **algae** (singular, alga), and single-celled, nonphotosynthetic protists are collectively known as **protozoa** (singular, protozoan).

In the following sections, we'll explore a small sample of protist diversity.

The Chromists Include Photosynthetic and Nonphotosynthetic Organisms

The *chromists* form a group whose members have fine, hairlike projections on their flagella (though in many chromists, flagella are present only at certain stages of the life cycle). Though chromists are united by shared evolutionary history, they display a wide range of different forms. Some are photosynthetic and some are not. Most are single-celled, but some are multicellular. Two major chromist groups are the diatoms and the brown algae.

Diatoms Encase Themselves within Glassy Walls

The *diatoms*, photosynthetic chromists found in both fresh and salt water, produce protective shells of silica (glass), some of exceptional beauty (Fig. 16-10). These shells consist of top and bottom halves that fit together like a pillbox or petri dish. Accumulations of diatoms' glassy walls over millions of years have produced fossil deposits of "diatomaceous earth" that may be hundreds of meters thick. This slightly abrasive substance is widely used in products such as toothpaste and metal polish.

Diatoms form part of the **phytoplankton**, the single-celled photosynthesizers that float passively in the upper layers of Earth's lakes and oceans. Phytoplankton play an immensely important ecological role. Marine phytoplankton account for nearly 70% of all the photosynthetic activity on Earth, absorbing carbon dioxide, recharging the atmosphere with oxygen, and supporting the complex web of aquatic life. Diatoms, as key components of the phytoplankton, are so important to marine food webs that they have been called the "pastures of the sea."

Figure 16-10 Some representative diatoms This photomicrograph illustrates the intricate beauty and variety of the glassy walls of diatoms.

(a)

(b)

Figure 16-11 Brown algae ***(a)*** *Fucus*, a genus found near shores, is shown here exposed at low tide. Notice the gas-filled floats, which provide buoyancy in water. ***(b)*** The giant kelp *Macrocystis* forms underwater forests off southern California.

Brown Algae Dominate in Cool Coastal Waters

Though most photosynthetic protists, such as diatoms, are single-celled, some form multicellular aggregations that are commonly known as *seaweeds*. Although some seaweeds seem to resemble plants, they are not closely related to plants and lack many of the distinctive features of the plant kingdom. For example, none of the seaweeds have roots or shoots, and none form embryos during reproduction.

The chromists include one group of seaweeds, the *brown algae*. The brown algae are named for the brownish yellow pigments that (in combination with green chlorophyll) increase the seaweed's light-gathering ability and produce a brown to olive-green color.

Almost all brown algae are marine. The group includes the dominant seaweed species that dwell along rocky shores in the temperate (cooler) oceans of the world, including the eastern and western coasts of the United States. Brown algae live in habitats ranging from nearshore, where they cling to rocks that are exposed at low tide, to far offshore. Several species use gas-filled floats to support their bodies (**Fig. 16-11a**). Some of the giant kelp found along the Pacific coast reach heights of 325 feet (100 meters) and may grow more than 6 inches (15 centimeters) in a single day. With their dense growth and towering height (**Fig. 16-11b**), kelp form undersea forests that provide food, shelter, and breeding areas for marine animals.

The Alveolates Include Parasites, Predators, and Phytoplankton

The *alveolates* are single-celled organisms that have distinctive, small cavities beneath the surface of their cells. Some alveolates are photosynthetic, some are parasitic, and some are predatory. The major alveolate groups are the dinoflagellates, apicomplexans, and ciliates.

Dinoflagellates Swim by Means of Two Whiplike Flagella

Most *dinoflagellates* are photosynthetic, although there are also some nonphotosynthetic species. Dinoflagellates are named for the motion created by their two whiplike flagella. One flagellum encircles the cell, and the second projects behind it. Some dinoflagellates are enclosed only by a cell membrane; others have cellulose walls that resemble armor plates (**Fig. 16-12**). Although some species live in fresh water, dinoflagellates are especially abundant in the ocean, where they are an important component of the phytoplankton and a food source for larger organisms. Many dinoflagellates are bioluminescent, producing a brilliant blue-green light when disturbed. Specialized dinoflagellates live within the tissues of corals, some clams, and even other protists, where they provide their hosts with

Figure 16-12 Dinoflagellates Two dinoflagellates covered with protective cellulose armor. Visible on each is a flagellum in a groove that encircles the body.

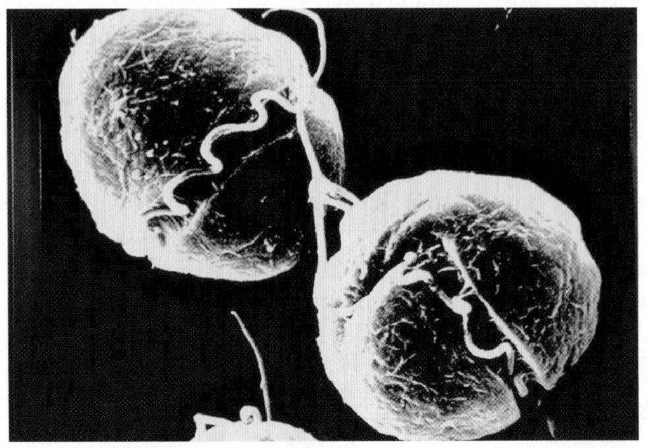

nutrients from photosynthesis and remove carbon dioxide. Reef-building corals live only in the shallow, well-lit waters in which their embedded dinoflagellates can survive.

Warm water that is rich in nutrients may bring on a dinoflagellate population explosion. Dinoflagellates can become so numerous that the water is dyed red by the color of their bodies, causing a "red tide" (Fig. 16-13). During red tides, fish die by the thousands, suffocated by clogged gills or by the oxygen depletion that results from the decay of billions of dinoflagellates. One type of dinoflagellate, *Pfisteria*, even eats fish directly, first secreting chemicals that dissolve the victim's flesh. But dinoflagellate explosions can benefit oysters, mussels, and clams, which have a feast, filtering millions of the protists from the water and consuming them. In the process, however, their bodies accumulate concentrations of a nerve poison produced by the dinoflagellates. Humans who eat these mollusks may be stricken with potentially lethal paralytic shellfish poisoning.

Figure 16-13 A red tide The explosive reproductive rate of certain dinoflagellates under the right environmental conditions can produce dinoflagellate concentrations so great that their microscopic bodies dye the seawater red or brown, as in this bay in Mexico.

Apicomplexans Are Parasitic and Have No Means of Locomotion

All *apicomplexans* (sometimes known as *sporozoans*) are parasitic, living inside the bodies and sometimes inside the individual cells of their hosts. They form infectious spores, resistant structures transmitted from one host to another through food, water, or the bite of an infected insect. As adults, apicomplexans have no means of locomotion. Many have complex life cycles, a common feature of parasites.

A well-known example is the malarial parasite *Plasmodium*. Parts of its life cycle are spent in the stomach, and later the salivary glands, of the female *Anopheles* mosquito. When the mosquito bites a human, it passes the *Plasmodium* to the unfortunate victim. The apicomplexan develops in the victim's liver, then enters the blood, where it reproduces rapidly in red blood cells. The release of large quantities of spores through the rupture of the blood cells causes the recurrent fever of malaria. Uninfected mosquitoes may acquire the parasite by feeding on the blood of a malaria victim, spreading the parasite when they bite another person.

Ciliates Are the Most Complex of the Alveolates

Ciliates, which inhabit fresh and salt water, represent the peak of unicellular complexity. They possess many specialized organelles, including **cilia** (singular, cilium), the short hairlike outgrowths after which they are named. The cilia may cover the cell or may be localized. In the well-known freshwater genus *Paramecium*, rows of cilia cover the organism's entire body surface. Their coordinated beating propels the cell through the water at a rate of 1 millimeter per second—a protistan speed record. Although only a single cell, *Paramecium* responds to its environment as if it had a well-developed nervous system. Confronted with a noxious chemical or a physical barrier, the cell immediately backs up by reversing the beating of its cilia and then proceeds in a new direction. Some ciliates, such as *Didinium*, are accomplished predators (Fig. 16-14).

Figure 16-14 A microscopic predator In this scanning electron micrograph, the predatory ciliate *Didinium* attacks a *Paramecium*. Note that the cilia of *Didinium* are confined to two bands, whereas *Paramecium* has cilia over its entire body. Ultimately, the predator will engulf and consume its prey. This microscopic drama could take place on a pinpoint, with room to spare.

Slime Molds Are Decomposers That Inhabit the Forest Floor

The *slime molds* are another distinctive protist group. The physical form of slime molds seems to blur the boundary between a colony of different individuals and a single, multicellular individual. The life cycle of the slime mold consists of two phases: a mobile feeding stage and a stationary reproductive stage called a *fruiting body*. There are two main types of slime mold: acellular and cellular. We focus here on the acellular slime molds.

Acellular Slime Molds Form a Multinucleate Mass of Cytoplasm

The acellular slime molds consist of a mass of cytoplasm that may spread thinly over an area of several square meters. Although the mass contains thousands of diploid nuclei, the nuclei are not confined in separate cells surrounded by plasma

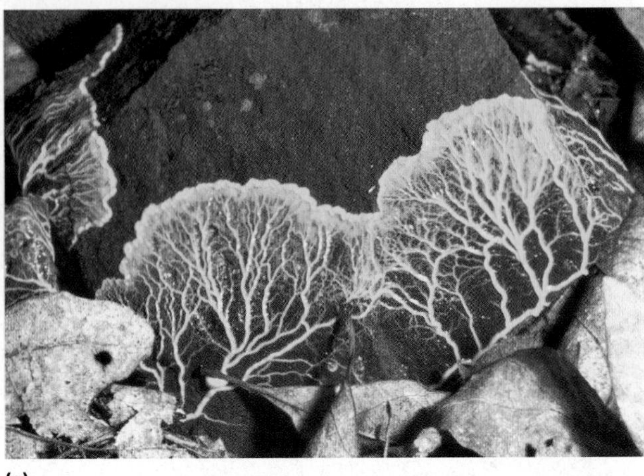
(a)

(b)

Figure 16-15 The acellular slime mold *Physarum* ***(a)*** *Physarum* oozes over a stone on the damp forest floor. ***(b)*** When food becomes scarce, the mass differentiates into black fruiting bodies in which spores are formed.

membranes, as in most multicellular organisms. This structure, called a plasmodium, explains why these protists are described as "acellular" (without cells). The plasmodium oozes through decaying leaves and rotting logs, engulfing food such as bacteria and particles of organic material. The mass may be bright yellow or orange; a large plasmodium can be rather startling (Fig. 16-15a). Dry conditions or starvation stimulate the plasmodium to form a fruiting body that produces haploid spores (Fig. 16-15b). The spores are dispersed and germinate under favorable conditions, eventually giving rise to a new plasmodium.

Various Protists Move by Means of Pseudopods

Like the slime molds, several other types of protist possess flexible plasma membranes that they can extend in any direction to form pseudopods, which are used for locomotion and for engulfing food (Fig. 16-16). Though these various protists all use pseudopods, they are probably not closely related to one another. But because a classification based on evolutionary history has not yet been developed for these groups, we'll discuss them together here.

Amoebas are common in freshwater lakes and ponds. Many amoebas are predators that stalk and engulf prey, but some species are parasites. One parasitic form causes amoebic dysentery, a disease that is prevalent in warm climates. The dysentery-causing amoeba multiplies in the intestinal wall, triggering severe diarrhea.

The *foraminiferans* and *radiolarians* are primarily marine protists that also produce beautiful shells. The shells of foraminiferans are constructed mostly of calcium carbonate (chalk; Fig. 16-17a), and those of radiolarians are of silica (Fig. 16-17b). These elaborate shells are pierced by myriad openings through which pseudopods extend. The chalky shells of foraminiferans, accumulating over millions of years, have resulted in immense deposits of limestone such as those that form the famous White Cliffs of Dover, England.

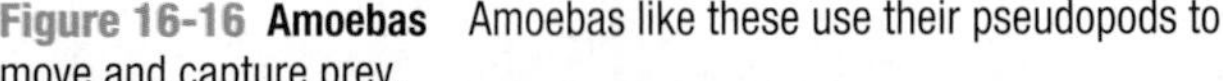
Figure 16-16 Amoebas Amoebas like these use their pseudopods to move and capture prey.

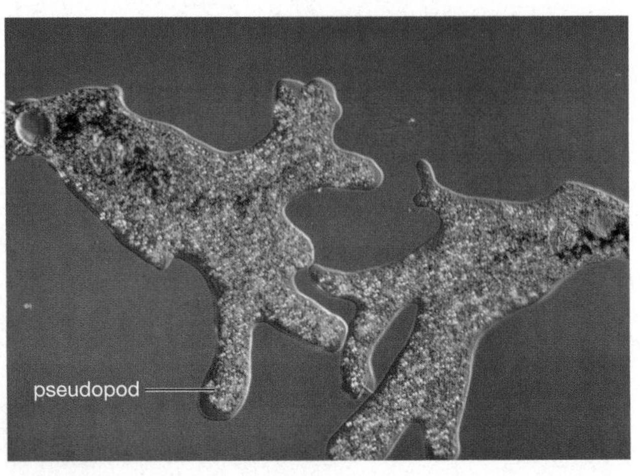

Green Algae Live Mostly in Ponds and Lakes

The *green algae*, a large and diverse group of photosynthetic protists, include both multicellular and unicellular species. Most species live in freshwater ponds and lakes, but some live in the seas. Some green algae, such as *Spirogyra*, form thin filaments from long chains of cells (Fig. 16-18). Other species of green algae form colonies containing clusters of cells that are somewhat interdependent and constitute a structure intermediate between unicellular and multicellular forms. These colonies range from a few cells to a few thousand cells, as in species of

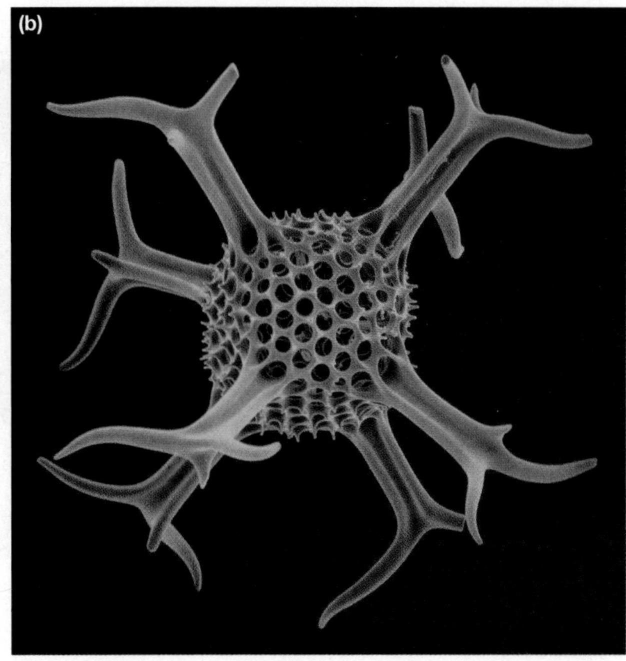

Figure 16-17 Foraminiferans and radiolarians ***(a)*** The chalky shells of foraminiferans show numerous interior chambers. ***(b)*** The delicate, glassy shell of a radiolarian. Pseudopods, which sense the environment and capture food, extend out through the openings in the shell.

Volvox. Most green algae are small, but some marine species are large. For example, the green alga *Ulva*, or sea lettuce, is similar in size to the leaves of its namesake.

The green algae are of special interest because, unlike other groups that contain multicellular, photosynthetic protists, green algae are closely related to plants. Plants share a common ancestor with some types of green algae, and many researchers believe that the very earliest plants were similar to today's multicellular green algae.

16.5 Fungi

When you think of a *fungus*, you probably picture a mushroom. Mushrooms, however, are just temporary reproductive structures that extend from the main bodies of certain kinds of fungus. The body of almost all fungi is a **mycelium** (Fig. 16-19a), which is an interwoven mass of one-cell-thick, threadlike filaments called **hyphae** (singular, hypha; Fig. 16-19b).

Fungi cannot move. They compensate for this lack of mobility with hyphae that can grow rapidly in any direction within a suitable environment. In this way, the fungal mycelium can quickly infuse itself into aging bread or cheese, beneath the bark of decaying logs, or into the soil. Periodically, the hyphae grow together and differentiate into reproductive structures that project above the surface beneath which the mycelium grows. These structures, including mushrooms, puffballs, and the powdery molds on unrefrigerated food, represent only a fraction of the complete fungal body, but are typically the only part of the fungus that we can easily see.

Like plant cells, fungal cells are surrounded by cell walls. Unlike plant cells, however, fungal cell walls are strengthened by chitin, the same substance found in the exoskeletons of arthropods.

Fungi Obtain Their Nutrients from Other Organisms

Like animals, fungi survive by breaking down nutrients stored in the bodies or wastes of other organisms. Some fungi digest the bodies of dead organisms. Others are parasitic, feeding on living organisms and causing disease. There are even a few predatory fungi, which attack tiny worms in soil.

Unlike animals, fungi do not ingest food. Instead, they secrete enzymes that digest complex molecules outside their bodies, breaking down the molecules into smaller subunits that can be absorbed. Fungal filaments can penetrate deeply

Figure 16-18 A green alga *Spirogyra* is a filamentous green alga composed of strands only one cell thick.

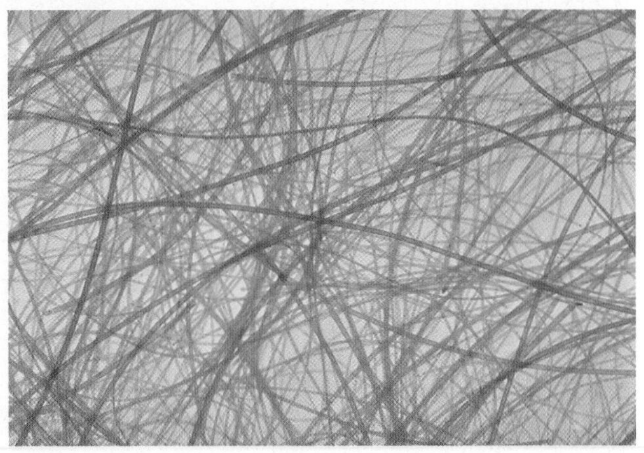

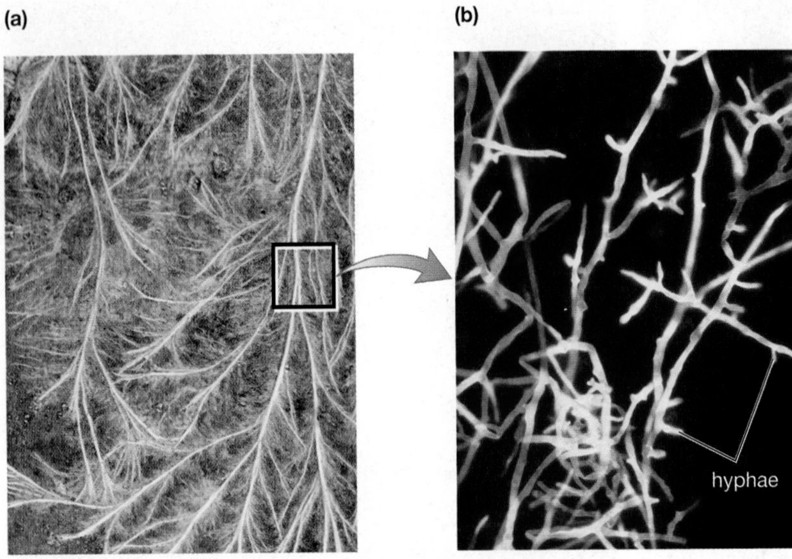

Figure 16-19 The filamentous body of a fungus *(a)* A fungal mycelium spreads over decaying vegetation. The mycelium is composed of *(b)* a tangle of microscopic hyphae, only one cell thick. **QUESTION:** *Which features of a fungus's body structure are adaptations related to its method of acquiring nutrients?*

into a source of nutrients and are only one cell thick, presenting an enormous surface area through which to secrete enzymes and absorb nutrients. This mode of securing nutrition serves fungi well. Almost every biological material can be consumed by at least one fungal species, so nutritional support for fungi is likely to be present in nearly every terrestrial habitat.

Most Fungi Can Reproduce Both Sexually and Asexually

Unlike plants and animals, fungi form no embryos. Instead, fungi propagate by means of tiny, lightweight reproductive packages known as **spores**, which are extraordinarily mobile, even though most lack a means for self-propulsion. Spores are distributed far and wide as hitchhikers on the outside of animal bodies, as passengers inside the digestive systems of animals that have eaten them, or as airborne drifters, cast aloft by chance or shot into the atmosphere by elaborate reproductive structures (Fig. 16-20). In addition, spores are often produced in great numbers (a single giant puffball may contain 5 trillion sexual spores). The fungal combination of prodigious reproductive capacity and highly mobile spores ensures that fungi are ubiquitous in terrestrial environments and accounts for the inevitable growth of fungi on every uneaten sandwich and container of leftovers.

In general, fungi are capable of both asexual and sexual reproduction. For the most part, asexual reproduction is the default mode under stable conditions, and sexual reproduction occurs mainly under conditions of environmental change or stress. Both asexual and sexual reproduction ordinarily involve the production of spores within special fruiting bodies that project above the mycelium.

Figure 16-20 Some fungi can eject spores A ripe earthstar mushroom, struck by a drop of water, releases a cloud of spores that will be dispersed by air currents.

Fungi Attack Plants That Are Important to People

Fungi cause the majority of plant diseases, and some of the plants that they infect are important to humans. For example, fungal pathogens have a devastating effect on the world's food supply. Especially damaging are the plant pests descriptively called rusts and smuts, which cause billions of dollars' worth of damage to grain crops annually (Fig. 16-21). Fungal diseases also affect the appearance of our landscape. The American elm and the American chestnut, two tree species that were once prominent in many of America's parks, yards, and forests, were destroyed on a massive scale by the fungi that cause Dutch elm disease and chestnut blight. Today, few people can recall the graceful forms of large elms and chestnuts, which are now almost entirely absent from the landscape.

The fungal impact on agriculture and forestry is not entirely negative, however. Fungal parasites that attack insects and other arthropod pests can be an important ally in pest control (Fig. 16-22). Farmers who wish to reduce their dependence on toxic and expensive chemical pesticides are increasingly turning to biological methods of pest control, including the application of "fungal pesticides." Fungal pathogens are currently used to control termites, rice weevils, tent caterpillars, aphids, citrus mites, and other pests.

Fungi Cause Human Diseases

The kingdom Fungi includes parasitic species that attack humans directly. Some of the most familiar fungal diseases are those caused by fungi that attack the skin, resulting in athlete's foot, jock itch, and ringworm. These diseases, though unpleasant, are not life threatening and can usually be treated with antifungal ointments. Prompt treatment can also usually control another common fungal disease, vaginal infections caused by the yeast *Candida albicans*. Fungi can also infect the lungs if victims inhale spores of disease-causing fungal species such as those that cause valley fever and histoplasmosis. Like other fungal infections, these diseases can, if promptly diagnosed, be controlled with antifungal drugs. If untreated, however, they can develop into serious, systemic infections. Singer Bob Dylan, for instance, became gravely ill with histoplasmosis when a fungus infected the pericardial membrane surrounding his heart.

Fungi Can Produce Toxins

In addition to their role as agents of infectious disease, some fungi produce toxins that are dangerous to humans. Of particular concern are toxins produced by fungi that grow on grains and other foodstuffs that have been stored in too-moist conditions. For example, molds of the genus *Aspergillus* produce highly toxic, carcinogenic compounds known as aflatoxins. Some foods, such as peanuts, seem especially susceptible to attack by *Aspergillus*. Since aflatoxins were discovered in the 1960s, food growers and processors have developed methods for reducing the growth of *Aspergillus* in stored crops, so aflatoxins have been largely eliminated from the nation's peanut butter supply.

Figure 16-21 Corn smut This fungal pathogen destroys millions of dollars' worth of corn each year. Even a pest such as corn smut has its admirers, though. In Mexico this fungus is known as *huitlacoche* and is considered to be a great delicacy.

Many Antibiotics Are Derived from Fungi

Fungi have also had positive impacts on human health. The modern era of lifesaving antibiotic medicines was ushered in by the discovery of penicillin, which is produced by a mold. Penicillin is still used, along with other fungi-derived antibiotics, to combat bacterial diseases. Other important drugs are also derived from fungi, including cyclosporin, which is used to suppress the immune response during organ transplants so that the body is less likely to reject the transplanted organs.

Figure 16-22 A helpful fungal parasite Some parasitic fungi are used by farmers to control insect pests. Here, a *Cordyceps* species has killed a grasshopper.

Fungi Make Important Contributions to Gastronomy

Fungi make important contributions to human nutrition. The most obvious components of this contribution are the fungi that we consume directly: wild and cultivated mushrooms and other fungi such as morels (Fig. 16-23) and the rare and prized truffle. The role of fungi in cuisine, however, also has less visible manifestations. For example, some of the world's most famous cheeses, including Roquefort, Camembert, Stilton, and Gorgonzola, gain their distinctive flavors from molds that grow on them as they ripen. Perhaps the most important and pervasive fungal contributors to our food supply, however, are the single-celled fungi known as yeasts.

Wine and Beer Are Made Using Yeasts

The discovery that yeasts could be harnessed to enliven our culinary experience is surely a key event in human history. Among the many foods and beverages that depend on yeasts for their production are bread, wine, and beer, which are consumed so widely that it is difficult to imagine a world without them. All derive their special qualities from fermentation by yeasts. Fermentation occurs when yeasts extract energy from sugar and, as by-products of the metabolic process, emit carbon dioxide and ethyl alcohol. As yeasts consume the fruit sugars in grape juice, the sugars are replaced by alcohol, and wine is the result. Eventually, the increasing concentration of alcohol kills the yeasts, ending fermentation. If the yeasts die before all available grape sugar is consumed, the wine will be sweet; if the sugar is exhausted, the wine will be dry.

Beer is brewed from grain (usually barley), but yeasts cannot effectively consume the carbohydrates that compose grain kernels. For the yeasts to do their work, the barley grains must have sprouted (recall that grains are actually seeds). Germination converts the kernels' carbohydrates to sugar, so the sprouted barley

Figure 16-23 A delicious fungus The morel, an edible delicacy. (Consult an expert before sampling any wild fungus—some are deadly!)

Figure 16-24 Bryophytes Both plants shown here are less than a half inch (about 1 centimeter) in height. ***(a)*** Liverworts grow in moist, shaded areas. This is the female plant, bearing umbrella-like structures, which hold the eggs. Sperm must swim up the stalks through a film of water to fertilize the eggs. ***(b)*** Moss plants, showing the stalks that carry spore-bearing capsules. **QUESTION:** *Why are all bryophytes short?*

Web Tutorial 16.3 Evolution of Plant Structure

provides an excellent food source for the yeasts. As with wine, fermentation converts sugars to alcohol, but beer brewers capture the carbon dioxide by-product as well, giving the beer its characteristic bubbly carbonation.

Yeasts Make Bread Rise

In bread making, carbon dioxide is the crucial fermentation product. The yeasts added to bread dough do produce alcohol as well as carbon dioxide, but the alcohol evaporates during baking. In contrast, the carbon dioxide is trapped in the dough, where it forms the bubbles that give bread its light, airy texture (and saves us from a life of eating sandwiches made of crackers). So the next time you're enjoying a slice of French bread with Camembert cheese and a nice glass of Chardonnay, or a slice of pizza and a cold bottle of your favorite brew, you might want to quietly give thanks to the yeasts. Our diets would certainly be a lot duller without the help we get from fungal partners.

Fungi Play a Crucial Ecological Role

No account of the fungi would be complete without mention of their fundamental ecological importance. The fungi are Earth's undertakers, consuming the dead of all kingdoms and returning their component substances to the ecosystems from which they came. The extracellular digestive activities of many fungi liberate nutrients such as carbon, nitrogen, and phosphorus compounds and minerals that can be used by plants. If fungi and bacteria were suddenly to disappear, the consequences would be disastrous. Nutrients would remain locked in the bodies of dead plants and animals, the recycling of nutrients would grind to a halt, soil fertility would rapidly decline, and waste and organic debris would accumulate. In short, ecosystems would collapse.

BACTERIA ARCHAEA EUKARYA
animals fungi plants protists

16.6 Plants

Two major groups of land plants arose from ancient algal ancestors. One group, the **bryophytes** (also called nonvascular plants), requires a moist environment to reproduce and thus straddles the boundary between aquatic and terrestrial life, much like the amphibians of the animal kingdom. The other group, the **vascular plants** (also called tracheophytes), has been able to colonize drier habitats.

Bryophytes Lack Conducting Structures

Bryophytes retain some characteristics of their algal ancestors. They lack true roots, leaves, and stems. They do possess rootlike anchoring structures called rhizoids that bring water and nutrients into the plant body, but bryophytes are nonvascular; they lack well-developed structures for conducting water and nutrients. They must instead rely on slow diffusion or poorly developed conducting tissues to distribute water and other nutrients. As a result, their body size is limited. Size is also limited by the absence of any stiffening agent in their bodies. Without such material, they cannot grow upward very far. Most bryophytes are less than 1 inch (2.5 centimeters) tall.

The most common bryophytes are the liverworts and mosses (**Fig. 16-24**), which are generally most abundant in areas where moisture is plentiful. Some mosses, however, have a waterproof covering that retains moisture, preventing water loss. These mosses can survive in deserts, on bare rock, and in far northern and southern latitudes where humidity is low and liquid water is scarce for much of the year.

The Reproductive Structures of Bryophytes Are Protected

Among the bryophytes' adaptations to terrestrial existence are their enclosed reproductive structures, which prevent the gametes from drying out. There are two types of structure, one in which eggs develop and one in which sperm are formed. In some bryophyte species, both egg-producing and sperm-producing structures are located on the same plant; in other species, each individual plant is either male or female. In all bryophytes, the sperm must swim to the egg (which emits a chemical attractant) through a film of water. Bryophytes that live in drier areas must time their reproduction to coincide with rains.

The Vascular Plants Have Conducting Vessels That Also Provide Support

Vascular plants are distinguished by specialized groups of conducting cells called vessels. The vessels are impregnated with the stiffening substance lignin and serve both supportive and conducting functions. Vessels allow vascular plants to grow taller than nonvascular plants, both because of the extra support provided by lignin and because the conducting cells allow water and nutrients absorbed by the roots to move to the upper portions of the plant. The vascular plants can be divided into two groups: the seedless vascular plants and the seed plants.

The Seedless Vascular Plants Include the Club Mosses, Horsetails, and Ferns

Like the bryophytes, *seedless vascular plants* have swimming sperm and require water for reproduction. As their name implies, they do not produce seeds but rather propagate by spores. The present-day seedless vascular plants—the club mosses, horsetails, and ferns—are much smaller than their ancestors, which dominated the landscape hundreds of millions of years ago. The bodies of ancient seedless vascular plants—transformed by heat, pressure, and time—are burned today as coal. The seedless vascular plants were once dominant, but today the more versatile seed plants have become dominant.

The club mosses are now limited to representatives a few inches in height. Their leaves are small and scale-like (Fig. 16-25a). Modern horsetails form a single genus, *Equisetum*, that contains only 15 species, most less than 3 feet tall. Horsetail leaves are reduced to tiny scales on the branches (Fig. 16-25b). The ferns, with 12,000 species, are the most diverse of the seedless vascular plants. In the tropics, tree ferns still reach heights reminiscent of their ancestors in the Carboniferous period. Ferns are the only seedless vascular plants that have broad leaves (Fig. 16-25c).

The Seed Plants Dominate the Land, Aided by Two Important Adaptations: Pollen and Seeds

The *seed plants* are distinguished from bryophytes and seedless vascular plants by their production of pollen and seeds. **Pollen** grains are tiny structures that carry sperm-producing cells. Pollen grains are dispersed by wind or by animal pollinators such as bees. In this way, sperm move through the air to fertilize egg cells. Thus, the distribution of seed plants is not limited by the need for water through which sperm can swim to the egg. Seed plants are fully adapted to life on dry land.

(a)

(b)

(c)

Figure 16-25 Seedless vascular plants Seedless vascular plants are found in moist woodland habitats. ***(a)*** The club mosses (sometimes called ground pines) grow in temperate forests. This specimen is releasing spores. ***(b)*** The giant horsetail (genus *Equisetum*) extends long, narrow branches in a series of rosettes. Its leaves are insignificant scales. ***(c)*** The leaves of this deer fern are emerging from coiled fiddleheads.

Analogous to the eggs of birds and reptiles, **seeds** consist of an embryonic plant, a supply of food for the embryo, and a protective outer coat. The *seed coat* maintains the embryo in a state of suspended animation or dormancy until conditions are proper for growth. The stored food helps sustain the emerging plant until it develops roots and leaves and can make its own food by photosynthesis. Some seeds possess elaborate adaptations that allow them to be dispersed by wind, water, and animals.

Seed plants are grouped into two general types: gymnosperms, which lack flowers, and angiosperms, the flowering plants.

Gymnosperms Are Nonflowering Seed Plants

Gymnosperms evolved earlier than the flowering plants. One group, the **conifers**, still dominates large areas of our planet. Conifers, whose 500 species include pines, firs, spruce, hemlocks, and cypresses, are most abundant in the cold latitudes of the far north and at high elevations where conditions are dry. Not only is rainfall limited in these areas, but water in the soil remains frozen and unavailable during the long winters. Conifers are adapted to dry, cold conditions in three ways. First, conifers retain green leaves throughout the year, enabling these plants to continue photosynthesizing and growing slowly during times when most other plants become dormant. For this reason, conifers are often called evergreens. Second, conifer leaves are actually thin needles covered with a thick, waterproof surface that minimizes evaporation. Finally, conifers produce an "antifreeze" in their sap that enables them to continue transporting nutrients in below-freezing temperatures. This substance gives them their fragrant piney scent. Other gymnosperms, such as the ginkgos and cycads (Fig. 16-26), have declined to a small remnant of their former range and abundance.

Figure 16-26 Two uncommon gymnosperms *(a)* The ginkgo, or maidenhair tree, has been kept from extinction by cultivation in China and Japan. Relatively resistant to pollution, these trees have become popular in U.S. cities. This ginkgo is female and bears fleshy seeds the size of large cherries, which are noted for their foul smell when ripe. *(b)* A cycad. Common in the age of dinosaurs, cycads are now limited to about 160 species living in warm, moist climates. Like ginkgos, cycads have separate sexes.

(a)

(b)

Angiosperms Are Flowering Seed Plants

Modern flowering plants, or **angiosperms**, have dominated Earth for more than 100 million years. The group is incredibly diverse, with more than 230,000 species. Angiosperms range in size from the diminutive duckweed (Fig. 16-27a) to the towering eucalyptus tree (Fig. 16-27b). From desert cactus to tropical orchids to grasses to parasitic mistletoe, angiosperms rule over the plant kingdom.

Flowers Attract Pollinators

Three major adaptations have contributed to the enormous success of angiosperms: flowers, fruits, and broad leaves. **Flowers**, the structures in which both male and female gametes are formed, may have evolved when gymnosperm ancestors formed an association with animals (most likely insects) that carried their pollen from plant to plant. According to this scenario, the relationship between these ancient gymnosperms and their animal pollinators was so beneficial that natural selection favored the evolution of showy flowers that advertised the presence of pollen to insects and other animals (Fig. 16-27b,e). The animals benefited by eating some of the protein-rich pollen, whereas the plant benefited from the animals' unwitting transportation of pollen from plant to plant. With this animal assistance, many flowering plants no longer needed to produce prodigious quantities of pollen and send it flying on the fickle winds to ensure fertilization. But there are also many wind-pollinated angiosperms (Fig. 16-27c,d).

Fruits Encourage Seed Dispersal

The ovary surrounding the seed of an angiosperm matures into a **fruit**, the second adaptation that has contributed to the success of angiosperms. Just as flowers encourage animals to transport pollen, so, too, many fruits entice animals to disperse seeds. If an animal eats a fruit, many of the enclosed seeds may pass through the animal's digestive tract unharmed, perhaps to fall at a suitable location for germination.

Figure 16-27 **Flowering plants** ***(a)*** The smallest angiosperm is the duckweed, found floating on ponds. These specimens are about $\frac{1}{8}$ inch (3 millimeters) in diameter. ***(b)*** The largest angiosperms are eucalyptus trees, which can reach 325 feet (100 meters) in height. Inconspicuous flowers and wind pollination are features of ***(c)*** grasses and many trees, such as ***(d)*** this birch, in which flowers are shown as buds (green) and blossom (brown). ***(e)*** More conspicuous flowers, such as those shown on this butterfly weed and eucalyptus tree [inset in part (b)], entice insects and other animals that carry pollen between individual plants. **QUESTION:** *What are the advantages and disadvantages of wind pollination? Of pollination by animals? Why do both types of pollination persist among the angiosperms?*

Not all fruits, however, depend on edibility for dispersal. Dog owners are well aware, for example, that some fruits (called burs) disperse by clinging to animal fur. Other fruits, such as those of maples, form wings that carry the seed through the air. The variety of dispersal mechanisms made possible by fruits has helped the angiosperms invade nearly all possible terrestrial habitats.

Broad Leaves Capture More Sunlight

The third feature that gives angiosperms an advantage in warmer, wetter climates is broad leaves. When water is plentiful, as it is during the warm growing season of temperate and tropical climates, broad leaves provide an advantage by collecting more sunlight for photosynthesis. In regions with seasonal variation in growing conditions, many trees and shrubs drop their leaves during periods when water is in short supply, because being leafless reduces evaporative water loss. In temperate climates, such periods occur during the fall and winter, at which time most temperate angiosperm trees and shrubs drop their leaves. In the tropics and subtropics, most angiosperms are evergreen, but species that inhabit certain tropical climates where periods of drought are common may drop their leaves to conserve water during the dry season.

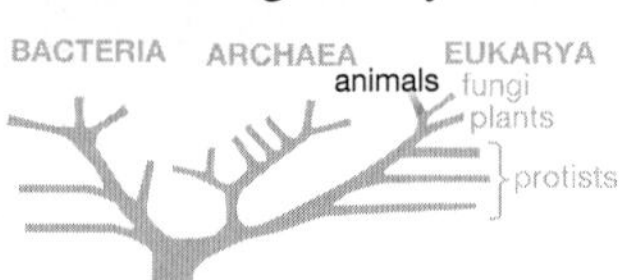

16.7 Animals

It is difficult to devise a concise definition of the term "animal." No single feature fully characterizes animals, so the group is defined by a list of characteristics. None of these characteristics is unique to animals, but together they distinguish animals from members of other kingdoms:

- Animals are multicellular.
- Animals obtain their energy by consuming the bodies of other organisms.
- Animals typically reproduce sexually. Although animal species exhibit a tremendous diversity of reproductive styles, most are capable of sexual reproduction.
- Animal cells lack a cell wall.
- Animals are motile (able to move about) during some stage of their life. Even the stationary sponges have a free-swimming larval stage (a juvenile form).
- Most animals are able to respond rapidly to external stimuli as a result of the activity of nerve cells, muscle tissue, or both.

Most Animals Lack Backbones

It's easy to overlook the differences among the multitude of small, boneless animals in the world. Even Carolus Linnaeus, the originator of modern biological classification, recognized only two phyla of animals without bones (insects and "worms"). Today, however, biologists recognize about 27 phyla of boneless animals.

For convenience, biologists often place animals in one of two major categories: **vertebrates**, those with a backbone (or vertebral column), and **invertebrates**, those lacking a backbone. The vertebrates—fish, amphibians, reptiles, birds, and mammals—are perhaps the most conspicuous animals from a human point of view, but less than 3% of all known animal species on Earth are vertebrates. The vast majority of animals are invertebrates.

The earliest animals probably originated from colonies of protists whose members had become specialized to perform distinct roles within the colony. In our survey of the kingdom Animalia, we will begin with the sponges, whose body plan most closely resembles the probable ancestral protozoan colonies.

Sponges Have a Simple Body Plan

Sponges lack true tissues and organs. In some ways, a sponge resembles a colony of single-celled organisms, and a few biologists believe that sponges should be classified as the protist group most closely related to animals, rather than within the animal kingdom.

All sponges have a similar body plan. The body is perforated by numerous tiny pores, through which water enters, and by fewer, large openings, through which it is expelled. Within the sponge, water travels through canals. As water passes through the sponge, oxygen is extracted, microorganisms are filtered out and taken into individual cells where they are digested, and wastes are released.

Sponges come in a variety of shapes and sizes. Some species have a well-defined shape, but others grow free-form over underwater rocks (Fig. 16-28). The largest sponges can grow to more than 3 feet (1 meter) in height. An internal skeleton composed of calcium carbonate (chalk), silica (glass), or protein spines provides support for the sponge's body. Natural household sponges, which are now largely replaced by factory-made cellulose imitations, are actually sponge skeletons.

All sponges live in water, mostly in saltwater environments. Adult sponges live attached to rocks or other underwater surfaces and do not move about. Sponges may reproduce asexually by budding, in which the adult produces miniature versions of itself that drop off and assume an independent existence. Alternatively, they may reproduce sexually through the fusion of sperm and eggs. Fertilized eggs develop inside the adult into active larvae that escape through the openings in the sponge body. Water currents disperse the larvae to new areas, where they settle and develop into adult sponges.

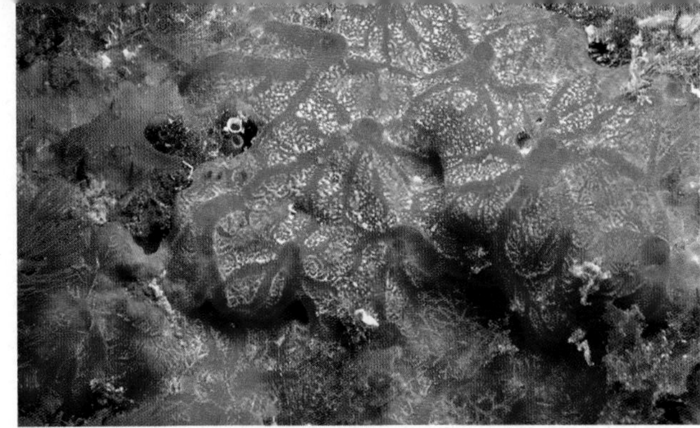

(a)

(b) (c)

Figure 16-28 Sponges Some sponges, such as ***(a)*** this fire sponge, grow in free-form pattern over undersea rocks. ***(b)*** Tiny appendages attach this tubular sponge to rocks, whereas ***(c)*** this reef sponge with flared tubular openings attaches to a coral reef. **QUESTION:** *Sponges are often described as the most "primitive" of animals. How can such a primitive organism have become so diverse and abundant?*

Cnidarians Are Well-Armed Predators

Cnidarians—jellyfish, sea anemones, corals, and hydrozoans—come in a bewildering and beautiful variety of forms (Fig. 16-29). In cnidarians, body parts are arranged in a circle around the mouth and digestive cavity. This arrangement of

Figure 16-29 Cnidarians ***(a)*** A red-spotted sea anemone spreads its tentacles to capture prey. ***(b)*** A small jellyfish. ***(c)*** A close-up of coral reveals bright yellow individuals in various stages of tentacle extension. At the lower right, areas where the coral has died expose the calcium carbonate skeleton that supports the animals and forms the reef. A crab with long banded legs sits on the coral, holding tiny white sea anemones in its claws. Their stinging tentacles help protect the crab. ***(d)*** A sea wasp, a cnidarian whose stinging cells contain one of the most toxic of all known venoms.

Figure 16-30 Annelids ***(a)*** A polychaete annelid projects brightly spiraling gills from a tube, made by the worm and attached to rock. When the gills retract, the tube is covered by a "trap door" visible as a reddish collar encircling the top of the tube. ***(b)*** The "fireworm" polychaete swims by using paddles on each segment. The bristles on each paddle can deliver a fiery sting. ***(c)*** This leech, a freshwater annelid, shows numerous segments. The whitish disk is a sucker encircling its mouth, allowing it to attach to its prey.

(a)

(b)

(c)

parts is well suited to these animals, which are either fixed in place or carried randomly by water currents, because it enables them to respond to prey or threats from any direction.

Cnidarian tentacles are armed with cells containing structures that, when stimulated by contact, explosively inject poisonous or sticky filaments into prey. The venom of some cnidarians can cause painful stings in humans unfortunate enough to come into contact with them, and the stings of a few jellyfish species can even be life threatening. The most deadly of these species is the sea wasp, *Chironex fleckeri*, which is found in the waters off northern Australia and Southeast Asia. The amount of venom in a single sea wasp could kill up to 60 people, and the victim of a serious sting may die within minutes of being stung.

The function of stinging cells, of course, is not to sting human swimmers but to capture prey. Although all cnidarians are predatory, none hunt actively. Instead, they wait for their victims to blunder, by chance, into the grasp of their enveloping tentacles. Stung and firmly grasped, the prey is forced through an expansible mouth into a digestive sac.

Like sponges, cnidarians are confined to watery habitats, and most species are marine. One group of cnidarians, the corals, is of particular ecological importance (see Fig. 16-29c). Coral individuals form large colonies, and each member of the colony secretes a hard skeleton of calcium carbonate. The skeletons persist long after the organisms die, serving as a base to which other individuals may attach themselves. The cycle continues until, after thousands of years, massive coral reefs are formed. Corals are restricted to the warm, clear waters of the tropics, where their reefs form undersea habitats that are the basis of an ecosystem of stunning diversity and unparalleled beauty (see Chapter 30).

Annelids Are Composed of Identical Segments

A prominent feature of the segmented worms, or *annelids*, is the division of the body into a series of repeating segments. Externally, these segments appear as ringlike depressions on the surface. Internally, most of the segments contain identical copies of nerves, excretory structures, and muscles. Segmentation is advantageous for locomotion, because the body compartments are each controlled by separate muscles and collectively are capable of more complex movement than could be achieved with only one set of muscles to control the whole body.

The annelid phylum includes three main subgroups: the oligochaetes, the polychaetes, and the leeches. The oligochaetes include the familiar earthworm and its relatives. Polychaetes live primarily in the ocean. Some polychaetes have paired fleshy paddles on most of their segments, used in locomotion. Others live in tubes from which they project feathery gills that both exchange gases and sift the water for microscopic food (Fig. 16-30a,b). Leeches (Fig. 16-30c) live in freshwater or moist terrestrial habitats and are either carnivorous or parasitic. Carnivorous leeches prey on smaller invertebrates; parasitic leeches suck the blood of larger animals.

Arthropods Are the Dominant Animals on Earth

In terms of both number of individuals and number of species, no other animal phylum comes close to the *arthropods*, which include insects, arachnids, and crustaceans. About 1 million arthropod species have been discovered, and scientists estimate that millions more remain undescribed.

All arthropods have an **exoskeleton**, an external skeleton that encloses the arthropod body like a suit of armor. This external skeleton protects against predators and is responsible for arthropods' greatly increased agility relative to their wormlike ancestors. The exoskeleton is thin and flexible in places, allowing movement of the paired, jointed appendages. By providing stiff but flexible appendages and rigid attachment sites for muscles, the exoskeleton makes possible the flight of the bumblebee and the intricate, delicate manipulations of the spider as it weaves its web. The exoskeleton also contributed enormously to the arthropod invasion of the land (arthropods were the earliest terrestrial animals) by providing a watertight covering for delicate, moist tissues such as those used for gas exchange.

Like a suit of armor, the arthropod exoskeleton poses some problems. First, because it cannot expand as the animal grows, the exoskeleton must be shed, or **molted**, periodically and replaced with a larger one. Molting uses energy and leaves the animal temporarily vulnerable until the new skeleton hardens. ("Soft-shelled" crabs are simply regular "hard-shelled" crabs that are caught just after molting.) The exoskeleton is also heavy, and its weight increases exponentially as the animal grows. It is no coincidence that the largest arthropods are crustaceans (crabs and lobsters), whose watery habitat supports much of their weight.

Arthropods are segmented, but their segments tend to be few and specialized for different functions such as sensing the environment, feeding, and movement. For example, in insects, sensory and feeding structures are concentrated on the front segment, known as the *head*, and digestive structures are largely confined to the *abdomen*, which is the rear segment. Between the head and the abdomen is the *thorax*, the segment to which structures used in locomotion, such as wings and walking legs, are attached.

Insects Are the Only Flying Invertebrates

The number of described *insect* species is about 850,000, roughly three times the total number of known species in all other classes of animals combined (Fig. 16-31). Insects have three pairs of legs, normally supplemented by two pairs of wings. Insects' capacity for flight distinguishes them from all other invertebrates and has contributed to their enormous success. As anyone who has pursued a fly can testify, flight helps in escaping from predators. It also allows the insect to find widely dispersed food. Swarms of locusts (Fig. 16-31d) have been traced moving all the way from Saskatchewan, Canada, to Texas on the trail of food.

During their development, insects undergo **metamorphosis**, a radical change from a juvenile body form to an adult body form. In insects with complete metamorphosis, the immature stage, called a **larva** (plural, larvae), is worm-shaped (for example, the maggot of a housefly or the caterpillar of a moth or butterfly; see Fig. 16-31e). The larva hatches from an egg, grows by eating voraciously and shedding its exoskeleton several times, and then forms a nonfeeding stage called a **pupa**. Encased in an outer covering, the pupa undergoes a radical change in body form, emerging in its adult winged form. The adults mate and lay eggs, continuing the cycle. Metamorphosis may include a change in diet as well as in shape, thereby eliminating competition for food between adults and juveniles and in some cases allowing the insect to exploit different foods when they are most available. For instance, a caterpillar that feeds on new green shoots in springtime metamorphoses into a butterfly that drinks nectar from the summer's blooming flowers. Some insects undergo a more gradual metamorphosis (called incomplete metamorphosis), hatching as young that bear some resemblance to the adult, then gradually acquiring more-adult features as they grow and molt.

(a)

(b)

(c)

(d)

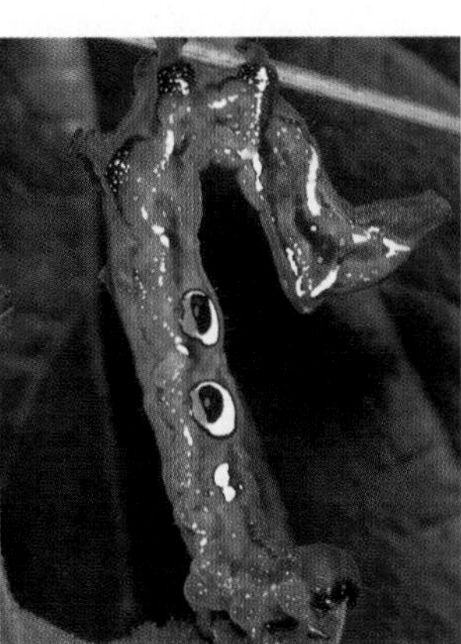

(e)

Figure 16-31 **Insects** ***(a)*** The rose aphid sucks sugar-rich juice from plants. ***(b)*** A mating pair of Hercules beetles. Only the male has the large "horns." ***(c)*** A June beetle displays its two pairs of wings as it comes in for a landing. The outer wings protect the abdomen and the inner wings, which are relatively thin and fragile. ***(d)*** Insects such as this locust can cause devastation of both crops and natural vegetation. ***(e)*** Caterpillars are larval forms of moths or butterflies. This caterpillar larva of the Australian fruit-sucking moth displays large eyespot patterns that may frighten potential predators, who mistake them for eyes of a large animal.

Most Arachnids Are Predatory Meat Eaters

The *arachnids* include spiders, mites, ticks, and scorpions (Fig. 16-32). All members of the class Arachnida have eight walking legs, and most are carnivorous. Many subsist on a liquid diet of blood or predigested prey. For example, spiders, the most numerous arachnids, first immobilize their prey with a paralyzing venom. They then inject digestive enzymes into the helpless victim (typically an insect) and suck in the resulting soup. Arachnid eyes are particularly sensitive to movement, and in some species they probably can form images. Most spiders have eight eyes placed in such a way as to give them a panoramic view of predators and prey.

(a)

(b)

(c)

Figure 16-32 Arachnids ***(a)*** The tarantula is among the largest of spiders but is relatively harmless. ***(b)*** Scorpions, found in warm climates, including deserts of the southwestern United States, paralyze their prey with venom from a stinger at the tip of the abdomen. A few species can harm humans. ***(c)*** Ticks before (*left*) and after feeding on blood. The uninflated exoskeleton is flexible and folded, allowing the animal to enlarge while feeding.

Most Crustaceans Are Aquatic

The *crustaceans*, including crabs, crayfish, lobster, shrimp, and barnacles, make up the only class of arthropods whose members live primarily in the water (Fig. 16-33). Crustaceans range in size from microscopic maxillopods that live in the spaces between grains of sand to the largest of all arthropods, the Japanese crab, with legs spanning nearly 12 feet (4 meters). Crustaceans have two pairs of sensory antennae, but the rest of their appendages are highly variable in form and number, depending on the habitat and lifestyle of the species.

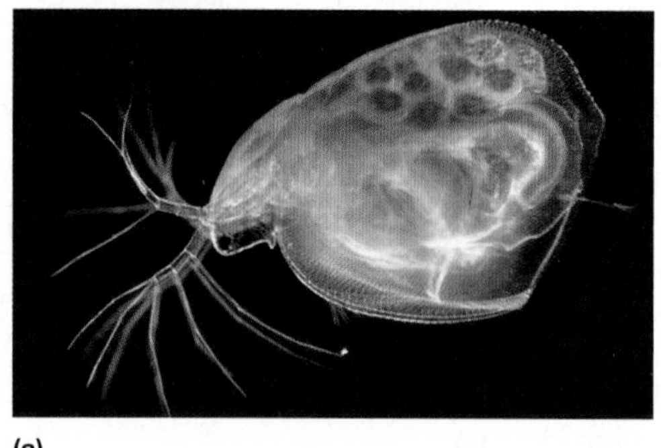

(a)

(b)

(c)

(d)

Figure 16-33 Crustaceans ***(a)*** The microscopic waterflea is common in freshwater ponds. Notice the eggs developing within the body. ***(b)*** The sowbug, found in dark, moist places such as under rocks, leaves, and decaying logs, is one of the few crustaceans whose ancestors successfully invaded the land. ***(c)*** The hermit crab protects its soft abdomen by inhabiting an abandoned snail shell. ***(d)*** The goose-neck barnacle uses a tough, flexible stalk to anchor itself to rocks, boats, or even animals such as whales. Other types of barnacles attach with shells that resemble miniature volcanoes (see Fig. 16-35b). Early naturalists thought barnacles were mollusks until the jointed legs, seen extending into the water, were observed.

(a)

(b)

Figure 16-34 Gastropod mollusks ***(a)*** A Florida tree snail displays a brightly striped shell and eyes at the tip of stalks that are retracted instantly if touched. ***(b)*** Spanish shawl sea slugs prepare to mate.

Most Mollusks Have Shells

In terms of number of species, the *mollusks* are second (albeit a distant second) only to the arthropods. Some mollusks protect their bodies with a shell of calcium carbonate, and others escape predation by moving swiftly or, if caught, by tasting terrible. With the exception of some snails and slugs, mollusks are water dwellers. Among the many classes of mollusks, we will discuss three in more detail: snails and their relatives, clams and their relatives, and octopuses and their relatives.

Gastropods Are One-Footed Crawlers

Snails and slugs, collectively known as *gastropods*, crawl on a muscular foot, and many are protected by shells that vary widely in form and color (**Fig. 16-34a**). Not all gastropods have shells, however. Sea slugs, for example, lack shells, but their brilliant colors warn potential predators that they are poisonous, or at least taste bad (**Fig. 16-34b**).

Gastropods feed with a radula, a flexible ribbon of tissue studded with spines that is used to scrape algae from rocks or to grasp larger plants or prey. Most snails use gills, typically enclosed in a cavity beneath the shell, for respiration. Gases can also diffuse readily through the skin of most gastropods, and most sea slugs rely on this mode of gas exchange. The few gastropod species (including the destructive garden snails and slugs) that live in terrestrial habitats use a simple lung for breathing.

Bivalves Are Filter Feeders

Included among the *bivalves* are scallops, oysters, mussels, and clams (**Fig. 16-35**). Not only do members of the class lend exotic variety to the human diet, but they are also important members of the nearshore marine community. Bivalves possess

Figure 16-35 Bivalve mollusks ***(a)*** This swimming scallop parts its hinged shells. The upper shell is covered with an encrusting sponge. ***(b)*** Mussels attach to rocks in dense aggregations exposed at low tide. White barnacles are attached to the mussel shells and surrounding rock.

(a)

(b)

(a)

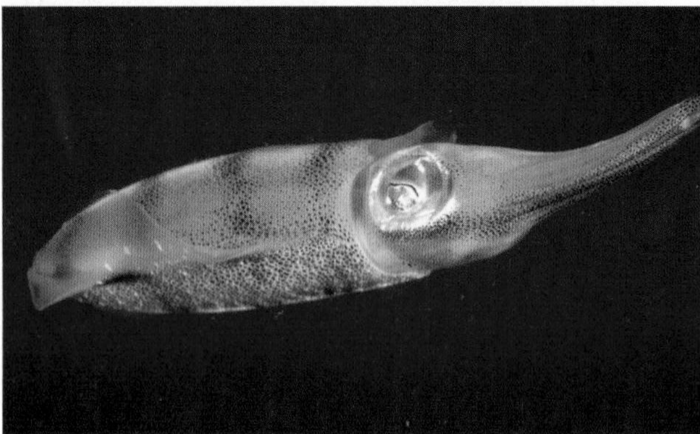

(b)

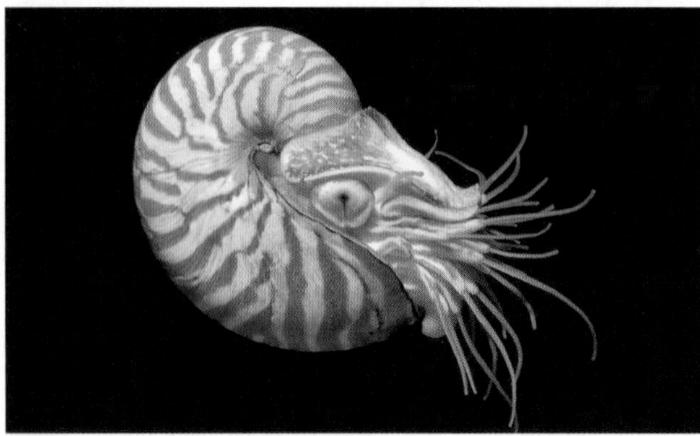

(c)

Figure 16-36 Cephalopod mollusks *(a)* An octopus can crawl rapidly by using its eight suckered tentacles. It can alter its color and skin texture to blend with its surroundings. In emergencies this mollusk can jet backward by vigorously contracting its mantle. *(b)* The squid moves by contracting its mantle to generate jet propulsion, which pushes the animal backward through the water. *(c)* The chambered nautilus secretes a shell with internal, gas-filled chambers that provide buoyancy. Note the well-developed eye and the tentacles used to capture prey.

two shells connected by a flexible hinge. A strong muscle clamps the shells closed in response to danger. This muscle is what you are served when you order scallops in a restaurant.

Clams use a muscular foot for burrowing in sand or mud. In mussels, which live attached to rocks, the foot is smaller and is used to help secrete threads that anchor the animal to the rocks. Scallops lack a foot and move by a sort of whimsical jet propulsion achieved by flapping their shells together.

Bivalves are filter feeders, using their gills as both respiratory and feeding structures. Water is circulated over the gills, which are covered with a thin layer of mucus that traps microscopic food particles. Food is conveyed to the mouth by the beating of cilia on the gills. Probably because they filter-feed and do not move extensively, bivalves "lost their heads" as they evolved.

Cephalopods Are Marine Predators

The *cephalopods* include octopuses, nautiluses, cuttlefish, and squids (Fig. 16-36). The largest invertebrate, the giant squid, which can be up to 60 feet (18 meters) long, belongs to this group. All cephalopods are predatory carnivores, and all are marine. In these mollusks, the foot has evolved into tentacles with well-developed sensory abilities and suction disks for detecting and grasping prey. Prey grasped by tentacles may be immobilized by a paralyzing venom in the saliva before being torn apart by beaklike jaws.

Cephalopods have highly developed brains and sensory systems. The cephalopod eye rivals our own in complexity and exceeds it in efficiency of design. The cephalopod brain, especially that of the octopus, is (for an invertebrate brain) exceptionally large and complex. It is enclosed in a skull-like case of cartilage and endows the octopus with highly developed capabilities to learn and remember. In the laboratory, octopuses can rapidly learn to associate certain symbols with food and to open a screw-cap jar to obtain food.

The Chordates Include Both Invertebrates and Vertebrates

It's hard to believe that humans have anything in common with a sea squirt. Nonetheless, we share the *chordate* phylum not only with the birds and the apes but also with the tunicates (sea squirts) and little fishy-looking creatures called lancelets. What characteristics do we share with these creatures that seem so different? All chordates are united by four features that all possess at some stage of their lives:

- A **notochord**. A stiff but flexible rod that extends the length of the body and provides an attachment site for muscles.
- A dorsal, hollow **nerve cord**. This hollow structure develops a thickening at one end that becomes a brain.
- **Pharyngeal gill slits**. Located in the pharynx (the cavity behind the mouth), these may form functional openings for gills (organs for gas exchange in water) or may appear only as grooves during an early stage of development.
- A **postanal tail**. An extension of the body past the anus.

This list may seem puzzling because, although humans are chordates, at first glance we seem to lack every feature except the nerve cord. But evolutionary relationships are sometimes seen most clearly during early stages of development, and it is during our embryonic life that we develop, and lose, our notochord, our gill slits, and our tails. Humans share these chordate features with all other vertebrates and with two invertebrate chordate groups, the lancelets and the tunicates.

The Invertebrate Chordates Live in the Seas

The invertebrate chordates lack the backbone that is the defining feature of vertebrates. These chordates comprise two groups, the lancelets and the tunicates. The small (2 inches, or about 5 centimeters, long) fishlike lancelet spends most of its time half-buried in the sandy sea bottom, filtering tiny food particles from the water. All four chordate features are present in the adult lancelet.

The tunicates form a larger group of marine invertebrate chordates that includes the sea squirts. It is difficult to imagine a less likely relative of humans than the immobile, filter-feeding, vaselike sea squirt (Fig. 16-37). Its ability to move is limited to forceful contractions of its saclike body, which can send a jet of seawater into the face of anyone who plucks it from its undersea home; hence the name sea squirt. Although adult sea squirts are immobile, their larvae swim actively and possess the four chordate features.

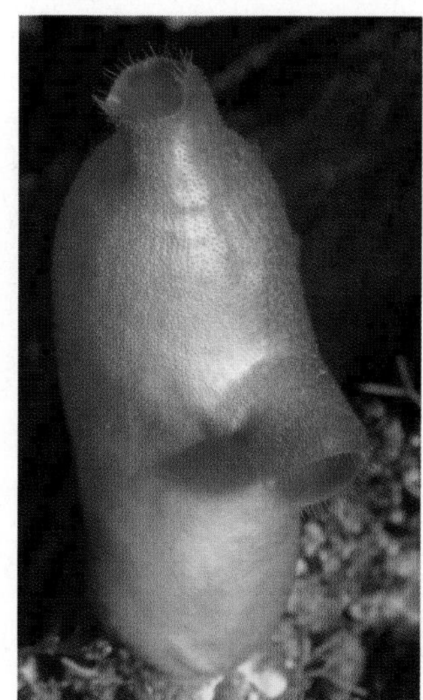

Figure 16-37 An invertebrate chordate This adult sea squirt (a type of tunicate) has lost its tail and notochord and has assumed a sedentary life.

Vertebrates Have a Backbone

In vertebrates, the embryonic notochord is normally replaced during development by a backbone, or **vertebral column**. The vertebral column is composed of bone or cartilage, a tissue that resembles bone but is less brittle and more flexible. This column supports the body, offers attachment sites for muscles, and protects the delicate nerve cord and brain. It is also part of a living internal skeleton that can grow and repair itself.

Vertebrates show other adaptations that have contributed to their successful invasion of most habitats. One is paired appendages. These first appeared as fins in fish and served as stabilizers for swimming. Over millions of years, some fins were modified by natural selection into legs that allowed animals to crawl onto dry land, and later into wings that allowed some to take to the air. Another adaptation that has contributed to the success of vertebrates is an increase in the size and complexity of their brains and sensory structures, which allow vertebrates to perceive their environment in detail and to respond to it in a great variety of ways.

The sequence of evolutionary events that led from invertebrate chordates to the first, fishlike vertebrates remains shrouded in mystery, because fossils of intermediate forms have never been discovered. Today, vertebrates are represented by several classes of fishes, and by amphibians, reptiles, birds, and mammals.

Bony Fishes Just as our size bias makes us overlook the most diverse invertebrate groups, our habitat bias makes us overlook the most diverse vertebrates. The most diverse and abundant vertebrates are not the birds or predominantly terrestrial mammals. The vertebrate diversity crown belongs instead to the lords of the oceans and fresh water, the bony fishes. From the snakelike moray eel to the bizarre, luminescent deep-sea forms, to the streamlined tuna, this enormously successful group has spread to nearly every possible watery habitat, both freshwater and marine (Fig. 16-38). Although about 17,000 species have been identified, scientists estimate that perhaps twice this number exist, including many in deep waters and remote areas.

(a)

(b)

(c)

(d)

Figure 16-38 Bony fishes Bony fishes have colonized nearly every aquatic habitat. ***(a)*** This female deep-sea angler fish attracts prey with a living lure that projects just above her mouth. The fish is ghostly white because in the 2000-meter depth where anglers live, no light penetrates so colors are superfluous. Male deep-sea angler fish are extremely small and remain as permanent parasites attached to the female, always available to fertilize her eggs. Two parasitic males can be seen attached to this female. ***(b)*** This tropical green moray eel lives in rocky crevices. A small fish (a banded cleaner goby) on its lower jaw eats parasites that cling to the moray's skin. ***(c)*** The tropical seahorse may anchor itself with its tail, which is adapted for grasping, while the animal feeds on small crustaceans. ***(d)*** A rare photo of a coelacanth, a fish once believed to be long extinct, in its natural habitat off the coast of South Africa.

(a)

(b)

(c)

Figure 16-39 Amphibians In many amphibian species, individuals make the transition from *(a)* the completely aquatic larval tadpole to *(b)* the adult animal, here a bullfrog, leading a semiterrestrial life. *(c)* The red salamander is restricted to moist habitats in the eastern United States. Salamanders hatch in a form that closely resembles the adult. **QUESTION:** *What advantages might amphibians gain from their two-part "double life"?*

The ocean's potential to conceal species was illustrated in dramatic fashion by the case of the coelacanth, a bony fish that had been known only from 75-million-year-old fossils. In 1939 coelacanths were found to be alive and well in deep water off the coast of South Africa (Fig. 16-38d). Since that initial discovery, coelacanths have also been found in other locations, including waters near Madagascar and Indonesia.

Amphibians The amphibians straddle the boundary between aquatic and terrestrial existence (Fig. 16-39). The limbs of amphibians show varying degrees of adaptation to movement on land, from the belly-dragging crawl of salamanders to the long leaps of frogs and toads. A three-chambered heart (in contrast to the two-chambered heart of fishes) circulates blood more efficiently, and lungs replace gills in most adult forms. Amphibian lungs, however, are poorly developed and must be supplemented by the skin, which serves as an additional respiratory organ. This respiratory function requires that the skin remain moist, a constraint that greatly restricts the range of amphibian habitats on land.

Amphibians are also tied to moist habitats by their breeding behavior, which requires water. Their fertilization is external and takes place in water, where the sperm can swim to the eggs. The eggs must remain moist, as they are protected only by a jellylike coating that leaves them vulnerable to water loss by evaporation. Different amphibian species keep their eggs moist in different ways, but many species simply lay their eggs in water. In some amphibian species, fertilized eggs develop into aquatic larvae such as the tadpoles of some frogs and toads. These aquatic larvae undergo a dramatic transformation into semiterrestrial

Figure 16-40 Reptiles *(a)* The mountain king snake has a color pattern very similar to that of the poisonous coral snake, which potential predators avoid. This mimicry helps the harmless king snake elude predation. *(b)* The outward appearance of the American alligator, found in swampy areas of the South, is almost identical to reconstructions of 150-million-year-old fossil alligators. *(c)* The tortoises of the Galapagos Islands, Ecuador, may live to be more than 100 years old.

(a)

(b)

(c)

EARTH WATCH

FROGS IN PERIL

Frogs and toads have lived in Earth's ponds and swamps for nearly 150 million years, somehow surviving the Cretaceous catastrophe that extinguished the dinosaurs and so many other species about 65 million years ago. Their evolutionary longevity, however, doesn't protect them from the environmental changes wrought by human activities. Over the past two decades, herpetologists (biologists who study reptiles and amphibians) from around the world have documented an alarming decline in amphibian populations. Thousands of species of frogs, toads, and salamanders are dramatically decreasing in number, and many have apparently gone extinct.

This is a worldwide phenomenon; population crashes have been reported from every part of the globe. Yosemite toads and yellow-legged frogs are disappearing from the mountains of California. Tiger salamanders have been nearly wiped out in the Colorado Rockies. Leopard frogs, eagerly chased by children, are becoming rare in the United States. Logging destroys the habitats of amphibians from the Pacific Northwest to the tropics, but even amphibians in protected areas are dying. In the Monteverde Cloud Forest Preserve in Costa Rica, the golden toad was common in the early 1980s but has not been seen since 1989. The gastric brooding frog of Australia fascinated biologists by swallowing its eggs, brooding them in its stomach, and later regurgitating fully formed offspring. This species was abundant and seemed safe in a national park. Suddenly, in 1980, the gastric brooding frog disappeared and hasn't been seen since.

The causes of the worldwide decline in amphibian diversity are not fully understood, but researchers have discovered that frogs and toads in many places are succumbing to infection by a pathogenic fungus. The fungus has been found in the skin of dead and dying frogs in widespread locations, including Australia, Central America, and the western United States. In those places, discovery of the fungus has coincided with massive frog and toad die-offs, and most herpetologists agree that the fungus is causing the deaths.

It seems unlikely, however, that the fungus alone is responsible for the worldwide decline of amphibians. For one thing, die-offs have occurred in many places where the fungus has not been found. In addition, many herpetologists believe that the fungal epidemic would not have arisen if the frogs and toads had not first been weakened by other stresses. So, if the fungus is not doing all of the damage on its own, what are the other possible causes of amphibian decline? All of the most likely causes stem from human modification of the biosphere—the portion of Earth that sustains life.

Habitat destruction, especially the draining of wetlands that are especially hospitable to amphibian life, is one major cause of the decline. Amphibians are also vulnerable to toxic substances in the environment. For example, researchers found that frogs exposed to trace amounts of atrazine, a widely used herbicide that is found in virtually all fresh water in the United States, suffered severe damage to their reproductive tissues. The unique biology of amphibians makes them especially susceptible to poisons in the environment. Amphibian bodies at all stages of life are protected only by a thin, permeable skin that pollutants can easily penetrate (**Fig. E16-1**). To make matters worse, the double life of many amphibians exposes their permeable skin to a wide range of aquatic and terrestrial habitats and to a correspondingly wide range of environmental toxins.

Amphibian eggs can also be damaged by ultraviolet (UV) light, according to research by Andrew Blaustein, an ecologist at Oregon State University. Blaustein demonstrated that the eggs of some species of frogs in the Pacific Northwest are sensitive to damage from UV light and that the most sensitive species are experiencing the most drastic declines. Unfortunately, many parts of Earth are subject to increasingly intense UV radiation levels, because atmospheric pollutants have caused a thinning of the protective ozone layer.

Many scientists believe that the troubles of amphibians signal an overall deterioration of Earth's ability to support life. According to this line of reasoning, the highly sensitive amphibians are providing an early warning of environmental degradation that will eventually affect more-resistant organisms as well. Equally worrisome is the observation that amphibians are not just sensitive indicators of the health of the biosphere but are also crucial components of many ecosystems. They may keep insect populations in check, in turn serving as food for larger carnivores. Their decline will further disrupt the balance of these delicate communities.

Margaret Stewart, an ecologist at the State University of New York, Albany, aptly summarized the problem: "There's a famous saying among ecologists and environmentalists: 'Everything is related to everything else.'... You can't wipe out one large component of the system and not see dramatic changes in other parts of the system."

Figure E16-1 Amphibians in danger The corroboree toad, shown here with its eggs, is rapidly declining in its native Australia. Tadpoles are developing within the eggs. The thin water-permeable and gas-permeable skin of the adult and the jellylike coating around the eggs provide little, if any, protection against air and water pollutants.

adults, a metamorphosis that gives the amphibians their name, which means "double life." Their double life and their thin, permeable skin have made amphibians particularly vulnerable to pollutants and to environmental degradation, as described in "Earth Watch: Frogs in Peril."

Reptiles and Birds The reptiles include the lizards and snakes (by far the most successful of the modern groups) and the turtles, alligators, and crocodiles (**Fig. 16-40**). Some reptiles, particularly desert dwellers such as tortoises and

(a)

(b)

(c)

Figure 16-41 Birds ***(a)*** The delicate hummingbird beats its wings about 60 times per second and weighs about 0.15 ounce (4 grams). ***(b)*** This young frigate bird, a fish-eater from the Galapagos Islands, has nearly outgrown its nest. ***(c)*** The ostrich is the largest of all birds, weighing more than 300 pounds (136 kilograms). Its eggs weigh more than 3 pounds (1500 grams). **QUESTION:** *Although the earliest birds could fly, many bird species—such as the ostrich—cannot. Why do you suppose flightlessness has evolved repeatedly among birds?*

lizards, are completely independent of their aquatic origins. This independence was achieved through a series of adaptations, of which three are outstanding: (1) Reptiles evolved a tough, scaly skin that resists water loss and protects the body. (2) Reptiles evolved internal fertilization, in which the male deposits sperm within the female's body. (3) Reptiles evolved a shelled egg, which can be buried in sand or dirt, far from water with its hungry predators. The shell prevents the egg from drying out on land. An internal membrane encloses the embryo in the watery environment that all developing animals require.

One very distinctive group of "reptiles" is the birds (Fig. 16-41). Although birds have traditionally been classified as a group separate from reptiles, biologists have shown that birds are really a subset of an evolutionary group that includes both birds and the groups that have been traditionally designated as reptiles. The first birds appear in the fossil record roughly 150 million years ago and are distinguished from other reptiles by feathers, which are essentially a highly specialized version of reptilian body scales. Modern birds retain scales on their legs—evidence of the ancestry they share with the rest of the reptiles.

Bird anatomy and physiology are dominated by adaptations that help them to fly. In particular, birds are exceptionally light for their size. Hollow bones reduce the weight of the bird skeleton to a fraction of that of other vertebrates, and many bones present in other reptiles have been lost in the course of evolution or fused with other bones. Reproductive organs shrink considerably during nonbreeding periods, and female birds possess only a single ovary, further minimizing weight. The shelled egg that contributed to the reptiles' success on land frees the mother bird from carrying her developing offspring internally. Feathers form lightweight extensions to the wings and the tail for the lift and control required for flight, and they also provide lightweight protection and insulation for the body.

Birds are also able to maintain body temperatures high enough to allow their muscles and metabolic processes to operate at peak efficiency, supplying the power to fly regardless of the outside temperature. This physiological ability to maintain an internal temperature that is usually higher than that of the surrounding environment is characteristic of both birds and mammals, which are therefore sometimes described as warm blooded. In contrast, the body temperature of invertebrates, fish, amphibians, and reptiles varies with the temperature of their environment, though these animals may exert some control of their body temperature by their behavior (such as basking in the sun or seeking shade).

Mammals One branch of the reptile evolutionary tree gave rise to a group that evolved hair and diverged to form the mammals. The mammals first appeared approximately 250 million years ago but did not diversify and become prominent on land until after the dinosaurs went extinct roughly 65 million years ago. In most mammals, fur protects and insulates the warm body. Legs designed for run-

Figure 16-42 Mammals ***(a)*** A humpback whale gives its offspring a boost. ***(b)*** A bat, the only mammal capable of true flight, navigates at night by using a kind of sonar. Large ears aid in detecting echoes as its high-pitched cries bounce off nearby objects. ***(c)*** Mammals are named after the mammary glands with which females nurse their young, as illustrated by this mother cheetah. ***(d)*** The male orangutan can reach 165 pounds (75 kilograms). These gentle, intelligent apes occupy swamp forests in limited areas of the tropics. They are endangered by hunting and habitat destruction.

ning rather than crawling make many mammals fast and agile. In contrast to birds, whose bodies reflect the requirements of flight, mammals have evolved a remarkable diversity of form. The bat, mole, impala, whale, seal, monkey, and cheetah exemplify the radiation of mammals into nearly all habitats, with bodies finely adapted to their varied lifestyles (Fig. 16-42).

Mammals are named for the milk-producing mammary glands used by all female members of this class to suckle their young (Fig. 16-42c). In addition to these unique glands, the mammalian body has sweat, scent, and sebaceous (oil-producing) glands, none of which are found in other vertebrates.

The mammalian nervous system has contributed significantly to the success of the mammals by making possible behavioral adaptation to changing and varied environments. The brain is more highly developed than in any other vertebrate group, giving mammals unparalleled curiosity and learning ability. Their highly developed brain allows mammals to alter their behavior on the basis of experience and helps them survive in a changing environment. Relatively long periods of parental care after birth allow some mammals to learn extensively under parental guidance. Humans and other primates are good examples. In fact, the large brains of humans have been the major factor leading to human domination of Earth.

AN UNSOLVED MYSTERY *REVISITED*

How did Terry Erwin arrive at his estimate that 30 million species of insects inhabit tropical rain forests? He began by arranging jars around the base of a tree in a Panamanian rain forest and fogging the tree's canopy with insecticide. As the insecticide took effect, dead insects rained down into the waiting jars. Erwin's team studied these insects and counted the number of beetle species.

Erwin repeated this procedure on dozens of trees, all of which were of the same species, *Luehea seemannii*. He found 1200 species of beetles, most of them previously unknown.

Erwin knew that many beetle species are highly specialized and live exclusively on a single species of tree. He estimated that about 13% of beetles have such specializations, so about 160 (13% of 1200) of the beetle species he collected were unique to *L. seemannii*. Erwin also knew that about 40% of all known insect species are beetles, so he reasoned that the 160 beetle species likewise represented 40% of a total of 400 insect species specialized to inhabit *L. seemannii*.

Erwin sampled only from the tree canopy, which, he assumed, holds about two-thirds of the insect species on a tree, with the remaining one-third living on the trunk and in the roots. For *L. seemannii*, that one-third would represent an additional 200 species, giving a total of 600 insect species specialized to live on it. If *L. seemannii* is a typical tropical tree, then each of the other rainforest tree species ought to also host about 600 specialized insect species. About 50,000 species of rainforest trees are known, so there must be at least $600 \times 50{,}000 = 30$ million species of rainforest insects altogether.

CONSIDER THIS: Each step in Erwin's chain of logic required an assumption. Make a list of those assumptions, and try recalculating the estimate using some different, but still reasonable, assumptions. What is the lowest estimate you come up with? The highest?

LINKS TO LIFE: Biological Vanity Plates

Looking for a special gift for a friend or loved one? Why not name a species after him or her? Or, for that matter, name one after yourself! Thanks to the BIOPAT project (**www.biopat.de**), anyone with $3000 can be immortalized in the Latin name of a newly discovered plant or animal.

Typically, the scientist who discovers and describes a new species is entitled to choose its Latin name. Scientists usually choose a name that describes a trait of the species or perhaps the location where it was found. Sometimes, however, more whimsical choices are made. For example, a recently discovered snail was named *Bufonaria borisbeckeri*, in honor of the German tennis player Boris Becker, and a frog was named *Hyla stingi* after the British rock star.

If you donate money to the BIOPAT project, the name of a new species will be entirely up to you. In return for a contribution that supports efforts to discover and conserve endangered species, the people at BIOPAT will offer you a selection of newly discovered but unnamed species. You can choose your species and pick a name, which is then given an appropriate Latin ending and published in a scientific journal. Your chosen name becomes the official, recognized scientific name of the new species.

Chapter Review

Summary of Key Concepts

16.1 How Are Organisms Named and Classified?

Organisms are classified and placed into hierarchical categories that reflect their evolutionary relationships. The eight major categories, in order of decreasing inclusiveness, are domain, kingdom, phylum, class, order, family, genus, and species. The scientific name of an organism, such as *Homo sapiens*, is its genus name followed by its species name. Anatomical and molecular similarities are used to classify organisms. Molecular similarities among organisms are a measure of evolutionary relatedness.

16.2 What Are the Domains of Life?

The three domains of life, each representing one of three main branches of the tree of life, are Bacteria, Archaea, and Eukarya. Each domain contains a number of kingdoms, but the details of kingdom-level classification are in a period of transition and remain unsettled. Within the domain Eukarya, however, the kingdoms Fungi, Plantae, and Animalia are universally accepted as valid groups.

16.3 Bacteria and Archaea

Members of the domains Bacteria and Archaea—the bacteria and archaea—are unicellular and prokaryotic. Archaea and bacteria are not closely related and differ in fundamental features, including cell wall composition and membrane structure. A cell wall determines the characteristic shapes of prokaryotes: round, rodlike, or spiral. Certain types of bacteria can form spores that disperse widely and withstand inhospitable environmental conditions. Prokaryotes obtain energy in a variety of ways. Some rely on photosynthesis. Others break down inorganic molecules to obtain energy. Many are anaerobic, able to obtain energy from fermentation when oxygen is not available.

Some bacteria are pathogenic, causing disorders such as pneumonia, tetanus, botulism, and the venereal diseases gonorrhea and syphilis. Most bacteria, however, are harmless to humans and play important roles in natural ecosystems. Bacteria and archaea have colonized nearly every habitat on Earth, including hot, acidic, very salty, and anaerobic environments. Some live in the digestive tracts of ruminants and break down cellulose. Nitrogen-fixing bacteria enrich the soil and aid in plant growth; many others live off the dead bodies and wastes of other organisms, liberating nutrients for reuse.

16.4 Protists

Most protists are single, highly complex eukaryotic cells, but some form colonies and some, such as seaweeds, are multicellular. Photosynthetic protists form much of the phytoplankton, which plays a key ecological role. Protists exhibit diverse modes of nutrition, reproduction, and locomotion. Protist groups include the chromists (diatoms and brown algae), the alveolates (dinoflagellates, apicomplexans, and ciliates), slime molds, several groups of pseudopod-using organisms (amoebas, radiolarians, and foraminiferans), and green algae (the closest relatives of plants). Some protists, especially types of apicomplexans, cause human diseases.

16.5 Fungi

Fungal bodies generally consist of filamentous hyphae that form large, intertwined mycelia. A cell wall of chitin surrounds fungal cells. Fungi obtain energy by secreting digestive enzymes outside their bodies and absorbing the liberated nutrients. Fungal reproduction depends on spores, and most species reproduce both sexually and asexually.

The majority of plant diseases are caused by parasitic fungi. Some parasitic fungi can help control insect crop pests. Others can cause human diseases, including ringworm, athlete's foot, and common vaginal infections. Some fungi produce toxins that can harm humans. Nonetheless, fungi add variety to the human food supply, and fermentation by fungi helps make wine, beer, and bread.

Fungi are extremely important decomposers in ecosystems. Their filamentous bodies penetrate rich soil and decaying organic material, liberating nutrients through extracellular digestion.

16.6 Plants

Two groups of plants, bryophytes and vascular plants, arose from ancient algal ancestors. Bryophytes, including the liverworts and mosses, are small land plants that lack conducting vessels. Although some have adapted to dry areas, most live in moist habitats. Reproduction in bryophytes requires water through which the sperm swims to the egg.

In vascular plants a system of vessels, stiffened by lignin, conducts water and nutrients absorbed by the roots into the upper portions of the plant and supports the body as well. Owing to this support system, seedless vascular plants, including the club mosses, horsetails, and ferns, can grow larger than bryophytes.

Vascular plants with seeds have two other major adaptive features: pollen and seeds. Seed plants are often classified into two categories: gymnosperms and angiosperms. Gymnosperms include ginkgos, cycads, and the highly successful conifers. Angiosperms, the flowering plants, dominate much of the land today. In addition to pollen and seeds, angiosperms produce flowers and fruits. The flower allows angiosperms to utilize animals as pollinators. In contrast to wind, animals can in some cases carry pollen farther and with greater accuracy and less waste. Fruits may attract animal consumers, which incidentally disperse the seeds in their feces.

16.7 Animals

Animals are multicellular, sexually reproducing, heterotrophic organisms. Most can perceive and react rapidly to environmental stimuli and are motile at some stage in their lives. Their cells lack cell walls.

The bodies of sponges are typically free-form in shape and cannot move. Sponges have relatively few types of cells. Digestion occurs exclusively within the individual cells.

Cnidarians (jellyfish, sea anemones, corals, and hydrozoans) use armed tentacles to capture prey.

The annelids, the most complex of the worms, include earthworms, marine tubeworms, and leeches.

Arthropods, which include the insects, arachnids, and crustaceans, are the most diverse and abundant organisms on Earth. They have invaded nearly every available terrestrial and aquatic habitat. Jointed appendages and well-developed nervous systems make possible complex, finely coordinated behavior. The exoskeleton (which conserves water and provides support) helps enable the insects and arachnids to inhabit dry land. The diversification of insects has been enhanced by their ability to fly. Crustaceans, which include the largest arthropods, are restricted to moist, usually aquatic habitats and respire by using gills.

The mollusks (gastropods, bivalves, and cephalopods) lack a skeleton. Some forms protect the soft, moist, muscular body with a single shell (many gastropods and a few cephalopods) or a pair of hinged shells (the bivalves). The lack of a waterproof external covering limits mollusks to aquatic and moist terrestrial habitats. The body plan of gastropods and bivalves limits the complexity of their behavior, but the cephalopod's tentacles are capable of precisely controlled movements. The octopus has the most complex brain and the best-developed learning capacity of any invertebrate.

The chordate phylum includes two invertebrate groups, the lancelets and tunicates, as well as the familiar vertebrates. All chordates possess a notochord, a nerve cord, pharyngeal gill slits, and a postanal tail at some stage in their development. Vertebrates have a backbone that is part of a living internal skeleton.

During the transitions from fishes to amphibians to reptiles, some vertebrates evolved adaptations that helped them colonize dry land. All amphibians have legs, and most have simple lungs for breathing in air rather than in water. Most are confined to relatively damp terrestrial habitats by their need to keep their skin moist, use of external fertilization, and requirement that their eggs and larvae develop in water. Reptiles are well adapted to the driest terrestrial habitats, thanks to well developed lungs, dry skin covered with relatively waterproof scales, internal fertilization, and shelled eggs with their own water supply. Birds and mammals are also fully terrestrial and have additional adaptations. The bird body is adapted for flight, with feathers and hollow bones. Mammals have insulating hair and give birth to live young that are nourished with milk from the mothers' mammary glands. The mammalian nervous system is the most complex in the animal kingdom, providing mammals with enhanced learning ability that helps them adapt to changing environments.

Key Terms

algae *p. 293*
anaerobe *p. 290*
angiosperm *p. 302*
Archaea *p. 287*
Bacteria *p. 287*
bryophyte *p. 300*
cilia *p. 295*
conifer *p. 302*
endospore *p. 289*
Eukarya *p. 287*
exoskeleton *p. 306*
flagella *p. 288*
flower *p. 302*
fruit *p. 302*
gymnosperm *p. 302*
hyphae *p. 297*
invertebrate *p. 304*
larva *p. 307*
metamorphosis *p. 307*
molt *p. 307*
mycelium *p. 297*
nerve cord *p. 310*
nitrogen-fixing bacteria *p. 290*
notochord *p. 310*
pathogenic *p. 291*
pharyngeal gill slit *p. 310*
phytoplankton *p. 293*
pollen *p. 301*
postanal tail *p. 310*
protist *p. 293*
protozoa *p. 293*
pupa *p. 307*
seed *p. 302*
spore *p. 298*
vascular plant *p. 300*
vertebral column *p. 311*
vertebrate *p. 304*

Thinking Through the Concepts

Multiple Choice

1. *Which of the following pairs is the most distantly related?*
  a. archaea and bacteria
  b. protists and fungi
  c. plants and animals
  d. fish and starfish
  e. fungi and plants

2. *Most pathogenic bacteria cause disease by*
  a. directly destroying individual cells of the host
  b. fixing nitrogen and depriving the host of this nutrient
  c. producing toxins that disrupt normal functions
  d. depleting the energy supply of the host
  e. depriving the host of oxygen

3. *Which of the following statements is true both of fungi and of animals?*
  a. Both photosynthesize.
  b. Both form embryos.
  c. They interact to form lichens.
  d. Both fix nitrogen.
  e. Both gain nutrition by consuming the bodies of other organisms.

4. *What is the function of a fruit?*
  a. It attracts pollinators.
  b. It provides food for the developing embryo.
  c. It stores excess food produced by photosynthesis.
  d. It helps ensure seed dispersal from the parent plant.
  e. It evolved so that people would cultivate the plant, ensuring its survival.

5. *Which of the following plants produces sperm that swim to the egg?*
  a. walnut tree
  b. Douglas fir tree, a conifer
  c. rattlesnake fern
  d. common dandelion
  e. none of the above

6. *Which of the following animals molts its exoskeleton, enabling it to grow larger?*
  a. the blue crab
  b. the earthworm
  c. the scallop
  d. the octopus
  e. the Venus clam

Review Questions

1. What features would you study to determine whether a dolphin is more closely related to a fish or to a bear?
2. Only a small fraction of the total number of species on Earth has been scientifically described. Why?
3. Describe some of the ways in which bacteria obtain energy and nutrients.
4. What are nitrogen-fixing bacteria, and what role do they play in ecosystems?
5. What is the major ecological role played by unicellular algae?
6. What portion of the fungal body is represented by mushrooms, puffballs, and similar structures? Why are these structures elevated above the ground?
7. List the adaptations that helped plants become successful terrestrial organisms. Which of these adaptations are possessed by bryophytes? By ferns? By gymnosperms and angiosperms?
8. The number of species of flowering plants is greater than the number of species in the rest of the plant kingdom. What feature(s) are responsible for the enormous success of angiosperms? Explain why.
9. Distinguish between vertebrates and invertebrates.
10. List four distinguishing features of chordates.
11. Describe the ways in which amphibians are adapted to life on land. In what ways are amphibians still restricted to a watery or moist environment?
12. List the adaptations that distinguish reptiles from amphibians and helped reptiles adapt to life in dry terrestrial environments.

Applying the Concepts

1. **IS THIS SCIENCE?** Biologist E. O. Wilson has estimated that 27,000 species go extinct each year. Some skeptics, however, point out that scientists have documented the disappearance of only 1100 species over the past 500 years. Does this observation demonstrate that Wilson's estimate is inaccurate? Explain your answer.

2. There are many areas of disagreement about the classification of organisms. For example, there is no consensus about whether the red wolf is a species distinct from the gray wolf or about how many kingdoms are within the domain Bacteria. What difference does it make whether biologists consider the red wolf a species, or into which kingdom a bacterial species falls? As Shakespeare put it, "What's in a name?"

3. The internal structure of the cells of many protists is much more complex than that of cells of multicellular organisms. Does this mean that the protist is engaged in more complex activities than the multicellular organism? If not, why should the protistan cell be much more complicated?

For More Information

Blaustein, A., and Johnson, P. T. J. "Explaining Frog Deformities." *Scientific American*, February 2003. Dramatic increases in the occurrence of deformed frogs are caused by a parasite epidemic exacerbated by environmental degradation.

Chadwick, D. H. "Planet of the Beetles." *National Geographic*, March 1998. The beauty and diversity of beetles, which constitute one-third of the world's insects, are described in text and photographs.

Hamner, W. "A Killer Down Under." *National Geographic*, August 1994. Among the most poisonous animals in the world is the box jellyfish, which lives off the coast of northern Australia.

Hudler, G. W. *Magical Mushrooms, Mischievous Molds*. Princeton, NJ: Princeton University Press, 1998. Engaging treatment of the fungi, focusing on their importance in human affairs.

Kaufman, P. B. *Plants—Their Biology and Importance.* New York: Harper & Row, 1989. Complete, readable coverage of all aspects of plant classification, physiology, and evolution.

Luoma, J. R. "Vanishing Frogs." *Audubon*, May–June 1997. Brief summary of declining amphibian populations, accompanied by striking photographs of some of the affected frog species.

Madigan, M., and Marrs, B. "Extremophiles." *Scientific American*, April 1997. Prokaryotes that prosper under extreme conditions, and potential industrial uses of the enzymes that they use to do so.

Mann, C., and Plummer, M. *Noah's Choice: The Future of Endangered Species*. New York: Knopf, 1995. A thought-provoking look at the hard choices we must make with regard to protecting biodiversity. Which species will we choose to preserve? What price are we willing to pay?

May, R. M. "How Many Species Inhabit the Earth?" *Scientific American*, October 1992. Although no one knows the precise answer to this question, an effective estimate is crucial to our effort to manage our biological resources.

Morell, V. "Life on a Grain of Sand." *Discover*, April 1995. The sand beneath shallow waters is home to an incredible range of microscopic creatures.

Pollan, M. *The Botany of Desire.* New York: Random House, 2001. A literate look at the mutually beneficial relationship between humans and plants.

Schaechter, E. *In the Company of Mushrooms*. Cambridge: Harvard University Press, 1997. An accessible account of the world of mushrooms, written in a warm, personal style.

Wilson, E. O. *The Diversity of Life*. Cambridge, MA: Harvard University Press, 1992. An outline of the processes that created the diversity of life and a discussion of the threats to that diversity and the steps required to preserve it.

Young, J., and Collier, R. J. "Attacking Anthrax." *Scientific American*, March 2002. A summary of recent research that could help develop new techniques for detecting and treating anthrax.

WEB TUTORIAL

To access a Web Tutorial visit http://www.prenhall.com/audesirk4. *Log in to the Web site selected by your instructor, navigate to this chapter, and select the appropriate Web Activity number.*

16.1 Taxonomic Classification
Estimated time: 5 minutes

Taxonomy is the branch of biology concerned with identifying, describing, and naming all organisms on Earth. This activity will help you understand the hierarchy of taxonomic groups used by biologists.

16.2 Tree of Life
Estimated time: 10 minutes

In this animation you will explore the Tree of Life, the type of evidence used to establish the three domains of life, and why domains are widely adopted today.

16.3 Evolution of Plant Structure
Estimated time: 10 minutes

This animation illustrates the evolutionary changes that have occurred in plant structures and how those changes have allowed plants to exploit different habitats.

16.4 Web Investigation: An Unsolved Mystery: How Many Species Are There?
Estimated time: 15 minutes

If you counted one know species of animal every second of an 8 hour workday (no lunch, please!), it would take more than 7 weeks (@ 5 days each week) just to count the animal species that scientists know about. Then, there are plants, fungi, and protists—don't even mention bacteria and archaea! So how many species are there and how could we possibly count them? This exercise will explain how one scientist came up with an estimate.

Appendices

Three appendices appear on the following pages:

Appendix I

Metric System Conversions

To Convert Metric Units:	Multiply by:	To Get English Equivalent:
	Length	
Centimeters (cm)	0.3937	Inches (in.)
Meters (m)	3.2808	Feet (ft)
Meters (m)	1.0936	Yards (yd)
Kilometers (km)	0.6214	Miles (mi)
	Area	
Square centimeters (cm^2)	0.155	Square inches ($in.^2$)
Square meters (m^2)	10.7639	Square feet (ft^2)
Square meters (m^2)	1.1960	Square yards (yd^2)
Square kilometers (km^2)	0.3831	Square miles (mi^2)
Hectare (ha) (10,000 m^2)	2.4710	Acres (a)
	Volume	
Cubic centimeters (cm^3)	0.06	Cubic inches ($in.^3$)
Cubic meters (m^3)	35.30	Cubic feet (ft^3)
Cubic meters (m^3)	1.3079	Cubic yards (yd^3)
Cubic kilometers (km^3)	0.24	Cubic miles (mi^3)
Liters (L)	1.0567	Quarts (qt), U.S.
Liters (L)	0.26	Gallons (gal), U.S.
	Mass	
Grams (g)	0.03527	Ounces (oz)
Kilograms (kg)	2.2046	Pounds (lb)
Metric ton (tonne) (t)	1.10	Ton (tn), U.S.
	Speed	
Meters/second (mps)	2.24	Miles/hour (mph)
Kilometers/hour (kmph)	0.62	Miles/hour (mph)
To Convert English Units:	**Multiply by:**	**To Get Metric Equivalent:**
	Length	
Inches (in.)	2.54	Centimeters (cm)
Feet (ft)	0.3048	Meters (m)
Yards (yd)	0.9144	Meters (m)
Miles (mi)	1.6094	Kilometers (km)
	Area	
Square inches ($in.^2$)	6.45	Square centimeters (cm^2)
Square feet (ft^2)	0.0929	Square meters (m^2)
Square yards (yd^2)	0.8361	Square meters (m^2)
Square miles (mi^2)	2.5900	Square kilometers (km^2)
Acres (a)	0.4047	Hectare (ha) (10,000 m^2)
	Volume	
Cubic inches ($in.^3$)	16.39	Cubic centimeters (cm^3)
Cubic feet (ft^3)	0.028	Cubic meters (m^3)
Cubic yards (yd^3)	0.765	Cubic meters (m^3)
Cubic miles (mi^3)	4.17	Cubic kilometers (km^3)
Quarts (qt), U.S.	0.9463	Liters (L)
Gallons (gal), U.S.	3.8	Liters (L)
	Mass	
Ounces (oz)	28.3495	Grams (g)
Pounds (lb)	0.4536	Kilograms (kg)
Ton (tn), U.S.	0.91	Metric ton (tonne) (t)
	Speed	
Miles/hour (mph)	0.448	Meters/second (mps)
Miles/hour (mph)	1.6094	Kilometers/hour (kmph)

Metric Prefixes

Prefix		Meaning	
giga-	G	10^9 =	1,000,000,000
mega-	M	10^6 =	1,000,000
kilo-	k	10^3 =	1000
hecto-	h	10^2 =	100
deka-	da	10^1 =	10
		10^0 =	1
deci-	d	10^{-1} =	0.1
centi-	c	10^{-2} =	0.01
milli-	m	10^{-3} =	0.001
micro-	μ	10^{-6} =	0.000001
nano-	n	10^{-9} =	0.000000001

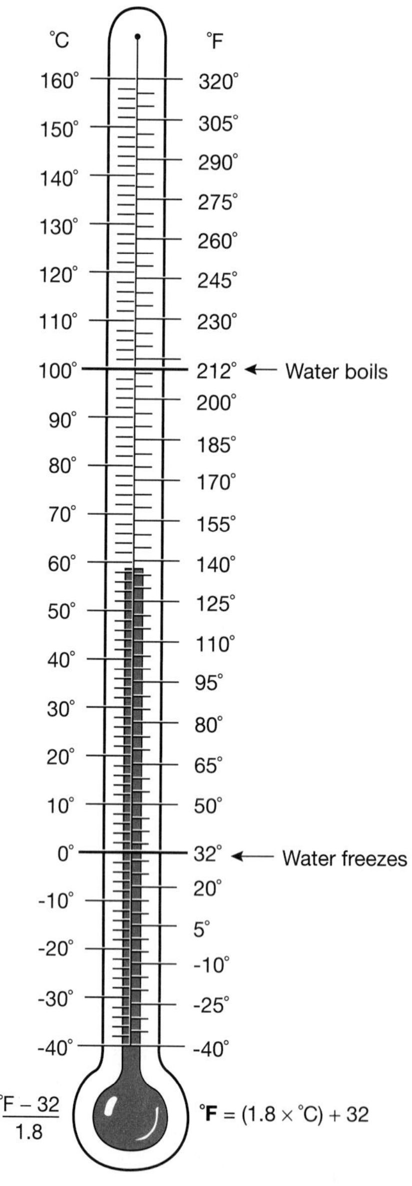

Appendix II

Classification of Major Groups of Organisms*

Domain	Kingdom	Phylum	Common Name
Bacteria (prokaryotic, peptidoglycan in cell wall)			bacteria
Archaea (prokaryotic, no peptidoglycan in cell wall)			archaeans
Eukarya (eukaryotic)		Rhodophyta	red algae
		Euglenophyta	euglenoids
		Myxomycota	plasmodial slime molds
		Acrasiomycota	cellular slime molds
		Sarcomastigophora	zooflagellates, amoebae
		Chlorophyta	green algae
	Alveolata		alveolates
		Apicomplexa	sporozoans
		Pyrrophyta	dinoflagellates
		Ciliophora	ciliates
	Chromista		chromists
		Oomycota	egg fungi
		Phaeophyta	brown algae
		Bacillariophyta	diatoms
	Fungi (multicellular, heterotrophic, absorb nutrients)		fungi
		Chytridiomycota	chytrids
		Zygomycota	zygote fungi
		Ascomycota	sac fungi
		Basidiomycota	club fungi
	Plantae (multicellular, photosynthetic)		plants
		Bryophyta	liverworts, mosses
		Pteridophyta	ferns
		Coniferophyta	evergreens
		Anthophyta	flowering plants
	Animalia (multicellular, heterotrophic, ingest nutrients)		animals
		Porifera	sponges
		Cnidaria	hydras, sea anemones, jellyfish, corals
		Ctenophora	comb jellies
		Platyhelminthes	flatworms
		Nematoda	roundworms
		Annelida	segmented worms
		Oligochaeta	earthworms
		Polychaeta	tube worms
		Hirudinea	leeches
		Arthropoda	arthropods ("jointed legs")
		Insecta	insects
		Arachnida	spiders, ticks
		Crustacea	crabs, lobsters
		Mollusca	mollusks ("soft-bodied")
		Gastropoda	snails
		Pelecypoda	mussels, clams
		Cephalopoda	squid, octopuses
		Echinodermata	sea stars, sea urchins, sea cucumbers
		Chordata	chordates
		Urochordata	tunicates
		Cephalochordata	lancelets
		Myxini	hagfishes
		Vertebrata	vertebrates
		Pertromyzontiformes	lampreys
		Chondrichthyes	sharks, rays
		Osteichthyes	bony fishes
		Actinopterygii	ray-finned fishes
		Sarcopterygii	lobe-finned fishes
		Amphibia	frogs, salamanders
		Anapsida	turtles
		Diapsida	
		Archosauria	birds, crocodiles
		Squamata	lizards, snakes
		Mammalia	mammals

**This table lists only those taxonomic categories described in the textbook.*

Appendix III

Periodic Table of the Elements

Main-group elements

- Metals
- Nonmetals
- Noble gases

Transition elements

Period	1 1A	2 2A	3 3B	4 4B	5 5B	6 6B	7 7B	8 8B	9	10
1	1 **H** 1.00794									
2	3 **Li** 6.941	4 **Be** 9.01218								
3	11 **Na** 22.9898	12 **Mg** 24.3050								
4	19 **K** 39.0983	20 **Ca** 40.078	21 **Sc** 44.9559	22 **Ti** 47.88	23 **V** 50.9415	24 **Cr** 51.9961	25 **Mn** 54.9381	26 **Fe** 55.847	27 **Co** 58.9332	28 **Ni** 58.693
5	37 **Rb** 85.4678	38 **Sr** 87.62	39 **Y** 88.9059	40 **Zr** 91.224	41 **Nb** 92.9064	42 **Mo** 95.94	43 **Tc** (98)	44 **Ru** 101.07	45 **Rh** 102.906	46 **Pd** 106.42
6	55 **Cs** 132.905	56 **Ba** 137.327	57 ***La** 138.906	72 **Hf** 178.49	73 **Ta** 180.948	74 **W** 183.84	75 **Re** 186.207	76 **Os** 190.23	77 **Ir** 192.22	78 **Pt** 195.08
7	87 **Fr** (223)	88 **Ra** 226.025	89 **†Ac** 227.028	104 **Rf** (261)	105 **Db** (262)	106 **Sg** (263)	107 **Bh** (262)	108 **Hs** (265)	109 **Mt** (266)	110 **Ds** (281)

*Lanthanide series	58 **Ce** 140.115	59 **Pr** 140.908	60 **Nd** 144.24	61 **Pm** (145)	62 **Sm** 150.36	63 **Eu** 151.965
†Actinide series	90 **Th** 232.038	91 **Pa** 231.036	92 **U** 238.029	93 **Np** 237.048	94 **Pu** (244)	95 **Am** (243)

** Not yet named

Notes: (1) Values in parentheses are the mass numbers of the most common or most stable isotopes of radioactive elements. (2) Some elements adjacent to the stair-step line between the metals and nonmetals have a metallic appearance but some nonmetallic properties. These elements are often called metalloids or semimetals. There is no general agreement on just which elements are so designated. Almost every list includes Si, Ge, As, Sb, and Te. Some also include B, At, and/or Po.

Main-group elements

11 1B	12 2B	13 3A	14 4A	15 5A	16 6A	17 7A	18 8A
							2 **He** 4.00260
		5 **B** 10.811	6 **C** 12.011	7 **N** 14.0067	8 **O** 15.9994	9 **F** 18.9984	10 **Ne** 20.1797
		13 **Al** 26.9815	14 **Si** 28.0855	15 **P** 30.9738	16 **S** 32.066	17 **Cl** 35.4527	18 **Ar** 39.948
29 **Cu** 63.546	30 **Zn** 65.39	31 **Ga** 69.723	32 **Ge** 72.61	33 **As** 74.9216	34 **Se** 78.96	35 **Br** 79.904	36 **Kr** 83.80
47 **Ag** 107.868	48 **Cd** 112.411	49 **In** 114.818	50 **Sn** 118.710	51 **Sb** 121.76	52 **Te** 127.60	53 **I** 126.904	54 **Xe** 131.29
79 **Au** 196.967	80 **Hg** 200.59	81 **Tl** 204.383	82 **Pb** 207.2	83 **Bi** 208.980	84 **Po** (209)	85 **At** (210)	86 **Rn** (222)
111 ** (272)	112 ** (285)		114 ** (289)		116 ** (292)		

64 **Gd** 157.25	65 **Tb** 158.925	66 **Dy** 162.50	67 **Ho** 164.930	68 **Er** 167.26	69 **Tm** 168.934	70 **Yb** 173.04	71 **Lu** 174.967
96 **Cm** (247)	97 **Bk** (247)	98 **Cf** (251)	99 **Es** (252)	100 **Fm** (257)	101 **Md** (258)	102 **No** (259)	103 **Lr** (260)

Glossary

abdomen: the body segment at the posterior end of an animal with segmentation; contains most of the digestive structures.

abiotic (ā-bī-ah′-tik): nonliving; the abiotic portion of an ecosystem includes soil, rock, water, and the atmosphere.

abortion: the procedure for terminating pregnancy; the cervix is dilated, and the embryo and placenta are removed.

abscisic acid (ab-sis′-ik): a plant hormone that generally inhibits the action of other hormones, enforcing dormancy in seeds and buds and causing the closing of stomata.

abscission layer: a layer of thin-walled cells, located at the base of the petiole of a leaf, that produces an enzyme that digests the cell wall holding leaf to stem, allowing the leaf to fall off.

absorption: the process by which nutrients are taken into cells.

accessory pigments: colored molecules other than chlorophyll that absorb light energy and pass it to chlorophyll.

acellular slime mold: a type of funguslike protist that forms a multinucleate structure that crawls in amoeboid fashion and ingests decaying organic matter; also called *plasmodial slime mold.*

acetylcholine (ah-sēt′-il-kō′-lēn): a neurotransmitter in the brain and in synapses of motor neurons that innervate skeletal muscles.

acid: a substance that releases hydrogen ions (H^+) into solution; a solution with a pH of less than 7.

acid deposition: the deposition of nitric or sulfuric acid, either dissolved in rain (acid rain) or in the form of dry particles, as a result of the production of nitrogen oxides or sulfur dioxide through burning, primarily of fossil fuels.

acidic: with an H^+ concentration exceeding that of OH^-; releasing H^+.

acquired immune deficiency syndrome (AIDS): an infectious disease caused by the human immunodeficiency virus (HIV); attacks and destroys T cells, thus weakening the immune system.

acrosome (ak′-rō-sōm): a vesicle, located at the tip of an animal sperm, that contains enzymes needed to dissolve protective layers around the egg.

actin (ak′-tin): a major muscle protein whose interactions with myosin produce contraction; found in the thin filaments of the muscle fiber; see also *myosin.*

action potential: a rapid change from a negative to a positive electrical potential in a nerve cell. This signal travels along an axon without a change in intensity.

activation energy: in a chemical reaction, the energy needed to force the electron shells of reactants together, before the formation of products.

active site: the region of an enzyme molecule that binds substrates and performs the catalytic function of the enzyme.

active transport: the movement of materials across a membrane through the use of cellular energy, normally against a concentration gradient.

adaptation: a trait that increases the ability of an individual to survive and reproduce compared with individuals without the trait.

adaptive radiation: the rise of many new species in a relatively short time as a result of a single species that invades different habitats and evolves under different environmental pressures in those habitats.

adenine: a nitrogenous base found in both DNA and RNA; abbreviated as *A.*

adenosine diphosphate (a-den′-ō-sēn dī-fos′-fāt; ADP): a molecule composed of the sugar ribose, the base adenine, and two phosphate groups; a component of ATP.

adenosine triphosphate (a-den′-ō-sēn trī-fos′-fāt; ATP): a molecule composed of the sugar ribose, the base adenine, and three phosphate groups; the major energy carrier in cells. The last two phosphate groups are attached by "high-energy" bonds.

adipose tissue (a′-di-pōs): tissue composed of fat cells.

adrenal cortex: the outer part of the adrenal gland, which secretes steroid hormones that regulate metabolism and salt balance.

adrenal gland: a mammalian endocrine gland, adjacent to the kidney; secretes hormones that function in water regulation and in the stress response.

adrenal medulla: the inner part of the adrenal gland, which secretes epinephrine (adrenaline) and norepinephrine (noradrenaline).

adrenocorticotropic hormone (a-drēn-ō-kor-tik-ō-trō′-pik; ACTH): a hormone, secreted by the anterior pituitary, that stimulates the release of hormones by the adrenal glands, especially in response to stress.

aerobic: using oxygen.

age structure: the distribution of males and females in a population according to age groups.

agglutination (a-gloo-tin-ā′-shun): the clumping of foreign substances or microbes, caused by binding with antibodies.

aggression: antagonistic behavior, usually among the same species, often from competition for resources.

aggressive mimicry (mim′ik-rē): the evolution of a predatory organism to resemble a harmless animal or part of the environment, thus gaining access to prey.

aldosterone: a hormone, secreted by the adrenal cortex, that helps regulate ion concentration in the blood by stimulating the reabsorption of sodium by the kidneys and sweat glands.

alga (al′-ga; pl., algae, al′-jē): any photosynthetic member of the eukaryotic Kingdom Protista.

alveolate (al-vē′-ō-lāt): a member of the Alveolata, a large assemblage of protists that is assigned kingdom status by many systematists. The alveolates, which are characterized by a system of sacs beneath the cell membrane, include ciliates, foraminiferans, dinoflagellates, and apicomplexans.

allantois (al-an-tō′-is): one of the embryonic membranes of reptiles, birds, and mammals; in reptiles and birds, serves as a waste-storage organ; in mammals, forms most of the umbilical cord.

allele (al-ēl′): one of several alternative forms of a particular gene.

allele frequency: for any given gene, the relative proportion of each allele of that gene in a population.

allergy: an inflammatory response produced by the body in response to invasion by foreign materials, such as pollen, that are themselves harmless.

allopatric speciation (al-ō-pat′-rik): speciation that occurs when two populations are separated by a physical barrier that prevents gene flow between them (geographical isolation).

allosteric regulation: the process by which enzyme action is enhanced or inhibited by small organic molecules that act as regulators by binding to the enzyme and altering its active site.

alternation of generations: a life cycle, typical of plants, in which a diploid sporophyte (spore-producing) generation alternates with a haploid gametophyte (gamete-producing) generation.

altruism: a type of behavior that may decrease the reproductive success of the individual performing it but benefits that of other individuals.

alveolus (al-vē′-ō-lus; pl., alveoli): a tiny air sac within the lungs, surrounded by capillaries, where gas exchange with the blood occurs.

amino acid: the individual subunit of which proteins are made, composed of a central carbon atom bonded to an amino group ($-NH_2$), a carboxyl group ($-COOH$), a hydrogen atom, and a variable group of atoms denoted by the letter *R.*

ammonia: NH_3; a highly toxic nitrogen-containing waste product of amino acid breakdown. In the mammalian liver, it is converted to urea.

amniocentesis (am-nē-ō-sen-tē′-sis): a procedure for sampling the amniotic fluid surrounding a fetus: A sterile needle is inserted through the abdominal wall, uterus, and amniotic sac of a pregnant woman; 10 to 20 milliliters of amniotic fluid is withdrawn. Various tests may be performed on the fluid and the fetal cells suspended in it to provide information on the developmental and genetic state of the fetus.

amnion (am′-nē-on): one of the embryonic membranes of reptiles, birds, and mammals; encloses a fluid-filled cavity that envelops the embryo.

amniote egg (am-nē-ōt′): the egg of reptiles and birds; contains an amnion that encloses the embryo in a watery environment, allowing the egg to be laid on dry land.

amoeba: a type of animal-like protist that uses a characteristic streaming mode of locomotion by extending a cellular projection called a *pseudopod.*

amoeboid cell: a protist or animal cell that moves by extending a cellular projection called a *pseudopod.*

amphibian: a member of the chordate class Amphibia, which includes the frogs, toads, and salamanders, as well as the limbless caecelians.

amplexus (am-plek′-sus): in amphibians, a form of external fertilization in which the male holds the female during spawning and releases his sperm directly onto her eggs.

ampulla: a muscular bulb that is part of the water-vascular system of echinoderms; controls the movement of tube feet, which are used for locomotion.

amygdala (am-ig′-da-la): part of the forebrain of vertebrates that is involved in the production of appropriate behavioral responses to environmental stimuli.

amylase (am′-i-lās): an enzyme, found in saliva and pancreatic secretions, that catalyzes the breakdown of starch.

anaerobe: an organism whose respiration does not require oxygen.

anaerobic: not using oxygen.
analogous structures: structures that have similar functions and superficially similar appearance but very different anatomies, such as the wings of insects and birds. The similarities are due to similar environmental pressures rather than to common ancestry.
anaphase (an′-a-fāz): in mitosis, the stage in which the sister chromatids of each chromosome separate from one another and are moved to opposite poles of the cell; meiosis I is the stage in which homologous chromosomes, consisting of two sister chromatids, are separated; meiosis II is the stage in which the sister chromatids of each chromosome separate from one another and are moved to opposite poles of the cell.
androgen: a male sex hormone.
androgen insensitivity: a rare condition in which an individual with XY chromosomes is female in appearance because the body's cells do not respond to the male hormones that are present.
angina (an-jī′-nuh): chest pain associated with reduced blood flow to the heart muscle, caused by the obstruction of coronary arteries.
angiosperm (an′-jē-ō-sperm): a flowering vascular plant.
angiotensin (an-jē-ō-ten′-sun): a hormone that functions in water regulation in mammals by stimulating physiological changes that increase blood volume and blood pressure.
annual ring: a pattern of alternating light (early) and dark (late) xylem of woody stems and roots, formed as a result of the unequal availability of water in different seasons of the year, normally spring and summer.
antagonistic muscles: a pair of muscles, one of which contracts and in so doing extends the other; an arrangement that makes possible movement of the skeleton at joints.
anterior: the front, forward, or head end of an animal.
anterior pituitary: a lobe of the pituitary gland that produces prolactin and growth hormone as well as hormones that regulate hormone production in other glands.
anther (an′-ther): the uppermost part of the stamen, in which pollen develops.
antheridium (an-ther-id′-ē-um): a structure in which male sex cells are produced, found in the bryophytes and certain seedless vascular plants.
antibiotic resistance: the ability of a mutated pathogen to resist the effects of an antibiotic that normally kills it.
antibody: a protein, produced by cells of the immune system, that combines with a specific antigen and normally facilitates the destruction of the antigen.
anticodon: a sequence of three bases in transfer RNA that is complementary to the three bases of a codon of messenger RNA.
antidiuretic hormone (an-tē-dī-ūr-et′-ik; ADH): a hormone produced by the hypothalamus and released into the bloodstream by the posterior pituitary when blood volume is low; increases the permeability of the distal tubule and the collecting duct to water, allowing more water to be reabsorbed into the bloodstream.
antigen: a complex molecule, normally a protein or polysaccharide, that stimulates the production of a specific antibody.
aphotic zone: the region of the ocean below 200 meters, where sunlight does not penetrate.
apical dominance: the phenomenon whereby a growing shoot tip inhibits the sprouting of lateral buds.
apical meristem (āp′-i-kul mer′-i-stem): the cluster of meristematic cells at the tip of a shoot or root (or one of their branches).
apicomplexan (ap-i-kum-plex′-un): a member of the protist phylum Apicomplexa, which includes mostly parasitic, single-celled eukaryotes such as *Plasmodium*, which causes malaria in humans. The apicomplexans are part of a larger group known as the alveolates.
appendicular skeleton (ap-pen-dik′-ū-lur): the portion of the skeleton consisting of the bones of the extremities and their attachments to the axial skeleton; the pectoral and pelvic girdles, the arms, legs, hands, and feet.
aqueous humor (ā′-kwē-us): the clear, watery fluid between the cornea and lens of the eye.
Archaea: one of life's three domains; consists of prokaryotes that are only distantly related to members of the domain Bacteria.
archegonium (ar-ke-gō′-nē-um): a structure in which female sex cells are produced; found in the bryophytes and certain seedless vascular plants.
arteriole (ar-tēr ′-ē-ōl): a small artery that empties into capillaries. Contraction of the arteriole regulates blood flow to various parts of the body.
artery (ar′-tuh-rē): a vessel with muscular, elastic walls that conducts blood away from the heart.
arthropod: a member of the animal phylum Arthropoda, which includes the insects, spiders, ticks, mites, scorpions, crustaceans, millipedes, and centipedes.
artificial selection: a selective breeding procedure in which only those individuals with particular traits are chosen as breeders; used mainly to enhance desirable traits in domestic plants and animals; may also be used in evolutionary biology experiments.
ascus (as′-kus): a saclike case in which sexual spores are formed by members of the fungal division Ascomycota.
asexual reproduction: reproduction that does not involve the fusion of haploid sex cells. The parent body may divide and new parts regenerate, or a new, smaller individual may form as an attachment to the parent, to drop off when complete.
association neuron: in a neural network, a nerve cell that is postsynaptic to a sensory neuron and presynaptic to a motor neuron. In actual circuits, there may be many association neurons between individual sensory and motor neurons.
atherosclerosis (ath′-er-ō-skler-ō′-sis): a disease characterized by the obstruction of arteries by cholesterol deposits and thickening of the arterial walls.
atom: the smallest particle of an element that retains the properties of the element.
atomic nucleus: the membrane-bound organelle of eukaryotic cells that contains the cell's genetic material.
atomic number: the number of protons in the nuclei of all atoms of a particular element.
atrial natriuretic peptide (ā′-trē-ul nā-trē-ū-ret′-ik; ANP): a hormone, secreted by cells in the mammalian heart, that reduces blood volume by inhibiting the release of antidiuretic hormone (ADH) and aldosterone.
atrioventricular (AV) node (ā′-trē-ō-ven-trik′-ū-lar nōd): a specialized mass of muscle at the base of the right atrium through which the electrical activity initiated in the sinoatrial node is transmitted to the ventricles.
atrioventricular valve: a heart valve that separates each atrium from each ventricle, preventing the backflow of blood into the atria during ventricular contraction.
atrium (ā′-trē-um): a chamber of the heart that receives venous blood and passes it to a ventricle.
auditory canal (aw′-di-tor-ē): a canal within the outer ear that conducts sound from the external ear to the tympanic membrane.
auditory nerve: the nerve leading from the mammalian cochlea to the brain, carrying information about sound.
autoimmune disease: a disorder in which the immune system produces antibodies against the body's own cells.
autonomic nervous system: the part of the peripheral nervous system of vertebrates that synapses on glands, internal organs, and smooth muscle and produces largely involuntary responses.
autosome (aw′-tō-sōm): a chromosome that occurs in homologous pairs in both males and females and that does not bear the genes determining sex.
autotroph (aw′-tō-trōf): "self-feeder"; normally, a photosynthetic organism; a producer.
auxin (awk′-sin): a plant hormone that influences many plant functions, including phototropism, apical dominance, and root branching; generally stimulates cell elongation and, in some cases, cell division and differentiation.
axial skeleton: the skeleton forming the body axis, including the skull, vertebral column, and rib cage.
axon: a long extension of a nerve cell, extending from the cell body to synaptic endings on other nerve cells or on muscles.
bacillus (buh-sil′-us; pl., bacilli): a rod-shaped bacterium.
Bacteria: one of life's three domains; consists of prokaryotes that are only distantly related to members of the domain Archaea.
bacterial conjugation: the exchange of genetic material between two bacteria.
bacteriophage (bak-tir′-ē-ō-fāj): a virus specialized to attack bacteria.
bacterium (bak-tir′-ē-um; pl., bacteria): an organism consisting of a single prokaryotic cell surrounded by a complex polysaccharide coat.
balanced polymorphism: the prolonged maintenance of two or more alleles in a population, normally because each allele is favored by a separate environmental pressure.
ball-and-socket joint: a joint in which the rounded end of one bone fits into a hollow depression in another, as in the hip; allows movement in several directions.
bark: the outer layer of a woody stem, consisting of phloem, cork cambium, and cork cells.
Barr body: an inactivated X chromosome in cells of female mammals, which have two X chromosomes; normally appears as a dark spot in the nucleus.
basal body: a structure resembling a centriole that produces a cilium or flagellum and anchors this structure within the plasma membrane.
base: (1) a substance capable of combining with and neutralizing H^+ ions in a solution; a solution with a pH of more than 7; (2) in molecular genetics, one of the nitrogen-containing, single- or double-ringed structures that distinguish one nucleotide from another. In DNA, the bases are adenine, guanine, cytosine, and thymine.
basic: with an H^+ concentration less than that of OH^-; combining with H^+.
basidiospore (ba-sid′-ē-ō-spor): a sexual spore formed by members of the fungal division Basidiomycota.
basidium (bas-id′-ē-um): a diploid cell, typically club shaped, formed by members of the fungal division Basidiomycota; produces basidiospores by meiosis.

basilar membrane (bas′-eh-lar): a membrane in the cochlea that bears hair cells that respond to the vibrations produced by sound.

basophil (bas′-ō-fil): a type of white blood cell that releases both substances that inhibit blood clotting and chemicals that participate in allergic reactions and in responses to tissue damage and microbial invasion.

B cell: a type of lymphocyte that participates in humoral immunity; gives rise to plasma cells, which secrete antibodies into the circulatory system, and to memory cells.

behavior: any observable activity of a living animal.

behavioral isolation: the lack of mating between species of animals that differ substantially in courtship and mating rituals.

bilateral symmetry: a body plan in which a single plane through the central axis will divide the body into mirror-image halves.

bile (bīl): a liquid secretion, produced by the liver, that is stored in the gallbladder and released into the small intestine during digestion; a complex mixture of bile salts, water, other salts, and cholesterol.

bile salt: a substance that is synthesized in the liver from cholesterol and amino acids and that assists in the breakdown of lipids by dispersing them into small particles on which enzymes can act.

binary fission: the process by which a single bacterium divides in half, producing two identical offspring.

binocular vision: the ability to see objects simultaneously through both eyes, providing great depth perception and accurate judgment of the size and distance of an object from the eyes.

biodegradable: able to be broken down into harmless substances by decomposers.

biodiversity: the total number of species within an ecosystem and the resulting complexity of interactions among them.

biogeochemical cycle: also called a *nutrient cycle*, the process by which a specific nutrient in an ecosystem is transferred between living organisms and the nutrient's reservoir in the nonliving environment.

biological clock: a metabolic timekeeping mechanism found in most organisms, whereby the organism measures the approximate length of a day (24 hours) even without external environmental cues such as light and darkness.

biological magnification: the increasing accumulation of a toxic substance in progressively higher trophic levels.

biomass: the dry weight of organic material in an ecosystem.

biome (bī′-ōm): a terrestrial ecosystem that occupies an extensive geographical area and is characterized by a specific type of plant community: for example, a desert.

biosphere (bī′-ō-sfēr): that part of Earth inhabited by living organisms; includes both living and nonliving components.

biotechnology: any industrial or commercial use or alteration of organisms, cells, or biological molecules to achieve specific practical goals.

biotic (bī-ah′-tik): living.

biotic potential: the maximum rate at which a population could increase, assuming ideal conditions that allow a maximum birth rate and minimum death rate.

birth control pill: a temporary contraceptive method that prevents ovulation by providing a continuing supply of estrogen and progesterone, which in turn suppresses luteinizing hormone (LH) release; must be taken daily, normally for 21 days of each menstrual cycle.

bladder: a hollow muscular storage organ for storing urine.

blade: the flat part of a leaf.

blastocyst (blas′-tō-sist): an early stage of human embryonic development, consisting of a hollow ball of cells, enclosing a mass of cells attached to its inner surface, which becomes the embryo.

blastopore: the site at which a blastula indents to form a gastrula.

blastula (blas′-tū-luh): in animals, the embryonic stage attained at the end of cleavage, in which the embryo normally consists of a hollow ball with a wall one or several cell layers thick.

blind spot: the area of the retina at which the axons of the ganglion cell merge to form the optic nerve; the blind spot of the retina.

blood: a fluid consisting of plasma in which blood cells are suspended; carried within the circulatory system.

blood–brain barrier: relatively impermeable capillaries of the brain that protect the cells of the brain from potentially damaging chemicals that reach the bloodstream.

blood clotting: a complex process by which platelets, the protein fibrin, and red blood cells block an irregular surface in or on the body, such as a damaged blood vessel, sealing the wound.

blood vessel: a channel that conducts blood throughout the body.

body mass index (BMI): a number derived from an individual's weight and height used to estimate body fat. The formula is: weight (in kg)/height^2 (in meters^2).

bone: a hard, mineralized connective tissue that is a major component of the vertebrate endoskeleton; provides support and sites for muscle attachment.

book lung: a structure composed of thin layers of tissue, resembling pages in a book, that are enclosed in a chamber and used as a respiratory organ by certain types of arachnids.

boom-and-bust cycle: a population cycle characterized by rapid exponential growth followed by a sudden massive die-off, seen in seasonal species and in some populations of small rodents, such as lemmings.

Bowman's capsule: the cup-shaped portion of the nephron in which blood filtrate is collected from the glomerulus.

bradykinin (brā′-dē-kī′-nin): a chemical, formed during tissue damage, that binds to receptor molecules on pain nerve endings, giving rise to the sensation of pain.

brain: the part of the central nervous system of vertebrates that is enclosed within the skull.

branch root: a root that arises as a branch of a preexisting root, through divisions of pericycle cells and subsequent differentiation of the daughter cells.

bronchiole (bron′-kē-ōl): a narrow tube, formed by repeated branching of the bronchi, that conducts air into the alveoli.

bronchus (bron′-kus): a tube that conducts air from the trachea to each lung.

bryophyte (brī′-ō-fīt): a simple nonvascular plant of the division Bryophyta, including mosses and liverworts.

bud: in animals, a small copy of an adult that develops on the body of the parent and eventually breaks off and becomes independent; in plants, an embryonic shoot, normally very short and consisting of an apical meristem with several leaf primordia.

budding: asexual reproduction by the growth of a miniature copy, or bud, of the adult animal on the body of the parent. The bud breaks off to begin independent existence.

buffer: a compound that minimizes changes in pH by reversibly taking up or releasing H^+ ions.

bulbourethral gland (bul-bō-ū-rē′-thrul): in male mammals, a gland that secretes a basic, mucus-containing fluid that forms part of the semen.

bulk flow: the movement of many molecules of a gas or fluid in unison from an area of higher pressure to an area of lower pressure.

bundle-sheath cell: one of a group of cells that surround the veins of plants; in C_4 (but not in C_3) plants, bundle-sheath cells contain chloroplasts.

C_3 cycle: the cyclic series of reactions whereby carbon dioxide is fixed into carbohydrates during the light-independent reactions of photosynthesis; also called *Calvin-Benson cycle*.

C_4 pathway: the series of reactions in certain plants that fixes carbon dioxide into oxaloacetic acid, which is later broken down for use in the C_3 cycle of photosynthesis.

calcitonin (kal-si-tōn′-in): a hormone, secreted by the thyroid gland, that inhibits the release of calcium from bone.

calorie (kal′-ō-rē): the amount of energy required to raise the temperature of 1 gram of water by 1 degree Celsius.

Calorie: a unit of energy, in which the energy content of foods is measured; the amount of energy required to raise the temperature of 1 liter of water 1 degree Celsius; also called a *kilocalorie*, equal to 1000 calories.

Calvin-Benson cycle: see *C_3 cycle*.

cambium (kam′-bē-um; pl., cambia): a lateral meristem, parallel to the long axis of roots and stems, that causes secondary growth of woody plant stems and roots. See *cork cambium; vascular cambium*.

camouflage (cam′-a-flaj): coloration and/or shape that renders an organism inconspicuous in its environment.

cancer: a disease in which some of the body's cells escape from normal regulatory processes and divide without control.

capillary: the smallest type of blood vessel, connecting arterioles with venules. Capillary walls, through which the exchange of nutrients and wastes occurs, are only one cell thick.

capsule: a polysaccharide or protein coating that some disease-causing bacteria secrete outside their cell wall.

carbohydrate: a compound composed of carbon, hydrogen, and oxygen, with the approximate chemical formula $(CH_2O)_n$; includes sugars and starches.

carbon fixation: the initial steps in the C_3 cycle, in which carbon dioxide reacts with ribulose bisphosphate to form a stable organic molecule.

cardiac cycle (kar′-dē-ak): the alternation of contraction and relaxation of the heart chambers.

cardiac muscle (kar′-dē-ak): the specialized muscle of the heart, able to initiate its own contraction, independent of the nervous system.

carnivore (kar′-neh-vor): literally, "meat eater"; a predatory organism that feeds on herbivores or on other carnivores; a secondary (or higher) consumer.

carotenoid (ka-rot′-en-oid): a red, orange, or yellow pigment, found in chloroplasts, that serves as an accessory light-gathering molecule in thylakoid photosystems.

carpel (kar′pel): the female reproductive structure of a flower, composed of stigma, style, and ovary.

carrier: an individual who is heterozygous for a recessive condition; displays the dominant phenotype but can pass on the recessive allele to offspring.

carrier protein: a membrane protein that facilitates the diffusion of specific substances across the membrane. The molecule to be transported binds to the outer surface of the carrier protein; the protein then changes shape, allowing the molecule to move across the membrane through the protein.

carrying capacity: the maximum population size that an ecosystem can support indefinitely; determined primarily by the availability of space, nutrients, water, and light.

cartilage (kar′-teh-lij): a form of connective tissue that forms portions of the skeleton; consists of chondrocytes and their extracellular secretion of collagen; resembles flexible bone.

Casparian strip (kas-par′-ē-un): a waxy, waterproof band, located in the cell walls between endodermal cells in a root, that prevents the movement of water and minerals into and out of the vascular cylinder through the extracellular space.

catalyst (kat′-uh-list): a substance that speeds up a chemical reaction without itself being permanently changed in the process; lowers the activation energy of a reaction.

catastrophism: the hypothesis that Earth has experienced a series of geological catastrophes, probably imposed by a supernatural being, that accounts for the multitude of species, both extinct and modern, and preserves creationism.

cell: the smallest unit of life, consisting, at a minimum, of an outer membrane that encloses a watery medium containing organic molecules, including genetic material composed of DNA.

cell body: the part of a nerve cell in which most of the common cellular organelles are located; typically a site of integration of inputs to the nerve cell.

cell cycle: the sequence of events in the life of a cell, from one division to the next.

cell division: splitting of one cell into two; the process of cellular reproduction.

cell-mediated immunity: an immune response in which foreign cells or substances are destroyed by contact with T cells.

cell plate: in plant cell division, a series of vesicles that fuse to form the new plasma membranes and cell wall separating the daughter cells.

cellular respiration: the oxygen-requiring reactions, occurring in mitochondria, that break down the end products of glycolysis into carbon dioxide and water while capturing large amounts of energy as ATP.

cellular slime mold: a funguslike protist consisting of individual amoeboid cells that can aggregate to form a sluglike mass, which in turn forms a fruiting body.

cellulase: an enzyme that catalyzes the breakdown of the carbohydrate cellulose into its component glucose molecules; almost entirely restricted to microorganisms.

cellulose: an insoluble carbohydrate composed of glucose subunits; forms the cell wall of plants.

cell wall: a layer of material, normally made up of cellulose or cellulose-like materials, that is outside the plasma membrane of plants, fungi, bacteria, and some protists.

central nervous system: in vertebrates, the brain and spinal cord.

central vacuole: a large, fluid-filled vacuole occupying most of the volume of many plant cells; performs several functions, including maintaining turgor pressure.

centriole (sen′-trē-ōl): in animal cells, a short, barrel-shaped ring consisting of nine microtubule triplets; a microtubule-containing structure at the base of each cilium and flagellum; gives rise to the microtubules of cilia and flagella and is involved in spindle formation during cell division.

centromere (sen′-trō-mēr): the region of a replicated chromosome at which the sister chromatids are held together until they separate during cell division.

cephalization (sef-ul-ī-zā′-shun): the tendency of sensory organs and nervous tissue to become concentrated in the head region over evolutionary time.

cerebellum (ser-uh-bel′-um): the part of the hindbrain of vertebrates that is concerned with coordinating movements of the body.

cerebral cortex (ser-ē′-brul kor′-tex): a thin layer of neurons on the surface of the vertebrate cerebrum, in which most neural processing and coordination of activity occurs.

cerebral hemisphere: one of two nearly symmetrical halves of the cerebrum, connected by a broad band of axons, the corpus callosum.

cerebrospinal fluid: a clear fluid, produced within the ventricles of the brain, that fills the ventricles and cushions the brain and spinal cord.

cerebrum (ser-ē′-brum): the part of the forebrain of vertebrates that is concerned with sensory processing, the direction of motor output, and the coordination of most bodily activities; consists of two nearly symmetrical halves (the hemispheres) connected by a broad band of axons, the corpus callosum.

cervical cap: a birth control device consisting of a rubber cap that fits over the cervix, preventing sperm from entering the uterus.

cervix (ser′-viks): a ring of connective tissue at the outer end of the uterus, leading into the vagina.

channel protein: a membrane protein that forms a channel or pore completely through the membrane and that is usually permeable to one or to a few water-soluble molecules, especially ions.

chaparral: a biome that is located in coastal regions but has very low annual rainfall.

chemical bond: the force of attraction between neighboring atoms that holds them together in a molecule.

chemical equilibrium: the condition in which the "forward" reaction of reactants to products proceeds at the same rate as the "backward" reaction from products to reactants, so that no net change in chemical composition occurs.

chemical reaction: the process that forms and breaks chemical bonds that hold atoms together.

chemiosmosis (ke-mē-oz-mō′-sis): a process of adenosine triphosphate (ATP) generation in chloroplasts and mitochondria. The movement of electrons down an electron transport system is used to pump hydrogen ions across a membrane, thereby building up a concentration gradient of hydrogen ions across the membrane; the hydrogen ions diffuse back across the membrane through the pores of ATP-synthesizing enzymes; the energy of their movement down their concentration gradient drives ATP synthesis.

chemoreceptor: a sensory receptor that responds to chemicals from the environment; used in the chemical senses of taste and smell.

chemosynthetic (kēm′-ō-sin-the-tik): capable of oxidizing inorganic molecules to obtain energy.

chemotactic (kēm-ō-tak′-tik): moving toward chemicals given off by food or away from toxic chemicals.

chiasma (kī-as′-muh; pl., chiasmata): a point at which a chromatid of one chromosome crosses with a chromatid of the homologous chromosome during prophase I of meiosis; the site of exchange of chromosomal material between chromosomes.

chitin (kī′-tin): a compound found in the cell walls of fungi and the exoskeletons of insects and some other arthropods; composed of chains of nitrogen-containing, modified glucose molecules.

chlamydia (kla-mid′-ē-uh): a sexually transmitted disease, caused by a bacterium, that causes inflammation of the urethra in males and of the urethra and cervix in females.

chlorophyll (klor′-ō-fil): a pigment found in chloroplasts that captures light energy during photosynthesis; absorbs violet, blue, and red light but reflects green light.

chloroplast (klor′-ō-plast): the organelle in plants and plantlike protists that is the site of photosynthesis; surrounded by a double membrane and containing an extensive internal membrane system that bears chlorophyll.

cholecystokinin (kō′-lē-sis-tō-kī′-nin): a digestive hormone, produced by the small intestine, that stimulates the release of pancreatic enzymes.

chondrocyte (kon′-drō-sīt): a living cell of cartilage. With their extracellular secretions of collagen, chondrocytes form cartilage.

chorion (kor′-ē-on): the outermost embryonic membrane in reptiles, birds, and mammals; in birds and reptiles, functions mostly in gas exchange; in mammals, forms most of the embryonic part of the placenta.

chorionic gonadotropin (CG): a hormone, secreted by the chorion (one of the fetal membranes), that maintains the integrity of the corpus luteum during early pregnancy.

chorionic villus (kor-ē-on-ik; pl., chorionic villi): in mammalian embryos, a fingerlike projection of the chorion that penetrates the uterine lining and forms the embryonic portion of the placenta.

chorionic villus sampling (CVS): a procedure for sampling cells from the chorionic villi produced by a fetus: A tube is inserted into the uterus of a pregnant woman, and a small sample of villi are suctioned off for genetic and biochemical analyses.

choroid (kor′-oid): a darkly pigmented layer of tissue, behind the retina, that contains blood vessels and pigment that absorbs stray light.

chromatid (krō′-ma-tid): one of the two identical strands of DNA and protein that forms a replicated chromosome. The two sister chromatids are joined at the centromere.

chromatin (krō′-ma-tin): the complex of DNA and proteins that makes up eukaryotic chromosomes.

chromosome (krō′-mō-sōm): a DNA double helix together with proteins that help to organize the DNA.

chromist (krō′-mist): a member of the Chromista, a large assemblage of protists that is assigned kingdom status by many systematists. The chromists include the diatoms, brown algae, and water molds.

chronic bronchitis: a persistent lung infection characterized by coughing, swelling of the lining of the respiratory tract, an increase in mucus production, and a decrease in the number and activity of cilia.

chyme (kīm): an acidic, souplike mixture of partially digested food, water, and digestive secretions that is released from the stomach into the small intestine.

ciliate (sil′-ē-et): a protozoan characterized by cilia and by a complex unicellular structure, including harpoonlike organelles called *trichocysts.* Members of the genus *Paramecium* are well-known ciliates.

cilium (sil′-ē-um; pl., cilia): a short, hairlike projection from the surface of certain eukaryotic cells that contains microtubules in a 9 + 2 arrangement. The movement of cilia may propel cells through a fluid medium or move fluids over a stationary surface layer of cells.
circadian rhythm (sir-kā′-dē-un): an event that recurs with a period of about 24 hours, even in the absence of environmental cues.
citric acid cycle: see *Krebs cycle*.
class: the taxonomic category composed of related genera. Closely related classes form a division or phylum.
cleavage: the early cell divisions of embryos, in which little or no growth occurs between divisions; reduces the cell size and distributes gene-regulating substances to the newly formed cell.
climate: patterns of weather that prevail from year to year and even from century to century in a given region.
climax community: a diverse and relatively stable community that forms the endpoint of succession.
clitoris: an external structure of the female reproductive system; composed of erectile tissue; a sensitive point of stimulation during sexual response.
clonal selection: the mechanism by which the immune response gains specificity; an invading antigen elicits a response from only a few lymphocytes, which proliferate to form a clone of cells that attack only the specific antigen that stimulated their production.
clone: offspring that are produced by mitosis and are therefore genetically identical to each other.
cloning: the process of producing many identical copies of a gene; also the production of many genetically identical copies of an organism.
closed circulatory system: the type of circulatory system, found in certain worms and vertebrates, in which the blood is always confined within the heart and vessels.
club fungus: a fungus of the division Basidiomycota, whose members (which include mushrooms, puffballs, and shelf fungi) reproduce by means of basidiospores.
clumped distribution: the distribution characteristic of populations in which individuals are clustered into groups; may be social or based on the need for a localized resource.
cnidocyte (nīd′-ō-sīt): in members of the phylum Cnidaria, a specialized cell that houses a stinging apparatus.
cochlea (kahk′-lē-uh): a coiled, bony, fluid-filled tube found in the mammalian inner ear; contains receptors (hair cells) that respond to the vibration of sound.
codominance: the relation between two alleles of a gene, such that both alleles are phenotypically expressed in heterozygous individuals.
codon: a sequence of three bases of messenger RNA that specifies a particular amino acid to be incorporated into a protein; certain codons also signal the beginning or end of protein synthesis.
coelom (sē′-lōm): a space or cavity that separates the body wall from the inner organs.
coenzyme: an organic molecule that is bound to certain enzymes and is required for the enzymes' proper functioning; typically, a nucleotide bound to a water-soluble vitamin.
coevolution: the evolution of adaptations in two species due to their extensive interactions with one another, such that each species acts as a major force of natural selection on the other.
cohesion: the tendency of the molecules of a substance to stick together.
cohesion–tension theory: a model for the transport of water in xylem, by which water is pulled up the xylem tubes, powered by the force of evaporation of water from the leaves (producing tension) and held together by hydrogen bonds between nearby water molecules (cohesion).
coleoptile (kō-lē-op′-tīl): a protective sheath surrounding the shoot in monocot seeds, allowing the shoot to push aside soil particles as it grows.
collagen (kol′-uh-jen): a fibrous protein in connective tissue such as bone and cartilage.
collar cell: a specialized cell lining the inside channels of sponges. Flagella extend from a sievelike collar, creating a water current that draws microscopic organisms through the collar and into the body, where they become trapped.
collecting duct: a conducting tube, within the kidney, that collects urine from many nephrons and conducts it through the renal medulla into the renal pelvis. Urine may become concentrated in the collecting ducts if antidiuretic hormone (ADH) is present.
collenchyma (kōl-en′-ki-muh): an elongated, polygonal plant cell type with irregularly thickened primary cell walls that is alive at maturity and that supports the plant body.
colon: the longest part of the large intestine, exclusive of the rectum.
colostrum (kō-los′-trum): a yellowish fluid, high in protein and containing antibodies, that is produced by the mammary glands before milk secretion begins.
commensalism (kum-en′-sal-iz-um): a symbiotic relationship in which one species benefits while another species is neither harmed nor benefited.
communication: the act of producing a signal that causes another animal, normally of the same species, to change its behavior in a way that is beneficial to one or both participants.
community: all the interacting populations within an ecosystem.
compact bone: the hard and strong outer bone; composed of osteons.
companion cell: a cell adjacent to a sieve-tube element in phloem, involved in the control and nutrition of the sieve-tube element.
competition: interaction among individuals who attempt to utilize a resource (for example, food or space) that is limited relative to the demand for it.
competitive exclusion principle: the concept that no two species can simultaneously and continuously occupy the same ecological niche.
competitive inhibition: the process by which two or more molecules that are somewhat similar in structure compete for the active site of an enzyme.
complement: a group of blood-borne proteins that participate in the destruction of foreign cells to which antibodies have bound.
complementary base pair: in nucleic acids, bases that pair by hydrogen bonding. In DNA, adenine is complementary to thymine and guanine is complementary to cytosine; in RNA, adenine is complementary to uracil, and guanine to cytosine.
complement reaction: an interaction among foreign cells, antibodies, and complement proteins that results in the destruction of the foreign cells.
complement system: a series of reactions in which complement proteins bind to antibody stems, attracting to the site phagocytic white blood cells that destroy the invading cell that triggers the reactions.
complete flower: a flower that has all four floral parts (sepals, petals, stamens, and carpels).
compound: a substance whose molecules are formed by different types of atoms; can be broken into its constituent elements by chemical means.
compound eye: a type of eye, found in arthropods, that is composed of numerous independent subunits called *ommatidia*. Each ommatidium apparently contributes a piece of a mosaic-like image perceived by the animal.
concentration: the number of particles of a dissolved substance in a given unit of volume.
concentration gradient: the difference in concentration of a substance between two parts of a fluid or across a barrier such as a membrane.
conclusion: the final operation in the scientific method; a decision made about the validity of a hypothesis on the basis of experimental evidence.
condensation: compaction of eukaryotic chromosomes into discrete units in preparation for mitosis or meiosis.
condom: a contraceptive sheath worn over the penis during intercourse to prevent sperm from being deposited in the vagina.
conducting portion: the portion of the respiratory system in lung-breathing vertebrates that carries air to the lungs.
cone: a cone-shaped photoreceptor cell in the vertebrate retina; not as sensitive to light as are the rods. The three types of cones are most sensitive to different colors of light and provide color vision; see also *rod*.
conifer (kon′-eh-fer): a member of a class of tracheophytes (Coniferophyta) that reproduces by means of seeds formed inside cones and that retains its leaves throughout the year.
conjugation: in prokaryotes, the transfer of DNA from one cell to another via a temporary connection; in single-celled eukaryotes, the mutual exchange of genetic material between two temporarily joined cells.
connective tissue: a tissue type consisting of diverse tissues, including bone, fat, and blood, that generally contain large amounts of extracellular material.
constant region: the part of an antibody molecule that is similar in all antibodies.
consumer: an organism that eats other organisms; a heterotroph.
contest competition: a type of intraspecific competition in which competing individuals interact directly with one another—for example, by fighting.
contraception: the prevention of pregnancy.
contractile vacuole: a fluid-filled vacuole in certain protists that takes up water from the cytoplasm, contracts, and expels the water outside the cell through a pore in the plasma membrane.
control: that portion of an experiment in which all possible variables are held constant; in contrast to the "experimental" portion, in which a particular variable is altered.
convergence: a condition in which a large number of nerve cells provide input to a smaller number of cells.
convergent evolution: the independent evolution of similar structures among distantly related organisms as a result of similar environmental pressures; see *analogous structures*.
convolution: a folding of the cerebral cortex of the vertebrate brain.
copulation: reproductive behavior in which the penis of the male is inserted into the body of the female, where it releases sperm.
coral reef: a biome created by animals (reef-building corals) and plants in warm tropical waters.

cork cambium: a lateral meristem in woody roots and stems that gives rise to cork cells.

cork cell: a protective cell of the bark of woody stems and roots; at maturity, cork cells are dead, with thick, waterproofed cell walls.

cornea (kor′-nē-uh): the clear outer covering of the eye, in front of the pupil and iris.

corona radiata (kuh-rō′-nuh rā-dē-a′-tuh): the layer of cells surrounding an egg after ovulation.

corpus callosum (kor′pus kal-ō′-sum): the band of axons that connect the two cerebral hemispheres of vertebrates.

corpus luteum (kor′-pus loo′-tē-um): in the mammalian ovary, a structure that is derived from the follicle after ovulation and that secretes the hormones estrogen and progesterone.

cortex: the part of a primary root or stem located between the epidermis and the vascular cylinder.

cotyledon (kot-ul-ē′don): a leaflike structure within a seed that absorbs food molecules from the endosperm and transfers them to the growing embryo; also called *seed leaf.*

coupled reaction: a pair of reactions, one exergonic and one endergonic, that are linked together such that the energy produced by the exergonic reaction provides the energy needed to drive the endergonic reaction.

covalent bond (kō-vā′-lent): a chemical bond between atoms in which electrons are shared.

crab lice: an arthropod parasite that can infest humans; can be transmitted by sexual contact.

creationism: the hypothesis that all species on Earth were created in essentially their present form by a supernatural being and that significant modification of those species—specifically, their transformation into new species—cannot occur by natural processes.

crista (kris′-tuh; pl., cristae): a fold in the inner membrane of a mitochondrion.

crop: an organ, found in both earthworms and birds, in which ingested food is temporarily stored before being passed to the gizzard, where it is pulverized.

cross-bridge: in muscles, an extension of myosin that binds to and pulls on actin to produce muscle contraction.

cross-fertilization: the union of sperm and egg from two individuals of the same species.

crossing over: the exchange of corresponding segments of the chromatids of two homologous chromosomes during meiosis.

cultural evolution: changes in the behavior of a population of animals, especially humans, by learning behaviors acquired by members of previous generations.

cuticle (kū′-ti-kul): a waxy or fatty coating on the exposed surfaces of epidermal cells of many land plants, which aids in the retention of water.

cyanobacterium: a photosynthetic prokaryotic cell that utilizes chlorophyll and releases oxygen as a photosynthetic by-product; sometimes called *blue-green algae.*

cyclic AMP: a cyclic nucleotide (adenosine monophosphate), formed within many target cells as a result of the reception of amino acid derivatives or peptide hormones, that causes metabolic changes in the cell; often called a *second messenger.*

cyclic nucleotide (sik′-lik noo′-klē-ō-tīd): a nucleotide in which the phosphate group is bonded to the sugar at two points, forming a ring; serves as an intracellular messenger.

cyst (sist): an encapsulated resting stage in the life cycle of certain invertebrates, such as parasitic flatworms and roundworms.

cystic fibrosis: an inherited disorder characterized by the buildup of salt in the lungs and the production of thick, sticky mucus that clogs the airways, restricts air exchange, and promotes infection.

cytokinesis (sī-tō-ki-nē′-sis): the division of the cytoplasm and organelles into two daughter cells during cell division; normally occurs during telophase of mitosis.

cytokinin (sī-tō-kī′-nin): a plant hormone that promotes cell division, fruit growth, and the sprouting of lateral buds and prevents the aging of plant parts, especially leaves.

cytoplasm (sī′-tō-plaz-um): the material contained within the plasma membrane of a cell, exclusive of the nucleus.

cytosine: a nitrogenous base found in both DNA and RNA; abbreviated as C.

cytoskeleton: a network of protein fibers in the cytoplasm that gives shape to a cell, holds and moves organelles, and is typically involved in cell movement.

cytotoxic T cell: a type of T cell that, upon contacting foreign cells, directly destroys them.

day-neutral plant: a plant in which flowering occurs as soon as the plant has grown and developed, regardless of daylength.

decomposer: an organism, normally a fungus or bacterium, that digests organic material by secreting digestive enzymes into the environment, in the process liberating nutrients into the environment.

deductive reasoning: the process of generating hypotheses about how a specific experiment or observation will turn out.

deforestation: the excessive cutting of forests, primarily rain forests in the Tropics, to clear space for agriculture.

dehydration synthesis: a chemical reaction in which two molecules are joined by a covalent bond with the simultaneous removal of a hydrogen atom from one molecule and a hydroxyl group from the other, forming water; the reverse of hydrolysis.

deletion mutation: a mutation in which one or more pairs of nucleotides are removed from a gene.

denature: to disrupt the secondary and/or tertiary structure of a protein while leaving its amino acid sequence intact. Denatured proteins can no longer perform their biological functions.

dendrite (den′-drīt): a branched tendril that extends outward from the cell body of a neuron; specialized to respond to signals from the external environment or from other neurons.

denitrifying bacterium (dē-nī′-treh-fī-ing): a bacterium that breaks down nitrates, releasing nitrogen gas to the atmosphere.

density-dependent: referring to any factor, such as predation, that limits population size more effectively as the population density increases.

density-independent: referring to any factor that limits a population's size and growth regardless of its density.

deoxyribonucleic acid (dē-ox-ē-rī-bō-noo-klā′-ik; DNA): a molecule composed of deoxyribose nucleotides; contains the genetic information of all living cells.

dermal tissue: plant tissue that makes up the outer covering of the plant body.

dermis (dur′-mis): the layer of skin beneath the epidermis; composed of connective tissue and containing blood vessels, muscles, nerve endings, and glands.

desert: a biome in which less than 25 to 50 centimeters (10 to 20 inches) of rain falls each year.

desertification: the spread of deserts by human activities.

desmosome (dez′-mō-sōm): a strong cell-to-cell junction that attaches adjacent cells to one another.

detritus feeder (de-trī′-tus): one of a diverse group of organisms, ranging from worms to vultures, that live off the wastes and dead remains of other organisms.

deuterostome (doo′-ter-ō-stōm): an animal with a mode of embryonic development in which the coelom is derived from outpocketings of the gut; characteristic of echinoderms and chordates.

development: the process by which an organism proceeds from fertilized egg through adulthood to eventual death.

diabetes mellitus (dī-uh-bē′-tēs mel-ī′-tus): a disease characterized by defects in the production, release, or reception of insulin; characterized by high blood glucose levels that fluctuate with sugar intake.

dialysis (dī-al′-i-sis): the passive diffusion of substances across an artificial semipermeable membrane.

diaphragm (dī′-uh-fram): in the respiratory system, a dome-shaped muscle forming the floor of the chest cavity that, when it contracts, pulls itself downward, enlarging the chest cavity and causing air to be drawn into the lungs; in a reproductive sense, a contraceptive rubber cap that fits snugly over the cervix, preventing the sperm from entering the uterus and thereby preventing pregnancy.

diatom (dī′-uh-tom): a protist that includes photosynthetic forms with two-part glassy outer coverings; important photosynthetic organisms in fresh water and salt water.

dicot (dī′-kaht): short for dicotyledon; a type of flowering plant characterized by embryos with two cotyledons, or seed leaves, modified for food storage.

differentially permeable: referring to a membrane through which some substances pass more readily than can other substances.

differential reproduction: differences in reproductive output among individuals of a population, normally as a result of genetic differences.

differentiated cell: a mature cell specialized for a specific function; in plants, differentiated cells normally do not divide.

differentiation: the process whereby relatively unspecialized cells, especially of embryos, become specialized into particular tissue types.

diffusion: the net movement of particles from a region of high concentration of that type of particle to a region of low concentration, driven by the concentration gradient; may occur entirely within a fluid or across a barrier such as a membrane.

digestion: the process by which food is physically and chemically broken down into molecules that can be absorbed by cells.

digestive system: a group of organs responsible for ingesting and then digesting food substances into simple molecules that can be absorbed and then expelling undigested wastes from the body.

dinoflagellate (dī-nō-fla′-jel-et): a protist that includes photosynthetic forms in which two flagella project through armorlike plates; abundant in oceans; can reproduce rapidly, causing "red tides."

dioecious (dī-ē′-shus): pertaining to organisms in which male and female gametes are produced by separate individuals rather than in the same individual.

diploid (dip′-loid): referring to a cell with pairs of homologous chromosomes.

direct development: a developmental pathway in which the offspring is born as a miniature version of the adult and does not radically change in body form as it grows and matures.

directional selection: a type of natural selection in which one extreme phenotype is favored over all others.
disaccharide (dī-sak′-uh-rīd): a carbohydrate formed by the covalent bonding of two monosaccharides.
disruptive selection: a type of natural selection in which both extreme phenotypes are favored over the average phenotype.
distal tubule: in the nephrons of the mammalian kidney, the last segment of the renal tubule through which the filtrate passes just before it empties into the collecting duct; a site of selective secretion and reabsorption as water and ions pass between the blood and the filtrate across the tubule membrane.
disulfide bridge: the covalent bond formed between the sulfur atoms of two cysteines in a protein; typically causes the protein to fold by bringing otherwise distant parts of the protein close together.
divergence: a condition in which a small number of nerve cells provide input to a larger number of cells.
divergent evolution: evolutionary change in which the differences between two lineages become more pronounced with the passage of time.
division: the taxonomic category contained within a kingdom and consisting of related classes of plants, fungi, bacteria, or plantlike protists.
DNA: see *deoxyribonucleic acid.*
DNA–DNA hybridization: a technique by which DNA from two species is separated into single strands and then allowed to re-form; hybrid double-stranded DNA from the two species can occur where the sequence of nucleotides is complementary. The greater the degree of hybridization, the closer the evolutionary relatedness of the two species.
DNA fingerprinting: the use of restriction enzymes to cut DNA segments into a unique set of restriction fragments from one individual that can be distinguished from the restriction fragments of other individuals by gel electrophoresis.
DNA helicase: an enzyme that helps unwind the DNA double helix during DNA replication.
DNA library: a readily accessible, easily duplicable complete set of all the DNA of a particular organism, normally cloned into bacterial plasmids.
DNA ligase: an enzyme that joins the sugars and phosphates in a DNA strand to create a continuous sugar-phosphate backbone.
DNA polymerase: an enzyme that bonds DNA nucleotides together into a continuous strand, using a preexisting DNA strand as a template.
DNA probe: a sequence of nucleotides that is complementary to the nucleotide sequence in a gene under study; used to locate a given gene within a DNA library.
DNA replication: the copying of the double-stranded DNA molecule, producing two identical DNA double helices.
DNA sequencing: the process of determining the chemical composition of a DNA molecule (in particular, the order in which the molecule's constituent nucleic acids are arranged).
domain: the broadest category for classifying organisms; organisms are classified into three domains: Bacteria, Archaea, and Eukarya.
dominance hierarchy: a social arrangement in which a group of animals, usually through aggressive interactions, establishes a rank for some or all of the group members that determines access to resources.
dominant: an allele that can determine the phenotype of heterozygotes completely, such that they are indistinguishable from individuals homozygous for the allele; in the heterozygotes, the expression of the other (recessive) allele is completely masked.
dopamine (dōp′-uh-mēn): a transmitter in the brain whose actions are largely inhibitory. The loss of dopamine-containing neurons causes Parkinson's disease.
dormancy: a state in which an organism does not grow or develop; usually marked by lowered metabolic activity and resistance to adverse environmental conditions.
dorsal (dor′-sul): the top, back, or uppermost surface of an animal oriented with its head forward.
dorsal root ganglion: a ganglion, located on the dorsal (sensory) branch of each spinal nerve, that contains the cell bodies of sensory neurons.
double covalent bond: a covalent bond in which two atoms share two pairs of electrons.
double fertilization: in flowering plants, the fusion of two sperm nuclei with the nuclei of two cells of the female gametophyte. One sperm nucleus fuses with the egg to form the zygote; the second sperm nucleus fuses with the two haploid nuclei of the primary endosperm cell, forming a triploid endosperm cell.
double helix (hē′-liks): the shape of the two-stranded DNA molecule; like a ladder twisted lengthwise into a corkscrew shape.
doubling time: the time it would take a population to double in size at its current growth rate.
douching: washing the vagina; after intercourse, an attempt to wash sperm out of the vagina before they enter the uterus; an ineffective contraceptive method.
Down syndrome: a genetic disorder caused by the presence of three copies of chromosome 21; common characteristics include mental retardation, distinctively shaped eyelids, a small mouth with protruding tongue, heart defects, and low resistance to infectious diseases; also called *trisomy 21.*
duct: a tube or opening through which exocrine secretions are released.
duplicated chromosome: a eukaryotic chromosome following DNA replication; consists of two sister chromatids joined at the centromeres.
echolocation: the use of ultrasonic sounds, which bounce back from nearby objects, to produce an auditory "image" of nearby surroundings; used by bats and porpoises.
ecological isolation: the lack of mating between organisms belonging to different populations that occupy distinct habitats within the same general area.
ecological niche (nitch): the role of a particular species within an ecosystem, including all aspects of its interaction with the living and nonliving environments.
ecology (ē-kol′-uh-jē): the study of the interrelationships of organisms with each other and with their nonliving environment.
ecosystem (ē′kō-sis-tem): all the organisms and their nonliving environment within a defined area.
ectoderm (ek′-tō-derm): the outermost embryonic tissue layer, which gives rise to structures such as hair, the epidermis of the skin, and the nervous system.
effector (ē-fek′-tor): a part of the body (normally a muscle or gland) that carries out responses as directed by the nervous system.
egg: the haploid female gamete, normally large and nonmotile, containing food reserves for the developing embryo.
electrolocation: the production of high-frequency electrical signals from an electric organ in front of the tail of weak electrical fish; used to detect and locate nearby objects.
electron: a subatomic particle, found in an electron shell outside the nucleus of an atom, that bears a unit of negative charge and very little mass.
electron carrier: a molecule that can reversibly gain or lose electrons. Electron carriers generally accept high-energy electrons produced during an exergonic reaction and donate the electrons to acceptor molecules that use the energy to drive endergonic reactions.
electron micrograph: a photographic image of an object viewed through an electron microscope.
electron shell: a region within which electrons orbit that corresponds to a fixed energy level at a given distance from the atomic nucleus of an atom.
electron transport system: a series of electron carrier molecules, found in the thylakoid membranes of chloroplasts and the inner membrane of mitochondria, that extract energy from electrons and generate adenosine triphosphate (ATP) or other energetic molecules.
element: a substance that cannot be broken down, or converted, to a simpler substance by ordinary chemical means.
embryo: in animals, the stages of development that begin with the fertilization of the egg cell and end with hatching or birth; in mammals in particular, the early stages in which the developing animal does not yet resemble the adult of the species.
embryonic disc: in human embryonic development, the flat, two-layered group of cells that separates the amniotic cavity from the yolk sac.
embryonic stem cell: a cell derived from an early embryo that is capable of differentiating into any of the adult cell types.
embryo sac: the haploid female gametophyte of flowering plants.
emergent property: an intangible attribute that arises as the result of complex ordered interactions among individual parts.
emigration (em-uh-grā-shun): migration of individuals out of an area.
emphysema (em-fuh-sē′-muh): a condition in which the alveoli of the lungs become brittle and rupture, causing decreased area for gas exchange.
endergonic (en-der-gon′-ik) reaction: pertaining to a chemical reaction that requires an input of energy to proceed; an "uphill" reaction.
endocrine disruptors: environmental pollutants that interfere with endocrine function, often by disrupting the action of sex hormones.
endocrine gland: a ductless, hormone-producing gland consisting of cells that release their secretions into the extracellular fluid, from which the secretions diffuse into nearby capillaries.
endocrine system: an animal's organ system for cell-to-cell communication, composed of hormones and the cells that secrete them and receive them.
endocytosis (en-dō-sī-tō′-sis): the process in which the plasma membrane engulfs extracellular material, forming membrane-bound sacs that enter the cytoplasm and thereby move material into the cell.
endoderm (en′-dō-derm): the innermost embryonic tissue layer, which gives rise to structures such as the lining of the digestive and respiratory tracts.
endodermis (en-dō-der′-mis): the innermost layer of small, close-fitting cells of the cortex of a root that form a ring around the vascular cylinder.
endogenous pyrogen: a chemical, produced by the body, that stimulates the production of a fever.
endometrium (en-dō-mē′-trē-um): the nutritive inner lining of the uterus.

endoplasmic reticulum (ER) (en-dō-plaz′-mik re-tik′-ū-lum): a system of membranous tubes and channels within eukaryotic cells; the site of most protein and lipid synthesis.
endorphin (en-dor′-fin): one of a group of peptide neuromodulators in the vertebrate brain that, by reducing the sensation of pain, mimics some of the actions of opiates.
endoskeleton (en′-dō-skel′-uh-tun): a rigid internal skeleton with flexible joints to allow for movement.
endosperm: a triploid food-storage tissue in the seeds of flowering plants that nourishes the developing plant embryo.
endospore: a protective resting structure of some rod-shaped bacteria that withstands unfavorable environmental conditions.
endosymbiont hypothesis: the hypothesis that certain organelles, especially chloroplasts and mitochondria, arose as mutually beneficial associations between the ancestors of eukaryotic cells and captured bacteria that lived within the cytoplasm of the pre-eukaryotic cell.
energy: the capacity to do work.
energy-carrier molecule: a molecule that stores energy in "high-energy" chemical bonds and releases the energy to drive coupled endothermic reactions. In cells, adenosine triphosphate (ATP) is the most common energy-carrier molecule.
energy level: the specific amount of energy characteristic of a given electron shell in an atom.
energy pyramid: a graphical representation of the energy contained in succeeding trophic levels, with maximum energy at the base (primary producers) and steadily diminishing amounts at higher levels.
entropy (en′-trō-pē): a measure of the amount of randomness and disorder in a system.
environmental estrogens: chemicals in the environment that mimic some of the effects of estrogen in animals.
environmental resistance: any factor that tends to counteract biotic potential, limiting population size.
enzyme (en′zīm): a protein catalyst that speeds up the rate of specific biological reactions.
eosinophil (ē-ō-sin′-ō-fil): a type of white blood cell that converges on parasitic invaders and releases substances to kill them.
epicotyl (ep′-ē-kot-ul): the part of the embryonic shoot located above the cotyledons but below the tip of the shoot.
epidermal tissue: dermal tissue in plants that forms the epidermis, the outermost cell layer that covers young plants.
epidermis (ep-uh-der′-mis): in animals, specialized epithelial tissue that forms the outer layer of skin; in plants, the outermost layer of cells of a leaf, young root, or young stem.
epididymis (e-pi-di′-di-mus): a series of tubes that connect with and receive sperm from the seminiferous tubules of the testis.
epiglottis (ep-eh-glah′-tis): a flap of cartilage in the lower pharynx that covers the opening to the larynx during swallowing; directs food down the esophagus.
epinephrine (ep-i-nef′-rin): a hormone, secreted by the adrenal medulla, that is released in response to stress and that stimulates a variety of responses, including the release of glucose from skeletal muscle and an increase in heart rate.
epithelial cell (eh-puh-thē′-lē-ul): a flattened cell that covers the outer body surfaces of a sponge.
epithelial tissue (eh-puh-thē′-lē-ul): a tissue type that forms membranes that cover the body surface and line body cavities and that also gives rise to glands.
equilibrium population: a population in which allele frequencies and the distribution of genotypes do not change from generation to generation.
erythroblastosis fetalis (eh-rith′-rō-blas-tō′-sis fē-tal′-is): a condition in which the red blood cells of a newborn Rh-positive baby are attacked by antibodies produced by its Rh-negative mother, causing jaundice and anemia. Retardation and death are possible consequences if treatment is inadequate.
erythrocyte (eh-rith′-rō-sīt): a red blood cell, active in oxygen transport, that contains the red pigment hemoglobin.
erythropoietin (eh-rith′-rō-pō-ē′-tin): a hormone produced by the kidneys in response to oxygen deficiency that stimulates the production of red blood cells by the bone marrow.
esophagus (eh-sof′-eh-gus): a muscular passageway that conducts food from the pharynx to the stomach in humans and other mammals.
essential amino acid: an amino acid that is a required nutrient; the body is unable to manufacture essential amino acids, so they must be supplied in the diet.
essential fatty acid: a fatty acid that is a required nutrient; the body is unable to manufacture essential fatty acids, so they must be supplied in the diet.
estrogen: in vertebrates, a female sex hormone, produced by follicle cells of the ovary, that stimulates follicle development, oogenesis, the development of secondary sex characteristics, and growth of the uterine lining.
estuary: a wetland formed where a river meets the ocean; the salinity there is quite variable but lower than in seawater and higher than in fresh water.
ethology (ē-thol′-ō-jē): the study of animal behavior in natural or near-natural conditions.
ethylene: a plant hormone that promotes the ripening of fruits and the dropping of leaves and fruit.
euglenoid (ū′-gle-noid): a protist characterized by one or more whiplike flagella that are used for locomotion and by a photoreceptor that detects light. Euglenoids are photosynthetic, but if deprived of chlorophyll, some are capable of heterotrophic nutrition.
Eukarya: one of life's three domains; consists of all eukaryotes (plants, animals, fungi, and protists).
eukaryote (ū-kar′-ē-ōt): an organism whose cells are eukaryotic; plants, animals, fungi, and protists are eukaryotes.
eukaryotic (ū-kar-ē-ot′-ik) cell: referring to cells of organisms of the domain Eukarya (plants, animals, fungi, and protists). Eukaryotic cells have genetic material enclosed within a membrane-bound nucleus and contain other membrane-bound organelles.
Eustachian tube (ū-stā′-shin): a tube connecting the middle ear with the pharynx; allows pressure between the middle ear and the atmosphere to equilibrate.
eutrophic lake: a lake that receives sufficiently large inputs of sediments, organic material, and inorganic nutrients from its surroundings to support dense communities; murky with poor light penetration.
evergreen: a plant that retains green leaves throughout the year.
evolution: the descent of modern organisms with modification from preexisting life-forms; strictly speaking, any change in the proportions of different genotypes in a population from one generation to the next.
excretion: the elimination of waste substances from the body; can occur from the digestive system, skin glands, urinary system, or lungs.
excretory pore: an opening in the body wall of certain invertebrates, such as the earthworm, through which urine is excreted.
exergonic (ex-er-gon′-ik) reaction: pertaining to a chemical reaction that liberates energy (either as heat or in the form of increased entropy); a "downhill" reaction.
exhalation: the act of releasing air from the lungs, which results from a relaxation of the respiratory muscles.
exocrine gland: a gland that releases its secretions into ducts that lead to the outside of the body or into the digestive tract.
exocytosis (ex-ō-sī-tō′-sis): the process in which intracellular material is enclosed within a membrane-bound sac that moves to the plasma membrane and fuses with it, releasing the material outside the cell.
exon: a segment of DNA in a eukaryotic gene that codes for amino acids in a protein; see also *intron*.
exoskeleton (ex′-ō-skel′-uh-tun): a rigid external skeleton that supports the body, protects the internal organs, and has flexible joints that allow for movement.
exotic/exotic species: a foreign species introduced into an ecosystem where it did not evolve; such a species may flourish and outcompete native species.
experiment: the third operation in the scientific method; the testing of a hypothesis by further observations, leading to a conclusion.
exponential growth: a continuously accelerating increase in population size.
extensor: a muscle that straightens a joint.
external ear: the fleshy portion of the ear that extends outside the skull.
external fertilization: the union of sperm and egg outside the body of either parent.
extinction: the death of all members of a species.
extracellular digestion: the physical and chemical breakdown of food that occurs outside a cell, normally in a digestive cavity.
extraembryonic membrane: in the embryonic development of reptiles, birds, and mammals, either the chorion (functions in gas exchange), amnion (provision of the watery environment needed for development), allantois (waste storage), or yolk sac (storage of the yolk).
eyespot: a simple, lensless eye found in various invertebrates, including flatworms and jellyfish. Eyespots can distinguish light from dark and sometimes the direction of light, but they cannot form an image.
facilitated diffusion: the diffusion of molecules across a membrane, assisted by protein pores or carriers embedded in the membrane.
fairy ring: a circular pattern of mushrooms formed when reproductive structures erupt from the underground hyphae of a club fungus that has been growing outward in all directions from its original location.
family: the taxonomic category contained within an order and consisting of related genera.
farsighted: the inability to focus on nearby objects, caused by the eyeball's being slightly too short.
fat (molecular): a lipid composed of three saturated fatty acids covalently bonded to glycerol; solid at room temperature.
fat (tissue): adipose tissue; connective tissue that stores the lipid fat; composed of cells packed with triglycerides.
fatty acid: an organic molecule composed of a long chain of carbon atoms, with a carboxylic acid (COOH)

group at one end; may be saturated (all single bonds between the carbon atoms) or unsaturated (one or more double bonds between the carbon atoms).

feces: semisolid waste material that remains in the intestine after absorption is complete and is voided through the anus. Feces consist of indigestible wastes and the dead bodies of bacteria.

feedback inhibition: in enzyme-mediated chemical reactions, the condition in which the product of a reaction inhibits one or more of the enzymes involved in synthesizing the product.

fermentation: anaerobic reactions that convert the pyruvic acid produced by glycolysis into lactic acid or alcohol and CO_2.

fertilization: the fusion of male and female haploid gametes, forming a zygote.

fetal alcohol syndrome (FAS): a cluster of symptoms, including retardation and physical abnormalities, that occur in infants born to mothers who consumed large amounts of alcoholic beverages during pregnancy.

fetus: the later stages of mammalian embryonic development (after the second month for humans), when the developing animal has come to resemble the adult of the species.

fever: an elevation in body temperature caused by chemicals (pyrogens) that are released by white blood cells in response to infection.

fibrillation: rapid, uncoordinated, and ineffective contractions of heart muscle cells.

fibrin (fī′-brin): a clotting protein formed in the blood in response to a wound; binds with other fibrin molecules and provides a matrix around which a blood clot forms.

fibrous roots: a root system, commonly found in monocots, characterized by many roots of approximately the same size arising from the base of the stem.

filament: in flowers, the stalk of a stamen, which bears an anther at its tip.

filtrate: the fluid produced by filtration; in the kidneys, the fluid produced by the filtration of blood through the glomerular capillaries.

filtration: within Bowman's capsule in each nephron of a kidney, the process by which blood is pumped under pressure through permeable capillaries of the glomerulus, forcing out water, dissolved wastes, and nutrients.

fimbria (fim′-brē-uh; pl., fimbriae): in female mammals, the ciliated, fingerlike projections of the oviduct that sweep the ovulated egg from the ovary into the oviduct.

first law of thermodynamics: the principle of physics that states that within any isolated system, energy can be neither created nor destroyed but can be converted from one form to another.

fission: asexual reproduction by dividing the body into two smaller, complete organisms.

fitness: the reproductive success of an organism, usually expressed in relation to the average reproductive success of all individuals in the same population.

flagellum (fla-jel′-um; pl., flagella): a long, hairlike extension of the plasma membrane; in eukaryotic cells, it contains microtubules arranged in a 9 + 2 pattern. The movement of flagella propel some cells through fluids.

flame cell: in flatworms, a specialized cell, containing beating cilia, that conducts water and wastes through the branching tubes that serve as an excretory system.

flexor: a muscle that flexes (decreases the angle of) a joint.

florigen: one of a group of plant hormones that can both trigger and inhibit flowering; daylength is a stimulus.

flower: the reproductive structure of an angiosperm plant.

flowering plant: member of a group of vascular plants whose seeds are formed inside protective chambers called ovaries; flowering plants are also known as angiosperms.

fluid: a liquid or gas.

fluid mosaic model: a model of membrane structure; according to this model, membranes are composed of a double layer of phospholipids in which various proteins are embedded. The phospholipid bilayer is a somewhat fluid matrix that allows the movement of proteins within it.

follicle: in the ovary of female mammals, the oocyte and its surrounding accessory cells.

follicle-stimulating hormone (FSH): a hormone, produced by the anterior pituitary, that stimulates spermatogenesis in males and the development of the follicle in females.

food chain: a linear feeding relationship in a community, using a single representative from each of the trophic levels.

food vacuole: a membranous sac, within a single cell, in which food is enclosed. Digestive enzymes are released into the vacuole, where intracellular digestion occurs.

food web: a representation of the complex feeding relationships (in terms of interacting food chains) within a community, including many organisms at various trophic levels, with many of the consumers occupying more than one level simultaneously.

foraminiferan (for-am-i-nif′-er-un): an aquatic (largely marine) protist characterized by a typically elaborate calcium carbonate shell.

forebrain: during development, the anterior portion of the brain. In mammals, the forebrain differentiates into the thalamus, the limbic system, and the cerebrum. In humans, the cerebrum contains about half of all the neurons in the brain.

fossil: the remains of a dead organism, normally preserved in rock—may be petrified bones or wood; shells; impressions of body forms, such as feathers, skin, or leaves; or markings made by organisms, such as footprints.

fossil fuel: a fuel such as coal, oil, and natural gas, derived from the remains of ancient organisms.

founder effect: a type of genetic drift in which an isolated population founded by a small number of individuals may develop allele frequencies that are very different from those of the parent population as a result of chance inclusion of disproportionate numbers of certain alleles in the founders.

fovea (fō′-vē-uh): in the vertebrate retina, the central region on which images are focused; contains closely packed cones.

free-living: not parasitic.

free nerve ending: on some receptor neurons, a finely branched ending that responds to touch and pressure, to heat and cold, or to pain; produces the sensations of itching and tickling.

free nucleotides: nucleotides that have not been joined together to form a DNA or RNA strand.

fruit: in flowering plants, the ripened ovary (plus, in some cases, other parts of the flower), which contains the seeds.

fruiting body: a spore-forming reproductive structure of certain protists, bacteria, and fungi.

functional group: one of several groups of atoms commonly found in an organic molecule, including hydrogen, hydroxyl, amino, carboxyl, and phosphate groups, that determine the characteristics and chemical reactivity of the molecule.

gallbladder: a small sac, next to the liver, in which the bile secreted by the liver is stored and concentrated. Bile is released from the gallbladder to the small intestine through the bile duct.

gamete (gam′-ēt): a haploid sex cell formed in sexually reproducing organisms.

gametic incompatibility: the inability of sperm from one species to fertilize eggs of another species.

gametophyte (ga-mēt′-ō-fīt): the multicellular haploid stage in the life cycle of plants.

ganglion (gang′-lē-un): a cluster of neurons.

ganglion cell: a type of cell, of which the innermost layer of the vertebrate retina is composed, whose axons form the optic nerve.

gap junction: a type of cell-to-cell junction in animals in which channels connect the cytoplasm of adjacent cells.

gas-exchange portion: the portion of the respiratory system in lung-breathing vertebrates where gas is exchanged in the alveoli of the lungs.

gastric inhibitory peptide: a hormone, produced by the small intestine, that inhibits the activity of the stomach.

gastrin: a hormone, produced by the stomach, that stimulates acid secretion in response to the presence of food.

gastrovascular cavity: a saclike chamber with digestive functions, found in simple invertebrates; a single opening serves as both mouth and anus, and the chamber provides direct access of nutrients to the cells.

gastrula (gas′-troo-luh): in animal development, a three-layered embryo with ectoderm, mesoderm, and endoderm cell layers. The endoderm layer normally encloses the primitive gut.

gastrulation (gas-troo-la′-shun): the process whereby a blastula develops into a gastrula, including the formation of endoderm, ectoderm, and mesoderm.

gel electrophoresis: a technique in which molecules (such as DNA fragments) are placed on restricted tracks in a thin sheet of gelatinous material and exposed to an electric field; the molecules then migrate at a rate determined by certain characteristics, such as length.

gene: a unit of heredity that encodes the information needed to specify the amino acid sequence of proteins and hence particular traits; a functional segment of DNA located at a particular place on a chromosome.

gene flow: the movement of alleles from one population to another owing to the migration of individual organisms.

gene pool: the total of all alleles of all genes in a population; for a single gene, the total of all the alleles of that gene that occur in a population.

generative cell: in flowering plants, one of the haploid cells of a pollen grain; undergoes mitosis to form two sperm cells.

genetic code: the collection of codons of messenger RNA (mRNA) each of which directs the incorporation of a particular amino acid into a protein during protein synthesis.

genetic drift: a change in the allele frequencies of a small population purely by chance.

genetic engineering: the modification of genetic material to achieve specific goals.

genetic equilibrium: a state in which the allele frequencies and the distribution of genotypes of a population do not change from generation to generation.

genetic recombination: the generation of new combinations of alleles on homologous chromosomes due to the exchange of DNA during crossing over.

genital herpes: a sexually transmitted disease, caused by a virus, that can cause painful blisters on the genitals and surrounding skin.

genital warts: a sexually transmitted disease, caused by a virus, that forms growths or bumps on the external genitalia, in or around the vagina or anus, or on the cervix in females or penis, scrotum, groin, or thigh in males.

genome (jē′-nōm): the entire set of genes carried by a member of any given species.

genotype (jēn′-ō-tīp): the genetic composition of an organism; the actual alleles of each gene carried by the organism.

genus (jē-nus): the taxonomic category contained within a family and consisting of very closely related species.

geographical isolation: the separation of two populations by a physical barrier.

germination: the growth and development of a seed, spore, or pollen grain.

germ layer: a tissue layer formed during early embryonic development.

gibberellin (jib-er-el′-in): a plant hormone that stimulates seed germination, fruit development, and cell division and elongation.

gill: in aquatic animals, a branched tissue richly supplied with capillaries around which water is circulated for gas exchange.

gizzard: a muscular organ, found in earthworms and birds, in which food is mechanically broken down before chemical digestion.

gland: a cluster of cells that are specialized to secrete substances such as sweat or hormones.

glial cell: a cell of the nervous system that provides support and insulation for neurons.

global warming: a gradual rise in global atmospheric temperature as a result of an amplification of the natural greenhouse effect due to human activities.

glomerulus (glō-mer′-ū-lus): a dense network of thin-walled capillaries, located within the Bowman's capsule of each nephron of the kidney, where blood pressure forces water and dissolved nutrients through capillary walls for filtration by the nephron.

glucagon (gloo′-ka-gon): a hormone, secreted by the pancreas, that increases blood sugar by stimulating the breakdown of glycogen (to glucose) in the liver.

glucocorticoid (gloo-kō-kor′-tik-oid): a class of hormones, released by the adrenal cortex in response to the presence of adrenocorticotropic hormone (ACTH), that make additional energy available to the body by stimulating the synthesis of glucose.

glucose: the most common monosaccharide, with the molecular formula $C_6H_{12}O_6$; most polysaccharides, including cellulose, starch, and glycogen, are made of glucose subunits covalently bonded together.

glyceraldehyde-3-phosphate (G3P): a three-carbon sugar that is an important component of the C_3 cycle in the light-independent reactions of photosynthesis.

glycerol (glis′-er-ol): a three-carbon alcohol to which fatty acids are covalently bonded to make fats and oils.

glycogen (glī′-kō-jen): a long, branched polymer of glucose that is stored by animals in the muscles and liver and metabolized as a source of energy.

glycolysis (glī-kol′-i-sis): reactions, carried out in the cytoplasm, that break down glucose into two molecules of pyruvic acid, producing two adenosine triphosphate (ATP) molecules; does not require oxygen but can proceed when oxygen is present.

glycoprotein: a protein to which a carbohydrate is attached.

goiter: a swelling of the neck caused by iodine deficiency, which affects the functioning of the thyroid gland and its hormones.

Golgi complex (gōl′-jē): a stack of membranous sacs, found in most eukaryotic cells, that is the site of processing and separation of membrane components and secretory materials.

gonad: an organ where reproductive cells are formed; in males, the testes, and in females, the ovaries.

gonadotropin-releasing hormone (GnRH): a hormone produced by the neurosecretory cells of the hypothalamus, which stimulates cells in the anterior pituitary to release follicle-stimulating hormone (FSH) and luteinizing hormone (LH). GnRH is involved in the menstrual cycle and in spermatogenesis.

gonorrhea (gon-uh-rē′-uh): a sexually transmitted bacterial infection of the reproductive organs; if untreated, can result in sterility.

gradient: a difference in concentration, pressure, or electrical charge between two regions.

Gram stain: a stain that is selectively taken up by the cell walls of certain types of bacteria (gram-positive bacteria) and rejected by the cell walls of others (gram-negative bacteria); used to distinguish bacteria on the basis of their cell wall construction.

granum (gra′-num; pl., grana): a stack of thylakoids in chloroplasts.

grassland: a biome, located in the centers of continents, that supports grasses; also called *prairie*.

gravitropism: growth with respect to the direction of gravity.

gray crescent: in frog embryonic development, an area of intermediate pigmentation in the fertilized egg; contains gene-regulating substances required for the normal development of the tadpole.

gray matter: the outer portion of the brain and inner region of the spinal cord; composed largely of neuron cell bodies, which give this area a gray color.

greenhouse effect: the process in which certain gases such as carbon dioxide and methane trap sunlight energy in a planet's atmosphere as heat; the glass in a greenhouse does the same. The result, global warming, is being enhanced by the production of these gases by humans.

greenhouse gas: a gas, such as carbon dioxide or methane, that traps sunlight energy in a planet's atmosphere as heat; a gas that participates in the greenhouse effect.

ground tissue: plant tissue consisting of parenchyma, collenchyma, and sclerenchyma cells that makes up the bulk of a leaf or young stem, excluding vascular or dermal tissues. Most ground tissue cells function in photosynthesis, support, or carbohydrate storage.

growth hormone: a hormone, released by the anterior pituitary, that stimulates growth, especially of the skeleton.

growth rate: a measure of the change in population size per individual per unit of time.

guanine: a nitrogenous base found in both DNA and RNA; abbreviated as G.

guard cell: one of a pair of specialized epidermal cells surrounding the central opening of a stoma of a leaf, which regulates the size of the opening.

gymnosperm (jim′-nō-sperm): a nonflowering seed plant, such as a conifer, cycad, or gingko.

gyre (jīr): a roughly circular pattern of ocean currents, formed because continents interrupt the currents' flow; rotates clockwise in the Northern Hemisphere and counterclockwise in the Southern Hemisphere.

habituation (heh-bich-oo-ā′-shun): simple learning characterized by a decline in response to a harmless, repeated stimulus.

hair cell: the type of receptor cell in the inner ear; bears hairlike projections, the bending of which causes the receptor potential between two membranes.

hair follicle: a gland in the dermis of mammalian skin, formed from epithelial tissue, that produces a hair.

halophile (hā′-lō-fīl): literally, "salt-loving"; a type of archaen that thrives in concentrated salt solutions.

haploid (hap′-loid): referring to a cell that has only one member of each pair of homologous chromosomes.

Hardy-Weinberg principle: a mathematical model proposing that, under certain conditions, the allele frequencies and genotype frequencies in a sexually reproducing population will remain constant over generations.

Haversian system (ha-ver′-sē-un): see *osteon*.

head: the anteriormost segment of an animal with segmentation.

heart: a muscular organ responsible for pumping blood within the circulatory system throughout the body.

heart attack: a severe reduction or blockage of blood flow through a coronary artery, depriving some of the heart muscle of its blood supply.

heartwood: older xylem that contributes to the strength of a tree trunk.

heat of fusion: the energy that must be removed from a compound to transform it from a liquid into a solid at its freezing temperature.

heat of vaporization: the energy that must be supplied to a compound to transform it from a liquid into a gas at its boiling temperature.

heliozoan (hē-lē-ō-zō′-un): an aquatic (largely freshwater) animal-like protist; some have elaborate silica-based shells.

helix (hē′-liks): a coiled, springlike secondary structure of a protein.

helper T cell: a type of T cell that helps other immune cells recognize and act against antigens.

hemocoel (hē′-mō-sēl): a blood cavity within the bodies of certain invertebrates in which blood bathes tissues directly; part of an open circulatory system.

hemodialysis (hē-mō-di-al′-luh-sis): a procedure that simulates kidney function in individuals with damaged or ineffective kidneys; blood is diverted from the body, artificially filtered, and returned to the body.

hemoglobin (hē′mō-glō-bin): the iron-containing protein that gives red blood cells their color; binds to oxygen in the lungs and releases it to the tissues.

hemophilia: a recessive, sex-linked disease in which the blood fails to clot normally.

herbivore (erb′-i-vor): literally, "plant-eater"; an organism that feeds directly and exclusively on producers; a primary consumer.

hermaphrodite (her-maf′-ruh-dīt′): an organism that possesses both male and female sexual organs.

hermaphroditic (her-maf′-ruh-dit′-ik): possessing both male and female sexual organs. Some hermaphroditic animals can fertilize themselves; others must exchange sex cells with a mate.

heterotroph (het′-er-ō-trōf′): literally, "other-feeder"; an organism that eats other organisms; a consumer.

heterozygous (het-er-ō-zī′-gus): carrying two different alleles of a given gene; also called *hybrid.*
hindbrain: the posterior portion of the brain, containing the medulla, pons, and cerebellum.
hinge joint: a joint at which one bone is moved by muscle and the other bone remains fixed, such as in the knee, elbow, or fingers; allows movement in only two dimensions.
hippocampus (hip-ō-kam′-pus): the part of the forebrain of vertebrates that is important in emotion and especially in learning.
histamine: a substance released by certain cells in response to tissue damage and invasion of the body by foreign substances; promotes the dilation of arterioles and the leakiness of capillaries and triggers some of the events of the inflammatory response.
homeobox (hō′-mē-ō-boks): a sequence of DNA coding for special, 60-amino-acid proteins, which activate or inactivate genes that control development; these sequences specify embryonic cell differentiation.
homeostasis (hōm-ē-ō-stā′sis): the maintenance of a relatively constant environment required for the optimal functioning of cells, maintained by the coordinated activity of numerous regulatory mechanisms, including the respiratory, endocrine, circulatory, and excretory systems.
hominid: a human or a prehistoric relative of humans, beginning with the Australopithecines, whose fossils date back at least 4.4 million years.
homologous structures: structures that may differ in function but that have similar anatomy, presumably because the organisms that possess them have descended from common ancestors.
homologue (hō-′mō-log): a chromosome that is similar in appearance and genetic information to another chromosome with which it pairs during meiosis; also called *homologous chromosome*.
homozygous (hō-mō-zī′-gus): carrying two copies of the same allele of a given gene; also called *true-breeding*.
hormone: a chemical that is synthesized by one group of cells, secreted, and then carried in the bloodstream to other cells, whose activity is influenced by reception of the hormone.
host: the prey organism on or in which a parasite lives; is harmed by the relationship.
host cell: a cell that has been invaded by a virus and is usually harmed by the invasion.
human immunodeficiency virus (HIV): a pathogenic retrovirus that causes acquired immune deficiency syndrome (AIDS) by attacking and destroying the immune system's T cells.
humoral immunity: an immune response in which foreign substances are inactivated or destroyed by antibodies that circulate in the blood.
Huntington disease: an incurable genetic disorder, caused by a dominant allele, that produces progressive brain deterioration, resulting in the loss of motor coordination, flailing movements, personality disturbances, and eventual death.
hybrid: an organism that is the offspring of parents differing in at least one genetically determined characteristic; also used to refer to the offspring of parents of different species.
hybrid infertility: reduced fertility (typically, complete sterility) in the hybrid offspring of two species.
hybrid inviability: the failure of a hybrid offspring of two species to survive to maturity.
hybridoma: a cell produced by fusing an antibody-producing cell with a myeloma cell; used to produce monoclonal antibodies.
hydrogen bond: the weak attraction between a hydrogen atom that bears a partial positive charge (due to polar covalent bonding with another atom) and another atom, normally oxygen or nitrogen, that bears a partial negative charge; hydrogen bonds may form between atoms of a single molecule or of different molecules.
hydrologic cycle: the water cycle, driven by solar energy; a nutrient cycle in which the main reservoir of water is the ocean and most of the water remains in the form of water throughout the cycle (rather than being used in the synthesis of new molecules).
hydrolysis (hī-drol′-i-sis): the chemical reaction that breaks a covalent bond by means of the addition of hydrogen to the atom on one side of the original bond and a hydroxyl group to the atom on the other side; the reverse of dehydration synthesis.
hydrophilic (hī-drō-fil′-ik): pertaining to a substance that dissolves readily in water or to parts of a large molecule that form hydrogen bonds with water.
hydrophobic (hī-drō-fō′-bik): pertaining to a substance that does not dissolve in water.
hydrophobic interaction: the tendency for hydrophobic molecules to cluster together when immersed in water.
hydrostatic skeleton (hī-drō-stat′-ik): a body type that uses fluid contained in body compartments to provide support and mass against which muscles can contract.
hydrothermal vent community: a community of unusual organisms, living in the deep ocean near hydrothermal vents, that depends on the chemosynthetic activities of sulfur bacteria.
hypertension: arterial blood pressure that is chronically elevated above the normal level.
hypertonic (hī-per-ton′-ik): referring to a solution that has a higher concentration of dissolved particles (and therefore a lower concentration of free water) than has the cytoplasm of a cell.
hypha (hī′-fuh; pl., hyphae): a threadlike structure that consists of elongated cells, typically with many haploid nuclei; many hyphae make up the fungal body.
hypocotyl (hī′-pō-kot-ul): the part of the embryonic shoot located below the cotyledons but above the root.
hypothalamus (hī-pō-thal′-a-mus): a region of the brain that controls the secretory activity of the pituitary gland; synthesizes, stores, and releases certain peptide hormones; directs autonomic nervous system responses.
hypothesis (hī-poth′-eh-sis): the second operation in the scientific method; a supposition based on previous observations that is offered as an explanation for the observed phenomenon and is used as the basis for further observations or experiments.
hypotonic (hī-pō-ton′-ik): referring to a solution that has a lower concentration of dissolved particles (and therefore a higher concentration of free water) than has the cytoplasm of a cell.
immigration (im-uh-grā′-shun): migration of individuals into an area.
immune response: a specific response by the immune system to the invasion of the body by a particular foreign substance or microorganism, characterized by the recognition of the foreign substance by immune cells and its subsequent destruction by antibodies or by cellular attack.
immune system: cells such as macrophages, B cells, and T cells and molecules such as antibodies that work together to combat microbial invasion of the body.
imperfect fungus: a fungus of the division Deuteromycota; no species in this division has been observed to form sexual reproductive structures.
implantation: the process whereby the early embryo embeds itself within the lining of the uterus.
imprinting: the process by which an animal forms an association with another animal or object in the environment during a sensitive period of development.
inclusive fitness: the reproductive success of all organisms that bear a given allele, normally expressed in relation to the average reproductive success of all individuals in the same population; compare with *fitness*.
incomplete dominance: a pattern of inheritance in which the heterozygous phenotype is intermediate between the two homozygous phenotypes.
incomplete flower: a flower that is missing one of the four floral parts (sepals, petals, stamens, or carpels).
independent assortment: see *law of independent assortment*.
indirect development: a developmental pathway in which an offspring goes through radical changes in body form as it matures.
induction: the process by which a group of cells causes other cells to differentiate into a specific tissue type.
inductive reasoning: the process of creating a generalization based on many specific observations that support the generalization, coupled with an absence of observations that contradict it.
inflammatory response: a nonspecific, local response to injury to the body, characterized by the phagocytosis of foreign substances and tissue debris by white blood cells and by the walling off of the injury site by the clotting of fluids that escape from nearby blood vessels.
inhalation: the act of drawing air into the lungs by enlarging the chest cavity.
inheritance: the genetic transmission of characteristics from parent to offspring.
inheritance of acquired characteristics: the hypothesis that organisms' bodies change during their lifetimes by use and disuse and that these changes are inherited by their offspring.
inhibiting hormone: a hormone, secreted by the neurosecretory cells of the hypothalamus, that inhibits the release of specific hormones from the anterior pituitary.
initiation complex: a structure that forms as the first step in assembly of the ribosome; includes a small ribosomal subunit, a methionine transfer RNA (tRNA), a methionine amino acid, and several proteins.
innate (in-āt′) behavior: behavior that can be performed on the first attempt, without the need for any experience or learning.
inner cell mass: in human embryonic development, the cluster of cells, on one side of the blastocyst, that will develop into the embryo.
inner ear: the innermost part of the mammalian ear; composed of the bony, fluid-filled tubes of the cochlea and the vestibular apparatus.
inorganic: describing any molecule that does not contain both carbon and hydrogen.
insertion: the site of attachment of a muscle to the relatively movable bone on one side of a joint.
insertion mutation: a mutation in which one or more pairs of nucleotides are inserted into a gene.
insight learning: a complex form of learning that requires the manipulation of mental concepts to arrive at adaptive behavior.
instinctive: innate; inborn; determined by the genetic makeup of the individual.

insulin: a hormone, secreted by the pancreas, that lowers blood sugar by stimulating the conversion of glucose to glycogen in the liver.
integration: in nerve cells, the process of adding up electrical signals from sensory inputs or other nerve cells to determine the appropriate outputs.
integument (in-teg′-ū-ment): in plants, the outer layers of cells of the ovule that surrounds the embryo sac; develops into the seed coat.
intensity: the strength of stimulation or response.
interferon: a protein released by certain virus-infected cells that increases the resistance of other, uninfected cells to viral attack.
intermediate filament: part of the cytoskeleton of eukaryotic cells that probably functions mainly for support and is composed of several types of proteins.
intermembrane compartment: the fluid-filled space between the inner and outer membranes of a mitochondrion.
internal fertilization: the union of sperm and egg inside the body of the female.
internode: the part of a stem between two nodes.
interphase: the stage of the cell cycle between cell divisions; the stage in which chromosomes are replicated and other cell functions occur, such as growth, movement, and acquisition of nutrients.
interspecific competition: competition among individuals of different species.
interstitial cell (in-ter-sti′-shul): in the vertebrate testis, a testosterone-producing cell located between the seminiferous tubules.
interstitial fluid (in-ter-sti′-shul): fluid, similar in composition to plasma (except lacking large proteins), that leaks from capillaries and acts as a medium of exchange between the body cells and the capillaries.
intertidal zone: an area of the ocean shore that is alternately covered and exposed by the tides.
intervertebral disk (in-ter-ver-tē′-brul): a pad of cartilage between two vertebrae that acts as a shock absorber.
intracellular digestion: the chemical breakdown of food within single cells.
intraspecific competition: competition among individuals of the same species.
intrauterine device (IUD): a small copper or plastic loop, squiggle, or shield that is inserted in the uterus; a contraceptive method that works by irritating the uterine lining so that it cannot receive the embryo.
intron: a segment of DNA in a eukaryotic gene that does not code for amino acids in a protein; see also *exon*.
invertebrate (in-vert′-uh-bret): an animal that never possesses a vertebral column.
ion (ī′-on): a charged atom or molecule; an atom or molecule that either has an excess of electrons (and hence is negatively charged) or has lost electrons (and is positively charged).
ionic bond: a chemical bond formed by the electrical attraction between positively and negatively charged ions.
iris: the pigmented muscular tissue of the vertebrate eye that surrounds and controls the size of the pupil, through which light enters.
islet cell: a cluster of cells in the endocrine portion of the pancreas that produces insulin and glucagon.
isolating mechanism: a morphological, physiological, behavioral, or ecological difference that prevents members of two species from interbreeding.
isotonic (ī-sō-ton′-ik): referring to a solution that has the same concentration of dissolved particles (and therefore the same concentration of free water) as has the cytoplasm of a cell.
isotope: one of several forms of a single element, the nuclei of which contain the same number of protons but different numbers of neutrons.
J-curve: the J-shaped growth curve of an exponentially growing population in which increasing numbers of individuals join the population during each succeeding time period.
joint: a flexible region between two rigid units of an exoskeleton or endoskeleton, allowing for movement between the units.
karyotype: a preparation showing the number, sizes, and shapes of all chromosomes within a cell and, therefore, within the individual or species from which the cell was obtained.
keratin (ker′-uh-tin): a fibrous protein in hair, nails, and the epidermis of skin.
keystone species: a species whose influence on community structure is greater than its abundance would suggest.
kidney: one of a pair of organs of the excretory system that is located on either side of the spinal column and filters blood, removing wastes and regulating the composition and water content of the blood.
kinetic energy: the energy of movement; includes light, heat, mechanical movement, and electricity.
kilocalorie: see *Calorie*.
kinetochore (ki-net′-ō-kor): a protein structure that forms at the centromere regions of chromosomes; attaches the chromosomes to the spindle.
kingdom: the second broadest taxonomic category, contained within a domain and consisting of related phyla.
kin selection: a type of natural selection that favors a certain allele because it increases the survival or reproductive success of relatives that bear the same allele.
Klinefelter syndrome: a set of characteristics typically found in individuals who have two X chromosomes and one Y chromosome; these individuals are phenotypically males but are sterile and have several female-like traits, including broad hips and partial breast development.
Krebs cycle: a cyclic series of reactions, occurring in the matrix of mitochondria, in which the acetyl groups from the pyruvic acids produced by glycolysis are broken down to CO_2, accompanied by the formation of adenosine triphosphate (ATP) and electron carriers; also called *citric acid cycle*.
kuru: a degenerative brain disease, first discovered in the cannibalistic Fore tribe of New Guinea, that is caused by a prion.
labium (pl., labia): one of a pair of folds of skin of the external structures of the mammalian female reproductive system.
labor: a series of contractions of the uterus that result in birth.
lactation: the secretion of milk from the mammary glands.
lacteal (lak-tēl′): a single lymph capillary that penetrates each villus of the small intestine.
lactose (lak′-tōs): a disaccharide composed of glucose and galactose; found in mammalian milk.
large intestine: the final section of the digestive tract; consists of the colon and the rectum, where feces are formed and stored.
larva (lar′-vuh): an immature form of an organism with indirect development before metamorphosis into its adult form; includes the caterpillars of moths and butterflies and the maggots of flies.
larynx (lar′-inks): that portion of the air passage between the pharynx and the trachea; contains the vocal cords.
lateral bud: a cluster of meristematic cells at the node of a stem; under appropriate conditions, it grows into a branch.
lateral meristem: a meristematic tissue that forms cylinders parallel to the long axis of roots and stems; normally located between the primary xylem and primary phloem (vascular cambium) and just outside the phloem (cork cambium); also called *cambium*.
law of independent assortment: the independent inheritance of two or more distinct traits; states that the alleles for one trait may be distributed to the gametes independently of the alleles for other traits.
law of segregation: Gregor Mendel's conclusion that each gamete receives only one of each parent's pair of genes for each trait.
laws of thermodynamics: the physical laws that define the basic properties and behavior of energy.
leaf: an outgrowth of a stem, normally flattened and photosynthetic.
leaf primordium (pri-mor′-dē-um; pl., primordia): a cluster of meristem cells, located at the node of a stem, that develops into a leaf.
learning: an adaptive change in behavior as a result of experience.
legume (leg′-ūm): a member of a family of plants characterized by root swellings in which nitrogen-fixing bacteria are housed; includes soybeans, lupines, alfalfa, and clover.
lens: a clear object that bends light rays; in eyes, a flexible or movable structure used to focus light on a layer of photoreceptor cells.
leptin: a peptide hormone. One of the functions of leptin, which is released by fat cells, is to help the body monitor its fat stores and regulate weight.
leukocyte (loo′-kō-sīt): any of the white blood cells circulating in the blood.
lichen (lī′-ken): a symbiotic association between an alga or cyanobacterium and a fungus, resulting in a composite organism.
life cycle: the events in the life of an organism from one generation to the next.
ligament: a tough connective tissue band connecting two bones.
light-dependent reactions: the first stage of photosynthesis, in which the energy of light is captured as adenosine triphosphate (ATP) and nicotinamide adenine dinucleotide phosphate (NADPH); occurs in thylakoids of chloroplasts.
light-harvesting complex: in photosystems, the assembly of pigment molecules (chlorophyll and accessory pigments) that absorb light energy and transfer that energy to electrons.
light-independent reactions: the second stage of photosynthesis, in which the energy obtained by the light-dependent reactions is used to fix carbon dioxide into carbohydrates; occurs in the stroma of chloroplasts.
lignin: a hard material that is embedded in the cell walls of vascular plants and provides support in terrestrial species; an early and important adaptation to terrestrial life.
limbic system: a diverse group of brain structures, mostly in the lower forebrain, that includes the thalamus, hypothalamus, amygdala, hippocampus, and parts

of the cerebrum and is involved in basic emotions, drives, behaviors, and learning.

limnetic zone: a lake zone in which enough light penetrates to support photosynthesis.

linkage: the inheritance of certain genes as a group because they are parts of the same chromosome. Linked genes do not show independent assortment.

lipase (lī′-pās): an enzyme that catalyzes the breakdown of lipids such as fats.

lipid (li′-pid): one of a number of organic molecules containing large nonpolar regions composed solely of carbon and hydrogen, which make lipids hydrophobic and insoluble in water; includes oils, fats, waxes, phospholipids, and steroids.

littoral zone: a lake zone, near the shore, in which water is shallow and plants find abundant light, anchorage, and adequate nutrients.

liver: an organ with varied functions, including bile production, glycogen storage, and the detoxification of poisons.

lobefin: a member of the fish order Sarcopterygii, which includes coelacanths and lungfishes. Ancestors of today's lobefins gave rise to the first amphibians, and thus ultimately to all tetrapod vertebrates.

locus: the physical location of a gene on a chromosome.

long-day plant: a plant that will flower only if the length of daylight is greater than some species-specific duration.

long-term memory: the second phase of learning; a more-or-less permanent memory formed by a structural change in the brain, brought on by repetition.

loop of Henle (hen′-lē): a specialized portion of the tubule of the nephron in birds and mammals that creates an osmotic concentration gradient in the fluid immediately surrounding it. This gradient in turn makes possible the production of urine more osmotically concentrated than blood plasma.

lung: a paired respiratory organ consisting of inflatable chambers within the chest cavity in which gas exchange occurs.

luteinizing hormone (LH): a hormone, produced by the anterior pituitary, that stimulates testosterone production in males and the development of the follicle, ovulation, and the production of the corpus luteum in females.

lymph (limf): a pale fluid, within the lymphatic system, that is composed primarily of interstitial fluid and lymphocytes.

lymphatic system: a system consisting of lymph vessels, lymph capillaries, lymph nodes, and the thymus and spleen; helps protect the body against infection, absorbs fats, and returns excess fluid and small proteins to the blood circulatory system.

lymph node: a small structure that filters lymph; contains lymphocytes and macrophages, which inactivate foreign particles such as bacteria.

lymphocyte (lim′-fō-sīt): a type of white blood cell important in the immune response.

lysosome (lī′-sō-sōm): a membrane-bound organelle containing intracellular digestive enzymes.

macronutrient: a nutrient needed in relatively large quantities (often defined as making up more than 0.1% of an organism's body).

macrophage (mak′-rō-fāj): a type of white blood cell that engulfs microbes and destroys them by phagocytosis; also presents microbial antigens to T cells, helping stimulate the immune response.

magnetotactic: able to detect and respond to Earth's magnetic field.

major histocompatibility complex (MHC): proteins, normally located on the surfaces of body cells, that identify the cell as "self"; also important in stimulating and regulating the immune response.

maltose (mal′-tōs): a disaccharide composed of two glucose molecules.

mammal: a member of the chordate class Mammalia, which includes vertebrates with hair and mammary glands.

mammary gland (mam′-uh-rē): a milk-producing gland used by female mammals to nourish their young.

mantle (man′-tul): an extension of the body wall in certain invertebrates, such as mollusks; may secrete a shell, protect the gills, and, as in cephalopods, aid in locomotion.

marsupial (mar-soo′-pē-ul): a mammal whose young are born at an extremely immature stage and undergo further development in a pouch while they remain attached to a mammary gland; includes kangaroos, opossums, and koalas.

mass extinction: the extinction of an extraordinarily large number of species in a short period of geologic time. Mass extinctions have recurred periodically throughout the history of life.

mast cell: a cell of the immune system that synthesizes histamine and other molecules used in the body's response to trauma and that are a factor in allergic reactions.

matrix: the fluid contained within the inner membrane of a mitochondrion.

mechanical incompatibility: the inability of male and female organisms to exchange gametes, normally because their reproductive structures are incompatible.

mechanoreceptor: a receptor that responds to mechanical deformation, such as that caused by pressure, touch, or vibration.

medulla (med-ū′-luh): the part of the hindbrain of vertebrates that controls automatic activities such as breathing, swallowing, heart rate, and blood pressure.

medusa (meh-doo′-suh): a bell-shaped, typically free-swimming stage in the life cycle of many cnidarians; includes jellyfish.

megakaryocyte (meg-a-kar′-ē-ō-sīt): a large cell type that remains in the bone marrow, pinching off pieces of itself that then enter the circulation as platelets.

megaspore: a haploid cell formed by meiosis from a diploid megaspore mother cell; through mitosis and differentiation, develops into the female gametophyte.

megaspore mother cell: a diploid cell, within the ovule of a flowering plant, that undergoes meiosis to produce four haploid megaspores.

meiosis (mī-ō′-sis): a type of cell division, used by eukaryotic organisms, in which a diploid cell divides twice to produce four haploid cells.

meiotic cell division: meiosis followed by cytokinesis.

melanocyte-stimulating hormone (me-lan′-ō-sīt): a hormone, released by the anterior pituitary, that regulates the activity of skin pigments in some vertebrates.

melatonin (mel-uh-tōn′-in): a hormone, secreted by the pineal gland, that is involved in the regulation of circadian cycles.

membrane: in multicellular organisms, a continuous sheet of epithelial cells that covers the body and lines body cavities; in a cell, a thin sheet of lipids and proteins that surrounds the cell or its organelles, separating them from their surroundings.

memory B cell: a type of white blood cell that is produced as a result of the binding of an antibody on a B cell to an antigen on an invading microorganism. Memory B cells persist in the bloodstream and provide future immunity to invaders bearing that antigen.

memory T cell: a type of white blood cell that is produced as a result of the binding of a receptor on a T cell to an antigen on an invading microorganism. Memory T cells persist in the bloodstream and provide future immunity to invaders bearing that antigen.

meninges (men-in′-jēz): three layers of connective tissue that surround the brain and spinal cord.

menstrual cycle: in human females, a complex 28-day cycle during which hormonal interactions among the hypothalamus, pituitary gland, and ovary coordinate ovulation and the preparation of the uterus to receive and nourish the fertilized egg. If pregnancy does not occur, the uterine lining is shed during menstruation.

menstruation: in human females, the monthly discharge of uterine tissue and blood from the uterus.

meristem cell (mer′-i-stem): an undifferentiated cell that remains capable of cell division throughout the life of a plant.

mesoderm (mēz′-ō-derm): the middle embryonic tissue layer, lying between the endoderm and ectoderm, and normally the last to develop; gives rise to structures such as muscle and skeleton.

mesoglea (mez-ō-glē′-uh): a middle, jellylike layer within the body wall of cnidarians.

mesophyll (mez′-ō-fil): loosely packed parenchyma cells beneath the epidermis of a leaf.

messenger RNA (mRNA): a strand of RNA, complementary to the DNA of a gene, that conveys the genetic information in DNA to the ribosomes to be used during protein synthesis; sequences of three bases (codons) in mRNA specify particular amino acids to be incorporated into a protein.

metabolic pathway: a sequence of chemical reactions within a cell, in which the products of one reaction are the reactants for the next reaction.

metabolism: the sum of all chemical reactions that occur within a single cell or within all the cells of a multicellular organism.

metamorphosis (met-a-mor′-fō-sis): in animals with indirect development, a radical change in body form from larva to sexually mature adult, as seen in amphibians (tadpole to frog) and insects (caterpillar to butterfly).

metaphase (met′-a-fāz): the stage of mitosis in which the chromosomes, attached to spindle fibers at kinetochores, are lined up along the equator of the cell.

methanogen (me-than′-ō-jen): a type of anaerobic archaean capable of converting carbon dioxide to methane.

microbe: a microorganism.

microevolution: change over successive generations in the composition of a population's gene pool.

microfilament: part of the cytoskeleton of eukaryotic cells that is composed of the proteins actin and (in some cases) myosin; functions in the movement of cell organelles and in locomotion by extension of the plasma membrane.

micronutrient: a nutrient needed only in small quantities (often defined as making up less than 0.01% of an organism's body).

microsphere: a small, hollow sphere formed from proteins or proteins complexed with other compounds.

microspore: a haploid cell formed by meiosis from a microspore mother cell; through mitosis and differentiation, develops into the male gametophyte.

microspore mother cell: a diploid cell contained within an anther of a flowering plant, which undergoes meiosis to produce four haploid microspores.

microtubule: a hollow, cylindrical strand, found in eukaryotic cells, that is composed of the protein tubulin; part of the cytoskeleton used in the movement of organelles, cell growth, and the construction of cilia and flagella.

microvillus (mī-krō-vi′-lus; pl., microvilli): a microscopic projection of the plasma membrane of each villus; increases the surface area of the villus.

midbrain: during development, the central portion of the brain; contains an important relay center, the reticular formation.

middle ear: the part of the mammalian ear composed of the tympanic membrane, the Eustachian tube, and three bones (hammer, anvil, and stirrup) that transmit vibrations from the auditory canal to the oval window.

middle lamella: a thin layer of sticky polysaccharides, such as pectin, and other carbohydrates that separates and holds together the primary cell walls of adjacent plant cells.

mimicry (mim′-ik-rē): the situation in which a species has evolved to resemble something else, typically another type of organism.

mineral: an inorganic substance, especially one in rocks or soil.

mitochondrion (mī-tō-kon′-drē-un): an organelle, bounded by two membranes, that is the site of the reactions of aerobic metabolism.

mitosis (mī-tō′-sis): a type of nuclear division, used by eukaryotic cells, in which one copy of each chromosome (already duplicated during interphase before mitosis) moves into each of two daughter nuclei; the daughter nuclei are therefore genetically identical to each other.

mitotic cell division: mitosis followed by cytokinesis.

molecule (mol′-e-kūl): a particle composed of one or more atoms held together by chemical bonds; the smallest particle of a compound that displays all the properties of that compound.

molt: to shed an external body covering, such as an exoskeleton, skin, feathers, or fur.

monoclonal antibody: an antibody produced in the laboratory by the cloning of hybridoma cells; each clone of cells produces a single antibody.

monocot: short for monocotyledon; a type of flowering plant characterized by embryos with one seed leaf, or cotyledon.

monoecious (mon-ē′-shus): pertaining to organisms in which male and female gametes are produced in the same individual.

monomer (mo′-nō-mer): a small organic molecule, several of which may be bonded together to form a chain called a *polymer*.

monophyletic: referring to a group of species that contains all the known descendants of an ancestral species.

monosaccharide (mo-nō-sak′-uh-rīd): the basic molecular unit of all carbohydrates, normally composed of a chain of carbon atoms bonded to hydrogen and hydroxyl groups.

monotreme: a mammal that lays eggs; for example, the platypus.

morula (mor′-ū-luh): in animals, an embryonic stage during cleavage, when the embryo consists of a solid ball of cells.

motor neuron: a neuron that receives instructions from the association neurons and activates effector organs, such as muscles or glands.

motor unit: a single motor neuron and all the muscle fibers on which it forms synapses.

mouth: the opening of a tubular digestive system into which food is first introduced.

mucous membrane: the lining of the inside of the respiratory and digestive tracts.

multicellular: many-celled; most members of the kingdoms Fungi, Plantae, and Animalia are multicellular, with intimate cooperation among cells.

multiple alleles: as many as dozens of alleles produced for every gene as a result of different mutations.

muscle fiber: an individual muscle cell.

muscle tissue: tissue that consists mainly of muscle fibers and whose function is accomplished when the tissue contracts.

mutation: a change in the base sequence of DNA in a gene; normally refers to a genetic change significant enough to alter the appearance or function of the organism.

mutualism (mū′-choo-ul-iz-um): a symbiotic relationship in which both participating species benefit.

mycelium (mū-sēl′-ē-um): the body of a fungus, consisting of a mass of hyphae.

mycorrhiza (mī-kō-rī′-zuh; pl., mycorrhizae): a symbiotic relationship between a fungus and the roots of a land plant that facilitates mineral extraction and absorption.

myelin (mī′-uh-lin): a wrapping of insulating membranes of specialized nonneural cells around the axon of a vertebrate nerve cell; increases the speed of conduction of action potentials.

myofibril (mū-ō-fī′-bril): a cylindrical subunit of a muscle cell, consisting of a series of sarcomeres; surrounded by sarcoplasmic reticulum.

myometrium (mī-ō-mē′-trē-um): the muscular outer layer of the uterus.

myosin (mī′-ō-sin): one of the major proteins of muscle, the interaction of which with the protein actin produces muscle contraction; found in the thick filaments of the muscle fiber; see also *actin*.

natural causality: the scientific principle that natural events occur as a result of preceding natural causes.

natural killer cell: a type of white blood cell that destroys some virus-infected cells and cancerous cells on contact; part of the immune system's nonspecific internal defense against disease.

natural selection: the unequal survival and reproduction of organisms due to environmental forces, resulting in the preservation of favorable adaptations. Usually, natural selection refers specifically to differential survival and reproduction on the basis of genetic differences among individuals.

nearshore zone: the region of coastal water that is relatively shallow but constantly submerged; includes bays and coastal wetlands and can support large plants or seaweeds.

nearsighted: the inability to focus on distant objects caused by an eyeball that is slightly too long.

negative feedback: a situation in which a change initiates a series of events that tend to counteract the change and restore the original state. Negative feedback in physiological systems maintains homeostasis.

nephridium (nef-rid′-ē-um): an excretory organ found in earthworms, mollusks, and certain other invertebrates; somewhat resembles a single vertebrate nephron.

nephron (nef′-ron): the functional unit of the kidney; where blood is filtered and urine formed.

nephrostome (nef′-rō-stōm): the funnel-shaped opening of the nephridium of some invertebrates such as earthworms; coelomic fluid is drawn into the nephrostome for filtration.

nerve: a bundle of axons of nerve cells, bound together in a sheath.

nerve cord: a paired neural structure in most animals that conducts nervous signals to and from the ganglia; in chordates, a nervous structure lying along the dorsal side of the body; also called *spinal cord*.

nerve net: a simple form of nervous system, consisting of a network of neurons that extend throughout the tissues of an organism such as a cnidarian.

nerve tissue: the tissue that make up the brain, spinal cord, and nerves; consists of neurons and glial cells.

net primary productivity: the energy stored in the autotrophs of an ecosystem over a given time period.

neural tube: a structure, derived from ectoderm during early embryonic development, that later becomes the brain and spinal cord.

neuromuscular junction: the synapse formed between a motor neuron and a muscle fiber.

neuron (noor′-on): a single nerve cell.

neuropeptide: a small protein molecule with neurotransmitter-like actions.

neurosecretory cell: a specialized nerve cell that synthesizes and releases hormones.

neurotransmitter: a chemical that is released by a nerve cell close to a second nerve cell, a muscle, or a gland cell and that influences the activity of the second cell.

neutral mutation: a mutation that has little or no effect on the function of the encoded protein.

neutralization: the process of covering up or inactivating a toxic substance with antibody.

neutron: a subatomic particle that is found in the nuclei of atoms, bears no charge, and has a mass approximately equal to that of a proton.

nicotinamide adenine dinocleotide phosphate (NADPH): an energy-carrier molecule produced by the light-dependent reactions of photosynthesis; transfers energy to the carbon-fixing reactions (light-independent) reactions.

nitrogen fixation: the process that combines atmospheric nitrogen with hydrogen to form ammonium (NH_4^+).

nitrogen-fixing bacterium: a bacterium that possesses the ability to remove nitrogen (N_2) from the atmosphere and combine it with hydrogen to produce ammonium (NH_4^+).

node: in plants, a region of a stem at which leaves and lateral buds are located; in vertebrates, an interruption of the myelin on a myelinated axon, exposing naked membrane at which action potentials are generated.

nodule: a swelling on the root of a legume or other plant that consists of cortex cells inhabited by nitrogen-fixing bacteria.

nondisjunction: an error in meiosis in which chromosomes fail to segregate properly into the daughter cells.

nonpolar covalent bond: a covalent bond with equal sharing of electrons.

nonpolar molecule: a molecule bound by covalent bonds in which electrical charge is symmetrically distributed, so that no portion of the molecule is electrically charged relative to other portions.

norepinephrine (nor-ep-i-nef-rin′): a neurotransmitter, released by neurons of the parasympathetic nervous system, that prepares the body to respond to stressful situations; also called *noradrenaline*.

northern coniferous forest: a biome with long, cold winters and only a few months of warm weather; populated almost entirely by evergreen coniferous trees; also called *taiga*.

notochord (nōt′-ō-kord): a stiff but somewhat flexible, supportive rod found in all members of the phylum Chordata at some stage of development.
nuclear envelope: the double-membrane system surrounding the nucleus of eukaryotic cells; the outer membrane is typically continuous with the endoplasmic reticulum.
nucleic acid (noo-klā′-ik): an organic molecule composed of nucleotide subunits; the two common types of nucleic acids are ribonucleic acid (RNA) and deoxyribonucleic acid (DNA).
nucleoid (noo-klē-oid): the location of the genetic material in prokaryotic cells; not membrane-enclosed.
nucleolus (noo-klē′-ō-lus): the region of the eukaryotic nucleus that is engaged in ribosome synthesis; consists of the genes encoding ribosomal RNA, newly synthesized ribosomal RNA, and ribosomal proteins.
nucleotide: a subunit of which nucleic acids are composed; a phosphate group bonded to a sugar (deoxyribose in DNA), which is in turn bonded to a nitrogen-containing base (adenine, guanine, cytosine, or thymine in DNA). Nucleotides are linked together, forming a strand of nucleic acid, as follows: Bonds between the phosphate of one nucleotide link to the sugar of the next nucleotide.
nucleotide substitution: a mutation that replaces one nucleotide in a DNA molecule with another; for example, a change from an adenine to a guanine.
nucleus (atomic): the central region of an atom, consisting of protons and neutrons.
nucleus (cellular): the membrane-bound organelle of eukaryotic cells that contains the cell's genetic material.
nutrient: a substance acquired from the environment and needed for the survival, growth, and development of an organism.
nutrient cycle: a description of the pathways of a specific nutrient (such as carbon, nitrogen, phosphorus, or water) through the living and nonliving portions of an ecosystem; also called a *biogeochemical cycle*.
nutrition: the process of acquiring nutrients from the environment and, if necessary, processing them into a form that can be used by the body.
observation: the first operation in the scientific method; the noting of a specific phenomenon, leading to the formulation of a hypothesis.
oil: a lipid composed of three fatty acids, some of which are unsaturated, covalently bonded to a molecule of glycerol; liquid at room temperature.
olfaction (ōl-fak′-shun): a chemical sense, the sense of smell; in terrestrial vertebrates, the result of the detection of airborne molecules.
oligotrophic lake: a lake that is very low in nutrients and hence clear with extensive light penetration.
ommatidium (ōm-ma-tid′-ē-um): an individual light-sensitive subunit of a compound eye; consists of a lens and several receptor cells.
omnivore: an organism that consumes both plants and other animals.
one gene, one protein rule: the premise that each gene encodes the information for the synthesis of a single protein.
oogenesis: the process by which egg cells are formed.
oogonium (ō-ō-gō′-nē-um; pl., oogonia): in female animals, a diploid cell that gives rise to a primary oocyte.
open circulatory system: a type of circulatory system found in some invertebrates, such as arthropods and mollusks, that includes an open space (the hemocoel) in which blood directly bathes body tissues.
operant conditioning: a laboratory training procedure in which an animal learns to perform a response (such as pressing a lever) through reward or punishment.
operculum: an external flap, supported by bone, that covers and protects the gills of most fish.
opioid (ōp′-ē-oid): one of a group of peptide neuromodulators in the vertebrate brain that mimic some of the actions of opiates (such as opium) and also seem to influence many other processes, including emotion and appetite.
optic nerve: the nerve leading from the eye to the brain, carrying visual information.
order: the taxonomic category contained within a class and consisting of related families.
organ: a structure (such as the liver, kidney, or skin) composed of two or more distinct tissue types that function together.
organelle (or-guh-nel′): a structure, found in the cytoplasm of eukaryotic cells, that performs a specific function; sometimes refers specifically to membrane-bound structures, such as the nucleus or endoplasmic reticulum.
organic molecule: a molecule that contains both carbon and hydrogen.
organism (or′-guh-niz-um): an individual living thing.
organogenesis (or-gan-ō-jen′-uh-sis): the process by which the layers of the gastrula (endoderm, ectoderm, mesoderm) rearrange into organs.
organ system: two or more organs that work together to perform a specific function; for example, the digestive system.
origin: the site of attachment of a muscle to the relatively stationary bone on one side of a joint.
osmosis (oz-mō′-sis): the diffusion of water across a differentially permeable membrane, normally down a concentration gradient of free water molecules. Water moves into the solution that has a lower concentration of free water from a solution with the higher concentration of free water.
osmotic pressure: the pressure required to counterbalance the tendency of water to move from a solution with a higher concentration of free water molecules into a solution with a lower concentration of free water molecules.
osteoblast (os′-tē-ō-blast): a cell type that produces bone.
osteoclast (os′-tē-ō-klast): a cell type that dissolves bone.
osteocyte (os′-tē-ō-sīt): a mature bone cell.
osteon: a unit of hard bone consisting of concentric layers of bone matrix, with embedded osteocytes, surrounding a small central canal that contains a capillary; also called the *Haversian system*.
osteoporosis (os′-tē-ō-por-ō′-sis): a condition in which bones become porous, weak, and easily fractured; most common in elderly women.
outer ear: the outermost part of the mammalian ear, including the external ear and auditory canal leading to the tympanic membrane.
oval window: the membrane-covered entrance to the inner ear.
ovary: in animals, the gonad of females; in flowering plants, a structure at the base of the carpel that contains one or more ovules and develops into the fruit.
oviduct: see *uterine tube*.
ovulation: the release of a secondary oocyte, ready to be fertilized, from the ovary.
ovule: a structure within the ovary of a flower, inside which the female gametophyte develops; after fertilization, develops into the seed.
oxaloacetate: a four-carbon molecule that participates in the Krebs cycle of glucose metabolism and also in the carbon-fixing pathway of C_4 photosynthesis.
oxytocin (oks-ē-tō′-sin): a hormone, released by the posterior pituitary, that stimulates the contraction of uterine and mammary gland muscles.
ozone layer: the ozone-enriched layer of the upper atmosphere that filters out some of the sun's ultraviolet radiation.
pacemaker: a cluster of specialized muscle cells in the upper right atrium of the heart that produce spontaneous electrical signals at a regular rate; the sinoatrial node.
pain receptor: a receptor that has extensive areas of membranes studded with special receptor proteins that respond to light or to a chemical.
palisade cell: a columnar mesophyll cell, containing chloroplasts, just beneath the upper epidermis of a leaf.
pancreas (pan′-krē-us): a combined exocrine and endocrine gland located in the abdominal cavity next to the stomach. The endocrine portion secretes the hormones insulin and glucagon, which regulate glucose concentrations in the blood. The exocrine portion secretes enzymes for fat, carbohydrate, and protein digestion into the small intestine and neutralizes the acidic chyme.
pancreatic juice: a mixture of water, sodium bicarbonate, and enzymes released by the pancreas into the small intestine.
parasite (par′-uh-sīt): an organism that lives in or on a larger prey organism, called a *host*, weakening it.
parasitism: a symbiotic relationship in which one organism (commonly smaller and more numerous than its host) benefits by feeding on the other, which is normally harmed but not immediately killed.
parasympathetic division: the division of the autonomic nervous system that produces largely involuntary responses related to the maintenance of normal body functions, such as digestion.
parathormone: a hormone, secreted by the parathyroid gland, that stimulates the release of calcium from bones.
parathyroid gland: one of a set of four small endocrine glands, embedded in the surface of the thyroid gland, that produces parathormone, which (with calcitonin from the thyroid gland) regulates calcium ion concentration in the blood.
parenchyma (par-en′-ki-muh): a plant cell type that is alive at maturity, normally with thin primary cell walls, that carries out most of the metabolism of a plant. Most dividing meristem cells in a plant are parenchyma.
parthenogenesis (par-the-nō-jen′uh-sis): a specialization of sexual reproduction, in which a haploid egg undergoes development without fertilization.
passive transport: the movement of materials across a membrane down a gradient of concentration, pressure, or electrical charge without using cellular energy.
pathogenic (path′-ō-jen-ik): capable of producing disease; refers to an organism with such a capability (a pathogen).
pedigree: a diagram showing genetic relationships among a set of individuals, normally with respect to a specific genetic trait.
pelagic (puh-la′-jik): free-swimming or floating.
penis: an external structure of the male reproductive and urinary systems; serves to deposit sperm into the

female reproductive system and delivers urine to the exterior.

peptide (pep′-tīd): a chain composed of two or more amino acids linked together by peptide bonds.

peptide bond: the covalent bond between the amino group's nitrogen of one amino acid and the carboxyl group's carbon of a second amino acid, joining the two amino acids together in a peptide or protein.

peptide hormone: a hormone consisting of a chain of amino acids; includes small proteins that function as hormones.

peptidoglycan (pep-tid-ō-glī′-kan): a component of prokaryotic cell walls that consists of chains of sugars cross-linked by short chains of amino acids called *peptides*.

pericycle (per′-i-sī-kul): the outermost layer of cells of the vascular cylinder of a root.

periderm: the outer cell layers of roots and a stem that have undergone secondary growth, consisting primarily of cork cambium and cork cells.

peripheral nerve: a nerve that links the brain and spinal cord to the rest of the body.

peripheral nervous system: in vertebrates, the part of the nervous system that connects the central nervous system to the rest of the body.

peristalsis: rhythmic coordinated contractions of the smooth muscles of the digestive tract that move substances through the digestive tract.

permafrost: a permanently frozen layer of soil in the arctic tundra that cannot support the growth of trees.

petal: part of a flower, typically brightly colored and fragrant, that attracts potential animal pollinators.

petiole (pet′-ē-ol): the stalk that connects the blade of a leaf to the stem.

phagocytic cell (fa-gō-sit′-ik): a type of immune system cell that destroys invading microbes by using phagocytosis to engulf and digest the microbes.

phagocytosis (fa-gō-sī-tō′-sis): a type of endocytosis in which extensions of a plasma membrane engulf extracellular particles and transport them into the interior of the cell.

pharyngeal gill slit (far-in′-jē-ul): an opening, located just posterior to the mouth, that connects the digestive tube to the outside environment; present (at some stage of life) in all chordates.

pharynx (far′-inks): in vertebrates, a chamber that is located at the back of the mouth and is shared by the digestive and respiratory systems; in some invertebrates, the portion of the digestive tube just posterior to the mouth.

phenotype (fēn′-ō-tīp): the physical characteristics of an organism; can be defined as outward appearance (such as flower color), as behavior, or in molecular terms (such as glycoproteins on red blood cells).

pheromone (fer′-uh-mōn): a chemical produced by an organism that alters the behavior or physiological state of another member of the same species.

phloem (flō′-um): a conducting tissue of vascular plants that transports a concentrated sugar solution up and down the plant.

phosphoenolpyruvate (PEP): a three-carbon molecule that reacts with carbon-dioxide in the carbon-fixing reaction of C_4 photosynthesis.

phosphoglyceric acid (PGA): a three-carbon sugar that is an important component of the C_3 cycle in the light-independent reactions of photosynthesis.

phospholipid (fos-fō-li′-pid): a lipid consisting of glycerol bonded to two fatty acids and one phosphate group, which bears another group of atoms, typically charged and containing nitrogen. A double layer of phospholipids is a component of all cellular membranes.

phospholipid bilayer: a double layer of phospholipids that forms the basis of all cellular membranes. The phospholipid heads, which are hydrophilic, face the water of extracellular fluid or the cytoplasm; the tails, which are hydrophobic, are buried in the middle of the bilayer.

photic zone: the region of the ocean where light is strong enough to support photosynthesis.

photon (fō′-ton): the smallest unit of light energy.

photopigment (fō′-tō-pig-ment): a chemical substance in photoreceptor cells that, when struck by light, changes in molecular conformation.

photoreceptor: a receptor cell that responds to light; in vertebrates, rods and cones.

photorespiration: a series of reactions in plants in which O_2 replaces CO_2 during the C_3 cycle, preventing carbon fixation; this wasteful process dominates when C_3 plants are forced to close their stomata to prevent water loss.

photosynthesis: the complete series of chemical reactions in which the energy of light is used to synthesize high-energy organic molecules, normally carbohydrates, from low-energy inorganic molecules, normally carbon dioxide and water.

photosystem: in thylakoid membranes, a light-harvesting complex and its associated electron transport system.

phototactic: capable of detecting and responding to light.

phototropism: growth with respect to the direction of light.

pH scale: a scale, with values from 0 to 14, used for measuring the relative acidity of a solution; at pH 7 a solution is neutral, pH 0 to 7 is acidic, and pH 7 to 14 is basic; each unit on the scale represents a tenfold change in H^+ concentration.

phycocyanin (fī-kō-sī′-uh-nin): a blue or purple pigment that is located in the membranes of chloroplasts and is used as an accessory light-gathering molecule in thylakoid photosystems.

phylogeny (fī-lah′-jen-ē): the evolutionary history of a group of species.

phylum (fī-lum): the taxonomic category of animals and animal-like protists that is contained within a kingdom and consists of related classes.

phytochrome (fī′-tō-krōm): a light-sensitive plant pigment that mediates many plant responses to light, including flowering, stem elongation, and seed germination.

phytoplankton (fī′-tō-plank-ten): photosynthetic protists that are abundant in marine and freshwater environments.

pilus (pil′-us; pl., pili): a hairlike projection that is made of protein, located on the surface of certain bacteria, and is typically used to attach a bacterium to another cell.

pineal gland (pī-nē′-al): a small gland within the brain that secretes melatonin; controls the seasonal reproductive cycles of some mammals.

pinocytosis (pi-nō-sī-tō′-sis): the nonselective movement of extracellular fluid, enclosed within a vesicle formed from the plasma membrane, into a cell.

pioneer: an organism that is among the first to colonize an unoccupied habitat in the first stages of succession.

pit: an area in the cell walls between two plant cells in which secondary walls did not form, such that the two cells are separated only by a relatively thin and porous primary cell wall.

pith: cells forming the center of a root or stem.

pituitary gland: an endocrine gland, located at the base of the brain, that produces several hormones, many of which influence the activity of other glands.

placenta (pluh-sen′-tuh): in mammals, a structure formed by a complex interweaving of the uterine lining and the embryonic membranes, especially the chorion; functions in gas, nutrient, and waste exchange between embryonic and maternal circulatory systems and secretes hormones.

placental (pluh-sen′-tul): referring to a mammal, possessing a placenta (that is, species that are not marsupials or monotremes).

plankton: microscopic organisms that live in marine or freshwater environments; includes phytoplankton and zooplankton.

plant hormone: the plant-regulating chemicals auxin, gibberellin, cytokinin, ethylene, and abscisic acid; somewhat resemble animal hormones in that they are chemicals produced by cells in one location that influence the growth or metabolic activity of other cells, typically some distance away in the plant body.

plaque (plak): a deposit of cholesterol and other fatty substances within the wall of an artery.

plasma: the fluid, noncellular portion of the blood.

plasma cell: an antibody-secreting descendant of a B cell.

plasma membrane: the outer membrane of a cell, composed of a bilayer of phospholipids in which proteins are embedded.

plasmid (plaz′-mid): a small, circular piece of DNA located in the cytoplasm of many bacteria; normally does not carry genes required for the normal functioning of the bacterium but may carry genes that assist bacterial survival in certain environments, such as a gene for antibiotic resistance.

plasmodesma (plaz-mō-dez′-muh; pl., plasmodesmata): a cell-to-cell junction in plants that connects the cytoplasm of adjacent cells.

plasmodial slime mold: see *acellular slime mold*.

plasmodium (plaz-mō′-dē-um): a sluglike mass of cytoplasm containing thousands of nuclei that are not confined within individual cells.

plastid (plas′-tid): in plant cells, an organelle bounded by two membranes that may be involved in photosynthesis (chloroplasts), pigment storage, or food storage.

platelet (plāt′-let): a cell fragment that is formed from megakaryocytes in bone marrow and lacks a nucleus; circulates in the blood and plays a role in blood clotting.

plate tectonics: the theory that Earth's crust is divided into irregular plates that are converging, diverging, or slipping by one another; these motions cause continental drift, the movement of continents over Earth's surface.

pleated sheet: a form of secondary structure exhibited by certain proteins, such as silk, in which many protein chains lie side by side, with hydrogen bonds holding adjacent chains together.

pleiotropy (plē′-ō-trō-pē): a situation in which a single gene influences more than one phenotypic characteristic.

pleural membrane: a membrane that lines the chest cavity and surrounds the lungs.

point mutation: a mutation in which a single base pair in DNA has been changed.

polar body: in oogenesis, a small cell, containing a nucleus but virtually no cytoplasm, produced by the first meiotic division of the primary oocyte.

polar covalent bond: a covalent bond with unequal sharing of electrons, such that one atom is relatively negative and the other is relatively positive.

polar molecule: a molecule bound by covalent bonds in which electrical charge is asymmetrically distributed, so that electrical charge differs in different portions of the molecule.

polar nucleus: in flowering plants, one of two nuclei in the primary endosperm cell of the female gametophyte; formed by the mitotic division of a megaspore.

pollen/pollen grain: the male gametophyte of a seed plant.

pollination: in flowering plants, when pollen grains land on the stigma of a flower of the same species; in conifers, when pollen grains land within the pollen chamber of a female cone of the same species.

polygenic inheritance: a pattern of inheritance in which the interactions of two or more functionally similar genes determine phenotype.

polymer (pah′-li-mer): a molecule composed of three or more (perhaps thousands) smaller subunits called *monomers*, which may be identical (for example, the glucose monomers of starch) or different (for example, the amino acids of a protein).

polymerase chain reaction (PCR): a method of producing virtually unlimited numbers of copies of a specific piece of DNA, starting with as little as one copy of the desired DNA.

polyp (pah′-lip): the sedentary, vase-shaped stage in the life cycle of many cnidarians; includes hydra and sea anemones.

polypeptide: a short polymer of amino acids; often used as a synonym for protein.

polyploidy (pahl′-ē-ploid-ē): having more than two homologous chromosomes of each type.

polysaccharide (pahl-ē-sak′-uh-rīd): a large carbohydrate molecule composed of branched or unbranched chains of repeating monosaccharide subunits, normally glucose or modified glucose molecules; includes starches, cellulose, and glycogen.

pons: a portion of the hindbrain, just above the medulla, that contains neurons that influence sleep and the rate and pattern of breathing.

population: all the members of a particular species within an ecosystem, found in the same time and place and actually or potentially interbreeding.

population bottleneck: a form of genetic drift in which a population becomes extremely small; may lead to differences in allele frequencies as compared with other populations of the species and to a loss in genetic variability.

population cycle: out-of-phase cyclical patterns of predator and prey populations.

population genetics: the study of the frequency, distribution, and inheritance of alleles in a population.

positive feedback: a situation in which a change initiates events that tend to amplify the original change.

post-anal tail: a tail that extends beyond the anus; exhibited by all chordates at some stage of development.

posterior: the tail, hindmost, or rear end of an animal.

posterior pituitary: a lobe of the pituitary gland that is an outgrowth of the hypothalamus and that releases antidiuretic hormone and oxytocin.

postmating isolating mechanism: any structure, physiological function, or developmental abnormality that prevents organisms of two different populations, once mating has occurred, from producing vigorous, fertile offspring.

postsynaptic neuron: at a synapse, the nerve cell that changes its electrical potential in response to a chemical (the neurotransmitter) released by another (presynaptic) cell.

postsynaptic potential (PSP): an electrical signal produced in a postsynaptic cell by transmission across the synapse; it may be excitatory (EPSP), making the cell more likely to produce an action potential, or inhibitory (IPSP), tending to inhibit an action potential.

potential energy: "stored" energy, normally chemical energy or energy of position within a gravitational field.

prairie: see *grassland*.

preadaptation: a feature evolved under one set of environmental conditions that, purely by chance, helps an organism adapt to new environmental conditions.

prebiotic evolution: evolution before life existed; especially, the abiotic synthesis of organic molecules.

precapillary sphincter (sfink′-ter): a ring of smooth muscle between an arteriole and a capillary that regulates the flow of blood into the capillary bed.

predation (pre-dā′-shun): the act of killing and eating another living organism.

predator: an organism that kills and eats other organisms.

premating isolating mechanism: any structure, physiological function, or behavior that prevents organisms of two different populations from exchanging gametes.

pressure-flow theory: a model for the transport of sugars in phloem, by which the movement of sugars into a phloem sieve tube causes water to enter the tube by osmosis, while the movement of sugars out of another part of the same sieve tube causes water to leave by osmosis; the resulting pressure gradient causes the bulk movement of water and dissolved sugars from the end of the tube into which sugar is transported toward the end of the tube from which sugar is removed.

presynaptic neuron: a nerve cell that releases a chemical (the neurotransmitter) at a synapse, causing changes in the electrical activity of another (postsynaptic) cell.

prey: organisms that are killed and eaten by another organism.

primary cell wall: cellulose and other carbohydrates secreted by a young plant cell between the middle lamella and the plasma membrane.

primary consumer: an organism that feeds on producers; an herbivore.

primary endosperm cell: the central cell of the female gametophyte of a flowering plant, containing the polar nuclei (normally two); after fertilization, undergoes repeated mitotic divisions to produce the endosperm of the seed.

primary growth: growth in length and development of the initial structures of plant roots and shoots, due to the cell division of apical meristems and differentiation of the daughter cells.

primary oocyte (ō′-ō-sīt): a diploid cell, derived from the oogonium by growth and differentiation, that undergoes meiosis, producing the egg.

primary phloem: phloem in young stems produced from an apical meristem.

primary root: the first root that develops from a seed.

primary spermatocyte (sper-ma′-tō-sīt): a diploid cell, derived from the spermatogonium by growth and differentiation, that undergoes meiosis, producing four sperm.

primary structure: the amino acid sequence of a protein.

primary succession: succession that occurs in an environment, such as bare rock, in which no trace of a previous community was present.

primary xylem: xylem in young stems produced from an apical meristem.

primate: a mammal characterized by the presence of an opposable thumb, forward-facing eyes, and a well-developed cerebral cortex; includes lemurs, monkeys, apes, and humans.

primitive streak: in reptiles, birds, and mammals, the region of the ectoderm of the two-layered embryonic disk through which cells migrate, forming mesoderm.

prion (prē′-on): a protein that, in mutated form, acts as an infectious agent that causes certain neurodegenerative diseases, including kuru and scrapie.

producer: a photosynthetic organism; an autotroph.

product: an atom or molecule that is formed from reactants in a chemical reaction.

profundal zone: a lake zone in which light is insufficient to support photosynthesis.

progesterone (prō-ge′-ster-ōn): a hormone, produced by the corpus luteum, that promotes the development of the uterine lining in females.

prokaryote (prō-kar′-ē-ōt′): an organism whose cells are prokaryotic; bacteria and archaea are prokaryotes.

prokaryotic (prō-kar-ē-ot′-ik) cell: cells of the domains Bacteria and Archaea. Prokaryotic cells have genetic material that is not enclosed in a membrane-bound nucleus; they lack other membrane-bound organelles.

prolactin: a hormone, released by the anterior pituitary, that stimulates milk production in human females.

promoter: a specific sequence of DNA to which RNA polymerase binds, initiating gene transcription.

prophase (prō′-fāz): the first stage of mitosis, in which the chromosomes first become visible in the light microscope as thickened, condensed threads and the spindle begins to form; as the spindle is completed, the nuclear envelope breaks apart, and the spindle fibers invade the nuclear region and attach to the kinetochores of the chromosomes. Also, the first stage of meiosis: In meiosis I, the homologous chromosomes pair up and exchange parts at chiasmata; in meiosis II, the spindle re-forms and chromosomes attach to the microtubules.

prostaglandin (pro-stuh-glan′-din): a family of modified fatty acid hormones manufactured by many cells of the body.

prostate gland (pros′-tāt): a gland that produces part of the fluid component of semen; prostatic fluid is basic and contains a chemical that activates sperm movement.

protease (prō′-tē-ās): an enzyme that digests proteins.

protein: polymer of amino acids joined by peptide bonds.

protist (prō′-tist): a eukaryotic organism that is not a plant, animal, or fungus. The term encompasses a diverse array of organisms and does not represent a monophyletic group. Algae, amoebas, slime molds, and ciliates are examples of protists.

protocell: the hypothetical evolutionary precursor of living cells, consisting of a mixture of organic molecules within a membrane.

proton: a subatomic particle that is found in the nuclei of atoms, bears a unit of positive charge, and has a relatively large mass, roughly equal to the mass of the neutron.

protonephridium (**prō-tō-nef-rid′-ē-um; pl., protonephridia**): an excretory system consisting of tubules that have an external opening but lack internal openings; for example, the flame-cell system of flatworms.

protostome (**prō′-tō-stōm**): an animal with a mode of embryonic development in which the coelom is derived from splits in the mesoderm; characteristic of arthropods, annelids, and mollusks.

protozoan (**prō-tuh-zō′-an; pl., protozoa**): a nonphotosynthetic or animal-like protist.

proximal tubule: in nephrons of the mammalian kidney, the portion of the renal tubule just after the Bowman's capsule; receives filtrate from the capsule and is the site where selective secretion and reabsorption between the filtrate and the blood begins.

pseudocoelom (**soo′-dō-sēl′-ōm**): "false coelom"; a body cavity that has a different embryological origin than a coelom but serves a similar function; found in roundworms.

pseudoplasmodium (**soo′-dō-plaz-mō′-dē-um**): an aggregation of individual amoeboid cells that form a sluglike mass.

pseudopod (**sood′-ō-pod**): an extension of the plasma membrane by which certain cells, such as amoebae, locomote and engulf prey.

Punnett square method: an intuitive way to predict the genotypes and phenotypes of offspring in specific crosses.

pupa: a developmental stage in some insect species in which the organism stops moving and feeding and may be encased in a cocoon; occurs between the larval and the adult phases.

pupil: the adjustable opening in the center of the iris, through which light enters the eye.

pyloric sphincter (**pī-lor′-ik sfink′-ter**): a circular muscle, located at the base of the stomach, that regulates the passage of chyme into the small intestine.

pyruvate: a three-carbon molecule that is formed by glycolysis and then used in fermentation or cellular respiration.

quaternary structure (**kwat′-er-nuh-rē**): the complex three-dimensional structure of a protein composed of more than one peptide chain.

queen substance: a chemical, produced by a queen bee, that can act as both a primer and a pheromone.

radial symmetry: a body plan in which any plane along a central axis will divide the body into approximately mirror-image halves. Cnidarians and many adult echinoderms have radial symmetry.

radioactive: pertaining to an atom with an unstable nucleus that spontaneously disintegrates, with the emission of radiation.

radiolarian (**rā-dē-ō-lar′-ē-un**): an aquatic protist (largely marine) characterized by typically elaborate silica shells.

radula (**ra′-dū-luh**): a ribbon of tissue in the mouth of gastropod mollusks; bears numerous teeth on its outer surface and is used to scrape and drag food into the mouth.

rain shadow: a local dry area created by the modification of rainfall patterns by a mountain range.

random distribution: distribution characteristic of populations in which the probability of finding an individual is equal in all parts of an area.

reactant: an atom or molecule that is used up in a chemical reaction to form a product.

reaction center: in the light-harvesting complex of a photosystem, the chlorophyll molecule to which light energy is transferred by the antenna molecules (light-absorbing pigments); the captured energy ejects an electron from the reaction center chlorophyll, and the electron is transferred to the electron transport system.

receptor: a cell that responds to an environmental stimulus (chemicals, sound, light, pH, and so on) by changing its electrical potential; also, a protein molecule in a plasma membrane that binds to another molecule (hormone, neurotransmitter), triggering metabolic or electrical changes in a cell.

receptor-mediated endocytosis: the selective uptake of molecules from the extracellular fluid by binding to a receptor located at a coated pit on the plasma membrane and pinching off the coated pit into a vesicle that moves into the cytoplasm.

receptor potential: an electrical potential change in a receptor cell, produced in response to the reception of an environmental stimulus (chemicals, sound, light, heat, and so on). The size of the receptor potential is proportional to the intensity of the stimulus.

receptor protein: a protein, located on a membrane (or in the cytoplasm), that recognizes and binds to specific molecules. Binding by receptor proteins typically triggers a response by a cell, such as endocytosis, increased metabolic rate, or cell division.

recessive: an allele that is expressed only in homozygotes and is completely masked in heterozygotes.

recognition protein: a protein or glycoprotein protruding from the outside surface of a plasma membrane that identifies a cell as belonging to a particular species, to a specific individual of that species, and in many cases to one specific organ within the individual.

recombinant DNA: DNA that has been altered by the recombination of genes from a different organism, typically from a different species.

recombination: the formation of new combinations of the different alleles of each gene on a chromosome; the result of crossing over.

rectum: the terminal portion of the vertebrate digestive tube, where feces are stored until they can be eliminated.

red blood cell: the most common type of cell in vertebrate blood; active in oxygen transport; contains the red pigment hemoglobin.

reflex: a simple, stereotyped movement of part of the body that occurs automatically in response to a stimulus.

regeneration: the regrowth of a body part after loss or damage; also, asexual reproduction by means of the regrowth of an entire body from a fragment.

releasing hormone: a hormone, secreted by the hypothalamus, that causes the release of specific hormones by the anterior pituitary.

renal artery: the artery carrying blood to each kidney.

renal cortex: the outer layer of the kidney; where nephrons are located.

renal medulla: the layer of the kidney just inside the renal cortex; where loops of Henle produce a highly concentrated interstitial fluid, important in the production of concentrated urine.

renal pelvis: the inner chamber of the kidney; where urine from the collecting ducts accumulates before it enters the ureter.

renal vein: the vein carrying cleansed blood away from each kidney.

renin: an enzyme that is released (in mammals) when blood pressure and/or sodium concentration in the blood drops below a set point; initiates a cascade of events that restores blood pressure and sodium concentration.

replacement-level fertility (**RLF**): the average birthrate at which a reproducing population exactly replaces itself during its lifetime.

replication bubble: the unwound portion of the two parental DNA strands, separated by DNA helicase, in DNA replication.

reproductive isolation: the failure of organisms of one population to breed successfully with members of another; may be due to premating or postmating isolating mechanisms.

reptile: a member of the chordate group that includes the snakes, lizards, turtles, alligators, and crocodiles; not a monophyletic group.

reservoir: the major source and storage site of a nutrient in an ecosystem, normally in the abiotic portion.

resource partitioning: the coexistence of two species with similar requirements, each occupying a smaller niche than either would if it were by itself; a means of minimizing their competitive interactions.

respiration: the process by which an organism exchanges gases with the environment.

respiratory center: a cluster of neurons, located in the medulla of the brain, that sends rhythmic bursts of nerve impulses to the respiratory muscles, resulting in breathing.

resting potential: a negative electrical potential in unstimulated nerve cells.

restriction enzyme: an enzyme, normally isolated from bacteria, that cuts double-stranded DNA at a specific nucleotide sequence; the nucleotide sequence that is cut differs for different restriction enzymes.

restriction fragment: a piece of DNA that has been isolated by cleaving a larger piece of DNA with restriction enzymes.

restriction fragment length polymorphism (**RFLP**): a difference in the length of restriction fragments, produced by cutting samples of DNA from different individuals of the same species with the same set of restriction enzymes; the result of differences in nucleotide sequences among individuals of the same species.

reticular formation (**reh-tik′-ū-lar**): a diffuse network of neurons extending from the hindbrain, through the midbrain, and into the lower reaches of the forebrain; involved in filtering sensory input and regulating what information is relayed to conscious brain centers for further attention.

retina (**ret′-in-uh**): a multilayered sheet of nerve tissue at the rear of camera-type eyes, composed of photoreceptor cells plus associated nerve cells that refine the photoreceptor information and transmit it to the optic nerve.

retrovirus: a virus that uses RNA as its genetic material. When it invades a eukaryotic cell, a retrovirus "reverse transcribes" its RNA into DNA, which then directs the synthesis of more viruses, using the transcription and translation machinery of the cell.

reverse transcriptase: an enzyme found in retroviruses that catalyzes the synthesis of DNA from an RNA template.

Rh factor: a protein on the red blood cells of some people (Rh-positive) but not others (Rh-negative); the exposure of Rh-negative individuals to Rh-positive blood triggers the production of antibodies to Rh-positive blood cells.

rhizoid (**rī′-zoid**): a rootlike structure found in bryophytes that anchors the plant and absorbs water and nutrients from the soil.

rhizome (**rī′-zōm**): an undergound stem, usually horizontal, that stores food.

rhythm method: a contraceptive method involving abstinence from intercourse during ovulation.
ribonucleic acid (rī-bō-noo-klā′-ik; RNA): a molecule composed of ribose nucleotides, each of which consists of a phosphate group, the sugar ribose, and one of the bases adenine, cytosine, guanine, or uracil; transfers hereditary instructions from the nucleus to the cytoplasm; also the genetic material of some viruses.
ribosomal RNA (rRNA): a type of RNA that combines with proteins to form ribosomes.
ribosome: an organelle consisting of two subunits, each composed of ribosomal RNA and protein; the site of protein synthesis, during which the sequence of bases of messenger RNA is translated into the sequence of amino acids in a protein.
ribozyme: an RNA molecule that can catalyze certain chemical reactions, especially those involved in the synthesis and processing of RNA itself.
ribulose biphosphate (RUBP): a six-carbon molecule that reacts with carbon dioxide in the carbon-fixing reaction of C_3 photosynthesis; an important participant in the Calvin-Benson C_3 cycle.
RNA polymerase: in RNA synthesis, an enzyme that catalyzes the bonding of free RNA nucleotides into a continuous strand, using RNA nucleotides that are complementary to those of a strand of DNA.
rod: a rod-shaped photoreceptor cell in the vertebrate retina, sensitive to dim light but not involved in color vision; see also *cone.*
root: the part of the plant body, normally underground, that provides anchorage, absorbs water and dissolved nutrients and transports them to the stem, produces some hormones, and in some plants serves as a storage site for carbohydrates.
root cap: a cluster of cells at the tip of a growing root, derived from the apical meristem; protects the growing tip from damage as it burrows through the soil.
root hair: a fine projection from an epidermal cell of a young root that increases the absorptive surface area of the root.
root system: the part of a plant, normally below ground, that anchors the plant in the soil; absorbs water and minerals; stores food; transports water, minerals, sugars, and hormones; and produces certain hormones.
rough endoplasmic reticulum: endoplasmic reticulum lined on the outside with ribosomes.
runner: a horizontally growing stem that may develop new plants at nodes that touch the soil.
sac fungus: a fungus of the division Ascomycota, whose members form spores in a saclike case called an *ascus.*
saprobe (sap′-rōb): an organism that derives its nutrients from the bodies of dead organisms.
sapwood: young xylem that transports water and minerals in a tree trunk.
sarcodine (sar-kō′-dīn): a nonphotosynthetic protist (protozoan) characterized by the ability to form pseudopodia; some sarcodines, such as amoebae, are naked, whereas others have elaborate shells.
sarcomere (sark′-ō-mēr): the unit of contraction of a muscle fiber; a subunit of the myofibril, consisting of actin and myosin filaments and bounded by Z lines.
sarcoplasmic reticulum (sark′-ō-plas′-mik re-tik′-ū-lum): the specialized endoplasmic reticulum in muscle cells; forms interconnected hollow tubes. The sarcoplasmic reticulum stores calcium ions and releases them into the interior of the muscle cell, initiating contraction.
saturated: referring to a fatty acid with as many hydrogen atoms as possible bonded to the carbon backbone; a fatty acid with no double bonds in its carbon backbone.
savanna: a biome that is dominated by grasses and supports scattered trees and thorny scrub forests; typically has a rainy season in which all the year's precipitation falls.
scientific method: a rigorous procedure for making observations of specific phenomena and searching for the order underlying those phenomena; consists of four operations: observation, hypothesis, experiment, and conclusion.
scientific name: the name of an organism formed from the two smallest major taxonomic categories—the genus and the species.
scientific theory: a general explanation of natural phenomena developed through extensive and reproducible observations; more general and reliable than a hypothesis.
sclera: a tough, white connective tissue layer that covers the outside of the eyeball and forms the white of the eye.
sclerenchyma (skler-en′-ki-muh): a plant cell type with thick, hardened secondary cell walls that normally dies as the last stage of differentiation and both supports and protects the plant body.
scramble competition: a free-for-all scramble for limited resources among individuals of the same species.
scrotum (skrō′-tum): the pouch of skin containing the testes of male mammals.
S-curve: the S-shaped growth curve that describes a population of long-lived organisms introduced into a new area; consists of an initial period of exponential growth, followed by decreasing growth rate, and, finally, relative stability around a growth rate of zero.
sebaceous gland (se-bā′-shus): a gland in the dermis of skin, formed from epithelial tissue, that produces the oily substance sebum, which lubricates the epidermis.
secondary cell wall: a thick layer of cellulose and other polysaccharides secreted by certain plant cells between the primary cell wall and the plasma membrane.
secondary consumer: an organism that feeds on primary consumers; a carnivore.
secondary growth: growth in the diameter of a stem or root due to cell division in lateral meristems and differentiation of their daughter cells.
secondary oocyte (ō′-ō-sīt): a large haploid cell derived from the first meiotic division of the diploid primary oocyte.
secondary phloem: phloem produced from the cells that arise toward the outside of the vascular cambium.
secondary spermatocyte (sper-ma′-tō-sīt): a large haploid cell derived by meiosis I from the diploid primary spermatocyte.
secondary structure: a repeated, regular structure assumed by protein chains held together by hydrogen bonds; for example, a helix.
secondary succession: succession that occurs after an existing community is disturbed—for example, after a forest fire; much more rapid than primary succession.
secondary xylem: xylem produced from cells that arise at the inside of the vascular cambium.
second law of thermodynamics: the principle of physics that states that any change in an isolated system causes the quantity of concentrated, useful energy to decrease and the amount of randomness and disorder (entropy) to increase.
second messenger: an intracellular chemical, such as cyclic AMP, that is synthesized or released within a cell in response to the binding of a hormone or neurotransmitter (the first messenger) to receptors on the cell surface; brings about specific changes in the metabolism of the cell.
secretin: a hormone, produced by the small intestine, that stimulates the production and release of digestive secretions by the pancreas and liver.
seed: the reproductive structure of a seed plant; protected by a seed coat; contains an embryonic plant and a supply of food for it.
seed coat: the thin, tough, and waterproof outermost covering of a seed, formed from the integuments of the ovule.
segmentation (seg-men-tā′-shun): an animal body plan in which the body is divided into repeated, typically similar units.
segmentation movement: a contraction of the small intestine that results in the mixing of partially digested food and digestive enzymes. Segmentation movements also bring nutrients into contact with the absorptive intestinal wall.
segregation: see *law of segregation.*
selectively permeable: refers to membranes across which some substances may pass freely while other substances cannot pass.
self-fertilization: the union of sperm and egg from the same individual.
selfish gene: the concept that genes promote their own survival in individuals through innate self-sacrificing behavior that enhances the survival of others that carry the same genes; helps explain the evolution of altruism.
semen: the sperm-containing fluid produced by the male reproductive tract.
semiconservative replication: the process of replication of the DNA double helix; the two DNA strands separate, and each is used as a template for the synthesis of a complementary DNA strand. Consequently, each daughter double helix consists of one parental strand and one new strand.
semilunar valve: a paired valve between the ventricles of the heart and the pulmonary artery and aorta; prevents the backflow of blood into the ventricles when they relax.
seminal vesicle: in male mammals, a gland that produces a basic, fructose-containing fluid that forms part of the semen.
seminiferous tubule (sem-i-ni′-fer-us): in the vertebrate testis, a series of tubes in which sperm are produced.
senescence: in plants, a specific aging process, typically including deterioration and the dropping of leaves and flowers.
sensitive period: the particular stage in an animal's life during which it imprints.
sensory neuron: a nerve cell that responds to a stimulus from the internal or external environment.
sensory receptor: a cell (typically, a neuron) specialized to respond to particular internal or external environmental stimuli by producing an electrical potential.
sepal (sē′-pul): the set of modified leaves that surround and protect a flower bud, typically opening into green, leaflike structures when the flower blooms.
septum (pl., septa): a partition that separates the fungal hypha into individual cells; pores in septa allow the transfer of materials between cells.
serotonin (ser-uh-tō′-nin): in the central nervous system, a neurotransmitter that is involved in mood, sleep, and the inhibition of pain.

Sertoli cell: in the seminiferous tubule, a large cell that regulates spermatogenesis and nourishes the developing sperm.

sessile (ses′-ul): not free to move about, usually permanently attached to a surface.

severe combined immune deficiency (SCID): a disorder in which no immune cells, or very few, are formed; the immune system is incapable of responding properly to invading disease organisms, and the individual is very vulnerable to common infections.

sex chromosomes: the pair of chromosomes that usually determines the sex of an organism; for example, the X and Y chromosomes in mammals.

sex-linked: referring to a pattern of inheritance characteristic of genes located on one type of sex chromosome (for example, X) and not found on the other type (for example, Y); also called X-linked. In sex-linked inheritance, traits are controlled by genes carried on the X chromosome; females show the dominant trait unless they are homozygous recessive, whereas males express whichever allele is on their single X chromosome.

sexually transmitted disease (STD): a disease that is passed from person to person by sexual contact.

sexual recombination: during sexual reproduction, the formation of new combinations of alleles in offspring as a result of the inheritance of one homologous chromosome from each of two genetically distinct parents.

sexual reproduction: a form of reproduction in which genetic material from two parent organisms is combined in the offspring; normally, two haploid gametes fuse to form a diploid zygote.

sexual selection: a type of natural selection in which the choice of mates by one sex is the selective agent.

shoot system: all the parts of a vascular plant exclusive of the root; normally aboveground, consisting of stem, leaves, buds, and (in season) flowers and fruits; functions include photosynthesis, transport of materials, reproduction, and hormone synthesis.

short-day plant: a plant that will flower only if the length of daylight is shorter than some species-specific duration.

sickle-cell anemia: a recessive disease caused by a single amino acid substitution in the hemoglobin molecule. Sickle-cell hemoglobin molecules tend to cluster together, distorting the red blood cell's shape and causing them to break and clog capillaries.

sieve plate: in plants, a structure between two adjacent sieve-tube elements in phloem, where holes formed in the primary cell walls interconnect the cytoplasm of the elements; in echinoderms, the opening through which water enters the water-vascular system.

sieve tube: in phloem, a single strand of sieve-tube elements that transports sugar solutions.

sieve-tube element: one of the cells of a sieve tube, which form the phloem.

simple diffusion: the diffusion of water, dissolved gases, or lipid-soluble molecules through the phospholipid bilayer of a cellular membrane.

single covalent bond: a covalent bond in which two atoms share one pair of electrons.

sink: in plants, any structure that uses up sugars or converts sugars to starch and toward which phloem fluids will flow.

sinoatrial (SA) node (sī′-nō-āt′-rē-ul): a small mass of specialized muscle in the wall of the right atrium; generates electrical signals rhythmically and spontaneously and serves as the heart's pacemaker.

skeletal muscle: the type of muscle that is attached to and moves the skeleton and is under the direct, normally voluntary, control of the nervous system; also called *striated muscle*.

skeleton: a supporting structure for the body, on which muscles act to change the body configuration; may be external or internal.

skin: the tissue that makes up the outer surface of an animal body.

slime layer: a sticky polysaccharide or protein coating that some disease-causing bacteria secrete outside their cell wall; helps the cells aggregate and stick to smooth surfaces.

small intestine: the portion of the digestive tract, located between the stomach and large intestine, in which most digestion and absorption of nutrients occur.

smooth endoplasmic reticulum: endoplasmic reticulum without ribosomes.

smooth muscle: the type of muscle that surrounds hollow organs, such as the digestive tract, bladder, and blood vessels; normally not under voluntary control.

sodium–potassium pump: in nerve cell plasma membranes, a set of active-transport molecules that use the energy of adenosine triphosphate (ATP) to pump sodium ions out of the cell and potassium ions in, maintaining the concentration gradients of these ions across the membrane.

solvent: a liquid capable of dissolving (uniformly dispersing) other substances in itself.

somatic nervous system: that portion of the peripheral nervous system that controls voluntary movement by activating skeletal muscles.

source: in plants, any structure that actively synthesizes sugar and away from which phloem fluid will be transported.

spawning: a method of external fertilization in which male and female parents shed gametes into the water, and sperm must swim through the water to reach the eggs.

speciation: the process of species formation, in which a single species splits into two or more species.

species (spē′-sēs): the basic unit of taxonomic classification, consisting of a population or series of populations of closely related and similar organisms. In sexually reproducing organisms, a species can be defined as a population or series of populations of organisms that interbreed freely with one another under natural conditions but that do not interbreed with members of other species.

specific heat: the amount of energy required to raise the temperature of 1 gram of a substance by 1°C.

sperm: the haploid male gamete, normally small, motile, and containing little cytoplasm.

spermatid: a haploid cell derived from the secondary spermatocyte by meiosis II; differentiates into the mature sperm.

spermatogenesis: the process by which sperm cells form.

spermatogonium (pl., spermatogonia): a diploid cell, lining the walls of the seminiferous tubules, that gives rise to a primary spermatocyte.

spermatophore: in a variation on internal fertilization in some animals, the males package their sperm in a container that can be inserted into the female reproductive tract.

spermicide: a sperm-killing chemical; used for contraceptive purposes.

spicule (spik′-ūl): a subunit of the endoskeleton of sponges that is made of protein, silica, or calcium carbonate.

spinal cord: the part of the central nervous system of vertebrates that extends from the base of the brain to the hips and is protected by the bones of the vertebral column; contains the cell bodies of motor neurons that form synapses with skeletal muscles, the circuitry for some simple reflex behaviors, and axons that communicate with the brain.

spindle microtubules: microtubules organized in a spindle shape that separate chromosomes during mitosis or meiosis.

spiracle (spi′-ruh-kul): an opening in the abdominal segment of insects through which air enters the tracheae.

spirillum (spi′-ril-um; pl., spirilla): a spiral-shaped bacterium.

spleen: an organ of the lymphatic system in which lymphocytes are produced and blood is filtered past lymphocytes and macrophages, which remove foreign particles and aged red blood cells.

spongy bone: porous, lightweight bone tissue in the interior of bones; the location of bone marrow.

spongy cell: an irregularly shaped mesophyll cell, containing chloroplasts, located just above the lower epidermis of a leaf.

spontaneous generation: the proposal that living organisms can arise from nonliving matter.

sporangium (spor-an′-jē-um; pl., sporangia): a structure in which spores are produced.

spore: a haploid reproductive cell capable of developing into an adult without fusing with another cell; in the alternation-of-generation life cycle of plants, a haploid cell that is produced by meiosis and then undergoes repeated mitotic divisions and differentiation of daughter cells to produce the gametophyte, a multicellular, haploid organism.

sporophyte (spor′-ō-fīt): the diploid form of a plant that produces haploid, asexual spores through meiosis.

sporozoan (spor-ō-zō′-un): a parasitic protist with a complex life cycle, typically involving more than one host; named for their ability to form infectious spores. A well-known sporozoan (genus *Plasmodium*) causes malaria.

stabilizing selection: a type of natural selection in which those organisms that display extreme phenotypes are selected against.

stamen (stā′-men): the male reproductive structure of a flower, consisting of a filament and an anther, in which pollen grains develop.

starch: a polysaccharide that is composed of branched or unbranched chains or glucose molecules; used by plants as a carbohydrate-storage molecule.

start codon: the first AUG codon in a messenger RNA molecule.

startle coloration: a form of mimicry in which a color pattern (in many cases resembling large eyes) can be displayed suddenly by a prey organism when approached by a predator.

stem: the portion of the plant body, normally located above ground, that bears leaves and reproductive structures such as flowers and fruit.

sterilization: a generally permanent method of contraception in which the pathways through which the sperm (vas deferens) or egg (oviducts) must travel are interrupted; the most effective form of contraception.

steroid: see *steroid hormone*.

steroid hormone: a class of hormone whose chemical structure (four fused carbon rings with various functional groups) resembles cholesterol; steroids, which are lipids, are secreted by the ovaries and placenta, the testes, and the adrenal cortex.

stigma (stig′-muh): the pollen-capturing tip of a carpel.

stoma (stō′-muh; pl., stomata): an adjustable opening in the epidermis of a leaf, surrounded by a pair of guard cells, that regulates the diffusion of carbon dioxide and water into and out of the leaf.

stomach: the muscular sac between the esophagus and small intestine where food is stored and mechanically broken down and in which protein digestion begins.

stop codon: a codon in messenger RNA that stops protein synthesis and causes the completed protein chain to be released from the ribosome.

strand: single polymer of nucleotides; DNA is composed of two strands.

striated muscle: see *skeletal muscle*.

stroke: an interruption of blood flow to part of the brain caused by the rupture of an artery or the blocking of an artery by a blood clot. Loss of blood supply leads to rapid death of the area of the brain affected.

stroma (strō′-muh): the semifluid material inside chloroplasts in which the grana are embedded.

style: a stalk connecting the stigma of a carpel with the ovary at its base.

subatomic particle: any of the particles of which atoms are made: electrons, protons, and neutrons.

subclimax: a community in which succession is stopped before the climax community is reached and is maintained by regular disturbances—for example, tallgrass prairie maintained by periodic fires.

substrate: the atoms or molecules that are the reactants for an enzyme-catalyzed chemical reaction.

subunit: a small organic molecule, several of which may be bonded together to form a larger molecule. See also *monomer*.

succession (suk-seh′-shun): a structural change in a community and its nonliving environment over time. Community changes alter the ecosystem in ways that favor competitors, and species replace one another in a somewhat predictable manner until a stable, self-sustaining climax community is reached.

sucrose: a disaccharide composed of glucose and fructose.

sugar: a simple carbohydrate molecule, either a monosaccharide or a disaccharide.

sugar-phosphate backbone: a major feature of DNA structure, formed by attaching the sugar of one nucleotide to the phosphate from the adjacent nucleotide in a DNA strand.

suppressor T cell: a type of T cell that depresses the response of other immune cells to foreign antigens.

surface tension: the property of a liquid to resist penetration by objects at its interface with the air, due to cohesion between molecules of the liquid.

survivorship curve: a curve resulting when the number of individuals of each age in a population is graphed against their age, usually expressed as a percentage of their maximum life span.

symbiosis (sim′-bī-ō′sis): a close interaction between organisms of different species over an extended period. Either or both species may benefit from the association, or (in the case of parasitism) one of the participants is harmed. Symbiosis includes parasitism, mutualism, and commensalism.

symbiotic: referring to an ecological relationship based on symbiosis.

sympathetic division: the division of the autonomic nervous system that produces largely involuntary responses that prepare the body for stressful or highly energetic situations.

sympatric speciation (sim-pat′-rik): speciation that occurs in populations that are not physically divided; normally due to ecological isolation or chromosomal aberrations (such as polyploidy).

synapse (sin′-aps): the site of communication between nerve cells. At a synapse, one cell (presynaptic) normally releases a chemical (the neurotransmitter) that changes the electrical potential of the second (postsynaptic) cell.

synaptic terminal: a swelling at the branched ending of an axon; where the axon forms a synapse.

syphilis (si′-ful-is): a sexually transmitted bacterial infection of the reproductive organs; if untreated, can damage the nervous and circulatory systems.

systematics: the branch of biology concerned with reconstructing phylogenies and with naming and classifying species.

taiga (tī′-guh): see *northern coniferous forest*.

taproot system: a root system, commonly found in dicots, that consists of a long, thick main root and many smaller lateral roots, all of which grow from the primary root.

target cell: a cell on which a particular hormone exerts its effect.

taste: a chemical sense for substances dissolved in water or saliva; in mammals, perceptions of sweet, sour, bitter, or salt produced by the stimulation of receptors on the tongue.

taste bud: a cluster of taste receptor cells and supporting cells that is located in a small pit beneath the surface of the tongue and that communicates with the mouth through a small pore. The human tongue has about 10,000 taste buds.

taxis (taks′-is; pl., taxes): an innate behavior that is a directed movement of an organism toward or away from a stimulus such as heat, light, or gravity.

taxonomy (tax-on′-uh-mē): the science by which organisms are classified into hierarchically arranged categories that reflect their evolutionary relationships.

Tay-Sachs disease: a recessive disease caused by a deficiency in enzymes that regulate lipid breakdown in the brain.

T cell: a type of lymphocyte that recognizes and destroys specific foreign cells or substances or that regulates other cells of the immune system.

T-cell receptor: a protein receptor, located on the surface of a T cell, that binds a specific antigen and triggers the immune response of the T cell.

tectorial membrane (tek-tor′-ē-ul): one of the membranes of the cochlea in which the hairs of the hair cells are embedded. In sound reception, movement of the basilar membrane relative to the tectorial membrane bends the cilia.

telophase (tēl′-ō-fāz): in mitosis, the final stage, in which a nuclear envelope re-forms around each new daughter nucleus, the spindle fibers disappear, and the chromosomes relax from their condensed form; in meiosis I, the stage during which the spindle fibers disappear and the chromosomes normally relax from their condensed form; in meiosis II, the stage during which chromosomes relax into their extended state, the nuclear envelopes re-form, and cytokinesis occurs.

temperate deciduous forest: a biome in which winters are cold and summer rainfall is sufficient to allow enough moisture for trees to grow and shade out grasses.

temperate rain forest: a biome in which there is no shortage of liquid water year-round and that is dominated by conifers.

template strand: the strand of the DNA double helix from which RNA is transcribed.

temporal isolation: the inability of organisms to mate if they have significantly different breeding seasons.

tendon: a tough connective tissue band connecting a muscle to a bone.

tendril: a slender outgrowth of a stem that coils about external objects and supports the stem; normally a modified leaf or branch.

tentacle (ten′-te-kul): an elongate, extensible projection of the body of cnidarians and cephalopod mollusks that may be used for grasping, stinging, and immobilizing prey, and for locomotion.

terminal bud: meristem tissue and surrounding leaf primordia that are located at the tip of the plant shoot.

territoriality: the defense of an area in which important resources are located.

tertiary consumer (ter′-shē-er-ē): a carnivore that feeds on other carnivores (secondary consumers).

tertiary structure (ter′-shē-er-ē): the complex three-dimensional structure of a single peptide chain; held in place by disulfide bonds between cysteines.

test cross: a breeding experiment in which an individual showing the dominant phenotype is mated with an individual that is homozygous recessive for the same gene. The ratio of offspring with dominant versus recessive phenotypes can be used to determine the genotype of the phenotypically dominant individual.

testis (pl., testes): the gonad of male mammals.

testosterone: in vertebrates, a hormone produced by the interstitial cells of the testis; stimulates spermatogenesis and the development of male secondary sex characteristics.

thalamus: the part of the forebrain that relays sensory information to many parts of the brain.

theory: in science, an explanation for natural events that is based on a large number of observations and is in accord with scientific principles, especially causality.

thermoacidophile (ther-mō-a-sid′-eh-fil): an archaean that thrives in hot, acidic environments.

thermoreceptor: a sensory receptor that responds to changes in temperature.

thick filament: in the sarcomere, a bundle of myosin that interacts with thin filaments, producing muscle contraction.

thin filament: in the sarcomere, a protein strand that interacts with thick filaments, producing muscle contraction; composed primarily of actin, with accessory proteins.

thorax: the segment between the head and abdomen in animals with segmentation; the segment to which structures used in locomotion are attached.

thorn: a hard, pointed outgrowth of a stem; normally a modified branch.

threshold: the electrical potential (less negative than the resting potential) at which an action potential is triggered.

thrombin: an enzyme produced in the blood as a result of injury to a blood vessel; catalyzes the production of fibrin, a protein that assists in blood clot formation.

thylakoid (thī′-luh-koid): a disk-shaped, membranous sac found in chloroplasts, the membranes of which contain the photosystems and ATP-synthesizing enzymes used in the light-dependent reactions of photosynthesis.

thymine: a nitrogenous base found only in DNA; abbreviated as T.

thymosin: a hormone, secreted by the thymus, that stimulates the maturation of cells of the immune system.

thymus (thī′-mus): an organ of the lymphatic system that is located in the upper chest in front of the heart

and that secretes thymosin, which stimulates lymphocyte maturation; begins to degenerate at puberty and has little function in the adult.
thyroid gland: an endocrine gland, located in front of the larynx in the neck, that secretes the hormones thyroxine (affecting metabolic rate) and calcitonin (regulating calcium ion concentration in the blood).
thyroid-stimulating hormone (TSH): a hormone, released by the anterior pituitary, that stimulates the thyroid gland to release hormones.
thyroxine (thī-rox′-in): a hormone, secreted by the thyroid gland, that stimulates and regulates metabolism.
tight junction: a type of cell-to-cell junction in animals that prevents the movement of materials through the spaces between cells.
tissue: a group of (normally similar) cells that together carry out a specific function; for example, muscle; may include extracellular material produced by its cells.
tonsil: a patch of lymphatic tissue consisting of connective tissue that contains many lymphocytes; located in the pharynx and throat.
trachea (trā′-kē-uh): in birds and mammals, a rigid but flexible tube, supported by rings of cartilage, that conducts air between the larynx and the bronchi; in insects, an elaborately branching tube that carries air from openings called *spiracles* near each body cell.
tracheid (trā′-kē-id): an elongated xylem cell with tapering ends that contains pits in the cell wall; forms tubes that transport water.
tracheophyte (trā′-kē-ō-fīt): a plant that has conducting vessels; a vascular plant.
transcription: the synthesis of an RNA molecule from a DNA template.
transducer: a device that converts signals from one form to another. Sensory receptors are transducers that convert environmental stimuli, such as heat, light, or vibration, into electrical signals (such as action potentials) recognized by the nervous system.
transfer RNA (tRNA): a type of RNA that binds to a specific amino acid by means of a set of three bases (the anticodon) on the tRNA that are complementary to the messenger RNA (mRNA) codon for that amino acid; carries its amino acid to a ribosome during protein synthesis, recognizes a codon of mRNA, and positions its amino acid for incorporation into the growing protein chain.
transformation: a method of acquiring new genes, whereby DNA from one bacterium (normally released after the death of the bacterium) becomes incorporated into the DNA of another, living, bacterium.
transgenic: referring to an animal or a plant that expresses DNA derived from another species.
translation: the process whereby the sequence of bases of messenger RNA (mRNA) is converted into the sequence of amino acids of a protein.
transpiration (trans′-per-ā-shun): the evaporation of water through the stomata of a leaf.
transport protein: a protein that regulates the movement of water-soluble molecules through the plasma membrane.
trial-and-error learning: a process by which adaptive responses are learned through rewards or punishments provided by the environment.
trichomoniasis (trik-ō-mō-nī′-uh-sis): a sexually transmitted disease, caused by the protist *Trichomonas*, that causes inflammation of the mucous membranes that line the urinary tract and genitals.
tricuspid valve: the valve between the right ventricle and the right atrium of the heart.
triglyceride (trī-glis′-er-īd): a lipid composed of three fatty-acid molecules bonded to a single glycerol molecule.
triple covalent bond: a covalent bond that occurs when two atoms share three pairs of electrons.
trisomy 21: see *Down syndrome*.
trisomy X: a condition of females who have three X chromosomes instead of the normal two; most such women are phenotypically normal and are fertile.
trophic level: literally, "feeding level"; the categories of organisms in a community, and the position of an organism in a food chain, defined by the organism's source of energy; includes producers, primary consumers, secondary consumers, and so on.
tropical deciduous forest: a biome with pronounced wet and dry seasons and plants that must shed their leaves during the dry season to minimize water loss.
tropical rain forest: a biome with evenly warm, evenly moist conditions; dominated by broadleaf evergreen trees; the most diverse biome.
true-breeding: pertaining to an individual all of whose offspring produced through self-fertilization are identical to the parental type. True-breeding individuals are homozygous for a given trait.
T tubule: a deep infolding of the muscle plasma membrane; conducts the action potential inside a cell.
tubal ligation: a surgical procedure in which a woman's oviducts are cut so that the egg cannot reach the uterus, making her infertile.
tube cell: the outermost cell of a pollen grain; digests a tube through the tissues of the carpel, ultimately penetrating into the female gametophyte.
tube foot: a cylindrical extension of the water-vascular system of echinoderms; used for locomotion, grasping food, and respiration.
tubular reabsorption: the process by which cells of the tubule of the nephron remove water and nutrients from the filtrate within the tubule and return those substances to the blood.
tubular secretion: the process by which cells of the tubule of the nephron remove additional wastes from the blood, actively secreting those wastes into the tubule.
tubule (toob′-ūl): the tubular portion of the nephron; includes a proximal portion, the loop of Henle, and a distal portion. Urine is formed from the blood filtrate as it passes through the tubule.
tumor: a mass that forms in otherwise normal tissue; caused by the uncontrolled growth of cells.
tundra: a biome with severe weather conditions (extreme cold and wind and little rainfall) that cannot support trees.
turgor pressure: pressure developed within a cell (especially the central vacuole of plant cells) as a result of osmotic water entry.
Turner syndrome: a set of characteristics typical of a woman with only one X chromosome: sterile, with a tendency to be very short and to lack normal female secondary sexual characteristics.
tympanic membrane (tim-pan′-ik): the eardrum; a membrane, stretched across the opening of the ear, that transmits vibration of sound waves to bones of the middle ear.
unicellular: single-celled; most members of the domains Bacteria and Archaea and the kingdom Protista are unicellular.
umbilical cord: structure that connects the circulatory systems of a mammalian mother and fetus.
uniform distribution: the distribution characteristic of a population with a relatively regular spacing of individuals, commonly as a result of territorial behavior.
uniformitarianism: the hypothesis that Earth developed gradually through natural processes, similar to those at work today, that occur over long periods of time.
unsaturated: referring to a fatty acid with fewer than the maximum number of hydrogen atoms bonded to its carbon backbone; a fatty acid with one or more double bonds in its carbon backbone.
upwelling: an upward flow that brings cold, nutrient-laden water from the ocean depths to the surface; occurs along western coastlines.
uracil: a nitrogenous base found in RNA; abbreviated as U.
urea (ū-rē′-uh): a water-soluble, nitrogen-containing waste product of amino acid breakdown; one of the principal components of mammalian urine.
ureter (ū′-re-ter): a tube that conducts urine from each kidney to the bladder.
urethra (ū-rē′-thruh): the tube leading from the urinary bladder to the outside of the body; in males, the urethra also receives sperm from the vas deferens and conducts both sperm and urine (at different times) to the tip of the penis.
uric acid (ūr′-ik): a nitrogen-containing waste product of amino acid breakdown; a relatively insoluble white crystal excreted by birds, reptiles, and insects.
urine: the fluid produced and excreted by the urinary system of vertebrates; contains water and dissolved wastes, such as urea.
uterine tube: the tube leading out of the ovary to the uterus, into which the secondary oocyte (egg cell) is released; also called the *oviduct*.
uterus: in female mammals, the part of the reproductive tract that houses the embryo during pregnancy.
vaccination: an injection into the body that contains antigens characteristic of a particular disease organism and that stimulates an immune response.
vacuole (vak′-ū-ol): a vesicle that is typically large and consists of a single membrane enclosing a fluid-filled space.
vagina: the passageway leading from the outside of a female mammal's body to the cervix of the uterus.
variable: a condition, particularly in a scientific experiment, that is subject to change.
variable region: the part of an antibody molecule that differs among antibodies; the ends of the variable regions of the light and heavy chains form the specific binding site for antigens.
vascular: describing tissues that contain vessels for transporting liquids.
vascular bundle (vas′-kū-lar): a strand of xylem and phloem in leaves and stems; in leaves, commonly called a *vein*.
vascular cambium: a lateral meristem that is located between the xylem and phloem of a woody root or stem and that gives rise to secondary xylem and phloem.
vascular cylinder: the centrally located conducting tissue of a young root, consisting of primary xylem and phloem.
vascular plant: a plant that has specialized structures (vessels) for transporting water and nutrients though its body. Horsetails, ferns, gymnosperms, and flowering plants are examples of vascular plants.
vascular tissue: plant tissue consisting of xylem (which transports water and minerals from root to shoot) and phloem (which transports water and sugars throughout the plant).

vas deferens (vaz de′-fer-enz): the tube connecting the epididymis of the testis with the urethra.

vasectomy: a surgical procedure in which a man's vas deferens are cut, preventing sperm from reaching the penis during ejaculation, thereby making him infertile.

vector: a carrier that introduces foreign genes into cells.

vein: in vertebrates, a large-diameter, thin-walled vessel that carries blood from venules back to the heart; in vascular plants, a vascular bundle, or a strand of xylem and phloem in leaves.

ventral (ven′-trul): the lower side or underside of an animal whose head is oriented forward.

ventricle (ven′-tre-kul): the lower muscular chamber on each side of the heart, which pumps blood out through the arteries. The right ventricle sends blood to the lungs; the left ventricle pumps blood to the rest of the body.

venule (ven′-ūl): a narrow vessel with thin walls that carries blood from capillaries to veins.

vertebral column (ver-tē′-brul): a column of serially arranged skeletal units (the vertebrae) that enclose the nerve cord in vertebrates; the backbone.

vertebrate: an animal that possesses a vertebral column.

vesicle (ves′-i-kul): a small, membrane-bound sac within the cytoplasm.

vessel: a tube of xylem composed of vertically stacked vessel elements with heavily perforated or missing end walls, leaving a continuous, uninterrupted hollow cylinder.

vessel element: one of the cells of a xylem vessel; elongated, dead at maturity, with thick, lignified lateral cell walls for support but with end walls that are either heavily perforated or missing.

vestigial structure (ves-tij′-ē-ul): a structure that serves no apparent purpose but is homologous to functional structures in related organisms and provides evidence of evolution.

villus (vi′-lus; pl., villi): a fingerlike projection of the wall of the small intestine that increases the absorptive surface area.

viroid (vī′-roid): a particle of RNA that is capable of infecting a cell and of directing the production of more viroids; responsible for certain plant diseases.

virus (vī′-rus): a noncellular parasitic particle that consists of a protein coat surrounding a strand of genetic material; multiplies only within a cell of a living organism (the host).

vitamin: one of a group of diverse chemicals that must be present in trace amounts in the diet to maintain health; used by the body in conjunction with enzymes in a variety of metabolic reactions.

vitreous humor (vit′-rē-us): a clear, jellylike substance that fills the large chamber of the eye between the lens and the retina.

vocal cord: one of a pair of bands of elastic tissue that extend across the opening of the larynx and produce sound when air is forced between them. Muscles alter the tension on the vocal cords and control the size and shape of the opening, which in turn determine whether sound is produced and what its pitch will be.

waggle dance: a symbolic form of communication used by honeybee foragers to communicate the location of a food source to their hivemates.

warning coloration: bright coloration that warns predators that the potential prey is distasteful or even poisonous.

water mold: a funguslike protist that includes some pathogens, such as the downy mildew, which attacks grapes.

water-vascular system: a system in echinoderms that consists of a series of canals through which seawater is conducted and is used to inflate tube feet for locomotion, grasping food, and respiration.

wax: a lipid composed of fatty acids covalently bonded to long-chain alcohols.

weather: short-term fluctuations in temperature, humidity, cloud cover, wind, and precipitation in a region over periods of hours to days.

Werner syndrome: a rare condition in which a defective gene causes premature aging; caused by a mutation in the gene that codes for DNA replication/repair enzymes.

white blood cell: cellular components of blood; not as numerous as red blood cells; most function as part of the immune system and help defend the body against invaders.

white matter: the portion of the brain and spinal cord that consists largely of myelin-covered axons and that give these areas a white appearance.

withdrawal: the removal of the penis from the vagina just before ejaculation in an attempt to avoid pregnancy; an ineffective contraceptive method.

working memory: the first phase of learning; short-term memory that is electrical or biochemical in nature.

xylem (zī-lum): a conducting tissue of vascular plants that transports water and minerals from root to shoot.

yolk: protein-rich or lipid-rich substances contained in eggs that provide food for the developing embryo.

yolk sac: one of the embryonic membranes of reptilian, bird, and mammalian embryos; in birds and reptiles, a membrane surrounding the yolk in the egg; in mammals, forms part of the umbilical cord and the digestive tract but is empty.

Z line: a fibrous protein structure to which the thin filaments of skeletal muscle are attached; forms the boundary of a sarcomere.

zona pellucida (pel-oo′-si-duh): a clear, noncellular layer between the corona radiata and the egg.

zooflagellate (zō-ō-fla′-jel-et): a nonphotosynthetic protist that moves by using flagella.

zooplankton: nonphotosynthetic protists that are abundant in marine and freshwater environments.

zoospore (zō′-ō-spor): a nonsexual reproductive cell that swims by using flagella; formed by members of the protistan division Oomycota.

zygospore (zī′-gō-spor): a fungal spore, produced by the division Zygomycota, that is surrounded by a thick, resistant wall and forms from a diploid zygote.

zygote (zī′-gōt): in sexual reproduction, a diploid cell (the fertilized egg) formed by the fusion of two haploid gametes.

zygote fungus: a fungus of the division Zygomycota, which includes the species that cause fruit rot and bread mold.

Selected Answers

CHAPTER 1

Answers to figure caption questions

Figure 1-6 Some examples:
Answerable at the cell level but not at the tissue level: How are signals transmitted in a neuron? How do white blood cells move to the site of wounds? How do chromosomes move during cell division? How do bacteria stick to surfaces?

Answerable at the tissue level but not at the cell level: Which part of the brain controls speech? How does the kidney help maintain the body's water balance? What are the functions of skin? How does water get from a plant's roots to its leaves?

Figure 1-8 The antibacterial chemicals produced by fungi probably evolved because they improved the fungi's ability to compete with bacteria for access to resources such as food and space (by excluding bacteria from areas where the fungi are present).

Figure E1-1 Redi's experiment demonstrated that the maggots were caused by something that was excluded by a gauze cover, but the possibility remained that some agent other than flies produced the maggots. An effective follow-up experiment might involve a series of closed, meat-containing systems that were identical in all respects other than the addition of a single possible causal element. Perhaps one container with flies added, one with roaches, one with dust or soot, and so on. And, of course, a control with nothing added.

Thinking Through the Concepts

Answers to Multiple-Choice Questions

1. d **2.** c **3.** c **4.** e **5.** e

Applying the Concepts

Hints

1. Include a clear, concise statement of your hypothesis and at least one testable prediction. Your experiment should include some control groups that are fed other diets. Specify the number of subjects in each group and precisely how you will measure variables.
2. One question that might be asked about the interaction between monarch butterfly caterpillars and milkweeds is, "Do feeding monarch caterpillars prefer milkweed leaves to the leaves of other species?" A hypothesis to answer the question might be, "Milkweed is the preferred food of monarch caterpillars." One prediction of this hypothesis is that caterpillars presented with a choice of leaves from two different plant species will consume a greater amount of milkweed leaves. An experiment might include tests in which single caterpillars are housed with milkweeds and various different alternative plant species.

CHAPTER 2

Answers to figure caption questions

Figure 2-2 Atoms with non-full outer shells are reactive, with a strong tendency to form bonds with other atoms, and thus are suitable for roles in the myriad chemical reactions of metabolism and for forming the complex molecules from which living things are constructed. Life's most prevalent molecules are notable for their tendency to participate in covalent bonds.

Figure 2-4 Oxygen's nucleus has eight protons, whereas hydrogen's has only a single proton. So the positive charge of the oxygen nucleus is far stronger than that of the hydrogen nucleus.

Figure 2-6 The energy that cells need to function is harvested from the chemical bonds in sugar molecules, so every cell must have access to the body's supply of sugars. Fortunately, blood, intercellular fluid, and cytoplasm consist largely of water, and sugars' easy solubility in water makes it possible for sugar molecules to be transported to and into all of the body's cells.

Figure 2-8 The pH of lemon juice is lower than the pH of tea, so the concentration of hydrogen ions in lemon juice is higher. Thus, the concentration of hydrogen ions in a mixture of tea and lemon juice is higher than the concentration in tea alone.

Figure 2-11 One the main drawbacks of widespread use of plastics is their resistance to natural degradation and consequent problems with long-term persistence of plastic waste in the environment. Starch is easily digestible by decomposer microbes, so starch-based plastics could potentially be far more biodegradable that plastics based on microbe-resistant molecules such as cellulose. The main challenge of designing starch-based plastics is making them sufficiently durable and strong.

Figure 2-14 As lipids, steroids are soluble in the lipid-based cell membrane and can cross it (and the nuclear membrane) to act inside the cell. Other types of hormones (mostly peptides) are not lipid soluble and cannot easily cross the cell membrane.

Figure 2-17 Heat energy can break chemical bonds, and the hydrogen bonds that account for secondary (and higher level) protein structure are especially susceptible to heat. Because a protein's functional ability usually depends on its shape, breaking shape-controlling bonds disrupts function.

Consider This

Hints: Both obesity and food additives pose risks to health, though the effects of obesity are better studied and less controversial. Think about how the two kinds of risks might be measured and compared. You may also wish to consider whether it is possible to promote behavior that avoids both kinds of risks.

Thinking Through the Concepts

Answers to Multiple-Choice Questions

1. c **2.** d **3.** a **4.** b **5.** b **6.** c

Applying the Concepts

Hints

1. Fat tissue consists mostly of fat molecules, and muscle tissue consists mostly of protein molecules. Think about the steps that would be required for conversion of fat to protein.
2. Bases combine with hydrogen ions, so adding a base to a solution tends to lower the solution's pH, unless a good buffer is also in the solution. Experiments might involve mixing fixed amounts of different antacids into acidic solutions (acidity in the range found in stomachs) and seeing how fast and how much the pH changes. Trials with different starting pH would help make the experiments more revealing.
3. If water molecules were nonpolar, there would be no hydrogen bonds between water molecules. As a result, we might hypothesize that water would be less effective as a solvent and that it would have less internal cohesion. Think about how such changes would affect such biological processes as transport of substances in blood and movement of water and nutrients in plants.

CHAPTER 3

Answers to figure caption questions

Figure 3-3 After some time has passed, compartment B will contain less water and less glucose than at the start of the experiment. The addition of sucrose to compartment A increases the total concentration of solutes there, so water flows by osmosis from B to A. The movement of water from compartment B to compartment A increases the concentration of glucose in B and decreases it in A, so glucose flows down its new concentration gradient from B to A.

Figure 3-4 In simple diffusion, the initial diffusion rate increases as the initial concentration gradient increases. In facilitated diffusion, the rate also increases with the concentration gradient, but eventually it reaches an upper limit when the carrier proteins are saturated.

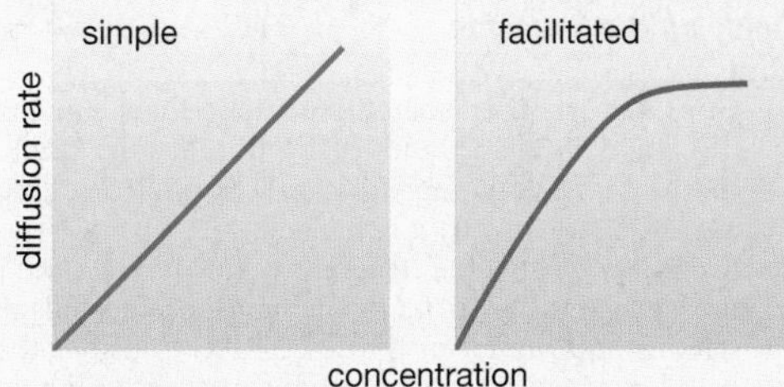

Figure 3-5 Freshwater fishes must have (and do have) physiological mechanisms that constantly export water to the environment to compensate for the water that flows into their bodies by osmosis.

Consider This

Hints: Think about how the relative health risks of obesity and high-fat diets might be compared. Consider whether weight might be reduced by diets other than low-carb and how one might determine the relative effectiveness of different weight-loss diets.

Thinking Through the Concepts

Answers to Multiple-Choice Questions

1. b **2.** e **3.** e **4.** a **5.** d **6.** d

Applying the Concepts

Hints

1. Blocking passage of cholesterol across all cell membranes would (as in people with FH) raise blood cholesterol. But ezetimibe affects mainly cells of the intestinal lining, preventing absorption of cholesterol from food in the intestine.
2. A *Paramecium* cell is surrounded by fresh water, but most of the cells of a large animal's body are surrounded by the body's extracellular fluid, which is rather salty. Think about the effects of osmosis on cells in these different circumstances. Also think about which molecules in a cell membrane have the most influence on the membrane's permeability to water.
3. Virtually all cells, including the cells of the immune system, can manufacture and secrete proteins. Some proteins form pores in membranes.
4. Energy is required to transport substances across membranes against a concentration gradient. Think about the membrane proteins required for active transport.

CHAPTER 4

Answers to figure caption questions

Figure 4-3 Of the four structures listed, only the ribosome is found in all of the main branches of life (i.e., found in bacteria, archaea, and all eukaryotes). Thus, ribosomes must have been present in the common ancestor of all living cells, and nuclei, mitochondria, and chloroplasts must have arisen later.
Figure 4-5 Key processes such as DNA replication and transcription require that enzyme molecules have access to the DNA strand. Condensation of the genetic material restricts that access because it leaves little space around individual strands.

Consider This

Hints: One potential advantage of growing stem cells outside the body is that they can produce tissues with a genotype different from that of the patient (useful for replacing tissues that are defective because of a genetic disease). One potential disadvantage is that the immune system may attack foreign cells introduced to a body.

Thinking Through the Concepts

Answers to Multiple-Choice Questions
1. a **2.** c **3.** d **4.** a **5.** a **6.** c

Applying the Concepts

Hints

1. Consider whether something other than a living cell could leave the kinds of traces that were found on the Martian meteor. Try to think of a trace that could be left only by a living cell.
2. Think about how the demands on a couch potato's muscle cells differ from the demands on the cells of a marathon runner or a weight lifter. Consider the cellular structures whose enhancement would help meet the different demands.
3. Think about how organelles and other materials are transported within cells.
4. Think about how substances enter and exit cells. Consider how a cell's need to acquire and eliminate substances increases as cell size increases, and consider the factors that limit the rate at which substances move into and out of a cell. Think about characteristics that would limit a cell's demand for substances.

CHAPTER 5

Answers to figure caption questions

Figure 5-2 Other possibilities include mechanical energy (e.g., shaking), electricity, and radiation.
Figure 5-3 The conversion breaks a phosphate bond, and the energy stored in that bond can be transferred to a molecule involved in a reaction.
Figure 5-7 No. A catalyst lowers a reaction's activation energy, but does not eliminate it. The reaction energy must still be overcome in order for the reaction to proceed.
Figure 5-8 The best bet would be to increase the concentration of enzyme, as reaction rate is most often limited by the number of available enzyme molecules. Other things that could be helpful include increasing the reaction temperature (but not so much as to denature the enzyme) and adjusting the pH to the level at which the enzyme's activity is highest (though this last modification would require specific knowledge about the enzyme).

Consider This

Hints: Criteria to consider might include how common a disorder is, how serious its consequences are, and the cost of testing for it. Think about how these and other criteria might be assessed and weighed.

Thinking Through the Concepts

Answers to Multiple-Choice Questions
1. d **2.** c **3.** b **4.** a **5.** c **6.** e

Applying the Concepts

Hints

1. The second law describes conditions inside a closed system. Consider outside inputs that life may be using to offset a tendency to increase disorder.
2. Consider the likely fate of enzymes in the acidic environment of the stomach. Think about how one might measure differences between experimental subjects who had or had not consumed enzyme supplements.
3. The laws of thermodynamics tell us that conversion of energy from one form to another is never completely efficient. Also, an organism's weight is determined in large measure by the amount of energy it consumes.
4. Consider what would happen if the body used only a single digestive enzyme to break down all types of nutrient molecules. Consider the advantages of an ability to control the time and location at which a given nutrient molecule is broken down.

CHAPTER 6

Answers to figure caption questions

Figure 6-2 Carotenoids are yellow, orange, or red.
Figure 6-3 Almost all of the ATP and NADPH produced in the chloroplast are used for the production of sugar in the Calvin-Benson cycle. Mitochondria are needed to extract the energy stored in the sugar molecules.
Figure 6-5 Yes, in principle the O_2 by-product of photosynthesis could be recycled within cells or transported within the plant for use in cellular respiration in the mitochondria.
Figure 6-6 The C_4 pathway is less efficient than the C_3 pathway; C_4 uses one extra ATP per CO_2 molecule (for regenerating PEP). Thus, when CO_2 is abundant and photorespiration is not a problem, C_3 plants produce sugar at lower cost and outcompete C_4 plants.

Consider This

Hints: Try to imagine a scenario in which a sudden decline in photosynthesis sets in motion a series of events that takes thousands of years and leads to thousands of extinctions over that period. Is your scenario plausible?

Thinking Through the Concepts

Answers to Multiple-Choice Questions
1. d **2.** c **3.** b **4.** e **5.** b

Applying the Concepts

Hints

1. Breath and body odors are generally caused by bacteria in the mouth and on the body surface. Consider how you might objectively assess the breath and body odors of experimental subjects who had and had not eaten chlorophyll tablets.
2. Carbon dioxide combines with RuBP to form PGA in the Calvin-Benson cycle (see Fig. 6-4). Water is split to provide electrons to photosystem II.
3. Consider the benefits that could be gained if photorespiration were prevented and plants could use C_3 photosynthesis even in hot, dry conditions.

CHAPTER 7

Answers to figure caption questions

Figure 7-5 In the absence of oxygen, ATP production halts. Oxygen is the final acceptor in the electron-transport chain, and if it is not present, electrons cannot proceed along the chain (they "stack up" in the chain) and production of ATP by chemiosmosis comes to a halt.
Figure 7-6 In oxygen-rich environments, both types of bacteria can survive, but aerobic bacteria prevail because their respiration is far more efficient (produces more ATP per glucose molecule). In oxygen-poor environments, however, aerobic bacteria are limited by the oxygen shortage, and anaerobes prevail despite their inefficiency.

Consider This

Hints: Consider the differences between an athlete's injecting an enzyme or drug and having it produced by genetically modified cells of the athlete's body. Consider the definition of "cheating" and "unfair competition."

Thinking Through the Concepts

Answers to Multiple-Choice Questions
1. a **2.** c **3.** d **4.** c **5.** a **6.** b

Applying the Concepts

Hints

1. If cellular respiration produces no ATP, cells will never receive the signal to halt respiration and will continuously consume glucose and other nutrient molecules. None of the body's tissues can function without ATP.
2. Yeasts respire when oxygen is present, and they ferment when oxygen is absent. Fermentation is much less efficient than aerobic respiration.
3. Consider the inputs required for the glucose activation stage of glycolysis.

CHAPTER 8

Answers to figure caption questions

Figure 8-2 It takes more energy to break apart a C-G base pair, because these are held together by three hydrogen bonds, compared with the two hydrogen bonds that bind A to T.
Figure 8-4 DNA polymerase can add nucleotides only to the free hydroxyl end of the DNA strand that it is synthesizing. The two strands of a DNA double helix are oriented in opposite directions, so one free hydroxyl end faces toward the replication fork and the other free hydroxyl end faces away from the fork. Therefore, DNA polymerase must move in opposite directions on the two strands.

Consider This

Hints: Not counting mutations, all people who share a common female ancestor have the same nucleotide sequence in their mitochondrial DNA. Most people lose track of ancestors more than three or four generations old. Therefore, it is possible that Billy the Kid, John Miller, and/or Brushy Bill Roberts, even if they were all different men, may have shared a common maternal ancestor (e.g., great-great grandmother) that nobody knows about. If so, their mitochondrial DNA would probably be identical. This scenario is probably quite unlikely, but it is not impossible.

Thinking Through the Concepts

Answers to Multiple-Choice Questions
1. d **2.** a **3.** d **4.** a **5.** d **6.** a

Applying the Concepts

Hints

1. In mammalian cells, DNA is found in the nucleus and directs the synthesis of mRNA and consequently proteins. Proteins are the principal structural molecule of cells, and protein enzymes perform most of the chemical reactions that are necessary to construct a cell and keep it alive. In general, DNA added to the *outside* of a mammalian cell doesn't get into the cell, much less get into the nucleus, become incorporated into a chromosome, and direct protein synthesis. Arbitrary DNA sequences, stored for months at room temperature in a shampoo, are even less likely to be able to direct the synthesis of useful proteins. Finally, the hair that actually sticks out of your scalp consists of dead cells, which aren't using DNA or synthesizing proteins anyway.
2. There isn't any single set of correct answers to these questions. Certainly, competition drives people to work long and hard to beat their competitors to an important discovery. On the other hand, competition, and the resulting secrecy, may cause people to duplicate each others' efforts and fail to

learn about, and avoid, other's mistakes. Collaboration avoids these pitfalls. Barriers to collaboration include the desire for fame and recognition, sometimes dislike for certain other researchers, and money (for example, patents for technologies or drugs may be very lucrative, and fewer collaborators means less sharing of revenues from patents).

3. The sequence of nucleotides on the "other" strand does not usually encode useful information. Using the "complementary English" analogy, the complementary phrase to "To be or not to be" would read "Gl xv li mlg gl xv," which is nonsense. Similarly, the protein synthesized from the instructions encoded in the nucleotide sequence of the "other" strand would probably have an arbitrary, useless amino acid composition. In all likelihood, DNA is double stranded because that structure provides the information needed to replicate DNA during cell division.

CHAPTER 9

Answers to figure caption questions

Figure 9-2 If the amino acid sequence of a useful protein is encoded in the sequence of nucleotides on one DNA strand, the sequence on the other strand is likely to encode an arbitrary, useless sequence of amino acids that would make a nonfunctional protein (see Applying the Concepts, question 3, in Chapter 8). Therefore, an RNA polymerase that transcribed both DNA strands simultaneously would waste nucleotides and energy by making one useful, and one useless, RNA strand.

Figure 9-3 A single mRNA molecule can direct the synthesis of only a small number of protein molecules. If the cell needs a lot of a certain protein (for example, a structural protein needed to make the plasma membrane, or a protein hormone that would be secreted into the bloodstream in large amounts), it would need many mRNA molecules.

Figure 9-5 The mRNA visible in **(i)** reads CGA AUC UAG UAA. Changing all Gs to Us would result in an mRNA with the sequence CUA AUC UAU UAA. As shown in the genetic code (Table 9-2), the codon CUA encodes the amino acid leucine rather than the arginine encoded by the original CGA codon. Further, the original stop codon UAG would be changed to UAU, which encodes tyrosine. Translation would not be terminated, but would add tyrosine to the growing protein chain. The next codon, UAA, happens to be a stop codon, so translation would stop there. The final protein would be met-val-his-arg-leu-ile-tyr instead of met-val-his-arg-arg-ile.

Consider This

Hints: There is no best answer to this question. A physician may withhold the chromosomal information from the patient for fear of causing psychological damage. But what happens if, later on, the patient somehow finds out anyway? Even if the physician is correct about the patient's psychological state, does a physician ever have the right to withhold medical information from a patient? On the other hand, what if the physician tells the patient that she is chromosomally XY, and she becomes severely depressed, or commits suicide? Who has the right and obligation to decide what to do?

Thinking Through the Concepts

Answers to Multiple-Choice Questions

1. c **2.** b **3.** c **4.** b **5.** c **6.** a

Applying the Concepts

Hints

1. The nucleotide sequences of RNA are encoded by genes. Except for rare mutations, gene sequences are not changed by the experiences of an organism, for example by learning. One could envision the possibility that, during millions of years of evolution, a species might evolve some very general types of "learning genes," such as a gene that would promote fear or anger responses. If such an organism experiences a frightening situation, it might selectively express "fear genes" and produce proteins that, for example, might make it more alert, more suspicious, or more likely to run away from strange or threatening objects. On the other hand, specific, novel learning, such as the letters of the genetic code, cannot possibly have been encoded in the genes during evolution, so specific "genetic code" RNAs cannot exist. One possible explanation for the *Planaria* results might be that the "normal" *Planaria* in the experiments did not receive enough RNA nucleotides in their diet, so feeding them trained worms would merely increase their supply of nucleotides, which in turn would allow them to make more RNA of all types and thereby synthesize more of whatever proteins are required for any learning to occur. To test this hypothesis, one could feed trained *Planaria* to one group of worms, untrained *Planaria* to a second group, RNA supplements to a third group, and no *Planaria* or RNA supplements to a fourth group, and then compare their learning abilities. Other experimental designs are also possible.
2. There are no single best answers to these questions. Factors to consider include: (1) whether the treatments (for example, steroids or growth hormones) are harmful to the person using them; (2) what one considers to be a biological or social "defect" (for example, is someone who stops growing at 4 feet "defective" or seriously handicapped? What about 5 feet?); (3) whether athletics is "fair" if some people but not others take treatments that enhance their abilities; and (4) whether athletics is "fair" if some people are genetically endowed with height, strength, or speed and others are not.

CHAPTER 10

Answers to figure caption questions

Figure 10-1 In this particular photo, the grove that dropped all its leaves is at a higher elevation than the grove with green leaves. Higher elevations are generally colder, so perhaps cold temperatures caused the grove at the higher elevation to drop its leaves earlier. There might also be differences in soil conditions or sunlight (if one grove faced south, for example, it would receive more direct sunlight). To test the hypothesis, one could dig up several trees from each grove and transplant them to the same place so they would have identical conditions. If leaf turn is (primarily) genetically determined, then the timing of when the leaves turn should still differ among trees from the three groves.

Figure 10-8 One daughter cell would have an extra copy of one chromosome and the other daughter cell would be missing one chromosome.

Figure 10-11 If one pair of homologues failed to separate at anaphase I, there would be two gametes with an extra copy of one chromosome and two gametes missing one chromosome. If one pair of sister chromatids failed to separate at anaphase II, there would be one gamete with an extra chromosome, one gamete missing a chromosome, and two normal gametes. If you can't figure out why these consequences would occur, try drawing meiosis with just one pair of chromosomes and see what happens. These failures of chromosomes or chromatids to separate during meiosis are called nondisjunction and can produce serious defects in the offspring (see Chapter 11).

Consider This

Hints: There is no "correct" answer. Factors to consider include whether you feel that a sperm fertilizing an egg is the only "right" way to produce a new human being and what you think about *in vitro* assisted reproduction in general. One should also remember that a cloned human being would *not* somehow "replace" the original person, or make the original person immortal. Further, the clone would be a baby. Because of differences in the environment (the surrogate mother, diet, and the enormous changes in society over the years), a clone might grow up to be very different from the original person, probably *much* less similar than identical twins that grow up in virtually identical environments.

Thinking Through the Concepts

Answers to Multiple-Choice Questions

1. c **2.** c **3.** b **4.** b **5.** d

Applying the Concepts

Hints

1. Generally, treatments that cure cancers, or at least greatly reduce their rate of growth, act by reducing the rate of cell division. Cell division is essential for many bodily functions, including hair growth, skin repair and replacement, and maintaining and repairing the linings of the lungs, stomach, and intestine. Any treatment that reduces cell division indiscriminantly will interfere with these functions and will therefore have "side effects." Any treatment that truly cures cancer without side effects must distinguish between cancerous cells and normal, dividing body cells. Cancer cells *are* different from normal cells, including often having different proteins on their cell surfaces, different responses to hormones, or the ability to promote the growth of blood vessels to nourish developing tumors. Experiments to test the specificity of cancer treatments might include adding the treatments to cultures of cancerous and normal (dividing) cells and seeing if there are differences, or treating animals and seeing if hair, skin, digestive tract, and so on functions are maintained.
2. Asexual reproduction produces offspring that are genetically identical to their parents. If the parents are well adapted to their environment, and if the environment does not change, then asexual reproduction should produce well-adapted offspring. Sexual reproduction produces offspring that differ from their parents. If the parents are well adapted to an environment that does not change, many, probably most, of their offspring will be less well adapted. (A few may be better adapted than their parents, but this outcome probably would be rare.) Therefore, asexual reproduction would generally produce a higher proportion of well-adapted offspring in a stable environment. In a changing environment, asexual reproduction would usually produce offspring that are not well adapted to the new conditions. Sexual reproduction, on the other hand, might produce some offspring that are fairly well adapted to new environmental conditions. In the long run, sexual reproduction has a better chance of producing at least a few offspring that are better adapted than their parents, even if the environment doesn't change very much.

CHAPTER 11

Answers to figure caption questions

Figure 11-4 The offspring phenotypes should be $^3/_4$ purple (450 plants) and $^1/_4$ white (150 plants). Finding 440 purple and 160 white-flowered plants is reasonably similar to the prediction. You could use statistical tests to determine if these numbers are significantly different from the prediction.

Figure 11-10 No. Pink-flowered snapdragons are always heterozygotes, and matings between heterozygotes will inevitably yield some homozygous (red- or white-flowered) offspring.

Consider This

Hints: There are no "correct" answers to these questions. In general, for cystic fibrosis or any other genetic disorder, factors to consider in these types of decisions include (1) the severity of the genetic disorder; (2) the current and likely future treatments available; and (3) your positions on abortion, adoption, and caring for a potentially seriously handicapped child.

Thinking Through the Concepts

Answers to Multiple-Choice Questions

1. c **2.** c **3.** e **4.** d **5.** a

Applying the Concepts

Hints

1. There is no correct answer to this question. If everyone ate a low-salt diet, then presumably no one would suffer from salt-induced hypertension. On the other hand, many people really like the taste of high-salt foods and are not harmed by salt. A few people even excrete too much salt and need to eat massive amounts to stay healthy. In a society in which a significant proportion of health care costs are paid by businesses and governments (and therefore customers and taxpayers), should the government produce guidelines, or even possibly enforce (for example, set legal limits on the amounts of salt in processed food)? Only if a large proportion of the population would benefit, if a "moderately" small proportion would benefit, or never? Can science provide solutions to such dilemmas, or only background information with which to inform political discussion?
2. These are issues with which science and society are likely to continue to struggle for some time. All scientific misconduct is reprehensible, but some has a much greater effect on society. For example, whether John Doe or Jane Roe receive royalties from a patent may not matter very much to the world at large. If, on the other hand, the patent produces a drug that people think will cure cancer, but instead makes it worse, this may cause misery and great financial loss to millions. Extensive, accurate record keeping can probably reduce the amount of scientific misconduct, but would also consume large amounts of time and money. How much oversight should government agencies require? In many cases, it's your money that's supporting the research—what do you think?

CHAPTER 12

Answers to figure caption questions

Figure 12-3 *Thermus aquaticus* lives in hot springs at temperatures as high as 80°C. Its DNA polymerase works faster at 72°C than at 50°C. Therefore, PCR works faster when the reaction temperature is raised to 72°C.

Figure 12-6 Each STR can be thought of as a gene, with multiple alleles. Because the different alleles have different numbers of tandem repeats, they have different sizes. Because they have different sizes, they move different distances during gel electrophoresis

As they do for other genes, each person normally has two copies of each STR gene, one on each of a pair of homologous chromosomes. A person may be homozygous (two copies of the same allele) or heterozygous (one copy of each of two alleles) for each STR. The bands on the gel represent individual alleles of an STR gene. Therefore, a single person can have one band (if homozygous) or two bands (if heterozygous). If a person is homozygous for an STR allele, then he has two copies of the same allele. The DNA from both (identical) alleles will run in the same place on the gel, and therefore that (single) band will have twice as much DNA as each of the two bands of DNA from a heterozygote. The more DNA, the brighter the band.

Consider This

Hints: There are no correct answers to these questions. Public funds support a great deal of scientific research, particularly in biology. The funds available are not unlimited. Who should decide which individual projects, or even which categories of projects, are worthy of funding—elected or appointed government officials, or scientists? Both groups may have their own biases, and neither can necessarily predict which research will "pay off" in improved public health, technological improvements, or environmental quality. Finally, should public funds be used for a project that you were fairly confident had no practical use (perhaps finding out about the colors of dinosaurs) but would be really fascinating?

Thinking Through the Concepts

Answers to Multiple-Choice Questions

1. e **2.** b **3.** c **4.** e **5.** d

Applying the Concepts

Hints

1. The vaccine banana and the vitamin C banana could be produced by the exact same genetic engineering techniques, with genes from the same organism. Therefore, biologically, they would be comparable. On the other hand, if a person feels that there is some danger, but not unlimited or catastrophic danger, to GM foods, then he or she may be more willing to risk the danger for a valuable vaccine than for vitamin C, which can be easily and cheaply obtained from other sources.
2. *Bacillus thuringiensis* can be used to control insects by spraying the bacteria on a field or by incorporating the Bt gene into crops. Some insects have developed resistance to the toxic effects of the Bt protein, usually because of spraying fields with *B. thuringiensis*. Therefore, it is reasonable to suppose that insects will also develop resistance to Bt crops (although there are methods that reduce the development of resistance, such as planting some nonresistant crops nearby or in alternate years). There is no single correct answer to the question of whether Bt crops should be planted anyway. One thing to consider is whether there is likely to be any significant, biological difference between Bt resistance that evolved because of spraying *B. thuringiensis* on a field or because of planting transgenic Bt crops.
3. There are no correct answers. Some people feel that all aspects of childbearing, including whether the children might inherit a genetic disorder, should not be tampered with, and therefore learning about their likelihood of passing on a genetic disorder is irrelevant or potentially depressing. Others feel that they would like to know if they carry a seriouly harmful allele. Their subsequent actions might be to choose not to reproduce, to prepare to care for an affected child, or to abort an affected embryo. In most societies, these decisions remain with the individual.

CHAPTER 13

Answers to figure caption questions

Figure 13-5 No. Mutations are the ultimate source of the phenotypic variation on which selection acts, and mutations occur in all organisms, including those that reproduce asexually.

Figure 13-6 No. Evolution can include changes in traits that are not revealed in morphology, such as physiological systems and metabolic pathways. More generally, evolution in the sense of changes in a species' gene pool is inevitable in all lineages; genetic evolution is not necessarily reflected in morphological change.

Figure 13-9 Analogous. Bird tails consist of feathers; dog tails consist of bone, muscle, and skin. Birds do have bones (fused to form the *pygostyle*) that are homologous to the dog's tail bones. (The tail feathers are attached to the pygostyle.)

Consider This

Hints: Try to describe a scenario in which natural selection causes the evolution of a structure that is nonessential at the time it first evolves. Is the scenario consistent with your understanding of how natural selection works?

Thinking Through the Concepts

Answers to Multiple-Choice Questions

1. e **2.** d **3.** e **4.** e **5.** a **6.** d

Applying the Concepts

Hints

1. Consider the likelihood that natural selection would cause the evolution and persistence of a large, complex structure with no function.
2. Review the description of science in Chapter 1. Think about whether the theory of evolution and the theory of special creation make predictions that can be tested with objective evidence.
3. Consider how the "best" traits might change when environmental conditions change.

CHAPTER 14

Answers to figure caption questions

Figure 14-1 The surviving colonies would be in different places on each treated plate, and there would be a different number of surviving colonies on each plate (because the antibiotic-caused mutations would arise unpredictably, depending on which bacteria happened to interact with the antibiotic such that mutations were caused). Another possibility would be for *all* of the colonies to survive (if the antibiotic *always* caused mutations in every colony).

Figure 14-2 Allele *A* should behave roughly as it does in the size 4 population, its frequency drifting to fixation or loss in almost all cases. But, because the population is a bit larger, the allele should, on average, take a longer time (greater number of generations) to reach fixation or loss. The longer period of drift should also allow for more reversals of direction (e.g., frequency drifting down, then up, then down again, etc.) than in the size 4 population.

Figure 14-3 Mutations inevitably and continually add variability to a population and, after the population becomes larger, the counteracting, diversity-reducing effects of drift decrease. The net result is an increase in genetic diversity.

Figure 14-5 Greater for males. A female's reproductive success is limited by her maximum litter size, but a male's potential reproductive success is limited only by the number of available females. When, as in bighorn sheep, males battle for access to females, the most successful males can impregnate many females, while unsuccessful males may not fertilize any females at all. Thus, the difference between the most and least successful males can be very large. In contrast, even the most successful female can have only one litter of offspring per breeding season, which is not that many more offspring than a female who fails to reproduce.

Consider This

Hints: Many species of bacteria and fungi have evolved chemical "weapons" that individuals secrete to help them compete for access to space and food.

Thinking Through the Concepts

Answers to Multiple-Choice Questions

1. a **2.** e **3.** d **4.** d **5.** b **6.** c

Applying the Concepts

Hints

1. After a speciation event, there are two or more descendant species. All of the descendant species continue to evolve.
2. Think about how one might determine whether two populations "interbreed freely under natural conditions."

3. Gene flow introduces new alleles to populations. Consider how this result might benefit an endangered population.

CHAPTER 15

Answers to figure caption questions

Figure 15-2 The presence of oxygen would prevent the accumulation of organic compounds by quickly oxidizing them or their precursors. All of the successful abiotic synthesis experiments used oxygen-free "atmospheres."
Figure 15-4 The bacterial sequence would be most similar to that of the plant mitochondrion, because (as the descendant of the immediate ancestor of the mitochondrion) the bacterium shares with the mitochondrion a more recent common ancestor than with the chloroplast or the nucleus.
Figure 15-8 No. The mudskipper merely demonstrates the plausibility of a hypothetical intermediate step in the proposed scenario for the origin of land-dwelling tetrapods. But the existence of a modern example similar in form to the hypothetical intermediate form does not provide information about the actual identity of that intermediate form.
Figure 15-10 On average, the rate at which new species arise has been larger than the rate at which species have gone extinct.
Figure E15-1 356.5 million years old. (3:1 ratio means that three-fourths of the original uranium-235 is left, so half of its half-life has passed.)

Consider This

Hints: Consider the likelihood that the fine details of a complex structure could evolve twice independently. Consider also whether this kind of evolutionary pattern is absolutely impossible. Think about how you would interpret a fossil bird that was older than the oldest dinosaur fossil.

Thinking Through the Concepts

Answers to Multiple-Choice Questions
1. a **2.** d **3.** c **4.** b, d **5.** e **6.** d

Applying the Concepts

Hints
1. Some meteors and comets contain organic molecules. Liquid water was probably present on Mars in the past and may currently be present elsewhere in our solar system.
2. We have inherited many physical characteristics from early hominids. Social behavior is controlled by the brain and nervous system.
3. Arguable possibilities include the origin of photosynthesis, the origin of aerobic respiration, the origin of multicellular bodies, and the invasion of land.

CHAPTER 16

Answers to figure caption questions

Figure 16-7 Protective structures such as endospores are most likely to evolve in environments in which protection is especially advantageous. Compared with other environments inhabited by bacteria, soils are especially vulnerable to drying out, which can be fatal to unprotected bacteria. Bacteria that could resist long dry periods would gain an evolutionary advantage.
Figure 16-8 The main advantage is efficiency. In binary fission, every individual produces new individuals. In sexual reproduction, only some individuals (e.g., females) produce offspring. So the average individual produces twice as many offspring by fission as it would by sexual reproduction.
Figure 16-19 Its filamentous shape helps the fungal body to penetrate and extend into its food sources and is also a shape that maximizes the ratio of surface area to interior volume (which maximizes the area available for absorbing nutrients). The extreme thinness of the filaments ensures that no cell is very far from the surface at which nutrients are absorbed.
Figure 16-24 Bryophytes lack lignin (which provides stiffness and support) and conducting vessels (which transport materials to distant parts of the body). Vessels and stiff stems seem to be required to achieve more than minimal height.
Figure 16-27

Type of pollination	Advantages	Disadvantages
Wind	Not dependent on presence of animals; no investment in nectar or showy flowers; pollen can disperse over large distances	Larger investment in pollen because most fails to reach an egg; higher chance of failure to fertilize any egg
Animal	Each pollen grain has much greater chance of reaching suitable egg	Depends on presence of animals; must invest in nectar and showy flowers

Both types of pollination persist in angiosperms because the cost-benefit balance, and therefore the most adaptive pollination system, differs depending on the ecological circumstances of a species.

Figure 16-28 Sponges are "primitive" only in the sense that their lineage arose early in the evolutionary history of animals and their body plan is comparatively simple. But early origin and simplicity do not determine effectiveness, and the sponge body plan and way of life are clearly suitable for excellent survival and reproduction in many habitats.
Figure 16-39 One advantage is that adults and juveniles occupy different habitats and therefore do not compete with one another for resources (the niche occupied by an individual over its lifetime is broadened).
Figure 16-41 Flight is a very expensive trait (consumes a lot of energy, requires many special structures). In circumstances in which the benefits of flight are low, such as in habitats without predators or in species whose size is very large, natural selection may favor individuals that forgo an investment in flight, and flightlessness can arise.

Consider This

Hints: One example of Erwin's assumptions is that 13% of beetle species are specialists that feed on only one plant species. You might assume that 5% (or some other proportion) are specialists.

Thinking Through the Concepts

Answers to Multiple-Choice Questions
1. a **2.** c **3.** e **4.** d **5.** c **6.** a

Applying the Concepts

Hints
1. Only a fraction of species has been discovered and identified. Many unknown species are small or rare, or both. It is difficult to document that no members of a species are still alive.
2. Naming and classification schemes reflect our understanding of evolutionary history. Much environmental policy depends on specifying species to be protected.
3. Consider likely differences between organisms in which different cells may be specialized for different functions and organisms in which all functions must be performed by a single cell.

CHAPTER 17

Answers to figure caption questions

Figure 17-3 Primary growth occurs at the tip of the primary root, at the tips of all lateral roots, at the terminal bud, at all lateral buds, and at the tips of lateral branches. Secondary growth will occur all along the margins of the primary and lateral roots and shoots.
Figure 17-8 Cortex and endodermis are ground tissues; xylem and phloem are vascular tissues. None of the cells shown are part of epidermal tissue (the root epidermis lies outside the area shown in the photo).
Figure 17-13 As can be seen in the figure, the phloem layer lies close to the outside of the stem, and is part of the bark. Removing a strip of bark entirely around the main stem (trunk) creates an unbridgeable break in the phloem vessels connecting the roots to the rest of the plant. Because phloem carries nutrients from their point of acquisition to their point of use, the plant cannot survive without a phloem connection between (for example) its sugar sources (e.g., leaves) and its sugar sinks (e.g., growing root tips).

Consider This

Hints: Try to think of some possible costs or disadvantages of producing anthocyanins and some circumstances in which the benefits of anthocyanins would be small or nonexistent.

Thinking Through the Concepts

Answers to Multiple-Choice Questions
1. a **2.** e **3.** a **4.** c **5.** e

Applying the Concepts

Hints
1. Your hypothesis should include a precise statement of the properties that differ in response to exposure to rock, classical, or no music. In designing your experiment, be sure to include appropriate controls and to specify the variables you will measure. Make sure that the measured variables accurately reflect your stated hypothesis.
2. Think about the advantages and disadvantages of deep roots versus shallow roots, assuming that the total amount of root tissue is equal in the two cases.
3. Consider the role of primary meristems in plant growth and the likely consequences if they are removed or destroyed.

CHAPTER 18

Answers to figure caption questions

Figure 18-4 Separate bloom times reduce the chances of inbreeding through self-fertilization. Individuals with lower levels of inbreeding generally have higher fitness (greater number of successful offspring), because inbred offspring are more likely to have low-fitness phenotypes (in large measure because inbreeding reduces heterozygosity and thus increases the odds that deleterious recessive alleles will be expressed).
Figure 18-6 Double fertilization ensures that no resources will be devoted to producing costly endosperm unless the egg in the same ovule is fertilized. (No endosperm develops unless the pollen tube reaches the ovule and both the egg and the polar nuclei are fertilized.)

Consider This

Hints: Ethylene helps initiate the formation of the abscission layer in plants that shed their leaves and also inhibits the growth of germinating seeds. These functions would be blocked if ethylene receptors were switched off.

Thinking Through the Concepts

Answers to Multiple-Choice Questions
1. d **2.** a **3.** d **4.** e **5.** c **6.** d

Applying the Concepts

Hints
1. Both apples and potatoes emit ethylene, but apples produce very high amounts and potatoes comparatively low amounts. Both potatoes and apples are

quite sensitive to ethylene. In designing an experiment, be sure to include appropriate controls.

2. Consider the consequences to a plant with a single, specialized pollinator species if the pollinator's population declines. Also consider the likelihood that members of the pollinator species will feed on other plant species.
3. Auxins stimulate growth. Consider what might happen if auxins were present in larger than normal concentration in plant tissues, or if auxins were present in tissues that normally lack them.

CHAPTER 19

Answers to figure caption questions

Figure 19-1 If the mammal's heat-sensing nerve endings were rendered nonfunctional, the nervous system would not send a signal to the hypothalamus when the body reached the set-point temperature. Consequently, the hypothalamus would send continuous "turn on" signals to the body's heat-generating and heat-retention mechanisms, which would continue to increase body temperature indefinitely (presumably until stored energy reserves were exhausted or the animal died). More generally, failure of the sensor in a negative feedback loop halts the flow of information to the control center, and the system remains in the action state it was in when the sensor failed.

Figure 19-6 In smaller bodies, the ratio of surface area to volume (and therefore to body mass) is greater than in larger bodies, so the rate of heat loss (per unit of body mass) is greater. Therefore, younger, smaller individuals require additional insulation to maintain body temperature. [Note: The relationship between volume and surface area is covered in Chapter 4.]

Figure 19-10 Skin is a functional unit composed of several tissue types (connective, epithelial, muscle, nerve), whereas blood is a more homogeneous functional unit composed of fairly similar cells and their surrounding matrix. One cannot identify two tissue types in blood.

Consider This

Hints: Consider the kinds of physiological responses to submersion in water that would result in a survival advantage to a mammal.

Thinking Through the Concepts

Answers to Multiple-Choice Questions

1. e **2.** d **3.** e **4.** d **5.** c **6.** e

Applying the Concepts

Hints

1. Consider the substances that are consumed or excreted by a body during vigorous activity. Think about the effects these changes would have on homeostasis. Also consider the effect that added flavors might have on consumers of sports drinks.
2. In the figure, note the locations of nerve cells and of epithelial cells (which can divide to give rise to new epithelial cells).
3. Make sure that your machine includes a positive feedback loop and a way to generate a signal that breaks the loop when the machine's task has been completed.

CHAPTER 20

Answers to figure caption questions

Figure 20-4 The right atrioventricular valve prevents backflow into the right atrium when the right ventricle contracts. If the valve were nonfunctional, less blood would be pumped from the right ventricle to the lungs (because some of the blood moved by the ventricle's contraction would flow back into the atrium), and flow from the vena cava to the right atrium would be impeded (because the atrium would be partly full of backwashed blood). Overall, each round of the cardiac cycle would circulate a smaller volume of blood.

Figure 20-5

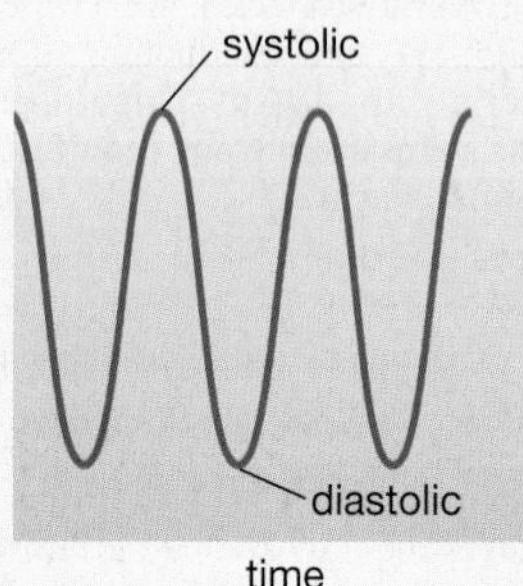

Figure 20-12 The formation of a fibrin mesh involves several different proteins, and a lack of any one of them can disrupt the formation of fibrin. The various proteins participate in several interlocking cascades, and this system of linear cascades means that one missing step can break the whole pathway (Fig. 6-7 illustrates this concept). The most common form of inherited hemophilia involves failure to produce Factor VIII, which is part of the pathway leading to thrombin production.

Figure 20-15 The principles of diffusion (see Chapter 3) tell us that molecules move down gradients from high to low concentration. Because the tissues through which capillaries pass generally consume oxygen, the concentration of oxygen is higher inside the capillaries than out, and oxygen diffuses along its concentration gradient from the inside of the capillary to the outside. Conversely, tissues generally produce carbon dioxide, so the concentration of carbon dioxide is higher outside the capillaries than in, and carbon dioxide diffuses along its concentration gradient from the outside of the capillary to the inside.

Figure 20-20 Convoluted nasal passages are an adaptation that helps conserve water and heat. (Convolution thus tends to be most extreme in species with especially great need to conserve water and/or heat, such as seals and kangaroo rats). Heat is conserved because body-warmed, exhaled air loses heat to surrounding tissue during its travel through the nasal passages. Water is conserved because the moisture in exhaled air condenses as the air cools during its travel through the nasal passages. Convoluted nasal passages may also help to warm and filter inhaled air.

Consider This

Hints: Try to predict the consequences of a cross-species disease transmission and weigh these against the potential benefits of xenotransplantation. Consider your view of how certain we ought to be about risks before proceeding with potentially rewarding new medical treatments.

Thinking Through the Concepts

Answers to Multiple-Choice Questions

1. c **2.** d **3.** d **4.** e **5.** b **6.** e

Applying the Concepts

Hints

1. Each red blood cell (RBC) can carry a certain number of oxygen molecules, so increasing the number of RBCs in a person's circulatory system increases the total amount of oxygen in the blood.
2. Nicotine is addictive, so cigarettes with less nicotine are less addictive. Most of the harmful effects of smoking arise from substances other than nicotine in tobacco smoke.
3. Consider the feedback mechanisms that determine breathing rate and think about the degree of conscious control that a person has over those mechanisms.

CHAPTER 21

Answers to figure caption questions

Figure 21-3 Yes. The label shows a product relatively low in calories, with almost no fat and no cholesterol. It has lots of fiber and a substantial percent of the recommended vitamins.

Figure 21-6 The muscular walls of the stomach generate the churning action that is an important part of the digestive process. An expandable stomach makes it possible to conserve space in the body cavity while still allowing for occasional consumption of large amounts of food. The expanded stomach serves to store such food until it can be processed by the rest of the digestive system.

Figure 21-9 A likely outcome is that the necessary large absorptive surface area would have been achieved by some other sort of adaptation; perhaps a much longer intestine and a correspondingly large abdominal cavity to hold it.

Figure 21-12 Excretion of nitrogenous wastes as ammonia is restricted to animals that live in water. Despite the toxicity of ammonia, animals that live surrounded by water can excrete it more or less continuously and thus escape its toxic effects (and also avoid using energy to convert ammonia to urea or uric acid). Ammonia excretion is especially common in freshwater animals, which must expel a lot of water to maintain osmotic balance.

Figure 21-13 Because the kidneys are responsible for fine-tuning blood composition and maintaining homeostasis, it is crucial that blood be filtered continuously. Each drop of blood in your body passes through a kidney about 350 times a day, giving the kidney ample opportunity to fine-tune the composition of the blood and help maintain homeostasis.

Figure 21-17 Consumption of alcohol tends to cause dehydration. Drinking large amounts of liquids (including alcoholic ones) decreases the osmolarity of blood below its set point, halting ADH release and stimulating urine flow and water loss. But with alcohol consumption, the effect of the resulting negative feedback (i.e., increased blood osmolarity) is dampened, because the effector (ADH release) is inhibited by alcohol. Thus, water loss continues even after blood osmolarity rises above its set point.

Consider This

Hints: A drug such as Xenical, which works entirely within the digestive system, preventing fats from being digested and entering the body cells, would be likely to have far fewer side effects. Neurotransmitters each have multiple effects within the nervous system, so a drug such as Meridia, which increases three different neurotransmitters, would be likely to have many side effects.

Thinking Through the Concepts

Answers to Multiple-Choice Questions

1. e **2.** d **3.** a **4.** d **5.** d **6.** b

Applying the Concepts

Hints

1. Homeostasis keeps the water content of the fluid that bathes our cells within very tight limits. Very small drops in blood water content trigger the urge to drink long before any danger occurs. Many animals roam far from reliable sources of water. A species that did not experience an urge to drink until it was in danger from dehydration would be unlikely to survive. Do you think that, to avoid starvation, you should eat before you feel hungry?
2. Most animals, including humans, evolved under conditions in which food was limited. Food preferences evolved because they helped animals survive to reproduce. What types of food are we and other animals likely to prefer?
3. Magnesium salts remaining in the large intestine cause water to move in by osmosis. The entry of

water stretches the large intestine and moistens the feces. Stretching the large intestine promotes peristalis, and moistening the feces makes them easier to move through the intestine and to expel.

4. The urine test takes advantage of the role of the kidneys in eliminating toxins and drugs from the body. If a person has taken drugs recently, traces of the drug itself or of chemicals that are produced when the drug is metabolized can be detected in urine.

CHAPTER 22

Answers to figure caption questions

Figure 22-4 Histamine causes inflammation by making capillary walls leaky and relaxing the smooth muscles surrounding arterioles, leading to increased blood flow. More blood flowing through leaky capillaries forces fluid to seep from capillaries into tissues surrounding the wound, which become swollen (and also warm and reddened).

Figure 22-5 The combination of constant and variable components allows an antibody to accomplish both functions that are unique to a particular antibody molecule (variable regions recognize and bind to a specific antigen) and functions that are common to all antibodies in a given class (constant regions carry out the appropriate effector response to attack the invader).

Figure 22-9 Cytotoxic T cells are activated by antigens on the target cell's surface, so one might expect that the membranes of cancer cells contain distinctive, abnormal molecules. This is in fact the case; cancer cell membranes bear antigens that are specific to cancerous cells.

Consider This

Hints: The extra experiment was to test ticks feeding on vaccinated mice for Lyme disease bacteria. It is rare for a vaccine to cause the death of a microbe outside of the vaccinated host's body. In the case of Lyme disease, however, the ticks were not merely injecting material into their hosts. They were also sucking blood (laden with antibodies to the Lyme bacterium) out. Because antibody-rich blood was entering the infected ticks, it was reasonable for scientists to hypothesize that the bacteria might be killed before they ever left the tick.

Thinking Through the Concepts

Answers to Multiple-Choice Questions

1. e **2.** c **3.** b **4.** a **5.** e **6.** e

Applying the Concepts

Hints

1. Before the advent of modern science, people relied on repeated observations of events that seemed to be related to one another. People who recovered from smallpox would never get it again, no matter how often they were exposed. People such as dairymaids who contracted cowpox never got smallpox, no matter how often they were exposed. Modern scientific inquiry also begins with observations, sometimes chance, sometimes resulting from careful experiments. In both prescientific and modern times, observations led to "experiments" such as deliberately exposing people to smallpox, and later to cowpox. Unlike modern experiments, these early experiments were not done on animals first, and they lacked specific controls (although one could consider that the people who did not expose themselves to mild cases of smallpox or cowpox served as an enormous control group). Of course, knowing exactly what causes infectious diseases allows modern scientists to work directly with the microbes (or even with their specific proteins) rather than relying on pus from an infected victim.
2. This is a thought question with no specific correct answer.
3. By blocking the activity of the chemical that stimulates helper T cells, cyclosporine will reduce the production of both B cells and cytotoxic T cells (see Figure 22-10). This reduction would suppress the immune response and reduce the likelihood of rejection of transplanted organs. However, it would also reduce the ability of the immune system to fight real threats, such as cancerous cells.

CHAPTER 23

Answers to figure caption questions

Figure 23-5 The hormone could only be obtained in tiny quantities from each cadaver. The cadaver had to be very fresh, and permission had to be obtained. Consequently, human growth hormone was enormously expensive and out of the reach of many who would have benefited from it. In addition, in rare instances, diseases could be transmitted from the deceased donor to the recipient.

Figure 23-6 Hunger in the child stimulates the suckling reflex, which triggers oxytocin release and milk letdown. As milk flows into the baby's stomach, hunger ceases, and suckling stops. Cessation of suckling in turn stops the release of oxytocin.

Consider This

Hints: (a) Blind studies prevent researchers from biasing the results on the basis of what they think should be happening in the different groups. (b) No, the study supported their hypothesis, but left room for other interpretations. Although rural men were exposed to higher levels of pesticides, and these were correlated with lower sperm quality, some other factor that wasn't tested could have produced the lowered sperm quality. (c) To conclude that the pesticides produced the lowered sperm count would require controlled studies in which all other variables were eliminated and the experimental groups had exactly the same level and duration of exposure to the same pesticides, while controls had no pesticide exposure at all. (d) For both ethical and logistical reasons, these studies could only be done on animals (such as laboratory rats) so that pesticides could be administered deliberately and all other variables could be controlled.

Thinking Through the Concepts

Answers to Multiple-Choice Questions

1. a **2.** e **3.** a **4.** b **5.** c

Applying the Concepts

Hints

1. "No" to all. For reading claims or purchasing supplements on the Internet or elsewhere, the rule is "Buyer beware!"
2. Note that growth hormone has potent and far-reaching effects throughout the body, which are far from fully understood. Think about how this situation compares with athletes' abusing steroids and other hormones.
3. These substances all have considerable value to society and to the companies that produce them, so it would not be practical to immediately ban them outright. A mandated phase-out, coupled with government-funded research into less harmful alternatives, would make substitutes both feasible and acceptable. Nontoxic substances that serve the same function but whose chemical structure does not interact with hormone receptors would be an ideal substitute. Alternatively, perhaps their chemical structure could be altered to cause them to break down quickly into harmless substances when dissolved in water. All new substances that might find their way into bodies of water should be required to pass rigorous tests to be sure they do not mimic hormones or do other environmental damage.

CHAPTER 24

Answers to figure caption questions

Figure 24-2 The IPSP would counteract the EPSP, and the neuron would not reach threshold and would not produce an action potential.

Figure 24-3 The evidence suggests that the toxin must act on the postsynaptic part of the synapse by blocking neurotransmitter receptors. If the toxin acted by blocking the release of neurotransmitters from the presynaptic neuron, then adding neurotransmitter to the synapse would have generated a postsynaptic potential.

Figure 24-4 Sensitive areas have a higher density of sensory neurons.

Figure 24-7 A severed or damaged spinal cord prevents the sensation of pain from being relayed to the brain but does not disrupt the reflex circuit, which requires only transmission within a small portion of the spinal cord.

Figure 24-14 For nearsightedness, the cornea is flattened. This reduces the convergence of light rays before they reach the lens. For farsightedness, the edges of the cornea are reduced, leaving the cornea with a more rounded, concave shape that converges incoming light rays.

Figure 24-19 Although large external ears would no doubt enhance porpoises' ability to detect ultrasound, the drag exerted in water by the structures would be costly. Strong selection for streamlining has forced the evolution of other means to enhance detection of ultrasound in animals such as whales and porpoises.

Consider This

Hints: There is no correct answer to this question, but as neuroimaging techniques become increasingly sophisticated and widely used, it is a question that society will need to address.

Thinking Through the Concepts

Answers to Multiple-Choice Questions

1. c **2.** c **3.** d **4.** e **5.** d **6.** c

Applying the Concepts

Hints

1. It may help to observe the PET scan in Figure 24-10b. Also recall that, at rest, the brain accounts for about 20% of your total energy expenditure. Think about this from an evolutionary standpoint: The human brain has increased in size over evolutionary time. Would natural selection favor such an increase if most of the brain went unused? The fact that you are using all of your brain does not mean that you can't continue to learn new facts and skills throughout life.
2. There is no correct answer to this question.
3. The "Links to Life" section describing synesthesia may help you to evaluate the implications.

CHAPTER 25

Answers to figure caption questions

Figure E25-1 Injecting the sperm avoids the complex chemical interaction between the sperm and egg that accompanies normal fertilization. It is difficult to predict the outcome of the interaction between a sperm and the egg's protective layers, or to know if a given sperm cell will be able to successfully penetrate an egg. So injecting the sperm will improve the odds of successful fertilization.

Figure 25-1 Mitosis. (In animals, meiosis occurs only as part of sexual reproduction.)

Figure 25-4 Courtship rituals help ensure that animals mate only with individuals of their own species.

Figure 25-10 The acrosome contains enzymes that dissolve protective layers around the egg. The rest of the head contains the haploid nucleus with genetic material. The midpiece contains mitochondria that provide energy to power swimming. The tail is a flagellum that whips back and forth to provide propulsion through the female reproducive tract to the egg.

Figure 25-17 Each sperm releases enzymes from its acrosome that help break down the barriers surrounding the egg, and the amount of enzyme released by many sperm increases the chances that one sperm will get through.

Figure 25-24 The placenta has evolved to provide easy diffusion of nutrients and oxygen to the fetus from the mother and wastes from the fetus to the mother. Because alcohol and many other drugs readily diffuse across cell membranes, the placenta could not exclude these substances without also providing a barrier to the diffusion of wastes and nutrients.

Figure 25-26 Estrogen and progesterone.

Consider This

Hints: Helping her find the information she needs might be more effective than giving instructions, but if she continues, her smoking and drinking could cause life-long problems for her child.

Thinking Through the Concepts

Answers to Multiple-Choice Questions

1. a **2.** d **3.** b **4.** e **5.** a **6.** b

Applying the Concepts

Hints

1. The Internet is an excellent source of information about the cloning process. Read more about cloning in Chapter 10 of this text.
2. In evaluating the popularity and features, consider side effects and convenience and ease of use.
3. There is no correct answer.

CHAPTER 26

Answers to figure caption questions

Figure 26-1 One possibility is that the variation necessary for selection has never arisen. (If no members of the species by chance gain the ability to discriminate between their own chicks and cuckoo chicks, then selection has no opportunity to favor the novel behavior.) Another possibility is that the cost of the behavior is relatively low. (If parasitism by cuckoos is rare, a parent that feeds any begging chick in its nest is, on average, much more likely to benefit than suffer.)

Figure 26-6 Because the crossbreeding experiment shows that differences in orientation direction between the two populations stem from genetic differences, birds from the western population should orient in a southwesterly direction, regardless of the environment in which they are raised.

Figure 26-19 If females do indeed gain any benefits from their mates, those benefits would necessarily be genetic (as the male provides no material benefits). Male fitness may vary, and to the extent that this fitness can be passed to offspring, females would benefit by choosing the fittest males. If a male's fitness is reflected in his ability to build and decorate a bower, females would benefit by preferring to mate with males that build especially good bowers.

Figure 26-22 Canids forage mainly by smell; apes are mostly visual foragers. Modes of sexual signaling are affected not only by the nature of the information to be encoded but also by sensory biases and sensitivities of the species involved. Communication systems may evolve to take advantage of traits that originally evolved for other functions.

Consider This

Hints: Consider how comparisons of the criteria for beauty in different cultures might be used to distinguish innate from learned preferences. Think about how to ensure that different cultures represent independent data points (i.e., have not influenced one another).

Thinking Through the Concepts

Answers to Multiple-Choice Questions

1. c **2.** d **3.** d **4.** e **5.** d **6.** a

Applying the Concepts

Hints

1. No human sex-attractant pheromones have been identified, though they have been found in other mammals such as rodents. Consider whether the company's research included appropriate controls and whether the treatment (the alleged pheromone) was assigned to subjects at random.
2. Consider the type of lure(s) most likely to attract mosquitoes and moths. Incorporate practical lures into your traps.
3. Wolves, lions, and chickens are species with well-known dominance hierarchies. Examples abound of human dominance behavior and territorial behavior.

CHAPTER 27

Answers to figure caption questions

Figure 27-5 Many variables interact in complex ways to produce real population cycles. Weather, for example, affects the lemmings' food supply and thus their ability to survive and reproduce. Predation of lemmings is influenced by both the number of predators and the availability of other prey, which in turn is influenced by multiple environmental variables.

Figure 27-8 Many scenarios are possible based on different assumptions about increases in technology, changes in birth and death rates, and the resiliency of the ecosystems that sustain us. (This is a subjective thought question.)

Figure 27-10 High birth rates in developing countries are sustained by cultural expectations, lack of health education, and lack of access to contraception. Lower birth rates in developed countries are encouraged by easy access to contraception, the relatively high cost of raising children, and more varied career opportunities for women. (Students should be able to expand on, or add to, these factors.)

Figure 27-11 U.S. population growth is in the rapidly rising "exponential" phase of the S-curve. Stabilization will require some combination of reduction in immigration rates and birth rates. An increase in death rates is less likely, but cannot be ruled out entirely in any future scenario. (Students should be encouraged to speculate about the timing and the reason for various time frames.)

Consider This

Hints: Consider that emigration is a typical "safety valve" for overcrowded populations, and consider factors that might restrict this option for Easter Island's humans.

Thinking Through the Concepts

Answers to Multiple-Choice Questions

1. a **2.** c **3.** d **4.** e **5.** b **6.** d

Applying the Concepts

Hints

1. Consider the likely outcome of a high death rate in a small portion of a very large population of mobile organisms. In designing your experiment, be sure to include appropriate controls and to specify how you will measure changes in population size.
2. Consider the outcome expected when death rates decline but birth rates do not. Consider the social and economic factors that affect birth rates in a country. Consider possible explanations of the link between affluence and birth rate.
3. Consider the ways in which human activities and technology can change the carrying capacity of an environment. Consider the effect on carrying capacity of the disappearance of nonhuman populations.

CHAPTER 28

Answers to figure caption questions

Figure 28-1 Exotic species, because they did not evolve in the habitat to which they were introduced, may occupy a niche that is nearly identical to that of a native species (for example, zebra mussels compete with other freshwater mussel and clam species). A successful exotic species may have adaptations (for example a higher growth rate or reproductive rate) that allow it to outcompete the native species. Further, all native species are likely to have local predators, while the introduced (exotic) species may not. The absence of predators would also help the exotic species outcompete a native species if they occupy very similar niches.

Figure 28-3 This is an open-ended question; a few examples follow: Keen eyesight of hawk and camouflage color of mouse. Forward-pointing eyes of predators allowing binocular vision and good localization of prey, and side-situated eyes of many prey that afford nearly 360-degree vision, thus allowing them to spot predators from almost any angle. Grasses have evolved tough silicon embedded in their leaves, and grazing herbivores have teeth that grow continuously throughout life so that they are not completely worn down by the abrasive grasses.

Figure 28-11 Whereas tasty prey has often evolved to blend in with its surroundings, poisonous prey (such as the monarch caterpillar that stores milkweed toxin in its body) frequently advertises its presence with bright, warning colors. These bright colors make it easy for predators to learn to actively avoid animals bearing them.

Consider This

Hints: Populations in healthy communities are generally stable or fluctuate within a narrow range. Think of the most common and likely causes of steady decline in the population of a species.

Thinking Through the Concepts

Answers to Multiple-Choice Questions

1. d **2.** c **3.** a **4.** c **5.** a **6.** c

Applying the Concepts

Hints

1. The concept of benefit to an ecosystem is difficult to define objectively. Biologists have documented many examples of exotic invaders harming particular native species. The effects of an exotic invader might be studied by comparing similar ecosystems with and without the invader present.
2. Think about possible evolutionary outcomes of competitive exclusion, such as character displacement.
3. Consider how you might manipulate field plots so that some contain kangaroo rats and others do not.

CHAPTER 29

Answers to figure caption questions

Figure 29-3 On land, high productivity is supported by optimal temperatures for plant growth, a long growing season, and plenty of moisture. In aquatic ecosystems, high productivity is supported by an abundance of nutrients and adequate light.

Figure 29-5 The primary consumers shown are a mouse, a ground squirrel, a bison, a prairie dog, a quail, a pheasant, a pronghorn, a rabbit, and a grasshopper.

Figure 29-6 Whenever energy is utilized, the second law of thermodynamics applies: When energy is converted from one form to another, the amount of useful energy decreases. Much of this energy is lost as heat. Because relatively little energy is captured in chemical bonds and thus available to maintain the tremendously organized state of life, animals must consume a large number of Calories from the trophic level below them to obtain the useful energy they need.

Figure 29-9 Humanity's need to grow crops to feed our growing population has led to the trapping of nitrogen using industrial processes; it is then used as fertilizer. Additionally, large-scale livestock feedlots generate enormous amounts of nitrogenous waste. Nitrogen oxides are also generated when fossil fuels are burned in power plants, vehicles, and factories, and when forests are burned. Consequences include the overfertilization of lakes, rivers, and portions of the ocean, which causes dead zones as decomposers deplete oxygen. Another important consequence is acid deposition, in which nitrogen oxides formed by combustion produce nitric acid in the atmosphere; this acid is then deposited on land.

Figure 29-15 In a greenhouse, solar energy enters as light that hits surfaces within the greenhouse and is converted to heat, which is trapped inside by the glass. In much the same way, the greenhouse effect occurs when greenhouse gases (including CO_2 and methane) absorb sunlight and convert it to heat, holding this heat in the atmosphere somewhat like the glass in a greenhouse.

Consider This

Hints: Compared with natural ecosystems, Biosphere 2 is much simpler and smaller. Also, it is unique, so experiments in it cannot be replicated. But Biosphere 2 is also especially well suited for controlled ecological experiments that would be impossible to do outdoors.

Thinking Through the Concepts

Answers to Multiple-Choice Questions

1. c **2.** b **3.** d **4.** e **5.** d

Applying the Concepts

Hints

1. Correlation between two trends does not prove that one caused the other. Consider methods and approaches that might be used to supplement data on recent worldwide trends in temperature and carbon dioxide levels. Possibilities include controlled experiments (what kinds?) and temperature and CO_2 data from the more distant past when the two were not rising in tandem (how might one get such data?).
2. Consider the stability of different substances and their relative solubility in fluids that are present in bodies. Think about energy pyramids and how they translate into biomass pyramids.
3. Consider that additional carbon dioxide and other pollutants are added to the environment by each additional person. Consider the comparative impact of births in different parts of the world.

CHAPTER 30

Answers to figure caption questions

Figure 30-1 The equator-to-pole temperature gradient would still remain, but there would be no change in daylength or seasons throughout the year.

Figure 30-9 Nutrients are indeed abundant in tropical rain forests, but they are not stored in the soil. The optimal temperature and moisture of tropical climates allow plants to make such efficient use of nutrients that nearly all nutrients are stored in plant bodies, and to a lesser extent, in the bodies of the animals they support. These growing conditions support such a vast array of plants that these, in turn, provide a wealth of habitats and food sources for diverse animals.

Figure 30-16 Tallgrass prairie is vulnerable to two major encroachments. First, if humans suppress natural wildfires, the adequate rainfall amounts allow forests to take over. Second, these biomes provide the world's most fertile soils and also have excellent growing conditions for farming, which has now displaced native vegetation.

Figure 30-26 Nearshore ecosystems have an abundance of the two limiting factors for life in water: nutrients and light to support photosynthetic organisms. Both upwelling from ocean depths and runoff from the land can provide nutrients, depending on the location of the ecosystem. The shallow water in these areas allows adequate light to penetrate to support rooted plants and anchored algae, which in turn provide food and shelter for a wealth of marine life.

Figure 30-27 Bleaching refers to the loss of symbiotic algae that normally inhabit the corals' tissues, providing them with energy captured during photosynthesis and calcium carbonate used in coral skeletons. Loss of these resources can eventually kill the coral. Bleaching is a common response to water that is excessively warm, and thus global warming may contribute to the demise of coral reefs.

Consider This

Hints: It is inevitable that the portion of Earth's surface devoted to agriculture will continue to grow, even if efforts to preserve some undisturbed habitat are successful.

Thinking Through the Concepts

Answers to Multiple-Choice Questions

1. a **2.** a **3.** c **4.** b **5.** e **6.** c

Applying the Concepts

Hints

1. Think of ways in which to identify and compare otherwise similar populations of frog and people in places that differ in the extent of ozone layer decay.
2. Consider which tropical biomes would be favored (or threatened) by a drier climate, and which by a moister climate. Think about how similar changes would (or would not) affect the distribution of temperate biomes.
3. Think about how the rate of decomposition in a biome affects soil fertility. Soils that are high in organic matter are often more fertile than soils with less organic matter.

Photo Credits

Chapter 1
Opener 1-1 NASA Headquarters
Opener 1-2 Dr. Yorgos Nikas/SPL/Photo Researchers, Inc.
1-2 Teresa and Gerald Audesirk
1-5 Corbis/Bettmann
1-7 Dr. Jeremy Burgess/Science Photo Library/Photo Researchers, Inc.
1-8 John Durham/Science Photo Library/Photo Researchers, Inc.
1-9a Andrew Syred/Science Photo Library/Photo Researchers, Inc.
1-9b Craig Tuttle/Corbis/Stock Market
1-9c Kim Taylor/Bruce Coleman Inc.
1-10 Johnny Johnson/DRK Photo
1-11 Kim Taylor/Bruce Coleman Inc.
1-12 Lawrence Livermore National Laboratory/Photo Researchers, Inc.
E1-2 Nigel J. Dennis/Photo Researchers, Inc.
E1-4 Luiz C. Marigo/Peter Arnold, Inc.

Unit 1
Opener 1 Manfred Kage/Peter Arnold, Inc.

Chapter 2
Opener 2-1 Roberto Brosan / Time Life Pictures / Getty Images, Inc.
Opener 2-2 Dennis MacDonald/PhotoEdit
2-7a Stephen Dalton/Photo Researchers, Inc.
2-7b Teresa and Gerald Audesirk
2-11uL Larry Ulrich Stock Photography, Inc./DRK Photo
2-11uM Dr. Jeremy Burgess/Science Photo Library/Photo Researchers, Inc.
2-11uR Biophoto Associates/Photo Researchers, Inc.
2-12a Jean-Michel Labat/Auscape International Pty. Ltd.
2-12b Donald Specker/Animals Animals/Earth Scenes
2-15a Robert Pearcy/Animals Animals/Earth Scenes
2-15b Jeff Foott/DRK Photo
2-15c Nuridsany et Perennou/Photo Researchers, Inc.
E2-1 Calvin Larsen/Photo Researchers, Inc.

Chapter 3
Opener 3-1 Oliver Meckes & Nicole Ottawa/Photo Researchers, Inc.
Opener 3-2 Don Fawcett/Photo Researchers, Inc.
3-5a-c Joseph Kurantsin-Mills, The George Washington University Medical Center
3-8b L.A. Hufnagel, Ultrastructural Aspects of Chemoreception in Ciliated Protists (Ciliophora), *Journal of Electron Microscopy Technique*, 1991. Photomicrograph by Jurgen Bohmer and Linda Hufnagel, University of Rhode Island.
3-9 Stephen J. Krasemann/DRK Photo

Chapter 4
Opener 4-1 David Young-Wolff/PhotoEdit
Opener 4-2 Dr. John P. Clare (www.caudata.org)
4-4b E. Guth, T. Hashimoto, and S.F. Conti
4-5 Carolina Biological Supply Company/Phototake NYC
4-6 Omikron/Science Source/Photo Researchers, Inc.
4-7UR, WR Barry F. King/Biological Photo Service
4-8M Don W. Fawcett/Photo Researchers, Inc.
4-9b Molecular Probes, Inc.
4-10aR Ellen R. Dirksen/Visuals Unlimited
4-10bR Yorgos Nikas/Getty Images Inc. - Stone Allstock
E4-1a1 Biophoto Associates/Photo Researchers, Inc.
E4-1a2 Cecil Fox/Science Source/Photo Researchers, Inc.
E4-1b1, b2 Brian J. Ford
E4-1c Jean-Claude Revy/Phototake NYC
E4-2a M.I. Walker/Photo Researchers, Inc.
E4-2b David M. Phillips/Visuals Unlimited
E4-2c Manfred Kage/Peter Arnold, Inc.
E4-2d National Library of Medicine

Chapter 5
Opener 5-1 Mike Powell/Getty Images, Inc. - Allsport Photography
Opener 5-2 Lockyer, Romilly/Getty Images Inc. - Image Bank
5-1 Photograph by Dr. Harold E. Edgerton. © Harold & Esther Edgerton Foundation, 204, courtesy of Palm Press, Inc.

Chapter 6
Opener 6-1 Joe Tucciarone
Opener 6-2 ©BIOS/Peter Arnold, Inc.
6-1a Ken W. Davis/Tom Stack & Associates, Inc.
E6-1 Fred Bruemmer/Peter Arnold, Inc.

Chapter 7
Opener 7-1 MIke Powell/Getty Images, Inc. - Allsport Photography
Opener 7-2 Agence Zoom/Stringer/Getty Images, Inc.
7-6aL-aR Teresa Audesirk
7-6b AP/Wide World Photos

Unit 2
Opener 2 Ron Kimball/Ron Kimball Photography

Chapter 8
Opener 8-1 Corbis/Bettmann
Opener 8-2 Michael Freeman/Phototake NYC
8-1a Tod Pitt/Image Direct/Getty Images/ImageDirect
8-1b Kevin Winter/Getty Images/ImageDirect
8-2c Michael Freeman/Phototake NYC
E8-1 A. Barrington Brown/Photo Researchers, Inc.

Chapter 9
Opener 9-1 Courtesy of Nautilus, Vancouver, Washington
E9-1 Howard W. Jones, Jr., M.D., Eastern Virginia Medical School

Chapter 10
Opener 10-1 AP Wide World Photos
Opener 10-2 Stuart Conway Photography
10-1a Michael Abbey/Photo Researchers, Inc.
10-1b John Durham/Science Photo Library/Photo Researchers, Inc.
10-1c Carolina Biological Supply Company/Phototake NYC
10-1d Teresa and Gerald Audesirk
10-5 Biophoto Associates/Photo Researchers, Inc.
10-6 CNRI/Science Photo Library/Photo Researchers, Inc.
10-7a-f Dr. Andrew S. Bajer, University of Oregon
10-8a-h M. Abbey/Photo Researchers, Inc.
10-9b T.E. Schroeder/Biological Photo Service
E10-1 Photograph courtesy of The Roslin Institute.

Chapter 11
Opener 11-1 Ronald Modra / Sports Illustrated
Opener 11-2 Corbis/Bettmann
11-2 Archiv/Photo Researchers, Inc.
11-8 Biophoto Associates/Photo Researchers, Inc.
11-11 Jane Burton/Bruce Coleman Inc.
11-13a Dennis Kunkel/Phototake NYC
11-13b Walter Reinhart/Phototake NYC
11-14a Hart-Davis/Science Photo Library/Photo Researchers, Inc.
11-15a CNRI/Science Photo Library/Photo Researchers, Inc.
11-15b Lawrence Migdale/Pix
E11-1 Abraham Menashe Inc.

Chapter 12
Opener 12-1 AP Wide World Photos
Opener 12-2 AP Wide World Photos
12-6 Courtesy of Orchid Cellmark, Inc., Germantown, Maryland.
12-8 Pharmacia Corporation
E12-1 Janet Chapple, author of *Yellowstone Treasures: The Traveler's Companion to the National Park.*
E12-2 Choy Hew, National University of Singapore, and Garth Fletcher, Memorial University of Newfoundland, St. John's, Newfoundland.

Unit 3
Opener 3 Francois Gohier/Photo Researchers, Inc.

Chapter 13
Opener 13-1 Johnny Jonson/DRK Photo
Opener 13-2 Carl Buell
13-2 Jim Steinberg/Photo Researchers, Inc.
13-3a © John Cancalosi / DRK PHOTO
13-3b David M. Dennis/Tom Stack & Associates, Inc.
13-3c Chip Clark
13-9a, b Stephen Dalton/Photo Researchers, Inc.
13-9c Douglas T. Cheeseman, Jr./Peter Arnold, Inc.
13-9d Gerard Lacz/Peter Arnold, Inc.
13-10a-c Photo Lennart Nilsson/Albert Bonniers Forlag AB
13-11a Stephen J. Krasemann/DRK Photo
13-11b Timothy O'Keefe/Tom Stack & Associates, Inc.
E13-1 © Bettmann / Corbis/Corbis/Bettmann
E13-2 Frans Lanting/Photo Researchers, Inc.

Chapter 14
Opener 14-1 RICHARD PRICE/Getty Images, Inc. - Taxi
Opener 14-2 David Scharf/Peter Arnold, Inc.
14-3b Luiz C. Marigo/Peter Arnold, Inc.
14-3c Y.R. Tymstra/Valan Photos
14-4a M.P.L. Fogden/Bruce Coleman Inc.
14-4b Tim Davis/Photo Researchers, Inc.
14-5 W. Perry Conway/Tom Stack & Associates, Inc.
14-6 D. Cavagnaro/DRK Photo
14-7a Wayne Lankinen/Valan Photos
14-7b Edgar T. Jones/Bruce Coleman Inc.
14-8a,b Pat & Tom Leeson/Photo Researchers, Inc.
14-9a Mark Smith/Photo Researchers, Inc.
14-9b Thomas Kitchin/Tom Stack & Associates, Inc.
14-9c Mark Smith/Photo Researchers, Inc.
14-10 Tim Laman/NGS Image Collection

14-11 Joy Spurr/Bruce Coleman Inc.
E14-1 Francois Gohier/Photo Researchers, Inc.
E14-2 Andrew J. Martinez/Photo Researchers, Inc.

Chapter 15

Opener 15-1 O. Louis Mazzatenta/NGS Image Collection
Opener 15-2 O. Louis Mazzatenta/National Geographics/Getty Images
15-3 Sidney Fox/Science VU/Visuals Unlimited
15-5 Michael Abbey/Visuals Unlimited
15-6a Milwaukee Public Museum, Photograph Collection
15-6b James L. Amos/Photo Researchers, Inc.
15-6c Carolina Biological Supply Company/Phototake NYC
15-6d Douglas Faulkner/Photo Researchers, Inc.
15-7 Illustration by Ludek Pesek/Science Photo Library/Photo Researchers, Inc.
15-8 Terry Whittaker/Photo Researchers, Inc.
15-9 Illustration by Chris Butler/Science Photo Library/Photo Researchers, Inc.
15-12a Tom McHugh/Chicago Zoological Park/Photo Researchers, Inc.
15-12b Frans Lanting/Minden Pictures
15-12c Nancy Adams/Tom Stack & Associates, Inc.
15-13 © Michel Brunet/M.P.F.T.
15-16 David Frayer, Dept. of Anthropology, University of Kansas
15-17 Jerome Chatin/Getty Images, Inc - Liaison

Chapter 16

Opener 16-1 Stephen Dalton/Photo Researchers, Inc.
Opener 16-2 Paul Lethaby/Bermuca Biological station for Research, Inc.
16-1a Wayne Lankinen/Bruce Coleman Inc.
16-1b M.C. Chamberlain/DRK Photo
16-1c Maslowski/Photo Researchers, Inc.
16-2a C. Steven Murphree/Biological Photo Service
16-2b Dr. Greg Rouse, Department of Invertebrate Zoology, National Museum of Natural History, Smithsonian Institution
16-2c Dr. Jeremy Burgess/Science Photo Library/Photo Researchers, Inc.
16-3a Hans Gelderblom/Getty Images Inc. - Stone Allstock
16-3b Reprinted by permission of Springer-Verlag from W.J. Jones, J.A. Leigh, F. Mayer, C.R. Woese, and R.S. Wolfe, Methanococcus jannaschii sp. nov., an extremely thermophilic methanogen from a submarine hydrothermal vent. *Archives of Microbiology* 136:254-261 (1983). © 1983 by Springer-Verlag GmbH & Co KG. Image courtesy of W. Jack Jones.
16-5a David M. Phillips/Visuals Unlimited
16-5b Karl O. Stetter, University of Regensburg, Germany
16-5c CNRI/Science Photo Library/Photo Researchers, Inc.
16-6 Karl O. Stetter, University of Regensburg, Germany
16-7 A.B. Dowsett/Science Photo Library/Photo Researchers, Inc.
16-8 CNRI/Science Photo Library/Photo Researchers, Inc.
16-9 Alan L. Detrick/Photo Researchers, Inc.
16-10 Manfred Kage/Peter Arnold, Inc.
16-11a D.P. Wilson/Eric and David Hosking/Photo Researchers, Inc.
16-11b Lawrence E. Naylor/Photo Researchers, Inc.
16-12 David M. Phillips/Visuals Unlimited
16-13 Kevin Schafer/Peter Arnold, Inc.
16-14 Oliver Meckes & Nicole Ottawa/Eye of Science/Photo Researchers, Inc.
16-15a P.W. Grace/Science Source/Photo Researchers, Inc.
16-15b Cabisco/Visuals Unlimited
16-16 M.I. Walker/Photo Researchers, Inc.
16-17a Ed Degginger/Color-Pic, Inc.
16-17b Manfred Kage/Peter Arnold, Inc.
16-18 Harry Rogers/Photo Researchers, Inc.
16-19a Robert & Linda Mitchell/Robert & Linda Mitchell Photography
16-19b Elmer Koneman/Visuals Unlimited
16-20 Jeff Lepore/Photo Researchers, Inc.
16-21 David M. Dennis/Tom Stack & Associates, Inc.
16-22 Michael Fogden/DRK Photo
16-23 David Dvorak, Jr.
16-24a John Gerlach/Tom Stack & Associates, Inc.
16-24b John Shaw/Tom Stack & Associates, Inc.
16-25a Dwight R. Kuhn/Dwight R. Kuhn Photography
16-25b Milton Rand/Tom Stack & Associates, Inc.
16-25c Larry Ulrich/DRK Photo
16-26a Maurice Nimmo/A-Z Botanical Collection, Ltd.
16-26b Teresa and Gerald Audesirk
16-27a Dwight R. Kuhn/Dwight R. Kuhn Photography
16-27b David Dare Parker
16-27b1 Matt Jones/Auscape International Pty. Ltd.
16-27c Dwight R. Kuhn/Dwight R. Kuhn Photography
16-27d Teresa and Gerald Audesirk
16-27e Larry West/Photo Researchers, Inc.
16-28a Larry Lipsky/DRK Photo
16-28b Brian Parker/Tom Stack & Associates, Inc.
16-28c Charles Seaborn/Odyssey Productions, Inc.
16-29a Gregory Ochocki/Photo Researchers, Inc.
16-29b Charles Seaborn/Odyssey Productions, Inc.
16-29c Teresa and Gerald Audesirk
16-29d David B. Fleetham/SeaPics.com
16-30a Kjell B. Sandved/Butterfly Alphabet, Inc.
16-30b David L. Bull/David Bull Photography
16-30c J.H. Robinson/Photo Researchers, Inc.
16-31a Carolina Biological Supply Company/Phototake NYC
16-31b Peter J. Bryant/Biological Photo Service
16-31c Stephen Dalton/Photo Researchers, Inc.
16-31d Werner H. Muller/Peter Arnold, Inc.
16-31e Stanley Breeden/DRK Photo
16-32a Dwight Kuhn/Dwight R. Kuhn Photography
16-32b Tim Flach/Getty Images Inc. - Stone Allstock
16-32c Teresa and Gerald Audesirk
16-33a Tom Branch/Photo Researchers, Inc.
16-33b Peter J. Bryant/Biological Photo Service
16-33c Carolina Biological Supply Company/Phototake NYC
16-33d Estate of Alex Kerstitch
16-34a Ray Coleman/Photo Researchers, Inc.
16-34b Estate of Alex Kerstitch
16-35a Fred Bavendam/Peter Arnold, Inc.
16-35b Ed Reschke/Peter Arnold, Inc.
16-36a Fred Bavendam/Peter Arnold, Inc.
16-36b Kjell B. Sandved/Photo Researchers, Inc.
16-36c Estate of Alex Kerstitch
16-37 Tom McHugh/Photo Researchers, Inc.
16-38a Peter David/Getty Images Inc. - Hulton Archive Photos
16-38b Mike Neumann/Photo Researchers, Inc.
16-38c Stephen Frink/Getty Images Inc. - Stone Allstock
16-38d Peter Scoones/Getty Images Inc. - Hulton Archive Photos
16-39a Breck P. Kent/Animals Animals/Earth Scenes
16-39b Joe McDonald/Tom Stack & Associates, Inc.
16-39c Cosmos Blank/National Audubon Society/Photo Researchers, Inc.
16-40a David G. Barker/Tom Stack & Associates, Inc.
16-40b Roger K. Burnard/Biological Photo Service
16-40c Frans Lanting/Minden Pictures
16-41a Walter E. Harvey/Photo Researchers, Inc.
16-41b Carolina Biological Supply Company/Phototake NYC
16-41c Ray Ellis/Photo Researchers, Inc.
16-42a Flip Nicklin/Minden Pictures
16-42b Jonathan Watts/Science Photo Library/Photo Researchers, Inc.
16-42c C. and M. Denis-Huot/Peter Arnold, Inc.
16-42d S.R. Maglione/Photo Researchers, Inc.
E16-1 Stanley Breeden/DRK Photo

Unit 4

Opener 4 John Shaw/Tom Stack & Associates, Inc.

Chapter 17

Opener 17-1 John Dudak/Phototake NYC
Opener 17-2 Peter E. Smith/The Natural Sciences Image Library (NSIL)
17-4 Dr. Jeremy Burgess/Science Photo Library/Photo Researchers, Inc.
17-7a Lynwood M. Chace/Photo Researchers, Inc.
17-7b Dwight R. Kuhn/Dwight R. Kuhn Photography
17-8UR Ed Reschke/Peter Arnold, Inc.
17-8WR E.R. Degginger/Animals Animals/Earth Scenes
17-9 Teresa and Gerald Audesirk
17-10 J. Robert Waaland, University of Washington/Biological Photo Service
17-11UR Ed Reschke/Peter Arnold, Inc.
17-14a Kenneth W. Fink/Photo Researchers, Inc.
17-14b Belinda Wright/DRK Photo
17-16 Kim Taylor/Bruce Coleman Inc.
17-17 Biophoto Associates/Science Source/Photo Researchers, Inc.
17-18R Hugh Spencer/National Audubon Society/Photo Researchers, Inc.
E17-1 Michael Fogden/DRK Photo

Chapter 18

Opener 18-1 Maximilian Stock Ltd/Animals Animals/Earth Scenes
Opener 18-2 Stone/Getty Images
18-1uL Jeff Foott/Bruce Coleman Inc.
18-3b Heather Audesirk
18-4 Teresa and Gerald Audesirk
18-5 Dennis Kunkel/Dennis Kunkel Microscopy, Inc.
18-6L Carolina Biological Supply Company/Phototake NYC
18-10 Teresa and Gerald Audesirk
18-11c Runk/Schoenberger/Grant Heilman Photography, Inc.
18-14 John Cancalosi/DRK Photo
18-15 Biophoto Associates/Photo Researchers, Inc.
18-16 Teresa and Gerald Audesirk
18-17a, b Thomas Eisner, Cornell University.
18-18 This item is reproduced by permission of The Huntington Library, San Marino, California.
18-19a Teresa and Gerald Audesirk
18-19b Carolina Biological Supply Company/Phototake NYC
18-20 Teresa and Gerald Audesirk

18-21 Scott Camazine/Photo Researchers, Inc.
E18-1 Myrleen Ferguson Cate/PhotoEdit

Unit 5
Opener 5 Randy Wells/Getty Images Inc. - Stone Allstock

Chapter 19
Opener 19-1 AP Wide World Photos
Opener 19-2 Norbert Wu/Peter Arnold, Inc.
19-4 Robert Brons/Biological Photo Service
19-5 Manfred Kage/Peter Arnold, Inc.
19-6a ©Susumu Nishinaga/SPL/Photo Researchers, Inc.
19-6b Fred Bruemmer/OKAPIA/Photo Researchers, Inc.
19-7 Yorgos Nikas/Science Photo Library/Photo Researchers, Inc.
19-8 Manfred Kage/Peter Arnold, Inc.

Chapter 20
Opener 20-1 Grant Heilman/Grant Heilman Photography, Inc.
Opener 20-2 Alix/Photo Researchers, Inc.
20-9 Copyright Dennis Kunkel Microscopy, Inc.
20-10 Photo Lennart Nilsson/Albert Bonniers Forlag
20-11 Kenneth Kaushansky, M.D., University of Washington Medical Center
20-12 CNRI/Science Photo Library/Photo Researchers, Inc.
20-15 Photo Lennart Nilsson/Albert Bonniers Forlag
E20-1b Lou Lainey/Getty Images/Time Life Pictures
E20-2 Bill Travis, M.D./National Cancer Institute
E20-3ab O. Auerbach/Visuals Unlimited

Chapter 21
Opener 21-1 LaCroix, Patrick J./Getty Images Inc. - Image Bank
Opener 21-2 John Henley/Corbis/Stock Market
21-1 Charles Krebs/Getty Images Inc. - Stone Allstock
21-2a Howard Brinton/General Board of Global Ministries
21-2b © 2005 A.D.A.M., Inc.
21-2c Biophoto Associates/Photo Researchers, Inc.
21-3 Robert and Beth Plowes Photography
21-4a E.R. Degginger/Photo Researchers, Inc.
21-10 Tom Adams/Visuals Unlimited
21-13b CNRI/Photo Researchers, Inc.
E21-1 De Keerle-Parker/Getty Images, Inc - Liaison
E21-2 Dan McCoy/Rainbow

Chapter 22
Opener 22-1 Volker Steiger/Science Photo Library/Photo Researchers, Inc.
Opener 22-2 © Scott Camazine/Photo Researchers, Inc.
22-1U David Scharf/Peter Arnold, Inc.
22-2 Photo Lennart Nilsson/Albert Bonniers Forlag
22-3a National Institute for Biological Standards and Control (U.K.)/Science Photo Library/Photo Researchers, Inc.
22-3b S.H.E. Kaufmann & J.R. Golecki/Science Photo Library/Photo Researchers, Inc.
22-8a Secchi-Lecaque/Roussel/CNRI/Science Source/Photo Researchers, Inc.
22-8b CNRI/Science Photo Library/Photo Researchers, Inc.
22-9 Oliver Meckes & Nicole Ottawa/Photo Researchers, Inc.
22-12b National Institute for Biological Standards and Control (U.K.)/Science Photo Library/Photo Researchers, Inc.
22-12c National Institute for Biological Standards and Control (U.K.)/Science Photo Library/Photo Researchers, Inc.
E22-2 Corbis/Bettmann

Chapter 23
Opener 23-1 © Juergen Berger, Max-Planck Institute/Science Photo Library/Photo Researchers, Inc.
Opener 23-2 Howard K. Suzuki
23-5 Corbis/Bettmann
23-7b Biophoto Associates/Photo Researchers, Inc.
23-10 Photo courtesy of The Jackson Laboratory, Bar Harbor, Maine.
E23-1 AP Wide World Photos

Chapter 24
Opener 24-1 The Warren Anatomical Museum, Francis A. Countway Library of Medicine, Harvard Medical School.
Opener 24-2 From H. Damasio, T. Grabowski, R. Frank, A.M. Galaburda, A.R. Damasio, The Return of Phineas Gage: Clues about the brain from the skull of a famous patient. *Science* 264:112-115, 1994. Department of Neurology and Image Analysis Facility, University of Iowa.
24-10a William J. Powers, M.D.
24-10b Monte S. Buchsbaum, M.D.
24-12a Robert S. Preston/Joseph E. Hawkins Jr.
24-15 Susan Eichorst
24-19a Charles E. Mohr/Photo Researchers, Inc.
24-19b Brian Parker/Tom Stack & Associates, Inc.
E24-1 Ken Graham/Bruce Coleman Inc.
E24-2 Ron Edmonds/AP Wide World Photos

Chapter 25
Opener 25-1 Bob Daemmrich/Bob Daemmrich Photography, Inc.
Opener 25-2 Sterling K. Clarren, M.D.
25-1 Biophoto Associates/Photo Researchers, Inc.
25-2 Peter J. Bryant/Biological Photo Service
25-3a Teresa and Gerald Audesirk
25-3b Peter Harrison
25-4 Anne et Jacques Six
25-5 Michael Fogden/Oxford Scientific Films/Animals Animals/Earth Scenes
25-6a Stephen J. Krasemann/DRK Photo
25-6b Tom McHugh/Photo Researchers, Inc.
25-6c Frans Lanting/Minden Pictures
25-10b CNRI/Science Photo Library/Photo Researchers, Inc.
25-14 C. Edelmann/La Villette/Photo Researchers, Inc.
25-17a Photo Lennart Nilsson/Albert Bonniers Forlag
25-17b Y. Nikas/Photo Researchers, Inc.
25-20b Photo Lennart Nilsson/Albert Bonniers Forlag
25-23a Photo Lennart Nilsson/Albert Bonniers Forlag
25-23b Petit Format/Nestle/Science Source/Photo Researchers, Inc.
25-23c Photo Lennart Nilsson/Albert Bonniers Forlag
E25-1 Hank Morgan/Photo Researchers, Inc.
E25-2 Brook Kraft/Corbis/Sygma

Chapter 26
Opener 26-1 Reuters NewMedia Inc./Corbis/Bettmann
Opener 26-2 Joe McDonald/DRK Photo
26-1a, b Eric and David Hosking/Frank Lane Picture Agency Limited
26-2a-e Boltin Picture Library
26-3 Frans Lanting/Minden Pictures
26-5 Thomas McAvoy/Getty Images/Time Life Pictures
26-7 Renee Lynn/Photo Researchers, Inc.
26-8 Ken Cole/Animals Animals/Earth Scenes
26-9 Richard K. LaVal/Animals Animals/Earth Scenes
26-10 Nuridsany et Perennou/Photo Researchers, Inc.
26-11 Robert & Linda Mitchell/Robert & Linda Mitchell Photography
26-12a Stephen J. Krasemann/Photo Researchers, Inc.
26-12b Hans Pfletschinger/Peter Arnold, Inc.
26-13a M.P. Kahl/DRK Photo
26-13b Marc Chamberlain
26-14 Ray Dove/Visuals Unlimited
26-15 William Ervin/Natural Imagery
26-16 Michael K. Nichols/National Geographic Image Collection
26-17 John D. Cunningham/Visuals Unlimited
26-18 Joe McDonald/Visuals Unlimited
26-19a © Konrad Wothe/Minden Pictures
26-19b Frans Lanting/Minden Pictures
26-20 Anne et Jacques Six
26-22a Dwight R. Kuhn Photography
26-22b Teresa and Gerald Audesirk
26-23 Fred Bruemmer/Peter Arnold, Inc.
26-25 Raymond A. Mendez/Animals Animals/Earth Scenes
26-26 Photo Lennart Nilsson/Albert Bonniers Forlag
26-27 William P. Fifer, New York State Psychiatric Institute, Columbia University
26-28a Frans Lanting/Minden Pictures
26-28b Daniel J. Cox/Getty Images, Inc - Liaison
26-28c Michio Hoshino/Minden Pictures
26-29 Reproduced by permission from Low-, normal-, high- and perfect-symmetry versions of a male face From Fig 1 page 235 in *Animal Behaviour*, 202, 64, 233-238 by Koehler N, Rhodes, G, & Simmons L.W. Copyright © 206 by Elsevier Science Ltd. Image courtesy of *Animal Behavior*.

Unit 6
Opener 6 *Paradise, Part One of The Trilogy of the Earth* by Suzanne Duranceau/Illustratrice, Inc.

Chapter 27
Opener 27-1 Fred Bruemmer/DRK Photo
Opener 27-2 Photo Researchers, Inc.
27-1a Robert & Linda Mitchell Photography
27-1b Tom and Susan Bean, Inc./DRK Photo
27-1c © 206 Gary Braasch
27-2a Prof. S. Cinti/University of Ancona, Italy/CNRI/Phototake NYC
27-2b © David G. Atfield, Sr.
27-7 Tom McHugh/Photo Researchers, Inc.
E27-3 Frans Lanting/Minden Pictures

Chapter 28
Opener 28-1 Schafer, Kevin/Getty Images Inc. - Image Bank
Opener 28-2 Kike Calvo/V&W/imagequestmarine.com
28-3a Henry Ausloos/Animals Animals/Earth Scenes
28-3b Stephen Dalton/Photo Researchers, Inc.
28-3c Frans Lanting/Minden Pictures
28-4a Art Wolfe, Inc.
28-4b Frans Lanting/Minden Pictures
28-5a Marty Cordano/DRK Photo
28-5b Paul A. Zahl/Photo Researchers, Inc.
28-5c Ray Coleman/Photo Researchers, Inc.
28-6a Jean-Paul Ferrero and Jean-Michel Labat/Auscape International Pty. Ltd.
28-6b Charles V. Angelo/Photo Researchers, Inc.
28-7 Dr. James L. Castner
28-8a Carmela Leszczynski/Animals Animals/Earth Scenes
28-8b Breck P. Kent
28-8c, d Ed Degginger/Color-Pic, Inc.

28-9a Robert P. Carr/Bruce Coleman Inc.
28-9b Monica Mather and Bernard Roitberg, Simon Fraser University
28-10a Zig Leszczynski/Animals Animals/Earth Scenes
28-10b Dr. James L. Castner
28-10c Jeff Lepore/Photo Researchers, Inc.
28-11a Thomas Eisner and Daniel Aneshansley, Cornell University.
28-11b Frans Lanting/Minden Pictures
28-12 Teresa Audesirk
28-13a Jim Zipp/Photo Researchers, Inc.
28-13b Stephen J. Krasemann/Photo Researchers, Inc.
28-14aL Dennis Oda/Getty Images Inc. - Stone Allstock
28-14aR Krafft-Explorer/Photo Researchers, Inc.
28-14bL, bR © 2002 Gary Braasch
28-14cL Raymond German/Corbis/Bettmann
28-14cR Douglas Faulkner/Corbis/Bettmann
28-17a, b Robert & Linda Mitchell Photography
28-18a, b Cristina Sandoval, Ph.D.
28-19a AL SATTERWHITE/Getty Images, Inc. - Taxi
28-19b Stephen G. Maka/DRK Photo
E28-1a Ron Peplowski/DTE Energy
E28-1b Chuck Pratt/Bruce Coleman Inc.
E28-1c Robert & Linda Mitchell Photography
E28-1d Jack Jeffrey/Photo Resource Hawaii Stock Photography

Chapter 29

Opener 29-1 Joseph Sohm/Chromosohm, Inc./Corbis/Bettmann
Opener 29-2 Biospheres Ventures/Corbis/Sygma
29-12 Tom Walker/Stock Boston
29-13 William E. Ferguson
29-14 Will McIntyre/Photo Researchers, Inc.
E29-1 Thomas A. Schneider/SchneiderStock Photography

Chapter 30

Opener 30-1 Vito Aluia/Index Stock Imagery, Inc.
Opener 30-2 Jungle Tech LLC
30-2b NASA/Johnson Space Center
30-6a Teresa and Gerald Audesirk
30-6b Tom McHugh/Photo Researchers, Inc.
30-9a Loren McIntyre
30-9b Schafer and Hill/Peter Arnold, Inc.
30-9c Michael Fogden/Oxford Scientific Films Ltd.
30-9d Shiell, Richard/Animals Animals/Earth Scenes
30-9e Frans Lanting/Minden Pictures
30-10 Jacques Jangoux/Peter Arnold, Inc.
30-11a W. Perry Conway/Aerie Nature Series, Inc.
30-11b Theo Allofs/Allofs Photography
30-11c Stephen J. Krasemann/DRK Photo
30-11d Aubry Lang/Valan Photos
30-12 James Hancock/Photo Researchers, Inc.
30-13a Don & Pat Valenti/DRK Photo
30-13b Brian Parker/Tom Stack & Associates, Inc.
30-13c Tom McHugh/Photo Researchers, Inc.
30-14 William H. Mullins/Photo Researchers, Inc.
30-15 Ken Lucas/Visuals Unlimited
30-16 Harvey Payne
30-17a Tom and Susan Bean, Inc./DRK Photo
30-17b W. Perry Conway/Tom Stack & Associates, Inc.
30-17c Nicholas DeVore III/Network Aspen
30-17d Bob Gurr/Valan Photos
30-17e Jim Brandenburg/Minden Pictures
30-18 Teresa and Gerald Audesirk
30-19a © 1993 Gary Braasch
30-19b Thomas Kitchin/Tom Stack & Associates, Inc.
30-19c Stephen J. Krasemann/DRK Photo
30-19d Joe McDonald/Animals Animals/Earth Scenes
30-20a Susan G. Drinker/Corbis/Stock Market
30-20b Teresa and Gerald Audesirk
30-20c Jeff Foott Productions
30-20d Richard Thom/Tom Stack & Associates, Inc.
30-21a Teresa and Gerald Audesirk
30-21b Marty Stouffer Productions/Animals Animals/Earth Scenes
30-21c Wm. Bacon III/National Audubon Society/Photo Researchers, Inc.
30-22 Gary Braasch/Gary Braasch Photography
30-23a Stephen J. Krasemann/DRK Photo
30-23b Jeff Foott/Tom Stack & Associates, Inc.
30-23c Michael Giannechini/Photo Researchers, Inc.
30-23d James Simon/Photo Researchers, Inc.
30-26a T.P. Dickinson/Comstock Images
30-26b Teresa and Gerald Audesirk
30-26b inset Walter Dawn/National Audubon Society/Photo Researchers, Inc.
30-26c Teresa and Gerald Audesirk
30-26c inset Thomas Kitchin/Tom Stack & Associates, Inc.
30-26d Robert R. Given, Marymount College
30-26d inset Stan Wayman/Photo Researchers, Inc.
30-27a Franklin J. Viola/Viola's Photo Visions, Inc.
30-27b Estate of Alex Kerstitch
30-27c Charles Seaborn/Odyssey Productions, Inc.
30-27d Estate of Alex Kerstitch
30-28a Stephen J. Krasemann/DRK Photo
30-28b Paul Humann/Jeffrey L. Rotman Photography
30-28c Fred McConnaughey/Photo Researchers, Inc.
30-28d James D. Watt/Getty Images Inc. - Hulton Archive Photos
30-28e Biophoto Associates/Science Source/Photo Researchers, Inc.
30-28f Robert Arnold/Getty Images Inc. - Hulton Archive Photos
30-29 J. Frederick Grassle, Institute of Marine and Coastal Sciences, Rutgers University
E30-1 NASA Headquarters
E30-2 Roy Morsch/Corbis/Stock Market

Index

Student Accelerator CD Instructions

To enhance the speed and performance of the website, simply

1. Insert the Student Accelerator CD into your CD-ROM drive
2. Access the website www.prenhall.com/audesirk, select your text, and log in

To view the Web Tutorials without an internet connection,

1. Insert the Student Accelerator CD into your CD-ROM drive
2. On a PC, double click "start.exe." On a Mac, double click "startOS9" or "startOSX." Then select a chapter to begin viewing the Web Tutorials.

Audesirk/Audesirk/Byers

***Life on Earth*, 4e Accelerator CD-ROM**

CD License Agreement

Pearson Prentice Hall

Pearson Education, Inc.

Upper Saddle River, NJ 07458

YOU SHOULD CAREFULLY READ THE TERMS AND CONDITIONS BEFORE USING THE CD-ROM PACKAGE. USING THIS CD-ROM PACKAGE INDICATES YOUR ACCEPTANCE OF THESE TERMS AND CONDITIONS.

Pearson Education, Inc. provides this program and licenses its use. You assume responsibility for the selection of the program to achieve your intended results, and for the installation, use, and results obtained from the program. This license extends only to use of the program in the United States or countries in which the program is marketed by authorized distributors.

LICENSE GRANT

You hereby accept a nonexclusive, nontransferable, permanent license to install and use the program ON A SINGLE COMPUTER at any given time. You may copy the program solely for backup or archival purposes in support of your use of the program on the single computer. You may not modify, translate, disassemble, decompile, or reverse engineer the program, in whole or in part.

TERM

The License is effective until terminated. Pearson Education, Inc. reserves the right to terminate this License automatically if any provision of the License is violated. You may terminate the License at any time. To terminate this License, you must return the program, including documentation, along with a written warranty stating that all copies in your possession have been returned or destroyed.

LIMITED WARRANTY

THE PROGRAM IS PROVIDED "AS IS" WITHOUT WARRANTY OF ANY KIND, EITHER EXPRESSED OR IMPLIED, INCLUDING, BUT NOT LIMITED TO, THE IMPLIED WARRANTIES OR MERCHANTABILITY AND FITNESS FOR A PARTICULAR PURPOSE. THE ENTIRE RISK AS TO THE QUALITY AND PERFORMANCE OF THE PROGRAM IS WITH YOU. SHOULD THE PROGRAM PROVE DEFECTIVE, YOU (AND NOT PEARSON EDUCATION, INC. OR ANY AUTHORIZED DEALER) ASSUME THE ENTIRE COST OF ALL NECESSARY SERVICING, REPAIR, OR CORRECTION. NO ORAL OR WRITTEN INFORMATION OR ADVICE GIVEN BY PEARSON EDUCATION, INC., ITS DEALERS, DISTRIBUTORS, OR AGENTS SHALL CREATE A WARRANTY OR INCREASE THE SCOPE OF THIS WARRANTY.

SOME STATES DO NOT ALLOW THE EXCLUSION OF IMPLIED WARRANTIES, SO THE ABOVE EXCLUSION MAY NOT APPLY TO YOU. THIS WARRANTY GIVES YOU SPECIFIC LEGAL RIGHTS AND YOU MAY ALSO HAVE OTHER LEGAL RIGHTS THAT VARY FROM STATE TO STATE.

Pearson Education, Inc. does not warrant that the functions contained in the program will meet your requirements or that the operation of the program will be uninterrupted or error-free.

However, Pearson Education, Inc. warrants the CD-ROM(s) on which the program is furnished to be free from defects in material and workmanship under normal use for a period of ninety (90) days from the date of delivery to you as evidenced by a copy of your receipt.

The program should not be relied on as the sole basis to solve a problem whose incorrect solution could result in injury to person or property. If the program is employed in such a manner, it is at the user's own risk and Pearson Education, Inc. explicitly disclaims all liability for such misuse.

LIMITATION OF REMEDIES

Pearson Education, Inc.'s entire liability and your exclusive remedy shall be: 1. the replacement of any CD-ROM not meeting Pearson Education, Inc.'s "LIMITED WARRANTY" and that is returned to Pearson Education, or 2. if Pearson Education is unable to deliver a replacement CD-ROM that is free of defects in materials or workmanship, you may terminate this agreement by returning the program.

IN NO EVENT WILL PEARSON EDUCATION, INC. BE LIABLE TO YOU FOR ANY DAMAGES, INCLUDING ANY LOST PROFITS, LOST SAVINGS, OR OTHER INCIDENTAL OR CONSEQUENTIAL DAMAGES ARISING OUT OF THE USE OR INABILITY TO USE SUCH PROGRAM EVEN IF PEARSON EDUCATION, INC. OR AN AUTHORIZED DISTRIBUTOR HAS BEEN ADVISED OF THE POSSIBILITY OF SUCH DAMAGES, OR FOR ANY CLAIM BY ANY OTHER PARTY.

SOME STATES DO NOT ALLOW FOR THE LIMITATION OR EXCLUSION OF LIABILITY FOR INCIDENTAL OR CONSEQUENTIAL DAMAGES,
SO THE ABOVE LIMITATION OR EXCLUSION MAY NOT APPLY TO YOU.

GENERAL

You may not sublicense, assign, or transfer the license of the program. Any attempt to sublicense, assign or transfer any of the rights, duties, or obligations hereunder is void.

This Agreement will be governed by the laws of the State of New York.

Should you have any questions concerning this Agreement, you may contact Pearson Education, Inc. by writing to:

ESM Media Development
Higher Education Division
Pearson Education, Inc.
1 Lake Street
Upper Saddle River, NJ 07458

Should you have any questions concerning technical support, you may write to:

New Media Production
Higher Education Division
Pearson Education, Inc.
1 Lake Street
Upper Saddle River, NJ 07458

YOU ACKNOWLEDGE THAT YOU HAVE READ THIS AGREEMENT, UNDERSTAND IT, AND AGREE TO BE BOUND BY ITS TERMS AND CONDITIONS. YOU FURTHER AGREE THAT IT IS THE COMPLETE AND EXCLUSIVE STATEMENT OF THE AGREEMENT BETWEEN US THAT SUPERSEDES ANY PROPOSAL OR PRIOR AGREEMENT, ORAL OR WRITTEN, AND ANY OTHER COMMUNICATIONS BETWEEN US RELATING TO THE SUBJECT MATTER OF THIS AGREEMENT.